Advanced Developments in Ultra-Clean Gasoline-Powered Vehicles

PT-104

Edited by

Fuquan (Frank) Zhao

Published by
Society of Automotive Engineers, Inc.
400 Commonwealth Drive
Warrendale, PA 15096-0001
U.S.A.
Phone: (724) 776-4841
Fax: (724) 776-5760
www.sae.org
January 2004

All rights reserved. No part of this publication may be reproduced, stored in a retrieval system, or transmitted, in any form or by any means, electronic, mechanical, photocopying, recording, or otherwise, without the prior written permission of SAE.

For permission and licensing requests contact:

SAE Permissions
400 Commonwealth Drive
Warrendale, PA 15096-0001-USA
Email: permissions@sae.org
Fax: 724-772-4891
Tel: 724-772-4028

Global Mobility Database®

All SAE papers, standards, and selected books are abstracted and indexed in the Global Mobility Database.

For multiple print copies contact:

SAE Customer Service
Tel: 877-606-7323 (inside USA and Canada)
Tel: 724-776-4970 (outside USA)
Fax: 724-776-1615
Email: CustomerService@sae.org

ISBN 0-7680-1420-4
Library of Congress Catalog Card Number: 2003115907
SAE/PT-104
Copyright © 2004 SAE International

Positions and opinions advanced in this publication are those of the author(s) and not necessarily those of SAE. The author is solely responsible for the content of the book.

SAE Order No. PT-104

Printed in USA

PREFACE

During the last several years, significant efforts have been directed toward the development of ultra-clean, gasoline-powered vehicles in the automotive industry to meet increasingly stringent emissions legislation. Several vehicle manufacturers have already succeeded in launching theses types of vehicles, which are powered by conventional internal combustion engines and are certified in California to meet the most difficult emissions standard of Partial Zero Emission Vehicle (PZEV) in the world. Accompanying this extensive development effort is the continuing generation of a very large volume of technical information, with a growing need for systematic organization, description of fundamental processes, sorting out insights on technical issues, and identification of key trends and future R&D directions. This volume was planned to serve this essential purpose.

This book is organized into 14 chapters. All the subjects related to the development and certification of ultra-clean gasoline-powered vehicles are covered in the book. Experts in the field were invited to address a broad spectrum of key research and development issues in the rapidly progressing area of ultra-clean gasoline-powered vehicles. Nearly all topics that a reader in the field may wish to comprehend are presented and the future technology directions and R&D needs are also outlined. The purpose is to provide the reader with a concise, brief introduction to the state-of-the-art of technology developments in ultra-clean gasoline-powered vehicles. At the end of each chapter, several carefully selected SAE technical papers are attached for advanced readings. Since many excellent papers are available in the literature, the selection of these papers has been a difficult task. It should be pointed out that each SAE technical paper in the book was carefully selected, based on the following criteria: 1. to provide the clearest explanation of a particular concept or development; 2. to be unique and ground breaking; or 3. to be the best representative of a group of similar papers.

It is evident that the compiled information in this book would constitute an excellent base for someone who either just started in the field, or for someone who has already been very familiar with the area and wants to hear opinions on certain issues from other experts in the field, or for someone who just wants to get a general idea about the recent technology developments in ultra-clean gasoline-powered vehicles. The selected SAE technical papers in the book are very convenient for those individuals who want to have a quick reference to well-selected technical information.

In closing, I would like to extend my thanks to all the chapter authors for their professional contributions that made the compilation of this book a great success. I would also like to express my gratitude to Ms. Martha Swiss of SAE for her excellent support in preparing this volume.

Fuquan (Frank) Zhao
DaimlerChrysler Corporation
November 17, 2003

Contents

CHAPTER 1

TRANSIENT ENGINE START-UP AND SHUT-DOWN PROCESSES

Wai K. Cheng
Massachusetts Institute of Technology

1.1 OVERVIEW

Start-up and shut-down processes are important engine transients that substantially affect overall vehicle emissions. In cold start, the catalyst is not effective until light-off. Since the conversion efficiency of a fully-warmed-up modern catalyst is in the upper ninety percents, the start-up emissions become a significant part of the trip total. Hydrocarbon (HC) emissions are of the most concern because of strict regulatory requirements. Precisely controlling the engine in the start-up period is difficult because there are significant uncertainties about the state of the engine (e.g. the amount of fuel left over inside the engine from the previous stop; the piston starting positions; the properties of the gasoline), and at low temperatures, the thermal environment for mixture preparation is not favorable. In addition, the engine should provide smooth operation in this period. Much effort has been made to understand the start-up process and improve the emissions of such.

The role of the engine shut-down process in emissions is largely due to its impact on the next start-up. If the ignition key-off is not synchronous with the engine cycle events, there could be several cylinders that receive fuel injections but have no ignition. For Port-Fuel-Injection (PFI) engines, there is also substantial fuel remaining in the cylinder and in the port. The unburned mixture in the non-firing cycles and the fuel vapor from the residual fuel contribute to the engine-out HC. The exhaust system, however, is usually large enough to trap all the engine effluent in the few cycles during which the engine coasts to a stop. This trapped gas will be pumped out in the next start-up and will contribute substantially to tailpipe emissions if the catalyst has cooled off.

The engine start-up behavior depends on the state of the engine from the last shut-down and the cranking/start-up strategy. Important considerations are the catalyst temperature, which determines whether the catalyst is effective in the start-up, and the coolant temperature (essentially the metal temperature of the engine), which affects the fuel evaporation process. The catalyst cool down time to ambient (i.e., time for the temperature difference to decrease by 1/e) is ~40 minutes. The coolant cool down time is ~60 minutes. Since the threshold temperature for effective catalyst operation is approximately 300°C, the catalyst remains effective for half an hour or so after shut down.

The engine start-up process may be divided into the following events:

(a) First round of firing of all the cylinders–In these firings, there is very little residual gas in the cylinders. The intake pressure is close to atmospheric. As such, the torque output is high and the engine accelerates quickly.

(b) The speed flare–The extent of the speed run up is determined by the process of drawing down the Manifold Air Pressure (MAP) to idle value by the engine. The engine accelerates up to 2000-plus rpm and then decelerates to the idle speed according to the engine brake torque, which is determined by the MAP. Care must be taken to limit the MAP undershoot in speed flare. The low MAP results in high residual gas fractions that could cause engine stability problems.

(c) The gradual warm-up of the engine—The speed flare usually takes place in a few seconds. After that, the engine idles at more or less constant speed and intake pressure until the driver demands the first acceleration. The thermal state of the engine, however, is far from equilibrium. Strategies such as high idle speed and retarded spark timing are usually used at start-up to create a high enthalpy flow to the catalyst to facilitate catalyst light-off.

The thermal environment in cold start is not favorable to mixture preparation. This fact is especially severe for the first round of firing for which the wall temperatures are cold and there is no reverse flow of hot burned gas at Intake Valve Open (IVO). With conventional PFI injectors, the drop size is on the order of 150 μm so that the drag force is small compared to the inertia force in the period of flight. Thus the majority of the fuel first lands on surfaces (port wall and valves). Fuel vapor is formed by the following processes:

(a) Direct evaporation from the injected droplets in flight
(b) Evaporation from the wall film in the intake port area
(c) Evaporation from the fuel drops strip-atomized by the reversed flow when the intake valve opens, and by the flow through the valve induced by the piston motion
(d) Evaporation from the cylinder and piston surfaces

As such, the factors that influence the mixture preparation in the start-up transient are:

(i) The port wall/intake valve temperatures—The port wall temperature does not change appreciably in the first 10 or so seconds. The intake valve temperature depends very much on the amount of liquid fuel that lands on the valve; it is typically on the order of 10°C per second. For a change of engine temperature from 80°C to 0°C, the minimum fuel amount required for stable combustion of the first firing may have to be increased by as much as a factor of 10.

(ii) The fuel properties—A more volatile fuel will facilitate mixture preparation in the start-up transient. The overall evaporative behavior of the fuel is usually characterized by a single number, the Driveability Index (DI), which is defined as $DI = 1.5 \times T_{10} + 3.0 \times T_{50} + 1.0 \times T_{90}$ where T10, T50 and T90 are the 10, 50 and 90% distillation temperatures in F. No direct relationship, however, has been established between DI (or the distillation temperatures) and the cold-start calibration requirement of an engine.

(iii) Fuel injection characteristics—The important parameters are the drop size, fuel targeting and injection timing. Since large drops typically land on the engine surfaces first, the mixture preparation process is not sensitive to drop size until the droplet diameters are substantially smaller than 100 μm. Injection is usually targeted at the back of the intake valve which is the hottest surface in the port and the fastest component to warm up. With conventional injectors, closed-valve injection usually produces lower HC emissions than open-valve injection.

(iv) Port flow characteristics–Strip-atomization of the surface fuel film by the port flow plays a significant role in the evaporation process. The reverse back flow at IVO can be enhanced by increasing the valve overlap. However, the large residual fraction at high overlap will be detrimental to idle quality. Increasing the local port flow velocity, usually done by a flap-type air control valve in the port, enhances the forward flow strip-atomization. The strategy has the additional benefit that it increases the burn rate and improves combustion stability.

(v) Intake valve timing–A late intake valve opening produces a cylinder pressure that is substantially below atmospheric when the valve opens. As such, there is no reversed blow-down, but the initial forward flow driven by the manifold/cylinder pressure difference is much stronger and thereby helps strip-atomization; the initial low cylinder pressure enhances evaporation; the temperature of the in-cylinder charge is higher due to the work done by the manifold/cylinder pressure difference on the gas flow during pressure equilibration. The overall effect, however, has not been sufficiently quantified because for an engine with a fixed cam, a later IVO would result in a later Intake Valve Close (IVC). Thus the volumetric efficiency would decrease (the effective compression ratio would also decrease). It is necessary to decouple the fuel evaporation effect and the air flow effect to assess the effectiveness.

Because of the low engine speed in the first few cycles, the combustion period is quite short in terms of crank angles. Thus the combustion phasing is effectively advanced and the gross indicated work is lower mainly because of the higher heat loss associated with the advanced timing.

For a properly calibrated engine, i.e., with no misfired or partial burn cycles, the engine-out HC emission mechanisms in the flare speed decay and engine gradual warm-up periods will be similar to those at steady state, although the engine temperature will be lower. Then the crevice volume will be larger, the solubility of the gasoline vapor in oil will be higher, and the extent of in-cylinder and exhaust HC oxidation will be reduced. All these factors contribute to a higher level of engine-out HC emissions.

There is, however, a significant amount of HC emissions associated with the cranking and run-up processes. Because of the difficulty in evaporating fuel at the cold start temperature, a large amount of fuel has to be injected in order to achieve a sufficiently combustible mixture. Then a significant amount of liquid fuel enters the combustion chamber. Part of the fuel that does not burn will escape the combustion chamber as HC emissions. (The other part will go into the oil sump.) Thus, there is substantial interest in bringing down the HC emissions in the cranking and run-up process.

The above remarks are pertinent mostly to PFI engines. For Direct Injection Spark Ignition (DISI) engines, the port fuel phenomena are eliminated; thus such engines could potentially have better cold start performance than PFI engines. A major difficulty in the DISI technology is the inability to quickly provide a high enough fuel pressure in the crank start process. A pressure of ~5-12 MPa is required to atomize the fuel to 25 μm or so droplets. The nominal start-up fuel pressure of about 0.4 MPa produces droplets of ~100 μm. These droplets land on the cylinder walls and result in start-up HC emissions which have been reported to be worse than those from a well-developed PFI system.

In conclusion, the engine start-up and shut-down transients are important processes that affect vehicle total emissions. Because of the difficulty in fuel evaporation at cold start, only a small amount of the injected fuel is vaporized and burned in the engine cycle. This fact is true for both the PFI and DISI engines. The large amount of residual fuel constitutes a major source of hydrocarbon emissions. An important area for development, therefore, is in improving the effectiveness of fuel delivery to form the combustible charge. Another area that requires attention is in the development of an effective mixture preparation model to be used for controlling the cranking and start-up process. Future engine calibration for low emissions will require sequential control of every injection event in the start-up sequence. Because of the highly dynamic nature of the process, model based fuel management is needed to minimize the calibration effort and to maximize the robustness of the fuel control procedure.

1.2 SELECTED SAE TECHNICAL PAPERS

Spark Ignition Engine Hydrocarbon Emissions Behaviors in Stopping and Restarting

Daniel Klein
RWTH-Aachen

Wai K. Cheng
Sloan Automotive Laboratory, Massachusetts Institute of Technology

Copyright © 2002 SAE International

ABSTRACT

Engine Hydrocarbon (HC) emissions behaviors in the shut down and re-start processes were examined in a production 4-cylinder 2.4 L engine. Depending on when the power to the ECU was cut off relative to the engine events, there could be two or three mis-fired cylinders (i.e. cylinders with fuel injected but no ignition). The total HC pumped out by the engine into the catalyst in the stopping process was ~4 mg (approximately equaled to the amount of one injection at idle condition). Because the size of the catalyst was larger than the total exhaust volume in the stopping process, this HC was not observed at the catalyst exit. The catalyst temperature was also not affected. When the engine was purged after shut down (by cranking the engine with the injectors and ignition disconnected), the total exit HC was 33 mg. In a restart 90 minutes after shut down, the integrated amount of HC emissions due to residual fuel from the stopping process was 16 mg. To assess the contribution of the mis-fired cylinders to the residual fuel, the engine was shut down by disabling the injectors but with ignition on at all time so that there was no mis-fired cylinders. The HC pumped out to the catalyst in the stopping process was a factor of two smaller (~2 mg). In the restarting process, the difference (compared to the normal shut down) in HC emissions due to the residual fuel was ~5mg. Thus the mis-fired cylinders contributed to a small but not insignificant amount of HC emissions in the restart process.

INTRODUCTION

The increasingly stringent emissions regulations has put tremendous pressure on automotive manufacturers to improve vehicle emissions. The restriction on Hydrocarbon (HC) emissions is especially severe. For the California Exhaust Emissions Standards, the HC level decreases from 0.25 g/mi for the 2001-2003 Tier 1 Vehicles to 0.01 g/mi for the 2004 and beyond LEV II – SULEV vehicles. To achieve this 25 fold reduction, there has been substantial development in the exhaust treatment technology. At the same time, the details of engine operation need to be attended to so as to devise appropriate control strategies. The basic engine HC mechanisms were outlined in Ref. [1]. Recent work has

focused on the engine cranking and start-up process [2,3] during which the mixture preparation environment is not favorable and the catalyst is ineffective. The subject of this paper is to investigate the engine behavior in the shut down and re-start process. We are interested in the amount of fuel left in the engine in engine shut down. This fuel is pumped out of the engine in the restart cranking process and contributes to the overall HC emissions.

There are two shut down/ restart scenarios: the normal vehicle ignition off and restart process and the engine-off and restart process in a hybrid vehicle. We are limiting our scope to the former case in this paper. The latter will be reported in a later publication.

ENGINE SHUT DOWN PROCESS

When the ignition key is turned off, the power off process may not be synchronized with the injection and ignition events of the engine. As a result, there may be "missed" cycles, which are the cycles with fuel delivery to the cylinder but without combustion.

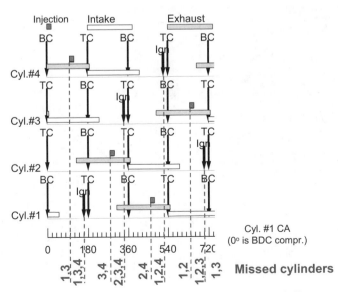

Fig.1   Engine event diagram at idle. The dash lines denote the boundaries on the crank angle axis for various cases of "missed" cylinders

The situation is illustrated in Fig. 1, which shows the event diagram for a 4-cylinder spark ignition engine at idle. (Engines are typically turned off at idle.) We are especially interested in the time when the engine is turned off relative to the injection and ignition events. For example if the power-off is at 400° CA (where 0° is the BDC compression position of cylinder #1; see bottom axis of Fig.1), then for cylinders #1, combustion is complete, and no more fuel is injected. For cylinder #3, combustion will be completed as the engine coasts to stop. For cylinders #2 and #4, however, fuel is injected but no spark was delivered. Thus these cylinders mis-fired. For power-off at different crank angle locations, there are two or three mis-fired cylinders as denoted by the dash lines in Fig. 1. (Since the power off may occur during an injection event or ignition coil charging time, "missed" conditions include those with partial pulse-width injections and non-fully dwelled ignitions.) The unburned mixture of the missed cycles will escape the cylinder as engine-out hydrocarbon.

APPARATUS AND PROCEDURE

The engine shut down and restart process was studied using a production DaimlerChrysler 2.4 L engine (Model year 1998) mounted on a Froude eddy-current dynamometer. The engine was a modern 4-cylinder, 4-valves per cylinder, port-fuel injection engine with a returnless fuel line. All the experiments were done with the OEM engine calibration. The cranking was provided by the production starter motor on the engine. The moment of inertia of the dynamometer rotor was approximately 25% of the flywheel and thus should not materially affect the engine cranking behavior. The engine was equipped with a 50K mile (dyno-aged equivalent) two-brick catalyst (for Model year 2001 ULEV). The fuel used was California Phase II calibration fuel from Philips Petroleum.

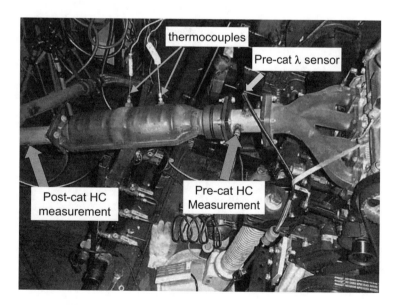

Fig.2 Exhaust system measurements

The procedure was to start the engine and run it at 2000 rpm and 0.5 bar intake pressure to a steady state condition (87° C coolant temperature). The catalyst temperature had also reached steady state. The process took about 15-20 minutes. Then the engine was idled for 0.5 to 1 minute before shut down.

The engine signals recorded were the injection and ignition pulses, the engine speed, the intake pressure and the cylinder #4 pressure. Measurements on the exhaust system are shown in Fig. 2. Two Fast-response Flame Ionization Detectors (FFID) were used to simultaneously monitor the HC levels at the catalyst entrance and exit. The catalyst temperatures at the front part of the first brick and in between the two bricks were measured by type K thermocouples. The Air/Fuel stoichiometric ratios (λ) were measured before and after the catalyst by Horiba λ sensors.

NORMAL ENGINE SHUT DOWN AND RESTART PROCESS

Normal Engine Shutdown
The engine behavior in a typical shut down process is shown in Fig. 3. The injection pulses and the ignition coil charging pulses are shown by the 12 to 0 V drop due to the pull down of the switching transistors. (Cylinders 1 and 4 share the same ignition; same is for cylinders 2 and 3.) The power shut-off to the ECU in the key-off process is indicated by the baseline voltage drop from 12V to 0 V in the injector coil and the ignition coil signals at approximately 2.2 s. The power-off occurred just after the ignition of cylinder #3. For cylinders 2 and 4, fuel was injected but there was no ignition. Therefore these cylinders mis-fired.

After the power-off, the engine took approximately 0.6 second (4 revolutions) to coast from the idle speed (800 rpm) to rest. The intake pressure, however, took longer; it did not recover to atmospheric until about 1.4 seconds from power-off due to the filling process of the manifold.

HC Emissions Behavior in Regular Shut down
To interpret the HC measurements from the FFID probes, the sampling geometry needs to be discussed. The time delay and transit time of the instrument (order of ms) is negligible compare to the time scale of interest (fraction of a second). Referring to Fig. 2, the volume of the exhaust system from the exhaust valve to the pre-catalyst HC measurement point for each individual cylinder was about 370 cc. The total volume was about 1 L. (The numbers did not add up because there was a large volume shared by all 4 runners.) Since the displaced volume of each cylinder was 600 cc., the FFID at the catalyst entrance should be able to sense the gas pumped out by the individual cylinder in each cycle. However, there was significant mixing of the gas flow from the different cylinders before the sampling point because (a) there was a significant shared exhaust system volume, and (b) during engine coast-down, the intake pressure was below atmospheric (see Fig. 3); as

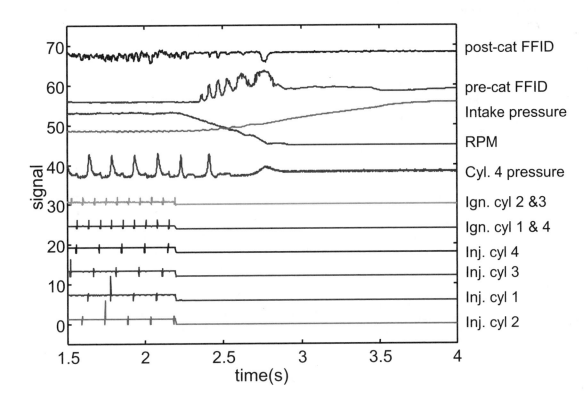

Fig. 3 Events in a normal engine shut-down process. The power-off is indicated by the drop of the baseline voltage from 12 to 0 V in the injector and the ignition signals. Last fired cycle: cylinder#3; missed cylinders: 2 and 4. Vertical scale for the various signals will be given in the subsequent figures.

a result, there was substantial backflow from the exhaust to the cylinder resulting in significant cycle-to-cycle and cylinder-to-cylinder mixing. Note that this back flow consisted of the normal valve-overlap backflow [4] plus the backflow resulting from the cylinder pressures at Exhaust-Valve-Open (EVO) for the non-firing cylinders were substantially below the exhaust manifold pressure.

The pre-cat FFID signal showed the pumping out of the unburned mixture from the cylinder as the engine coasted to a stop. In the first revolution (of the 4 revolutions leading to stop) immediately after power-off, the HC level did not increase because the exhaust was from cylinders 1 and 3 which had complete combustion. The subsequent 6 peaks corresponded to the displacement of the unburned mixtures in the remaining 3 revolutions (in the order of cylinders 4-2-1-3-4-2). When the engine stopped, the FFID was sampling from a stagnant gas; the signal decayed slowly (time scale ~minutes) due to diffusion.

The gradual build-up of the pre-cat FFID signal during the coast-down process was due to the incremental filling of the exhaust volume (portion of the exhaust system before the sampling point) by the HC coming out of the cylinders. The source of HC included both the unburned mixture from the mis-fired cycles of cylinders 2 and 4 and the subsequent delivery of the residual fuel from the intake port after the injectors were cut off. Because of the substantial exhaust mixing as described previously, the signal could not be unscrambled to determine the individual HC flow from each cylinder.

The increase in HC shown in the pre-cat FFID measurement was not observed in the post-cat FFID signal, which was obtained simultaneously. This fact was a result of the significant volume of the catalyst – the exhaust system volume between the pre- and post-catalyst measurement points was approximately 5 L (see Fig. 2). The total exhaust flow of the three non-firing revolutions was approximately 3.6 L (not taking into account of complications such as backflow and temperature corrections etc.). Thus in the engine stopping process, there was not enough of a displacement flow to deliver the unburned HC out of the catalyst.

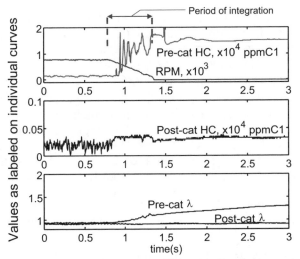

Fig. 4 Pre-cat, post-cat HC mole fractions and A/F equivalence ratios in the regular engine shut down process.

The details of the HC emissions and the pre- and post-cat A/F equivalence ratios are shown in Fig. 4. The data for this figure were deliberately chosen from a different run than from that of Fig. 3 to show the run-to-run variations in the signals. The HC values for all the runs were similar both quantitatively and qualitatively. The peak values of HC were ~ 18kppmC$_1$. These values were significantly lower than those of stoichiometric mixtures (125 kppmC$_1$, excluding residual gas dilution). This low value was due to the mixing with the burned gas in the exhaust volume.

Since the exhaust outflow never reached the post-cat FFID sampling location, the post-cat HC level remained constant at the engine idle level of 500 ppmC$_1$. The change in the 'noise' level of the signal was due to the change in vibration level of the engine as it coasted down.

The pre-cat λ value showed the dilution due to the air pumped out by the cylinder during the coast-down process. The post-cat λ value, however, remained at stoichiometric. This result confirms our earlier observation from the post-cat HC signal that all the exhaust gas pumped out in the coasting process were trapped in the catalyst volume.

The total HC (m_{HC}) pumped into the catalyst in the shut down process was obtained by integrating the HC flow measured by the pre-cat FFID:

$$m_{HC} = \int \dot{m}_{exhaust} \; x_{HC} \frac{w_{HC}}{\overline{W}} \, dt \qquad (1)$$

where $\dot{m}_{exhaust}$ is the exhaust flow rate, x_{HC} is the HC mole fraction measured by the FFID, and W_{HC} and $\overline{W}$ are the molecular weight of the HC (as C$_1$) and average molecular weight of the exhaust gas. The integration period is over the range where the engine speed coasted to zero. See Fig. 4.

Because of the complex nature of the multi-cylinder exhaust flow during the coast-down process, it was difficult to make an exact correspondence of the HC signal to the cylinder-by-cylinder exhaust flow. Instead, a simple estimate for an average $\dot{m}_{exhaust}$ was obtained by using the engine volumetric efficiency calibration, the instantaneous speed and instantaneous intake pressure. For a normal running engine and an engine coasting to a stop with non-firing cycles, there are differences between the thermal states of the residual gas and between the exhaust process. That the residual gas in the non-firing cycles has a lower temperature (than the normal firing cycles) will lead to an underestimate of $\dot{m}_{exhaust}$. The presence of the backflow due to the sub-atmospheric cylinder pressure at EVO in a non-firing engine will lead to an overestimate of the average $\dot{m}_{exhaust}$. These two errors roughly cancelled each other out.

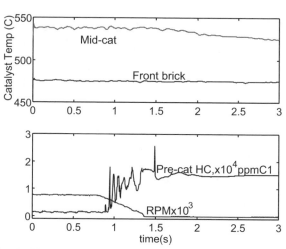

Fig 5 Front-brick and mid-cat temperatures in regular engine shut down. See text for description of measurement points.

For the regular engine shut down, the integrated HC emissions at the catalyst entrance in the period of the engine coasting to stop was 4.1 ± 0.2mg (4 trials). This amount corresponds to approximately the fuel mass of one injection (4.5 mg) at idle.

Catalyst Temperature Behavior in Regular Shut Down

It is of interest to see whether the HC pumped out by the engine in the coast-to-stop process would adversely affect the catalyst temperature. The catalyst temperatures are shown in Fig. 5. Referring to Fig. 2, the "front brick" temperature was measured at 5 cm from the leading edge of the front brick (of length 14.5 cm); the "mid-cat" temperature was measured at the air space between the front and the back brick. There was no observable impact of the coast-down HC emissions on the catalyst temperatures in the shut down process.

After the engine stopped, the front brick temperature remained constant because the measuring thermocouple, being very close to the catalyst brick, was reading the substrate temperature. This temperature changed very little in the time scale of seconds because of the thermal inertia of the materials. (In a long time record of the catalyst temperature, the rate of cooling of the front brick was approximately 10° C per minute.)

There was a gradual drop off in the "mid-brick" temperature after the engine stopped. This drop was due to the lead-wire heat loss from the thermocouple junction which was no longer convectively heated by the exhaust flow. (The measurement was done by an exposed-junction thermocouple.)

That the high exhaust HC flow during coast-down (compared to steady state flow at idle) did not affect the catalyst temperature may be explained by the following. At 10^4 ppmC1 HC concentration, the heating value of the HC is significant – it is about 35% of the sensible exhaust enthalpy (referenced to the catalyst temperature) at idle condition. However, the period of the flow was very brief – if the integrated HC of 4 mg was deposited evenly in the front brick, the corresponding temperature rise would be 0.1° C. Thus no noticeable temperature change was observed.

Contribution of Residual HC From Shut Down to Restarting After Regular Shut Down

To assess the contribution of the residual HC from engine shut down to restarting, the following two experiments were performed.

ENGINE PURGING EXPERIMENT

In the first experiment, the engine was purged by cranking the engine with the injectors disconnected 1 minute after shut down. The objective was to assess for a brief shut down period, how much of the left-over residual HC from shut down would be pumped out by the cranking process, and would the catalyst, which was still warm (front brick at 550° C; mid-cat at 400° C), be able to remove them.

The results are shown in Fig. 6. In this figure, cranking started at 0.45 s. At time zero (before cranking), the pre-cat HC trace started at a small non-zero value which varied from run to run. This value depended on the details of the mixing of the residual HC in the exhaust system in the time between engine shut down and restart.

The residual HC from the stopping process was observed as the engine was purged by cranking. It was, however, found that the HC trace settled to a non-zero value. This offset was believed to be a result of the HC released from the engine oil in the cranking process. A procedure was established to correct for this effect. See Appendix. After the correction, the integrated amount of residual HC (obtained using eq.(1)) which was delivered to the catalyst in the purge process was 33 mg.

The post-cat HC level was substantially lower than the pre-cat level, showing that the catalyst was effective in removing the HC. The efficiency was greater than 80%.

At time zero (before cranking) in Fig. 6, both the pre-cat and post-cat λ values were essentially stoichiometric, since the λ sensors were reading the stagnant burned gas. As the engine cranked and delivered air to the exhaust, the pre-cat λ value went up. The post-cat λ value followed much later because of the large catalyst volume (~5L) between the two sensors.

ENGINE RESTART EXPERIMENT

In the second experiment, the engine was restarted after a cool down period of 90 minutes. The catalyst had come down to non-active temperature (~80° C). The fuel metering was controlled by the OEM engine controller.

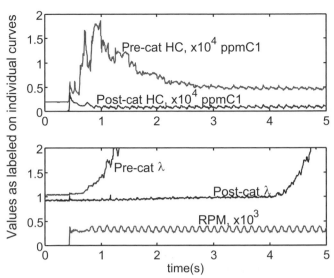

Fig. 6   Purge of engine HC by cranking with injectors disconnected. Procedure performed 1 minute after regular shut down.

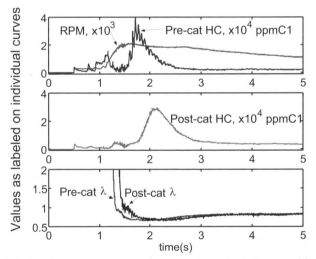

Fig 7   Engine restart after regular shut down with 90 minutes of cool down period. OEM engine calibration used.

The engine emissions behaviors are shown in Fig. 7. The engine was cranked at ~200 rpm and started firing after 0.5 s at λ~1.2. Then the speed accelerated to ~2000 rpm in the next 0.5 s before gradually decreasing to idle speed.

The pre-cat HC trace showed two peaks. The first one (with a magnitude of ~15kppmC1) was attributed to the purging of the residual HC from shut down by the cranking process. This attribution is illustrated in Fig. 8 in which the pre-cat HC traces from the purge and from the restart experiments were aligned and displayed. In the restart experiment, this residual HC pulse was shortened by the acceleration of the engine.

The second peak in the pre-cat HC trace was due to the fueling strategy of the starting procedure. Fuel was

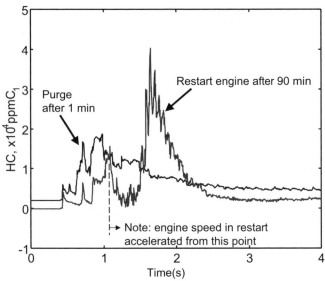

Fig 8 Pre-cat HC traces from engine purging (1 minute after shut down) and from engine restarting (90 minute after shut down). The two traces were aligned to start at the same cranking time.

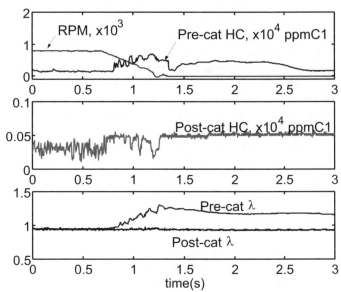

Fig.9 Engine behavior in shut-down with fuel cut-off. Ignition was on until engine coasted to stop. There was no missed cylinders.

simultaneously injected into all four cylinders a short time (0.2 second) after cranking commenced. Some of the cylinders did not fired in the first cycle and the unburned mixture was pumped out by the engine to the catalyst. This peak varied significantly from run to run (from 1.5 to 6 x 10^4 ppmC1) depending on the details of cranking process. The integrated HC from the first peak using Eq.(1) was 16 mg. That of the second peak was 55 mg for the data in Fig.8.

Because the catalyst did not lighted off, the HC trace at the catalyst exit was essentially the same as that at the catalyst entrance.

ENGINE SHUT DOWN BY FUEL CUT-OFF

To eliminate the mis-fired cylinders in the shut down process, the following procedure was used. The power to the ECU and the ignition was left on; the injectors were disconnected so that the engine coasted to stop. (Care was taken to ensure that the disconnection did not happen during injection nor during charging of an ignition coil.) Thus all the cylinders with fuel injection fired before engine came to rest.

HC emissions in shut down by fuel cut-off
The engine behavior by cutting out the injectors is shown in Fig. 9. There was no qualitative difference in behavior compared to the regular shut down (see Fig. 4). The HC emissions during the coast down process was, however, substantially lower. The integrated HC emissions at the catalyst entrance in the period of the engine coasting to stop was 2.2 mg± 0.1mg (7 trials)., which was approximately half of that (4.1 mg) in the regular shut down process.

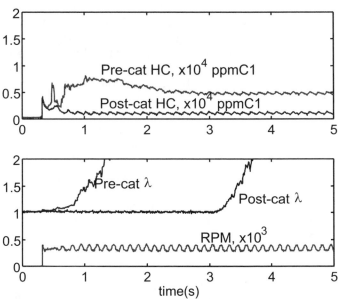

Fig 10 Purge of engine HC by cranking with injectors disconnected. Procedure performed 1 minute after engine shut down by injector cut-off.

Engine purging after shut down by fuel cut-off
The residual HC in the engine in the shut down process was purged by cranking the engine 1 minute after engine shut down. The results, shown in Fig. 10, were qualitatively the same as those for the normal shut down (shown in Fig. 6) except that the pre-cat HC level was substantially lower. Again, because the catalyst had not cooled off, removal of this HC was effective (>80%). Note that the pre-cat HC level settled into the same level as that of the regular shut down case in Fig.6. This observation supports the explanation that this background HC level was due to the HC from the engine oil in the cranking process.

After the contribution of the engine oil was corrected (see Appendix), the integrated HC emissions in the purging process was 14 mg

Engine Restart After Shut Down By Injector Cut-Off
The engine was restarted after being shut down by injector cut-off with a 90 minutes of cool down period. The catalyst had cooled down to inactive temperatures. The behavior is shown in Fig. 11. There was no qualitative difference between this behavior and that for restarting after normal shut down except that the first pre-cat HC peak (due to purging of the residual fuel in the engine by cranking) was lower. The second peak, which was the emissions from the fueling of the new start, had run-to-run variations similar to that of the restart after regular shut down.

The integrated HC from the first peak was 11 mg. That of the second peak was 43 mg.

DISCUSSIONS

A summary of the HC flow to the catalyst in the regular shut down and in the injector cut-off shut down is shown in Table 1.

Because the shut down by injector cut-off had no mis-fired cylinders, the mole fraction and the total HC flow to the catalyst in the shut down process were lower. In the normal shut down, the integrated flow was 4 mg, which corresponded to approximately the fuel amount of one injection at idle (4.5 mg). For the injector cut-off shut down, the value was a factor of two lower (2 mg).

Because of the large catalyst volume (~5L), the HC pumped out by the engine in the coasting down process never reached the exit of the catalyst. Thus the HC emissions during the engine coast down was contained.

Because the heating value of the integrated amount of HC was modest, the catalyst temperatures were not affected by the flow of HC in the engine coast down process.

The total HC out in purging the engine by cranking with injectors disconnected after shut down were 33 and 14 mg respectively for the two cases. This HC consisted of the unburned mixture from the mis-fired cylinders, and that from the purging of the residual fuel in the intake manifold. It is perplexing that the difference of 19 mg is more than the amount of a maximum of three injections (3 x 4.5 = 13.5 mg) for the mis-fired cylinders in the shut down process. The lack of closure may be a result of the uncertainties in estimating the $\dot{m}_{exhaust}$ term in Eq.(1), and the errors associated with the oil layer contribution correction described in the Appendix.

When the catalyst temperature was high, the purged HC was effectively removed by the catalyst. Using the initial catalyst temperature of 550° C (measured), a cooling rate of ~10° C per minute (see Appendix), and assuming

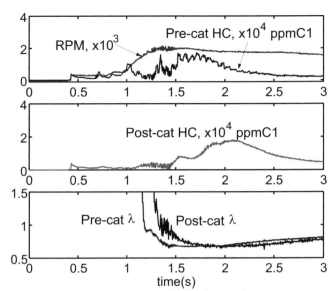

Fig. 11 Engine restart behavior after shut down by injector cut-off. Crank start 90 minute after shut down using OEM engine calibration.

Table 1 Comparison of HC emissions with regular shut down and injector cut-off shut down

	Normal shut down	Shut down by Injector cut-off
HC flow to catalyst	1-2% (C1) (mole fraction)	Below 1% C1
Total flow to catalyst	4mg – correspond to about 1 fuel injection (4.5mg) at idle	2 mg
Catalyst temperature	Not affected	Not affected
Total HC out by purging after shut down	33 mg	14 mg
Among of residual HC pumped out in the restart process	16 mg	11 mg

Difference of 5 mg (~1 injection at idle)

a catalyst deactivation temperature of 350° C, It would take 20 minutes of cool down period before the catalyst is ineffective.

In restarting the engine, two HC peaks were observed in the flow to the catalyst. The first peak corresponded to the residual HC from shut down being pumped out by the cranking process. The second peak was a result of the fuel delivery in the restart cranking process. The residual HC obtained by integrating over the first peak were 16 and 11 mg respectively for the restart after a regular shut down, and after an injector cut-off shut down. If the catalyst was not active, this HC would be the contribution of the shut down residual HC to the tailpipe emissions. The difference of 5 mg (corresponded to about 1 injection at idle) may be attributed to the mis-fired cylinders. To put this number in perspective, the LEVII-SLEV regulation of 0.01g/mi for the 11 mile FTP test would amount to 110 mg of total HC emissions for the test. Thus the 5 mg value is a modest but not insignificant amount[1].

CONCLUSIONS

In an unmanaged engine shut down, depending on the timing of the key-off relative to the engine events, there could be 2 or 3 mis-fired cylinders (i.e. cylinders with fuel injected but no ignition) in a four cylinder SI engine. The unburned HC in the shut off process is contained in the engine and exhaust system in the coast down process. In restarting the engine, the unburned fuel from the mis-fired cylinders contributes to an exhaust HC mass equivalent to approximately one fuel injection at idle.

REFERENCES

1. W.K. Cheng, D. Hamrin, J.B. Heywood, S. Hochgreb, K.D. Min, M. Norris, "An Overview of Hydrocarbon Emissions Mechanisms in Spark-Ignition Engines," SAE Transaction, 102, Paper 932708, 1207-1220, 1993.
2. N.A. Henein, M.K. Tagomori, M.K. Yassine, T.W. Asmus, C.P. Thomas, and P.G. Hartman, "Cycle by-cycle Analysis of HC Emissions During Cold Start of Gasoline Engines," SAE Paper 952402, 1995.
3. B.M. Castaing, J.S. Cowart, W.K. Cheng, "Fuel Metering Effects on Hydrocarbon Emissions and Engine Stability During Cranking and Start-up in a Port Fuel Injected Spark Ignition Engine," SAE Paper 2000-01-2836, 2000.
4. J.W. Fox, W.K. Cheng, J.B. Heywood, "A Model for Predicting Residual Gas Fraction in Spark-Ignition Engines," SAE Transaction, 102, No.3, Paper 931025, 1538-1544, 1993.

ACKNOWLEDGEMENT

The authors would like to thank Mr. Yuetao Zhang of the MIT Sloan Automotive Lab for repeating some of the engine data. This work was supported by an Industrial Consortium on Engine and Fuels Research. The member companies are: DaimlerChrysler, Ford Scientific Research Lab, Ford/Volvo, GM and ExxonMobil.

APPENDIX

Correction of HC contribution due to engine oil in the purging process

When the engine was purged by cranking with injectors disconnected, the exhaust HC mole fraction settled to a background level g_∞ after a long cranking time. It was believed that this background was a result of the release of HC from the engine oil (due to dissolved gasoline) in the cranking process. To eliminate this background, the

mole fraction (g) of this HC was assumed to be delivered to the sensor with the same time constant τ (due to the filling and emptying of the exhaust manifold volume) as the decay of the HC trace after the peak; see Fig. A1. The algorithm was to measure g_∞ from the HC trace in the figure and guessed a value τ' for the time constant. Then for the portion of the HC curve after the peak, a time constant τ was found from the the slope of the semilog plot (least square fit) of the quantity $y(t) = [x_{HC}(t) - g_\infty (1-e^{-t/\tau'})]$ versus time. The process was repeated until $\tau' = \tau$. Then y is the corrected HC mole fraction from purging of the engine without the engine oil contribution.

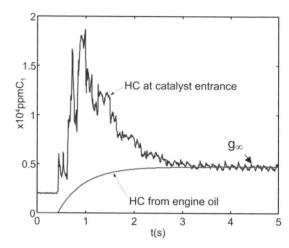

Fig. A1 HC emissions in purging of engine by cranking. Procedure done 90 minutes after regular engine shut-down. Engine cranking started at ~0.45 second on the x axis.

Catalyst cool down characteristics

The measured catalyst and engine coolant temperatures after shut down are shown in Fig. A2. The catalyst cooled at approximately $10°$ C per minute. Note that it took 30 minutes for the catalyst to cool to inactive temperature (below $300°$ C).

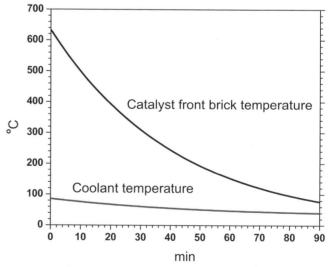

Fig. A2 Catalyst and coolant temperatures after engine shut down.

---

[1] The regulation is on Non-Methane Organic Gas. Methane, however, only comprises 2-6% of the engine-out HC. Therefore the correction to the 5 mg value should be inconsequential.

Mixture Preparation and Hydrocarbon Emissions Behaviors in the First Cycle of SI Engine Cranking

Halim Santoso and Wai K. Cheng
Sloan Automotive Laboratory, Massachusetts Institute of Technology

Copyright © 2002 SAE International

ABSTRACT

The mixture preparation and hydrocarbon (HC) emissions behaviors for a single-cylinder port-fuel-injection SI engine were examined in an engine/dynamometer set up that simulated the first cycle of cranking. The engine was motored continuously at a fixed low speed with the ignition on, and fuel was injected every 8 cycles. Unlike the real engine cranking process, the set up provided a well controlled and repeatable environment to study the cranking process. The parameters were the Engine Coolant Temperature (ECT), speed, and the fuel injection pulse width. The in-cylinder and exhaust HC were measured simultaneously with two Fast-response Flame Ionization Detectors. A large amount of injected fuel (an order of magnitude larger than the normal amount that would produce a stoichiometric mixture in a warm-up engine) was required to form a combustible mixture at low temperatures. That was because fuel delivery efficiency (the fraction of the injected fuel that constituted the combustible charge) decreased significantly with temperature, and that this efficiency also decreased with an increase in the injection amount. A thermodynamic model for mixture preparation based on representing the fuel by its major components, and on an equilibrium between the fuel liquid, vapor and part of the charge air was developed. The model agreed well with the observed dependence of the delivery efficiency on temperature and on the amount of fuel injection. The first cycle HC emissions as a function of the in-cylinder fuel equivalence ratio were bracketed by incomplete combustion on both the lean and the rich side. At stoichiometric condition, the HC emissions increased with lower ECT and with lower speed.

INTRODUCTION

With the development of fast light-off catalyst (typical light-off time < 10 s), the Hydrocarbon (HC) emissions in the start-up process in a Spark Ignition (SI) engine equipped vehicle become a significant part of the total trip emissions. Therefore there is considerable interest to understand and quantify the engine HC emissions behavior in the start-up process so that appropriate strategy can be devised to control it [1,2]. This paper addresses the engine behavior for a port-fuel-injection SI engine in the first cranking cycle.

To ensure a robust and quick start, a large amount of fuel (many times the amount required to prepare a stoichiometric mixture) is usually injected for the first cranking cycle. This strategy is necessary because of the unfavorable mixture preparation environment – both the cold port wall temperature and the atmospheric port pressure are detrimental to fuel evaporation. As such, a substantial quantity of the injected fuel does not contribute to the combustible mixture of the first cycle. Part of this fuel ends up in the engine lubrication system; part remains in the intake port and affects the subsequent mixture preparation process. The objective of this study is to quantify the effectiveness of fuel delivery (i.e. the fraction of the injected fuel that forms the combustible mixture) as a function of temperature, engine cranking speed and the amount of fuel injection.

This work is a continuation of the engine start-up study in Ref. [2] in which a production multi-cylinder engine was crank-started by the starter motor. As such, however, it was difficult to control the engine conditions precisely; for example, the engine speed could not be adjusted independently. Thus while the real engine behavior was captured in that study, systematic variations of the engine variables were not performed. Here, the measurement method was to motor a single cylinder engine on a dynamometer at various speeds, and to supply fuel once every so many cycles to simulate the first cranking cycle behavior. Then the engine conditions were precisely controlled and statistics over many injection cycles could be collected easily.

TYPICAL ENGINE START-UP BEHAVIOR

It is useful to examine the typical engine start-up behavior to assess the magnitude of the HC emissions problem. The engine variables and the HC entering and exiting the catalyst are shown in Fig. 1. The engine was a production four-cylinder, four-valve per cylinder, engine calibrated for 1998 Model Year LEV. The HC mole fractions were measured with Fast-response Flame Ionization Detectors (FFID). The crank position before

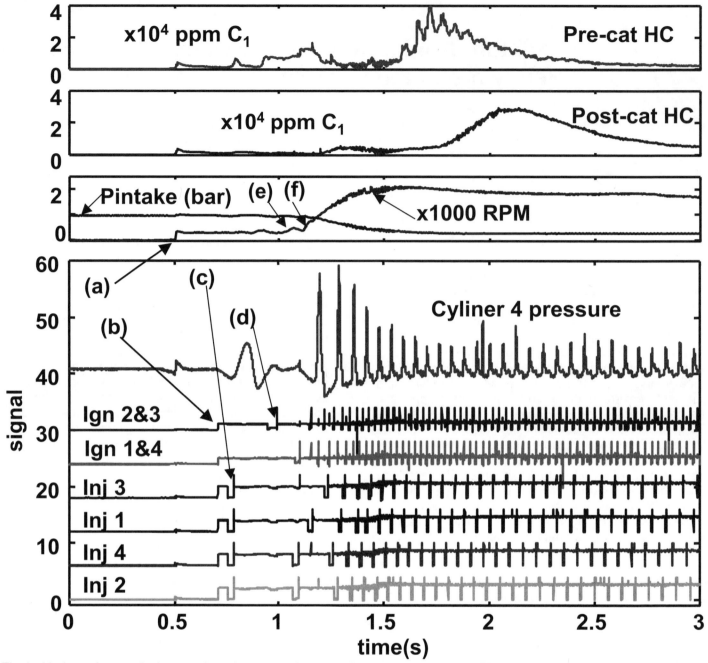

Fig 1. Hydrocarbon emissions and engine events in an engine start-up process. The events marked by labels (a)-(f) are discussed in the text.

starting was at the mid stroke of compression of cylinder 2. (The firing order was 1-3-4-2.)

Crank started at clock time of 0.5 s (label (a) in the figure); the starter motor accelerated the engine speed to ~250 rpm almost instantaneously. At (b) the ECU was powered up (0.2 s after crank start). The first injection was at (c) – fuel was supplied by all four injectors simultaneously. Injection started at crank location of 300° CA (with 0° defined to be the Bottom-Dead-Center (BDC) compression location of Cylinder #1). Thus for Cylinders #1-3, the first injection occurred with valve closed; for Cylinder #4, the first

occurred with valve opened. At clock time of 1.1s, a second injection was applied simultaneously to Cylinders #2 and 4. The subsequent injection events were then in-sync with the engine operation.

At (d), ignition was first applied to Cylinder #2. The torque resulted from combustion accelerated the engine speed (e). This acceleration, however, was not as pronounced as that of the second firing cycle (of Cylinder #1, at (f)). The observation may be explained by the difference in the mixture preparation process of the two cylinders. For Cylinder #2, the first injection pulse ended just before Intake-Valve-Open (IVO). For

Cylinder #1, the injected fuel had the time of a full stroke in the port before it was inducted into the cylinder. Evidently the longer time for evaporation produced a better mixture for combustion. The subsequent firing accelerated the engine to 2000 rpm and the increase in engine speed caused the intake pressure to come down.

It was noted that in the first injection (simultaneously to all the cylinders), substantial amount of fuel (~70 mg per injection) was introduced. This amount was about twice of that required to produce a stoichiometric mixture.

The mole fractions of the HC flows into (Pre-cat HC) and out of (Post-cat HC) the catalyst are shown at the top of Fig. 1. Because the catalyst had not reached light-off temperature, the value for the Post-cat was essentially the same at that for the Pre-cat (with the appropriate flow delay).

The Pre-cat HC values showed two peaks. The first peak was due to the flushing out of the residual HC from the previous engine shut-down – this peak could be observed when the engine was crank-started with the injectors disconnected; see Ref. [3]. The second peak was identified to be the HC emission due to the current crank-start process.

The integrated HC mass values from the two peaks were 16 and 55 mg respectively. (These values were obtained by integrating the product of the HC mass fraction and an estimate of the instantaneous exhaust mass flow rate.) To put these values in perspective, the California Standard for the 2004 and beyond LEVII – SULEV HC emissions is 0.01 g/mi. For the equivalent of 11 miles of in FTP testing, the total allowable exhausted HC mass is 110 mg. Thus the first two HC peaks in cranking with a HC emissions sum of 71 mg (16 + 55) will contribute to 65% of the total allowance. (Note: the regulation is on NMOG. Since methane only comprises 2-6% of the engine-out HC, the correction to the above value of 71 mg should be immaterial.)

Furthermore, since the two peaks occurred within 2 seconds from crank start, it is unlikely that the catalyst will be effective in this short period of time without external pre-heating which is costly. Therefore there is substantial interest in devising appropriate start-up strategy to lower the cranking HC emissions.

APPARATUS AND PROCEDURE

The experiments were performed on a single cylinder engine which was modified from a production 2L, 4-valves per cylinder, SI Engine. The engine specifications are tabulated in Table I.

The port fuel injection was done with the OEM fuel injector with 4 nozzles which created two colliding jets and two jets aiming respectively at the back of the two intake values. California Phase II calibration gasoline from Philips Petroleum was used as fuel.

Table I : Engine specification (Nissan SR 20E)

Bore/Stroke	86 mm /86 mm
Displacement	497 cc per cylinder
Compression Ratio	9.5
IVO/IVC	13° BTC / 55° ABC
EVO/EVC	57° BBC / 3° ATC

The engine was motored on a dynamometer with a pulley speed reduction system to run in the 300 – 900 rpm speed range. The operation was to have the ignition on at all time and to inject fuel once every 8 to 10 cycles. The number of skipped cycles were chosen to make sure that all the port fuel was flushed out of the system (as confirmed by exhaust FFID measurements) while maintaining a reasonable data taking efficiency. (It is recognized that in a real engine starting, there may be residual fuel left over from the previous shut-down process (see previous discussion), but to include such effect in this experiment would confound the accounting of fuel delivery.)

For the measurement of the fraction of the injected fuel that was delivered to the combustible mixture in the cylinder, two series of experiments were done. Since the intake pressure would decrease as the engine speed increased, the intake Manifold Air Pressure (MAP) was set as a function of engine speed in the first set of experiments. The dependence was obtained from the time history of speed and MAP in a normal crank-start process in a 4-cylinder engine of the same displacement per cylinder. This dependence is shown in Fig. 2. The data obtained from this set of experiments will be useful for examining the effect of cranking speed on fuel delivery.

In the second set of experiments, the intake pressure and the speed were varied independently to generate a

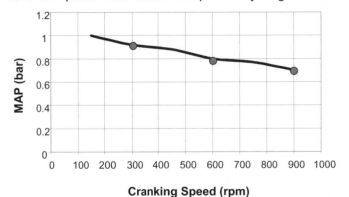

Fig.2 Intake manifold pressure as a function of engine speed in an actual start-up process. In the simulated start, the intake pressure was set at 0.92, 0.8 and 0.7 bar at speeds of 300 600 and 900 rpm

set of data for validation of a mixture preparation model.

For determining the combustion properties on the HC emissions of the first cycle, the spark timing was varied from 25° CA BTC to 10° CA TDC.

FUEL DELIVERY TO THE CHARGE

The fuel mass delivered to the cylinder as the combustible charge was measured by the in-cylinder FFID with the sampling inlet located at the ground electrode of the spark plug. A typical signal trace is shown in Fig. 3. Although there may be fuel-air non-uniformity in the charge, there was a relatively flat in-cylinder HC mole fraction prior to combustion (signified by the sharp drop in the HC signal). The indicated that the charge was reasonably well mix and the HC in the combustible mixture may be represented by the average HC level just prior to combustion.

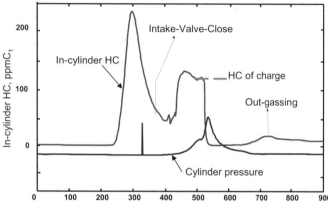

Fig 3 In-cylinder HC measurement and pressure trace. Horizontal scale is crank angle with arbitrary offset.

From the above HC mole fraction measurement, the fuel mass m_f in and the equivalence ratio Φ of the mixture may be calculated. Typically 40 to 180 cycles of injection data were collected. Typical statistics of the Φ values are show in Fig. 4. The frequency distribution was well behaved (single peak with all the data within approximately $\pm 10\%$ of the mean) so that an averaged Φ value is meaningful.

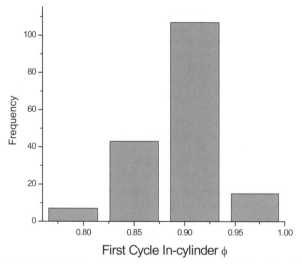

Fig.4 Statistics of the first cycle in-cylinder Φ. Test condition: 300 rpm, 0.92 bar MAP, 38 mg fuel injected. Average $\Phi=0.91$; averaged in-cylinder fuel mass=28mg

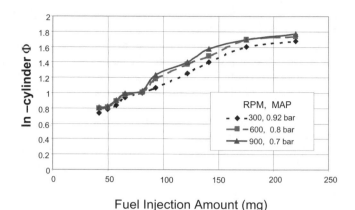

Fig.5 The first crank cycle in-cylinder Φ value as a function of amount of fuel injection and engine speed at ECT of 40° C. The MAP was set as a function of speed according to Fig. 2.

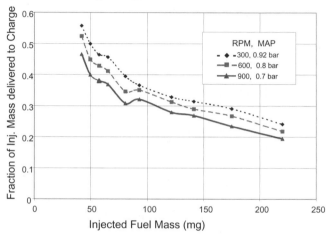

Fig. 6 The first crank cycle fuel delivery efficiency as a function of amount of fuel injection and engine speed at ECT of 40° C. The MAP was set as a function of speed according to Fig. 2.

Fuel Delivery to the First Cycle

The Φ values of the first crank cycle as a function of the amount of fuel injected at 40° C ECT for different cranking speeds are shown in Fig. 5. For this data, the MAP at the different speeds were set according to Fig. 2. A richer charge resulted with increase of m_{inj}. The sensitivity, however, was modest: a more than four fold increase in m_{inj} (from 50 to 220 mg) only resulted in an increase of Φ from 0.8 to 1.8. The speed change of 300 to 900 RPM had only a small effect, with a slight increase of Φ at the higher speed for the same value of m_{inj}.

The corresponding fuel delivery efficiency (ε_f) which is defined as the ratio of the fuel mass in the combustible mixture to the injected fuel mass ($\varepsilon_f = m_f / m_{inj}$) is shown in Fig. 6. The dependence on m_{inj} shows a diminishing return – the more was the fuel injected, the less was the fraction that went into the combustible mixture. At the same value of m_{inj}, there was a small but definite increase in the delivery efficiency with lower speed. Note that the speed dependence was the reverse of that shown in Fig. 5. This was because of the MAP setting of the experiments. At a higher speed, the higher value of Φ in Fig. 5 (at the same m_{inj}) was offset by the lower charge mass at the lower MAP; the net result was a lower m_f and therefore lower ε_f .

The first cycle in-cylinder Φ results for the test series with the MAP adjusted according to the engine speeds are summarized in Fig.7. In addition, the results at 0° C and 20° C for a 4-cylinder engine (Ford Ztec) crank-started by the starter motor [4] are also included.

The general results show a similar dependence of Φ on m_{inj} as which was discussed in connection with the 40° C ECT case (Fig. 5). The Φ values increase with increase of m_{inj}. At the same m_{inj}, there is a slight increase in Φ with a higher engine speed. The most sensitive parameter, however, is the ECT.

A significant feature of this graph is that the slope of Φ versus m_{inj} decreases significantly with decrease in ECT; the slopes are especially small at temperatures below ~20° C.

Also drawn in Fig. 7 is the lean threshold (approximately at Φ =0.7-0.9) for robust firing in the first cranking cycle. It has been shown [4] that this threshold in terms of the in-cylinder Φ is independent of the ECT. Because of the insensitivity of Φ to m_{inj}, it takes a tremendous amount of fuel to achieve robust firing at low temperatures. For example, at 0°C, 300 mg is needed. This amount is almost an order of magnitude larger than what would be required to prepare a stoichiometric charge if all the fuel had gone to the combustible mixture.

The corresponding first cycle fuel delivery efficiency values (ε_f) are shown in Fig. 8. The general trend is that the efficiency decreases with temperature and there is a diminishing return: the larger is the minj, the smaller is the ε_f. At 0° C, the delivery efficiency is below 10%.

The combined effect of the significantly lowered delivery efficiency with temperature and the diminishing return with increase of fuel mass injected explains why that a large amount of fuel is required for a robust engine start under cold condition. This fuel delivery behavior will be examined by a thermodynamics based mixture preparation model later.

Fuel Delivery to the Second Cycle

Since a substantial amount of the first cycle injected fuel does not contribute to the combustible mixture of the first cycle, it is of interest to see how much does this residual fuel contribute to the mixture of the second cycle. The results are shown in Fig. 9. The data were obtained from the in-cylinder FFID measurement of the second cycle. It should be noted that there was no fuel injection in this cycle; therefore the intake port environment was

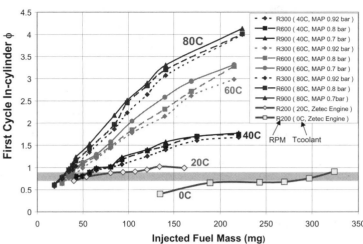

Fig. 7 First cycle in-cylinder Φ as a function of injected fuel mass, ECT and engine speed. The MAP values were set as a function of engine speed according to Fig. 2. The horizontal band at Φ=0.7-0.9 marks the lean limit of robust firing

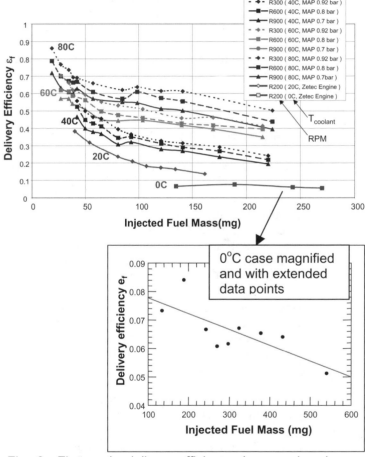

Fig. 8 First cycle delivery efficiency ($\varepsilon_f = m_f / m_{inj}$) as a function of fuel mass (m_f), ECT and engine seed. The MAP values were set as a function of engine speed according to Fig. 2.

different from that of an actual engine in which fuel would be injected.

The results show that the residual fuel could contribute to a significant Φ value in the second cycle. For the cases with high fuel injection in the first cycle, the second cycle mixture was combustible even without further fuel injection. Positive IMEP values were indeed observed in these cases since the ignition was on.

ENGINE OUT HYDROCARBON

The Engine-Out HC (EOHC) values as a function of m_{inj} at different engine speeds and at a fixed ECT of 40° C are shown in Fig. 10. These values were the average over 40 to 150 cycles. At low injection mass, there was significant misfired and partial burned cycles, and the EOHC values rose sharply. At high injection mass, the EOHC values increased because the overly rich mixture reduced the post-combustion HC oxidation, and there may be partial burn cycles .

The EOHC at different ECT, but at the same speed of 300 RPM are shown in Fig. 11. The behaviors at the low and high injection mass are similar to that of Fig. 10.

The data presented in Fig.10 and 11 are confounded by effect of the mixture preparation process on the different cases. To minimize this effect, all the EOHC data were re-plotted versus the corresponding in-cylinder Φ values. (It should be noted that the EOHC mechanism due to the residual in-cylinder unburned liquid fuel is still there.) The results are shown in Fig. 12, which were obtained with a fixed spark timing (at TDC).

The general behaviors of the EOHC in Fig. 12 are:

(a) On the lean side (Φ < 0.75), there is a sharp increase due to presence of misfired and partial burned cycles.
(b) On the rich side (Φ > 1.5), the rapid increase of EOHC to >10 kppmC1 which is a significant level compared to that of the unburned mixture indicates the presence of partial burned cycles. (For reference, the HC mole fraction of the stoichiometric mixture was 120 kppmC1.)
(c) In the range of 1 < Φ < 1.5, there is a gradual increase of EOHC due to the reduced post-combustion HC oxidation under the rich condition.
(d) In the 'normal behavior' range (0.8< Φ <1.5), the EOHC increases rapidly with decrease in ECT. For example, at 300 rpm and Φ = 1, the EOHC were 3, 4 and 8 kppmC1 at ECT of 80, 60 and 40° C. This trend is the result of several effects. (i) With a decrease in ECT, the crevice volume is larger (volume increases by ~4% with every 10° C decrease in ECT); (ii) the post-combustion oxidation processes in both the cylinder wall boundary layer and the port are slower; (iii) there is more unburned liquid fuel in the cylinder (since more fuel has to be injected to achieve the same in-cylinder Φ) and this residual fuel may exit the cylinder as EOHC.
(e) Again in the 'normal behavior' range (0.8< Φ <1.5), the EOHC increases with decrease in engine speed. The major factor here is the temperature in the post-combustion HC oxidation. Several factors influence this behavior. (i) At a lower speed, the engine MAP

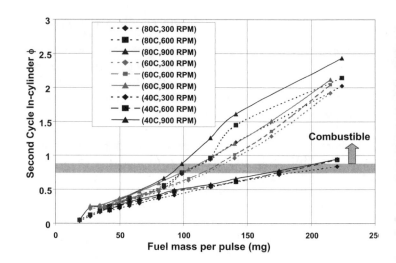

Fig. 9 Second cycle in-cylinder Φ values. There was no fuel injection after the first cycle. Combustion was observed for the cases with Φ above the limit as shown.

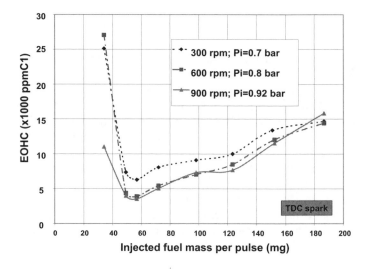

Fig. 10 Engine-out HC as a function of injected fuel mass at 40° C ECT.

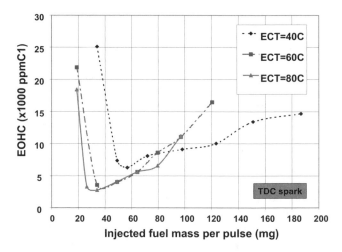

Fig. 11 Engine-out HC as a function of injected fuel mass at 300 rpm (0.92 bar MAP); various ECT.

is higher (see Fig. 2). The burned charge temperature should then be higher since heat lost as a fraction of the charge energy is lower. (ii) The heat loss per crank angle of the charge in the expansion process is higher at lower speed[1], leading to a lower temperature. (iii) At a fixed spark timing, combustion is faster (relative to piston motion) at the lower engine speeds. Therefore the charge temperature is lower in the later part of the expansion. Factor (iii) is the most prominent one. The overall effect is a lowering of burned charge temperature with decrease of speed; hence the post-combustion HC oxidation is reduced.

Combustion Phasing Effect on EOHC

To examine the combustion phasing issues at different engine speed, cycles were selected from the data with the same in-cylinder equivalence ratios so that the mixture preparation effects can be decoupled. The burned mass fraction (defined as the indicated energy release normalized by the fuel energy of the charge as measured by the in-cylinder FFID) was obtained from a heat release analysis of the pressure trace. By using a mass fraction, the effects of the different MAP at the different engine speeds were normalized.

The burned mass fraction per unit CA and the burned mass fraction history are shown in Fig. 13 for a stoichiometric charge at 300, 600 and 900 rpm. Spark timing was fixed at 10° BTC. On a crank angle basis, combustion at 300 RPM is faster than that at 900 RPM by approximately 5 degree-CA (based on either the shift of the location of the peak pressure or that of the 50% mass fraction burned; the two values were about the same.)

The faster combustion (with respect to piston motion) at the lower speed implies that the burned gas has a higher effective expansion ratio and thus a lower temperature at the later part of the expansion process. This lower temperature reduces both the level of HC oxidation in-cylinder and in the exhaust port and leads to a higher EOHC.

This effect is illustrated in Fig. 14. The data in this figure were obtained at ECT of 60° C and at 42 mg of fuel injection. Under this condition, the mixture Φ was at ~0.9, and was not sensitive to engine speed. The EOHC increases both with the decrease in speed and with spark retard. Because of the non-linear dependence of the HC oxidation to temperature, it is difficult to establish a direct connection between the data in Fig. 14 and 13. The agreements in the directional dependence of the EOHC on speed and spark timing, however, give support to the above explanation.

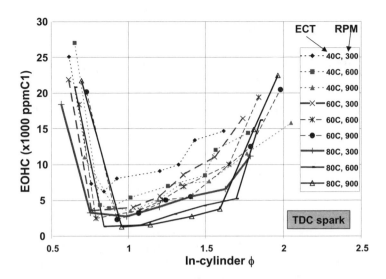

Fig. 12 First cycle EOHC as a function of the in-cylinder Φ; various ECT and engine speeds.

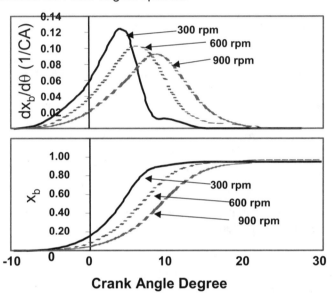

Fig. 13 Mass fraction burn rate ($dx_b/d\theta$) and burned mass fraction (x_b) as a function of crank angle at various engine speeds. The in-cylinder Φ values were all stoichiometric. Spark was at 10° BTC.

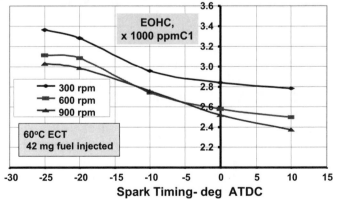

Fig. 14 Dependence of EOHC on engine speed and spark timing. The charge Φ was at ~0.9, independent of engine speed.

---

[1] Heat loss per unit time $\propto$ speed$^{1/2}$; time per CA $\propto$ speed^{-1}; overall scaling : heat loss per CA $\propto$ speed$^{-1/2}$

MODEL OF THE FIRST CYCLE FUEL DELIVERY

A thermodynamics model was developed to explain the observed behavior of the first cycle fuel delivery efficiency ε_f in Fig. 8. In particular, we would like to reproduce the large sensitivity of ε_f to the engine coolant temperature, and the decrease of this efficiency with an increase in injected fuel mass.

Because it is a thermodynamics model, geometric factors of the mixture preparation process such as the fuel spray pattern and the port geometry are not explicitly represented. Instead, these factors are embedded in a few calibrated parameters (two) that will be engine specific. The utility is that for specific engines (or typical class of engines), the model can be exercised to obtain the dependence of ε_f on fuel properties and engine operating conditions (speed, MAP and temperature).

Fuel Model

The sensitivity of ε_f to temperature suggests that the evaporative properties of the fuel must be important. The Major Component Model [5] of the fuel is used. The more than 100 species of fuel components are divided into several groups according to the number of carbon atoms in the molecules and to the type of organic compounds (paraffins, olefins, aromatics). Then all the compounds in each group is represented by the most abundant species of that group which takes on the weight percentage of the whole group. It has been demonstrated that the distillation curve of the representation matches well with the actual distillation curve.

The representation of the California Phase II gasoline is shown in Table II.

Table II
Major Component Model of the California Phase II gasoline

Species	Mol. Frac.
n-Butane	0.020
Iso-Pentane	0.166
Iso-Hexane	0.106
23 Di-methyl Pentane	0.119
224 Tri-Methyl Pentane	0.161
22 Di-Methyl Heptane	0.026
n-Decane	0.007
n-Undecane	0.004
n-Dodecane	0.003
Benzene	0.012
Toluene	0.074
m-Xylene	0.101
Iso-Propyl Benzne	0.047
Iso-Butyl Benzene	0.024
MTBE	0.130

Partial Air Equilibrium Model

The preparation of the first cycle combustible mixture from the injected fuel is a complex process that involves the evaporation of the fuel and the mixing of the vapor with the air. To simplify this complex non-equilibrium process to a thermodynamic model, the conceptual picture of Fig. 15 is used.

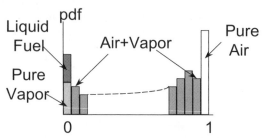

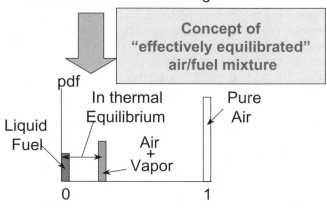

Fig. 15 concept of the Partial Air Equilibrium Model for modeling the fuel delivery of the injected fuel to the combustible mixture of the charge.

The conceptual procedure is to first divide the fluid mass in the intake process into mass elements. Then these elements are sorted according to the mass fraction of air in them. The probability distribution of these elements as a function of the air mass fraction (y_a) is shown in the upper figure of Fig. 15. Thus mass elements with $y_a = 1$ has only air and no fuel in it; with $y_a = 0$ has only fuel (liquid and vapor) in it and no air. Elements with values of y_a in between are fuel/air mixture elements.

In the second step of the conceptual procedure, the elements with non-zero fuel vapor fraction and the vapor mass in the elements with fuel only (i.e. $y_a = 0$) are lumped together to one group. This group is then assumed to be in thermodynamics equilibrium with the liquid fuel in the charge. The situation is illustrated in the lower figure of Fig. 15. Thus the pure air elements are segregated; the elements with fuel/air mixtures and the liquid fuel are effectively equilibrated. This model is termed the Partial Air Equilibrium Model because only part of the charge air is included in the thermodynamic equilibration with the fuel.

24

So far we have not specified the physical locations of the mass elements. It is now assumed that all the vapor obtained in this procedure goes into the combustible mixture. The liquid fuel could be partly inside the cylinder and partly in the port. Further evaporation of the residual liquid will contribute to the mixture preparation of the second cycle or the exhaust emissions of the first cycle. Thus the fuel delivery efficiency ε_f is the total fuel vapor mass divided by m_{inj}.

The air mass m_a which effectively equilibrated with the fuel is a model parameter. This mass is physically the amount of air in the boundary layer between the liquid fuel and the air. As such, the engine speed dependence may be modeled by an inverse square root term (see final model later).

The model algorithm is to perform, for the mass of injected liquid fuel and specified value of m_a, a flash boiling calculation (using the code provided by Ref. [6]) of the vapor mass at the given ECT and MAP. Then ε_f is the total fuel vapor mass divided by m_{inj}.

In the process of the model development, it was found that the model over-predicted the value of ε_f at high temperatures. It was decided that an isothermal flash boiling at ECT would not represent the actual process well since there would be substantial evaporation directly from the fuel drops without wall contact. As such, the evaporative cooling will lower the droplet temperature and reduce the amount of evaporation. To remedy this effect, the flash boiling process was changed from isothermal to a process with limited heat transfer. A new parameter, f, defined as the fraction of the heat of vaporization supplied externally to evaporate the fuel, was introduced. For example, f=0.8 means that 80% of the heat of vaporization was supplied externally (through the contact of the liquid fuel to the walls). Then the flash boiling process was carried out with the initial temperature of the fuel equal to the ECT; the final temperature was adjusted according to energy balance.

The value of f should be approaching one at low temperatures (with reference to the fuel distillation curve) since then almost all of the fuel lands on the port walls before evaporation, and approaching one at high temperatures. A complimentary error function model based on the distillation temperatures is used. See final model.

Final Model Details

The inputs to the model for calculating the fuel delivery efficiency ε_f are m_{inj}, ECT, MAP, engine speed, and fuel composition as represented by the Major Component Model (see Table II for the California Phase II Calibration Gasoline).

The two model parameters are m_a, the air mass which effectively equilibrated with the fuel, and f, the fraction of the heat of vaporization which is supplied externally. Note that these parameters are empirical and

specific to a particular engine geometry and injection details (although parameters for engines with similar geometry and injector behaviors should be similar). Models for these parameters for the particular engine (see Table I) used in this study are as follows.

The air mass m_a is given by:

$$m_a = 22\sqrt{\frac{1000}{RPM}}\frac{MAP^{0.5}}{(T/300)^{0.25}} \qquad (1)$$

where ma is in mg, MAP is in bar, and T is the ECT in $^\circ K$. The logic behind Eq.(1) is that since ma is the air mass in the boundary layer between the liquid fuel and the air, it should scale as

$$m_a \propto \rho_a\sqrt{Dt}\, Sh \qquad (3)$$

Since the air density ρ_a scales as P/T, the mass diffusivity D scales as $T^{3/2}/P$, the diffusion time t scales as the inverse of the engine speed, and assuming that the Sherwood number Sh to be approximately constant, Eq.(1) follows.

It is noted that in Eq.(1), m_a is not dependent on m_{inj}. The logic here is that for normal injectors with droplets $>\sim 100\ \mu m$, almost all the fuel is deposited on the port wall. The area of mass exchange with air is essentially the foot print of the injection. Therefore, for a fixed port geometry and spray pattern, m_a should be independent of m_{inj} to the first order of approximation.

The fraction f of heat of vaporization that is supplied externally is given by:

$$f = 0.5\, erfc\left(\frac{ECT - T_{50}}{0.5(T_{90} - T_{10})}\right) \qquad (2)$$

where ECT is the Engine Coolant Temperature in C, and T_{10}, T_{50} and T_{90} are the 10, 50 and 90% distillation temperatures respectively in C.

Comparison with Experimental Data

In the experiments simulating the first cranking cycle, the MAP was set as a function of the engine speed according to the typical MAP/speed dependence in a real cranking process. To extensively test the fuel delivery model, however, a new set of data was obtained in which the speed and the MAP were varied independently.

A specific purpose of varying the speed and MAP independently is to clarified the dependence of ε_f on these variables respectively. In particular, thermodynamics indicates that fuel evaporation, and therefore ε_f should go up with decrease of MAP. The results of the first series of experiment, however had MAP and speed coupled, with the apparent effect of a decrease in ε_f with the decrease of MAP associated with the speed increase.

The delivery efficiency ε_f as a function of MAP for constant engine speeds is shown in Fig. 16. The data show that indeed ε_f increases with decrease of MAP at constant speed, and increases with decrease of speed at constant MAP. With the cranking MAP/speed relation of Fig. 2, the net result for the coupled MAP/speed variation is shown in Fig. 16 by the arrow. In this coupled variation, the effect on ε_f by the change in MAP was over powered by that in speed. The results of this second series of experiments were indeed consistent with the first series.

The calculated delivery efficiency is plotted against all the experimental data in Fig. 17 for this second series of experiment. The ECT was varied from 40 to 80° C, the speed was varied from 300 to 600 RPM, MAP was varied from 0.5 to 1 bar, and m_{inj} was varied from 55 to 163 mg.. The agreement over the wide range of data is excellent.

The model was then applied to the delivery efficiency data of the first series of experiment (see Fig. 8). The results are shown in Fig. 18. The overall agreement is very good.

DISCUSSION

The model may be used to interpret the nature of the mixture preparation process in the first cranking cycle. That a large amount of injected fuel is required to prepare a combustible mixture at low temperatures (see Fig. 7 and 8) is due to (a) the fuel delivery ratio ε_f decreases significantly with decrease in temperature, and (b) the diminishing return of increasing the amount of fuel injected – ε_f decreases with increase of m_{inj}. The Partial Air Equilibrium Model for Mixture Preparation explains quantitatively that the temperature dependence (a) is essentially a fuel effect related to the distillation properties of fuel. Factor (b) is a fuel air mixing effect and it is based on the facilitating of fuel evaporation when air is present in the liquid/vapor/air equilibrated mixture. Evaporation is favored by the increase of air presence. Since to the first order of approximation, the fuel air interface is limited to the injection foot print in the intake region, mixing does not increase substantially with increase in m_{inj}. This mixing limitation leads to the observed diminishing return.

With the above discussion, the control of the first cycle fuel delivery may be decoupled into a fuel part which is governed by the volatility of the fuel components, and a mechanical part which controls the effectiveness of the air mixing process. There are, therefore, opportunity for improvement both in the engineering of the fuel properties and in the design of injection hardware to facilitate the fuel air mixing process. An example of the latter is the injector design to substantially decrease the droplet size and increase the fuel dispersion.

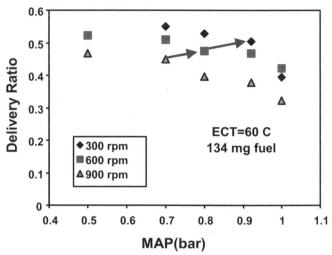

Fig. 16   First cranking cycle fuel delivery ratio ε_f as a function of MAP and engine speed respectively. The arrows indicate the coupled change of the cranking speed/MAP from 900 rpm/0.7 bar to 300 rpm/0.92 bar.

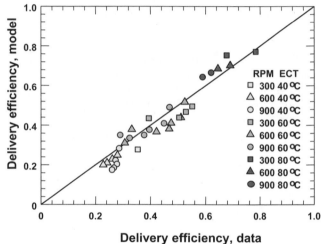

Fig. 17   Comparison of the fuel delivery efficiency calculated by the model and obtained from the experiments.

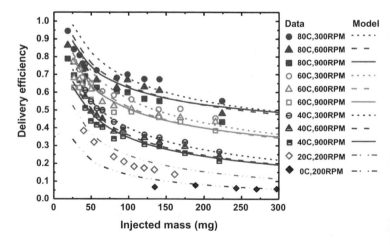

Fig. 18   Comparison of Partial Air Equilibrium Model for Mixture Preparation in the first cranking cycle with data. Data are represented by points, model results by lines.

CONCLUSIONS

The HC emissions from the cranking process could contribute to a substantial part (upwards of 60%) of the total regulation allowance in the FTP cycle. The mixture preparation and HC emissions behaviors for a single-cylinder port-fuel-injection SI engine were examined in an engine/dynamometer set up that simulated the first cycle of cranking. The results show that a large amount of injected fuel (an order of magnitude larger than the amount that would produce a stoichiometric mixture in a warm-up engine) is needed to form a combustible mixture at low temperatures. This behavior is due to a substantial decrease of the fuel delivery efficiency at low temperatures, and to a diminishing return in the efficiency when more fuel is injected. The first cycle Engine-Out Hydrocarbon (EOHC) is primarily a function of the in-cylinder fuel equivalence ratio (Φ). At low Φ (<0.75) and at high Φ (>1.5), the EOHC increases sharply due to misfired and partial burn cycles. A simple thermodynamics model based on the equilibration of the liquid fuel, vapor and a limited amount of air (the Partial Air Equilibrium Model) is developed to explain the mixture preparation behavior. There is good agreement between the model and the data over an extensive range of Engine Coolant Temperature, engine speed and mass of fuel injected.

ACKNOWLEDGEMENT

The authors would like to thank Daniel Klein of the MIT Sloan Automotive Lab for the data used in Fig. 1. This work was supported by the Engine and Fuels Research Consortium; the members were: DaimlerChrysler, Ford Research Lab, Ford/Volvo car, General Motors, and ExxonMobil.

CONTACT

Prof. Wai K. Cheng: 31-165 MIT Cambridge, MA 02139-4307. Email: wkcheng@mit.edu.

REFERENCES

1. N.A. Henein, M.K. Tagomori, M.K. Yassine, T.W. Asmus, C.P. Thomas, and P.G. Hartman, "Cycle by-cycle Analysis of HC Emissions During Cold Start of Gasoline Engines," SAE Paper 952402, 1995.
2. B.M. Castaing, J.S. Cowart, W.K. Cheng, "Fuel Metering Effects on Hydrocarbon Emissions and Engine Stability During Cranking and Start-up in a Port Fuel Injected Spark Ignition Engine," SAE Paper 2000-01-2836, 2000.
3. D. Klein, W.K. Cheng, "Spark Ignition Engine Hydrocarbon Emissions Behaviors in Stopping and Restarting," paper submitted to SAE Fuels and Lubricant Meeting, October, Paper 02FFL-69, 2002.
4. Brigitte Castaing, "Fuel Metering Effects on Hydrocarbon Emissions and Engine Stability During Cranking and Start-up in a Port Fuel Injected Spark Ignition Engine," MS Thesis, Dept. of Mech. Eng., MIT, 2000.
5. K.C. Chen, K. DeWitte, W.K. Cheng, "A Species-Based Multi-Component Volatility Model for Gasoline," SAE Paper 941877, 1994.
6. J.F. Fly, M.L. Huber, "NIST Thermophysical Properties of Hydrocarbon Mixtures Database (Supertrapp)," National Institute of Standards and Technology Standard Reference Data Base 4, Gaithersburg, MD 20899, 1992.

NOMENCLATURE

CA	Crank angle
D	Mass diffusivity
$dx_b/d\theta$	Mass fraction burned per unit CA
ECT	Engine Coolant Temperature
EOHC	Engine-out Hydrocarbon emissions
f	Fraction of the heat of vaporization supplied by external heat transfer in the mixture preparation model
IMEP	Indicated Mean Effective Pressure
MAP	Intake Manifold Air Pressure
m_a	Air mass which effectively equilibrated with the fuel in the Partial Air Equilibrium Model
m_{inj}	Injected fuel mass per cylinder per cycle
m_f	Fuel mass of the in-cylinder combustible mixture
NMOG	Non-Methane Organic Gas
P	pressure
Sh	Sherwood number
t	time
T	Temperature
x_b	Burned mass fraction
y_a	Mass fraction of air in charge mixture mass element
ε_f	Fuel delivery efficiency; $\varepsilon_f = m_f / m_{inj}$
Φ	In-cylinder fuel equivalence ratio Φ
ρ_a	Air density

941872

Liquid Gasoline Behavior in the Engine Cylinder of a SI Engine

Younggy Shin, Wai K. Cheng, and John B. Heywood
Massachusetts Institute of Technology

ABSTRACT

The liquid fuel entry into the cylinder and its subsequent behavior through the combustion cycle were observed by a high speed CCD camera in a transparent engine. The videos were taken with the engine firing under cold conditions in a simulated start-up process, at 1,000 RPM and intake manifold pressure of 0.5 bar. The variables examined were the injector geometry, injector type (normal and air-assisted), injection timing (open- and closed-valve injection), and injected air-to-fuel ratios.

The visualization results show several important and unexpected features of the in-cylinder fuel behavior: 1) strip-atomization of the fuel film by the intake flow; 2) squeezing of fuel film between the intake valve and valve seat at valve closing to form large droplets; 3)deposition of liquid fuel as films distributed on the intake valve and head region. Some of the liquid fuel survives combustion into the next cycle. The time evolution of the in-cylinder liquid film is influenced by the injection geometry, injection timing, injected air-to-fuel ratio, and port surface temperature. Photographs showing the liquid fuel features and an explanation of the observed phenomena are given in the paper.

INTRODUCTION

A plausible mechanism for hydrocarbon (HC) emissions is the entry of liquid fuel into the engine cylinder. This is especially critical in the engine starting process due to poor fuel evaporation in the cold engine. To achieve smooth starting, fuel enrichment is used so that there is sufficient vaporized fuel to form a combustible mixture in the engine cylinder. A substantial amount of the injected fuel, especially the less volatile components, does not vaporize [1,2][1], and enter into the cylinder as liquid. This fuel may be stored somewhere in the cylinder (such as in the deposit, oil layer or crevices in the combustion chamber) and comes out into the bulk gas in the expansion and exhaust processes. The

---

[1] Numbers enclosed in square brackets designate references.

incomplete oxidation of this fuel would contribute to the exhaust HC emissions.

In spite of the significant contribution of the liquid fuel effect to the HC emissions during engine starting and warm-up, liquid fuel behavior in the engine cylinder has been poorly understood. This is because of the difficulty associated with observing the liquid fuel in the cylinder of a practical engine due to the lack of optical access; and that the phenomenon cannot be easily reproduced in a bench scale flow rig because processes such as the back flow of the hot burned gas from the cylinder to the intake port in a firing engine could not be reproduced easily.

The purpose of this paper is to show the important features of the liquid fuel entry into the engine cylinder and the behavior of the liquid fuel deposited on the combustion chamber wall through the engine cycle. The liquid fuel is visualized in a firing transparent engine using a high-speed CCD camera.

EXPERIMENTAL APPARATUS

To visualize the liquid fuel behavior within the cylinder, it is desirable to use a transparent engine with the following features:

1) the windows for visualization in the engine should be easily taken apart for frequent cleaning since unvaporized liquid fuel tends to make the window dirty in a short period of time;

2) the windows should be wide enough to admit adequate lighting and to accommodate various camera view angles;

3) the windows should withstand the thermal loading of engine firing.

In order to satisfy the above requirements, the following experimental apparatus was used.

TRANSPARENT ENGINE - Fig. 1 shows a schematic of the single-cylinder transparent visualization engine [3] that was used in this study.

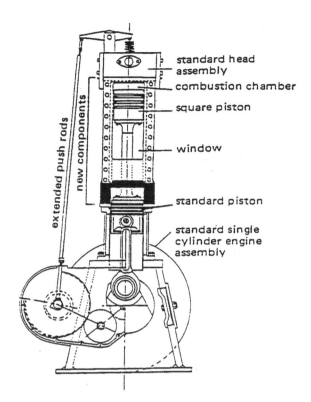

Fig. 1 Schematic of square piston engine

The engine has a square cross-section cylinder assembly with two opposing glass walls which serve as windows for ease of optical access. The glass walls could easily be taken apart from the enigne without disassembling other parts. The square piston is sealed by graphite seal bars that overlap at the corners and these bars are pressed against the cylinder walls by springs from behind. The graphite seals provide a dry means of sealing the combustion chamber so that the windows are not obscured by the lubrication oil. Details of the engine configuration are listed in Table 1.

Table 1. Geometrical data of square piston engine

Displacement Volume	785 cm^3
Stroke	114.3 mm
Side	82.6 mm
Connecting Rod Length	254 mm
Compression Ratio	6.07

Valve Timing:

IVO	10 deg after TC
IVC	25 deg after BC
EVO	45 deg after BC

EVC 10 deg after TC

HIGH SPEED CAMERA - The high Speed CCD camera used in this study was a Kodak Ektapro system. The sensor consists of an 192x239 pixel NMOS array. The recording rates used for the current set of experiments were 500, 1,000 and 2,000 frames/sec. A 90 mm f/2.5 camera lens was used for imaging. The field of view was illuminated by a collimated 650W candescent light source which was carefully placed to avoid direct reflection of the light into the camera. The images were recorded in image files and transferred onto video tapes.

TEST VARIABLES

ENGINE OPERATING CONDITIONS - In the simulated starting experiments, the engine was motored at 1,000 rpm and the throttle position was fixed at an intake manifold pressure of 0.5 - 0.6 bar. Then fuel injection (with fixed pulse width) started. Since the engine was uncooled, the engine was only fired for approximately 2 minutes until the cylinder wall temperature reached 100 °C. After each run, the engine was cooled down before the next run to keep the initial temperatures of the engine and intake system at 20 °C.

AIR-TO-FUEL RATIO - Two air-to-fuel ratios (based on the amount of fuel injected) of 14.6 and 11.0 were selected to represent stoichiometric and enriched conditions. The amount of fuel injected was calibrated by the measured values using a Horiba A/F meter and a laminar flow sensor (for the air flow) taken at the end of the engine warm-up test at which there was complete vaporization of the liquid fuel.

FUEL INJECTION METHOD - Two types of fuel injectors were used to compare the effects of intake port wetting area due to fuel injection geometry on in-cylinder liquid fuel behavior. Fig. 2(a) shows the fuel injection geometry in the intake port equipped with a normal injector (Bosch EV1.1A). The Sauter Mean Diameter for this injector is about 130 μm. It is noted that the intake port geometry in this engine (which was constructed using the head of a CFR engine) is substantially different from that of modern engines. As seen in the figure, the injected fuel spray did not hit the back of the intake valve directly and, as a result, there was substantial fuel wetting over a wide area of the port. Figure 2(b) is a drawn-to-scale drawing showing the fuel injection geometry in the intake port equipped with an air-assisted injector provided by the Ford Motor company [4]. It was designed to minimize wall wetting and to improve atomization by shattering the fuel drops with an air jet supplied by compressed air at about 0.68 bar gauge pressure. Although the SMD (Sauter Mean Diameter) of the fuel spray was reported to be about 40 μm [4], the improvement in atomization was believed to have much less impact in this port configuration since the distance between the injector tip and the impinging surface of the intake valve was rather short so that most of the fuel impinged on the back of the intake valve. Therefore, the main purpose of using the air-assisted injector in this test was to change the injection geometry and to minimize wall wetting area.

FUEL INJECTION TIMING - The fuel Injection timing relative to Intake-Valve-Open (IVO) influences the liquid fuel behavior in terms of the residence time and transport of the injected fuel in the intake port. Two cases were selected for this study: (i) open-valve-injection (injection started at 50 degree after TDC), and (ii) closed-valve-injection (injection started at 205 degrees after TDC). The relative phasing of the injection periods with respect to valve lift is shown in Fig. 3.

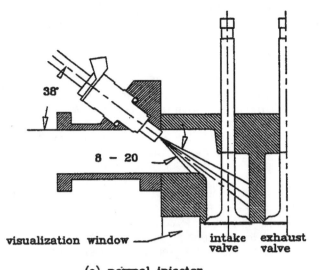

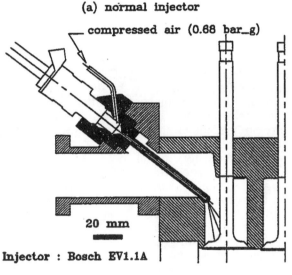

(a) normal injector

compressed air (0.68 bar_g)

20 mm

Injector : Bosch EV1.1A

(b) air-assisted injector

Fig. 2 Fuel injection method; (a) normal injector (b) air-assisted injector (The drawing is to-scale)

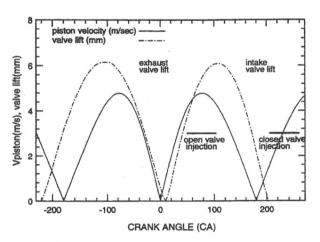

Fig. 3 Fuel injection timing

Also shown in this figure is the piston velocity. The open valve injection period was selected to be at the maximum piston velocity so that the injection took placed at the highest port flow.

VISUALIZATION RESULTS: DESCRIPTION OF LIQUID FUEL BEHAVIOR OVER ONE CYCLE

We will now describe the visualization results obtained in a typical (but not the first) engine cycle during the warm-up process. (The warm-up process and the different behaviors under different test conditions will be discussed later.) The specific case under discussion is for closed-valve injection, at air-fuel ratio of 11.0, and for the normal injector. The results, however, are typical, and were generally observed under all the test conditions. The sequence of events that governs the liquid fuel behavior in-cylinder is illustrated in Fig. 4.

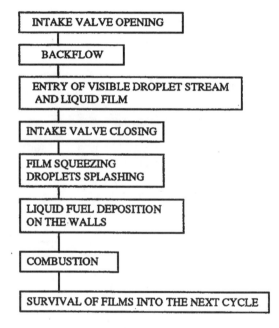

Fig. 4 In-cylinder liquid fuel behavior in time sequence

When the intake valve opens, no fuel droplet stream could be observed initially. This is because the cylinder pressure then is higher than the pressure in the intake manifold (see Fig. 5), and there is a back flow of the charge into the intake port [5]. The back flow stops when the cylinder pressure equilibrates with that of the intake manifold; then visible droplet stream begins to enter the cylinder as the intake forward flow commences. (There is a delay between the time of pressure equilibration which is at 30 deg. ATC in Fig. 5, and the first appearance of the fuel drops at 60 deg. ATC. The delay will be discussed in the next section.) Flow of liquid into the cylinder is also observed. The liquid manifests as substantial puddles with wavy surfaces residing around the intake valve seat and on the valve face.

Figure 6 is a sequence of video frames showing the entry of the droplet mist into the cylinder. The mist is observed as a fuzzy obscuration of the image. Note also that around tile Periphery of the valve, there is a substantial liquid film that survives combustion in the previous cycle.

When the intake valve closes, the puddles at the valve seat are squeezed between the valve lip and valve seat. The squeezed film is broken up into fairly large droplets (estimated to be ~1mm diameter from the video) which splashes into the cylinder. When these large drops land on the head and liner walls, they form significant isolated puddles. These wall puddles stay in place due to surface tension.

To clarify the features of liquid fuel behavior within the cylinder, the following topics will be described in detail:

1)strip-atomization of liquid fuel by the intake flow;

2)fuel film formation on the valve surface and seat area;

3)liquid film squeezing by the closing valve; and

4)the in-cylinder liquid film distribution

STRIP-ATOMIZATION OF LIQUID FUEL BY THE INTAKE FLOW - The observed dominant atomization process for the liquid fuel is the stripping off of droplets from the fuel film on the port, intake valve stem, and back-of-intake-valve

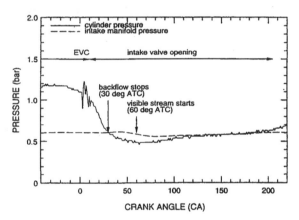

Fig - 5 Cylinder and intake manifold pressures during intake process

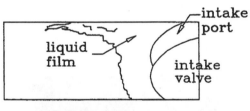

(a) Reference sketch

(b) 57 deg ATC (intake stroke)

(c) 60 deg ATC (intake stroke)

(d) 63 deg ATC (intake stroke)

(e) 66 deg ATC (intake stroke)

(f) 69 deg ATC (intake stroke)

Fig. 6 Sequence of pictures showing visible droplet stream in the intake process right after start-up

surfaces by the shear stress of the intake flow. Because in the intake-geometry used, most of the fuel spray from the injector landed on the port surface, the above process was dominant both for closed-valve and open-valve injections. The process is illustrated in Fig.7. When the air flow velocity is high (when the valve is not fully lifted and when the piston is at ~mid-stroke), a fine and dense mist of droplets is formed (Fig. 7a). Since the droplet size is inversely proportional to the Weber number of the flow, the fine mist wanes into a stream of large and sparsely distributed droplets at the later part of the intake process when the intake air velocity is small, i.e. when the valve is fully lift and the piston is close to the end stroke (Fig. 7b).

The delay (see Fig. 5) between the stop of the back flow and the observed start of the droplet stream may be attributed

is a sufficiently high intake air velocity, which is dependent on the valve lift and the piston velocity, see Fig. 3. (Except for the first few cycles, the delay is not due to the fuel film transport at the port because a liquid film from the previous cycles is always present).

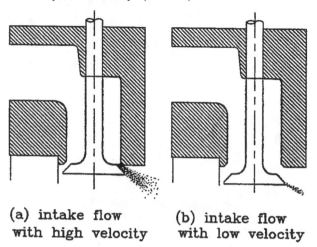

(a) intake flow with high velocity (b) intake flow with low velocity

Fig. 7 Strip-atomization by the intake flow

FUEL FILM FORMATION ON THE VALVE SURFACE AND SEAT AREA - The fuel deposited in the valve stem, back of the intake valve and the port surfaces is pushed towards the cylinder as a film-flow by the shear stress of the intake flow and gravity. This film flow stops, however, at the valve exit because of the diverging flow field as the intake air enters the cylinder (see Fig. 8). The recirculation region induced by the air flow at the valve face and at the head area next to the valve seat traps the film at the lip of the valve face and at the valve seat and its vicinity. The film thickness of the film would be determined by a balance between the surface tension, the shear force produced by the air flow and gravity. The accumulation of liquid film is especially significant towards the end of the intake process because of the reduced amount of shear-atomization, and because in most practical cases, the intake valve closes at part way into the compression stroke so that there is a reverse displacement flow then.

When the wall temperature is cold, a good fraction of the liquid in many of the in-cylinder fuel films, especially

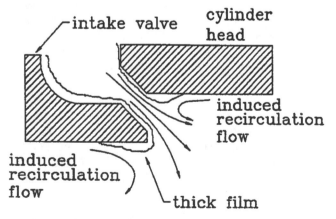

Fig. 8 Formation of thick liquid film at the vicinity of the valve lip and seat

those on the valve surface and around the valve seat on the head, survive combustion into the next cycle in the early cycles of the warm-up process.

LIQUID FILM SQUEEZED BY THE CLOSING VALVE - As discussed in the above, there is substantial accumulation of liquid films on the valve lip and the valve seat at the end of the intake process. When the valve closes, these fuel films are squeezed by the impact of the lip on the seat. The liquid is squeezed into large droplets which splash into the cylinder (see Fig. 9).

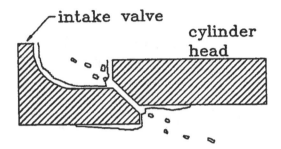

Fig. 9 Squeezing of liquid film when valve is closed. Large droplets are splashed into the cylinder

The squeezing event is shown in the sequence of video images in Fig. 10. The splashing of the fuel drops are clearly visible. The drop size estimated from the images is of the order of 1 min. These big drops land on the head and the cylinder wall to form substantial puddles.

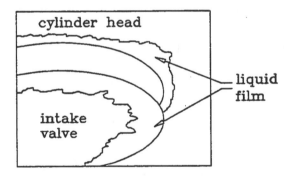

(a) Reference sketch

(b) 40 deg BBC (compression)

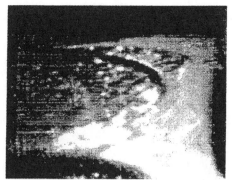

(c) 25 deg ABC (compression); right after film squee

(d) 30.4 deg ABC (compression)

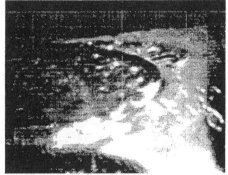

(e) 41.2 deg ABC (compression)

Fig. 10 Sequence of pictures showing film squeezing phenomenon

IN-CYLINDER LIQUID FILM DISTRIBUTION - Based on the flow visualization results, the liquid film distribution may be divided into three types, according to the process by which they are formed, see Fig. 11.

(1) A thick film at the valve surface and around the valve seat - This film is formed by the liquid film flow from the back of the valve and the port surfaces.

(2) A thin film on the combustion chamber surfaces formed by the impingement of the droplet stream from strip-atomization.

(3) Isolated puddles formed by the landing of the splashed drops from the intake valve closing process.

The latter two types of liquid film disappear quickly in the combustion process (within one cycle) even at the very early stage of the warm-up process. The thick film around the valve area, however, survives combustion into the next cycle well into the warm-up period (of the order of 1 minute after firing).

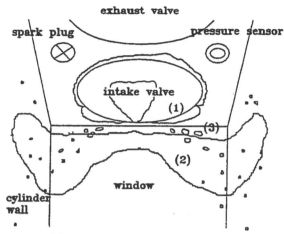

(1) film around intake valve
(2) film from droplet stream
(3) film from splashing droplets

Fig. 11 Typical in-cylinder film distribution

ENGINE WARM-UP BEHAVIOR

We will now discuss the engine warm-up behavior. The warm-up experiments were done using indolene as fuel at speed fixed at 1,000 rpm and at intake pressure of ~0.5 bar. The fuel injection was fixed at stoichiometric (A/F=14.6) and at fuel rich condition (A/F=11.0). Three injection configurations were tested for each A/F ratio: (A) using the normal injector in closed-valve injection; (B) using the air-assisted injector in closed-valve injection; and (C) using the air-assisted injector in open-valve injection.

DIFFERENCES DUE TO THE INJECTOR CONFIGURATION - Figure 12 shows the first cycle for the appearance of the liquid fuel and the first firing cycle for the three test configurations under stoichiometric injection condition. For each configuration, the three data points represent observations from repeating the experiment. The cycle at which liquid fuel appears is very repeatable. There is, however a significant spread in the first firing cycle with the same configuration.

With open-valve injection (C), there is appearance of liquid fuel in the first cycle of injection. With closed-valve injection, that happens at the second cycle for the air-assisted injector, and at the fifth cycle for the normal injector. This difference is attributed to the injection geometry rather than the atomization characteristics of the injector: for the air-assisted injector (Fig. 2b), most of the injected fuel lands on the back of the intake valve, which is a short distance from the entrance of the cylinder. For the normal injector (Fig. 2a), most of the fuel lands on the valve stem and on the portion of the port wall quite far away from the cylinder

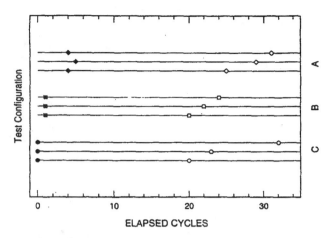

A: Normal injector / Closed valve injection
B: Air-assisted injector / Closed valve injection
C: Air-assisted injector / Open valve injection
solid symbols : First appearance of liquid fuel in the cylinder
open symbols : First firing

Fig. 12 Engine starting behavior at injected A/F=14.6

entrance. The delay in appearance of liquid entry (both in terms of liquid film flow and of strip-atomized droplets) into the cylinder of the latter case is due to the transport time of the fuel film from the fuel spray impingement points to the cylinder entrance.

The first firing cycle happens a little bit sooner for the air-assisted injector than the normal injector. There is no significant difference between open valve and closed valve injection using the air-assisted injector, although there is a larger variance in the first firing cycle with the former.

The in-cylinder liquid film behavior during warm up for closed-valve injection is shown in Fig. 13. The significant in-cylinder liquid films are those around the intake valve. The sketches in Fig. 13 were based on the video images at 30, 60 and 90 seconds since injection started Compared to the

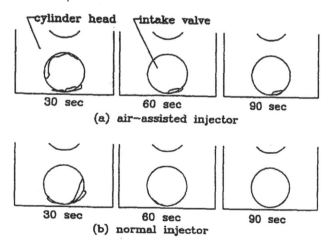

Fig. 13 Fuel film behavior with respect to fuel injector configuration during warm-up (A/F=14.6, closed-valve injection)

normal injector results, there is substantially more liquid in the cylinder and the liquid film persists longer with the air-assisted injector. This difference, again, may be attributed to the injection geometry. In the case of the normal injector (Fig. 2a), the foot Print of the fuel spray in the port has a large area which facilitates vaporization in the port (The facilitation is due to two (related) factors. When the fuel is spread over a larger area, the energy required per unit area to evaporate the fuel is less. The surfaces are less cooled by the fuel so that the temperature is higher which facilitates heat transfer and the evaporation of the heavier fuel components.) As a result, the film flow is reduced. For the air-assisted injector, because of the proximity of the injector tip to the back of the valve (Fig.2b), most of the fuel land on a small area, and there is a substantial film flow into the cylinder.

The comparison between open-valve and closed-valve injection (with the air-assisted injector) is shown in Fig.14. For the open-valve injection, the liquid film on the valve is distributed more on the side of the location of the fuel spray (the bottom of Fig. 14). For the closed-valve injection case, the film is distributed more uniformly around the valve. This observation could be attributed to the redistribution of the liquid on the back of the valve during the reverse flow process when the intake valve first opens.

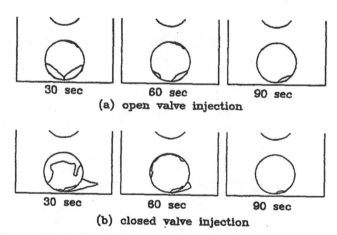

Fig. 14 Fuel film behavior with respect to fuel injection timing during warm-up (A/F=14.6, air-assisted injector)

FUEL ENRICHMENT EFFECTS - The results for fuel injection at A/F of 11.0 are shown in Fig. 15. For the normal injector with closed-valve injection, the first cycle for the appearance of liquid fuel happens sooner compared to the stoichiometric case (Fig. 12). This observation may be attributed to a larger amount of liquid impinged on the wall (thicker film) so that there is a faster film flow (both the shear stress induced flow and the gravitational flow increase with the film thickness).

The first firing cycle occurs much sooner than the stoichiometric case, and is more repeatable. This is due to the larger amount of fuel vapor present with enrichment. As in the stoichiometric case, the first firing cycle occurs slightly sooner in the air-assisted injector than in the normal injector.

The fuel film behaviors for all the configurations under stoichiometric and fuel rich conditions are shown in Fig. 16. For all the configurations, the pattern of the fuel film is

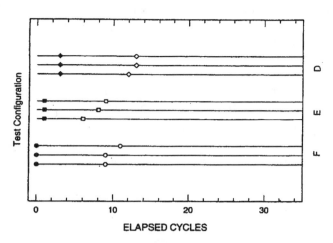

D: Normal injector / Closed valve injection
E: Air-assisted injector / Closed valve injection
F: Air-assisted injector / Open valve injection
solid symbols : First appearance of liquid fuel in the cylinder
open symbols : First firing

Fig. 15 Engine starting behavior at injected A/F=11.0

similar for both the stoichiometric and the fuel rich condition; except that the quantity is larger in the latter.

MEASUREMENT OF TEMPERATURES IN THE INTAKE SYSTEM

Because the fuel evaporation process is highly temperature dependent, the temperatures of selected components of the intake system were measured in the warm-up process. These measurements will also

elucidate the differences between the fuel injection configurations.

The apparatus is shown in Fig. 17. The measurement consisted of the port surface temperature at 43.0 min from the cylinder entrance, and the temperatures at the back of the intake valve which was pinned so that the valve orientation was fixed. For the latter, two measurement locations were used. They were opposite each other along a diameter which was parallel to the line connecting the centers of the intake and exhaust valves. The two measurements will be referred to as the temperature of the intake valve at the intake side (denoted by A) and at the exhaust side (denoted by B), see Fig 17b.

The major factor affecting the temperature of the intake system is the back flow of the burned gas from the cylinder to the intake when the valve opens. To study this phenomenon, the engine was operated first with propane fuel (which was injected into the intake port) so that complication due to the cooling of the valve by the impingement of the liquid was absent. The temperature measurement results for closed valve injection are shown in the first portion (at time t less than 60 seconds, with start of fueling at time zero) of Fig. 18. The temperatures at the back of the intake valve are higher than that of the port wall which has a much higher thermal inertia.

The difference in temperature between the two locations on the intake valve was a result of the pattern of the back flow. At the location A, the valve opening was close to the cylinder wall, and therefore, the flow path was restricted by

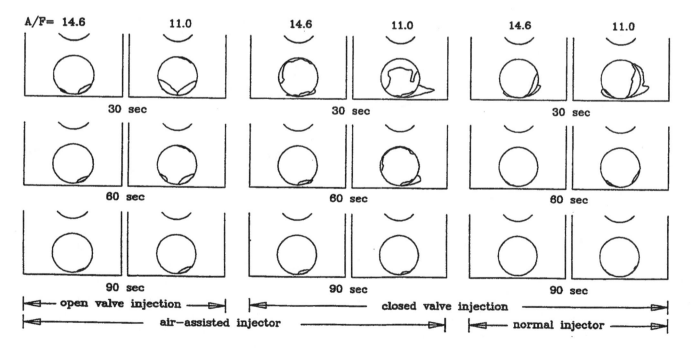

Fig. 16 Fuel film behavior with respect to air-to-fuel ratio during warm-up

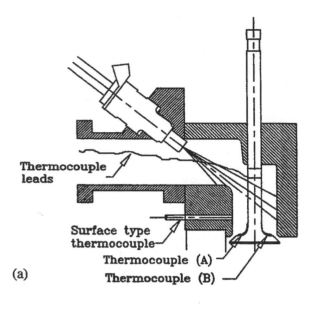

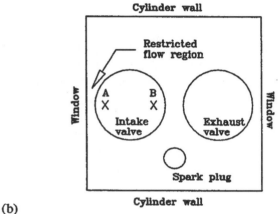

(a)

(b)

Fig. 17 Temperature measurements in the intake "system; (a) section side-view (b) top view

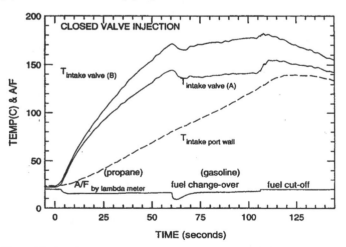

Fig. 18 Intake system component temperatures during warm-up (p_i--0.5 bar, λ=1.0, 1,000 rpm). Liquid fuel was delivered by the normal injector in closed-valve injection.

the wall. Thus the back flow was stronger at location B than location A, leading to a higher heat transfer rate at B.

The propane injection was switched off at t=60 sec. in Fig. 18, and gasoline injection using the normal injector with closed-valve injection was switched on for t=60 to 120 sec. The "step" drop of temperatures at the back of the intake valve was due to the liquid fuel cooling effect [6]. There was no step-decrease in the port wall temperature because the measurement location was in a shadow outside the footprint of the fuel spray.

The intake system component temperatures during warm-up using the air-assisted injector and the normal injector with closed-valve injection are compared in Fig 19. For both configurations, the temperatures in location A were lower than those in location B due to the back flow pattern as described above. For the air-assisted injector, however, there was the extra factor due to the significant fuel spray impingement at location A; thus a considerably lower temperature was measured at location A.

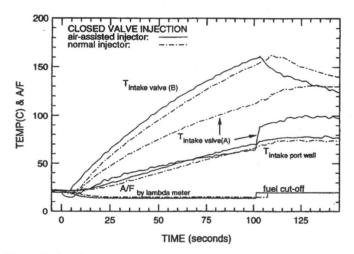

Fig. 19 Intake system component temperatures during warm-up for different injector configurations.(A/F=14.6, closed-valve injection)

CONCLUSIONS

The liquid fuel entry into the cylinder and its subsequent behavior through the combustion cycle were visualized in a transparent engine using a high speed CCD camera. There is significant liquid fuel entry into the cylinder, in particular, during the early cycles in the warm-up process. The major liquid fuel transport processes are: the strip atomization of the fuel film on the surfaces of the intake system into droplet streams by the intake air flow; the film flow which forms significant liquid puddles at the valve surface and at the vicinity of the intake valve; and the squeezing of the liquid film at the valve lip and seat into large droplets in the valve closing process. A substantial fraction of the liquid in the thick film formed at the valve surface and the seat area survives combustion into the next cycle. The time evolution of the in-cylinder liquid film is influenced by the injection geometry, whether the fuel is introduced by open- or closed-valve injection and the amount of fuel enrichment.

surface and the seat area survives combustion into the next cycle. The time evolution of the in-cylinder liquid film is influenced by the injection geometry, whether the fuel is introduced by open- or closed- valve injection and the amount of fuel enrichment.

ACKNOWLEDGMENTS

This work has been supported by a Consortium for Engine Research (members are the Chrysler Corp., Ford Motor Co., General Motors Corp., Peugeot S.A., Regie Nationale des Usines Renault, and the Volvo Car Corp.). We would like to thank Prof. Kim Vandiver, Anthony Caloggero and Steve Kim of the Edgerton Center at MIT for making the high speed video equipment available and technical assistance.

REFERENCES

1. Cheng, W.K., Hanmrin, D., Heywood, J.B., Hochgreb, S., Min, K-D., and Norris, M., "An Overview of Hydrocarbon Emissions Mechanisms in Spark-Ignition Engines," SAE Paper 932078, 1993.

2. Fox, J.W., Min, K-D., Cheng, W.K., and Heywood, J.B., "Mixture Preparation in a SI Engine with Port Fuel Injection During Starting and Warm-up," SAE Paper 922170, 1992.

3. Namazian, M., Hansen, S.P., Lyford-Pike, E.J., Sanchez-Barsse, J., Heywood, J.B., and Rife, J., "Schlieren Visualization of the Flow and Density Fields in the Cylinder of a Spark-Ignition Engine," SAE Paper 800044, 1980.

4. Yang, J., Kaiser, E.W., Siegl, W.O., and Anderson, R.W., "Effects of Port-Injection Timing and Fuel Droplet Size on Total and Speciated Exhaust Hydrocarbon Emissions," SAE Paper 930711, 1993.

5. Cheng, C.O., Cheng, W.K., Heywood, J.B., Maroteaux, D., and Collings, N., "Intake Port Phenomena in a SI Engine at Part Load," SAE Paper 912401, 1991.

6. Martins, J.J.G., Finlay, I.C., "Fuel Preparation in Port Injected Engines," SAE Paper 920518, 1992.

CHAPTER 2

MIXTURE FORMATION PROCESSES

Ron Matthews and Matt Hall
The University of Texas

2.1 OVERVIEW

Cold start and warm-up pose special problems for hydrocarbon emissions control over the FTP cycle. When the engine's surface temperatures are cold, the sources of engine-out hydrocarbons (HCs) are high, resulting in high tailpipe hydrocarbons until the catalyst reaches light-off temperature. Due to their temperature-dependency, engine-out HCs due to the flame quench, oil film, deposit, and crevice sources are inherently higher during cold start and warm-up than after the surfaces have reached normal operating temperatures. However, the most important source during this period results from the need to over-fuel the engine due to the low efficiency of fuel evaporation off cold surfaces. This results in a rich fuel vapor/air mixture for at least the first several seconds of the FTP. In turn, this increases engine-out HCs from the exhaust valve leakage source. More importantly, the liquid fuel source is extremely high during cold start and warm-up. It has been shown that the liquid fuel source can persist well after catalyst light-off, yielding high tailpipe HCs during the first three accelerations of Phase 1 of the FTP.

When gasoline is injected onto the port sides of intake valves that are initially cold, little of the gasoline will evaporate – only the most volatile (light) components. Because fuel vapor is required for ignition and combustion, the engine starts and warms up on the lightest fractions of the gasoline, and the heavier components evaporate later after the surface temperatures have increased appreciably. Thus, 8-15 times the stoichiometric amount of gasoline is typically injected during the first several cycles of the cold start. An in-cylinder fuel vapor/air equivalence ratio, ϕ, of 0.7-0.9 (1.1<λ<1.4) is required on the first cycle to obtain a robust first fire. The minimum in engine-out HCs occurs for a fuel vapor/air ϕ of about 0.9. Generally, a rich fuel vapor/air mixture is required during starting and warm-up to ensure rapid cold start and stable idle during warm-up. Even if a richer in-cylinder mixture is used for the first fire, there is obviously a lot of liquid fuel remaining in films and these films are enriched in the heavier, more difficult-to-evaporate components.

The fuel films on the port surfaces and on the port side of the intake valves are important because they lead to wetting of in-cylinder surfaces. The four mechanisms that produce in-cylinder wall wetting are 1) fuel stripping from the intake valve during intake valve opening, 2) the direct flow of liquid fuel into the combustion chamber from the injector if open valve injection (OVI) timing is used, 3) transport during the middle of the intake stroke driven by the high speed flow of intake air, and 4) droplets formed when the fuel in the film on the face of the valve and the valve seat is squeezed during intake valve closing. Difficulty in evaporating the heavier components of the gasoline can result not only in in-cylinder wall wetting, but also in charge inhomogeneity. In turn, the charge inhomogeneity could increase HC emissions via rich pockets. Also, in-cylinder fuel films can lead to pool fires that can lead to emissions of both HCs and particulates. However, the slow evaporation of in-cylinder fuel films appears to be the most important reason that engine-out HC emissions are high during cold start and warm-up.

The key factors that affect fuel film formation and thus engine-out HCs include injection timing, injector targeting, droplet size, and in-cylinder bulk flow.

Steady-state experiments--which do not capture all of the phenomena that are important during starting and warm-up--have shown that OVI yields higher engine-out HC emissions than closed valve injection (CVI). For CVI, the HC emissions for steady state operation are largely insensitive to injection timing unless the start of injection is within ~80°CA of intake valve opening (IVO), when HCs begin to increase for steady-state operation. If batch fire is used while awaiting synchronization, some cylinders will have OVI while others will have CVI, but possibly too near IVO. In general, OVI yields liner wetting under the exhaust valves and head wetting around the intake valves and between the exhaust valves. CVI produces more port and valve wetting than OVI plus some piston wetting and head wetting between the intake valves. In-cylinder fuel films persist and build up over successive cycles when the surfaces are cold. Some of the liner film is scraped into the crankcase as "lost fuel" and some of the liner film is scraped up closer to the exhaust valves. The in-cylinder wetting location is important because the closer that location is to the exhaust port, the less likely evaporated fuel from that location will undergo in-cylinder oxidation or be retained within the cylinder. That is, wetting locations closer to the exhaust port contribute more to engine-out HCs. Also, the head films from OVI begin to form on earlier cycles than for CVI. Even if injection is delayed until after synchronization such that CVI can be used exclusively, and even if the start of injection for CVI is right after IVC to provide the maximum time for fuel evaporation, little fuel will evaporate when the port and intake valve are cold. The reverse flow right after IVO is advantageous because it redistributes the fuel in the port and aids evaporation. Variable valve timing systems, even though they cannot normally be activated until the oil pressure increases and the oil viscosity decreases sufficiently, can be used to provide an "inactive" valve overlap that promotes this reverse flow during starting and warm-up. However, port films can persist for over 10 minutes even with the very small droplets obtained via air forced injection. Again, port wetting is important because it leads to in-cylinder wall wetting.

Injector targeting is important because the rate of evaporation is dictated by the surface temperature. The intake valve heats much faster than the port wall, for which the temperature is very close to that of the coolant outlet throughout warm-up. Similarly, flame passage does not completely evaporate in-cylinder fuel films because the in-cylinder surface temperatures increase relatively slowly. In-cylinder fuel films, especially in the vicinity of the exhaust, contribute to exhaust HCs because evaporation off cold surfaces is slow. Some of these in-cylinder films will evaporate late in the expansion stroke, when in-cylinder gas temperatures are too low for complete oxidation, or during the exhaust stroke.

The beginning of the FTP, and the corresponding calibration, can be divided into stages. The first stage is cranking/starting. No fuel compensation can be used for the first injection because there is no liquid fuel left from the prior cycle.

The engine starts on the lighter (more volatile) ends of the gasoline, leaving the heavier ends behind in fuel films. Also, the crank pulse must be calibrated to compensate for high Driveability Index (DI) gasolines, which are less abundant in light ends. Thus, 8-15 times the stoichiometric amount of gasoline is typically injected during the first several cycles of the cold start. If sufficient fuel is injected on the first cycle, enough vapor may be generated between cycles that the extreme of fuel compensation--no injection--can be used for the second cycle. However, a higher mass injected results in thicker port and valve films, and consequently thicker in-cylinder films, which evaporate more slowly.

The second stage is warm-up at idle in neutral (from crank to 15 s of the FTP). The fuel films from cranking/starting persist and build. Because it takes about one minute for the intake valves to complete the first (rapid) stage of warm-up, fuel evaporation remains a problem. Also, the port and in-cylinder surface temperatures change little from the temperature before cranking. Fuel compensation can be used during this and subsequent stages. However, the consumer expects a stable idle, and this normally requires a rich vapor/air mixture with little spark retard. It has been shown that crankshaft speed fluctuations can be used to compensate for high DI gasolines during this period. Although this results in more mass injected for the high DI gasolines, it eliminates the need for calibration enrichment beyond that required for certification gasoline.

The third stage is warm-up at idle in drive (15-20 s of the FTP). The increase in manifold absolute pressure (MAP) impedes evaporation of fuel films both in the intake and in-cylinder. The intake valves are still in the first stage of heating and the port and in-cylinder surface temperatures have changed little. Evaporation from these films remains a problem.

The next stage is the first accel and cruise (20-114 s of the FTP). The intake valves complete the rapid stage of heating perhaps 40-50 s into the FTP but do not reach normal operating temperature. The port and liner continue to heat slowly. The increased MAP impedes vaporization. The first accel generates high engine-out HCs due to transfer of fuel film from the intake to in-cylinder surfaces as a result of the increased flow velocities past the valves. Typically, the catalyst will light-off during this stage.

The key to minimizing engine-out HCs during cold start and warm-up is to minimize in-cylinder surface wetting, especially at locations near the exhaust valves. Both tumble and swirl, especially in conjunction with small droplets, can decrease in-cylinder wetting, allow warm-up with mixtures that are not as rich (perhaps even lean), and allow more spark retard. Air flow control valves have been shown to be effective in increasing both tumble and swirl, not only for idle speeds but also during cranking. Air forced injectors produce the smallest droplets, followed by air assist and then by 12-hole injectors. In fact, multiple hole injectors with a high rail pressure can produce droplets nearly as small as via air forced injectors. However, several PZEV vehicles are in production that use droplets that have an SMD on the order of 70 μm, which is larger than air forced (~10 μm) or air assist (~25 μm). Open valve injection minimizes port wetting (with proper targeting) but must be used in conjunction with small droplets and a strong bulk flow that minimizes impaction on in-cylinder surfaces, especially those near the exhaust valves. Two other technologies are being developed that either eliminate or greatly minimize fuel wetting in the intake and, thereby, in-cylinder. These reformer and on-board distillation technologies merit continued development.

2.2 SELECTED SAE TECHNICAL PAPERS

1999-01-3661

Further Experiments on the Effects of In-Cylinder Wall Wetting on HC Emissions from Direct Injection Gasoline Engines

Jianwen Li and Ronald D. Matthews
The University of Texas at Austin

Rudolf H. Stanglmaier and Charles E. Roberts
Southwest Research Institute

Richard W. Anderson
Ford Motor Company

Copyright © 1999 Society of Automotive Engineers, Inc.

ABSTRACT

A recently developed in-cylinder fuel injection probe was used to deposit a small amount of liquid fuel on various surfaces within the combustion chamber of a 4-valve engine that was operating predominately on liquefied petroleum gas (LPG). A fast flame ionization detector (FFID) was used to examine the engine-out emissions of unburned and partially-burned hydrocarbons (HCs). Injector shut-off was used to examine the rate of liquid fuel evaporation. The purpose of these experiments was to provide insights into the HC formation mechanism due to in-cylinder wall wetting. The variables investigated were the effects of engine operating conditions, coolant temperature, in-cylinder wetting location, and the amount of liquid wall wetting.

The results of the steady state tests show that in-cylinder wall wetting is an important source of HC emissions both at idle and at a part load, cruise-type condition. The effects of wetting location present the same trend for idle and part load conditions. Wetting the cylinder liner under the exhaust port yields the highest increase in HC emissions, wetting the cylinder liner under the intake port yields the smallest HC emissions increase, and wetting the top of the piston yields an increase in HC emissions that lies between these two extremes. The coolant temperature has a stronger effect on HCs due to liner wetting for part load than for idle, but little effect for piston wetting for either operating condition. Depositing 10% of the total fuel mass as a liquid on the top of the piston results in approximately 30% and 70% increases in HC emissions for idle and part load conditions, respectively. The percentage increase in HC emissions relative to the LPG-only baseline initially increases rapidly with increasing liquid percentage of the total fuel mass flow but eventually appears to approach a constant value.

The FFID results show that liquid fuel on in-cylinder surfaces produced increased HC concentrations during the entire exhaust process. Wetting the liner under the exhaust valves had a very pronounced effect on the middle peak relative to operation at the same equivalence ratio with a gaseous fuel. Wetting the piston most strongly affected the final peak and also produced a small peak just before the final peak. Wetting the liner under the intake valves had the least effect on the FFID results, increasing the HC concentration of the first peak most, but yielding a somewhat larger percentage increase in the middle peak. Physical explanations for these characteristics are proposed.

Injector shut-off tests indicate that the process of evaporation of liquid fuel from in-cylinder surfaces is slow in comparison to engine cycle times. For the four combinations of operating condition and coolant temperature examined, the 1/e time constant for fuel evaporation from the piston was about 3-6 engine cycles, with the engine continuing to operate on the gaseous fuel after injector shut-off.

INTRODUCTION

The present study is relevant to hydrocarbon (HC) emissions due to in-cylinder wall wetting for both port fuel injected (PFI) and direct injection spark ignition (DISI) engines. A brief background is presented in this section, followed by the objectives of the present research.

It is well known that 60-80% of tailpipe HC emissions from port fuel injected (PFI) vehicles are emitted during the first 1-2 minutes of operation (Refs. 1-4) when the catalyst is reaching light-off temperature; the in-cylinder crevices are relatively large and cold; both liquid fuel effects and oil layer adsorption/desorption may play important roles; the engine is over-fueled to assure rapid start-up, stable idle, and good drivability during warm-up; and - due to the lack of initial synchronization - some cylinders have open valve injection while others have closed valve injection. This, along with low wall temperatures, results in the presence of liquid fuel in the intake manifold and combustion chamber during cold start and warm-up. Many researchers have shown that the presence of liquid fuel in the combustion chamber is an important cause of high engine-out unburned HCs during cold start and warm-up conditions for PFI engines (5-9). In these studies, the liquid fuel is injected into the intake port and is subsequently forced into the combustion chamber as a liquid/vapor mixture during the intake stroke. A portion of the liquid fuel entering the combustion chamber ultimately exists as fuel droplets suspended in the bulk gases but another portion is deposited as a fuel film located on the combustion chamber walls. The fuel droplets vaporize very quickly due to strong air motion and heat transfer in the chamber, whereas the fuel film on the wall vaporizes more slowly. Stanglmaier and coworkers (10) and Kelly-Zion and coworkers (11) showed that open valve injection yields fuel wetting on the liner under the exhaust valves whereas closed valve injection produces wetting of the piston top. However, no studies have directly quantified how both the amount and location of liquid fuel on in-cylinder surfaces influences HC emissions.

Direct injection gasoline engines have been the subject of attention because of their potential fuel economy benefits. However, a recent comparison (12) of the FTP emissions from a DISI production (Japanese market) vehicle to those from a comparable PFI vehicle showed that the emissions barriers impeding the use of this technology in the United States are, in decreasing order of severity, oxides of nitrogen (NOx), HCs, and particulates. The high tailpipe NOx occurred in spite of a factor of two decrease in engine-out NOx mass emissions, because of the low efficiency of the lean NOx trap/catalyst. Although the 3-way catalysts on this DISI vehicle produced a very high efficiency for HC oxidation, the tailpipe HCs were higher than for the PFI vehicle because the engine-out HCs were significantly higher. Furthermore, unlike PFI vehicles, the tailpipe HC emissions are not dominated by the first 1-2 minutes of operation. This is partially due to the fact that the DISI vehicle operates with early injection timing, using a near-stoichiometric mixture and with a throttled intake for 250 seconds at the beginning of the FTP to light-off the underfloor NOx trap/catalyst, and partially due to very high engine-out HCs during late injection, which is used for most of the remainder of the FTP. In a related study (13), it was shown that the engine-out Emissions Index for Hydrocarbons (EIHCs) is about a factor of two higher for early injection compared to the PFI vehicle, and at least a factor of four higher for late injection. Since the fuel is injected directly into the combustion chamber, liquid fuel impingement on the walls is largely unavoidable. The extent of wall wetting during late injection depends, in part, on how the stratified charge is formed - wall guided, spray guided, or air motion guided (14). It has been shown that liquid fuel can impact both the liner and the piston in DISI engines, with the location and amount of wetting dependent upon bulk motion, injection timing, and other factors (e.g., 15,16)

The present study is the second in a series of papers examining the effects of in-cylinder wall wetting on HC emissions. In this set of studies, a 4 cylinder 4-valve SI engine is operated on liquefied petroleum gas (LPG) to eliminate liquid fuel effects. A much smaller amount of liquid fuel is then injected into cylinder #4 using a fuel injection probe developed specifically for this purpose. This probe can deposit liquid fuel on any radial location on the liner just below the head, or on the top of the piston. This probe yields a stream of fuel rather than a spray, so that the injected liquid impacts a surface with little, if any, evaporation during the injection process. This technique allows isolation of one of the three sources of HC emissions from DISI engines. The three sources are bulk flame extinction due to the flame encountering an overly lean mixture, low in-cylinder temperatures for oxidation of fuel that escapes combustion, and wall wetting accompanied by slow evaporation. Operation of the engine predominantly on gaseous LPG at an overall excess air ratio (λ) of about 1.1 eliminates flame extinction and low in-cylinder temperatures, thereby isolating the wall wetting source. The results are also relevant to PFI engines because, as noted above, open valve injection yields wetting of the liner under the exhaust valves, whereas closed valve injection (for a cold engine) yields wetting of the piston top. The initial study (17) examined idle-type operation with California Phase 2 reformulated gasoline (RFG) as the liquid fuel. Among the conclusions of this study were:

- wetting of any in-cylinder surface yields a large increase of HC emissions, with wetting the liner under the exhaust valves being the worst case, wetting the liner under the intake valves being the best case, and wetting the piston top being not quite as bad as wetting the liner under the exhaust valves, and

- the HC emissions due to in-cylinder wetting are almost independent of injection timing (ranging from injection during the exhaust stroke to during early compression, as illustrated in Figure 1), indicating that the HC increase may be due to evaporation during the subsequent exhaust stroke or during a later cycle.

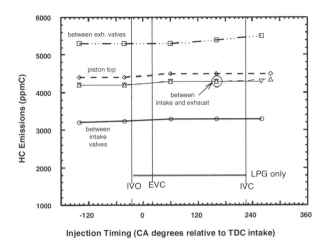

Figure 1. Effects of liquid fuel injection timing on HC emissions (17).

Further experimental studies are continuing and results of part of this effort are presented in this paper.

The primary objective of the present study was to explore the details of the HC formation mechanism due to in-cylinder wall wetting. Crank angle resolved HC emissions, measured using a Combustion Fast Flame Ionization Detector (FFID), were used to ascertain why the HC emissions depend upon wetting location. Other objectives were to determine the characteristic time for liquid fuel evaporation from in-cylinder surfaces and to investigate how engine operating conditions and liquid fuel quantity affect the engine-out HC emissions.

APPARATUS AND PROCEDURE

The experimental apparatus, test procedure, and operating conditions examined in this study are discussed in the following subsections.

TEST APPARATUS – Table 1 provides specifications of the 1992 2.3L GM Quad-4 engine used for the present experiments.

Table 1. Specifications of the Test Engine

Cylinders	4
Valves/cylinder	4
Displacement (L)	2.259
Bore (mm)	92
Stroke (mm)	85
Comp. Ratio	10.0:1
IVO	22° BTDC
IVC	45° ABDC
EVO	120° ATDC
EVC	20° ATDC

An Accurate Technologies Motec M8 engine management system was used to control fuel injection and ignition. The engine was operated on pre-vaporized LPG to eliminate liquid fuel effects from the predominant fuel. A much smaller amount of liquid fuel was injected onto a chosen surface in the fourth cylinder using a directional fuel injection probe that has been described previously (17). As shown in Figure 2, the exhaust manifold was modified to separate the exhaust for the fourth cylinder from that for the other three cylinders. A Combustion FFID probe (HFR400) and a thermocouple were placed into the exhaust port of the fourth cylinder. Additionally, a sampling port for a traditional Flame Ionization Detector (Horiba FIA-34A-2) and a wide range lambda sensor (Horiba Model 101λ) were used to monitor steady state HC emissions and excess air ratio. A DSP Technology data acquisition system was used to record signals from the Fast FID, lambda sensor, and fuel injector. This DSP included a 4012A system controller, 4325 real time processor, 2904 spincoder, and 6001 crate controller.

As in the initial study (17), the liquid fuel used for the present study was California Phase 2 reformulated gasoline. The atmospheric pressure distillation curve for this fuel is provided in Figure 3; the curve shifts upward for higher pressures (e.g., 18). The end boiling point is almost 180 °C (~450 K). In comparison, simulations for an engine that was also fueled with propane and running at this same operating condition (including a coolant temperature of 90 °C and an oil temperature of 73 °C) revealed a liner temperature of 395K and a piston temperature of 434 K (19). However, the simulated engine was a 2-valve engine with a somewhat smaller swept volume and a compression ratio of 9.

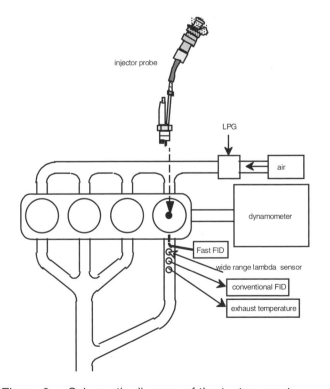

Figure 2. Schematic diagram of the test apparatus.

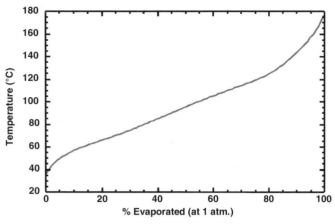

Figure 3. Distillation curve for California Phase 2 RFG.

The injector probe is illustrated in Figure 4. It is mounted on a Champion 304-802 offset electrode spark plug through the hole that is normally used for mounting a cylinder pressure transducer. The probe used for most of the experiments was that used in the initial study and had an ID of 0.51 mm and a 0.15 mm orifice. Validation of the performance of this probe in the optical version of this engine was discussed in the initial paper (17). Two additional probes were tested for some of the piston wetting cases. These were hypodermic needles with IDs of 0.20 and 0.15 mm. The steady state HC emissions were not affected by the probe used, in spite of the longer injection pulse width required due to the increased flow resistance of the needles.

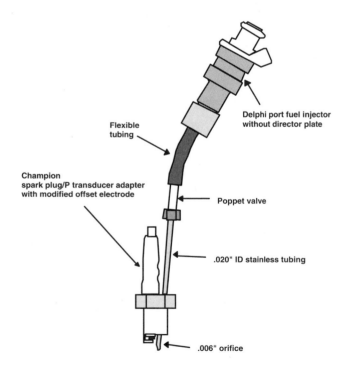

Figure 4. Schematic of the injector probe used to deposit liquid fuel on in-cylinder surfaces (17).

TEST PROCEDURE – The aim of this study was to examine the influence of in-cylinder wall wetting on HC emissions. To minimize the sensitivity of the HC emissions to minor variations in stoichiometry, the excess air ratio was maintained at a constant and slightly lean value before and after depositing the California Phase 2 RFG ($\lambda = 1.1$, corresponding to an equivalence ratio $\phi \approx 0.9$). Additionally, constant coolant and lubrication oil temperatures were also maintained (two coolant temperatures: 40 °C and 90 °C, and an oil temperature of 60-70 °C). The test procedure shown in Figure 5 was adopted for all tests. During the initial LPG-only operation, the LPG flow rate was adjusted to hold the excess air ratio slightly lean at $\lambda=1.1$. In the second stage of the procedure, a chosen mass of liquid fuel was injected onto the desired in-cylinder surface using the injection probe. Upon liquid fuel introduction, the overall excess air ratio decreases (the mixture approaches stoichiometric or becomes rich). In Stage 3 of the procedure, the overall excess air ratio is returned to $\lambda = 1.1$ by reducing the LPG flow rate. When the excess air ratio returns to this new steady state, the HC emissions are recorded using both the Fast FID and the conventional FID.

To investigate the evaporation of the liquid fuel film, the transient excess air ratio signal from the wide range lambda sensor and the HC emissions were recorded before and after shutting-off the injection probe. The step-change in liquid fueling is followed by a period during which the excess air ratio and HC emissions approach new steady states during Stage 4 of the procedure.

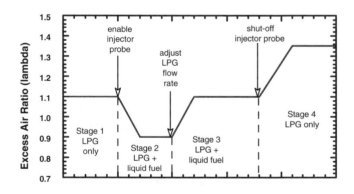

Figure 5. Illustration of the test procedure.

TEST CONDITIONS – Engine speed and load effect film-generated HC emissions because of their influence on cylinder wall temperatures, air motion, and in-cylinder gas temperatures. To minimize fluctuations in engine speed, all idle tests were conducted at 1000 rpm with a manifold air pressure of 32.5-34.5 kPa, depending upon the coolant temperature. Part-throttle conditions of 1500 rpm and 262 kPa BMEP (Ford's "world wide mapping point", WWMP) were also tested. The WWMP represents a typical cruise condition on the FTP driving cycle.

The ignition timing was fixed at MBT for both idle (31° BTDC for 40 °C coolant, 29° BTDC for 90 °C coolant, both with λ=1.1) and part load conditions (27° BTDC for 40 °C coolant, 25° BTDC for 90 °C coolant, again with I=1.1). The injection timing for injecting the liquid fuel was fixed at 160° ATDC during the intake stroke. It should be noted again that our previous study (17) showed that injection timing has little effect on HC emissions for these experiments.

RESULTS AND DISCUSSION

In the following subsections, the results are divided into: effects of operating conditions and wetting location, effects of liquid mass, Fast FID results, and results of the injector shut-off tests.

EFFECTS OF ENGINE OPERATING CONDITIONS AND WETTING LOCATIONS – The effects of wetting location on HC emissions was discussed in our previous paper (17) when operating the engine at idle conditions. The results obtained for part load conditions (Ford's WWMP of 1500 rpm, 262 kPa BMEP) are compared with the idle results in Figures 6a and 6b. For the tests illustrated in Figures 6a and 6b, the liquid fuel amount injected was maintained constant at 1.5 mg/cycle for both operating conditions. For idle conditions, about 82% of the fuel was LPG and 18% liquid fuel. For operation at the WWMP, about 89% of the fuel was LPG and about 11% liquid fuel. Three wetting locations were tested and are denoted as "I-I" for the cylinder liner under the intake ports, "E-E" for the cylinder liner under the exhaust ports, and "Piston Top" for fuel deposited directly onto the piston crown.

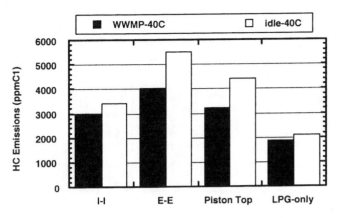

Figure 6b. Effects of wetting location and operating conditions on HC emissions with 40 °C coolant.

It is apparent from Figures 6a and 6b that the wetting location within the combustion chamber has a substantial impact on the resulting HC emissions at both idle and part load. All wetting locations resulted in significant increases in HC emissions from the baseline (LPG-only) levels. The difference in HC emissions for the three wetting locations was explained for the idle condition in our previous paper (17) by considering the physical distance from the wetting location to the exhaust port. The HC variation from one wetting location to another at part load is similar to the idle condition. Namely, the smallest increase in HC emissions was recorded for wetting the liner under the intake valves, while the highest increase in HC emissions was recorded for wetting the liner under the exhaust valves. Wetting the piston crown resulted in an HC increase between these two extremes. The increases in the HC emissions due to wall wetting at part load appear to be less pronounced than those at idle for the same wetting location. However, for 1.5 mg of liquid fuel injected per cycle, only 11% of the fuel is liquid at the WWMP while 18% is liquid at idle. The effects of the amount of wetting are explored in the next subsection.

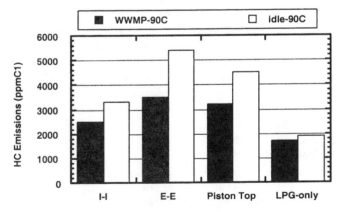

Figure 6a. Effects of wetting location and operating conditions on HC emissions during fully warmed up operation (90 °C coolant).

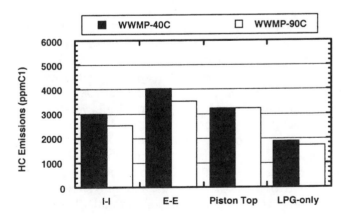

Figure 7a. Effects of coolant temperature on H emissions at the World Wide Mapping Point part load operating condition.

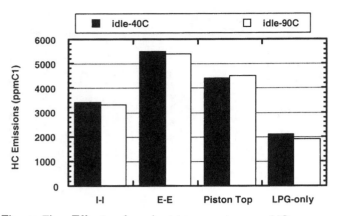

Figure 7b. Effects of coolant temperature on HC emissions at idle.

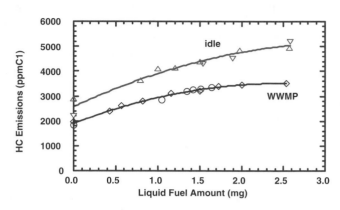

Figure 8. Effects of mass of liquid fuel injected on HC emissions. Over the ranges examined, the trends are second order with a correlation coefficient (R) > 0.98.

In our previous study of HC emissions due to in-cylinder fuel wetting at idle (17), it was shown that increasing the coolant temperature produced lower HC emissions for all wetting locations except the piston top, for which the HC emissions increased with increasing coolant temperature. The piston top differs from the liner locations in several ways: the piston top is hotter than the liner, the temperature of the piston crown is less sensitive to the coolant temperature than is the liner temperature, and little, if any, of the liquid impacting the piston can get into the topland where it can either add to the crevice source or escape into the crankcase. Figures 7a and 7b indicate that the coolant temperature also affects the HC emissions due to wall wetting at part load. For the WWMP condition compared to idle, the wall wetting HCs are more sensitive to coolant temperature except for wetting the piston top. It is surmised that as the liner temperature increases, the amount of fuel vaporized during the exhaust stroke increases as well. Because the piston temperature is not as sensitive as the liner temperature to the coolant temperature at the WWMP, the HC emissions due to piston wetting are essentially independent of the coolant temperature at this operating condition.

EFFECTS OF AMOUNT OF LIQUID FUEL INJECTED – The results from our prior study (17) indicated that DISI design and/or operating conditions that produce more piston wetting should yield increased HC emissions. To quantify the effect of the amount of liquid fuel injected on HC emissions, the mass of the liquid fuel was changed by adjusting the injection pulse width. The liquid fuel flow rate was also monitored using a fuel flow meter. The excess air ratio was maintained constant at 1.1, the coolant temperature was 40 °C, and only piston wetting was examined for these experiments. Figure 8 shows the HC emissions as a function of the amount of liquid fuel injected for both idle and part load conditions.

Obviously, the HC emissions increase proportional to the mass of liquid fuel injected for both operating conditions; i.e., the greater the liquid impingement, the higher the HC emissions. Therefore, liquid fuel impingement on in-cylinder surfaces should be minimized in direct injection gasoline engines. Figure 8 also indicates that, for the same mass of liquid fuel injected, the increase in the concentration of HC emissions due to wall wetting for idle conditions is higher than that for part load conditions. However, any given liquid mass injected is a greater percentage of the total (LPG plus liquid fuel) for the idle condition. Also, the baseline HC emissions due to LPG are different between these two operating conditions (lower HC emissions on LPG-only at the WWMP). Figure 9 presents the results from Figure 8 in terms of the percent increase in HC emissions (relative to LPG-only) as a function of the liquid fuel fraction (percentage of the liquid fuel to the total fuel flow). As shown in Figure 9, the percentage increase in HC emissions due to wall wetting at part load is higher than that at idle for the same liquid fuel fraction. For instance, 10% liquid fuel impingement causes approximately a 30% increase in HC emissions at idle and approximately a 70% HC increase at the World Wide Mapping Point. Numerical models of direct injection gasoline engines have predicted as much as 50% liquid fuel impingement (20,21). Figure 9 illustrates an apparent leveling-off of the percentage increase in the HC emissions with increasing liquid fuel fraction. However, it is not clear that the decreasing rate of rise of the HC emissions with increasing liquid fuel fraction shown in Figure 9 will extrapolate to a DISI engine. Because the present experiments were intended to isolate the wall wetting HC source from the bulk flame quench and low oxidation temperature sources via operation predominately on LPG, these experiments include a crevice source that may be negligible for a DISI engine operating with late injection (stratified charge). As the liquid fuel fraction increases in Figure 9, the LPG fraction necessarily decreases and, thus, the strength of the crevice source also decreases. This may partially explain the slower rate of increase of HC emissions with increasing liquid fuel in Figures 8 and 9.

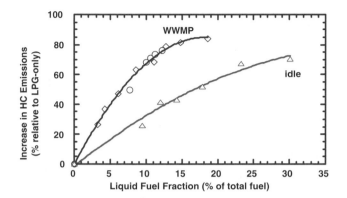

Figure 9.   Effects of the percentage of liquid fuel injected (of the total fuel) on the percentage increase in HC emissions. Over the ranges examined, the trends are second order with a correlation coefficient (R) > 0.99.

The results in Figure 9 imply that wall wetting plays a more important role at part load than that at idle, even though increased engine speed and load cause increased heat transfer and wall temperatures, effects which should promote vaporization of fuel films from the wall. Apparently, the effects of increased speed and load on promoting vaporization are not strong enough to compensate for the decreased time available for evaporation and oxidation at the higher speed. At both idle and the WWMP, it appears that a portion of the liquid fuel remains on the wall into subsequent cycles. This effect is discussed in more detail in a later section of this paper.

FAST FID RESULTS – The Fast Flame Ionization Detector (FFID) has been used by several investigators to measure time-resolved HC concentrations in the exhaust port [e.g., 22-26], the combustion chamber, and in the intake port of SI engines. In general, the results from exhaust port FFID investigations have shown that the HC concentration history has a peak near exhaust valve opening and another peak near exhaust valve closing. A middle peak is also occasionally observed. The first peak is believed to be due to unburned HCs contained in the crevice volume of the exhaust valve seat and the quench layers in its vicinity. The peak near EVC is believed to be due to HCs in the roll-up vortex that originated from the piston crevices and from oil layer absorption/desorption. Thompson and Wallace (25) observed a middle peak for an engine operating on a gaseous fuel, and postulated that it might be due to the interaction between the "jet" outflow from the inter-ring zone with the bulk flow leaving the cylinder. By conducting Fast FID tests using various pure fuels with different volatilities in a PFI engine, Stache and Alkidas [9] found that the HCs attributed to liquid fuel stored in the piston topland crevice exit the cylinder near the end of the exhaust process, along with the HCs due to fuel vapor storage in the topland. Additionally, the HCs attributed to liquid fuel deposited on the cylinder liner under the exhaust valves - due to open valve injection - exit the cylinder as a middle peak that is roughly in the middle of the exhaust process. The findings by Stache and Alkidas have led to better understanding and interpretation of Fast FID results obtained when liquid fuel is deposited on different locations within the cylinder of a PFI engine.

For the present study, FFID measurements were performed at the WWMP condition with 90 °C coolant and λ=1.1. Measurements were obtained during operation solely on LPG and also with 1.5 mg of California Phase 2 RFG injected onto the piston, the liner under the exhaust valves, or the liner under the intake valves. The results presented are averages over 100 cycles of steady state operation. The FFID probe was located ~25 mm from the exhaust valve and close to the septum.

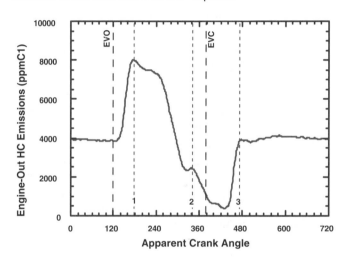

Figure 10.   Transient HC concentration for the baseline condition (LPG only) at the WWMP with 90 °C coolant. Three peaks are easily identified. These results have not been corrected for the transport delay, thus yielding the final peak after apparent exhaust valve closing.

Figure 10 presents the measured HC concentration histories at the WWMP part load condition for the baseline fuel (LPG only). It is well known that the FFID trace depends upon the location of the FFID probe (9, 24). Seabrook and coworkers (27) showed that it is also dependent upon the intake and exhaust valve timing events. The trace illustrated in Figure 10 is similar to that by Seabrook and coworkers for standard valve timing with the probe 25 mm from the exhaust valve and the engine operating at 1500 rpm and 2.62 bar BMEP (the WWMP that is also examined in the present study). Figure 10 is also similar to the baseline case by Stache and

Alkidas at 1300 rpm, 2.75 bar BMEP with the probe 15 mm from the exhaust valve stem. Specifically, there is a peak right after EVO, after which the HC concentration falls until a small middle peak is encountered, after which the HC concentration again decreases until the final peak occurs. Additionally, without correction for the effect of the low exit velocities near EVC, the final peak appears to occur after EVC (27).

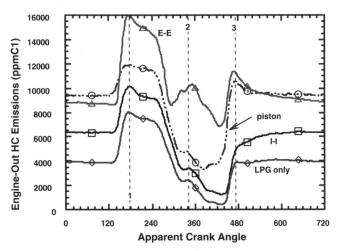

Figure 11. Transient HC concentrations for wetting the piston top, the liner under the intake valves, and the liner under the exhaust valves at the WWMP with 90 °C coolant.

Figure 11 shows the influence of wetting different locations in the cylinder on the HC concentration trace at the WWMP conditions. Compared to the baseline case (LPG only), the HC concentration for each test case is increased over the entire exhaust period due to the presence of liquid fuel directed onto an in-cylinder surface, even though the equivalence ratio was the same for all liquid injection cases as for the LPG-only baseline. Similarly, Stache and Alkidas (9) found that the entire trace shifted to higher concentrations for a PFI engine when the injection timing was changed from closed to open valve injection or the coolant temperature was decreased. However, as with the Stache and Alkidas results, some peaks are more strongly affected than others. This dependence is quantified in Table 2.

Table 2 shows the hydrocarbon concentration for the first, middle, and last peaks (points 1, 2, and 3 in Figures 10 and 11) for the three wetting locations and the baseline, LPG-only case. The concentration difference between the liquid wetting and LPG-only data is also shown in Table 2, as is the percent difference relative to the baseline for each of the three wetting locations and each of the three peaks.

Table 2. Effects of Wetting Location on Peaks 1, 2, and 3

Case	1	2	3
HC concentration (ppmC1)			
LPG-only (baseline)	8005	2415	3869
piston top	11875	4635	10569
liner under exhaust valves	15980	10118	11307
liner under intake valves	10125	3403	4867
Difference (ppmC1)			
piston-baseline	3870	2220	6700
exhaust-baseline	7975	7703	7438
intake-baseline	2120	988	998
Difference (%)			
(piston-baseline)/baseline	48.3	91.9	173.2
(exhaust-baseline)/baseline	99.6	319.0	192.2
(intake-baseline)/baseline	26.5	40.9	25.8

Wetting the cylinder liner under the exhaust valves increased the HC concentration, in ppmC1, for all three peaks by almost the same amount: 7400-8000 ppm. However, in terms of percentage increase, this wetting location increased the first peak by a factor of two and the final peak by a factor of ~3, but had the strongest effect on the middle peak which increased by more than a factor of 4. This predominant effect on the middle peak of liquid on the liner under the exhaust valves was also identified by Stache and Alkidas (9) for a PFI engine when they changed from closed to open valve injection, which yields liquid on the liner under the exhaust valves, as discussed previously. This effect of wetting the liner under the exhaust valves on the middle peak is the most easily noticeable feature of Figure 11. The strong increase in the HC concentration for the first peak is probably due to the short distance between the exhaust port and this wetting location, which gives rise to less time available for oxidizing unburned fuel that evaporates early but remains within the thermal boundary layer, or evaporates after the end of combustion. As noted above, the processes leading to the middle peak are the least well understood. As the piston ascends, much of the liquid on the liner will get scraped into the topland, from which a portion may flow into the inter-ring zone. When the piston descends, some of the liquid in the piston/liner crevices will get smeared on the liner as a thin film. The middle peak may be due to evaporation of this film during the exhaust stroke. Alternatively, or perhaps additionally, following the arguments of Thompson and Wallace (25) this middle peak may be due to liquid that evaporates within the inter-ring zone and emerges with the "jet" from the top ring end gap. Here, it should be noted that the liner temperature within the inter-ring zone is probably not much lower than the temperature near the top of the liner, but the pressure is lower and this promotes vaporization. Furthermore, Roberts and Matthews (19) showed that, for these operating conditions, this jet emerges during the exhaust stroke, when in-cylinder conditions are least favorable for oxidation. However, if vaporization within the inter-ring zone is an important contributor to the strength of the middle peak for this case, one would expect that it should depend upon the position of the top ring end gap with respect to the exhaust valves (19). Because the HC emissions for wetting the liner under the exhaust valves has not varied significantly over several months of performing these tests, it must be concluded that evaporation within the inter-ring zone is not a significant contributor and that the strength of the middle peak is due to vaporization of the wall film. The increase in the HC concentration for the final peak is due to the increased fuel loading of the topland by the liquid fuel that is scrapped into the topland during the compression stroke.

Wetting the piston top also increased the middle peak, but in this case by "only" a factor of ~2, whereas piston wetting had the strongest effect on the final peak, which increased by a factor of ~2.7. Additionally, as shown in Figure 11, the minimum that occurs just before the rise to the final peak for all of the other cases is replaced by a small peak in the case of piston wetting. This indicates that fuel vapor due to piston wetting exits shortly before as well as along with the roll-up vortex.

Wetting the liner under the intake valves affected the HC concentration of the first peak most strongly (in ppm), but the percentage effect on the middle peak was somewhat larger. Overall, wetting the liner under the intake valves had the least effect on the FFID results compared to operation on the gaseous fuel. This location is farthest from the exhaust valves, thus providing the maximum amount of in-cylinder transit time for unburned hydrocarbons to undergo oxidation.

These differences in the FFID histories between the wall wetting and LPG-only baseline must be due to liquid fuel that either evaporated before combustion but was protected from the combustion process or evaporated later in the cycle when the potential for in-cylinder oxidation was low. In either case, the vaporized fuel can affect different portions of the FFID trace depending upon the wetting location. Liquid that is deposited farther from the exhaust valves and evaporates early in the cycle has a greater probability of in-cylinder oxidation than fuel that is deposited closer to the exhaust valves due to the difference in distances to the exhaust port. Similarly, liquid that is deposited farther from the exhaust valves and evaporates later in the cycle has a lower probability of escaping the cylinder than fuel that is deposited closer to the exhaust valves. Thus the greater effect on the FFID trace for wetting the liner under the exhaust valves than of wetting the liner under the intake valves is to be

expected independent of when the liquid evaporates. However, there are several reasons to believe that evaporation occurs during the exhaust stroke, when conditions do not favor in-cylinder oxidation, as discussed in the next section.

INJECTOR SHUT-OFF RESULTS – To understand the lifetime of residual liquid fuel on in-cylinder walls, the transient signals from the wide range lambda sensor, the conventional FID, and the injector probe were recorded during a step-change shut-off of the injector probe while the engine continued to operate on LPG (stage 4 in Figure 5). Data were recorded every 0.05 seconds. The wide range lambda sensor was located ~1/3 of the swept volume downstream from the exhaust valve. The excess air ratio was held constant at 1.1 before stopping injection, and the mass injected per cycle was 1.5 mg. The liquid was injected onto the piston for all of the shut-off experiments.

As expected, there was a short delay in the response of the wide range lambda sensor and a much longer delay in the HC response. The HC response is dominated by a transport delay to the conventional FID (the ~4 mm ID sampling line was ~3 m long). After accounting for the delays, the 1/e time constant was determined for both signals. The time constant from the conventional FID was slightly longer than that for the wide range lambda sensor because the instrument response time for the FID is much longer than that for the lambda sensor. Therefore, only the results obtained using the wide range lambda sensor are discussed below.

Three injector probes were used for the initial injector shut-off experiments, which were conducted at idle with 90 °C coolant. Although the size of the probe did not affect steady state emissions, the probe volume did affect the transient results from the shut-off tests. The original probe holds more than 10 times the injected mass (1.5 mg), the 0.20 mm ID needle holds 59% more mass than the injected mass, and the 0.15 mm ID needle holds 11% less mass than the 1.5 mg injected. Due to the present interest in time constants, the results are presented in terms of the relative excess air ratio:

$$\text{relative } \lambda = \frac{\lambda_t - \lambda_\infty}{\lambda_0 - \lambda_\infty} \qquad (1)$$

Figure 12 illustrates that there is a delay between injector shut-off and the resulting increase in lambda (which produces a decrease in the relative excess air ratio). This delay was ~3 seconds long for the original probe but only ~0.25 seconds for the two needles. Furthermore, the original probe produced a rich spike before the decay, whereas the two needles did not. It appears that the excess fuel in the original probe continued affecting the combustion process for these several seconds. From the beginning of the decay, the 1/e time constant for the original probe was ~1.2 s (~10 cycles) but only ~0.7s (~6 cycles) for the two needles. Thus, the dominant effect of the probe volume is on the period between injector shut-off and the beginning of the decay in the relative excess air ratio. The 0.15 mm ID needle was used for the remaining shut-off tests.

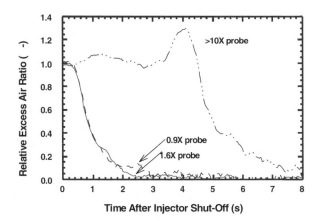

Figure 12. Effect of the injector probe volume on the relative excess air ratio following injector shut-off during idle with 90° C coolant and 1.5 mg injected per cycle.

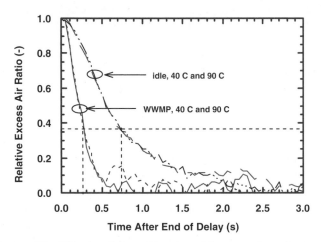

Figure 13. Effects of operating conditions and coolant temperature on the relative excess air ratio following injector shut-off with 1.5 mg injected per cycle using the 0.15 mm ID needle.

Figure 13 presents the results of the shut-off tests for operation at idle and at the WWMP part load condition, with both 90 °C and 40 °C coolant temperatures used at each operating condition. As shown in Figure 13, the coolant temperature has no effect on the 1/e time constant at either the WWMP or idle. This is probably due to the relatively weak effect of the coolant temperature on the piston temperature. In contrast, the operating condition has a significant effect on the 1/e time constant. At the WWMP, the 1/e time constant is less than half that at idle (~0.3 s or ~3 cycles compared to ~0.7 s or ~6 cycles at idle).

These injector shut-off experiments indicate that the liquid fuel survives for several cycles while the engine continues to operate on LPG. However, because the probe volume is finite, other interpretations of the data are possible. The fact that the 0.2 mm ID needle has a 78% greater volume than the 0.15 mm ID needle but produced the same results supports the hypothesis that the decay rate is not due to the probe but, instead, is due to the rate of evaporation of the liquid fuel from the piston. The weak effect of injection timing on the HC emissions (Figure 1) supports the hypothesis that liquid fuel survives for more than one cycle. Other experimental data also support this hypothesis. For example, Xiong and Yuen (28) examined the evaporation rates of droplets impinging on hot surfaces, including single component fuels and diesel fuel. They found that these fuels, including the multicomponent blend (diesel), exhibited the characteristics first observed by Leidenfrost in 1756. Specifically, the evaporation and heat transfer rates increase as the surface temperature approaches and surpasses the boiling temperature, but reach a maximum at a transition temperature that is 50 – 60 °C above the boiling temperature for the single component fuels studied (n-heptane, hexadecane, octanol, and 1,3,5 trimethyl benzene). The evaporation and heat transfer rates decrease as the temperature increases further, up to the Leidenfrost point, which is ~120 °C above the boiling temperature for these single component fuels. For diesel fuel, the transition temperature is ~350 °C and the Leidenfrost temperature is ~450 °C. At the transition temperature, diesel fuel evaporates in some tens of microseconds, but about 1 second is required to evaporate a droplet at either ~200 °C or ~400 °C. Of course, gasoline is more volatile than diesel fuel, but California Phase 2 RFG does contain components that have a boiling temperature (at atmospheric pressure) that is above the estimated piston temperature (434 K). As importantly, at 434 K many of the components in gasoline are above their Leidenfrost temperature. Additionally, the experiments by Xiong and Yuen (28) were performed at atmospheric pressure, and the distillation curve in Figure 3 is also for atmospheric pressure. As the pressure increases, the boiling temperature of pure hydrocarbons increases and the distillation curve moves up to higher temperatures. For the less volatile components in gasoline, this implies that the piston temperature is even further below their boiling point (slower evaporation) and the less volatile components may be closer to their Leidenfrost point (again, slower evaporation). From the perspective of the present experiments, this indicates that liquid fuel that does not evaporate during the intake stroke is generally less prone to evaporate during compression and expansion and more prone to evaporate during the subsequent exhaust stroke or during a later cycle.

All of these injector shut-off experiments were performed for injection onto the piston. An even longer time constant might be expected for liner wetting for two reasons.

Any liquid on the liner that gets into the inter-ring zone will continue to get smeared on the liner until the liquid is depleted. Also, the less volatile fuel fragments have a greater tendency to absorb in the oil and this could also contribute to the lambda histories after injector shut-off.

SUMMARY AND CONCLUSIONS

The present study is a continuation of an investigation of the effects on HC emissions of wetting in-cylinder surfaces with liquid fuel. Because the engine is fueled predominantly with LPG, the wall wetting/slow evaporation source is isolated from the other sources of HCs for DISI engines: bulk flame extinction and low in-cylinder post-combustion oxidation temperatures. The influences of engine operating conditions and the amount of liquid fuel have been investigated. Fast FID measurements have been used to clarify the reasons that fuel wetting of different in-cylinder surfaces yields differences in the HC emissions. Injector shut-off tests have been performed to characterize the rate of fuel evaporation. The principal conclusions of this study are:

1. Liquid fuel wetting of any surface within the combustion chamber is a very important source of HC emissions, even for part load conditions that are representative of an FTP cruise. The coolant temperature has a stronger effect on HCs due to liner wetting for part load than for idle, but little effect for piston wetting for either operating condition.

2. The increase in HC emissions due to wall wetting is proportional to the amount of liquid fuel impacting an in-cylinder surface. Depositing 10% of the fuel as a liquid on the top of the piston results in approximately 30% and 70% increases in HC emissions for idle and the WWMP part load condition, respectively.

3. The results from Fast FID tests showed that 1.5 mg of liquid fuel on in-cylinder surfaces produced increased HC concentrations during the entire exhaust process. Wetting the liner under the exhaust valves had a pronounced effect on the middle peak but increased the HC concentration of all three peaks almost equally relative to operation at the same equivalence ratio with a gaseous fuel. Wetting the piston most strongly affected the final peak and also produced a small peak just before the final peak. Wetting the liner under the intake valves had the least effect on the FFID results, increasing the HC concentration of the first peak most, but yielding a somewhat larger percentage increase on the middle peak. Physical explanations for these characteristics have been proposed.

4. There are several reason to conclude that the process of evaporation of the liquid film from in-cylinder surfaces is slow:

   a. The effect of liquid fuel injection timing on the HC emissions is weak.

   b. Heat transfer considerations, including the Leidenfrost effect, dictate that surface tempera-

tures that are not much higher or lower than the boiling point can decrease the evaporation rate by two orders of magnitude, even for multicomponent fuel blends.

   c. Increased cylinder pressure during compression and expansion inhibits evaporation by shifting the distillation curve up to higher temperatures and by increasing the transition and Leidenfrost temperatures.

   d. The 1/e time constant is ~3-6 cycles for 1.5 mg of Cal. Phase 2 RFG impinging on the piston: ~3 cycles at the WWMP condition and ~6 cycles at idle. The 1/e time constant is independent of the coolant temperature due to the relatively weak dependence of the piston temperature on the coolant temperature.

The present results have been compared to a gaseous fuel baseline. This LPG-only baseline undoubtedly yields lower HC emissions than would be observed with port injection of California Phase 2 RFG at $\lambda=1.1$, especially with a coolant temperature of 40° C. In the next paper in this series, we will explore the liquid-fueled PFI baseline, examine a range of liquid fuels including pure hydrocarbons of various volatilities, and speciate the exhaust hydrocarbons.

ACKNOWLEDGMENTS

This work was supported by a grant from Ford Motor Company and conducted using the facilities of the General Motors Foundation Combustion Sciences and Automotive Research Labs on the campus of The University of Texas.

REFERENCES

1. Kim, C., and D.E. Foster (1985), "Aldehyde and unburned fuel emissions measurements from a methanol-fueled Texaco stratified-charge engine", SAE Paper 852120; also in: *SAE Trans.* Vol. 94.

2. Guillemot, P., B. Gatellier, and P. Rouveirolles (1994), "The influence of coolant temperature on unburned hydrocarbon emissions from spark ignition engines", SAE Paper 941962.

3. Crane, M.E., R.H. Thring, D.J. Podnar, and L.G. Dodge (1997), "Reduced cold-start emissions using rapid exhaust port oxidation (REPO) in a spark-ignition engine", SAE Paper 970264; also in *Journal of Fuels and Lubricants*.

4. Witze, P.O., and R.M. Green (1997), "LIF and flame-emission imaging of liquid fuel films and pool fires in an SI engine during a simulated cold start", SAE Paper 970866; also in *Journal of Engines*.

5. Fulcher, S.K., B.F. Gajdeczko, P.G. Felton, and F.V. Bracco (1995), "The effects of fuel atomization, vaporization, and mixing on the cold-start UHC emissions of a contemporary SI engine with intake manifold injection", SAE Paper 952482; also in *Journal of Engines*.

6. Saito, K., K. Sekiguchi, N. Imatake, K. Takeda, and T. Yaegashi (1995), "A new method to analyze fuel behavior in a spark ignition engine," SAE Paper 950044.

7. Alkidas, A.C., and R.J. Drews (1996), "Effects of mixture preparation on HC emissions of an S.I. engine running under steady-state cold conditions", SAE Paper 961958; also in *Journal of Fuels and Lubricants*.

8. Chen, K C., W.K. Cheng, J.M. van Doren, J.P. Murphy III, M.D. Hargus, and S.A. McSweeney (1996), "Time-resolved, speciated emissions from an SI engine during starting and warm-up", SAE Paper 961955; also in *Journal of Fuels and Lubricants*.

9. Stache, I., and A.C. Alkidas (1997), "The influence of mixture preparation on the HC concentration histories from an SI engine running under steady-state conditions", SAE Paper 972981; also in *Journal of Fuels and Lubricants*.

10. Stanglmaier, R.H., M.J. Hall, and R.D. Matthews (1997), "In-cylinder fuel transport during the first cranking cycles in a port injected 4-valve engine", SAE Paper 970043; also in *Journal of Engines* .

11. Kelly-Zion, P.L., J.P. Styron, C.F. Lee, R.P. Lucht, J.E. Peters, and R.A. White (1997), "In-cylinder measurements of liquid fuel during the intake stroke of a port-injected spark-ignition engine", SAE Paper 972945.

12. Stovell, C., R.D. Matthews, B.E. Johnson, H. Ng, and B. Larsen (1999), "Emissions and fuel economy of 1998 Toyota with a direct injection spark ignition engine", SAE Paper 1999-01-1527; also in: Emissions Formation Processes in SI and Diesel Engines, SAE Special Publication SP-1462.

13. Matthews, R.D., C. Stovell, H. Ng, B. Larsen, and B.E. Johnson (1999), "Effects of load on emissions and NOx trap/catalyst efficiency for a direct injection spark ignition engine", SAE Paper 1999-01-1528; also in: Emissions Formation Processes in SI and Diesel Engines, SAE Special Publication SP-1462.

14. Fraidl, G.K., W.F. Piock, and M. Wirth (1996), "Gasoline direct injection: actual trends and future strategies for injection and combustion systems", SAE Paper 960465; also in *Journal of Engines*.

15. Han, Z., L. Fan, and R.D. Reitz (1997), "Multidimensional modeling of spray atomization and air-fuel mixing in a direct-injection spark-ignition engine", SAE Paper 970884; also in *Journal of Engines*.

16. Alger, T., M.J. Hall, and R.D. Matthews, (1999), "Fuel spray dynamics and fuel vapor concentration near the spark plug in a direct-injected 4-valve SI engine", SAE Paper 1999-01-0497.

17. Stanglmaier, R.H., J.W. Li, and R.D. Matthews (1999), "The effect of in-cylinder wall wetting on HC emissions from SI engines", SAE Paper 1999-01-0502; also in: Direct Injection SI Engine Technology 1999, SAE Special Publication SP-1416.

18. Lenz, H.P. (1992), Mixture Formation in Spark-Ignition Engines, Springer Verlag, Vienna.

19. Roberts, C.E., and R.D. Matthews (1996), "Development of an improved ringpack model for hydrocarbon emissions studies", SAE Paper 961966; also in: *Journal of Fuels and Lubricants* **105**:1480- .

20. Glaspie, C.R., J.R. Jaye, T.G. Lawrence, T.H. Lounsberry, L.B. Mann, J.J. Opra, D.B. Roth, and F.-Q. Zhao (1999), "Application of design and development techniques for direct injection spark ignition engines", SAE Paper 1999-01-0506; also in: Direct Injection SI Engine Technology 1999, SAE Special Publication SP-1416.

21. Suh, E.S., and C.J. Rutland (1999), "Numerical study of fuel/air mixture preparation in a GDI engine", SAE Paper 1999-01-3657.

22. Collings, N., and J. Willey (1987), "Cyclically resolved HC emissions from a spark ignition engine", SAE Paper 871691.

23. Collings, N., and D. Eade (1988), "An improved technique for measuring cyclic variations in the hydrocarbon concentration in an engine exhaust", SAE Paper 880316.

24. Finlay, I.C., D.J. Boam, J. F. Bingham,.and T.A. Clark (1990), "Fast response FID measurement of unburned hydrocarbons in the exhaust port of a firing gasoline engine", SAE Paper 902165.

25. Thompson, N.D., and J.S. Wallace (1994), "Effect of engine operating variables and piston and ring parameters on crevice hydrocarbon emissions", SAE Paper 940480.

26. Min, K., W.K. Cheng, and J.B. Heywood (1994), "The effects of crevices on the engine-out hydrocarbon emissions in SI engines", SAE Paper 940306.

27. Seabrook, J., C. Nightingale, and S.H. Richardson (1996), "The effect of engine variables on hydrocarbon emissions - an investigation with statistical experiment design and fast response FID measurements", SAE Paper 961951.

28. Xiong, T.Y., and M.C. Yuen (1991), "Evaporation of a liquid droplet on a hot plate", *International Journal of Heat and Mass Transfer* **34**(7): 1881-1894.

930711

Effects of Port-Injection Timing and Fuel Droplet Size on Total and Speciated Exhaust Hydrocarbon Emissions

Jialin Yang, Edward W. Kaiser, Walter O. Siegl, and Richard W. Anderson
Ford Research Laboratory

ABSTRACT

The requirement of reducing HC emissions during cold start and improving transient performance has prompted a study of the fuel injection process. Port-fuel-injection with the Intake-valve open using small droplets is a potentially feasible option to achieve the goals. To gain a better understanding of the injection process, the effects of droplet size, injection timing, and coolant temperature on the total and speciated HC emissions were tested In a Single-cylinder engine. It was found that droplet size plays an important role in the total HC emission increase during open-valve injection, especially with cold operation. Large droplets (300 μm SMD) produced a substantial HC increase while small droplets (14 μm SMD) produced no observable increase. Increase In the total HC emissions was always accompanied by an increase in the heavy fuel components in the exhaust gases. Fuel loss during open-valve injection with cold operation was also observed when using large droplets, while no fuel loss was present with smaller droplets. The test results imply that cylinder-liner-wetting by large fuel droplets during open-valve injection is the main reason for the HC emission increase, the change in the HC species distribution in the exhaust gases, and the fuel losses. The effects of droplet size on wall-wetting were also studied through modeling using particle dynamics. The modeling results confirmed that large droplets can result in cylinder-liner-wetting, and small droplets can substantially avoid both Intake-port-wetting and cylinder-liner-wetting.

INTRODUCTION

Hydrocarbon emisslons from gasoline engines have received more attention recently due to stricter legislation on clean air. Studies have shown the fuel/air mixture preparation to be an important factor in engine HC emissions. It is known that during cold start, gasoline operation. For a conventional port-fuel-injection (PFI) engine, relatively large fuel droplets are usually injected when the intake valve is closed. During warmed-up operation, fuel vaporization and fuel/air mixing rely on the hot surface of the intake valve/port and the back flow of burned gases [1]. Fuel vaporization at cold conditions, however, is insufficient, and more fuel has to be injected to make the mixture around the spark plug rich enough to ignite, resulting in a low overall air/fuel ratio (fuel-rich condition) in the combustion chamber. A similar condition occurs during transient operation when there is Intake-port wall-wetting from fuel injection. The response of the air/fuel ratio to a variation of load/throttle will be slow. There will be a temporary air/fuel ratio increase and a torque loss at the moment of throttle opening [2, 3, 4], except when extra fuel is injected, which will cause larger HC and CO emissions. Therefore, it is desirable to avoid intake-port wall-wetting and the fuel/air mixing process should ideally not rely on fuel vaporization on hot wall surfaces. This can be reduced by injecting fuel when the intake valve opens.

When fuel is injected into an open intake-valve, the fuel droplets can enter the cylinder with the intake air if the droplets are sufficiently small. Compared to closed-valve injection, the injection quality in this case needs to be improved and the fuel-spray droplet size needs to be reduced. On the other hand, smaller droplet size usually requires more costly injection systems. Also, depending on temperature, very fine droplets may not benefit from a higher volumetric efficiency provided by larger droplets, which may not be vaporized completely during the intake process and have a higher density than that of vapor. Therefore, there must be an optimal droplet size for open-valve Injection. To determine this droplet size, the effects on hydrocarbon emissions have to be studied.

Another issue relative to open-valve injection is injection timing. Since the intake air flow varies periodically, the fuel/air mixing process will depend on injection timing. The effects of injection timing on hydrocarbon emissions also need to be studied, especially during the open-valve event.

Exhaust HC emissions consist of unburned and partially-burned fuel. When the droplet size and injection time are varied, fuel/air mixing, combustion, and HC emission formation can be different. Hence, the composition of the exhaust HC might also be different, resulting in a change in the atmospheric reactivity of the exhaust emissions. For this reason, the composition of the exhaust HC emissions were determined by gas chromatographic (GC) analysis.

EXPERIMENT

Tests were conducted on the same single-cylinder research engine used in an earlier study [5]. This engine was built on the crank case of a CFR engine which was modified so that the combustion system was similar to a production 1.9L engine with a bore of 82 mm, a stroke of 88 mm, and a compression ratio of 9. The intake port and the combustion-chamber geometry of the two-valve engine were so designed that a strong swirl flow, with an "AVL" swirl ratio of 3.0, was created. The tip of the injector was about 100 mm from the base of the intake valve stem at which it was aimed, as shown in Fig.1.

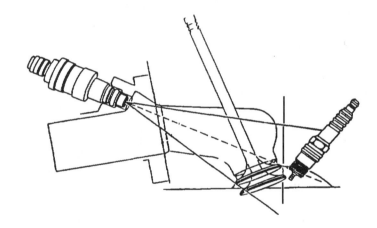

Fig.1   Schematic of intake port and fuel injection.

Three different injectors were used during the test. A production pintle gasoline injector [6] created the largest fuel droplets for this test. The Sauter mean diameter (SMD) of the fuel droplets from this injector was about 300 μm. An air-assisted injector (AAI) with fuel droplets of 40 μm SMD was also used. The third injector was an air-forced injector (AFI) [7]. It created the smallest fuel droplets, 14 μm SMD, for this test. Two spray angles, 80° and 15°, were used for the AFI injector.

The fuel used in these experiments is a multi-component gasoline with a research octane of 91 and a T_{90} distillation point of 157°C. The fuel contains 68% saturates, 31% aromatics, and 1% olefins by weight. GC analysis of the liquid fuel shows that it contains 11%

the fuel. Thus, this fuel has a somewhat more limited distribution of species than might be encountered in a typical gasoline in the market place. In particular, the aromatic content is skewed toward lighter aromatic compounds, notably toluene.

The tests were conducted under steady-state conditions, providing several advantages. Any specified parameter, such as coolant temperature, can remain constant during one test and can have a different value in a second test. Meanwhile, all other parameters that do not interact remain constant during both tests. This minimizes the interaction of different parameters. Therefore, the major effects of a single parameter on HC emissions can be identified.

The test conditions and major variables are listed in Table 1:

Table 1. Operating conditions.

speed	1500 rpm
imep	3.78 bar
EGR	0%
spark timing	MBT*
A/F ratio	16.2; 12.7 (Φ= 0.9; 1.15)
coolant T**	30 ± 4°C; 89 ± 1°C
oil T***	43 ± 3°C; 76 ± 1°C
inlet T	28 ± 2°C
droplet size	300; 40; 14 µm SMD
inj. timing	variable
fuel	gasoline (91)****

* location of peak pressure at 13 CAD ATDC
** at the outlet of cylinder block
*** lower T correrponds to lower coolant T
**** as described above

The amount of fuel injected was controlled by the electronic pulse width. The pulse widths of injection for the three injectors are listed in Table 2.

Table 2. Injection pulse width.

injector:	A/F 16.2	A/F 12.7
production	4.6 ms 41. CAD	5.2 ms 47. CAD
air-assisted	6.4 ms 58. CAD	7.9 ms 71. CAD
air-forced* (p.w. of air)	6.0 ms 54. CAD	6.0 ms 54. CAD

*Air-forced injector relies on two electric pulses. One pulse controls the amount of fuel into the injector, the other controls the compressed air, which blows the fuel out of the injector [7].

A standard emission bench was used to analyze dried exhaust gases for HC, NO, CO, CO_2 and O_2. To avoid hydrocarbon loss during the exhaust-gas-dehydration process, a hot FID (HFID), Horiba FIA-34A-2, was also employed. The air/fuel ratio in this test was determined by three different methods. The first method involved measuring intake air and fuel mass flows. The other two methods depended on exhaust gas measurements. One was based on a calculation of the exhaust-gas composition, chemical Spindt [8]; the other was based on a NTK universal exhaust gas oxygen sensor (UEGO).

In addition to measuring total HC emissions, samples for GC analysis were taken from the exhaust pipe approximately 6 feet downstream of the exhaust port. The samples were withdrawn through a heated sample line to minimize HC sample loss and were stored in an evacuated Pyrex flask at a pressure of approximately 70 torr. After dilution to 1 atm with nitrogen, the samples were analyzed by a GC technique described elsewhere [9]. These analyses determined the concentrations of C_1 to C_{12} hydrocarbon species with little mutual interference. No analysis of oxygenated organic species was performed.

TEST RESULTS AND ANALYSIS
Total HC Emissions

The exhaust HC emissions measured by the HFID under the fuel-lean condition are plotted in Fig.2 versus the start-of-injection (SOI) timing at two different coolant temperatures using three different injectors. Similar results but under the fuel-rich condition are plotted in Fig.3. Analysis of the effects of these parameters can provide insight into the HC emission mechanism. It was found that the HC emissions measured on different days can vary by ±10% from the mean. This variation can result from slight changes in the operating parameters, combustion chamber deposits, and other reasons. However, the effect of injection timing on HC emissions during an injection timing scan is quite consistent. In other words, the shape of the curves in Figs.2 and 3 is quite reproducible even though the curve itself might be shifted up or down. In the following discussion, therefore, the HC emissions for each data point are divided into two parts: the HC emissions corresponding to closed-valve injection, HC_{closed}, and the HC emissions increase during open-valve injection, ΔHC_{open}. The total HC emissions will be

$$HC_{total} = \Delta HC_{open} + HC_{closed}.$$

We will focus on the ΔHC_{open} since it is more reproducible.

Figures 2 and 3 show that the droplet size has a significant impact on HC emissions during open-valve injection, particularly at the lower coolant temperature. The largest droplets (300μm SMD) caused the largest increase in HC emissions, while the smallest droplets (14μm SMD) produced no HC emission increase within experimental error. The effects of droplet size on

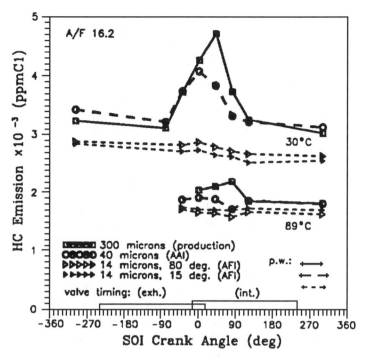

Fig.2 Total HC emissions at fuel-lean operation.

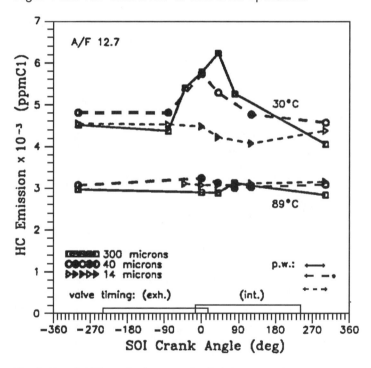

Fig.3 Total HC emissions at fuel-rich operation.

ΔHC_{open} were much smaller at high coolant temperature. Both droplet size and coolant temperature play an important role in this HC emission increase during open-valve injection.

When using the largest droplets (300μm SMD) at low coolant temperature (30°C), the HC emission level during open-valve injection was 50% higher than that of closed-valve injection as shown in Fig.2. Since the conventional sources of HC emissions, such as the crevice volume and oil film thickness, do not change with injection timing, this HC emission increase must result from effects caused by the droplets entering directly into the cylinder. When large droplets enter the cylinder with the intake air during open-valve injection, some of these fuel droplets might be centrifugally separated from the swirl flow and impinge on the combustion chamber surfaces, including the cylinder liner. Fuel impingement on the cylinder liner is especially undesirable since some of the fuel can stick on the liner surface resulting in cylinder-liner wetting. This fuel film can be trapped in the piston-liner crevice when the piston moves up. Some of the trapped liquid fuel might be vaporized in the crevice, and eventually flow out of the crevice with the trapped gases when the cylinder pressure deceases. Some of the remaining liquid fuel might be left on the liner surface when the piston moves down and be vaporized later in the expansion process. Components of this fuel may also be absorbed/desorbed by the oil layer on the liner. The vaporization rate of the fuel on the cylinder liner is the slowest since the average temperature on this surface is the lowest in the whole combustion chamber. In contrast, other combustion-chamber surfaces, such as the piston and cylinder head surfaces, have higher temperatures which help to vaporize the fuel film. A fuel film on these surface cannot be trapped in the piston-liner crevice. it can be vaporized and burned during the main combustion process. Wetting of these surfaces may not be as important. This speculation is supported by the conclusions of Frank and Heywood [10] from tests in a spark-ignited direct-injection engine, in which fuel was injected on the piston surface. They claimed that a 50°C change of piston temperature had small effect on HC emissions. Therefore, it is speculated that the HC emission increase for the production injector in Fig.2 which represents 1.33% of the total intake fuel charge results mainly from cylinder-liner-wetting by the fuel.

It is also possible that some of this HC emission increase comes from incomplete oxidation of local rich regions generated by the large droplets, as speculated by Matthes and McGill [11], and Quader [12]. Incomplete oxidation of local rich regions should increase CO emission, because the CO concentration is very sensitive to air/fuel ratio under fuel-rich conditions. However, the increase of measured CO emission during open-valve injection was not significant and consistent. The CO emission during open-valve injection under fuel-lean condition and low coolant temperature increased by 8% above that observed during closed-valve injection as shown in Fig.4. This is a much smaller increase than the 50% increase in the HC emissions shown in Fig.2 Under the fuel-rich condition (A/F 12.7), the CO emission actually decreased by 5.5% during open-valve Injection as shown in Fig.5 Therefore, incomplete oxidation of local rich regions generated by the large droplets might not be as important a contributor to the HC emission increase as compared with the effects of cylinder-liner-wetting.

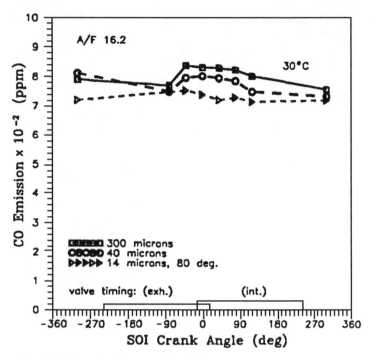

Fig.4 CO emissions at fuel-lean operation.

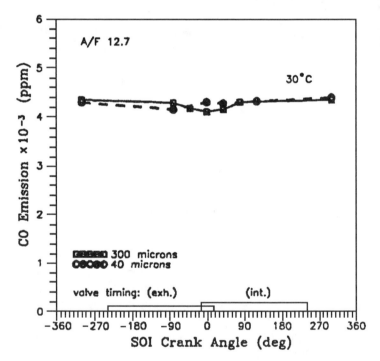

Fig.5 CO emissions at fuel-rich operation.

ber with a higher swirl ratio resulted in higher HC emissions with port fuel injection. Besides other possible reasons for this phenomena, droplet trajectory may play an important role. Higher swirl motion of the gases can result in more fuel droplets being separated centrifugally from the gases and more cylinder-liner-wetting. Another effect of higher swirl includes higher bum rate. Higher bum rate can result in lower exhaust temperature and greater crevice loading, which may also lead to higher HC emissions. Alternatively, gas swirl motion will generate turbulence near the chamber walls which promotes vapor/air mixing. Therefore, if locally rich regions produced by large droplets exist, they will tend to be homogenized by the stronger swirl motion, and the associated HC emission increase will tend to be reduced. However, the HC emissions in Quader's tests actually increased with increased swirl flow and port injection. This implies that incomplete oxidation of focally rich regions may not play an important role in the HC emission process for engines with strong swirl.

For high coolant temperature (89°C), Δ HC for the large droplets was reduced to 0.32% of the total intake fuel charge. The decrease of ΔHC_{open} from 1.33% to 0.32% of the total intake fuel caused by increasing coolant temperature probably results from a higher vaporization rate of the fuel film, even though a comparable amount of fuel impinges on the combustion chamber surface.

For the 40μm droplets, ΔHC_{open} was smaller (0.88% of the total fuel) at 30°C coolant temperature. For the smallest droplets (14μm SMD), no measurable HC emission increase at the tow coolant temperature was observed. At the high coolant temperature, the smaller droplets, both 14μm and 40μm SMD, might produce slightly lower HC emissions during open-valve injection, although the significance of this decrease is marginal.

It is believed that the dependence of ΔHC_{open} on droplet size occurs because smaller droplets are less likely to be separated from the flow, resulting in fess liner-wetting. The dependence of cylinder-liner-wetting on droplet size will be discussed later.

Changes in injection timing during open-valve injection produced different HC emission increases for the same injector and coolant temperature. As shown in Figs.2 and 3, the largest HC emission increases always occur when fuel injection starts early in the intake stroke. In fact, measurable HC emission increases ΔHC_{open} are observed only when injection starts before, or at, 80 CAD ATDC (crank-angle degrees after top-dead-center of gas exchange). When injection starts late in the intake stroke, i,e., at or after 120 CAD ATDC, even the largest droplets produce similar HC emissions to those during the closed-valve injection.

The effect of injection timing on HC emissions while the intake valve is open might result from the following two observations. First, altering the injection timing can result in changes in cylinder-liner-wetting. When fuel is injected late during the intake stroke, the piston has already moved away from the intake valve, and a large percentage of the inducted air has already entered the cylinder. The intake swirl flow is getting weak. Therefore, the droplets will take a longer time of travel to reach the walls, resulting in more time for vaporization. Fewer fuel droplets will be separated centrifugally from the gas flow and impinge on the liner surface.

Second, changes in injection timing can result in charge stratification in the axial direction. At 90 CAD ATDC of gas exchange, probably more than half of the total mass has already been trapped In the cylinder. This fraction of the total mass contains all of the residuals and tends to be located in the lower part of the cylinder at bottom-dead-center (BDC). When fuel is injected late during the intake stroke, the droplets enter the cylinder with fresh air, and will tend to remain In the top part of the combustion chamber if they are sufficiently small. The in-cylinder swirl motion promotes stronger mixing in the radial direction than in the axial direction. Therefore, axial charge stratification, at least at the beginning of the compression stroke, is most likely present for some open-valve injection timings [14, 15]. Axial stratification may, therefore, result in a tower hydrocarbon density near the piston/liner crevice region which can manifest itself in lower HC emissions from the crevice.

Some level of charge stratification appears to exist in this engine. Experiments show that the burning rate and maximum cylinder pressure vary during the injection-timing scan with large droplets (300 μm and 40 μm).

When fuel injection started just before the intake Valve opened, i.e., injecting slightly before TDC, ΔHC_{open} was moderate. It was higher than that of closed-valve injection but tower than the peak value during open-valve Injection, as shown in Figs.2 and 3 for large droplets. Fuel/air mixing in this case is a transition from closed to open-valve injection.

As mentioned earlier, because of experimental variability, less significance can be attached to the differences between the absolute HC emissions during closed-valve injection, HC_{closed}, for the three injectors. Therefore, no quantitative comparison is made for the HC_{closed}. However, in general, these HC emissions are closer to each other, even at tow coolant temperature, than is shown during open-valve injection. This is because the air/fuel mixing in this case depends mainly on hot-surface vaporization and the back flow of burned gases into the intake port during the valve overlap period. As long as the droplets can reach the surfaces of the intake port/valve and form a liquid fuel film, the remaining mixing and vaporization process will be similar. The only differences will be the residence time for fuel vaporization between injection and intake valve opening and the geometry (area and location) of the liquid fuel film.

HC Species Disributions

For all three injectors, exhaust-gas samples were withdrawn and analyzed by gas chromatography. During each series of experiments, a sample was first taken with the start of injection at 300 CAD ATDC of gas exchange (intake valve dosed), followed by samples at 0° (open), 40° (open), and -300° (closed). This provided two closed intake valve and two open valve data points at each operating condition, permitting a check of the reproducibility of the experiments over a period of approximately 4 hours. Comparing data over a relatively short period of time is important to the determination of small changes in HC species distributions resulting from changes in engine operating conditions. Both fuel-rich and fuel-lean equivalence ratios (Φ = [F/A]/[F/A]$_{stoic}$) were tested at the two wall temperatures.

Because well over 100 species can be identified in the engine-out emissions from gasoline fuel, we choose

63

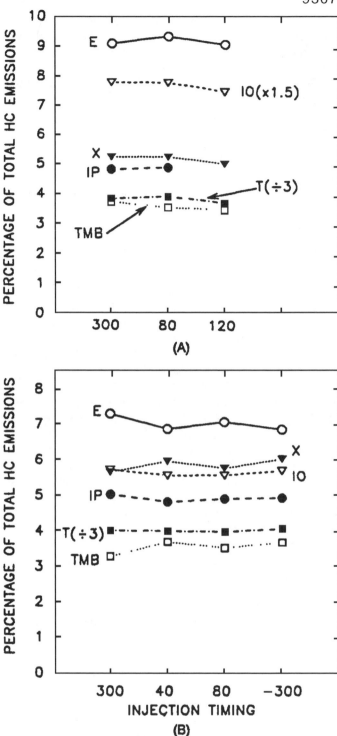

in the interest of clarity to present data on only selected species. These are representative of the classes of hydrocarbons present in the exhaust gas and in sum constitute approximately 40% of the total HC emissions. The species selected are: ethylene, a major combustion product; isopentane, a light aliphatic fuel component; isooctane, a heavier aliphatic fuel component; toluene, a light aromatic fuel component; the sum of all xylenes and the sum of all trimethyl benzenes, which are representative of the heavier aromatic components in the fuel. In, all experiments, the trends reported for ethylene are representative of these for methane, acetylene, and propylene to within the experimental error of ~ ±5%.

As discussed above, with an engine coolant temperature of 89°C, ΔHC_{open} is small for the production injector and zero for both the AAI and AFI injector. The percentage contribution of each of the selected species to the total HC emissions for each injector also shows no significant change at this coolant temperature when the injection timing is varied. This is illustrated in Fig.6, which presents the percentage contribution of the selected HC species to the total HC emissions at four injection timings for the AAI injector during fuel-rich operation and three timings during fuel-lean operation. The notation of the species in this and the following figures are listed in Table 3.

Table 3. Notation of the species shown in figures.

E	ethylene
IP	isopentane
IO	isooctane (2,2,4-trimethyl pentane)
T	toluene
X	sum of xylenes
TMB	sum of trimethyl benzenes

The abscissa in these figures is not drawn to scale: for 300 and -300 CAD, injection occurs with the intake valve closed; for -40, 0, 40, 80 and 120 CAD, injection occurs with the intake valve open.

At 30° C coolant temperature, however, significant differences are observed. Figures 7A and 7B present results obtained with the production injector during fuel-lean and fuel-rich operation, respectively. At both equivalence ratios, the contributions of the combustion product, ethylene, and the light fuel component, isopentane, to the total HC emissions decrease during open-valve injection. The percentage contributions of toluene and isooctane remain essentially constant, while the heavier aromatics, xylene and trimethyl benzene, show a substantial increase in importance. The injection timing scan in Fig.7 for each equivalence ratio was obtained on a single day as described above. The fuel-lean condition was repeated on a second day to test the experimental reproducibility. While the total HC emis-

Fig.6 Percentage contribution of selected HC species to the total HC emissions as a function of the start of injection timing for the AAI injector at 89° C and A/F of 16.2 (A) or A/F of 12.7 (B).

distribution remained similar to those presented in Figs.2 and 7A.

Qualitatively, the same effect is observed for the AAI injector as illustrated in Fig.8B for a rich fuel-air ratio using low coolant temperature. However, the magnitude of the change is less as might be expected because the increase in total HC emission during open-valve injection, ΔHC_{open}, is approximately 1/2 of that observed for the production injector. During fuel lean operation with the

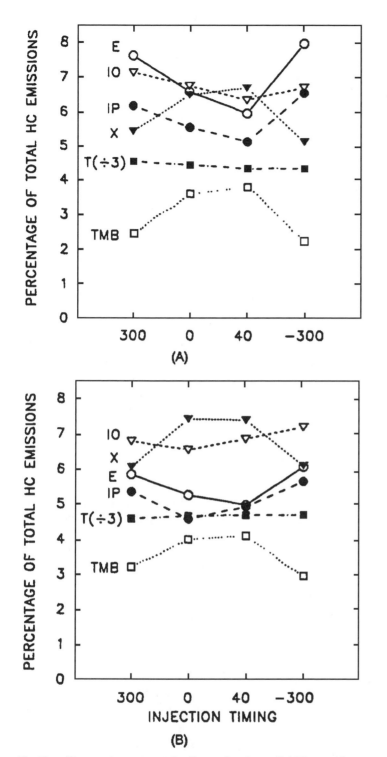

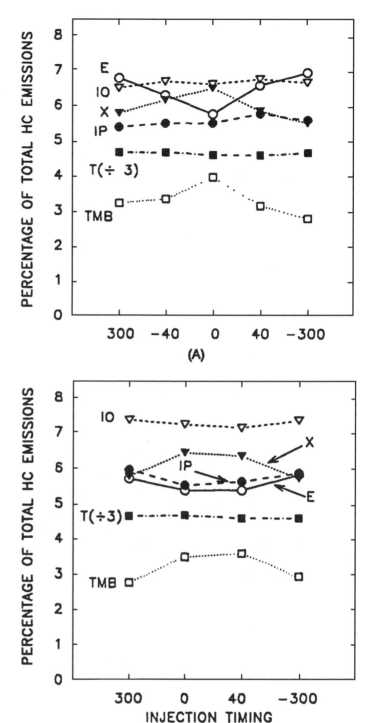

Fig.7 Percentage contribution of selected HC species to the total HC emissions as a function of the start of injection timing with 300 μ m droplets at 30° C and A/F of 16.2 (A) or A/F of 12.7 (B).

Fig.8 Percentage contribution of selected HC species to the total HC emissions as a function of the start of injection timing with 40 μm droplets at 30°C and A/F of 16.2 (A) or A/F of 12.7 (B).

AAI injector, similar trends in species distributions are observed as shown in Fig.8A. However, the effect of injection timing is localized within a somewhat narrower region near TDC. Experiments were repeated at 300, 0, and -300 CAD under the conditions of Fig.6A on a second day to verify the reproducibility of these data Essentially identical trends in the distribution of HC species and in the total HC emission were observed in both data sets.

Limited species data were obtained for the AFI injector. One injection timing scan was run both at Φ = 0.9 and at Φ = 1 .15 with no repeat run of either data set. As shown in Fig.9, particularly under fuel-rich conditions, the trend in species distribution seems to follow a pattern opposite to that obtained from the production and AAI injectors. For the API injector, the percentage contributions of the heavier aromatics tends to <u>decrease</u> while the lighter components increase during open-valve

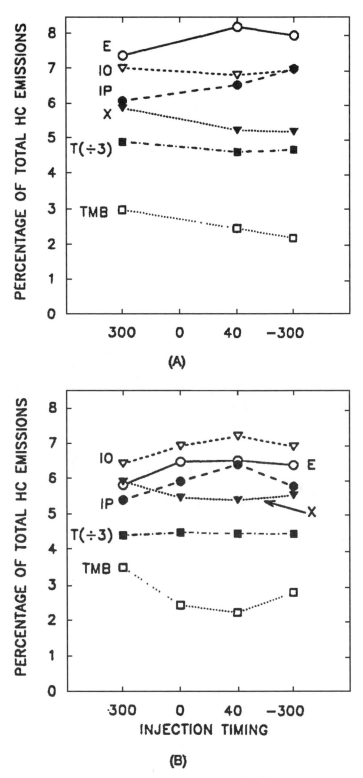

Fig.9 Percentage contribution of selected HC species to the total HC emissions as a function of the start of injection timing with 14 μm droplets at 30° C and A/F of 16.2 (A) or A/F of 12.7 (B).

injection. We caution, however, that neither of the AFI experiments was repeated in contrast to the production and AAI injectors, and the data are, therefore, very limited.

These results demonstrate clearly that fuel injection timing can affect the distribution of HC species in the

exhaust gas. For both the production and AAI injectors, open valve injection at 30° C coolant temperature produces increased emissions of the heavier fuel components relative to the fighter ones and to the light combustion products. This observation is consistent with the earlier suggestion that the increased emission during open-valve injection for both these injectors results from impingement of fuel droplets onto the cylinder wall. Wetting the wall with fuel might be expected to result in an increase in the exhaust emissions of less volatile components, because these components will be slower to vaporize and will not, therefore, be burned as efficiently by the flame. Operating at higher wail temperature reduces the effect for the production injector and eliminates it for the AAI. The effect of wall temperature is also consistent with the above wall wetting hypothesis since hotter wall temperatures will produce higher gas temperatures, promoting the vaporization of fuel droplets during injection and assisting vaporization of any fuel deposited on the walls.

However, the above explanation cannot account for the seemingly opposite trends noted from the AFI injector, which should have the feast wall wetting. We have no certain explanation for these fatter observations at present although increased charge stratification during open-valve injection of the very small droplets from the AFI might cause the lower vapor pressure components to remain closer to the canter of the chamber because they will vaporize more slowly. This might reduce the amount of these species stored in crevice volumes and oil films. However, this explanation is speculative and additional data confirming these initial results are needed.

Because a change in the HC species distribution could also alter the atmospheric reactivity of the engine-out emissions, a calculation of the exhaust gas reactivity was performed for the production injector data in Fig.7 at 300 and 40 CAD injection timing. In carrying out these calculations, all species measured by the GC were included, and the most recent Maximum Incremental Reactivities (MIR) provided by the California Air Resources Board were used. These calculations indicate reactivities of 4.2 and4.2 gO_3/gHC for 300 and 40 CAD at Φ= 0.9; for Φ = 1 .15, the calculated reactivities are 3.5 and 3.6, respectively. These results indicate that for this injector essentially no reactivity difference is observed even though significant changes in the species distributions occur. However, because the species distribution does change, the potential for change in the exhaust gas reactivity certainly exists and should be considered for any injector-engine combination.

As discussed above, the fuel used In these experiments contains relatively large percentages of isopentane, isooctane, and toluene. This is not characteristic of typical broad-cut gasolines, which will have larger percentages particularly of heavier aromatics, and the results observed in these experiments could be accentuated in fuels containing larger amounts of heavier aromatic species.

Fuel Consumption Effects

A comparison of the air/fuel ratio from intake mass flow measurement to that from exhaust gas composition measurement can be used to determine whether there is a loss of fuel within the engine, as has been reported by Quader et al. [12] in their study. Shown in Fig.10 are the air/fuel ratios obtained by three different measurement techniques at the high coolant temperature (89°C) using the production injector (300μm SMD droplets) under the fuel-lean condition. Generally speaking, the three values are within a 0.1 A/F ratio of each other which is the measurement capability of the different techniques.

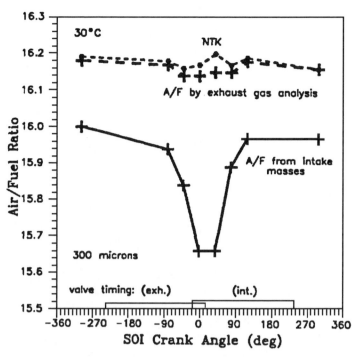

Fig.11. Air/fuel ratios measured by three methods at cold, fuel-lean, condition with 300 μm SMD droplets.

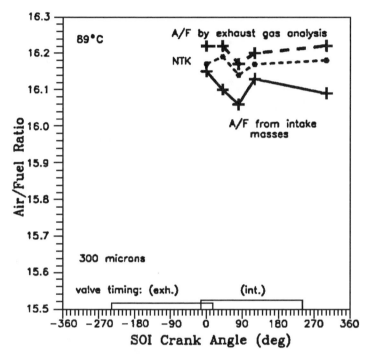

Fig.10 Air/fuel ratios measured by three methods at warmed-up, fuel-lean, condition with 300μm SMD droplets.

When the coolant temperature is low (30°C), however, significant differences are observed, as shown in Fig.11. Similar results can be seen in Fig.12 for the fuel-rich condition. The air/fuel ratio determined from the intake mass flows is tower than that from the two exhaust-gas methods, chemical Spindt and NTK. This difference suggests that some fuel is lost in the cylinder, resulting in leaner operation than is expected from the mass flow measurement. The fraction of input fuel that is lost can be calculated by

$$\frac{\Delta m_{fuel}}{m_{fuelin}} = \frac{\Delta A/F}{A/F_{exh}}$$

The air/fuel ratios in the exhaust were controlled during the test; therefore, they are relatively invariant during the injection timing scan, as shown in Fig.11. However, the air/fuel ratio from the intake mass flow varied during the injection timing scan. There is a valley in the curve, indicating that additional fuel is lost from the cylinder

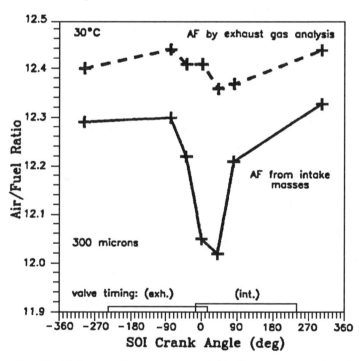

Fig.12 Air/fuel ratios measured by three methods at cold, fuel-rich, condition with 300 μm SMD droplets.

during open-valve injection. This air/fuel ratio difference can be divided into two parts: the difference $\Delta A/F_{closed}$ between the exhaust and intake analyses at closed-valve injection; and the difference $\Delta A/F_{open}$ between open and closed valve injection in the curve of air/fuel ratio from the intake-mass flows. Thus, the total difference in air/fuel ratio can be expressed as

$$\Delta A / F_{total} = \Delta A/F_{open} + \Delta A/F_{closed}$$

similar to the expression presented for HC emissions. Correspondingly, the fuel loss can be divided into two parts,

$$\Delta m_{fuel} = m_{f.open} + \Delta m_{f.closed}$$

If there were system errors involved, such as air leakage, set point repeatability, or calibration error, the A/F curves would be shifted from one another. Therefore, the air/fuel ratio difference $\Delta A/F_{closed}$ might result from systematic inaccuracies in the measurements. This type of error, however, would not affect the shape of the curves since the data for each injection timing scan were taken on the same day, and any system errors should have been constant. Thus, the values of $\Delta A/F_{open}$ and $\Delta m_{f.open}$ should be reliable and are used in the following analyses.

The depth of the valley, $\Delta A/F_{open}$, in the air/fuel ratio curve of Fig. 11 indicates that about 1.9% of the total fuel flow was lost at the cold condition. About 1% of the total fuel was lost using medium sized droplets (40μm SMD) at the same low coolant temperature, as shown in Fig.13. No fuel loss was found with the smallest droplets (14μm SMD), as shown in Fig.14. The shift between the results of the NTK and chemical Spindt methods, shown in Fig.14, is systematic and larger than that in other tests. But, it is still within the error range of the NTK system.

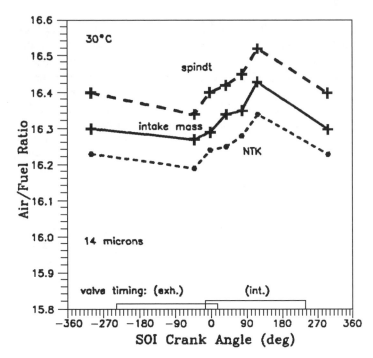

Fig.14. Air/fuel ratios measured by three methods at cold, fuel-lean, condition with 14μm SMD droplets.

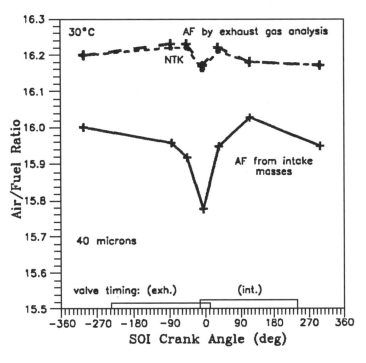

Fig.13. Air/fuel ratios measured by three methods at cold, fuel-lean, condition with 40μm SMD droplets.

The loss of fuel resulted in increased fuel consumption. Figure 15 shows the indicated specific fuel consumption (ISFC) of the three injectors at the same low coolant temperature. The peaks in the curves indicate

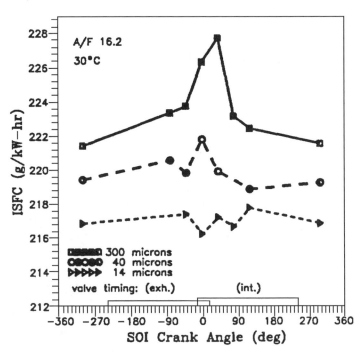

Fig.15. Measured ISFC at cold, fuel-lean, condition using three injectors.

that more fuel, $\Delta m_{f.extra}$, has been added during the open-valve injection operation.

The relationships between the amount of fuel lost $\Delta m_{f.open}$, the increase of fuel consumption $\Delta m_{f.extra}$ and the increase of HC emission ΔHC_{open} can be identified by studying the magnitudes and phases of the different systems. To study their magnitudes, the values of these three parameters for each of the injectors are shown in Table 4.

Table 4. Comparisons of three parameters at open-valve injection as percent of total fuel under fuel-lean and cold operating condition.

droplet size	$300 \mu m$ SMD	$40 \mu m$ SMD	$14 \mu m$ SMD
ΔHC_{open}	1.33%	0.8%	$0 \pm 0.2\%$
$\Delta m_{f.open}$	1.9%	1.0%	$0 \pm 0.3\%$
$\Delta m_{f.extra}$	2.7%	1.3%	$0 \pm 0.5\%$

For each injector, the fuel loss $\Delta m_{f.open}$ is the same order as the HC emission increase ΔHC_{open} The ratios of ΔHC_{open} to $\Delta m_{f.open}$ of the first two injectors are about the same: slightly smaller than unity. The summation of the HC emission increase and the fuel loss should be roughly equal to the increase of the fuel consumption $\Delta m_{f.extra}$ However, in Tab. 4, this conclusion is not exactly right, which might result from their measurement accuracy range, ±0.2%, ±0.3% and ±0.5% of total fuel, respectively.

The phases of these parameters were also consistent. Figures 2, 10, 11 and 15 show that the maximum (or minimum) for each injector occurred at the same injection timing. These results indicate that the three phenomena are related.

It was suggested earlier that the increase in HC emissions during open-valve injection results mainly from the cylinder-liner-wetting. The fuel lost during open-valve injection must also be related to the liner-wetting. This explanation is supported by the observation that liner-wetting can lead to fuel loss to the crank case, which is the most feasible destination of the lost fuel. The unvaporized fuel film on the cylinder liner will be trapped in the piston/liner crevice during the compression stroke. In addition, some of the liquid fuel can be left on the liner surface below the rings. If the piston rings move up and wn relative to their grooves, some of liquid fuel in the crevice can be pumped down to the crank case through the back of the rings. Some of the fuel in the crevice can be blown down to the crank case with the blowby gases as well. All these phenomena will cause loss of fuel. When the wall temperature is higher, less fuel will be trapped in the crevice and less fuel will be lost, consistent with the experimental results.

Next, we consider the difference in air/fuel ratios obtained by different measurement approaches for closed-valve injection, i.e., $\Delta A/F_{closed}$ As mentioned before, this value, as well as the fuel loss $\Delta m_{f.closed}$ is less reliable than $\Delta A/F_{open}$ since there is a possibility that system errors are involved. However, we believe that the

ture using large droplets. Therefore, analyses of the air/fuel ratio difference $\Delta A/F_{closed}$ and the fuel lost, $\Delta m_{f.\ closed}$ under these conditions might still be meaningful. Figures 11 and 13 imply that about the same amount of fuel, 1.% of the total fuel, are lost for both $300 \mu m$ and $40 \mu m$ SMD droplets at a low coolant temperature with closed-valve injection. This suggests that the fuel evaporation and the air/fuel mixing in this case might still be insufficient and cylinder-liner-wetting might still be present.

DROPLET DYNAMICS

The behavior of droplets in the intake port and the cylinder can be analyzed by particle dynamics. The phenomena involved are quite complicated. They include the transient air motion in a geometrically compii-cated container, the interaction between air motion and droplet motion which results in air entrainment, jet/droplet breakup and droplet coalescence [16], and the evaporation process of the droplets with interaction between the large amount of droplets. Therefore, it is impossible to simulate all the phenomena without multi-dimensional modelling. in the following, however, a rather simple model will be presented which shows the basic behavior of a single droplet moving in air. All of the conditions and parameters in the simulation are consistent with the experiment. The calculation improves the understanding of the effect of droplet size on cylin-der-liner-wetting and HC emissions.

The drag force of air on a droplet can be calculated by [17]

$$F_d = CA_p \rho u^2/2$$

where Fd is drag force, C is drag coefficient, A is projected area of the droplet in the direction of m&on, p is surrounding gas density, and u is the relative velocity between droplet and gas. For the range of Reynolds numbers (Re < 100) calculated in this study, the drag coefficient C can be determined by [17]

$$C = \frac{24}{Re}(1+0.14Re^{0.7})$$

which describes the intermediate region between Stoke's law and Newton's law regions.

To study the response of a droplet which has a velocity different from the surrounding air, calculated droplet speeds versus time (start from stationary) are plotted in Fig.16. As shown, small droplets (<20 μm) can catch up with the air flow quickly (in less than 2 ms or 18 CAD), while the large droplets (300 μm and larger) can hardly follow the gas motion. In the intake port, this implies that the large droplets can reach the walls after injection without significant disturbance by the gases. During open-valve injection, many of the large droplets can impinge on the valve or port surfaces first, rebound, then enter the cylinder with the intake air. In contrast, the small droplets are entrained in the air quickly after

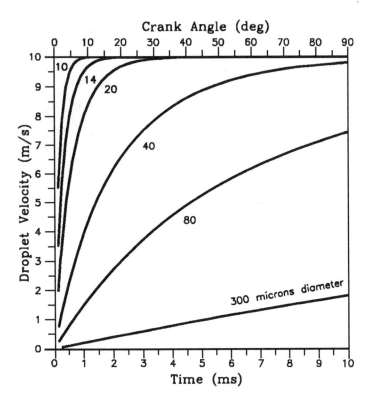

Fig.16. Calculated droplet velocity after emerging into an air flow of 10 m/s.

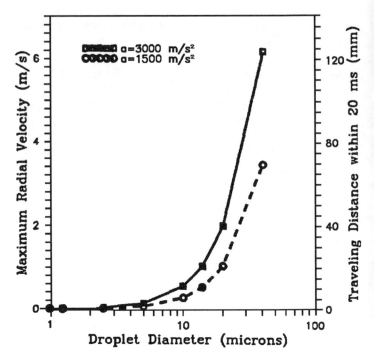

Fig.17. Terminal radial velocities of droplets at two different radial accelerations generated by centrifugal force of swirl flow.

injection and move with it, without significant wall wetting. In the combustion chamber, this means that small droplets can move with the air flow - subject to centrifugal force. However, the large droplets tend to move along their initial direction. They need more time to catch up with the in-cylinder fluid motion.

Then, the effect of centrifugal force on droplet motion is studied. Assuming the droplets can follow the air swirl motion, the calculated droplet radial velocity component generated by the centrifugal force, or the corresponding traveling distance in the radial direction if it moves at this velocity for 20 ms (i.e., 180 CAD), is plotted in Fig.17 for different droplet sizes. To estimate the centrifugal force, the droplets are assumed to be located at a 35 mm radius from the cylinder center line, and the air swirl speeds there are assumed to be 10.26 or 7.25 m/s, which results in a local radial acceleration of 3000 or 1500 m/s^2 respectively. The higher speed is estimated for this engine at 1500 rpm, while the tower speed is to simulate the phenomena at low speed or in low-swirl-ratio engines. Generally speaking, the calculated results confirmed the speculation that large droplets can result in cylinder-liner-wetting. The radial velocity of small droplets (10 μm and smaller), as shown is so small that the air swirl motion cannot cause significant cylinder-liner-wetting. However, the radial motion of large droplets caused by the centrifugal force can be significant, if they can follow the air swirl flow. In this ease, cylinder-liner-wetting can become a significant problem. Fortunately, larger droplets take a longer time to catch up with

wetting by the large droplets, though large, cannot be as significant as implied by Fig.17.

The size of fuel droplets In an engine will reduce though fuel evaporation and droplet breakup [16]. The evaporation rate of a droplet depends on the velocity difference between the droplet and surrounding air, the droplet and air temperatures, as well as the fuel-vapor concentration in the surrounding air which represents the Interaction between droplets. The droplet breakup depends mainly on the velocity difference between the droplet and surrounding air. Droplet s&e can affect both the evaporation rate and the droplet breakup. Larger droplets take a longer time to catch up with the air motion, as shown in Fig.16. Therefore, they have more time for droplet breakup and have a higher evaporation rate per droplet surface area But, smaller droplets can have a faster evaporation rate per unite mass due to 1) a larger surface/volume ratio

$$\frac{A}{V} \propto \frac{1}{d}$$

where A is the surface area, V is the volume, and d is the diameter of a droplet, and 2) a faster heating rate per unit mass is the surrounding temperature is higher

$$\frac{q}{m} \propto \frac{1}{d}\Delta T$$

where q is heat transfer rate and m is mass of the droplet.

The analysis shows that wall-wetting in both the intake port and the cylinder liner can be avoided substantially for droplets smaller than 10 μm, or completely avoided for droplets smaller than 5 μm. Since the fuel droplets will continuously reduce their size through evaporation, the desirable droplet size for an injector can be larger than that indicated by Fig.17, as 10 μm or 5 μm. The injector with 14 μm SMD droplets might be quite sufficient, though a further small reduction of HC emissions from the results of the 14 μm SMD droplets of these tests may still be possible.

The analysis also shows a strong influence of droplet size on cylinder-liner-wetting. The largest droplets in a spray are more pertinent to cylinder-liner-wetting. Therefore, the size distribution in a spray should also be known to characterize the injector for its effect on HC emissions.

SUMMARY AND CONCLUSIONS

Total hydrocarbon emissions, HC species distribution, air/fuel ratio, and fuel consumption were measured In a single cylinder engine under steady-state operation using three different injectors. The effect of injection timing and coolant temperatures were investigated. The test results show that:

1. Fuel droplet size played an important role in the HC emissions during open valve injection for the cold coolant temperature (30°C). Large droplets (300μm) produced 50% higher HC emissions, about 1.33% of the total fuel, at open-valve injection than at closed-valve injection. Small droplets (14μm), however, produced no HC emission increase compared with closed-valve injection;

2   For 89°C coolant temperature, the effects of droplet size on the HC emissions were much smaller. For small and medium-size droplets (14μm and 40μm) there was no HC emission increase during open-valve injection;

3. The HC emission increase during open-valve injection occurred only when the injection started in the first half of the intake stroke. When injection started later in the intake stroke, even the largest droplets did not cause a measurable HC emission increase;

4. For the production and AAI injectors, increases in the total HC emissions during open-valve injection were accompanied by changes In the distribution of HC species in the exhaust gas. The percentage contribution of the heavier aromatics, toluene, xylene, and trimethyl benzenes increased while the percentage of lighter combustion products and fuel components decreased;

5. Essentially no reactivity difference (gO₃/gHC) in the exhaust hydrocarbons was observed even though significant changes in the species distributions occurred during open-valve injection at the cold condition;

6. Open-valve injection with large droplets resulted in fuel loss. The amount of the fuel lost during open-valve injection was found to be of the same order, and a little bit higher than, the HC emission increase. The maximum values of the lost fuel and the HC emission increase were also found to be in phase during the injection timing scan;

7. Fuel consumption also increased during open-valve injection with large droplets. The fuel-consumption increase was close to the summation of the lost fuel and the HC emission increases. It's maximum value was also in phase with the other two during the injection-timing scan

The movement of a single droplet in air was simulated using particle dynamics. It was shown:

1. Using droplets smaller than 5 - 10 μm can sub stantially avoid both cylinder-liner-wetting and intake-port-wetting;

2. Larger droplets have higher radii velocity generated by the centrifugal force when they follow the gas swirl motion. This radial movement of the droplets can result in cylinder-liner-wetting:

3. Large droplets (300 μm and larger) cannot catch up with the air motion during the available time in the intake port or in the combustion chamber. They can produce intake-port-wetting;

4. Since the behavior of a droplet is so sensitive to its size, the size distribution is also an important parameter to characterize a spray, in addition to the average size (e.g. SMD).

The test and modeling results can help to understand the HC emissions during open-valve Injection. It is speculated that:

1. All of the changes during open-valve injection with large droplets, such as the HC emission increase, the increased contribution of the heavier HC species, the fuel toss, and the increase in fuel consumption, mainly result from cylinder-liner-wetting by the fuel droplets and trapping of this liquid fuel in the piston/liner crevice;

2. Wetting of other combustion-chamber surfaces and incomplete oxidation of locally rich regions which are generated by the large droplets may play some role in the HC emission increases. But their effects probably are much smaller than the cylinder-liner-wetting:

3. A droplet size of 14 μm SMD is probably sufficient for open-valve injection during cold and transient operation. Droplets smaller than 10 μm may not be needed.

ACKNOWLEDGEMENT

The authors would like to thank Dave Cotton for working in the test cell and taking data, Fred Trinker for discussions and suggestions related to this study, Bill Clemons for providing the air-assisted injector, and Michael Schechter and Levin for providing the air-forced injector.

REFERENCES

1. C.-O. Cheng, W.K Cheng, J.B. Heywood, D. Maroteaux, and N. Collings, "Intake port phenomena in a spark-ignition engine at part load," SAE paper 912401, 1991.

2. S.R. Fozo and C.F. Aquino, "Transient A/F characteristics for cold operation of a 1.6 liter engine with sequential fuel injection," SAE paper 886691, 1988.

3. D.R. Hamburg and D. Klick, "The measurement and improvement of the transient A/F characteristics of the electric fuel Injection system," SAE paper 620766, 1982.

4. M. Iwata, M. Furuhashi, and M. Ujihashi, "Two-hole injector improves transient performance and exhaust emissions of 4-valve engines," SAE paper 870125, 1987.

5. F.H. Trinker, R.W. Anderson, Y.I. Henig, W.O. Siegl, and E.W. Kaiser, "The effect of fuel-oil solubility on exhaust HC emissions," SAE paper 912349, 1991,

6. M. Greiner, P. Romann, and U. Steinbrenner, "Bosch fuel injectors - new developments," SAE paper 870124, 1987.

7. M.M. Schechter and M.B. Levin, "Air-forced fuel injection system for 2-stroke D.I. gasoline engine," SAE paper 910664, 1991.

8. J.B. Heywood, Internal Combustion Engine Fundamentals, McGraw-Hill Book CO., New York, 1988.

9. E.W. Kaiser, J.M. Andino, W.O. Siegl, R.H. Hammerle,and J.W. Butler, "Hydrocarbon and aldehyde emissions for an engine fueled with ethyl-t-butyl ether," J. Air & Waste Management Assoc., 41, 195-7, 1991.

10. R.M. Frank and J.B. Heywood, "The effect of piston temperature on hydrocarbon emissions from a spark-ignited direct-injection engine," SAE paper 910558, 1991.

11. W.R. Matthes and R.N. McGill, "Effects of the degree of fuel atomization on single-cylinder engine performance,", SAE paper 760117, 1976.

12. A.A. Quader, T.M. Sloane, R.M. Sinkevitch, and K.L. Olson, "Why gasoline 90% distillation temperature affects emissions with port fuel injection and premixed charge," SAE paper 912430, 1991.

13. A.A. Quader, "How injector, engine, and fuel variables impact smoke and hydrocarbon emissions with port fuel injection," SAE paper 890623, 1989.

14. A.A. Quader, "The axially-stratified-charge engine," SAE paper 820131,1982.

15. K. Horie, K. Nishizawa, T. Ogawa, S. Akazaki, and K. Miura, "The development of a high fuel economy and high performance four-valve lean burn engine," SAE paper 920455,1992.

16. G. Bower, S.K. Chang, M.L Corradini, M. El-Beshbeeshy, J.K. Martin, and J. Krueger, "Physical mechanisms for atomization of a jet spray: A comparison of models and experiments," SAE paper 881318, 1988.

17. B.C. Sakiadis, Perry's Chemical Engineer's Handbook, Sect. 5, Edited by R.H. Perry, D.W. Green, and J.O. Maloney, McGraw-Hill Book Co., New York 1984

Quantitative Analysis of Fuel Behavior in Port-Injection Gasoline Engines

Nobuo Imatake, Kimitaka Saito, and Shingo Morishima
Nippon Soken, Inc.

Shunji Kudo and Akira Ohhata
Toyota Motor Co.

Copyright 1997 Society of Automotive Engineers, Inc.

ABSTRACT

We have studied the fuel behavior in Port-injection gasoline engines as the following :

1. We have developed a 100%-sampling quantitative analysis method where fuel is sealed up in the intake port and cylinder at a specific point during firing operation, using an engine with intake and exhaust valves that are opened and closed by electronic control.

2. As a result of our analysis of steady and transient state characteristics of fuel behavior using this method, it was verified that the amount of wall-wetting fuel in the port and cylinder is apparently different before and after the warm-up process.

As for transient fuel behavior, a delay in fuel transfer has been acknowledged in the amount of wall-wetting fuel not only in the port but also in the cylinder.

Different from the existing indirect analysis, this method enables direct measurement of fuel behavior even during the actual firing operation. This allows the separation of the amount of fuel behavior between the intake port and the cylinder to occur, thus making it clear that the amount of wall-wetting fuel not only in the port but also in the cylinder should be taken into consideration when evaluating the fuel transfer delay in the transient performance. By using this method, we were able to gain a complete picture of the fuel behavior, which has always been rather ambiguously defined.

INTRODUCTION

In attempting to improve Port-injection gasoline engines combustion while achieving a reduction in tail-pipe emissions and improving drivability, an important subject is optimizing control of the fuel injection system.

During transient state, there is a difference between (a) the amount of wetting fuel in the intake port and on the valve surfaces and (b) the amount of vaporization fuel from wall-wetting. That means there is a discrepancy between the fuel injection amount and the amount of fuel which intakes the cylinder. As a result, research on fuel transfer in the intake port and cylinder is extremely important.There have been reports on research on physical and chemical models which were created to clarify the behavior of fuel spray and wall-wetting to carry out analysis of phenomena [1] ~ [3] .

More recently, there has been abundant research on wall-wetting fuel for the purpose of improving the ability of the fuel injection system, especially with Port-injection gasoline engines [4] , [5] . There have been various suppositions made in calculating from the thickness of the fuel film of the port wall-wetting, as well as the indicated mean effective pressure (Pi) or from the exhaust air/fuel ratio for engine transient conditions.

A model of fuel behavior was created as shown in Fig. 1 in order to obtain the coefficients of the direct intake rate and the vaporization rate of wall-wetting fuel, etc. by means of experiments. This was used to correct the fuel injection amounts during transient state and provide a method of controlling the fuel amount entering the cylinder [6] .

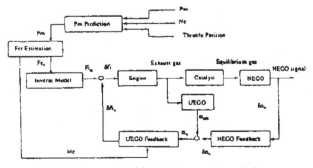

Fig. 1 Concept of fuel transport model

This paper involved a survey of fuel behavior comprising the following three steps:

(1) Survey on the relationship of the intake port wall-wetting amount during single and serial injection.

(2) Observation of intake port wall-wetting conditions during single injection to survey the effect that the injection equipment and air flow in the intake port has on the injection spray.

(3) Direct measurement of fuel behavior in the different cycles in firing operation.

When deducing from the exhaust A/F value under transient state during warm-up when the wall-wetting is high, it is not possible to clarify the total amount of wall-wetting. This makes it difficult to correct the amount of the injection fuel. The authors used independently developed analysis equipment to carry out qualitative and quantitative analysis of fuel behavior in the intake port and cylinder. This was done to clarify the effect that wall-wetting fuel during warm-up have on fuel transport properties during acceleration transient state.

Model of fuel behavior and test parameter

Model of fuel behavior- Fig. 2 shows the changes in conditions during a single cycle of Port-injection gasoline engines.

In response to the engine drive conditions, an electronic fuel control device injects the Fi amount of fuel from the injector by means of the injection signals

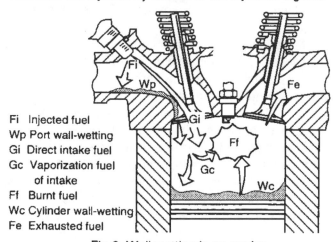

Fi Injected fuel
Wp Port wall-wetting
Gi Direct intake fuel
Gc Vaporization fuel
     of intake
Ff Burnt fuel
Wc Cylinder wall-wetting
Fe Exhausted fuel

Fig.2 Wall-wetting in an engine

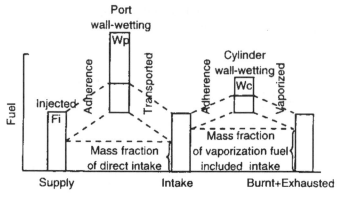

Fig.3 Relationship of wall-wetting and fuel behavior
at steady condition

corresponding to the intake air mass and injection timing. During air intake, part of the port wall-wetting Wp adhering to the intake port and valve up to the prior cycle either vaporizes or becomes wall surface flow which flows into the cylinder. The remainder is left over for the next cycle. As a result, the fuel taken into the cylinder is the sum of the direct intake fuel amount (Gi) of Fi which was injected in this cycle as well as the intake fuel amount from Wp. The fuel Ff which can contribute to combustion during compression is the sum of the vaporization fuel Gc of the intake fuel to the cylinder as well as the vaporization fuel amount of the cylinder wall-wetting (Wc) which was adhered in the cylinder wall, piston surface and combustion chamber up to the prior cycle. In addition, in the exhaust process, part of the unburned fuel undergoes exhaust Fe and the fuel (Wc) remaining in the cylinder is left over for the next cycle.

As Fig. 3 shows, in this model the amount of wall-wetting fuel in the port and cylinder during steady state is a constant amount. Thus, if there is injection of a fuel amount corresponding to the amount of intake air, this agrees with the required A/F value. However, because there are fluctuations in the amount of wall-wetting fuel that flows from the intake port to the cylinder during transient state, there is a corresponding change in the amount of wall-wetting fuel in the port and cylinder. In the transient state there is correction of the amount of fuel injected in carrying out complex controls to obtain the required A/F value. As conditions for evaluation, the authors concentrated on conditions in which the amount of wall-wetting fuel was minimal and the elements of the injection and intake systems which become more numerous. This was done to clarify differences in fuel behavior.

Test parameter

Relationship of the intake port wall-wetting amount during single and serial injection- When examining the elements of the injection and intake systems, it is easier to determine the engine elements in single injection since visible evaluation is possible. In order to determine whether it is possible to represent steady state with single injection, there was a study on tendencies for wall-wetting fuel in the port in two situations when operating the engine under the required conditions (Table 1): (a) When there was only single cycle injection, (b) When there was serial injection under steady state.

Table 1 Parameter of experimental

Swirl control valve (SCV)	Close vs. Open
Fuel injection timing	Open-valve injection (TDC) Closed-valve injection (BTDC180℃A)
Connection hole diameter	10mm vs. 6mm
Angle between two sprays	26˚ vs. 19˚

There is a comparison of differences of fuel behavior in two situations: (a) when the port walls have dried after single cycle injection, (b) when there is moistening of the same walls with fuel to an extent corresponding to an actual vehicle. If there isn't correlation, it was decided that it was necessary to make an evaluation with serial injection.

Observation of intake port wall-wetting conditions-
In order to understand tendencies for wall-wetting fuel regarding the fuel injection and air flow control devices, there was observation of wall-wetting fuel. The method of observation is shown in Fig. 4. The engine was motored at constant speed under the required conditions (Table 1), injected once, stopped after closing the intake valve, and removed the intake manifold. And then it was possible to take the pictures of the wall-wetting fuel's traces in the port.

A fluorescent agent was mixed with the injection fuel and illuminated with ultraviolet light for photography. If the port wall-wetting fuel was vaporized, the fluorescent agent was remained.

Even if normal fuel is injected, it is not possible to distinguish it from the lubrication oil wetting. As a result, it is not possible to determine the wall-wetting position on the port wall. The authors thus mixed a fluorescent agent in the fuel and illuminated it with ultraviolet light to take pictures. Xanthene dye was used for the fluorescent agent. The fluorescent agent receives the influence of the ultraviolet light to emit yellow visible light with a median wavelength of 5400 Å. This makes it possible to distinguish from the lubrication oil section on the port wall and observe only the wall-wetting fuel. The two ultraviolet lamps (ultra-high pressure mercury lamps) emit ultraviolet light with a median wavelength of 3650 Å. This makes it possible to observe only the illuminated section of the visible fluorescent agent and thus increase the visibility of the wall-wetting fuel.

Direct measurement of fuel behavior in different cycles during firing operation- In observing fuel behavior during firing operation, there was a division into port wall-wetting, cylinder wall-wetting, burnt fuel and exhausted fuel amounts in developing a new method of quantitative measurement of the behavior for each cycle [7] , [8] .

Under the required conditions, there are fuel transfer cycles shown in Fig. 5 in the port and cylinder for the different processes during engine operation. As a result, if it is possible during an arbitrary cycle and an arbitrary process to quantify the state of fuel in the port and cylinder under sealed conditions, it is possible to make a division into the following equations for calculating the port wall-wetting, direct intake fuel, vaporization fuel of intake, cylinder wall-wetting and exhaust fuel.

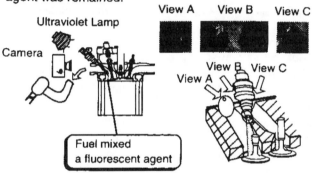

Fig.4 Observation method of wall-wetting in the port

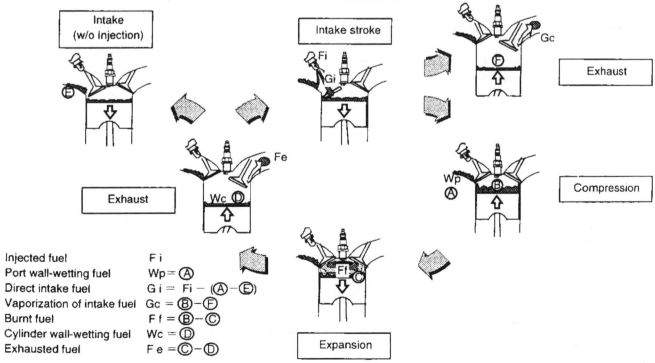

Injected fuel F_i
Port wall-wetting fuel $W_p = Ⓐ$
Direct intake fuel $G_i = F_i - (Ⓐ - Ⓔ)$
Vaporization of intake fuel $G_c = Ⓑ - Ⓕ$
Burnt fuel $F_f = Ⓑ - Ⓒ$
Cylinder wall-wetting fuel $W_c = Ⓓ$
Exhausted fuel $F_e = Ⓒ - Ⓓ$

Fig.5 Measurement procedure for fuel behavior

(1) Port wall-wetting: Port wall-wetting amount Ⓐ after intake stroke.

(2) Direct intake fuel: This is obtained as follows: The port wall-wetting amount Ⓔ obtained when stopping injection in the final cycle is subtracted from the port wall-wetting amount Ⓐ after intake stroke. Then the amount adhering to the port in the final cycle Ⓐ-Ⓔ is subtracted from the injection amount Fi.

(3) Vaporization fuel of intake: The cylinder wall-wetting amount Ⓕ exhausted during compression stroke is subtracted from the cylinder wall-wetting amount Ⓑ after intake stroke.

(4) Burnt fuel: The cylinder wall-wetting amount Ⓒ after combustion stroke is subtracted from the cylinder wall-wetting amount Ⓑ after intake stroke.

(5) Cylinder wall-wetting: Cylinder wall-wetting amount Ⓓ after exhaust stroke. This adhered fuel is carried over to next cycle.

(6) Exhausted fuel: The cylinder wall-wetting amount Ⓓ after exhaust stroke is subtracted from cylinder wall-wetting amount Ⓒ before exhaust stroke.

Using the different fuel amounts, the mass fraction of direct intake and vaporization fuel included intake were calculated in terms of percentages divided by the injected fuel amount Fi.

It is possible with the above calculations to achieve quantitative measurement of the wall-wetting volumes in the port and cylinder during steady state and transient state. It is thus possible with this measurement method to carry out direct total mass sampling of the wall-wetting amount to achieve accurate measurement. However, it requires new technology for sealing the fuel conditions of the engine during firing operation.

Engine with hydraulic control of intake and exhaust valves- As a means of sealing the fuel conditions during firing operation, there was use of an engine allowing electronic freely control of the intake and exhaust valves (Fig. 6) [9] . There was elimination of the engine cam-shaft which is mounted on standard vehicles. This was replaced with a valve drive unit that allows independent hydraulic drive of the intake and exhaust valves of each cylinder. The valve drive unit is composed of a hydraulic passage, a spool valve for switching the passageway, a piezo-actuator, and a hydraulic cylinder. These are housed in the cylinder head of the base engine. The time required for opening and closing the intake and exhaust values is 2 msec. That means the valve lift characteristics are rectangular during low revolution. For this reason, a diaphragm was added to the hydraulic passage in response to engine revolution and thus match the valve lift characteristics found during use of a standard cam drive. It was possible with this engine to independently and freely control the opening and closing timing of the intake and exhaust valves for each cylinder and also is stopped quickly.

Fuel wall-wetting measurement made use of the special feature of the engine of being able to carry out swift stop of intake and exhaust valve drive. This was done to seal the fuel conditions in the engine. Other than eliminating the cam-shaft on the base engine and using hydraulic drive, the engine was not modified. This made it possible to measure fuel behavior in the market-base engines.

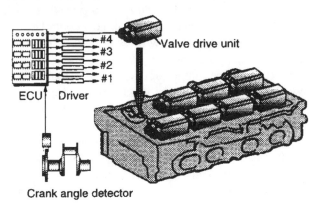

Fig.6 Cylinder head with electro-hydraulic actuated valve

Method of measuring port wall-wetting amount- Fig. 7 shows the port wall-wetting measurement device that was used with this engine. Simultaneous with swift stop of drive of the intake valve during operation under the required conditions, the intake shutter is closed. As a result, the area inside the intake port is hermetically sealed up and it is possible to seal up the fuel conditions inside the intake port during operation. After sealing, there is insertion in the intake port of high temperature air heated to 200℃ (end point of distillation amount of test fuel). This is done to achieve total vaporization of the wall-wetting fuel. At the same time a sampling pipe is used to extract vaporized fuel inside the intake port and then measure HC concentration by the FID analyzer.

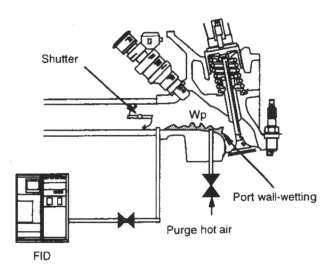

Fig.7 Measure the amount of port wall-wetting

Method of measuring cylinder wall-wetting amount-
Fig. 8 shows the cylinder wall-wetting measurement device for use with this engine. Simultaneous with swift stop of drive of the intake and exhaust valves during operation under the required conditions, the engine is stopped. As a result, the cylinder becomes hermetically sealed and it is possible to seal up the fuel conditions inside the cylinder during operation. After sealing, there is insertion in the cylinder of high temperature air heated to 200℃ (end point of distillation amount of test fuel). This is done to achieve total vaporization of the wall-wetting fuel. At the same time a sampling pipe is used to extract vaporized fuel inside the cylinder and then measure HC concentration by the FID analyzer.

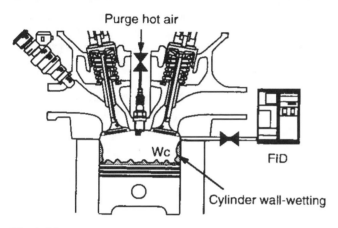

Fig.8 Measure the amount of cylinder wall-wetting

Wall-wetting quantification- As Fig. 9 shows, there is quantification of the wall-wetting amount by adding the integer value of the FID analyzer and the sampling gas amount.

In order to verify the accuracy of wall-wetting amount measurement when using this method, an arbitrary amount of fuel is inserted in the intake port and cylinder and measured. This confirmed that it was possible to detect up to an accuracy of approximately ±2 mg of fuel.

Even when there was sending of heated air (200℃) after cutting fuel injection and carrying out motoring, because there is no measurement as HC it was

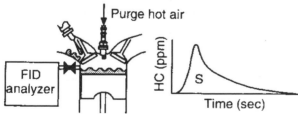

Measured fuel amount
$= Q \times S \times M \times 10^{-6}$ (g)
Q : Sampling gas flow (mol / sec)
S : HC concentration (ppm · sec)
M : Mean molecular weight of fuel (g / mol)
Fig.9 Experimental apparatus measuring method for quantitative analysis

determined that there was no influence of the lubrication oil. Regarding fuel for which there is oil dilution to the cylinder or where the fuel has entered the piston top land, it was confirmed that, if there is sampling after lowering the piston position, it is possible to measure almost all the fuel in the cylinder.

Test engine and fuel properties- As Fig. 10 shows, the engine used in the experiment was a port-injection gasoline engine with four valve system. The engine has two separate intake ports per one cylinder (one side is helical port with helical protrusion, and the other side is straight port). In the straight port side, it has Swirl Control Valve (SCV). In response to the SCV angle, it can control swirl power in the cylinder. The engine specifications are shown in Table 2. The engine control system involves independent control of cylinders for the fuel injection and ignition systems. There was use of circuits allowing start and stop of injection and ignition at an arbitrary cycle and arbitrary process.

Table 3 shows the test fuel properties. There was comparison of three types of fuel for which the distillation characteristics differ. Other than when investigating the influence of the fuel conditions, there was use of fuel A.

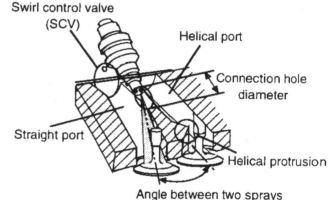

Fig.10 Characteristics of test engine's intake port

Table 2 Test engine specifications

Number of cylinder	4
Displacement (cc)	1587
Bore × Stroke (mm)	81 × 77
Compression ratio	9.5
Valves per cylinder	4

Table 3 Test fuel properties

		Fuel A	Fuel B	Fuel C
Specific gravity (g/cm³) @15℃		0.7484	0.7445	0.7621
Distillation characteristics (℃)	Initial point	30.0	31.5	31.5
	10%	45.5	53.5	56.5
	50%	91.0	105.0	119.5
	90%	143.0	155.5	169.0
	End point	181.5	176.5	204.5

Results

Correlation between single and serial injections-
Fig. 11 shows the results of measurement under the measurement conditions listed in Table 1. The correlation in the mass fraction of direct intake between single and serial injections was within 10 percent. There was also a division into groups according to water temperature and whether there was with ignition or without.

(Influence of water temperature):

at the constant water temperature, it was compared with mass fraction of direct intake single and serial injection. When the water temperature is low, mass fraction of direct intake for single injection is higher then serial injection. This is reversed when the water temperature is high. When the water temperature is low during single injection, the adhered fuel in the port is remained. During serial injection, the adhered fuel flows together with the wall-wetting fuel of a liquid film. When the water temperature is high during single injection, the wall-wetting fuel in the port readily vaporizes due to the temperature of the walls. During serial injection, there is vaporization cooling by the wall-wetting fuel which causes the temperature of the walls to drop so that the wall-wetting fuel does not vaporize easily.

(With or without ignition):

With and without ignition, it was compared with mass fraction of direct intake for single and serial injection. When there is with ignition, mass fraction of direct intake for single injection is larger then serial injection. When there is without ignition, there is the reverse situation. This is because, when there is ignition for serial injection, the exhaust pressure rises due to combustion so that there is an increase in flowing backward inside the cylinder to the intake port. Though port wall-wetting amount with ignition is less than without ignition. Because there is by firing operation with ignition which causes the temperature of the intake valves to rise so that the wall-wetting fuel vaporizes easily.

Observation of wall-wetting fuel during single injection- (Table 4 *next page)
(1) During SCV open, as opposed to close, there are less wall-wetting on the helical protrusion and connection hole between helical port and straight port.
(2) Closed-valve injection, as opposed to open-valve injection, there are less wall-wetting on the helical protrusion and connection hole.
(3) When the connection hole diameter between straight port and helical port are small (6mm) as opposed to large (10mm), because there is a stronger air current, there are more wall-wetting on the connection hole and the outside surface of the straight port side.
(4) When the angle between two sprays is small (19 degrees) as opposed to large (26 degrees), there are more wall-wetting on the helical protrusion but less wall-wetting on the outside surface of the straight port side.

Direct measurement of fuel behavior for individual cycles during firing operation with actual equipment- The measurement of the steady state is wall-wetting after firing operation for 3 minutes (1800 cycles) under the required conditions.
Effect of engine load- Fig. 12 shows effect between the engine load and wall-wetting amount (unit: mg) during steady state and warm-up. As the load increases, the wall-wetting amounts in the port and cylinder increase. With the increased load, the pressure in the intake port increases so that the vapor pressure increases. In addition, due to an increase in the injection amount, the wall temperature decreases on the areas where the spray impinges due to vaporization cooling. Closed-valve injection, as opposed to open-valve injection, there is slightly more port wall-wetting under all conditions, although there is major less cylinder wall-wetting during high loads.

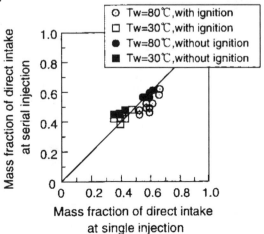

Fig.11 Correlation between single and serial injections for mass fraction of direct intake

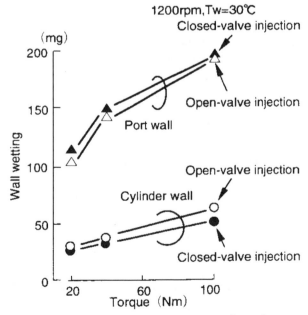

Fig.12 Effect of engine torque on wall wetting

78

Table 4 Observation of wall-wetting in the port

SCV / Injection timing / Connection hole diameter / Angle between two sprays	View A	View B	View C
	Connection hole	Helical protrusion	Connection hole
Close / Open valve / 10 mm / 26 °			
Open / Open valve / 10 mm / 26 °			
Close / Closed valve / 10 mm / 26 °			
Close / Open valve / 6 mm / 26 °			
Close / Open valve / 10 mm / 19 °			

Effect of warm-up conditions- Fig. 13 shows effect of the results of measuring the wall-wetting amounts of individual sections during and after warm-up at the steady state.

During warm-up the injected fuel (15 mg for a single cylinder) is the sum of the direct intake fuel (7 mg) as well as the adhering fuel to the intake port wall (8 mg). And then the intake fuel into the cylinder (15 mg) is the sum of the direct intake fuel (7 mg) as well as the transporting fuel from the port wall-wetting (8 mg). And then the taken fuel in the cylinder (15 mg) is the sum of the vaporization fuel (8 mg) as well as the adhering fuel to the cylinder wall (7 mg). And then burnt and exhausted fuel (15 mg) is the sum of the vaporization fuel included intake (8 mg) as well as the vaporization fuel of the cylinder wall-wetting (7 mg).

The port wall-wetting amount during warm-up is 135 mg for a single cylinder which corresponds to 9 times the injection amount for a single cycle. The cylinder wall-wetting amount is 37 mg. This corresponds to 2.5 times the injection amount for a single cycle.

The port wall-wetting amount after warm-up is 45 mg for a single cylinder or one-third the amount for warm-up. The cylinder wall-wetting amount for a single cylinder is 12 mg or one-third the amount for warm-up.

The port and cylinder wall-wetting amount during operation of the engine is at a level that is several times the injection amount. That means the warm-up state of the engine is a particularly influential factor.

Ne=1200rpm,40Nm,Open-valve injection

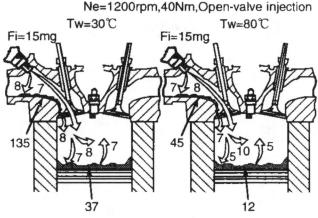

Fig.13 An illustration to measure
the amount of fuel wall-wetting

Effect of the injection timing- Fig. 14 shows the effect that the injection timing under steady state during warm-up has on the direct intake amount and wall-wetting amount. Open-valve injection, as opposed to closed-valve injection, mass fraction of direct intake is 18% more for each cycle, although the port wall-wetting is 7% less. Mass fraction of vaporization fuel included intake is 12% less, while the cylinder wall-wetting amount is 16% more. This is believed to occur for the following reasons: Open-valve injection is transferred to the flow of intake air so that it flows directly to the cylinder. As a result, the time for vaporization of the injection fuel is shorter and mass fraction of

vaporization fuel included intake is lower so that the cylinder wall-wetting increases. In case of closed-valve injection, the wall-wetting which is retained temporarily on the intake valve surface is vaporized due to the high-speed air flow in the space of the intake valve seat during the initial stages of intake, after which it flows to the cylinder. As a result, the vaporization rate of intake fuel is higher and the cylinder wall-wetting decreases. In addition, the reason why there is less difference in port wall-wetting compared to direct intake rate is because the port wall-wetting spreads throughout the port and not just in the area around the intake valve. As a result, the total mass is more dependent on the port configuration and temperature than the injection timing.

Cylinder wall-wetting is exhausted as engine-out HC emissions from the cylinder wall surfaces during the exhaust stroke.

As a result, under conditions with a large cylinder wall-wetting (starting, warm-up, open-valve injection, spray with poor vaporization etc.) the exhaust HC concentration tends to increase.

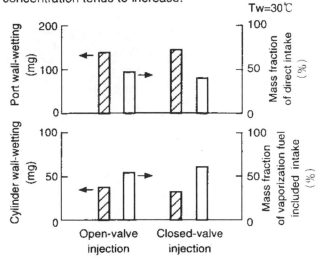

Fig.14 Effect of injection timing on wall-wetting

Effect of the fuel properties- Fig. 15 shows effect of the fuel properties. The port wall-wetting amount is highest for fuel C and then for fuel B and A in that order. Likewise, the cylinder wall-wetting amount is highest for fuel C and then for fuel B and A in that order.

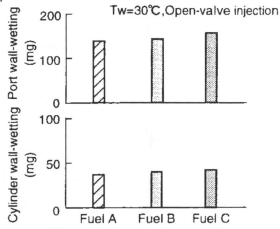

Fig.15 Effect of fuel properties on wall-wetting

Compared to the intake port wall, the temperature on the cylinder wall is higher, and the time for fuel vaporization is longer. As a result effects of the fuel properties is minimal. As for the exhaust amount of unburned HC, in this case as well it corresponded to an increase in the cylinder wall-wetting amount. The 50% distillation temperature shown in Table 3 corresponds well to increases or decreases in the port wall-wetting amount. The higher the 50% distillation temperature for the fuel, the more a method is required for reducing the wall-wetting amount.

Varies in wall-wetting amount during transient state– The authors verified that it was possible to deduct the vary in the wall-wetting amount during transient state in terms of time series very during measurement of wall-wetting amount under steady state. As Fig. 16 shows, if there is acceleration at 100 msec from 20 Nm (Injected fuel amount is 10 mg) to WOT (Injected fuel amount is 30 mg) without transient correcting the injection amount, the cylinder wall-wetting increases in correspondence with intake fuel amount, as is the case under steady state.

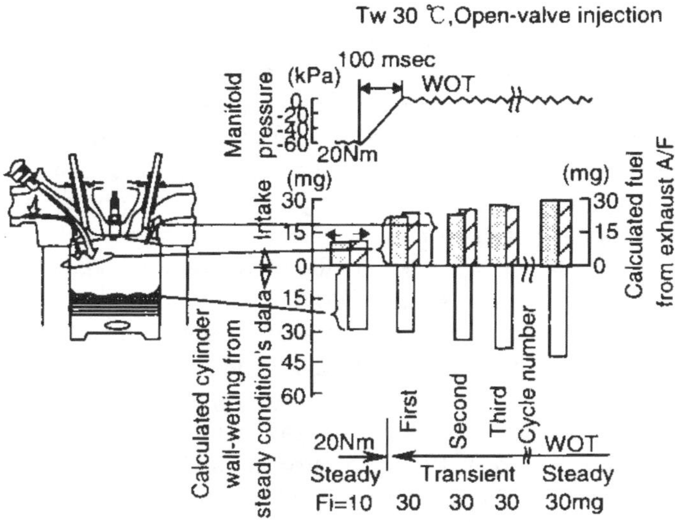

As Fig. 17 shows, by deducing the intake fuel mass is calculated from the mass fraction of direct intake and the mass fraction of the transported fuel from the port wall-wetting, the cylinder wall-wetting amount should increase as a time constant in relation to the amount of intake fuel and then become steady. However the authors discovered that there is a difference between the intake fuel and amount of fuel calculated from the exhaust A/F value using an A/F sensor.

Fig. 18 shows the relationship between the wall-wetting amounts of individual cycles under transient state during warm-up and the fuel amount for burnt + exhaust (exhaust A/F). During 2-3 cycles after transient state the cylinder wall-wetting amount decreases and then continues to increase for several tens of cycles until a constant mass is reached. In case of open-valve injection, the cylinder wall-wetting which was 30 mg at 20 Nm under steady state decreases to 27 mg in the first cycle during transient state, and then to 25 mg in the second cycle. Then it gradually increases to 62 mg which is the WOT's steady amount. This reduction

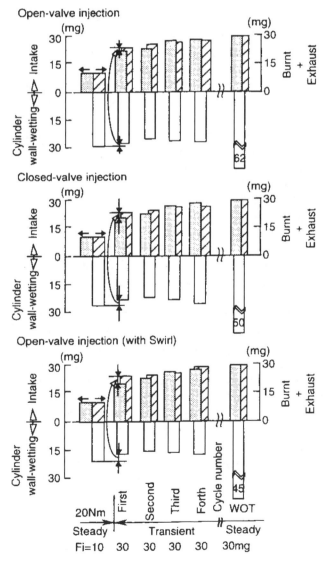

Fig.18 Cycle variation of wall-wetting
at transient condition

Fig.16 Expect of cycle variation of wall-wetting
at transient condition

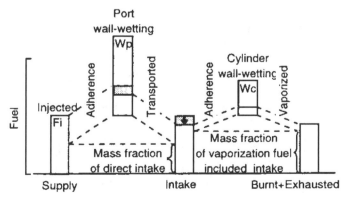

Fig.17 Relationship of wall-wetting and fuel behavior
at transient condition

amount undergoes burnt and exhausted so that the lean spike is not as large as the increase in the port wall-wetting. The reduction amount of the cylinder wall-wetting in the first 2-3 cycles after transient state becomes larger under conditions where the vaporization fuel included intake rate is large and the cylinder wall-wetting is small (closed-valve injection, with swirl).

The reason for the reduction in cylinder wall-wetting is as follows: During steady state, when the piston is moving up in the compression stroke, the cylinder wall-wetting is scraped up by the top ring groove of the piston. When the piston is moving down in the expansion stroke, the fuel captured by the top land crevice and the top ring groove is spread thinly on the cylinder wall again.. This thin fuel wall-wetting does not burn and carries over next cycle. Immediately after acceleration, due to an increase in the cylinder wall temperature the thermal boundary layer and quench layer become thinner so that there is a lean state in the cylinder with a corresponding decrease in the rich amount. Also, the reason why there is an increase in the amount by which the cylinder wall-wetting decreases by addition of a swirl is because the eddies in the cylinder make the thermal boundary layer thinner. In addition, there is promotion of the lean burn.

The reason for the small difference between open- and closed-valve injection is as follows: Open-valve injection, as opposed to closed-valve injection, mass fraction of direct intake is more, although mass fraction of vaporization fuel included intake is less. Therefore open-valve injection, as opposed to closed-valve injection, a difference between the amount of fuel supplied and the amount of fuel which intakes to the cylinder is less, although the amount of reduction in the cylinder wall-wetting is less, too.

There have been reports on the importance of the port wall-wetting amount in correcting the injection amount during transient state by means of the exhaust A/F value. After acceleration during warm-up, the cylinder wall-wetting contributes to combustion. If it continues constant increase in fuel injection amount, there is a possibility of a slip to lean when the cylinder wall-wetting increases. As a result, instead of simply controlling the amount of fuel entering the cylinder in terms of the exhaust A/F value, it is necessary to control the A/F in the cylinder while considering changes in the cylinder wall-wetting. That means it is necessary to deduct the amount of change in the cylinder wall-wetting from the cylinder temperature during and after warm-up to achieve detailed injection control during transient state.

Relationship between fuel behavior and engine performance

The port wall-wetting amount is higher closed-valve injection than open-valve injection. The opposite is true for the cylinder wall-wetting amount. Fig. 19 shows the influence that the wall-wetting amount has on the injection fuel amount, the amount of fuel flowing to the cylinder, and the burnt + exhausted fuel amount.

The port wall-wetting amount has a major influence on the transfer of fuel. However, the port wall-wetting does not concern directly the exhaust HC. On the other hand, the cylinder wall-wetting amount has an influence directly on the exhaust HC.

Under steady state, the relationship between the wall-wetting amount in the port and cylinder and the fuel injection, vaporization, transfer and flow to the cylinder for each cycle is in a state of balance. It is equal to the amount of fuel supplied and the amount of fuel that intakes to the cylinder. However, under transient state, the balanced state of the port wall-wetting amount is destroyed so that a difference appears between the amount of fuel supplied and the amount of fuel which intakes to the cylinder. In addition, the balanced state of the cylinder wall-wetting amount is destroyed so that a difference appears between the intake fuel amount and the burnt + exhausted fuel amount. In an engine where the port wall-wetting amount is large, there is a major fluctuation in the A/F ratio during transient state. During acceleration, an A/F ratio lean occurs. This in turn leads to a leaner exhaust emission, sluggish torque and backfires. Conversely, during deceleration, there is worsening of the exhaust emission and misfiring due to an overly rich A/F ratio. On the other hand, in the case of engines with a high cylinder wall-wetting amount there is a major increase in the exhaust HC, regardless of whether it is steady or transient state.

In deciding on engine elements (injection system, intake system, etc.) it is necessary to make a judgment on which of the following to carry out: (a) Visualizing the fuel wall-wetting and carrying out improvements so that the fuel reaches the target position in relation to the intake air flow under various engine conditions.

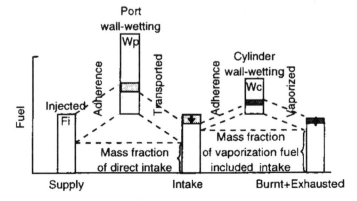

Fig.19 Relationship of wall-wetting and fuel behavior at transient condition

(b) Giving priority to correcting transient state instead of emission and combustion under steady state.

In this report an engine with hydraulically driven valves was used for all measurements. However, during investigation of the influence of the injection elements (attachment position, injection target position, atomization level, injection timing, etc.) and the intake port elements (configuration, air flow configuration, valve open/close timing, etc.) it was possible to carry out quantitative evaluation of the port wall-wetting amount under fixed conditions using a standard engine with cam-driven valves. As a result, after determining the optimum elements of the injection and intake systems by means of simulation involving visualization of port wall-wetting fuel and single injection by means of an engine with cam-driven valves, a method is useful for determining the coefficients for correcting transient state by means of a study on fuel behavior regarding wall-wetting amounts using an engine in which one can freely seal up the fuel in the port and cylinder, as is the case in this report.

Summary

(1) The difference in the direct intake rate for single and serial injection is within 10%. However, the rate for single injection is slightly higher during warm-up, and the rate for serial injection is slightly higher after warm-up. This is believed to be the result of whether there is a liquid film flow on the intake port wall and differences in the vaporization amount of the wall-wetting fuel.

(2) It was possible to clarify differences in the port wall-wetting condition due to the air flow, injection timing and injection elements by observation of wall-wetting fuel into which a fluorescent agent had been mixed.

(3) In order to analyze the fuel behavior during firing operation with a port-injection gasoline engine, a method was established for measurement of wall-wetting quantities using an engine on which the intake and exhaust valves are driven hydraulically by means of electronic control.

The measurement of the port and cylinder wall-wetting amount was used to for quantitative clarification of the engine load, warm-up conditions, injection timing and the varies in the cycle period during transient state. (For example, the port wall-wetting amount is 3 to 9 times the injection amount for a single cycle and the cylinder wall-wetting amount is 1 to 2.5 times.)

As a result, it was possible to clarify the mechanism for varies in the exhaust A/F during warm-up. As a method of reducing varies in the exhaust A/F value, it is effective to carry out control of the fuel injection mass during transient state while considering the reduction in the wall-wetting amount and the relationship between the intake fuel and wall-wetting amount.

References

[1] Ather A. Quadwe, "How Injector, Engine, and Fuel Variables Impact Smoke and Hydrocarbon Emissions With Port Fuel Injection", SAE Paper 890623,1989

[2] Yoshishige Ohyama, Toshiharu Nogi and Mamoru Fujieda, "Study on Variable Injection Pattern Control System in a Spark Ignition Engine", SAE Paper 910080, 1991

[3] Ryuji Morishima and Katsutoshi Asai, "Mixture Strength at Cranking Cycles of Gasoline Engine Starting", SAE Paper 920235, 1992

[4] Hatsuo Nagaishi, Hiromichi Miwa, Yoshihisa Kawamura and Masaaki Saitoh, "An Analysis of Wall Flow and Behavior of Fuel in Induction Systems of Gasoline Engines", SAE Paper 890837, 1989

[5] Goran Almkvist and Soren Eriksson, "An analysis of Air to Fuel Ratio Response in a Multi Point Fuel Injected Engine Under Transient Conditions", SAE Paper 932753, 1993

[6] Akira Ohata, Michihiro Ohashi, Masahiro Nasu, and Toshio Inoue, "Model Based Air Fuel Ratio Control for Reducing Exhaust Gas Emissions", SAE Paper 950075, 1995

[7] Kimitaka Saito, Kiyonori Sekiguti, Nobuo Imatake, Keiso Takeda and Takehisa Yaegashi, "A New Method to Analyze Fuel Behavior in a Spark Ignition Engine", SAE Paper 950044, 1995

[8] Keiso Takeda , Takehisa Yaegashi , Kimitaka Saito, Kiyonori Sekiguti, and Nobuo Imatake, "Mixture Preparation and HC Emissions of a 4-Valve Engine with Port Fuel Injection During Cold Starting and Warm-up", SAE Paper 950074, 1995

[9] Ken-ichiro Shindoh,Daisaku Sawada,Taiichi Mori, Hideo Saruhashi and Kenji Oshima,"Development of Hydraulically Actuated Valve Train",Proc. of the 13th Internal Combustion Engine Symposium,1995, (in Japanese).

2001-01-3587

Contribution of Liquid Fuel to Hydrocarbon Emissions in Spark Ignition Engines

Gary B. Landsberg, John B. Heywood and Wai K. Cheng

Massachusetts Institute of Technology

Copyright © 2001 Society of Automotive Engineers, Inc.

ABSTRACT

The purpose of this work was to develop an understanding of how liquid fuel transported into the cylinder of a port-fuel-injected gasoline-fueled SI engine contributes to hydrocarbon (HC) emissions. To simulate the liquid fuel flow from the valve seat region into the cylinder, a specially designed fuel probe was developed and used to inject controlled amounts of liquid fuel onto the port wall close to the valve seat. By operating the engine on pre-vaporized Indolene, and injecting a small amount of liquid fuel close to the valve seat while the intake valve was open, we examined the effects of liquid fuel entering the cylinder at different circumferential locations around the valve seat. Similar experiments were also carried out with closed valve injection of liquid fuel at the valve seat to assess the effects of residual blowback, and of evaporation from the intake valve and port surfaces.

The amount and location of liquid fuel entering the cylinder was found to have a significant impact on engine-out hydrocarbon emissions. Around the intake valve seat periphery, liquid fuel delivered closest to the exhaust valve resulted in the highest engine-out HC's, while delivery farthest from the exhaust valve had the lowest HC's. Closed valve (CV) probe liquid fuel injection resulted in lower engine-out HC emissions than did the same amount of liquid fuel with open-valve (OV) injection. This difference between OV and CV probe fuel injection was about equal in magnitude for all probe locations, indicating that the blowback during valve overlap has a similar impact on fuel redistribution and vaporization at all circumferential locations. Overall, liquid fuel flow into the cylinder produces 3 to 7 times more HC

emissions per unit mass of fuel than does vapor fuel flow, depending of flow location.

SOURCES OF ENGINE HYDROCARBON EMISSIONS

The sources of hydrocarbons in the SI engine cylinder are crevices, oil layers, deposits, liquid fuel, flame quenching, and exhaust valve leakage. Cheng et al. [1] developed a flowchart which quantified the relative magnitude of these HC sources during steady-state engine operation. The largest source of HC under steady-state conditions is crevices in the combustion chamber: narrow regions connected to the combustion chamber where unburned mixture can be stored and escape combustion. The piston lands, ring pack crevices, the volumes contained in the piston ring grooves and clearances, account for some 80% of the total crevice volume [2]. However, HC emissions in a cold engine following start-up, appear to be largely independent of crevice volume [3].

Under warm operating conditions, the oil layer HC mechanism accounts for less than 10% of the total HC emissions [4], because 80% of the fuel desorbed from the oil is oxidized as it mixes with the hot in-cylinder gases [5] [6]. The fuel-in-oil contribution to HC emissions is expected to be higher at lower temperatures [1] [4], when a larger amount of fuel is absorbed into the oil [7]. Under steady-state operation, HC emissions resulting from flame quenching at the cylinder walls are expected to be small. During cold engine operation when wall temperatures are cooler, the quench layer is expected to increase in thickness, increasing its contribution to HC emissions.

Several researchers have studied the effects of liquid fuel entering the cylinder during steady-state warm engine operation, and during warm-up. Min et al. [8] compared engine-out HC with pre-vaporized Indolene and normal port injection of Indolene, and found that HC levels with port injection were four times higher during the first 50 seconds of engine operation. However, Kaiser et al. [9] in a similar experiment found that fuel preparation with a cold engine had a much more modest effect on engine-out HC. Imatake et al. [10] developed a special engine where the intake and exhaust valves could be sealed at a specific point during firing operation. By measuring the trapped liquid fuel, they found that during warm-up, the amount of liquid in the intake port corresponded to about ten injection pulses, and the amount of liquid fuel in the cylinder was 2-3 times that of each injected pulse. Under steady-state warmed-up conditions these liquid fuel amounts were about three times lower. With open-valve injection, they found that 15-20% more of the injected fuel enters the cylinder as liquid resulting in some 16% more in-cylinder wall wetting. Shayler et al. [11] performed a fuel audit on a four cylinder engine during starting and warm-up and found similar results.

Fry and Nightingale [12] found that liquid fuel can collect in the crevices around the inlet and exhaust valves and in the apex of the pent-roof head during induction. It remains unburned throughout the combustion process, and can be pulled out of the cylinder during the exhaust blowdown process.

Stanglmaier et al. studied the effect of the location of the in-cylinder wall wetting region on HC emissions using a specially designed spark-plug-mounted injector probe [13][14]. The highest increase in HC emissions was found to come from wall wetting on the exhaust side of the cylinder. The next largest HC increase came from liquid fuel wetting the piston top. The lowest HC increase came from wetting the cylinder wall on the intake valve side. Stanglmaier et al. suggest that the physical location of the liquid fuel wall film within the cylinder, relative to the exhaust valves, is the critical factor in these differences.

Despite the above cited work, the effects of liquid fuel entering the cylinder on engine-out HC emission are not well understood. These effects are especially important during cold engine operation when the intake valve, port, and cylinder wall temperatures are lower. The relationship between liquid fuel location in the port, and subsequent entry into and behavior in the cylinder on HC emissions is also unclear.

APPROACH

The purpose of this study was to develop a data-based description of liquid fuel buildup and transport into the engine and thereby quantify the liquid fuel contribution to HC emissions. A novel liquid-fuel injector probe was developed that deposits controlled amounts of liquid fuel onto the intake port wall, just upstream of the intake valve seat, at three specific circumferential locations. By controlling the amount and location of the liquid fuel injected at the valve seat during open valve injection, and measuring the engine-out hydrocarbon (HC) emissions, an estimate of the contribution to these emissions due to liquid fuel directly entering the cylinder, could be made.

Liquid fuel which is injected into the intake port can be transported into the cylinder in three forms; fuel vapor, fuel droplets, or as a fuel film. Various researchers have examined this fuel transport from the port to the cylinder (e.g. [15]-[17]). During normal closed-valve injection, the injected fuel is largely deposited as films on the port walls and intake valve prior to valve opening. Open-valve injection results in a smaller mass of liquid fuel being deposited on the port wall and valve, and more liquid fuel is transported directly into the cylinder than occurs with closed-valve injection [10]. The exhaust gas backflow at intake valve opening causes the injected liquid fuel distribution in the port to be significantly different for closed and open-valve injection. This difference in port fuel distribution has a direct impact on the amount and distribution of liquid fuel entering the cylinder [16].

An approximate assessment of the initial liquid fuel film distribution in the intake can be made using the fuel injector location and spray angle, airflow velocity, and port geometry. However, the liquid distribution set-up by the injection process is substantially changed by the high-velocity residual-gas backflow that occurs at inlet valve opening, especially under throttled conditions [18], [19]. Thus, our experimental approach to measuring the effect on HC of liquid fuel entering the cylinder was to simplify this complex situation.

A special liquid fuel injector probe which deposits controlled amounts of liquid fuel at the intake valve seat was used. The remainder (majority) of the fuel was put into the intake port in vapor form (as propane, or as vaporized Indolene using a prevaporizing injector). By controlling the amount and location of the injected liquid fuel and measuring the changes in engine-out hydrocarbon emissions we can estimate the liquid fuel contribution to these emissions. These different injector probe locations allowed us to evaluate the impact of liquid fuel entering the cylinder from different positions around the valve seat.

EXPERIMENTAL SET-UP

A Nissan SR20DE, 2-liter production four-cylinder spark-ignition engine was used for all experiments. This 1991 Nissan Sentra engine has four valves per cylinder and a pent-roof head design with direct overhead camshafts.

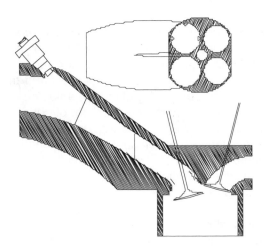

Fig. 1: Intake runner and intake port geometry for the test engine.

The engine had been modified so that only one cylinder is fired. Its intake manifold and exhaust manifold had been separated from those of the other cylinders. The engine's operating condition was chosen to be 1500 rev/min with intake manifold pressure equal to 0.5 bar. This is a typical part-load operating test condition used by the auto industry. The engine was operated with a stoichiometric mixture. The spark timing was optimized for maximum brake torque (MBT). Figure 1 shows the engine's intake port and cylinder geometry.

The hydrocarbon emissions were measured using a flame ionization detector. The engine exhaust was sampled from a damping tank located 1.7m downstream of the exhaust port. A heated line was used to prevent HC condensation in the sampling line. At this fixed operating condition, the engine-out hydrocarbon emissions variation, day-to-day, was found to be approximately 200 ppmC1.

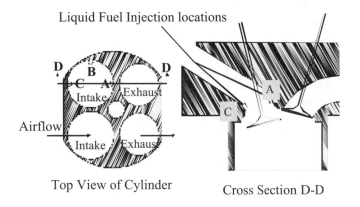

Fig. 2: Liquid fuel injector probe locations.

To simulate the liquid fuel transfer from the valve seat region into the cylinder, a timed liquid fuel delivery system was constructed to inject controlled amounts of liquid fuel onto the port wall close to the valve seat. This fuel delivery system or probe consisted of a small, standard port fuel injector connected to a 1.6 mm inner

diameter Teflon tube. Three fuel probe tubes were installed, approximately 35 cm in length, and each was epoxied to the intake runner wall. The end of each fuel probe tube was mounted about 6 mm back from the intake valve seating surface, aligned perpendicular to the valve seat. The probe tube locations are shown in Fig. 2.

The standard fuel injector for this engine is a four-hole injector, located about 22 cm upstream of the valve. Each of the liquid fuel probe tubes was fueled, in turn, by a lower flow-rate pulsed port-fuel-injector. Visualization of the probe liquid fuel flow with a strobe light indicated that a probe injector pulse width of 25° crank angle (CA) results in an effective liquid flow pulse out of the probe tube of 100°CA duration, with a delay of 185°CA as shown in Fig. 3.

The prevaporizing injector used to inject the majority of the Indolene in vapor form consists of a smaller than standard injector mounted at one end of a heated brass tube. A small swirling airflow (about 15% of total airflow) is supplied at the injector end of the tube to throw the liquid fuel onto the tube surface where it fully vaporizes. The design is based on the prevaporizing injector developed at the National Engineering Laboratory in Scotland [20].

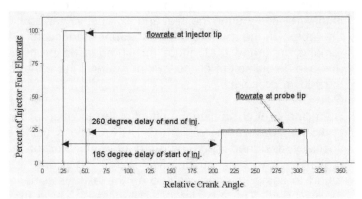

Fig. 3: Liquid fuel injection delay through fuel delivery probes, based on probe visualization study.

EXPERIMENTAL RESULTS AND THEIR INTERPRETATION

The HC emissions variations with spark timing, relative air/fuel ratio, coolant temperature, and fuel injection timing were as expected. At the standard test conditions, stoichiometric, MBT spark, 0.5 bar intake manifold pressure, 1500 rev/min, HC emissions with propane were 1100 ppm C_1, about half those with standard closed-valve port-injected Indolene (2250 ppm C_1). HC emissions with Indolene with the prevaporized injector (2000 ppm C_1) were about 10% lower than the standard port-injection values [21]. This latter difference is attributed primarily to a modest fraction of the fuel entering the cylinder as liquid in the standard case.

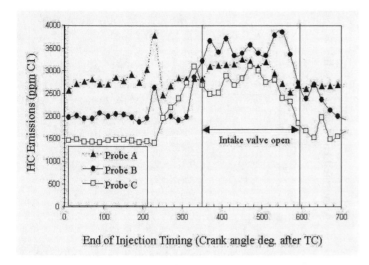

Fig. 4: Comparison of engine-out HC emissions for probes A, B, and C. 36% Indolene through probe, balance Propane into intake.

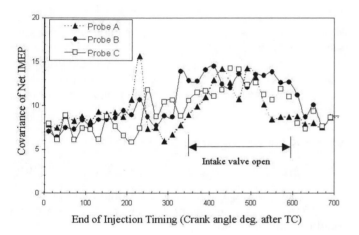

Fig. 5: Comparison of covariance in net imep for probes A, B, and C. 36% Indolene through probe, balance Propane into intake.

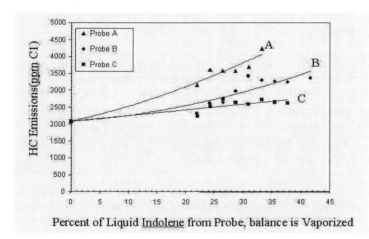

Fig. 6: Engine-out HC emissions as a function of amount of liquid fuel going through probe; balance of fuel is prevaporized Indolene. Liquid fuel injected while intake valve is open, end of injection 410 ATC, 60 after IVO.

The impact of the liquid fuel mass entering the cylinder was studied by injecting different amounts of liquid fuel and measuring the corresponding effect on the engine-out HC emissions at each probe location. Since the remaining fuel was prevaporized, and was the same fuel (Indolene), the effect of known amounts of liquid fuel entering the cylinder could be determined.

Figure 4 shows the impact of the liquid fuel probes, as a function of probe liquid pulse timing, when 36% of the total fuel mass was Indolene injected through one of the three probes and the remaining 64% was propane flowing steadily into the intake. With probe location A (see Fig. 2), HC emissions show only modest differences between the liquid fuel pulse when the intake valve is closed and when it is open. With probe location C, a much higher impact on HC emissions is observed if the probe fuel reaches the valve seat when the intake valve is open. Results for probe B location lie in-between. The HC levels with all three probe locations, during the first two-thirds of the intake valve open period are essentially independent of timing. The coefficients of variation of net imeps for this same set of experiments are shown in Fig. 5. These values are higher when the probe liquid fuel pulse reaches the valve seat when the intake valve is open. The magnitude of COV of net imep varies approximately with the magnitude of the HC emissions. Note that these net imep COV values are relatively high (5-15%) due to the substantial direct flow of liquid fuel into the cylinder (36%). It is presumed that the mixing of this liquid fuel in the cylinder with the lean propane-air mixture is incomplete.

Figure 6 shows the HC emissions as a function of percent Indolene as liquid from the probe, for probe locations A, B and C, with open-valve probe injection. (The remaining percentage of the Indolene goes through the prevaporizing injector.) The rate of increase with probe A, on the top of the intake port on the exhaust-valve side of the intake valve seat (see Fig. 2), is greatest; the rate of increase for probe location C (farthest from the exhaust valve) is lowest. It proved difficult to control liquid fuel probe injections below 20% of the total fuel flow. For closed-valve probe injections, the HC emissions increase less, as expected, since the residual gas backflow at intake valve opening redistributes the probe liquid fuel, and more of it will evaporate before entry into the cylinder. Probe A is higher than probe C as before; the increase in HC emissions for probe C is small. See Fig. 7. The increases in HC emissions due to the liquid fuel from the probes are given in Figs. 8 and 9, where linear fits to the data have been added.

These results indicate that for all amounts of liquid fuel injected at the valve seat, the fuel delivered closest to the exhaust valve (probe A) results in the highest increase in engine-out HC emissions, while the location farthest from the exhaust valve (probe C) has the lowest increase in HC emissions.

DATA ANALYSIS

These experiments, where an incremental amount of liquid fuel is introduced into the cylinder of an engine operating on pre-vaporized fuel, allow us to estimate the fraction (ψ_{liq}) of the in-cylinder liquid fuel that will escape the engine as unburned hydrocarbon (HC), in comparison with the HC fraction from prevaporized fuel (ψ_{vap}).

In the experiment, a mass fraction y of the total fuel (of mass m_{fuel}) is introduced as liquid by injecting the fuel close to the intake-valve seat; pre-vaporized fuel comprises the remainder, fraction $1-y$. Of the liquid that enters the cylinder, there is further vaporization within the cylinder so that only a fraction z remains as liquid. Thus the mass of exhausted HC (m_{HC}) may be expressed as:

$$m_{HC} = m_{fuel} [\ yz\psi_{liq} + (1 - yz)\psi_{vap}) \ ] \qquad (1)$$

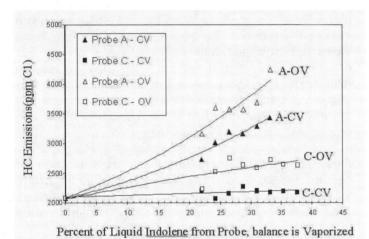

Fig. 7: Comparison of HC emissions with open valve and closed valve probe injection, as a function of amount of liquid fuel going through probe. Probe locations A and C.

where

ψ_{liq} = Mass fraction of the in-cylinder liquid fuel that escapes as exhaust HC

ψ_{vap} = Mass fraction of the in-cylinder fuel vapor that escapes as exhaust HC

y = Mass fraction of the total fuel injected as liquid

z = Mass fraction of the injected liquid fuel that is not vaporized

Equation (1) may be rewritten as:

$$x_{HC} \equiv m_{HC} / m_{fuel} = \ yz \left(\psi_{liq} - \psi_{vap} \right) + \ \psi_{vap} \qquad (2)$$

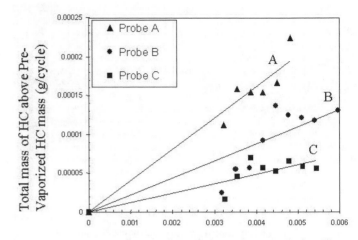

Fig. 8: Mass of HC emissions above prevaporized HC emissions, as a function of probe liquid fuel mass and probe location, with open valve probe injection.

where x_{HC} is mass fraction of the engine-out HC with respect to the fuel mass. The value of ψ_{vap} is known from the measured value of x_{HC} at $y = 0$ (no liquid fuel); its value for this experiment is 1.45%. Thus the value of ψ_{liq} may be obtained from the slope of the plot of x_{HC} versus y, if z is known.

An equivalent way of expressing Eq.(1) is:

$$m_{HC} = (m_{fuel} \ y) \ z \ (\psi_{liq} - \psi_{vap}) + \ m_{fuel}\psi_{vap} \qquad (3)$$

Since $m_{fuel}y$ is equal to the mass of the injected liquid fuel m_{inj}, and $m_{fuel} \ \psi_{vap}$ is the engine-out HC measured with no liquid fuel injection, then

$$m_{HC} - m_{HC} (y = 0) = \ m_{inj} \ z \ (\psi_{liq} - \psi_{vap}) \qquad (4)$$

Thus, the plot of the increase in HC emission (from the all-pre-vaporized-fuel level) versus the amount of liquid fuel injected (m_{inj}) has a slope of $z(\psi_{liq} - \psi_{vap})$. From this slope, ψ_{liq} may be calculated for a given value of z.

The plots of Eq. (4) for open-valve (OV) probe liquid Injection and closed-valve (CV) probe injection are shown in Figs. 8 and 9. In each case, the slope is higher from position A (injection on the exhaust valve side) and lower at position C (injection away from the exhaust valve side). For the same injection location, the slope is higher for OV than CV probe injection.

Previous work by Meyer and Heywood [22] suggests that for a warmed-up engine, approximately 50% of the liquid fuel entering the cylinder vaporized prior to combustion. Thus a value of z of 0.5 was assumed for the open valve liquid injection case. Using this value, with $\psi_{vap} = 1.45\%$, the values of ψ_{liq} are shown in Table 1 and Fig. 10 for the different cases. It should be noted that the analysis here assumes the z value to be independent of the liquid injection location around the valve seat. (Values of ψ_{liq} obtained with different values of z indicated that the precise value assumed was not that critical [21].)

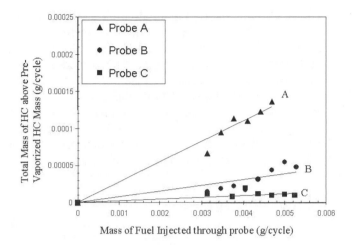

Fig. 9: Mass of HC emissions above prevaporized HC emissions as a function of probe liquid fuel location, with closed valve probe injection.

These results (Fig. 10) show that for OV injection, introducing the liquid fuel at location A (closest to the exhaust valve side) has a higher fraction of this fuel contributing to the exhaust HC than introducing liquid fuel at location C (furthest from the exhaust valve side.

Table 1 Fraction (ψ_{liq}) of fuel mass that escapes engine as engine-out HC, as a function of injection location (using $z = 0.5$ in Eq. (4)).

Open-Valve Injection

	Location A	B	C	Average
ψ_{liq}	0.098	0.065	0.039	0.067
ψ_{liq}/ψ_{vap}	7	5	3	5

The ratios ψ_{liq}/ψ_{vap} (see Table 1) indicate that with introduction of liquid fuel at locations A, B and C using OV injection, the respective masses of exhaust HC derived from unit mass of the in-cylinder liquid fuel are 7, 5 and 3 times those derived from unit mass of the in-cylinder fuel vapor. An average value for the multiplication factor is 5.

For the CV probe liquid-fuel-injection cases, because of the evaporation in the vicinity of the valve seat prior to valve opening and the redistribution of the liquid due to the backflow when the valve opens, the value of z should be lower and may not be uniform around the valve circumference. The data from our experiment can only determine the values for $z(\psi_{liq} - \psi_{vap})$, which are the slopes of the curves in Figs. 8 and 9. To analyze the CV probe injection cases, we assume that the values for ψ_{liq}

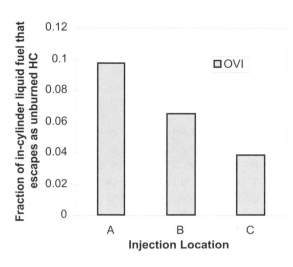

Fig. 10 Fraction of in-cylinder liquid fuel that escapes as hydrocarbon emissions for open-valve probe injection (OVI); $z = 0.5$ in Eq. (4) See Fig. 2 for definition of probe injection locations.

are the same as those in the OV cases at the same location. Then z is adjusted so that the $z(\psi_{liq} - \psi_{vap})$ values match the slopes of Fig. 9. These values of z are 0.35, 0.3 and 0.2 respectively for probe injection at locations A, B and C. The average of these values is 0.28, which is 40% less than the corresponding value assumed for the OV case ($z = 0.5$). Thus of the liquid fuel injected on the valve seat region with CV injection about 60% enters the cylinder as liquid. This suggests that with probe injection during CV injection, the backflow of hot residual, and the consequent redistribution and evaporation of the liquid fuel in the intake region, significantly reduces the amount of liquid fuel flow into the cylinder.

We can use these data to estimate the amount of liquid fuel entering the cylinder in this Nissan engine with its standard port-fuel-injection system. At the same operating condition (1500 rpm, 0.5 bar intake pressure, stoichiometric mixture), the engine-out HC mass was 1.57% of the fuel mass. Using an average ψ_{liq} value of 0.067 and $z = 0.5$ (Table 1) which is appropriate for the liquid fuel entering the cylinder, Eq. (1) gives $y = 0.05$. Thus, using the normal fuel injection system at warmed-up conditions, 5% of the injected mass enters the cylinder as liquid.

SUMMARY AND CONCLUSIONS

An experimental technique and data analysis procedure to examine the increase in HC emissions resulting from liquid gasoline entering the cylinder has been developed and used. Probes--tubes depositing known amounts of liquid fuel close to the intake valve seat during the first half of the intake process, were combined with a prevaporizing gasoline injector for the rest (majority) of the fuel, thereby allowing the HC emissions that result from liquid fuel and from vapor fuel present in the cylinder to be estimated separately. The results obtained are self-consistent, and provide insight into the significantly larger contribution of liquid fuel in the cylinder to engine out HC emissions. The specific conclusions from this study are the following:

1. As the amount of liquid fuel entering the cylinder during the intake process (assumed equal to the liquid fuel delivered to the valve seat by each probe during the intake valve open period) increases, so do the engine-out HC emissions.

2. The increase in HC emissions depends on the probe liquid fuel deposition location. A given amount of liquid fuel entering one intake valve of this 4-valve-per-cylinder engine on the exhaust valve side of the seat (probe location A) caused three times the increase in HC than did liquid fuel entering on the other side of the intake valve (probe location C), farthest from the exhaust valve. The probe location midway between (B) showed intermediate HC increases.

3. Liquid fuel probe injection at the valve seat while the intake valve was closed (not the primary operating mode for these experiments) showed a similar but smaller (about half) linear increase in HC with amount injected. This would be expected since the vigorous hot residual gas backflow at intake valve opening at the throttled engine operating condition studied (intake pressure 0.5 atm) redistributes liquid fuel in the vicinity of the valve seat on the upstream port walls and increases fuel vaporization within the intake port. The difference in HC emissions between open valve and closed valve probe liquid deposition, for a given amount of probe liquid fuel, was about the same at each probe location suggesting this backflow results in a roughly equal increase in fuel evaporation at all three probe locations.

4. A model was developed which, when fitted to the data, estimates the HC emissions indices, ψ_{liq} and ψ_{vap}, for liquid and vapor fuel, respectively, in the cylinder at time of combustion (emission index is the mass of HC/mass of liquid or vapor fuel). With open valve probe liquid deposition at the valve seat, the ratio ψ_{liq} / ψ_{vap} varied from about 3 to 7, depending on probe location.

5. With probe liquid deposition when the intake valve is closed, the ψ_{liq} values are lower due to evaporation in the intake of some of the liquid fuel deposited as a result of redistribution and heating by the hot residual gas backflow that occurs at intake valve opening. Our data and model indicate that this additional evaporation is substantial: some 40% of the liquid fuel deposited by the probe evaporates in the port with closed valve probe liquid fuel deposition.

6. The model was used to estimate the amount of liquid fuel in the cylinder with this production engine operating with its standard port fuel injection system (with closed valve injection), at this warmed-up light-load low-speed condition typical of urban driving. About 5% of the injected fuel enters the cylinder as liquid, and this increases the HC emissions by about 11% above the fully vaporized gasoline HC emission level assuming half this liquid vaporizes in the cylinder and does not impinge on the walls.

ACKNOWLEDGMENTS

An Industrial Consortium on Engine and Fuels Research supported this work. The members of the consortium are: DaimlerChrysler Corp., Ford Motor Co., General Motors, ExxonMobil Corp., Shell Development Co., and Volvo Car Corp.

REFERENCES

1. Cheng, W.K., D. Hamrin, J.B. Heywood, S. Hochgreb, M. Min, and M. Norris, "An Overview of Hydrocarbon Emissions Mechanisms in Spark-Ignition Engines," SAE paper 932708, 1993.

2. Min, K., W.K. Cheng, and J.B. Heywood, "The Effects of Crevices on the Engine-Out Hydrocarbon Emissions in SI Engines," SAE 940306, 1994.

3. Sterlepper, J. U. Spicher, and H. Ruhland, "Flame Propagation into Top Land Crevice and Hydrocarbon Emissions from a SI Engine During Warm-Up," SAE paper 932648, 1993.

4. Linna, J., Malberg, H., Bennett, P.J., Palmer, P.J., Tian, T., and Cheng, W.K., "Contribution of Oil Layer Mechanism to the Hydrocarbon Emissions from Spark-Ignition Engines," SAE paper 972892, 1997.

5. Kaiser, E.W., Seigel, W.O., and Russ, S.G., "Effect of Fuel Dissolved in Crankcase Oil on Engine-Out Hydrocarbon Emissions from Spark-Ignited Engine," SAE 972891, 1997.

6. Norris, M.G., and Hochgreb, S., "Extent of Oxidation of Hydrocarbons Desorbing from the Lubricant Oil Layer in Spark-Ignition Engines," SAE 960069, 1993.

7. Parks, J., Armfield, J., Storey, J., Barber, and T., Wachter, E., "In Situ Measurement of Fuel

Absorption into the Cylinder Wall Oil Film During Engine Cold Start," SAE 981054, 1998.

8. Min, K., "The Effects of Crevices on the Engine-Out Hydrocarbon Emissions in Spark Ignition Engines," Ph.D. Thesis Department of Mechanical Engineering, MIT, 1993.

9. Kaiser, E.W., Siegl, W.O., Lawson, G.P., Connolly, F.T., Cramer, C.F., Dobbins, K.L., Roth, P.W., Smokovitz, M., ,"Effect of Fuel Preparation on Cold-Start Hydrocarbon Emissions from a Spark-Ignited Engine," SAE 961957, 1997.

10. Imatake, N., Saito, K., Morishima, S., Kudo, S., and Ohhata, A., "Quantitative Analysis of Fuel Behavior in Port-Injection Gasoline Engines," SAE 971639, 1997.

11. Shayler, P.J., Davies, M.T., "Audit of Fuel Utilization During the Warm-Up of SI Engines," SAE 971656, 1997.

12. Fry, M., Nightingale, C., Richardson, S., "High-Speed Photography and Image Analysis Techniques Appolied to Study Droplet Motion Within the Porting and Cylinder of a 4-Valve SI Engine," SAE 952525, 1995.

13. Stanglmaier, R.H., Li, J., and Matthews, R.D., "The Effect of In-Cylinder Wall Wetting Location on the HC Emissions from SI Engines," SAE 1999-01-0502, 1999.

14. Li, J., Matthews, R.D., Stanglmaier, R.H.,, Roberts, C.E., and Anderson, R.W., "Further Experiments on the Effects of In-Cylinder Wall Wetting on HC Emissions from Direct Injection Gasoline Engines," SAE 1999-01-3661, 1999.

15. Meyer, R., and Heywood, J.B., "Liquid Fuel Transport Mechanisms into the Cylinder of a Firing Port-Injected SI Engine During Start-Up," SAE paper 970865, presented at the SAE International Congress & Exposition, Detroit, MI, February 24-27, 1997.

16. Meyer, R., Yilmaz, E., and Heywood, J.B., "Liquid Fuel Flow in the Vicinity of the Intake Valve of a Port-Injected SI Engine," SAE paper 982471, presented at the SAE International Fall Fuels and Lubricants Meeting and Exposition, San Francisco, CA, Oct. 19-22, SAE Trans., Vol. 107, 1998.

17. Meyer, R., and Heywood, J.B., "Effect of Engine and Fuel Variables on Liquid Fuel Transport into the Cylinder in Port-Injected SI Engines," SAE paper 1999-01-0563, presented at the 1999 SAE International Congress and Exposition, Detroit, MI, March 1-4, 1999.

18. Cheng, C., Cheng, W., Heywood, J.B., Maroteaux, D., Collings, N., "Intake Port Phenomena in a Spark-Ignition Engine at Part Load," SAE paper 912401, 1991.

19. Shin, Y., Min, K., Cheng, W., "Visualization of Mixture Preparation in a Port Fuel Injection Engine During Engine Warm-Up, SAE paper 952481, 1995.

20. Boyle, R.J., Boam, D.J., and Finlay, I.C., "Cold Start Performance of an Automotive Engine Using Prevaporized Gasoline," SAE paper 93710, 1993.

21. Landsberg, G., "Liquid Fuel Hydrocarbon Emissions Mechanisms in Spark Ignition Engines," S.M. Thesis, Department of Mechanical Engineering, June 2000.

22. Meyer, R., and Heywood, J.B., "Evaporation of In-Cylinder Liquid Fuel Droplets in an SI Engine: A Diagnostic-Based Modeling Study," SAE paper 1999-01-0567, presented at the 1999 SAE International Congress and Exposition, Detroit, MI, March 1-4, 1999.

CHAPTER 3

COLD-START HYDROCARBON EMISSION MECHANSIMS

James A Eng
General Motors Research

3.1 OVERVIEW

As the hydrocarbon emission standards are lowered to ULEV and SULEV levels, a larger fraction of the total hydrocarbon (HC) emissions over the FTP test are emitted during the initial cold-start transient. At the ULEV emissions level, 80 to 90% of the tailpipe HC are emitted during the cold-start and first cycle of the FTP test. Tailpipe HC emissions during the cold-start are high because the catalyst is not at its light-off temperature (typically around 300°C) to efficiently oxidize the HC. Thus, tailpipe HC emissions during the cold-start are essentially equal to engine-out HC emissions until the catalyst lights-off. In order to shorten the time to obtain catalyst light-off, advanced engine control strategies are used, such as retarding the spark to rapidly increase the exhaust gas temperature. However, since the rate of catalyst heating is limited by the thermal stresses produced within the catalyst, it will not be possible to obtain the required emissions reductions through fast catalyst light-off and control strategies alone. Engine-out HC emission during the cold-start will also have to be reduced to meet the SULEV emission standards.

Engine-out hydrocarbon emissions from spark ignition (SI) engines are the result of two processes. The first is storage of raw fuel within the cylinder that escapes combustion. The second process is the oxidation of these unburned HC during the expansion and exhaust strokes. After the primary combustion process has consumed the major portion of the combustible (i.e., vaporized) fuel within the cylinder, the unburned HC can either be completely consumed into CO_2 and H_2O, or be partially oxidized into lower molecular weight species that exit the engine and contribute to engine-out HC emissions.

The most significant mechanisms by which fuel can be stored within the cylinder and escape combustion during a cold-start are: 1) storage of fuel in combustion chamber crevices, 2) liquid fuel within the cylinder that is too rich to burn, and 3) partial burns. Additional mechanisms that may contribute to HC emissions are quench layers and oil layer absorption. However, both experimental and computational results indicate that the contribution of these mechanisms to HC emissions under steady-state conditions is less than 5 to 10%. During a cold-start the engine is operated fuel-rich in order to avoid misfire, and the rich air/fuel ratio leads to increased HC emissions. It is important to note that while the HC storage mechanisms have been identified, the relative contribution of an individual mechanism is not known at present. Given the significant impact of engine design and control strategies on HC emissions, it seems clear that the relative importance of any specific mechanism is very dependent on the particular engine in which the research was performed. With this caveat in mind, the major HC emission mechanisms for an engine cold-start will be discussed below.

Storage in Crevices--During compression unburned fuel enters crevices into which the flame cannot propagate and escapes combustion during the main combustion process. Crevices within the combustion chamber in which fuel can be stored are in the exposed threads around the spark plug, in the crevices around the circumference of the intake and exhaust valves, in the cylinder head gasket, and in the piston ring-pack crevices. The largest crevice volume in the engine is the piston ring-pack. In modern engines the ring-pack can account for up to 80% of the total crevice volume. Of the ring-pack volume, the largest volumes are the top-land volume and behind the top compression ring. It is generally recognized that piston ring-pack crevices are the largest source of HC emissions at steady-state warmed-up operation. Estimates for the contribution of ring-pack crevices to HC emissions at steady-state warmed-up conditions range from 50% up to 90%.

As the coolant temperature is reduced the contribution of the ring-pack crevices becomes more significant due an increase in gas density within the crevice. Additionally, the piston top-land side clearance increases due to thermal contraction of the piston. With a constant cylinder head temperature, experimental results show a 4% increase in HC emissions per 10°C change in cylinder block coolant temperature. Roughly the same changes in HC emissions were obtained from experiments performed with iso-pentane (which vaporizes at 28°C at atmospheric pressure) and gasoline. These results indicate that HC emissions would increase by roughly 25% as the coolant temperature is reduced from 90°C to 20°C.

Liquid Fuel – When liquid fuel impacts a surface within the intake port or cylinder that is below the vaporization temperature of the higher molecular weight compounds of the fuel it will remain a liquid and can be a significant source of HC emissions. The in-cylinder fuel films are very thin (on the order of 50 to 300 μm) such that the fuel is at the temperature of the surface. Since the higher boiling point components in gasoline vaporize over a range of temperatures from 100 to 200°C, much of the fuel remains unburned within the cylinder for a significant number of engine cycles following a cold-start. The effects of liquid fuel on emissions at steady-state conditions can be readily observed by performing an injection timing sweep. Experiments performed with injection timing sweeps show that HC emissions increase significantly when the fuel is injected during the intake stroke. Speciated exhaust emissions obtained when the fuel was injected during the open intake valve show an increased percentage of xylenes and methyl-benzenes in the exhaust gas. These results are consistent with the physical model that the increase in HC emissions is due to wall impaction of the liquid fuel on a surface that is below the vaporization temperature of the heavy components of the fuel. After combustion these high molecular weight fractions are vaporized by the high temperature burned gas. The temperature-time history during the expansion stroke is not sufficient to vaporize all of these HC and consume them before exhaust valve opening, and thus they are a source of HC emissions.

It is known that large amounts of liquid fuel enter the cylinder during engine warm-up. Fuel films are formed in both the intake port and cylinder. Models of fuel film dynamics during a warm-up transient have indicated that the mass of the fuel in the cylinder can rapidly increase to ~120 to 150 mg per cylinder. This in-cylinder fuel film is predicted to decrease to 20 mg over a period of 100 sec. It is generally believed that liquid fuel is a significant, if not the largest, contributor to cold-start HC emissions. More research is needed in order to determine the importance of liquid fuel to HC emissions.

Partial Burns - A partial burn takes place when the fuel in the cylinder is not completely consumed before the exhaust valve opens. Post-flame reactions are quickly quenched after the cylinder pressure and temperature decrease during exhaust blowdown. The vaporized fuel that remains within the cylinder when the flame extinguishes will exit the engine and be a significant source of HC emissions. At the SULEV emissions level the vehicle will fail to meet the emissions standard if the engine has even one misfire during a cold-start. At the ultra-low emissions levels it is extremely important that the engine and control strategy are developed to make the engine robust to operating temperatures and fuel composition changes so that the engine does not misfire or have a significant number of partial burns during a cold-start. The cold-start calibration is typically developed with the minimum required fueling levels and spark timings to ensure that the engine does not misfire. Identifying a partial burn during a cold-start is difficult because the combustible (vaporized) fuel within the cylinder is not known well. In addition, the rapid engine speed and manifold absolute pressure (MAP) changes make it difficult to determine the fresh air delivered per cycle. As a practical matter, the indicated mean effective pressure (IMEP) can often be used to determine when a misfire or partial burn occurs. A complete misfire is easy to identify since the IMEP is negative. When the engine speed is not changing rapidly, a good metric that can be used to identify a partial burn is when the IMEP of a cycle is more than 20% lower then the previous and successive cycle.

Rich Air/Fuel Ratio - The effect of air/fuel ratio on HC emissions from SI engines is well known. Hydrocarbon emissions increase more or less linearly on either side of the stoichiometric air/fuel ratio. At steady-state conditions the minimum HC emissions occurs near air/fuel equivalence ratios of 0.90-0.95. The variation in HC emissions with air/fuel ratio is strictly one of chemistry. On the rich side there is not sufficient oxygen present to complete the oxidation of all of the fuel. On the lean side there is sufficient oxygen present, however the maximum temperatures are reduced due to the increased dilution and a lower amount of post-flame oxidation due to depletion of the oxygen. The primary reason the engine is over-fueled during the cold-start is the poor vaporization of the fuel. Additionally, a large percentage of the injected fuel from the first few engine cycles is stored in fuel films in the intake port and cylinder and does not participate in combustion. The engine must be over-fueled to account for both of these effects so that the fuel that does vaporize and enter the cylinder will form a combustible mixture. Significant reductions in HC emissions could be achieved if the engine could be operated closer to stoichiometry during the cold-start. The effects of fuels and fuel reforming on HC emissions is discussed in Chapter 7.

Post-Flame HC Consumption - After the fuel escapes the primary combustion process it must also avoid being oxidized both within the cylinder and the exhaust port in order to contribute to engine-out HC emissions. At steady-state warmed-up conditions there is a significant amount of post-flame oxidation taking place within the cylinder. Under these conditions, experimental and computational results indicate that 50 to 90% of the unburned HC are consumed within the cylinder before the exhaust valve opens.

The post-flame HC oxidation processes in engines are extremely complex. The unburned HC are distributed non-uniformly throughout the combustion chamber and there are large temperature and concentration gradients near the cylinder walls. The available experimental data on post-flame HC consumption in engines indicate that HC consumption (complete conversion of the fuel into H_2O and CO_2) takes place at temperatures higher than 1300 to 1500 K. For normal operating conditions with spark advances near the minimum for best torque (MBT), the post-flame HC oxidation is essentially completed within the cylinder before exhaust valve opening (EVO). Simultaneous in-cylinder imaging of OH radicals and unburned fuel concentrations in an optical engine has also indicated that post-flame HC oxidation takes place at temperatures higher than 1400 K. When the temperature decreased below this following EVO the OH signal rapidly extinguished, and unburned fuel was detected near the top of the piston only after the disappearance of OH signal. During the expansion stroke the unburned fuel is layered along the cylinder wall boundary layer and diffuses into the high-temperature burned gas. The fuel rapidly breaks down by thermal pyrolysis (C-C bond cleavage) and a radical pool is quickly established that consumes the smaller fuel fragments. At temperatures higher than 1400 K the HC are able to initiate and replenish their own radical pool without the presence of radicals within the burned gas.

Computational studies of post-flame HC consumption in engines have resulted in similar conclusions to those obtained experimentally. The model calculations are for a simplified one-dimensional diffusion and reaction of wall layer HC into the high temperature core gas. The computations have been performed using detailed chemical reaction mechanisms for both propane and iso-octane. The results of these calculations showed that the chemical reactions are effectively quenched at temperatures below 1400 to1600 K for both fuels. The computational studies were extended to cold engine operation, and specifically cold-start operation. For these simulations a thin layer of liquid fuel along the cylinder wall was modeled using the same one-dimensional approach. The results indicated that regardless of the initial thickness of the liquid fuel layer, 95% to 97% of the fuel that was vaporized within a given engine cycle was fully consumed by EVO.

Secondary air injection into the exhaust has been used for many years to obtain reduced cold-start HC emissions. The principle behind using secondary air is to operate the engine overall fuel-rich and then inject secondary air into the exhaust manifold to provide oxygen to react CO and HC in the catalyst. While the main objective of secondary air is to obtain faster catalyst light-off, it is also possible to obtain HC consumption within the exhaust port. The chemical timescales for HC consumption increase significantly when the temperature is decreased below 1300-1500 K. For temperatures lower than this range there simply is not enough time available to consume the HC with the timescales available in an engine. The temperature of the exhaust gas changes significantly during an engine cycle, and immediately following EVO and exhaust blowdown the exhaust gas temperature is within this critical 1300 to 1500 K window for a short period of time. Hydrocarbon oxidation can take place if the air can be mixed with the exhaust gas without significantly reducing the temperatures within the exhaust port during blowdown. Since the chemical timescale increase exponentially as the temperature decreases, it is expected that the mixing time between the secondary air and the exhaust gas will be critical to achieving HC consumption within the exhaust port.

The effects of secondary air injection location, exhaust manifold design and engine operation on engine-out (or equivalently, converter-in) HC emissions have been recently extensively investigated as an effective means to obtain reduced HC emissions. In general, experiments performed with secondary air injection directed towards the exhaust valve have not shown any evidence of exhaust port oxidation. This is due to poor mixing between the secondary air and the exhaust gas and the reduced temperatures caused by the secondary air addition. The pressure pulse in the exhaust gas caused by the blowdown pulse shuts off the secondary air momentarily when the airstream is directed towards the exhaust valve. However, results obtained with a sparger-type design, where the air was injected perpendicularly to the exhaust flow, has shown evidence of HC consumption taking place within the exhaust port. With the sparger design the secondary air is not shut off by the strong blowdown flow and the air is entrained into the exhaust gas. To gain additional time for post-engine reactions to take place the residence time within the exhaust manifold can be increased by changing the manifold volume. Experimental results have shown that it is possible to obtain significantly increased exhaust gas temperature 10 sec after the cold-start due to sensible energy release within the exhaust port, which leads to a significant reduction in catalyst light-off time. Achieving HC oxidation within the exhaust port requires optimization of the mixing rate between the secondary air and the exhaust gas, and the exhaust port and manifold design.

3.2 SELECTED SAE TECHNICAL PAPERS

950074

Mixture Preparation and HC Emissions of a 4-Valve Engine with Port Fuel Injection During Cold Starting and Warm-up

Keiso Takeda and Takehisa Yaegashi
Toyota Motor Corp.

Kiyonori Sekiguchi, Kimitaka Saito, and Nobuo Imatake
Nippon Soken, Inc.

ABSTRACT

In order to reduce tail-pipe hydrocarbon emissions from SI gasoline engines, rapid catalyst warm-up and improvement of catalyst conversion efficiency are important. There are many reports which have been published by manufacturers and research institutes on this issue. For further reduction of tail-pipe hydrocarbon emissions, it is necessary to reduce engine-out hydrocarbon emissions and to improve after treatment, during the time the catalyst is not activated.

This paper quantitatively analyzed the fuel amount of intake port and cylinder wall-wetting, burned fuel and engine-out hydrocarbon emissions, cycle by cycle in firing condition, utilizing a specially designed analytical engine.

The effect of mixture preparation and fuel properties for engine-out hydrocarbon emissions, during the cold engine start and warm-up period, were quantitatively clarified.

INTRODUCTION

In SI gasoline engines, mixture preparation in the cylinders during cold engine start and warm-up period is closely related not only to the engine starting performance but also to the exhaust emissions [1] ~ [3][*].

Clarification of these relationships is an important issue. Recently, a high catalyst conversion efficiency of exhaust gases has been achieved with enhanced Air/Fuel (A/F) ratio control by electronic fuel injection (EFI) and catalyst technologies. However, from the view point of environmental protection; we must further decrease exhaust emissions. To achieve this, it is very important to reduce unburned hydrocarbon, which is emitted much during cold engine start and warm-up period when the catalyst is not activated.

Fig. 1 and 2 show the hydrocarbon (HC) emissions

during LA#4 mode driving. Over 60% of the non-methane hydrocarbon (NMHC) emissions are produeced just after engine cold start and warm-up period. As shown in Fig. 2, most of HCs are not converted during the first sixty seconds following start, due to inadequate catalyst activity. To reduce these emissions during cold operation, we must analyze the fuel behavior in the intake port and cylinder.

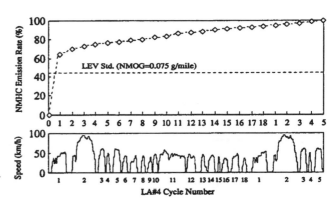

Fig. 1 NMHC Emission Rate of Toyota LS400

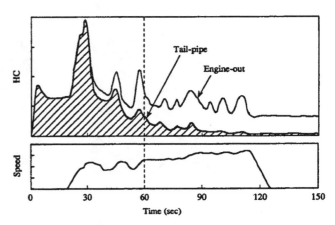

Fig. 2 HC Concentration during the First Cycle

---

[*] Numbers in parentheses designate references at the end of the paper.

A large number of studies have been conducted on fuel atomization and injecting direction, which greatly affect fuel behavior in engines [4] ~ [8].

However, fuel behavior during cold operation is complicated, because injected fuel flows onto the intake port wall and cylinder wall.

The comprehensive relationships between HC emissions and fuel behavior, during cold operation including intake port and cylinder wall-wetting, still remains unreported. The authors have clarified the effects of intake port and cylinder wall-wetting on engine-out HC emissions during cold engine start and warm-up by analyzing the fuel behavior qualitatively and quantitatively with our specially designed analytical engine and equipments.

EXPERIMENTAL APPARATUS AND PROCEDURES

TEST ENGINE AND FUEL PROPERTIES - A 4-cylinder, 4-valve, port fuel injection engine was utilized for the experiment. Table 1 shows its specifications, and Table 2 shows the properties of the test fuels. We compared three types of fuels with different distillation characteristics.

Fuel B is the refference, and always used except as noted.

Table 1. Test Engine Specifications

Number of cylinders	4
Displacement (cc)	2164
Bore x Stroke(mm)	87 x 91
Compression ratio	9.5
Valves per cylinder	4

Table 2. Fuel Properties

		Fuel A	Fuel B	Fuel C
Specific (g/cm^3) gravity (Tf = 15°C)		0.7409	0.7445	0.7635
Distillation characteristic(°C)	Initial point	39.0	31.5	33.5
	10%	61.5	53.5	56.5
	50%	96.0	105.0	119.5
	90%	143.5	155.5	169.0
	End point	187.0	176.5	204.0

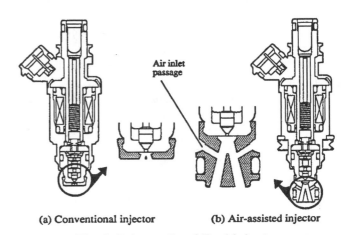

Air inlet passage

(a) Conventional injector   (b) Air-assisted injector

Fig. 3 Schematic of Fuel Injector

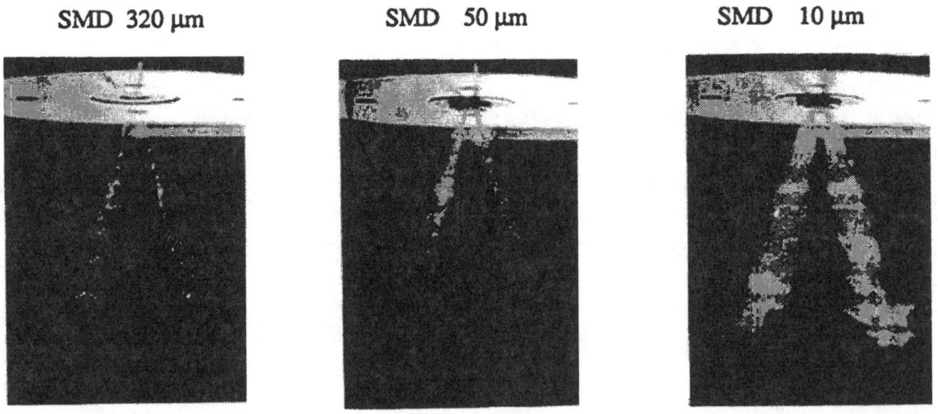

SMD 320 μm   SMD 50 μm   SMD 10 μm

Aspirated air-assist (ΔP = 50kPa)   Pressurized air-assist (ΔP = 200 kPa)

(a) Conventional Injector   (b) Air-assisted Injector

Fig. 4 Spray Shapes

FUEL INJECTORS - Fig. 3 shows schematics of the two types of fuel injectors used for the experiment. (a) is a conventional fuel injector, and (b) is an air-assisted fuel injector designed for fuel atomization with assist air. For the latter, the following two types of air assistance were compared.

(1) "Aspirated air-assist type"; utilizing the pressure difference between atmospheric and intake port.

(2) "Pressurized air-assist type"; using pressurized pump air

Conventional fuel injector is the reference, and always used except as noted.

Fig. 4 shows photographs of injected fuel comparing the conventional and air-assisted fuel injectors. In all cases, the fuel was injected into two directions avoiding the partition between the siamese intake ports specific to 4-valve engines.

The fuel droplet diameter (SMD) from the conventional fuel injector, the aspireted air-assist type and the pressurized air-assist type, were about 320μm, 50 μm and 10μm, respectively.

EXPERIMENTAL APPARATUS FOR QUANTITATIVE ANALYSIS AND MEASUREING METHOD - Fig. 5 shows a specially designed analytical engine, for quantitative analysis of intake port and cylinder wall-wetting, and engine-out HC emissions, under firing conditions.

The intake and exhaust valves were modified from the conventional cam-driven into the hydraulic controlled type, so that the valve opening and closing timing could be independently controlled. To determine the amount of fuel, in case of the "cylinder wall-wetting", the engine is started and the valves are operated normaly until the measuring point of the end of exhaust stroke. Then, the engine is turned off, and both the intake and exhaust valves are kept closed, to trap the fuel inside the cylinder. To vaporize the fuel inside the cylinder, purge air heated up to 200°C is fed through the spark plug hole. The vaporized fuel is analyzed by FID and its total mass is determined by integration. The intake port wall-wetting is also determined with the same technique. The analytical engine is reported in detail in our another SAE paper 950044 [9].

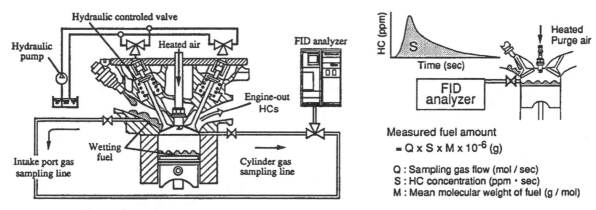

Measured fuel amount
= Q x S x M x 10⁻⁶ (g)

Q : Sampling gas flow (mol / sec)
S : HC concentration (ppm · sec)
M : Mean molecular weight of fuel (g / mol)

Fig. 5 Experimental Apparatus and Measuring Method for Quantitative Analysis of Wall-Wetting and Engine-Out HC Emissions

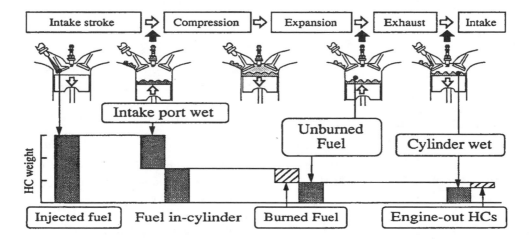

Fig. 6 Measurement Procedure for Quantitative Analysis of Wall-Wetting and Engine-Out HC Emissions

MEASUREMENT PROCEDURE FOR QUANTITATIVE ANALYSIS - Fig. 6 shows the measurement procedure for quantitative analysis of intake port wall-wetting, cylinder wall-wetting, burned fuel and engine-out HC emissions.

Injected fuel amount is calibrated to the minimum amount needed for stable combustion.

Fuel injected is separated to intake port wall-wetting and fuel in-cylinder, which is separated into burned and unburned fuel. Unburned fuel is separated into cylinder wall-wetting and engine-out HC emissions. In Fig. 6, gray portions show actual measured values and hatched portions show calculated values.

The amounts of intake port wall-wetting, unburned fuel and cylinder wall-wetting are measured at these measuring points [↑] shown in Fig. 6 respectively, utilizing the method described in previous section.

Each measurement begins at engine start and ends at the determined measuring point. It is impossible to measure continuously, because of the need to purge the wall wetting fuel as a sample.

By repeating this test sequence, the each portion of the fuel amounts for each cycle, are quantified, during the cold start and warm-up period.

RESULTS

GENERAL FUEL BHAVIOR DURING COLD ENGINE START AND WARM-UP PERIOD - Fig. 7 shows required fuel amount for stable combustion, during cold engine start and warm-up period.

During the period, the engine requires much more fuel compared with the warmed-up condition. It was observed that the 2nd cycle requires minimal amount of injection fuel.

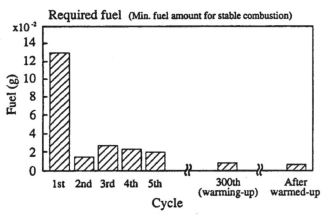

Fig. 7 Required Fuel during Engine Start and Warm-Up Period

Fig. 8 shows the required, out-going and remaining fuel amount during the first three cycles. Injected fuel is separated into intake port wall-wetting, cylinder wall-wetting, burned and engine-out HC emissions. Most of

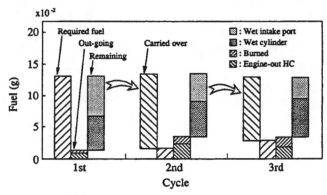

Fig. 8 Required, Out-going and Remaining Fuel for First Three Cycles

injected fuel adheres to the intake port and cylinder wall. As the adhered fuel is carried over to next cycle, the required amount of fuel needed for the following cycle is reduced. The large amounts of remaining fuel which are carried over to next cycle, makes precise air fuel ratio control difficult, and results in high engine-out HC emissions.

Fig. 9 shows the remaining fuel and engine-out HC during cold engine start and warm-up period. In this figure, "1st-3rd cycles" means the average value of three cycles after the cold start. "300th cycle" means warming-up condition, around 30 seconds after cold start.

Intake port wall-wetting increases until about the 300th cycle and then decreases gradually as the engine warms up, because of the slow increase in the port wall temperature during the first 300 cycles allows an increase of wetting area.

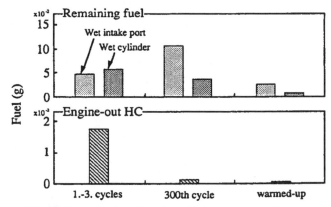

Fig. 9 Remaining Fuel and Engine-out HC Emissions in Warming-up Period

As shown in Fig. 9. cylinder wall-wetting decreases gradually with time. High engine-out HC emitted just after cold engine start and decreases gradually as the engine warms up. This tendency is similar to cylinder wall-wetting. Indicating that the engine-out HC emissions are due to vaperization of the fuel remaining on the cylinder wall, during the expansion stroke.

To reduce engine-out HC during cold engine start and warm-up period, it is important that to reduce the intake port wall-wetting and cylinder wall-wetting simultaneously.

EFFECTS OF INJECTION TIMING - Fig. 10 shows the remaining fuel and engine-out HC for the first cycle after starting. Injected fuel was adjusted to the minimum amount for stable combustion.

The open intake valve injection (End of injection timing: 90°ATDC) had greater cylinder wall-wetting and engine-out HC emissions but less intake port wall-wetting than the closed intake valve injection (End of injection timing: 90°BTDC).

This indicates that the injected fuel goes directly into the cylinder, colliding with and adhering to the cylinder wall. This shows the difficulty of simultaneously reducing the intake port wall-wetting and the cylinder wall-wetting by only optimizing the injection timing.

Open intake valve injection is the reference, and always used except as noted.

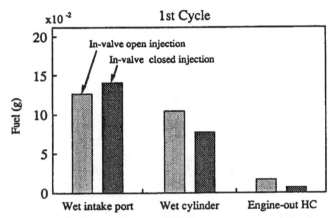

Fig. 10 Remaining Fuel and Engine-out HC Emissions for First Cycle

EFFECTS OF FUEL PROPERTIES - Fig. 11 shows the effects of fuel volatility on the amount of fuel in the following categories, during three cycles after the cold start.

    (a) minimum required fuel
    (b) intake port wall-wetting
    (c) cylinder wall-wetting
    (d) burned fuel
    (e) engine-out HC emissions

Fuel requirement is significantly affected by the 50% distillation temperature. The data shows that a lower 50% distillation temperature requires the less fuel to maintain stable combustion. This is because a lower distillation temperature fuel will vaporize easier; providing more fuel for combustion.

Fuel C, which has the highest 50% distillation temperature, showed the largest intake port wall-wetting,

cylinder wall-wetting and engine-out HC emissions, followed by fuel B, and then by fuel A.

These results show that the 50% distillation temperature is one of the essential factors for the reduction of engine-out HC emissions during cold start and warm-up.

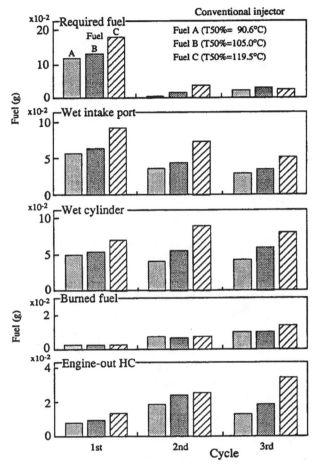

Fig. 11 Required, Remaining and Out-going Fuel for First Three Cycles

EFFECTS OF FUEL SPRAY CHARACTERISTICS - Fig. 12 shows the results of a similar study on the effects of fuel spray with different SMD values (shown in Fig. 3 and 4).

Compared with the conventional injector, these five categories of fuel amounts were improved in both the aspirated and pressurized air-assist injectors, due to better fuel atomization. In the first cycle, the amounts of required fuel, intake port wall-wetting and cylinder wall-wetting were about 40% less with the pressurized air-assist injector than those with conventional injector. Also, in the first cycle, the amount of engine-out HC emissions were reduced by about 80% compared with the conventional injector due to the reduced cylinder wall-wetting fuel and homogenized fuel mixture in the cylinder.

With a pressurized air-assist injector, the fuel

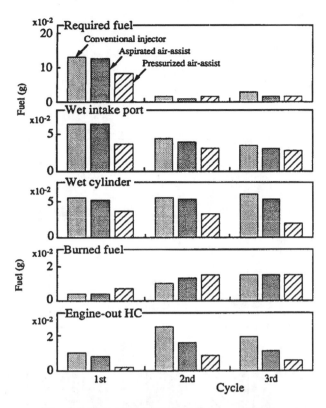

Fig. 12 Required, Remaining and Out-going Fuel for First Three Cycles

requirement reduces linearly with each cycle and allows more precise air-fuel ratio control. The amount of combusted fuel increaces becasuse of better mixture preparation, allowing a reduction in the amount of fuel required for stable combustion.

The combination of fine atomization and the open intake valve injection can reduce both intake port and cylinder wall-wetting of every cycle.

The 1st cycle engine-out HCs are less than that of 2nd or 3rd cycle in these three cases, because the lower cylinder wall temperature of 1st cycle causes less vaporized fuel at the end of expansion stroke than that of the 2nd or 3rd cycle.

In the case of an aspirated air assist injector, for the first cycle, lack of pressure diference causes poor atomization characteristics. For all categories of fuel amounts are little improvement is shown. The second cycle generated pressure diference improves atomization and all categories of fuel amounts are improved, compaired with the conventional injector.

Fig. 13 shows the comparison of three types of fuels with the pressurized air-assist injector. Compared with the conventional injector (Fig. 11), the pressurized air-assist injector improved all categories of fuel amounts, and also reduced these fuel amount differences between fuel A and fuel C, due to fuel atomization. Even though pressurized air-assist injector reduces the difference

caused by fuel volatility; a difference still remains. The 50% distillation temperature is still the key factor for the reduction of engine-out HC emissions during cold start and warm-up period.

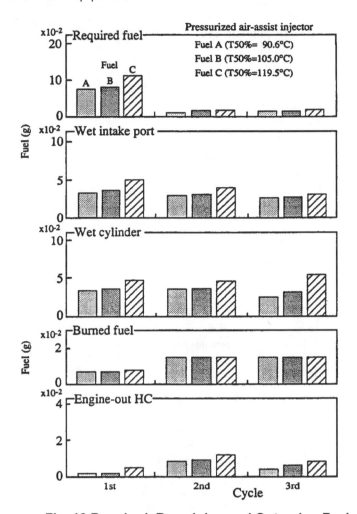

Fig. 13 Required, Remaining and Out-going Fuel for First Three Cycles

HYDROCARBON EMISSIONS - Fig. 14 shows the engine-out HC emissions difference, after engine cold start between the pressurized air-assist injector and conventional injector.

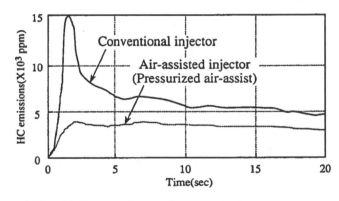

Fig. 14 Comparison of HC Emissions Traces

Improved fuel atomization associted with open intake-valve injection reduced engine-out HC emissions by 50%, during the 20 sec period immediately after a cold start.

CONCLUSION

(1) Utilizing a specially designed analytical engine, it was possible to quantify the following categories of fuel amounts cycle by cycle, on firing conditions.
   (a) minimum required fuel
   (b) intake port wall-wetting
   (c) cylinder wall-wetting
   (d) burned fuel
   (e) engine-out HC emissions

(2) Utilizing the quantification methods, transitional characteristics were clarified for each category of fuel, and suggested a direction for air-fuel ratio control, during cold engine start and warm-up period was determined.

(3) For fuel properties, distillation characteristics, particularly the 50% distillation temperature, affects engine-out HC emissions significantly.

(4) Improved fuel atomization associated with open intake-valve injection, reduced both intake port wall-wetting and cylinder wall-wetting, and resulted in much lower engine-out HC emissions, during a cold engine start and warm-up period.

(5) The pressurized air-assist injector reduced engine-out HC emissions differences between fuel A and fuel C, due to fuel atomization.

REFERENCES

[1] Jordan A. Kaplan, and John B. Heywood, "Modeling the Spark Ignition Engine Warm-Up Process to Predict Component Temperatures and Hydrocarbon Emissions", SAE Paper 910302

[2] R. Morishima, and K. Asai, "Mixture Strength at Cranking Cycles of Gasoline Engine Starting", SAE Paper 920235

[3] J. W. Fox, K. D. Min, W. K. Cheng, and J. B. Heywood, "Mixture Preparation in a SI Engine with Port Fuel Injection During Starting and Warm-Up", SAE Paper 922170

[4] A. A. Quader, "How Injector, Engine, and Fuel Variable Impact Smoke and Hydrocarbon Emissions with Port Fuel Injection" , SAE Paper 890623

[5] M. Kashiwaya, T. Kosuge, K. Nakagawa, and Y. Okamoto, "The Effect of Atomization of Fuel Injectors on Engine Performance" , SAE Paper 900261

[6] Y. Ohyama, T. Nogi, and M. Fujieda, "Study on Variable Injection Pattern Control System in a Spark Ignition Engine", SAE Paper 910080

[7] T. Sugimoto, K. Takeda, and H. Yoshizaki, "Toyota Air-Mix Type Two-Hole Injector for 4-Valve Engines", SAE Paper 912351

[8] K. Harada, R. Shimizu, K. Kurita, and M. Muramatsu, "Development of Air-Assisted Injector System", SAE Paper 920294

[9] K. Saito, K. Sekiguchi, N. Imatake, K. Takeda, and T. Yaegashi, "A New Method to Analyze Fuel Behavior in a Spark Ignition Engine", SAE Paper 950044

1999-01-0220

Emissions and Fuel Utilisation After Cold Starting Spark Ignition Engines

P. J. Shayler and C. Belton
University of Nottingham

A. Scarisbrick
Ford Motor Co.

Copyright © 1999 Society of Automotive Engineers, Inc.

ABSTRACT

A model has been developed to interpret experimental results for emissions and air/fuel ratio variations recorded during warm-up from cold starts at temperatures down to −20°C. The model describes fuel transport and utilisation after injection to its exhaust as fuel products or loss to the crankcase, and allows for the storage of fuel in films on the intake port surface, in-cylinder surfaces and in the piston "crevice". Engine-out emissions of unburned hydrocarbons are treated as being comprised of contributions from the bulk charge, fuel returning from in-cylinder wetted surfaces and from fuel stored in the piston crevices. The model characterises engine-out emissions and air/fuel ratio variations successfully under both quasi-steady and transient engine operating conditions during warm-up. Good agreement between experimental data and model predictions has been achieved for a wide range of engine operating conditions. The model equations set has been derived from phenomenological arguments. Despite the complexity of the processes described, empirical constants/functions depend on engine coolant temperature only. The model has been used to investigate implied variations of in-cylinder fuel film mass, fuel loss to the crankcase, exhaust air/fuel ratio variations and engine-out emissions.

INTRODUCTION

Poor mixture ratio control and high emissions of unburned hydrocarbons (HC) and carbon monoxide (CO) typify problems arising during the warm-up of a spark ignition engine after a cold start. High engine-out emissions of HC and CO emissions during the early minutes of engine running contribute disproportionately to totals emitted during short vehicle journeys. The weighting is higher still for tail-pipe emissions because three-way-catalytic converters are ineffective before they reach light-off temperatures, which can take tens of seconds or more, and then for high converter efficiencies, mixture ratio must be maintained very close to stoichiometric values. The lower the ambient temperature, the greater do differences between cold and fully-warm engine behaviour become. At extreme cold start temperatures it is difficult to ensure reliable starting and stable running without high levels of mixture supply enrichment, and lubricating oil can be diluted with fuel washed into the crankcase.

The importance of engine performance during cold start and warm-up is now widely recognised and emissions test procedures are being revised to reflect this. When European Stage III specifications become mandatory in the year 2000, gas sampling will be required to commence at key-on, and the introduction of a −7°C cold-start test for HC and CO emissions is under discussion [1]. In the USA, and California in particular, similar or more stringent regulatory measures are in place or planned. In recent years, there has been a growth of research activity directed at understanding and defining the various mechanisms which effect cold-running performance [2]-[13]. In the work reported here, fuel transport and utilisation, and the connection between these and engine-out emissions have been investigated for a range of quasi-steady and transient engine conditions. Conditions following cold starts at temperatures down to -20°C have been examined, for a representative 4 cylinder in-line 16 valve engine with electronically controlled, port fuel injection.

The processes of interest give rise to a discrepancy between fuel injected and fuel products exhausted in the exhaust gas flow, and strongly influence emissions of HC and CO. The path of fuel transport through the engine starts at the point of fuel injection. The fuel injectors deliver a spray of atomised fuel into the intake port and, under cold running conditions, a significant part of the spray is deposited on the intake port surfaces and valve backs forming a thin film. Fuel induced into the cylinder is comprised of vapour and liquid which has remained airborne after injection or been sourced by fuel transfer from the port surface films. Liquid fuel reaching the valve seat can accumulate in sufficient quantity to splash back into the intake port as the intake valve closes, as observed by [2], [3] and [4]. As the intake valves open, liquid feeds from the valve seat over the upper surface of the combustion chamber [5]. At the same time airborne droplets and droplets generated by the shear flow across the valve are carried into the cylinder and can strike the cylinder walls and piston crown [5], [6] and [7]. These processes, and further deposition encouraged by charge motion during the intake stroke, wet the entire exposed surface area. During the compression stroke fuel is carried into the clearance gap around the piston by blowby and, as the piston rises, surface fuel on the liner will be swept into the piston land and ring grooves (referred to hereafter as the piston crevices) [6]. This fuel either remains trapped in the crevices, re-wets the liner during subsequent down strokes and is exposed to the bulk mixture in the cylinder, or it is lost to the lubricating oil. In the last case, this occurs by absorption into oil collected by the scraper ring and by wetting out from the blowby gases as these pass through the crankcase [8].

In the following, experimental facilities and test procedures are described and representative examples of experimental results are presented. A phenomenological model of fuel transport paths from injection into the intake port to exhaust flow or crankcase is then defined. Of particular interest has been the behaviour of fuel deposited on cylinder walls and what happens to this subsequently, and the model has been extended to cover the description of engine-out hydrocarbon and carbon monoxide emissions. The model embodies a small number of empirical coefficients. These have been determined to best match predictions to a broad range of observed features. The result of these developments is a model defining fuel behaviour and the generation of emissions which greatly assists understanding the influence and consequences of the processes involved. These are discussed.

EXPERIMENTAL STUDIES

All the experimental data used in this investigation has been acquired on Ford 1.8 ℓ, Zetec engines. Fuel injection was sequential and timed to end before intake valve opening. The fuel used was European unleaded Class 8 pump-grade gasoline drawn from a single source. The specified minimum to maximum ranges of percent evaporated are 20-50% at 70°C, 43-70% at 100°C, and a minimum of 85% at 180°C. The test facility is comprised of a compact cold test cell with an external motor/regenerator. The test engine is installed in the cell and coupled to the motor/regenerator using an in-line transmission. The cold cell provides engine cooling by circulating a refrigerated ethylene glycol/water mixture through the block and head

coolant passages and the circulation of dry, refrigerated air over the external surfaces of the engine. Additional refrigerant loops through the engine sump to promote oil cooling, and through radiators in the cell to assist ambient temperature control, allow a uniform engine temperature of down to -29°C from 20°C to be achieved in around 3 hours.

Individual test runs were carried out to a standard procedure. Before soaking down the engine to the target temperature it was run fully-warm under load for 30 minutes. After shutting down, the engine was motored for a period of five to ten minutes to dry out port surfaces, as part of the conditioning process. The lubricating oil was changed after typically 30 tests. Before each test, after the engine temperature has been lowered to a uniform target value, the engine was motored to a target speed setting before fuel injection was enabled. This speed was varied between 1000rpm and 2500rpm in the test matrix. The motoring procedure followed during starting avoids variations in test conditions associated with cold start characteristics at low temperatures and allowed a constant mixture supply ratio to be used from key-on. Within the main test matrix the intake air/fuel ratio was varied from 11:1 to 17:1. The manifold absolute pressure (MAP) was varied between 0.3 bar and 0.9 bar, and cold soak temperatures were varied from 20°C down to -20°C. In total some 150 tests were carried out in the main test array.

Each warm-up test was allowed to progress from key-on through to fully-warm engine operating conditions. Typically data logging continued between 400 to 1200 seconds. The key parameters recorded were fuel injected, intake air mass flow rate, values of exhaust concentrations of CO and unburned HC, together with the air/fuel ratio recorded by a UEGO sensor located at the exhaust manifold flange of cylinder one. This is referred to as the exhaust air/fuel ratio. Exhaust gas analysers sampled the exhaust flow at the confluence point of flows from all cylinders. A Signal 3000 analyser with a heated sample line provided FID measurements of HC. CO was recorded using a Horiba Mexa-554JE NDIR gas analyser. The sample transport and response times are close to 2 seconds in each case.

In the first instance, the experimental data was processed using the method described in [9], to infer values for the apparent masses of fuel in intake port films, in-cylinder surface films and piston crevices, and fuel lost to the crankcase. An example of processed results for a warm-up after a motored start at -20°C is given in Figure 1. The measured exhaust air/fuel ratio variation (Figure 1a) follows a pattern of deviation from the mixture supply air/fuel ratio which is typical for starts at very low temperatures. The exhaust air/fuel ratio is lean of the supply value for several minutes of engine running.

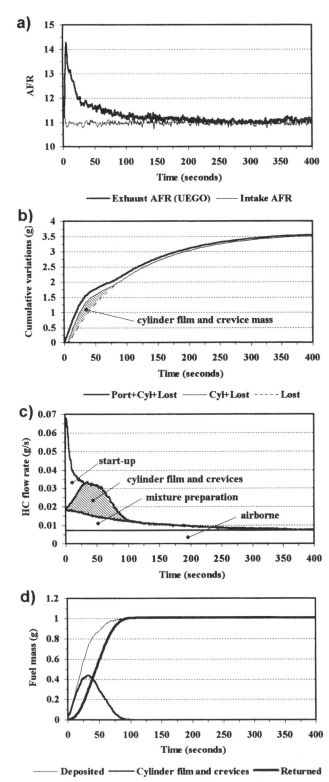

Figure 1. Variation of (a) measured exhaust and intake AFR (b) inferred cumulative port film, cylinder film and lost fuel (c) measured exhaust hydrocarbon flow rate (d) inferred cumulative fuel deposited to and returned from the cylinder film during engine warm-up at 2000rpm, 0.6 bar MAP, 11:1 AFR supply, -20°C from a motored start with carbon deposits present. All values shown are on a per cylinder basis.

After the first 2-3 seconds required for firing conditions to be established, the discrepancy between the exhaust and supply air/fuel ratio values is more than three air/fuel ratio units and is still more than half an air/fuel ratio after 50 seconds. The cumulative discrepancy between the fuel injected and fuel exhausted as unburned or burned products is shown in Figure 1b. The discrepancy is attributed to the effects of fuel being lost past the piston into the crankcase, or retained in films and crevices in the intake port and cylinder. The mass flow rate of unburned hydrocarbons in the exhaust flow is given in Figure 1c. In this example the hydrocarbon flow rate peaks within a few seconds of engine starting and decays monotonically throughout engine warm-up. Characteristically the decay in flow rate is arrested for a time between 25 and 100 seconds after starting. The total flow rate is shown as being made up of contributions from four sources. The baseline is the HC flow rate corresponding to fully-warm operating conditions. The second contribution is associated with the deterioration in mixture preparation which occurs during cold operation. The remaining two contributions are associated with the return of liquid fuel from combustion chamber surfaces and crevices, and the exhaust of unburned mixture supplied before firing conditions are established. The connection between cylinder wall films generated under cold start conditions and as unburned hydrocarbons was described in [9], [10] and [11]. In Figure 1d the inferred variations of cumulative fuel mass deposited and returned is given together with the difference between these which dictate the fuel mass in the film at any given time.

FUEL DISTRIBUTION AND TRANSPORT MODEL

The method of inferring quantities of fuel stored, returned, and lost from experimental data has limitations pointed out in [9]. These include the choice of a function describing the loss of fuel to the crankcase. The selected function satisfies constraints on the initial value of cumulative fuel lost, and the variation of this after cylinder films have dried-out. There remains uncertainty as to the variation during the first 100 seconds of engine running, and the effects of fuel stored in the piston crevices. Additionally, when engine operating conditions change rapidly, a more detailed model is required to describe the complex response of exhaust air/fuel ratio and emissions. This is defined in the following section, in the first instance for the description of fuel transport processes. Figure 2 is a schematic representation of these processes and the parameters used in the model to describe the fuel behaviour.

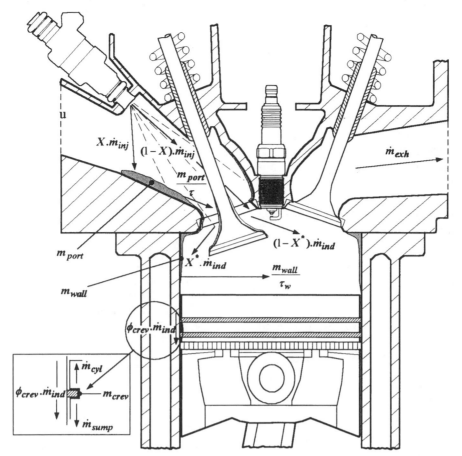

Figure 2. Parameters used to describe the fuel passage from the point of fuel injection to exhaust as fuel products or loss to crankcase.

INTAKE PORT FUEL TRANSPORT – One of the simplest and most widely applied models of fuel transport in the intake port is the τ-X model described in [14] and [15]. They assumed that a fraction X of the fuel injected is deposited on the port surfaces and that fuel leaves the surface film at a rate proportional to film mass and inversely proportional to a time constant τ. These assumptions lead to two first order equations for the mass flow rate of fuel induced and the rate of change of film mass on the port surface, respectively:

$$\dot{m}_{ind} = (1-X).\dot{m}_{inj} + \frac{m_{port}}{\tau}$$

(1)

$$\dot{m}_{port} = X.\dot{m}_{inj} - \frac{m_{port}}{\tau}$$

(2)

CYLINDER WALL FILM – A similar approach is taken here to describe fuel behaviour after induction. Of the fuel induced, a fraction X^* is assumed to be deposited as liquid on surfaces exposed during the induction stroke. The fraction $(1-X^*)$ remains airborne. Some of the fuel deposited will be swept into the piston crevices as the piston rises during the compression stroke. This is taken to be a fraction ϕ_{crev} of the induced mass flow rate such that the rate of deposition of fuel stored in the wall film above the piston is given:

$$\dot{m}_{dep} = (X^* - \phi_{crev}).\dot{m}_{ind}$$

(3)

Assuming, as in the case of the intake port, that fuel leaves the wall film at a rate proportional to film mass and inversely proportional to a time constant τ_w such that:

$$\dot{m}_{ret} = \frac{m_{wall}}{\tau_w}$$

(4)

then equations (3) and (4) can be combined to derive an expression for the rate of change of film mass:

$$\dot{m}_{wall} = \dot{m}_{dep} - \dot{m}_{ret}$$

(5)

The two parameters X^* and ϕ_{crev} are not independent. Initially, X^* will be greater than ϕ_{crev}, but these must become equal at some time during warm-up when surface films have dried out. The rate at which X^* falls to ϕ_{crev} has a time scale similar to τ_w, which is of the order of seconds. In addition, exhaust air/fuel ratio and emissions variations following rapid changes in engine operating conditions indicate that X^* is also a function of induced fuel flow rate, cylinder wall film mass and engine speed. When these parameters are combined as a dimensionless group then, consistent with data, the variation of X^* can be described by:

$$X^* = \phi_{crev} \cdot \left\{ 1 + \frac{C \cdot \dot{m}_{ind}^2 \cdot (\tau_w - \tau_{w_{fw}})}{N \cdot (b_1 + m_{wall})^2} \right\} \tag{6}$$

In this, b_1 is a constant which avoids a singularity in equation (6) before deposition creates a significant cylinder film mass on engine start-up. τ_{wfw} is the value of τ_w at fully-warm conditions and ensures X^* and ϕ_{crev} are the equal. The value of C depends upon the level of deposit build-up in the combustion chamber. During the experimental investigations, data were acquired on an engine both before and after carbon deposits on combustion chamber surfaces were removed. Comparing the two sets of data showed that the effect of the deposits was to reduce the value of C such that:

$$C = \frac{b_2}{(b_3 + m_d)} \tag{7}$$

It should be noted that mean-value variables are used throughout the equations (3) to (7). These account for the net effect of variations within a cycle. Thus, for example, at fully-warm conditions fuel may still be deposited onto the cylinder walls during induction but within one cycle this has been collected into the piston crevices or been transferred to the bulk gas mixture so that the residual cylinder wall film mass is zero. The process of deposition and immediate return to the bulk mixture, within a cycle, will also occur during warm-up. X^* is therefore the net, apparent, rate of deposition over a cycle. The true fraction of induced fuel which is deposited will always be higher. The difference between these, the intra-cycle deposition and return rate, will depend upon the way fuel is absorbed and desorbed from the deposits and provides an indication of why the value of C depends upon the mass of deposits present.

PISTON LAND AND RING GROOVE STORAGE – Implicit in equation (3), the rate at which fuel is transported into the piston land and ring groove crevices is $\phi_{crev} \cdot \dot{m}_{ind}$. Conditions in the crevices are different to those of the exposed surfaces in the combustion chamber. Assuming again that fuel leaves these crevices at a rate proportional to the fuel stored and inversely proportional to a new time constant τ^*, then the rate of change of liquid fuel stored in the crevice volume is given by:

$$\dot{m}_{crev} = \phi_{crev} \cdot \dot{m}_{ind} - \frac{m_{crev}}{\tau^*} \tag{8}$$

Fuel leaving the crevices must either be returned to the bulk cylinder charge or be lost to the crankcase. If the rate of loss to the crankcase is given by:

$$\dot{m}_{sump} = a \cdot \left(\frac{m_{crev}}{\tau^*} \right) \tag{9}$$

then for conservation of mass, the rate returning to the cylinder will be given by:

$$\dot{m}_{cyl} = [1 - a] \left(\frac{m_{crev}}{\tau^*} \right) \tag{10}$$

The ratio of these:

$$\frac{\dot{m}_{cyl}}{\dot{m}_{sump}} = \frac{1 - a}{a} \tag{11}$$

depends upon the value of a, which is determined as described later.

EXHAUST FLOW RATE OF FUEL AND FUEL PRODUCTS – The exhaust flow rate of fuel and fuel products is given by the sum of the contributions from (a) the fraction of fuel entering the cylinder that is airborne and (b) fuel returning from the cylinder wall film and piston crevice storage. Summing these gives:

$$\dot{m}_{exh} = (1 - X^*) \cdot \dot{m}_{ind} + \frac{m_{wall}}{\tau_w} + (1 - a) \cdot \frac{m_{crev}}{\tau^*} \tag{12}$$

HC EMISSIONS MODEL

When the engine is running cold and in the absence of misfires, the main sources of engine-out hydrocarbon emissions are airborne fuel escaping combustion in the bulk of the mixture, fuel returning from the cylinder wall film and fuel returning from the piston crevices. These sources are summarised in Figure 3. The rate of fuel returning from the film and crevices is defined by the equations developed in the preceding section. Only part of these returning fuel flow rates escapes combustion and contribute to the engine-out flow of unburned hydrocarbons. The fraction is assumed to be a constant in each case. Therefore, from equation (4) the contribution to engine-out emissions made by fuel returning from the cylinder wall film is given by:

$$\dot{m}_{wall_{HC}} = c_1 \cdot \dot{m}_{ret} = c_1 \cdot \frac{m_{wall}}{\tau_w} \tag{13}$$

where c_1 is a constant. Similarly the contribution of fuel returning from the piston land and ring groove source is proportional to the mass flow rate given by equation (10), such that:

$$\dot{m}_{crev_{HC}} = c_2 \cdot \dot{m}_{cyl} = c_2 \cdot (1 - a) \cdot \frac{m_{crev}}{\tau^*} \tag{14}$$

where c_2 is a constant.

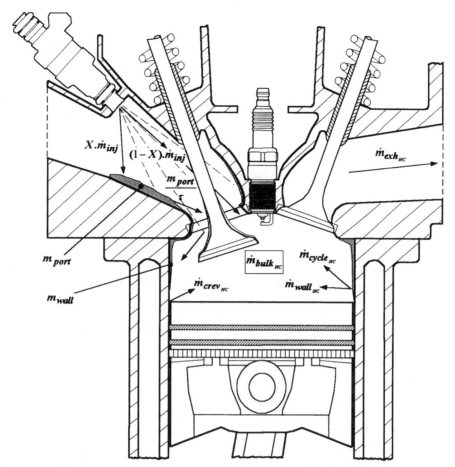

Figure 3. The contribution of various sources to the unburned hydrocarbons in the exhaust flow.

The contributions defined by equations (13) and (14) are mean-value flow rates based on net changes over the engine cycle. Fuel deposited and then returned to the bulk mixture within one cycle also contributes to engine-out hydrocarbons and requires an addition to the quantity defined by equation (13). The fuel returned within one cycle does not contribute to the apparent value of fuel deposited, but a fraction of this remains uncombusted and passes into the exhaust as unburned hydrocarbons. If the additional contribution is proportional to the net rate of change of cylinder film mass such that, with reference to equation (6), the contribution to hydrocarbon flow rate is:

$$\dot{m}_{cycle_{HC}} \propto (X^* - \phi_{crev}).\dot{m}_{ind} \qquad (15)$$

or

$$\dot{m}_{cycle_{HC}} = e.\left\{\frac{\phi_{crev}.C.\dot{m}_{ind}^2.(\tau_w - \tau_{w_{fw}})}{N.(b_1 + m_{wall})^2}\right\}.\dot{m}_{ind} \qquad (16)$$

where e is the constant of proportionality.

The value of C given by equation (7) is reduced by increasing carbon deposits. This reduces emissions of HC in line with experimental data obtained during the current work. There have been a number of recent investigations which indicate, collectively, that combustion chamber deposits can produce either an increase or a decrease in HC emissions, depending on engine design and operating condition. In the study reported in [16], deposits were observed to increase HC emissions. Four mechanisms which might account for this were described and a physics-based model was formulated. However, in [17], combustion chamber deposits were reported to produce increases in HC emissions for one engine design and decreases for a second. References cited in [17] give other examples of both trends. It is clear that the problem is complex and not yet fully understood. Here, no attempt has been made to model the fundamental mechanisms involved.

BULK MIXTURE CONTRIBUTION – The remaining contribution to engine-out hydrocarbon emissions is associated with fuel which remains airborne in the bulk of the mixture, minor crevice contributions and absorption/desorption in regions of the combustion system not wetted by the cylinder wall fuel film. At fully-warm engine operating conditions these and the piston crevices are the sources of the total engine-out hydrocarbon flow. After cold-starting and during warm-up the bulk mixture contribution is raised by poor mixture preparation and the possibility of partial burning and/or misfires. This will be proportional to the mass flow rate of fuel induced which is not deposited or deposited and returned within the engine cycle as the intra-cycle contribution to HC. The HC contribution associated with the bulk mixture is then defined by:

$$\dot{m}_{bulk_{HC}} = d.\left[(1 - X^*).\dot{m}_{ind} - \dot{m}_{cycle_{HC}}\right] \quad (17)$$

where d is the fraction of the remaining fuel which escapes combustion. The value of d depends upon the quality of mixture preparation in the bulk of the charge and the probability of a partial burn or complete misfire occurring.

During cold start-up and for a few seconds thereafter the value of d changes rapidly. At times later than t_0 when first fire occurs, the variation of d is taken to be:

$$d = d_1 + d_0.e^{-\left(\frac{(t-t_0)}{\tau_{airborne}}\right)} \quad (18)$$

The time constant $\tau_{airborne}$ is 5 seconds. This is representative of observed times for the initial hydrocarbon decay. The constant d_1 matches the total engine-out hydrocarbon mass flow rate to measurements at fully-warm engine operation conditions. This is then equal to the emission index of HC. The sum of d_0 and d_1 represents the value of d at the time of first fire. This value of d at time t_0 is less than unity (this is illustrated in Figure 4c). Prior to first fire no combustion takes place and all of the airborne fuel not deposited is exhausted as unburned hydrocarbons. In this case and at any time during operation when a complete misfire occurs, the contribution to the engine-out flow rate of hydrocarbons is given by:

$$\dot{m}_{bulk_{HC}} = (1 - X^*).\dot{m}_{ind} \quad (19)$$

First fire is assumed to occur when the air/fuel ratio of the airborne charge induced falls below the lean misfire limit. Similarly a misfire is assumed to occur when the air/fuel ratio exceeds this limit. In both cases the air/fuel ratio of the airborne charge is calculated from:

$$AFR_{airborne} = \frac{\dot{m}_{air}}{(1 - X^*).\dot{m}_{ind} - \dot{m}_{cycle_{HC}} + (1 - c_1).\dot{m}_{ret} + (1 - c_2).\dot{m}_{cyl}} \quad (20)$$

In equation (20), the HC returned from the cylinder wall and piston crevice which persists as unburned HC in the exhaust is treated as not contributing to the airborne mixture ratio.

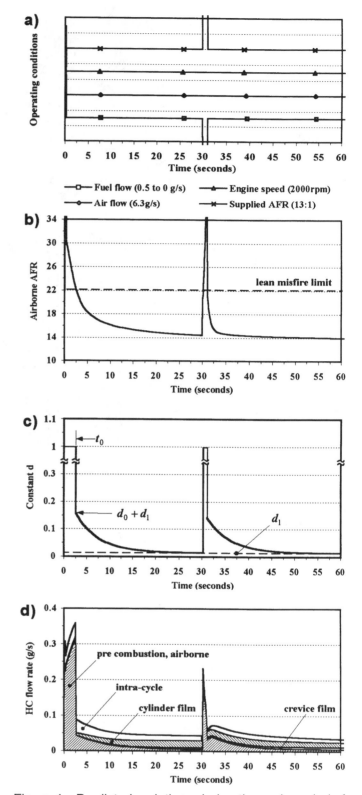

Figure 4. Predicted variations during the early period of engine operation from key-on under the operating conditions shown in (a) from a −20°C start. The variation of the airborne AFR is shown in (b), the variation of parameter d is shown in (c) and the exhaust mass flow rate of HC is given in (d). All values shown are on a per cylinder basis.

116

Table 1.　Coefficients for temperature function

MODEL PARAMETERS	COEFFICIENTS					
	f_1	f_2	f_3	g_1	g_2	g_3
ϕ_{crev}	1	-0.023	1.32×10^{-4}	7.42	-0.167	0.00109
ϕ_{tw}	3.36	-0.067	4.8×10^{-4}	1	0.055	9.5×10^{-4}
ϕ^{t^*}	16.8	-0.368	0.0021	1	0	0

Table 2.　List of constants for the model

CONSTANTS	a	b_1	b_2	b_3	c_1	c_2	d_0	d_1	e
VALUES	0.23	0.60	1000	0.75	0.40	0.11	0.16	0.015	0.40

The total engine-out mass flow rates of unburned hydrocarbons can now be derived by summing the contributions given by equations (13), (14), (16) and (17) or (19) which gives:

$$\dot{m}_{exh_{HC}} = \dot{m}_{wall_{HC}} + \dot{m}_{crev_{HC}} + \dot{m}_{bulk_{HC}} + \dot{m}_{cycle_{HC}} \quad (21)$$

An illustrative example of how these are predicted to vary during the first seconds of engine operation from key-on is given in Figure 4. The lean misfire limit is taken to be at an air/fuel ratio of 22:1. The engine operating conditions are shown in Figure 4a. The fuel supply has been interrupted after 30 seconds to induce a misfire. The variation of the parameter d is shown in Figure 4c, and the exhaust mass flow rate of HC is given in Figure 4d. The values of coefficients used in the model are as summarised in Table 1 and Table 2. The determination of these is discussed later.

CO EMISSIONS MODEL

It is well known that exhaust CO concentrations effectively depend only on equivalence ratio for cases when the burned mixture is homogeneous. This is supported by data from numerous investigations [18]. At fully-warm, steady engine operating conditions, volumetric concentrations of CO measured in the current study were consistent with these results. For values of equivalence ratio ϕ greater than unity, the CO concentration can be determined to an accuracy of within about half a percentage point from the simple fit:

$$CO_{\%vol} = 41.9 \left(\frac{\phi - 1}{\phi} \right) \quad (22)$$

For equivalence ratios of unity or less, CO concentrations are 0.5% or lower and are neglected.

During warm-up after a cold-start, the mixture in the cylinder is inhomogeneous and it is necessary to distinguish between the overall air/fuel ratio of the mixture induced and the air/fuel ratio of the mixture combusted since the second of these will dictate the CO mass flow rate in the exhaust gas stream. The air/fuel ratio of the combusted mixture, AFR_{comb}, is the same as $AFR_{airborne}$, given by equation (20), provided the lean misfire limit is not exceeded. The denominator takes account of fuel available for combustion in the bulk charge. This excludes the HC sourced by fuel returning from the cylinder wall and piston crevice which remains unburned into the exhaust. The air/fuel ratio given by equation (20) is substantially different from the air/fuel ratio of the mixture supplied throughout start-up and warm-up. At fully-warm engine operating conditions the two values are essentially equal. Experimental results taken during warm up show that the CO concentration by volume in the exhaust gas stream is given by equation (22) provided that the equivalence ratio ϕ is calculated from $AFR_{stoich}/AFR_{airborne}$. For both warm-up and fully-warm operating conditions, the mass flow rate of CO in the exhaust stream is given by:

$$\dot{m}_{CO} = \dot{m}_{exh} \cdot \frac{\tilde{m}_{CO}}{\tilde{m}_{exh}} \cdot [CO_{\%vol}] \quad (23)$$

MODEL COEFFICIENTS AND COMPARISON WITH DATA

The set of equations described contain a total of nine constants and five parameters. A summary of values for the constants and function coefficients used is given in Tables 1 and 2. These have been determined empirically. In most cases, the values effect specific features of behaviour making independent identification possible. The τ and X parameters of the intake port fuel transfer model form one sub-set. Methods of determining these have been described elsewhere. For the engines used in the current work, expressions relating τ and X to engine speed and coolant temperature were derived in [19]. These have been used without modification here. The constant b_1 is required simply to avoid a singularity in equation (6) and has no physical meaning. The constants

b_2 and b_3 in equation (7) essentially recalibrate the model in keeping with observed shifts in behaviour after carbon deposits were removed from the surfaces of the combustion chamber within the test engine. As noted earlier, investigators have reported mixed trends with regard to the effect of carbon deposits on emissions, and equation (7) must therefore be considered specific to the engine design used in the work reported here. The constants d_0 and d_1 define the bulk mixture contribution to HC emissions after first fire. The value of d_1 is matched to the emissions index for HC at fully-warm conditions. The value of d_0 has been chosen to match the predicted and experimental variation in HC emissions over the first few seconds of engine firing. The influence of d_0 is limited to this period and is an approximate value. The constant e, in equation (16), has been determined by examining the response of HC emissions to step changes in engine operating conditions. The contribution to the total HC flow is substantial under these conditions and the constant can be determined with reasonable confidence. The constants c_1 and c_2, in equations (13) and (14) respectively, define the proportion of fuel returning from the cylinder and crevice films that remains unburned. The remaining constant a, in equation (9), defines how fuel leaving the piston crevice is split between returning to the cylinder and lost to the crankcase. This, along with the functions describing variations of ϕ_{crev}, τ_w and τ^*, has been determined from the variations of exhaust air/fuel ratio, and HC and CO flow rates recorded throughout cold start and warm-up tests covering some 150 combinations of test conditions.

The parameters ϕ_{crev}, τ_w and τ^* are associated with the behaviour of fuel films on the cylinder walls and in the piston crevice region. These parameters depend on the thermal state of the engine, particularly surface temperatures. The assumption made here is that to a first approximation, they can be defined as functions of engine coolant temperature only. This has proved adequate to match model predictions and measured exhaust air/fuel ratio, CO and HC emissions characteristics as later examples illustrate. In each case, the function was taken to have the form:

$$\frac{f_1 + f_2.T + f_3.T^2}{g_1 + g_2.T + g_3.T^2}$$

where T is engine coolant temperature in degrees Celcius. The empirical constants f_1 to f_3 and g_1 to g_3 for each parameter are given in Table 1.

The variations of ϕ_{crev}, τ_w and τ^* with engie coolant temperature are plotted in Figure 5. The parameter values fall with increasing temperature in a well-behaved manner. Although temperature varies over the combustion chamber, piston and liner surfaces wetted by fuel films, no direct account has been taken of this. The main spatial variations about a mean value are quickly established after engine start-up. Distributions continue to evolve during warm-up but the effect of history is relatively weak.

The mean surface temperature generally, and the liner temperature particularly, are regulated by coolant temperature.

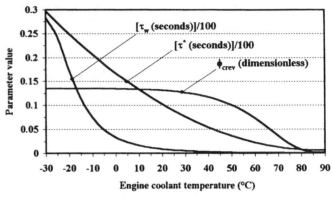

Figure 5. Variation of parameter values as defined by coefficients given in Table 1.

Two example sets of predicted and experimental results are given in Figures 6 and 7. The experimental results are taken from the test database and the agreement between experimental and model predictions is typical. Figure 6 illustrates results for a cold-start at -20°C and warm-up conditions of 2000rpm, an air/fuel ratio of 11:1, and a manifold absolute pressure of 0.6 bar. This test was carried out on the test engine before removing carbon deposits built up on the combustion chamber surfaces during previous service. Figure 6a shows the variation of operating conditions, Figure 6b shows the agreement between predicted and measured exhaust air/fuel ratio variations, Figure 6c shows agreement for the exhaust mass flow rate of unburned hydrocarbons and finally Figure 6d shows the predicted and experimental data for the exhaust CO concentration variation. The test results shown in Figure 7 were recorded on the same engine after the combustion chamber surfaces were cleaned and the engine re-built. During this test the engine was cold-started at -10°C and allowed to run for 30 seconds at 1000rpm, a supply air/fuel ratio of 11:1 and a manifold absolute pressure of 0.4 bar. After 30 seconds the throttle was opened to raise manifold pressure to 0.7 bar. Thereafter, this was then maintained constant. The change in operating conditions part way through warm-up has a substantial influence on exhaust air/fuel ratio, the engine-out unburned hydrocarbon flow rate, and the CO oncentration. This can be clearly seen in Figures 7b, c and d respectively. In each case the variation is accurately predicted by the model. As an indication of the performance of the model over a much wider range of operating conditions predicted and measured values for the total cumulative unburned hydrocarbons exhausted between cold-start and fully-warm engine conditions being reached is shown in Figure 8.

118

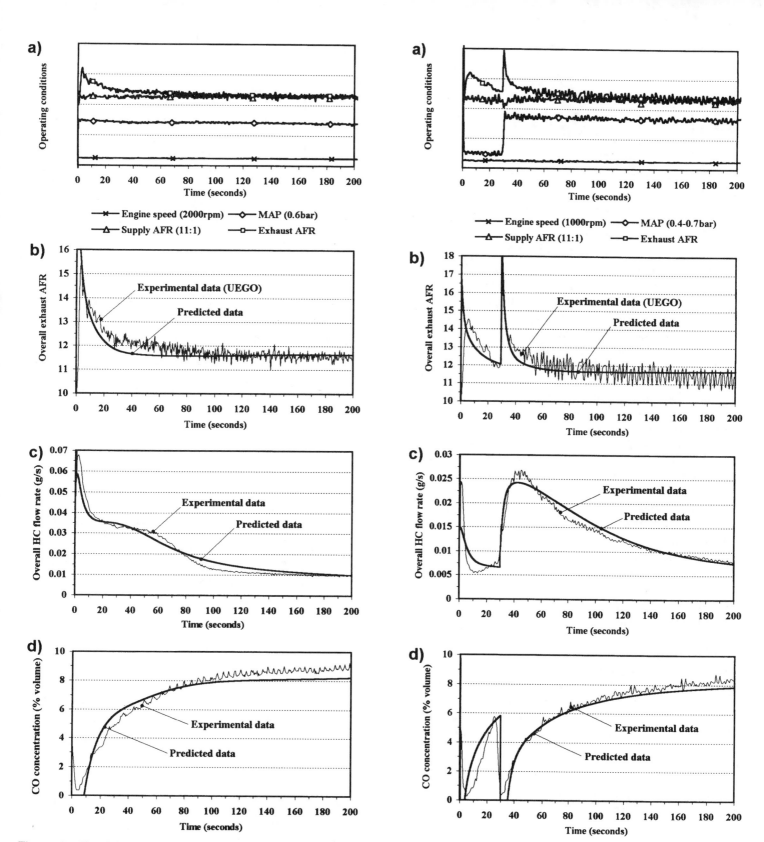

Figure 6. For the operating conditions given in (a) during engine warm-up from a motored start at –20°C with carbon deposits present, comparison between experimental data and model predictions for (b) exhaust AFR (c) exhaust unburned hydrocarbon flow rate (d) exhaust CO concentration. All values shown are on a per cylinder basis.

Figure 7. For the operating conditions given in (a) during engine warm-up from a motored start at –10°C without carbon deposits present, comparison between experimental data and model predictions for (b) exhaust AFR (c) exhaust unburned hydrocarbon flow rate (d) exhaust CO concentration. All values shown are on a per cylinder basis.

119

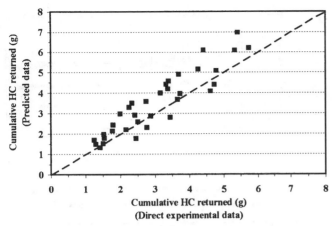

Figure 8. Comparison of predicted and experimental results for the cumulative unburned hydrocarbons exhausted during warm-up. The results cover quasi-steady state tests at various start temperatures, AFRs, manifold pressure and engine speeds. All values shown are on a per cylinder basis.

Figure 9 illustrates how the various hydrocarbon sources contribute to the overall engine-out mass flow rate of hydrocarbons throughout warm-up. In this case the start temperature was -20°C and warm-up was carried out at quasi-steady conditions of 2000rpm, 11:1 air/fuel ratio and a manifold pressure of 0.6 bar absolute. Figure 9c shows that in the early seconds of engine running the main contributors to the total hydrocarbon flow rate are the cylinder fuel film, the start-up and intra-cycle contribution, and the bulk mixture contribution. Fuel returning from the piston crevices as unburned hydrocarbons increases relatively slowly and subsequently decays to a near constant proportion of the total as fully-warm conditions are approached. The corresponding cumulative unburned hydrocarbons returned is shown as a function of time since start in Figure 9d. Figure 10 presents sets of data and predictions for the more complicated, varying pattern of operating conditions shown in Figure 10a. In this case the overall hydrocarbon flow rate is a maximum 100 seconds after start-up, when operating conditions are changed. The large spike only occurs when the operating conditions are changed early in the warm-up and, as can be seen from the breakdown given in Figure 10c, is associated largely with the cylinder film and intra-cycle contributions. A breakdown of the individual contributions to the total cumulative unburned hydrocarbons exhausted is shown in Figure 10d.

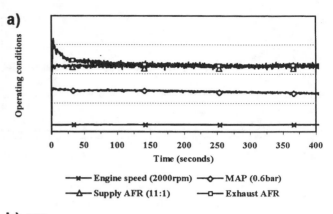

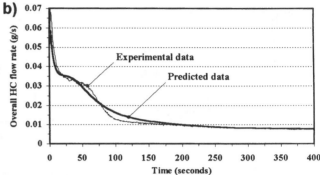

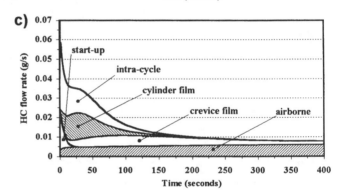

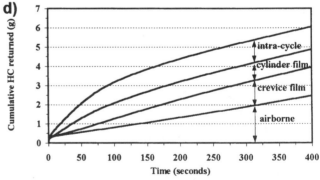

Figure 9. For the operating conditions in (a) during engine warm-up from –20°C with carbon deposits present, comparison between experimental and predicted data for the total unburned hydrocarbons is given in (b), individual unburned hydrocarbon sources is given in (c) and cumulative variations of the individual unburned hydrocarbons returning in the exhaust from various sources are given in (d). All values shown are on a per cylinder basis.

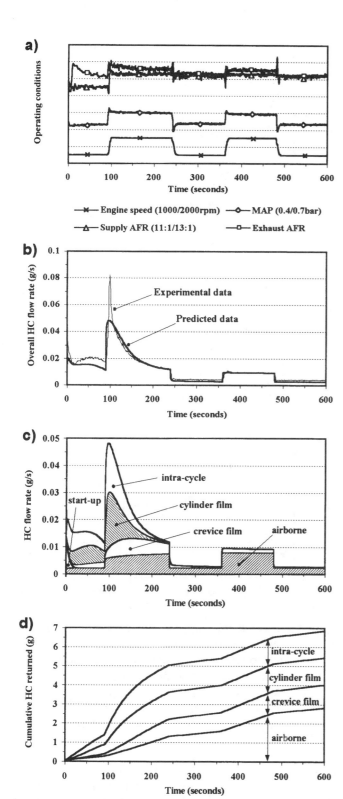

a)

b) Experimental data / Predicted data

c) intra-cycle / cylinder film / start-up / crevice film / airborne

d) intra-cycle / cylinder film / crevice film / airborne

Engine speed (1000/2000rpm) ◇ MAP (0.4/0.7bar)
△ Supply AFR (11:1/13:1) □ Exhaust AFR

Figure 10. For the operating conditions in (a) during engine warm-up from −20°C without carbon deposits present, comparison between experimental and predicted data for the total unburned hydrocarbons is given in (b), individual unburned hydrocarbon sources is given in (c) and cumulative variations of the individual unburned hydrocarbons returning in the exhaust from various sources are given in (d). All values shown are on a per cylinder basis.

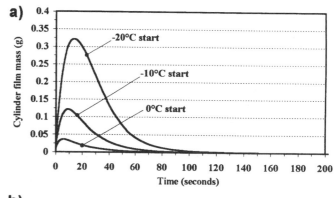

a) -20°C start / -10°C start / 0°C start

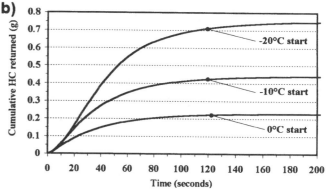

b) -20°C start / -10°C start / 0°C start

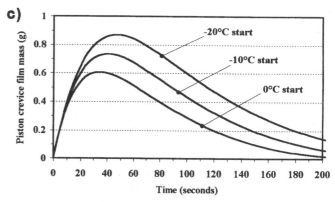

c) -20°C start / -10°C start / 0°C start

Figure 11. The effect of start temperature on (a) cylinder wall film mass (b) cumulative unburned hydrocarbons returned from the cylinder wall film (c) piston land and ring crevice film mass. All values shown are on a per cylinder basis for operating conditions of 2000rpm, 0.6 bar MAP, 13:1 AFR supply from a motored start with carbon deposits present.

121

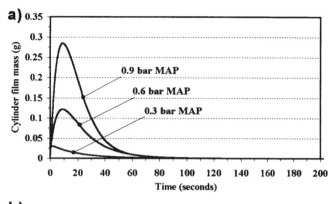

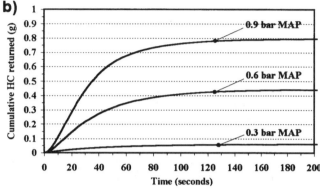

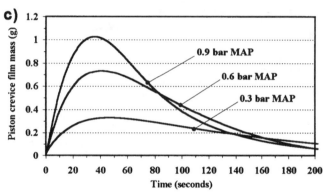

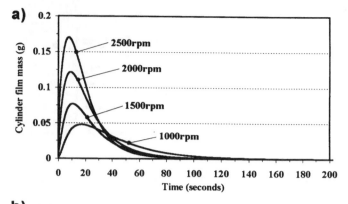

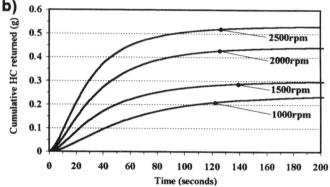

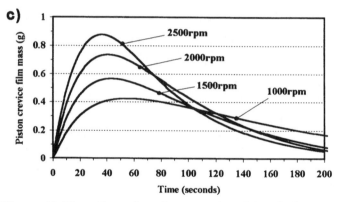

Figure 12. The effect of manifold absolute pressure on (a) cylinder wall film mass (b) cumulative unburned hydrocarbons returned from the cylinder wall film (c) piston land and ring crevice film mass. All values shown are on a per cylinder basis for operating conditions of 2000rpm, 13:1 AFR supply, -10°C motored start with carbon deposits present.

Figure 13. The effect of engine speed on (a) cylinder wall film mass (b) cumulative unburned hydrocarbons returned from the cylinder wall film (c) piston land and ring crevice film mass. All values shown are on a per cylinder basis for operating conditions of 0.6 bar MAP, 13:1 AFR supply, -10°C motored start with carbon deposits present.

The magnitude and variation of three of the most significant dependent parameters in the model are plotted in Figures 11, 12, 13 and 14. These show the influence of start temperature, manifold absolute pressure, engine speed and supply air/fuel ratio respectively, on the cylinder film mass, the cumulative unburned hydrocarbons exhausted, and the variation of fuel mass stored in the piston crevices.

The base conditions for all of the predictions are an engine speed of 2000rpm, an absolute manifold pressure of 0.6bar, an air/fuel ratio of 13:1 and a cold-start temperature of −10°C. The cylinder film mass peaks typically after 20 seconds and decays to near negligible values after 100 seconds. The peak film mass increases rapidly with falling cold-start temperature. Start temperature is the strongest single influence. The influence of increasing MAP, increasing speed and reducing air/fuel ratio is in each case to raise the peak cylinder film mass stored. These dependencies are reflected in the cumulative unburned hydrocarbons exhausted over the first two hundred seconds shown in (b) of each figure set. The fuel mass stored in the piston crevices is also sensitive to start temperature, MAP, engine speed, and supply air/fuel ratio. The crevice fuel mass peaks later than the cylinder film mass and, at the conditions illustrated, is generally larger by a factor of 3 or more. The crevice fuel mass also takes a much longer time to decay and does not dry out completely. The corresponding variations of the cumulative mass of fuel lost to the crankcase are shown in Figure 15, over an extended time. The magnitude and variation of this depends on the rate at which fuel is transferred to the piston crevices as well as the current fuel mass stored. The cumulative lost appears least sensitive to start temperature over the first two hundred seconds of engine running. However, over a longer time scale, the effect of start temperature is very significant and for start temperatures above 0°C the total lost diminishes rapidly.

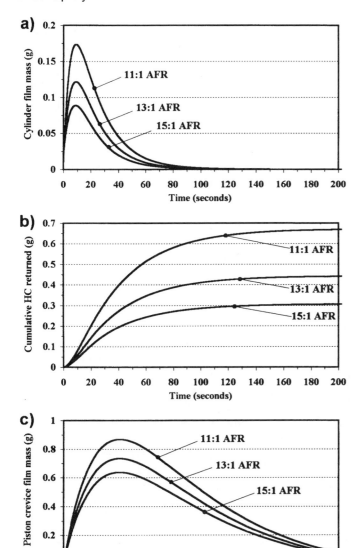

Figure 14. The effect of supply AFR on (a) cylinder wall film mass (b) cumulative unburned hydrocarbons returned from the cylinder wall film (c) piston land and ring crevice film mass. All values shown are on a per cylinder basis for operating conditions of 2000rpm, 0.6 bar MAP, -10°C motored start with carbon deposits present.

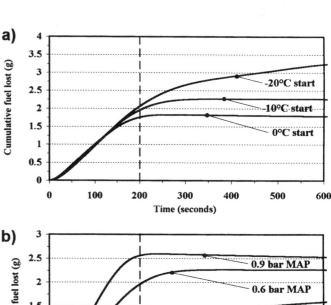

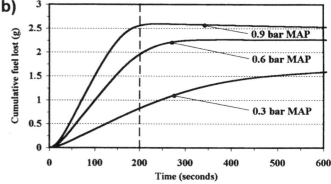

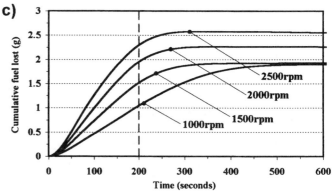

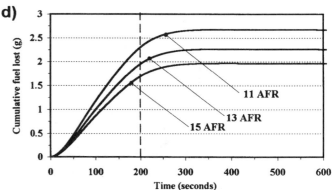

Figure 15. Variation of cumulative fuel lost to crankcase due to the effect of (a) start temperature (b) manifold absolute pressure (c) engine speed (d) air-to-fuel ratio during warm-up. All values are on a per cylinder basis. Operating conditions are as for Fig.10 to Fig.13.

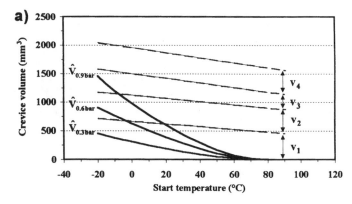

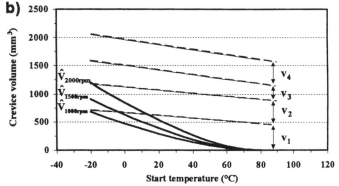

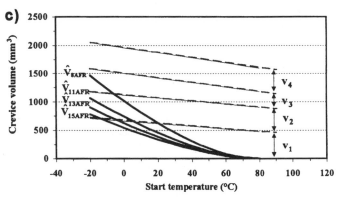

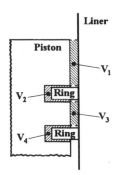

Figure 16. For any given start temperature, the variation in peak quantities of the predicted fuel stored within the available piston crevice volumes due to the effect of (a) manifold absolute pressure (b) engine speed (c) supply air/fuel ratio for base conditions of 1500rpm, 0.6bar MAP and 13:1 AFR supply from a motored start. All values are on a per cylinder basis.

DISCUSSION AND CONCLUSIONS

The model defined in this work, to the authors' knowledge, is the first capable of predicting the wide range of observed characteristics associated with exhaust air/fuel ratio and engine-out emissions variations during warm-up after a cold-start. Although the model contains coefficients which have been determined empirically, these are constants or simple functions of engine coolant temperature. The model and the experimental data on which the formulation of this has been based show how fuel behaviour dictates the rate of fuel loss to the crankcase and engine-out emissions during cold-running. The formulation has been very successful in characterising the relative influence of engine operating conditions, the influences of transient changes in these, and the relationship between various features in engine-out hydrocarbon emissions and the causes.

There remains considerable scope for further development of the model. The role of fuel absorption and desorption in the oil film on the liner above and below the ring pack, and the influence of blowby, are taken into account only implicitly. This is a weakness which requires further work. It appears probable that the oil film plays an important role as a mechanism of transporting fuel 'lost' into the sump oil reservoir and in the return of fuel from the piston crevice as unburned hydrocarbons. Gas flow into and out of the piston crevice will also influence this. The net blowby flow will encourage the wash-down of liquid past the piston and additionally contains unburned mixture from the bulk of the cylinder contents. Only part of the unburned mixture is returned to the combustion system via the crankcase breather route, particularly under cold operating conditions. In the model as presented, these processes are subsumed primarily into the description of piston crevice fuel behaviour and the empirical coefficients associated with this sub-model. The peak quantities of fuel predicted to be stored in the crevice occupy a substantial proportion of the volume available, at low temperatures, as illustrated in Figure 16. There is no direct experimental evidence to confirm the accuracy of the fuel mass prediction. The possible influence of the oil film and blowby processes outlined suggest this should be viewed as an open issue, notwithstanding the consistency of the model with experimental data where comparisons are currently possible.

The influence of basic design parameters such as bore, stroke and ring pack details have not been investigated. The evidence in the literature suggests that within the mainstream of designs, cold start characteristics are quite similar and that the model predictions are representative of these. The magnitudes of fuel stored in cylinder wall films and trends with operating conditions described in [9] are generally consistent with those described in [10]. Experimental measurements of HC emissions presented in [20] and [21] are compared to model predictions in Figures 17a and 17b. The agreement is reasonably good in each case.

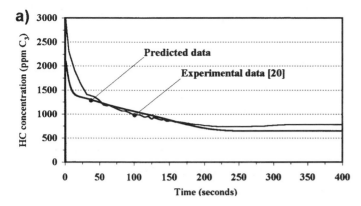

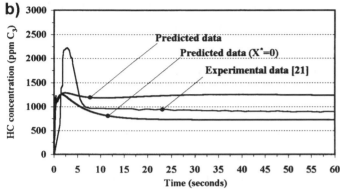

Figure 17. Comparison between experimental measurements of HC concentration and predictions for operating conditions of (a) 2000rpm, 0.44 bar MAP, 14.6:1 AFR supply and 10°C start {modern production engine} [20] (b) 1200rpm, 0.54 bar MAP, 14.6:1 AFR supply and 23°C start {8 cylinder, Vee, 4.6 litre engine} [21]. The latter shows the effect of pre-vaporised fuel.

All the experimental work reported here was carried out using a representative pump-grade fuel, and clearly the model reflects this. The effect of fuel composition has not been investigated. Fuel volatility has a significant influence on intake port fuel transport characteristics and emissions during warm-up [4]. The effects on in-cylinder fuel behaviour and how these might be taken into account in the model are open issues. It is interesting to note, however, how limiting cases can be explored as in Figure 17b. A pre-vaporised fuel supply was used in the experimental study. A prediction made with X^*=0 corresponds to the limiting case of no wall-wetting. The original and this modified prediction bracket the experimental data for times later than 5 seconds after engine start-up.

In summary, the main sources of engine-out HC emissions after cold-starting at low temperatures are inferred to be fuel deposited on cylinder walls and returned intracycle, a persisting film of fuel on the walls, fuel trapped in the piston land and ring grooves, and airborne fuel which remains in the bulk mixture throughout until its exhaust. The deposition of fuel onto cylinder surfaces is the most important cause of high HC emissions after start-up and when operating conditions are changed. The loss of fuel to the crankcase can account for 2g to 3g per cylinder at

sub-zero start temperatures. This increases with increasing engine speed, manifold pressure and mixture air/fuel ratio. The latter influence is primarily due to increasing fuel supply. The fuel loss is sensitive to start temperature, and diminishes rapidly as this is raised towards normal ambient and higher temperatures. At very low start temperatures, the initial rate of loss is limited by the volume and restriction offered by the piston crevice path.

ACKNOWLEDGEMENT

The authors wish to express their thanks to the Ford Motor Company for support and permission to publish this work which was carried out within the Department of Mechanical Engineering at the University of Nottingham.

REFERENCES

1. Greening, P, "The future of European Emissions Regulation", Proc. AVL Conf. Engine and Environment, Gratz, 1997, Vol II, pp.3-26.
2. Meyer, R. and Heywood, J.B., "Liquid Fuel Transport Mechanisms into the Cylinder of a Firing Port-Injected SI Engine during Start Up, SAE Paper No. 970865, 1997.
3. Shin, Y., Cheng, W.K. and Heywood, J.B., "Liquid Gasoline Behaviour in the Engine Cylinder of a SI Engine", SAE Paper No. 941872, 1994.
4. Shayler, P.J., Davies, M.T., Colechin, M.J.F. and Scarisbrick A., "Intake Port Fuel Transport and Emissions: The Influence of Injector Type and Fuel Composition", SAE Paper No. 961996, 1996.
5. Fry, M., Nightingale, C. and Richardson, S., "High-Speed Photography and Image Analysis Techniques Applied to Study Droplet Motion within the Porting and Cylinder of a 4-valve SI Engine", SAE Paper No. 952525, 1995.
6. Saito, K., Sekiguchi, K., Imatake, N., Takeda, K. and Yaegashi, T., "A New Method to Analyse Fuel Behaviour in a Spark Ignition Engine", SAE Paper No. 950044, 1995.
7. Witze, P.O. and Green, R.M., "LIF and Flame-Emission Imaging of Liquid Fuel Films and Pool Fires in an SI Engine during a Simulated Cold Start", SAE Paper No. 970866, 1997.
8. Frottier, V., Heywood, J.B. and Hochgreb, S., "Measurement of Gasoline Absorption into Engine Lubricating Oil", SAE Paper No. 961229, 1996.
9. Shayler, P.J., Davies, M.T. and Scarisbrick, A., "Audit of Fuel Utilisation during the Warm-Up of SI Engines", SAE Paper No. 971656, 1997.
10. Imatake, N., Saito, K., Morishima, S., Kudo, S. and Ohhata, A., "Quantitative Analysis of Fuel Behaviour in Port-Injection Gasoline Engines", SAE Paper No. 971639, 1997.
11. Takeda, K., Yaegashi, T., Sekiguchi, K., Saito, K. and Imatake, N., "Mixture Preparation and HC Emissions of a 4-Valve Engine with Port Fuel Injection during Cold Starting and Warm-Up", SAE Paper No. 950074, 1995.
12. Sampson, M.J. and Heywood, J.B., "Analysis of Fuel Behaviour in the Spark Ignition Start-Up Process", SAE Paper No. 950678, 1995.
13. Shayler, P.J., Colechin, M.J.F. and Scarisbrick, A., "Intra-Cycle Resolution of Heat Transfer to Fuel in the Intake Port of an SI Engine", SAE Paper No. 961995, 1996.
14. Hires, S.D. and Overington, M.T., "Transient Mixture Strength Excursions – An Investigation of their Causes and the Development of a Constant Mixture Strength Fuelling Strategy", SAE Paper No. 810495, 1981.
15. Aquino, C.F., "Transient A/F Control Characteristics of the 5 Litre Central Fuel Injection Engine", SAE Paper No. 810494, 1981.

125

16. Haidar, H.A. and Heywood, J.B., "Combustion Chamber Deposit Effects on Hydrocarbon Emissions from a Spark-Ignition Engine", SAE Paper No. 972887, 1997.
17. Kalghatgi, G.T., "Effects of Combustion Chamber Deposits, Compression Ratio and Combustion Chamber Design on Power and Emissions in Spark-Ignition Engines", SAE Paper No. 972886, 1997.
18. Heywood, J.B., "Internal Combustion Engine Fundamentals", McGraw-Hill International Edition, 1988.
19. Shayler, P.J., Teo, Y.C. and Scarisbrick, A., "Fuel Transport Characteristics of Spark Ignition Engines for Transient Fuel Compensation", SAE Paper No. 950067, 1995.
20. Edwards, J., Tidmarsh, D. and Willcock, M. "Evaluation of Total HCs and Individual Species for SI Engine Cold Start and Comparison with Predictions", IMechE Paper No. C517/040/96, 1996.
21. Kasier, E.W., Siegl, W.O., Lawson G.P., Connolly, F.T., Cramer, C.F., Dobbins, K.L., Roth, P.W. and Smokovit, M. "Effect of Fuel Preparation on Cold-Start Hydrocarbon Emissions from a Spark-Ignited Engine", SAE Paper No. 961957, 1996.

NOMENCLATURE

$AFR_{airborne}$: Air/fuel ratio of the airborne mixture

AFR_{comb} : Air/fuel ratio of the combusted mixture

AFR_{stoich} : Stoichiometric air/fuel ratio of gasoline

$CO_{\%vol}$: Carbon monoxide concentration (% volume)

$\tilde{m}_{CO}$: Molar mass of carbon monoxide (kg/kmol)

$\tilde{m}_{exh}$: Molar mass of exhaust products (kg/kmol)

m_{crev} : Mass of fuel within the piston crevices (g)

m_d : Mass of deposits in the combustion chamber (g)

m_{port} : Mass of fuel within the port film (g)

m_{wall} : Mass of fuel within the cylinder wall film (g)

$\dot{m}_{air}$: Air mass flow rate (g/s)

$\dot{m}_{bulk_{HC}}$: Mass flow rate of unburned HC from bulk mixture (g/s)

$\dot{m}_{CO}$: Mass flow rate of carbon monoxide (g/s)

$\dot{m}_{crev}$: Rate of change of fuel mass in piston crevices (g/s)

$\dot{m}_{crev_{HC}}$: Mass flow rate of unburned HC from piston crevices (g/s)

$\dot{m}_{cycle_{HC}}$: Mass flow rate of unburned HC from intra-cycle source (g/s)

$\dot{m}_{cyl}$: Rate of fuel returning to the cylinder from the piston crevices (g/s)

$\dot{m}_{dep}$: Rate of fuel deposition to the cylinder wall film (g/s)

$\dot{m}_{exh}$: Mass flow rate of fuel products (g/s)

$\dot{m}_{exh_{HC}}$: Mass flow rate of unburned HC in the exhaust (g/s)

$\dot{m}_{inj}$: Rate of fuel delivery by the injectors (g/s)

$\dot{m}_{ind}$: Rate of fuel induction to the cylinder (g/s)

$\dot{m}_{port}$: Rate of change of the port film mass (g/s)

$\dot{m}_{ret}$: Rate of fuel returning from the cylinder film (g/s)

$\dot{m}_{sump}$: Rate of fuel loss to the oil sump from the piston crevices (g/s)

$\dot{m}_{wall}$: Rate of change of the cylinder wall film mass (g/s)

$\dot{m}_{wall_{HC}}$: Mass flow rate of unburned HC from cylinder wall film (g/s)

N: Engine speed (rpm)

t_0: Time when first fire occurs (s)

T : Engine coolant temperature (°C)

X: Fraction of injected fuel deposited on port surface

X^*: Fraction of the induced fuel flow deposited on cylinder surface

τ : Time constant for fuel return from port film (s)

τ_w: ime constant for fuel return from cylinder wall film (s)

τ^*: Time constant for fuel return from piston crevices (s)

ϕ_{crev}: Fraction of the induced fuel swept into the piston crevices

950678

Analysis of Fuel Behavior in the Spark-Ignition Engine Start-Up Process

Michael J. Sampson and John B. Heywood
Massachusetts Institute of Technology

ABSTRACT

An analysis method for characterizing fuel behavior during spark-ignition engine starting has been developed and applied to several sets of start-up data. The data sets were acquired from modern production vehicles during room temperature engine start-up. Two different engines, two control schemes, and two engine temperatures (cold and hot) were investigated. A cycle-by-cycle mass balance for the fuel was used to compare the amount of fuel injected with the amount burned or exhausted as unburned hydrocarbons. The difference was measured as "fuel unaccounted for". The calculation for the amount of fuel burned used an energy release analysis of the cylinder pressure data. The results include an overview of starting behavior and a fuel accounting for each data set

Overall, starting occurred quickly with combustion quality, manifold pressure, and engine speed beginning to stabilize by the seventh cycle, on average. To facilitate rapid starting at cold engine conditions, approximately five times the amount of fuel required for a stoichiometric mixture is injected during the first one or two cycles. A large portion of this fuel, equivalent to nearly ten injections at stoichiometric idle conditions, remains "unaccounted for" after ten cycles of this analysis. About 10% of the fuel injected during the initial overfueling that is "unaccounted for" at first, shows up later in underfueled cycles as burned fuel or as hydrocarbon emissions. Similar trends occurred with both engines, and start-up strategies, although, during warm engine start-up conditions the overfueling is only 130% of stoichiometric and the mass "unaccounted for" after ten cycles represents only one injection at idle. The most successful start-up strategies that were analyzed injected close to the stoichiometric requirement for each cycle after the initial overfueling. The stoichiometric requirement for a particular cycle is directly proportional to the manifold pressure at a given temperature. It is recommended that methods for using manifold pressure in start-up strategies be investigated.

INTRODUCTION

Recent Clean Air Act amendments have placed very strict limits on the amount of hydrocarbons that can be emitted. The engine starting and warm-up contribute disproportionately highly to engine unburned hydrocarbon emissions. The work described in this paper was motivated by the belief that an improved understanding of fuel behavior during engine start-up, an extreme transient process under open loop control, will aid the development of systems and strategies that reduce emissions.

Current engine starting strategies inject approximately five times the required fuel during the first cycle or first two cycles at room temperature to get the engine started as quickly as possible. During this time the air flow is low and fuel evaporation is poor. Therefore, a large fraction of this fuel ends up as a liquid film, or "puddle", on the port walls, on the intake valve, and in the cylinder. According to Shayler et. al. [1], at 16 deg. C, only 25-30% of the injected fuel is in vapor form inside the cylinder during the first few cycles of cranking. This liquid fuel "puddle also exists during many normal, warmed-up engine operating conditions. If the extent and transient behavior of this puddle were better understood then the overfueling and higher hydrocarbon emissions that can occur during transients might be reduced. Additionally, when liquid fuel from this puddle enters the cylinder and impinges on the liner, it reduces the effectiveness of the oil film and increases hydrocarbon emissions. Over a time this oil dilution contributes to the breakdown of the entire oil supply.

The objective of the research reported here was to characterize engine start-up behavior, focusing especially on the fuel transport. Several sets of start-up data have been analyzed, incorporating runs with two different engines and both room temperature soaked and warm-engine starts. The results obtained are: 1) a characterization of the combustion behavior during starting, and 2) a fuel "accounting" for each run based on fuel injected, engine-out hydrocarbon measurements and estimates of fuel burned using an energy

---

[1]Michael Sampson is currently with the Advanced Powertrain Controls group at General Motors Corp..

release analysis of the cylinder pressure traces. This paper summarizes the analysis method and explains these results. An overview of engine starting, including the experimental setup, engine specifications, measurements, and operating conditions, will be given first. The fuel accounting method is then explained and results for several starts are presented and discussed.

CHARACTERIZATION OF ENGINE STARTUP

Data from two different modern engines, a 3.8 liter V6, and a 4.6 liter V8, were used in this study. The tests were conducted with the engines in cars on chassis dynamometers. A description of the major engine parameters is given in Table 1. These engines and vehicles were production units except for alterations to the starting strategy as noted below.

Table 1 Engine Parameters

	4.6L V8	3.8L V6
Bore [mm]	90.2	96.8
Stroke [mm]	90.0	86.0
Con. Rod Length [mm]	150.7	150.2
Compression Ratio	9.02	9.00
Valves/Cylinder	2	2
Valve Events:		
IVO [BTC]	12	18
IVC [ABC]	64	56
EVO [BBC]	63	70
EVC [ATC]	21	20
Firing Order	1-3-7-2-6-5-4-8	1-4-2-5-3-6
Approx. vehicle mileage	2,000	8,000

The tests represent the first portion of the Environmental Protection Agency's Federal Test Procedure driving cycle. The test conditions reproduced an "in-service typical" start as closely as possible. Each start-up test was conducted at room temperature and atmospheric pressure (approximately 1 bar and 23°C) with the vehicles in park on a chassis dynamometer. The engine was either at room temperature or close to warmed-up operating temperature, depending on the test. Both cars had low mileage and were fueled with indolene. Starting was initiated by a driver in the vehicle turning the ignition key, and the data acquisition was triggered automatically by the turning of the crankshaft.

The tests include starts with two different engines, two different control schemes, and two different engine temperature conditions. The test matrix is outlined in Table 2. The "Cold" condition refers to the engine at room temperature. This was achieved by allowing the car to sit for an extended period, or by force cooling with external heat exchangers and fans. The "Hot" start was done after the engine had reached normal operating temperature. The car was shut down, and started again before any significant cooling had occurred.

Table 2 Experimental Test Matrix

Test	Engine	Strategy	Temperature
1	V6	Sequential	Cold
2	V8	Simultaneous	Cold
3	V8	Sequential	Cold
4	V8	Simultaneous	Hot

The two starting strategies differed in the way that fuel injection was synchronized with the crankshaft during start-up. The "simultaneous" scheme is representative of the production strategy, where all the injectors are fired simultaneously until the control unit can determine the absolute position of the crankshaft. At that point each injection is gradually shifted, over the next few cycles, to the correct relative location for each cylinder (i.e. to sequential injection during normal running). The "sequential" injection strategy provides additional position information to the controller. This allows each injector to start delivering fuel at the same relative point for its cylinder from the beginning of start. All sequential injections were done onto closed intake valves. There are other, more subtle, changes in the start-up strategy such as the exact point where the controller changes modes. These changes do not effect the outcome of this study, and are not included in this analysis.

A closer look at the starting process allows a better understanding of the difficulties and tradeoffs involved. Figure 1 shows several of the outputs for a typical start. The in-cylinder pressure for cylinder number two is shown as a reference. For this cylinder the first cycle is a non-firing cycle, as indicated by a low maximum pressure relative to the intake manifold pressure. The second cycle is clearly a firing cycle with a high intake pressure. The subsequent cycles have a peak pressure that follows the same trend as the manifold pressure. The intake manifold is initially at atmospheric pressure and starts to decrease after the first firing cycle. The intake pressure falls rapidly as the engine utilizes the air in the manifold while the throttle remains nearly closed. By the sixth cycle for cylinder number two, the manifold pressure is approximately one half of atmospheric pressure and the rate of pressure drop has decreased. By 12 cycles the intake pressure is close to the minimum value. After approximately 20 cycles the minimum of 1/3 atmosphere has been reached and the pressure starts to increase slightly as engine speed continues to decrease.

The crankshaft speed increases rapidly during the first several cycles. The transition from the cranking speed of 150 rev/min occurs before every cylinder has completed its first cycle since one of the cylinders late in the sequence fires on its first pass. Within three cycles for cylinder number two, the speed has reached 1000 rev/min. After this point the engine's acceleration is less pronounced and the maximum speed of 1500 rev/min occurs after eight cycles for cylinder two. Engine speed then decreases almost linearly until it levels out at approximately 900 rev/min.

Figure 2 shows the amount of fuel injected per cycle and the amount of fuel required for a stoichiometric air/fuel ratio during a typical room temperature start. The injection strategy consists of three distinct phases: the initial, very large injections where there is substantial excess fuel, the following injections which are below the stoichiometric level,

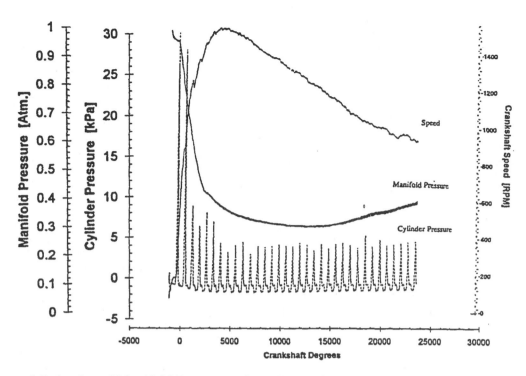

Figure 1 Behavior of Manifold Pressure, Speed, and Cylinder Pressure versus crankshaft degrees during engine start-up. Peak cylinder pressure approximately follows the trends of manifold pressure. Engine speed rises rapidly during the first five cycles.

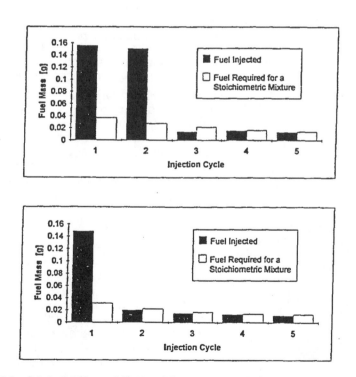

Figure 2 Mass of fuel injected vs. Mass of fuel required for a stoichiometric mixture with the amount of air inducted. Examples of the first five injection cycles of the 4.6L V8 sequential start-up showing that while most cylinders receive two large injection pulses, some have only one.

and the subsequent injections which are close to the amount of fuel required for that cycle. Injected amounts for two different cylinders are displayed, showing that occasionally a cylinder will only get one large injection pulse. For the data analyzed in this study, the overwhelming majority of cylinders (more than 80%) received two large injections.

Figure 3 presents the relative work output for each cycle of each cylinder for the V-8 engine, normalized by the manifold pressure during the intake process for each cycle. Work is expressed as indicated mean effective pressure (IMEP), which is the indicated work per cycle divided by cylinder volume. Dividing IMEP by manifold pressure gives insight into the stability of the combustion process since the manifold pressure is proportional to the mass of charge in the cylinder. The first engine cycle can be seen to be a cranking cycle, since all the cylinders show a negative work output. During the next several cycles not all cylinders produce their full output; some are very slow and/or late burns. By the seventh cycle the balance between the cylinders has improved and the variation remains almost constant. Note that the work output scales with the manifold pressure since the IMEP/Pman ratios are all the same magnitude.

Figure 4 summarizes the above discussion with an overview of several engine parameters based on data from numerous starts. The process can be divided into several distinct phases: Cranking, Unstable Combustion, Combustion Stabilization, and Steady Idle. The cranking phase commonly lasts for less than two complete engine cycles before several of the cylinders fire. Unstable combustion occurs in the following three to five cycles, as the manifold pressure drops rapidly, and the fuel and air flows change significantly. Combustion stabilization starts as the engine reaches its peak speed, the manifold pressure begins to level out, and each cylinder receives a similar charge of air and fuel. Approximately 15 to 20 cycles after the first firing, the engine is idling with only minor changes in manifold pressure, speed, and cylinder work output.

MODELS USED FOR FUEL ACCOUNTING

The primary objective of this work was quantifying the fuel behavior during start-up. This fuel accounting was performed for each cycle of each cylinder, and involved calculating the amount of fuel that is injected, the amount that is burned, and the amount that leaves the exhaust manifold as unburned hydrocarbons. This is done for the first ten cycles of each cylinder. This section describes the models used, their assumptions, and the estimated accuracy of these calculations.

FUEL INJECTED - The quantity of fuel injected into the engine can be accurately determined by recording the signal to the injector solenoid. The duration or pulse width of this signal is essentially the time that the injector is open. The injectors were calibrated by measuring fuel flow for a given pulse width and battery voltage. Using the calibration and the solenoid signal the fuel mass injected can be calculated. Changing battery voltages, variations between injectors, and the resolution of the data acquisition system results in an estimated uncertainty of 5% for the amount of fuel injected.

FUEL BURNED - A single-zone burn-rate model based on in-cylinder pressure was used to determine the amount of fuel oxidized during the combustion process. This model [2] uses an energy release approach based on the First Law of Thermodynamics for the gasses in the cylinder. The model inputs include: cylinder pressure data, engine geometry, the amount of air inducted each cycle, the air/fuel ratio, the residual gas fraction, and several thermodynamic parameters.

The in-cylinder charge estimate was performed in two steps: first, the total mass of charge in the cylinder at intake valve closing was estimated using the ideal gas law; second, the makeup of the charge was estimated based on a residual gas model developed by Fox et.al. [3] and by assuming a stoichiometric air/fuel ratio. The central part of the model then calculates the mass fraction burned and burning rate profiles. The outputs consist of the burn profiles, as well as statistics for these parameters and for the pressure data. To determine the mass of fuel burned during each cycle, the maximum mass fraction burned value was multiplied by the estimate for the mass of fuel inducted that cycle.

The assumption of stoichiometric mixture composition was justified as follows. For a lean mixture, the product of calculated maximum mass fraction burned (for the lean relative air/fuel ratio) and the amount of fuel inducted (for the lean air/fuel ratio) is essentially the same as this product using the stoichiometric assumption. Therefore the amount of fuel burned is obtained within 5% accuracy. For fuel-rich mixtures, this precision is lost since combustion energy release is then primarily air limited. However, the requirement of low carbon monoxide emissions in modern engines precludes running under Federal Test Procedure conditions with mixtures richer than "slightly rich", so the error introduced should still be of order 10%. Further details of the energy release model and the related uncertainties can be found in the appendix.

HYDROCARBON EMISSION - The mass of engine-out hydrocarbons per cycle (before the catalyst) was estimated from the mass of charge in the cylinder and the feedgas hydrocarbon measurement. The measurement was made by a fast response flame-ionization hydrocarbon detector. To convert the measured ppm hydrocarbon concentration to a mass of fuel exhausted for that cycle, the molar fraction (ppm) value was converted to a mass fraction (assuming a stoichiometric composition) and multiplied by the mass of mixture for that cycle. The measured value at top center of the exhaust stroke was the HC concentration used. Due to the time delay in the sampling system this corresponded approximately to the middle of the exhaust process. Changes in HC concentration over several cycles were larger than the variation within one cycle and the uncertainty of the HC mass calculation is estimated to be approximately 15%. Overall, the injected fuel accounted for by hydrocarbon emission is a small fraction compared to the other components.

FUEL UNACCOUNTED FOR - For each cycle, the mass of fuel burned and the mass exhausted as hydrocarbons are subtracted from the mass injected. The balance is labeled "fuel unaccounted for", and represents fuel that remains in

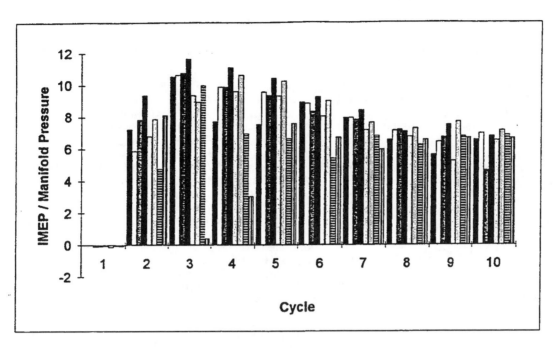

Figure 3 Normalized work output for each individual cylinder in sequence during a V8 engine start-up. The first cycle is a cranking cycle.

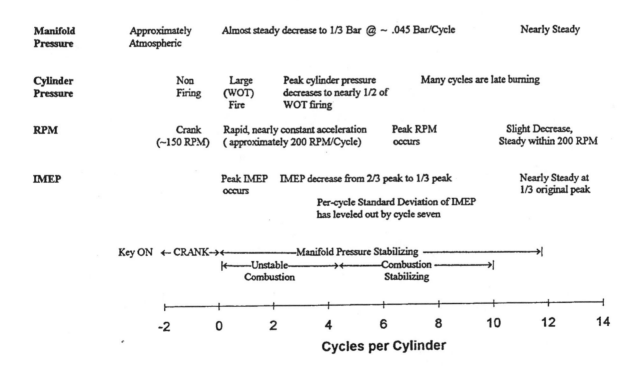

Manifold Pressure	Approximately Atmospheric	Almost steady decrease to 1/3 Bar @ ~ .045 Bar/Cycle		Nearly Steady
Cylinder Pressure	Non Firing	Large (WOT) Fire	Peak cylinder pressure decreases to nearly 1/2 of WOT firing	Many cycles are late burning
RPM	Crank (~150 RPM)	Rapid, nearly constant acceleration (approximately 200 RPM/Cycle)	Peak RPM occurs	Slight Decrease, Steady within 200 RPM
IMEP		Peak IMEP occurs	IMEP decrease from 2/3 peak to 1/3 peak Per-cycle Standard Deviation of IMEP has leveled out by cycle seven	Nearly Steady at 1/3 original peak

Figure 4 Overview of engine behavior during start-up compiled from observations of many raw data sets. Average behavior is displayed.

the intake manifold, is blown by the piston rings, is absorbed into the oil on the cylinder walls, is in liquid form in the cylinder or is otherwise retained or lost from the cylinder. It also includes the errors in the fuel injected and fuel accounting computations.

The "fuel unaccounted for" can have a negative value due to fuel retention in the intake and cylinder. For example, liquid fuel retained in the intake port that was injected during previous cycles may evaporate and make its way into the cylinder to supplement the current injection. In this case, the mass of fuel burned could be larger than the mass injected, resulting in a negative "unaccounted for" value for that cycle. By summing the fuel unaccounted for over each cycle, the cumulative mass of fuel unaccounted for is obtained. This representation gives insight to the fuel transport behavior over several cycles.

RESULTS AND DISCUSSION

The fuel accounting analysis was applied to several sets of start-up data to provide a more quantitative overview of the engine start-up process and to observe differences in the starting strategies. The engine start-up cases reviewed in this section include a simultaneous injection strategy, a sequential injection strategy, and a warm start using the sequential strategy. Each data set analyzed was chosen to be representative of data taken at similar conditions.

SIMULTANEOUS INJECTION - The majority of automobiles currently in production use the simultaneous injection strategy during starting. This provides quick starting, but there is no flexibility for different fuel needs between cylinders as the manifold pressure changes rapidly, since all injectors are fired with one command.

Figure 5 shows the amount of fuel injected for each cylinder during the first ten cycles of a room temperature start of the 4.6 liter V8 engine using the simultaneous strategy. The first eight injections occur in the same point in time, but because of the timing of this injection in the firing order, the injection happens during either cycle one or cycle two depending on the cylinder. One cylinder shows a slightly lower fuel mass for this injection, probably due to inaccuracies in the data acquisition and measurement process. The second eight injections are also simultaneous, and at about 0.13 grams of fuel, have a slightly lower mass. The third injection for each cylinder is distinctly different than the previous two. The fuel mass drops to less than 0.02 grams, almost one seventh the previous value, and the injections become sequential, although not every cylinder yet has injections occurring at the same relative point in the cycle. The transition to full sequential injection has occurred by the end of cycle five. From cycle four to ten the fuel mass injected decreases by less than 0.01 grams. Figure 5 clearly shows the fuel injection strategy: large initial injections followed by an immediate transition to much smaller fuel pulses. Figure 6 shows how these pulses compare to the fuel required for a relative air/fuel ratio equal to one (a stoichiometric mixture) for cylinder three. Significant overfueling in the first two cycles is followed by moderate underfueling in the following seven cycles.

The normalized work outputs (IMEP/Pman) for each cylinder of this 4.8L V8 engine start-up are presented in Fig. 7. Cranking cycles are seen as negative work output. Combustion is intermittent for the first cycle and a half, followed by the highest relative levels of work output. This high output is likely due to the fact that there is no burned residual in the first firing cycle and the excess fuel from the first one or two injections has full air utilization. The variation of work output between cylinders generally decreases with the exception of the group of high output cycles and one cylinder in cycles eight and nine. Figure 8 displays the coefficient of variation (COV) for IMEP computed on an individual cycle basis. COVimep is the standard deviation of IMEP for all cylinders for that engine cycle, divided by the average IMEP value for that cycle, expressed as a percentage. The first cycle is omitted since the majority are cranking cycles and COVimep then has little meaning. The second cycle shows a high value because not every cylinder has fired yet. The COVimep is close to 30% for cycles three to five, and decreases thereafter. A COVimep above 10% is considered problematic for a warmed-up engine, but for the early part of start-up a value below 20% by cycle five is typical.

Figure 9 displays the results of the fuel accounting analysis for cylinder number seven. The remaining cylinders show closely similar trends. The majority of the fuel injected during the initial large injections is unaccounted for. A small fraction of this "lost" fuel is accounted for in the following cycles, as evidenced by a negative "fuel unaccounted for" mass. The mass of fuel burned per cycle increases slightly from cycle two to three, and then gradually decreases to about half of its peak value of 0.025 grams by cycle ten. Except for the first two injections, the burned mass accounts for the largest fraction of fuel injected, and by cycle eight the other portions are negligible. The mass of hydrocarbon emissions increases to its largest values by the second and third injection (cycles three and four), and by cycle seven the amount is not visible in the figure.

The cumulative mass of fuel "unaccounted for" is shown for each cylinder in Fig. 10. This is plotted against injection number (where number one is the cycle before the first injection) instead of by engine cycle to avoid confusion due to the phasing of the first injection. Cylinders one and four are omitted because of problems encountered by the burn-rate program in analyzing one or more of the cycles for these cylinders. For all cylinders, the data show the large unaccounted for mass resulting from the first two large injections. The unaccounted mass for each cylinder peaks between 0.2 grams and 0.24 grams, or approximately ten to twelve times the mass of one injection for cycle five (a "normal" sized injection). After the peak is reached, the unaccounted for mass begins to decrease, first by 0.01 grams, then leveling out. The mass "regained" in later cycles is 0.02 grams by the end of ten cycles, equivalent to 10% of the mass originally unaccounted for, or about one normal injection.

This regained fuel is likely liquid fuel that has evaporated after being stored in the intake port or in the cylinder for one or more cycles [4]. There are several possibilities for the fuel that remains unaccounted for after ten cycles. Since the slope of the unaccounted for mass is

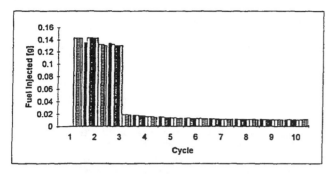

Figure 5 Mass of fuel injected for each cylinder in sequence for the 4.6L V-8 simultaneous start-up strategy showing two full cycles of large fuel pulses.

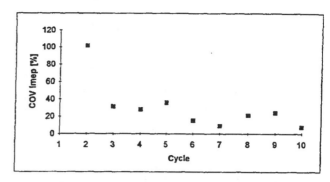

Figure 8 Coefficient of Variation for Indicated Mean Effective Pressure (COVimep) computed based on all cylinders for each cycle for the 4.6L V8 simultaneous start. The first cycle is omitted because of the majority of cranking cycles. Cycle two is high due to the presence of misfiring cycles.

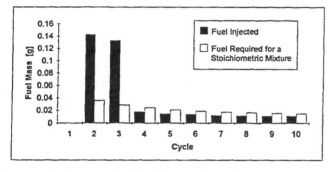

Figure 6 Mass of fuel injected and mass of fuel required for a stoichiometric mixture for cylinder three of the 4.6L V8 simultaneous start-up strategy showing the initial overfueling and subsequent underfueling.

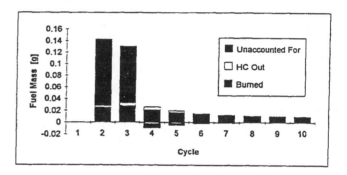

Figure 9 Fuel Accounting result for one typical cylinder (#7) of the 4.6L V8 during the simultaneous start-up test. By cycle seven the hydrocarbon out and "unaccounted for" portions are too small to be visible. Cycle one received no injection for this cylinder.

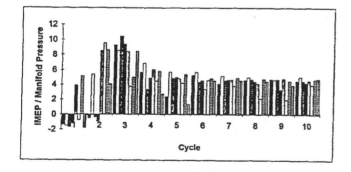

Figure 7 Normalized work output for the 4.6L V-8 simultaneous start-up. Negative values represent cranking or misfiring cycles.

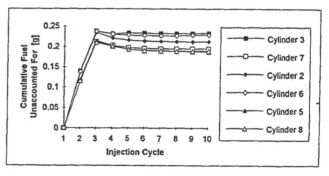

Figure 10 Cumulative mass of fuel "unaccounted for" for the 4.6L V8 simultaneous start-up test vs. injector cycle. Cylinders one and four have been omitted due to problems that the burn rate program had with these cylinders.

close to zero, and the engine conditions are almost steady after ten cycles, it appears unlikely that the remaining portion will be accounted for in subsequent idling cycles. Part of this mass could represent a steady-state mass of fuel that remains in the port, with a portion vaporizing and entering the cylinder while the current injection deposits more liquid. Other likely fuel sinks are blowby of fuel vapor, liquid fuel flow into the crankcase, and fuel absorption into the oil. During starting the piston and ring clearances are large, the oil is viscous and the engine speed is low. These conditions promote fuel loss in a cold engine.

Overall, this engine start-up could be characterized as "fair", with unsteady combustion (COVimep > 20%) occurring past cycle eight, and with a relatively high hydrocarbon output (0.156 grams) over the first 10 cycles. This is likely due to the underfueling condition that continues after cycle five where there is significant fuel "re-accounted for" from fuel puddling in the intake port.

SEQUENTIAL INJECTION - The sequential strategy is a more advanced method of injection that requires information on crankshaft position at the beginning of the start. This scheme is used on some vehicles in an attempt to improve the fuel delivery and the resulting in-cylinder air/fuel ratio in order to reduce the hydrocarbon emissions while maintaining the same startability.

Figure 11 indicates the fuel injected for each cylinder during a sequential start-up of the 4.6 L V8. Figure 12 compares the fuel injected to the fuel required for a stoichiometric in-cylinder mixture for cylinder number one. The overall trend is similar to the simultaneous start: very large fuel delivery for the first two cycles followed by close-to-stoichiometric injection. However, several important differences should be noted. First, from Fig. 11 it can be seen that the transition to the smaller fuel pulses has a short stage where two cylinders inject an intermediate value. Second, except for the first cycle, the amount of underfueling is significantly less than in the simultaneous case, even though the injected fuel mass is almost identical. In addition, the initial injections are slightly lower.

Figures 13 and 14 display the normalized work output (IMEP/P_{man}) for each cylinder and the cycle to cycle COVimep, respectively. The work output quickly transitions from cranking to a high output level with little variation among cylinders. Generally, the work output remains high for the balance of the cycles although the variation is changing. COVimep doubles during cycle three due to one misfire, and drops below 20% by cycle five. By cycle seven COVimep is close to 10%. These are indications of a good start-up.

Figure 15 gives the cumulative "fuel unaccounted for" during the sample sequential start-up. Again, the result of the initial overfueling can clearly be seen. After the initial overfueling, approximately 10% of the peak unaccounted for fuel is re-accounted for over the next eight cycles. The significant result shown here is that the cylinders with transition injections have a lower mass of "fuel unaccounted for" than the cylinders with two large injections. The lower fuel requirement due to the falling manifold pressure has been exploited and the last three cylinders in the firing order (5,4,8) received less fuel, resulting in a better start with a rapid

speed increase, low COVimep, and low emissions (0.074 grams HC total).

WARM START - The third set of data analyzed came from the 4.6 liter V8 engine started using the simultaneous strategy while close to warmed up operating temperature. The warm start condition increases the evaporation of the fuel and decreases the density of air in the intake manifold. The injection enrichment level is therefore decreased and the engine starts more easily.

Figure 16 displays a comparison of fuel injected to fuel required for one typical cylinder (#3). The initial overfueling injections are between 70 to 75 percent lower than the corresponding injections during cold start, and the injections are much closer to that required for a stoichiometric mixture (only 1.30 times instead of 6.5 times). The injections following overfueling are about half that required for a stoichiometric in-cylinder mixture during cycles 4 and 5. The fuel needed for a stoichiometric incylinder mixture at this point is approximately 25% less than for a cold start. The manifold pressure drops and the fuel requirement drops, while the injections actually become larger. However, the injected fuel does not rise above 90% of the stoichiometric requirement until cycle 10.

Figures 17 and 18 present the work output for each cylinder and the COVimep for each cycle for the warm start. The work output is high for the first two cycles. The high output from the first cycle with injected fuel indicates that there is enough fuel evaporation for high air utilization, likely due to the fact that the air has been sitting in the warm intake manifold. However, as this warm air supply is depleted, the cylinder outputs become quite low and the COVimep actually increases to unacceptably high levels (40-60%) before falling again for cycle nine. By this time the fuel injections are approaching the requirements for a stoichiometric mixture in the cylinder.

The fuel accounting results are displayed using a typical cylinder's output (cylinder # 1) in Fig. 19. In contrast to the cold starts, the vast majority of the mass injected during the initial overfueling is burned, and only slightly more than 10% is unaccounted for. A large fraction of this unaccounted for mass is re-accounted for during the first underfueled injection. As previously indicated by the work output, the results are quite variable after cycle four. Overall, the mass of fuel released as hydrocarbons is more consistent from cycle to cycle than during the cold starts, although the total mass released is close to the amount released in the sequential cold start example.

Figure 20 shows the cumulative mass of fuel unaccounted for during the warm start-up for each cylinder. The peak mass unaccounted for is one order of magnitude less than on the cold starts (0.025 grams vs. 0.25 grams). However, the trend is similar for the first three cycles: two increases followed by a decrease. Two cylinders do not follow this trend, and it is clear that there are some errors in the traces since one cylinder shows a negative "cumulative unaccounted for" value. This is likely due to the uncertainty in these mass levels. The final fuel level shows a wide scatter. The variation in the final mass of "fuel unaccounted for" is nearly the same as in the cold starts, but the percentage variation is much higher. The overall amount of

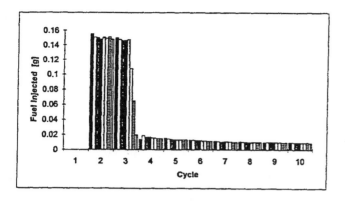

Figure 11 Mass of fuel injected for each cylinder in sequence for the 4.6L V8 sequential start-up strategy showing two cycles with intermediate injection masses. Cycle one received no injections.

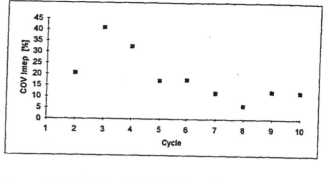

Figure 14 Coefficient of Variation for Indicated Mean Effective Pressure (COVimep) computed based on all cylinders for each cycle for the 4.6L V8 sequential start. The first cycle is omitted because of the majority of cranking cycles. Cycle two is low (good combustion) for such an early cycle.

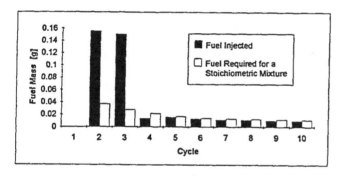

Figure 12 Mass of fuel injected and mass of fuel required for a stoichiometric mixture for cylinder one of the 4.6L V8 sequential start-up showing the initial overfueling and subsequent underfueling. The underfueling is substantially less than the simultaneous start-up case.

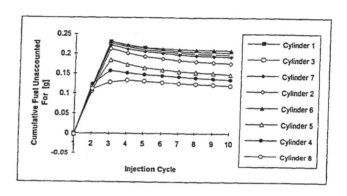

Figure 15 Cumulative mass of fuel "unaccounted for" for the 4.6L V8 sequential start-up test vs. injection cycle. Note the wide variation in final unaccounted for mass due to the difference in the second injection pulse.

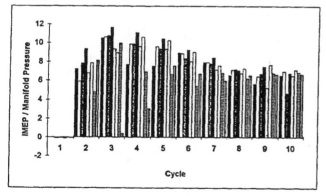

Figure 13 Normalized work output for the 4.6L V8 sequential start-up. Negative values represent cranking or misfiring cycles. Note the high output levels and small amount of variation between the cylinders.

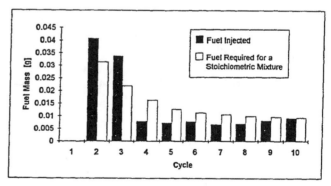

Figure 16 Mass of fuel injected and mass of fuel required for a stoichiometric mixture for cylinder three of the 4.6L V8 warm start-up showing the reduced initial overfueling and subsequent underfueling. The underfueling is substantially greater than the cold start-up cases.

135

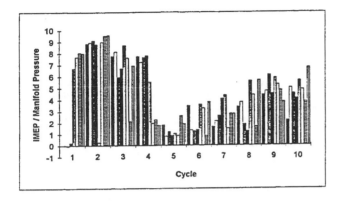

Figure 17 Normalized work output for the 4.6L V8 warm start-up. Negative values represent cranking or misfiring cycles. Note the high output levels during the first three and a half cycles followed by much lower levels.

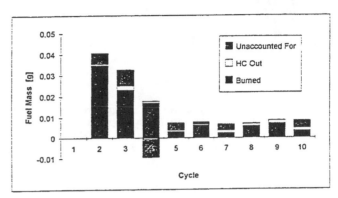

Figure 19 Fuel Accounting result for one typical cylinder (#1) of the 4.6L V8 during the warm start-up test. Cycle one received no injection for this cylinder. The unaccounted for mass portion is much smaller than in the cold start cases and the variation of the proportions is greater after cycle five.

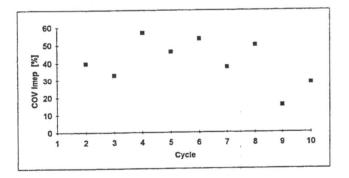

Figure 18 Coefficient of Variation for Indicated Mean Effective Pressure (COVimep) computed based on all cylinders for each cycle for the 4.6L V8 warm start. The first cycle is omitted because of the majority of cranking cycles. Cycle two would be much lower except for one misfiring cylinder. The variation remains quite high until cycle nine.

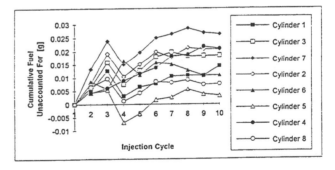

Figure 20 Cumulative mass of fuel "unaccounted for" for the 4.6L V8 warm start-up test vs. injection cycle. Note the wide variation in final unaccounted for mass. The trends are similar to the cold cases for the first four cycles. Cylinder five shows a negative value which is likely an error due to the burn analysis in cycle three.

fuel unaccounted for is lower by an order of magnitude, and the final mass unaccounted for after ten cycles is between 0.005 and 0.025 grams, which represents approximately one half to two-and-one-half "normal" injections; significantly fewer than in the cold start cases. The variation in the traces is likely due to a combination of the uncertainty of the analysis and the variability of the process. In spite of its good beginning, overall this start is poor, emitting slightly more HC than the sequential cold start (0.096 grams) and showing a high COVimep (approximately 50%) after cycle two. We do not know how typical this warm start is because only one data set was available for analysis.

DISCUSSION

The intention of this work was to develop a fuel accounting method for engine start up studies and apply this analysis to several sets of engine starting data to describe the fuel behavior during start-up and investigate differences in start-up strategies. Another objective was to identify areas for improvement. It has been seen that a number of phenomena can be identified and quantified through a combination of fuel measurement, charge mixture calculation, burn rate analysis, and observation of work output measurements.

Current cold start-up schemes inject nearly five times the normal fuel requirement per cycle during the first one or two cycles. A small portion of this fuel is recovered over the next ten or so cycles, but the majority remains "unaccounted for" during this period. A period of two or three cycles of underfueling immediately follows the first injections as the engine quickly reaches stable idle conditions. During warm starts the overfueling is reduced since favorable conditions exist in the manifold and cylinder for good fuel evaporation.

The data samples that were analyzed were chosen because the engine parameters contained no obvious deviations from the norm of the supplied data set. Many raw data sets were examined to be certain that the examples chosen were representative. It is unclear how typical the starts are at a detailed level, since only a limited number of sets were investigated in depth. However, there are consistent trends and the analysis demonstrates which start-up sets perform well. The accuracy of the analysis is sufficient to define these common trends during the cold starts, but during the warm start the unaccounted fuel mass is so small that the uncertainty limit of the analysis is approached.

CONCLUSIONS

The conclusions from this work can be summarized as follows:

1. During cold starting, approximately five times the amount of fuel required for a stoichiometric mixture is injected during the first one or two cycles. A large portion of this fuel (approximately 0.2 grams per cylinder or nearly ten times the mass of fuel injected during a normal idle cycle) is "unaccounted for" by this analysis. Only 10% of this mass (or approximately one "normal" injection) is then "re-accounted for" over the following eight cycles.

2. During the warm start example analyzed, the mass of fuel "unaccounted for" is an order of magnitude lower than in the cold starts (0.02 grams), representing the mass of approximately one injection at normal idle conditions. The uncertainty limit of the analysis is approached with the warm start data due to the smaller masses involved.

3. Likely sources for this "unaccounted for" fuel include storage in a liquid fuel puddle in the intake port (and cylinder) and loss to blowby, liquid fuel flow into the crankcase and absorption into oil layers in the cylinder.

4. The initial large overfueling is followed by much smaller underfueling. The transition to this stage has an impact on hydrocarbon emissions. The level and duration of underfueling is important to the combustion stability in the subsequent cycles. Some starts are poor after this transition point because of lean operation due to excessive underfueling.

5. The amount of fuel recovered from the initial overfueling must be considered in scheduling the degree of underfueling. The best start inducted very close to the desired stoichiometric fuel amount. This fuel inducted was a combination of fuel injected for that cycle and fuel from previous injections.

6. Best starting and emissions occurred with the sequential injection start strategy, which had a more gradual transition from the overfueling phase to the normal fueling phase. It is likely that as soon as the fuel puddle has been established, this strategy provides injections that are close to the optimal level. By more closely following the decreasing trend in manifold pressure, these injections would better match the mass of air being inducted into the cylinder.

7. The quality of the start can be predicted by looking at how closely the injections follow the stoichiometric requirement after the second cycle (while allowing for a small, decreasing portion of "re-accounted form fuel). An optimal strategy might start with a large injection to create the fuel puddle, and then base injection on intake manifold pressure (which is directly proportional to the stoichiometric requirement at a given temperature).

8. Initially, in a warm engine start, the air in the manifold is hot, which promotes fuel evaporation, leading to immediate combustion and work output from the injections and little unaccounted for fuel. However, once this hot air supply has been depleted (after approximately four cycles) the injected mass should be increased to compensate for the decreased evaporation rate and increased air density. A larger injection pulse at this point may aid the formation of a required fuel puddle.

9. The order of magnitude difference in the fuel mass "unaccounted for" between warm and cold starts may be a direct consequence of fuel loss or storage during cold start. If the other fuel loss mechanisms are minimized during warm starting, and the puddle formation mechanism is similar for both conditions, then the difference between the mass "unaccounted for" may represent the mass lost from or stored in the cylinder during cold start, and the mass unaccounted for during the warm start may represent the steady-state puddle mass. Future experiments, where the manifold is cooled and flushed of the hot air immediately before a warm start, in order to make the manifold evaporation similar to that of a cold start, could confirm this hypothesis.

REFERENCES

[1] Shayler, P.J., Isaacs, R.M., and Ma, T.H., "*The Variation of In-Cylinder Mixture Ratios During Engine Cranking at Low Ambient Temperatures*," IMechE, Journal of Automobile Engineering, 1992.

[2] Cheung, H. M., and Heywood, J.B., "*Evaluation of a One-Zone Burn-Rate Analysis Procedure Using Production SI Engine Pressure Data*" S.A.E. Paper 932749, S.A.E. Transactions, vol. 102, 1993.

[3] Fox, J.W., "*A Model for Predicting Residual Gas Fraction in Spark-Ignition Engines*," SAE Paper 931025, 1993.

[4] Shin,Y., Cheng, W.K., and Beywood, J.B., "*Liquid Gasoline Behavior in the Engine Cylinder of a SI Engine*, S.A.E. Paper 941872, 1994.

[5] Gonzalez, M.A., Boman, G.L., and Reitz, R.D., "A Study of Diesel Cold Starting Using Both Cycle Analysis and Multidimensional Calculations," S.A.E. 910180, 1991.

ACKNOWLEDGMENTS

Special thank are due to Bill Kryska, Steve Smith and John Hoard from Ford's Advanced Powertrain Group for their assistance with this research project. This research was made possible by funding from the National Science Foundation and the M.I.T. Sun Jae Professorship Fund.

APPENDIX

DETAILS OF FUEL BURNED CALCULATION- The energy release model used in the fuel burned calculation uses several assumptions to obtain the parameters for the ideal gas law used to calculate the in-cylinder mixture mass. The cylinder pressure was determined by averaging the manifold pressure between the times of bottom center piston position (BC) and intake valve closing (IVC). Cylinder volume was calculated at intake valve closing using the engine geometry provided. The valve closing volume was used due to the slow speeds encountered and the absence of any charging or tuning effects at these speeds. One of two values was used for incylinder temperature, depending on the cycle conditions. For the first cycles, where the previous cycle did not contain any significant combustion, the temperature was assumed to be room temperature (295.9 K). For cycles following burning cycles, a higher temperature (318 K) was used due to the fraction of hot combustion products that remain in the cylinder. These values were used to calculate charge mass. A residual model [3] gave a residual mass, and a stoichiometric air/fuel mixture was assumed to comprise the remaining fraction of the cylinder charge. Although the fuel is injected at levels much richer than stoichiometric, only a limited fraction of this fuel is in vapor form in the cylinder. Since the injection strategy is optimized through calibration, it is likely that the gas phase air/fuel ratio is close to stoichiometric for most cycles during the start-up. Fuel that has entered the cylinder in liquid form is only accounted for by the fact that it may evaporate to form a stoichiometric mixture.

The results of the fraction burned analysis are dependent on a number of assumptions in addition to the incylinder composition. Inputs are required for the inlet temperature, atmospheric pressure, spark timing, inlet pressure, engine speed, wall temperature, swirl ratio, heat transfer calibration constant and crevice volume. The model does not account for the effect of blowby because "the significance of blowby is small in modern engines." [2] However, at start-up, blowby is significant. Up to 16% of the cylinder charge is lost to blowby at cranking speeds, and approximately 5% at 400 RPM[5]. This is likely a major source of "lost fuel" during the early part of the start-up analysis.

Most of the uncertainties in the parameters in the burned fuel calculation result in a minimal change in the maximum fraction burned result. However, the initial mass estimate, the mixture stoichiometry and the accuracy of the pressure data have a direct effect on the peak fraction burned result Unfortunately, these are also areas that have potentially large uncertainties. The accuracy of the pressure data is tied to the accuracy of the intake manifold pressure and the assumption that the two pressures are equal during the period shortly after bottom center during the intake process. The accuracy of the mass estimate is directly proportional to the accuracy of the pressure measurement and the temperature estimate. The inaccuracy due to the pressure signal is likely less than 7%, but the uncertainty of the temperature value may be much greater (close to 15%), depending on the cycle. The uncertainty of the total mass and fuel/air mixture estimates are large, but these uncertainties are less important to the final analysis since the product of fuel mass and maximum fraction burned is the important quantity: the product of fuel mass and fraction burned will be almost constant. As explained in the text, the mass of fuel burned is robust to errors in the estimated mass of mixture inducted. Most cycles during start-up appear to have lean vaporized Fuel/Air mixtures, even though the injected fuel/air ratio is rich, due to blowby and liquid fuel transport. If this is correct then the overall effect of the uncertainties in the mixture composition estimate is minimal.

Flames and Liquid Fuel in an SI Engine Cylinder During Cold Start

S. Campbell, E. Clasen, C. Chang, and K. T. Rhee
Rutgers, The State University of New Jersey

Copyright 1996 Society of Automotive Engineers, Inc.

Abstract

The flame propagations in the very first firing and subsequent cycles in an SI engine during cold start were studied to gain a better understanding of reaction fronts associated with liquid fuel (regular unleaded) in the cylinder. This work was performed using the Rutgers high-speed spectral infrared digital imaging system on a single-cylinder engine with optical access. The engine was mounted with a production engine cylinder-head mated with a conventional port fuel injection (PFI) system.

In the study, four images in respective spectral bands were simultaneously obtained at successive instants of time during the combustion period, which was done for eight sequential cycles. This multiple-band successive-imaging was repeated in intervals of about two minutes over a period of more than twenty-five minutes after the engine start. During this experiment, the temperature changes at the intake port, the water jacket and the exhaust gas were monitored. In addition, pressure-time data was obtained from individual cycles in order to gain some insight into the overall in-cylinder reactions. Note that the fuel rate by the PFI for the first set of successive images was about 3.5 times stoichiometric and that for others was near-stoichiometric.

The first firing cycle exhibited almost invariably weak flame propagation, which was followed by very intense flame fronts in the next cycle. Note that the flame propagation in the first cycle seems to only indicate consumption of the fuel vapor available in the cycle. The flames in the third cycle were also intense in some cases, but mostly weaker than those in the second. Upon formation of the flame front in the beginning of combustion, some exceedingly strong local reactions started to grow, but no earlier than 15CA after TDC. The reactions appeared to be diffusion reaction fronts around liquid fuel layered over the chamber surfaces. The scale of these local reaction zones decreased with time and exhibited some significant transient changes. This variation continued to occur even though the engine was relatively well warmed. Results from some parametric studies are also reported.

Introduction

The large amount of unburned hydrocarbon (UHC) emitted by spark ignition (SI) engines during cold start basically stems from the fact that the liquid fuel introduced into the engine is poorly vaporized. Because of the fuel's low vapor pressure at this time, extra fuel is injected into the engine in order to produce enough vapor to achieve successful development of the flame propagation (with the throttle valve closed). This extra fuel causes accordingly large amounts of liquid fuel layered at the intake port and in the combustion chamber.

Let's consider what may occur during cold start in a typical modern SI engine with a conventional port injection fuel system (PIF). The present discussion concerns reactions during two different time periods from the start: They are: (1) the first period of several seconds, and (2) the early period before attaining a well warmed engine condition. The former is of interest because, in a typical modern SI engine, the over-rich fuel injection ceases immediately after the start. The latter is separately discussed because, during the warm up period, the in-cylinder formation of UHC continues to be high even with the mixture near stoichiometric.

Engine Operation with Extra Fuel. The very first fuel injected, although in an amount sufficient to produce a rich mixture of several times the stoichiometric, will probably not be connected to an immediate flame propagation, but will mostly wet the intake port and the combustion chamber surface. With no ignition occurring in this cycle, both a great amount of fuel vapor and probably even some liquid fuel would be wasted "raw" to the exhaust. In the next cycle, the fuel at the intake port will be added by the next fuel delivery to form a thicker liquid layer of fuel and an increased amount of vapor. When the intake valve opens, more fuel from this accumulation will be combined with the trapped fuel (in the cylinder from the previous cycle), which will also produce thicker layers of liquid fuel over the chamber surface and, of course, an increased amount of vapor. In spite of this, the cycle may not achieve a flame propagation either, and then the same will be repeated. The process involving wasting and accumulation of the fuel will continue until a sufficiently rich vapor-air mixture is produced near the spark plug for a successful fire ball formation. Even the first firing cycle may not be followed by the same, and then the above process will be repeated as before. The number and mode of such unsuccessful cycles during the start will depend on various factors, e.g. the amount of fuel injection per cycle; fuel distillation characteristics; temperature of the cold start; amount of the residual fuel trapped in the cylinder and the intake port after the previous engine operation. Note that the last is affected by the piston locations, the time period after the previous engine operation, and more.

The combustion, then, will produce the first hot (rushing) back-flow of combustion products, which will alter the thermal condition and fluid flows at the intake port to increase the atomization and vaporization of the liquid fuel deposit. The combustion will also change the transport process over the liquid fuel layers in the cylinder, which increases vapor formation and decreases the amount of liquid fuel loaded over the surface. The waste gas at this time is expected to contain a greater portion of incomplete combustion products for several reasons. For example, there will be over-rich diffusion flame fronts off the liquid layer, and fuel vapor leaving the layer (after the reaction fronts disappear) will not be well oxidized. Note that since the combustion chamber surface is at low temperatures, the wall quenching effects will be highly significant. In addition, the UHC diffused out of the quenched layer will be poorly consumed by the bulk combustion products. Such strong quenching effects and poor post-flame oxidation, plus the raw fuel mentioned above, will become main sources of the engine-out UHC during the cold start. The high complexity of the formation processes may befurther realized as considered in the following.

In order to minimize the amount of UHC during this period, several possible injection strategies may be implemented in the very early cycles. As one such strategy, a minimum amount of fuel may be injected to produce a lean vapor-air mixture for a marginal ignition followed by a completing flame propagation, or a large amount of fuel could be injected to produce a near-stoichiometric vapor-air mixture. The first flame propagation by the former strategy is expected to be weak but leave a small amount of unconsumed liquid fuel over the chamber surface. Such a weak (lean-mixture) flame would be subjected to strong quenching effects to cause more incomplete products, but the small amount of liquid fuel is expected to produce less over-rich diffusion reactions and a smaller amount of wasteful fuel vapor before the exhaust-valve-open (EVO). The first flame by the latter method, if properly achieved in a well controlled manner, which depends on the cold start temperature, fuel characteristics and others, will be strong to help the engine to attain a high temperature sooner. This strategy will need fewer cycles of operation than the former before attaining a high temperature, which will additionally bring beneficial aspects such as an active post flame oxidation, but leaves a larger amount of liquid fuel over the surface, a negative factor in achieving low UHC emissions.

Regardless of strategies, there will be a significant amount of liquid fuel deposit over the surfaces causing UHC, which would exhibit severe transient and cyclic variations.

The variations would be dependent on the amount of extra fuel, the temperature of the cold start, the engine speed (and change) and more. The massive amount of liquid fuel layers formed over the surface during the cold start may be disposed of in several routes, including: (1) consumed in the following cycles; (2) washed down to the crankcase; (3) vaporized (in the absence of reaction fronts) and wasted during EVO; and (4) carbonized over or within the deposit formation. It is desirable to obtain a better understanding of such in-cylinder events during this over-rich combustion period, particularly the flames (of consuming fuel vapor) and (diffusion) reactions over the liquid layers. In order to minimize negative consequences on not only the UHC but also on others as considered above, a delicate compromise will have to be made among various factors for the PFI control strategy in a prompt manner during the transient period.

The present discussion also is concerned with the number of cycles from the start when the over-rich fuel injection is taken off and a near-stoichiometric mixture is provided. It's importance can be seen from the fact that if the fuel is fully vaporized (upon the over rich injection of several times an amount of fuel in a near stoichiometric mixture) the mixture will be simply too rich to ignite. As the combustion cycle continues, the engine temperature rises to promote the vaporization of the fuel at both the intake port and in the combustion chamber. With the mixture preparation shifted to near stoichiometric, sooner or later, the catalytic converter becomes lit up.

Liquid Layers during the Warm-up Period.

When the engine attains a warm condition, the fuel introduced for producing a near stoichiometric mixture may be well vaporized achieving predictable flame propagations. However, only recently, some new evidence came to light suggesting that the regular unleaded gasoline at ignition may not be well vaporized in the combustion chamber even after the engine is relatively well warmed. This poor vaporization may be a significant source of UHC emissions [1]. This study reports a new discovery of locally reacting centers in the combustion chamber, in which successive images were captured by using our high-speed multispectral IR imaging system. Note that when the engine was warm, these local reactions were mainly found around the intake valve and that they were not found when the same engine was operated by gaseous fuels (namely propane and natural gas). In addition, when the engine was started at room temperature, reaction centers were found at multiple locations, even including the exhaust valve and spark plug. Also, there is some observation suggesting that the formation of local reaction centers may be affected by the distillation characteristics of fuel.

The above observation of local reaction centers, which were considered to occur due to liquid layers formed

*Numbers in parentheses designate references at end of paper.

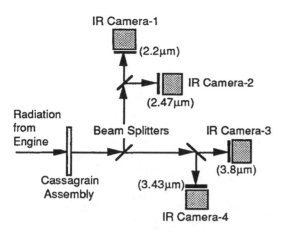

Fig. 1. Schematic Presentation of Rutgers SIS.

Table-I. Engine Dimensions

Bore × Stroke (mm),	101.6 × 101.6
Compression Ratio,	9:1
Spark Ignition,	6 BTDC
Valve Timing:	
EVO	135 ATDC
EVC	10 ATDC
IVO	10 BTDC
IVC	135 BTDC
Fuel Rail Pressure	200 kPa

over the cylinder head during the intake period, may not be surprising in view that many research individuals report liquid fuel layers or puddles observed at the intake port (in warm SI engines). In addition, the local reaction centers, according to the above study, remain initially a surface phenomenon, which later become a volume phenomenon, i.e., around 70 after top-dead-center (ATDC). The radiation from the local centers was similar to that expected in diffusion flames. Note that an SI engine operated by a gaseous fuel, which did not exhibit any sign of local reactions, produces much lower engine-out UHC than the same engine fueled by regular gasoline. As to the phase of fuel in the mixture when it rushes through the intake port, it is actually desirable to have it only partially vaporized in order to achieve a high brake mean effective pressure (bmep). If, instead, an SI engine is designed to achieve a nearly full vaporization of the fuel at this time, at the expense of bmep, (for example, by excessively heating the port) the liquid fuel layer in a cylinder may not be formed. After finding this liquid burning, a strong need for obtaining better understanding of the phenomenon associated with many issues was pronounced, such as its transient variation and effects of engine-fuel factors.

If the formation of liquid layers is significantly responsible for UHC even in a warm SI engine, it should be reasonable to infer that the same would be a main process of the engine-out UHC during the remaining warm-up period after the over-rich fuel injection ceases at the intake port. It is desirable to investigate the behaviors of flames and diffusion reactions over the liquid layers during this transient period, while the engine is run by a near stoichiometric mixture.

Experiment

Since the present investigation was performed based on findings from the earlier work [1], the same apparatus were used. They include (1) high-speed multispectral IR imaging system and (2) SI engine with optical access, which will be only briefly described here. More details of these apparatus may be found elsewhere [1-3].

Multispectral IR Imaging System. This is a one-of-a-kind system designed and fabricated at Rutgers University, which is referred to as the Rutgers System or Super Imaging System (SIS). As shown in Fig. 1, this SIS has four high-speed IR digital camera units connected to a single optical train. The radiation passing through the optical access of the engine is collected by a cassegrain assembly consisting of two reflective mirrors, and is then relayed through three different spectral beam splitters. This arrangement produces four geometrically identical (pixel-to-pixel matching) images in respective spectral domains. A narrow-band filter installed in front of each camera further specifies, within the corresponding domain, the spectral nature of the image for the camera.

Some of the performance features of the SIS having Pt-Si imagers (64×128 pixels each) are: imaging rate over 1,800frames/sec per camera; independently variable exposure period as short as 20μsec; spectral range of 1.5- 5.5 μm; and total 256 images to be captured (in each experiment) per camera. The cameras in the SIS are simultaneously operated according to the predetermined setting, including: the exposure period; the total number of images to be obtained per cycle; the start of imaging (in crank angle, CA) with respect to a reference marker (here top-dead-center, TDC); the interval between successive images in CA.

Four spectrally distinct images simultaneously obtained by the SIS at successive instants of time implies that, ideally, distributions of four different pieces of information may be obtained at the CA of imaging, such as temperature, and concentrations of water vapor and soot. While new data processing methods for achieving such quantitative imaging are being developed at Rutgers, it was found that raw images produced by the SIS permit us to collect new pieces of in-cylinder information, which are difficult to obtain using conventional diagnostic devices. Some of them are reported here.

Engine Apparatus. The apparatus was built on a single-cylinder engine base by mounting a new Ford 302 cylinder head and a matching port-injection fuel system (PIF). Construction of the set-up was performed in an attempt to preserve the representative characteristics of the real-world SI engines. Figure 2 shows the arrangement of the optical access in the engine. Since an in-depth description was made in previous papers, some relevant engine information is summarized in Table-I.

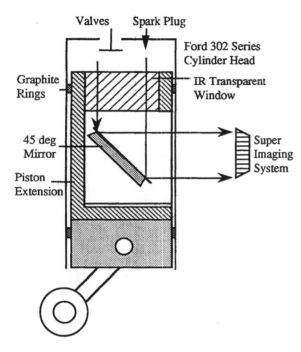

Fig. 2. Spark-Ignition Engine with Optical Access

Fig. 3. A Visible-Ray Photograph of the Cylinder Head Exhibiting the Imaging View.

The engine apparatus was sufficiently instrumented in order to obtain consistent results from the experiment. Particularly to monitor the thermal condition of the intake port, thermocouples were installed where deemed appropriate. In addition, cylinder pressure-time (p-t) history was recorded for the corresponding sets of instantaneous images. This was achieved by synchronizing both pressure-time data acquisition and imaging to the same engine encoder.

Figure 3 is included here in order to indicate the imaging view, which shows the intake valve on the left, the exhaust valve on the right and the spark plug in between. This figure will be referred to when the instantaneous spectral images are presented later.

Results and Discussion

The main results from the study are a set of four separate spectral IR digital images simultaneously taken at successive instants of time during the combustion period. This high-speed imaging from a cold engine was done by starting from the very first cycle until the engine was fully warmed up with the intake manifold vacuum maintained at 24kPa. While the imaging was performed, the corresponding pressure-time data was obtained to gain some idea of the overall combustion condition. Temperature changes at the intake-port, the water-jacket (the cylinder heat) and the exhaust gas were recorded. The experiment was performed by using regular unleaded gasoline with ignition time at 6BTDC.

In each imaging, four sets (in spectral bands of 2.2, 2.47, 3.43 and 3.8µm) of thirty-two digital images per cycle, were simultaneously obtained from eight successive cycles, which filled up the memory in the SIS (over four megabytes in 12 bit dynamic resolution). Note that, in a separate run, the imaging was also carried out by obtaining (in each spectral band) eight images per cycle from 32 sequential cycles. In yet another experiment, the same was made to obtain a single image per cycle at 35 ATDC from 256 consecutive cycles. These data were transferred to hard drive to free the SIS memory for the next imaging, a process which typically took about two minutes. That is, in this experiment, such a batch of images was repeatedly obtained in about two-mimute intervals during the warm-up period,

The imaging of combustion events during the cold start was done in two steps in sequence: (1) over-rich mixture imaging; and (2) stoichiometric mixture imaging. In the first step, after the engine was motored to attain a speed of 350rpm. The injection started in an amount of about 3.5times that of a stoichiometric mixture, which seemed to produce a satisfactory start. Note that when a smaller amount of fuel was injected, the first successful firing cycle was not always followed by the same. The SIS was synchronized to capture images from the very first firing and subsequent cycles, which took only a few seconds and were immediately followed by a fuel-injection-rate change to provide a stoichiometric mixture. At this time, the external load was engaged on the engine in such a way that the speed did not exceed 500rpm. As soon as the SIS was ready again for imaging after transferring the first batch of images, the second batch was obtained in the same way using a stoichiometric mixture at constant engine speed of 500rpm

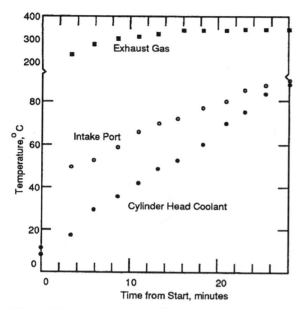

Fig. 4. Temperature-time Data at each Imaging.

1	-5	TDC	5	10	15	20	25	30
2	35	40	45	50	55	60	65	70
3	75	80	85	90	95	100	105	110
	a	b	c	d	e	f	g	h

Fig. 7. Look-up-table indicating the Time of Imaging.

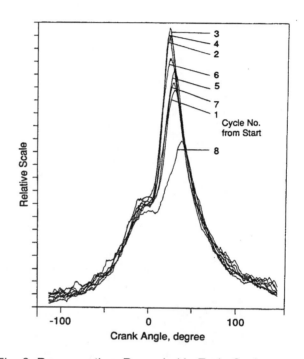

Fig. 8. Pressure-time Recorded in Early Cycles.

Figure 4 summarizes the typical sequence of imaging and temperature measurements in the engine, which, although self-explanatory, still deserves some attention. During the warming period, the temperature at the intake port is consistently higher than at the cylinder head coolant, which is caused by the back-flow of the combustion products. As soon as the imaging from the over-rich starting was done, the heater in the coolant loop was turned on for a while in order to shorten the warm-up period. In this experiment, when the measured temperature at the coolant outlet was near 90°C, the coolant was almost boiling, which may be a condition a bit more over-heated than in a typical engine operation.

It should be added that the engine running for such a long period of time produced a layer of deposit over the optical window, which would have been impossible for visible ray to pass through. But the quality of spectral IR images in the present experiment did not appear to be significantly degraded by the deposit.

A large number of imaging trials were performed. Typical sets of spectral IR images obtained from the very first firing and the following seven cycles are presented in Fig. 5 (A) 3.43μm and (B) 3.8μm. The results obtained thereafter in about two-minute intervals are shown in Fig. 6 by indicating the start time of imaging. They are displayed in pseudo-color in order to enhance the presentation of more local variations. Those images in the former band were expected to indicate radiation from mainly water vapor (and soot if formed), exhibiting the consumption of fuel vapor. Note that this band was also observed to capture radiation from some intermediate species in the preignition zone [1, 2]. Since the latter band is transparent to radiation from main combustion products including water vapor and carbon dioxide, it was considered to mostly exhibit those from the cylinder head surface, and since the soot formation was expected in very rich or diffusion

reactions, it would be also reflected by the results in this band. The sequence of images captured in a 5CA intervals is from the left to the right and goes downward, like in a calendar. In order to indicate the individual times of imaging in CA, a look-up table (LUT) for the images is included in Fig. 7. In addition, Fig. 8 displays the pressure-time (p-t) histories to match with respective imaging cycles.

While in-cylinder events with the over-rich fuel injection (Fig. 5) were meaningful even after 110CA ATDC, as indicated by 3-h of Fig. 7, those with a near-stoichiometric mixture (Fig. 6) did not show such after 70ATDC, as included in the results. Because of the difference in strength of the radiation, imaging was made with the exposure period set to 170μsec for those with the over-rich fuel, and 310μsec for the near-stoichiometric mixture.

Over-Rich Mixture for Start. Discussing the results from the over-rich starting, the images of flame propagation (via 3.43μm) in the very first firing cycle are

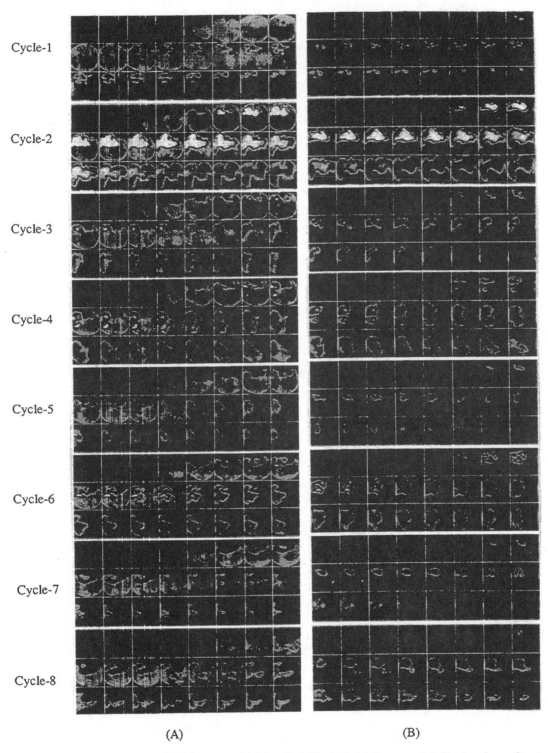

Cycle-1

Cycle-2

Cycle-3

Cycle-4

Cycle-5

Cycle-6

Cycle-7

Cycle-8

(A)   (B)

Fig. 5 High-speed Spectral IR Images Obtained at First Eight Sequential Cycles from Start
via Bands of: (A) 3.43mm; and (B) 3.8mm.

144

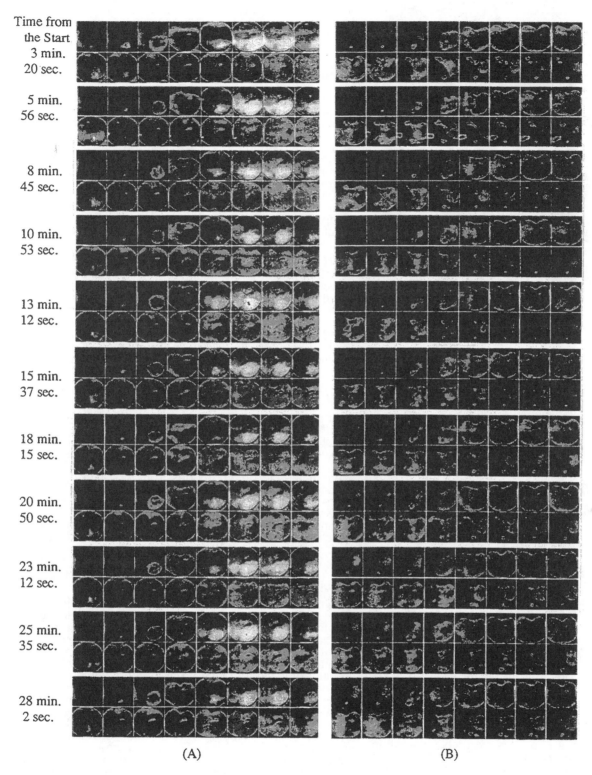

Fig. 6. High-speed Spectral IR Images Obtained during the Warm-up Period via Bands of: (A) 3.43mm; and (B) 3.8mm.

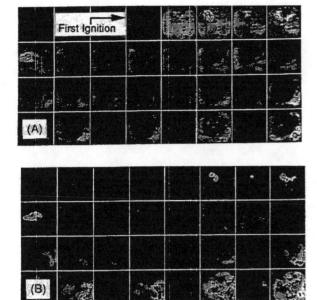

Fig. 9. Images obtained from Successive Cycles:
One per cycle at 35CA ATDC: (A) 3.43μm and
(B) 3.8μm.

relatively weak (Fig. 5-(A)). This fact seems to suggest that a relatively small amount of vapor was produced prior to ignition in the cycle. This weak flame front, however, became somewhat intense at the later stage of combustion in the cycle, exhibiting a sluggish consumption of fuel, which can be also seen from the p-t history (Fig. 8). Although the actual amount of accumulated fuel in the cylinder prior to the ignition is expected to be large, the energy release in the first cycle is small, which suggests that the unused fuel was either layered over the surface or partially lost (e.g. via wash-down). It is noted that, in the first ignition cycle, the flame fronts are observed over the entire imaging view of the chamber, which suggests that the fuel vapor was accumulated in the zone during the previous (no-ignition) cycles. This is in contrast with the following cycles having zones without emitting any significant mount of radiation by combustion products, particularly near the squish area. The radiation in 3.8#m (Fig.5-(B)), which is expected to reveal radiation from soot indicating diffusion reactions, is also rather insignificant, which indicates possibilities either that the amount of liquid fuel layered over the surface was small or that the flame was too weak (due to low temperatures) to produce a strong diffusion reaction.

The flame propagations in the 2nd, 3rd and 4th cycles are more intense and exhibit a remarkable difference in strength, i.e., it is extremely intense in 2nd and comparably weak in 3rd and 4th cycles. (In some trials, the difference of intensity between these cycles was small.) The radiation intensity, however, does not seem to properly represent the heat release according to the corresponding p-t data. That is, in spite of the remarkable difference in radiation among these cycles, which is most notably strong in the 2nd cycle, their p-t data look about the same. While the intensities of the flame propagations (via 3.43μm band) are somewhat weak but yet continue to be similar to each other, the radiation via 3.8μm (from a high soot formation to reflect rich diffusion reactions) became strong in 4th cycle, which was followed by weak radiation in 5th cycle.

Such a cyclically varying mode of diffusion reactions repeated in the remaining cycles in the over-rich operation. It is noted that in some cases, two consecutive cycles exhibited strong radiation, followed by weak radiation and so forth, and vice versa. During these early cycles, the local diffusion reactions started to grow after 15-20ATDC, which is at least a period of 20CA after the flame fronts started. This time interval appears to represent the heating period required for the liquid fuel layered over the surface to commence/support the local diffusion reactions, which continues in an increasing intensity even long after the flame fronts disappeared. The continuing local (diffusion) reactions in the absence of flame fronts is further discussed later.

The p-t data indicates that the combustion started to deteriorate after the 5th cycle, which also seems to be reflected by the imaging results. The p-t data from the 8th cycle is particularly suggestive of what has been mentioned earlier, i.e., the vapor formation becomes sufficient enough within several cycles, and thereafter the vapor-air mixture becomes too rich to burn. In order to confirm this observation, an additional experiment was performed: One image per cycle was obtained in each band at the same CA (i.e., 35ATDC) from 256 successive cycles, but those from the first 32 cycles are displayed here (Fig. 9). It is clear that beyond several cycles after the start, the over-rich operation was no longer useful in achieving satisfactory flame propagation. This supports a fuel injection strategy of switching off the over-rich mixture within a few or several cycles from the start. The results also suggest that it would be meaningless to obtain images from over-rich cycles other than those already included in Fig. 5. Also, there was a considerable amount of liquid fuel flowing down along the cylinder surface as the engine was operated by the extended over-rich fuel injection. This observation plus the weak radiation of both flame propagation and diffusion reactions seen from images of the very first firing cycle (Fig. 5) lead us to expect that some of fuel washes down along the cylinder liner.

Continuously changing diffusion reactions (Fig. 5-(B)) strongly suggest a possibility of accordingly varying amount of fuel layers over the chamber surface. Such a variation in the cylinder may be dictated by the amount of fuel accumulated in the intake port during this early period. That is, the amount of liquid fuel layered at the port would affect the activities in the cylinder in a similar way. On the other hand, changes in fuel vaporization may be another possible factor affecting the variation, which would be related to the thermal condition of the cylinder entity, including temperatures of back-flow and the residual gas. Reviewing the images, however, particularly when comparing those in the 5th through 8th cycles with each other, a strong diffusion flame is not necessarily found with correspondingly intense flame propagation. The observations suggest that the variation in vaporization could

of local diffusion reactions. That is, the varied amount of liquid fuel present in the cylinder seems to be the most probable factor in determining the subsequent in-cylinder reactions.

It is further noted that the activities exhibiting the local (diffusion) reactions started no earlier than 15ATDC (1-e in Fig. 7), which coincides with the time after the entire imaging view is covered by the flame fronts indicating near depletion of vapor-air mixtures (seen via 3.43μm band). This indicates that the liquid fuel layered over the (low-temperature) surface begins an active consumption only after the chamber temperature is elevated. Since the piston location becomes relatively far from the TDC thereafter, the reaction around the layers would be increasingly sluggish.

A few last observations to be mentioned about the images from the over-rich start are after the fuel vapor-air is consumed, some regions previously with no reaction front began to exhibit a new center of flame as observable in cycles 4, and 6, particularly in the squish area. The only possible explanation for this finding seems to be the slowly evaporating liquid fuel landed in those regions, which was not consumed by the flame fronts in the beginning, and the liquid layer upon receiving heat transfer produced enough vapor to support diffusion flames around. It is also noted that it is not unusual to find a cycle having flames similar to those in cycle-5, which has only small amounts of (diffusion) flames around the liquid layers. It seems most probable that such severe variations during those early cycles can be attributed to the intake-port condition dictating the fuel-air preparation in the cylinder. It is not conceivable that the in-cylinder thermal condition, which would affect the fuel vapor formation, changes so swiftly to cause such cyclic variations.

Images with Stoichiometric Mixture. Recall that immediately after the first batch of imaging was done with an over-rich fuel injection, the engine was fed with a near stoichiometric mixture for the next imaging (see Fig. 4). Since each batch of imaging contains results from eight consecutive cycles, it was attempted to choose a representative set of images in preparing Fig. 6.

In general, the radiation by gaseous mixture (as seen via 3.43μm band) gradually decreases with time particularly at the later stage of combustion, which may be seen by comparing those taken at 70ATDC with each other. This observation may be explained by the possibility of reducing postflame oxidation as the engine warms up. Since the early flame growths appear to be similar to each other (by looking at those obtained at 5ATDC, for example), the accelerating flame propagation before the piston goes away too far from the TDC may also explain the phenomena. As expected, the images captured in 3.8μm band show an increasing strength of radiation by the surface until the coolant temperature attained about 60°C (at about 20minutes under the present coolant system) and thereafter they seem not to vary considerably. Unlike the images taken right after the start, which show some parts of the chamber having almost no reaction (Fig. 5), those included in Fig. 6 indicate the flames propagating through out the entire reaction volume. According to the present experiment, once the engine ran for a few minutes from the start, the in-cylinder reactions become relatively predictable to expect a reasonable power output, except for a new finding as discussed below.

Although the difference in exposure period of imaging is taken into consideration, it was quite obvious that the amount of liquid fuel dramatically decreased within the period of about two minutes after the start. During the interval, there was a change in intake-port temperature from about 10 to 50°C. In general, the scale of the diffusion flame decreased with time. In turn, the temperature increase in the engine, most notably at the intake port (Fig. 4) and presumably over the chamber surface.

Let us review the sequential images in Fig. 6-(A) in reference to the imaging view (Fig. 3). For example examine those taken at 3.3 minutes after the start. There are intense local centers at the upper portion of the intake valve clearly visible starting from 35ATDC. (Note that in typical SI engines the flame fronts become no longer visible after about 30ATDC.) According to imaging results from the first eight cycles, the local reaction center started as early as 15ATDC, which was observable late in a warm engine. The reason for the difference was due to the radiation from combustion product masking that from the local centers over the surface. The dynamic nature of the local reaction is clearly observable when they are reviewed in a video animation display: It is restated that the centers stay as a stationary surface phenomena in the beginning, which become a moving volume reaction around.

The local centers continued to react even far after the flame propagation disappeared but in gradually decreasing intensity as the piston moves away from the TDC. According to observation by a video display, the diminishing image of the local centers seemed most likely due to the depletion of liquid fuel off the surface. Note also that imaging results obtained by a longer exposure period indicate that the local radiation continue to be remarkable even after the EVO. The weakening radiation may be also due to the lower piston location, which will cause the cylinder temperature to be low, resulting in both a low oxidation and low radiation. The above consideration brings up an expectation that the chamber surface is considerably cooled down during the non-combustion period before the next liquid fuel is delivered.

Without displaying a bulky volume of results, it is noted that the liquid layers producing such diffusion reaction centers showed remarkable cyclic variations. In some cycles, the local center was hardly observable, even when the engine was not at a high temperature, which can be seen in the sample results obtained at different times after the start (Fig. 6). For example, there were cycles exhibiting almost no local burning such as those captured at 8.25 minutes after the start, while the same obtained at 20.8 minutes (and also 25.6 minutes) indicated some remarkable local reactions. Again, it remains to be further studied that these reactions centers were often found in a rather remarkable strength even after the engine was relatively warmed, which observation was reported earlier [1].

Significance of Diffusion Reactions after Flames Disappeared. Regardless of the engine temperature, it is reasonable to expect that the continuing local reaction centers would be a significant source of UHC emissions. As mentioned earlier, the local diffusion reactions started to grow after around 15ATDC and continued far after the flame fronts disappeared (e.g. after EVO). It is again noted that the flames here mean the reaction fronts of consuming fuel-vapor and air mixtures.

Since the slowly reacting local centers are considered to occur over the liquid layers, it is quite appropriate to analyze the nature of the reaction environment. Many issues come into consideration. They are: (1) the location of liquid layers; (2) the flow of the layer over the surface; (3) the reaction anchored over the layers, (4) the stability of reactions; and (4) deposit effects on the reaction centers. The formation of layers is dictated by various factors such as the geometric configurations, and thermal and fluid flow characteristics of the engine. In this experiment, the most significant layers are found over the surface which perpendicularly divides between the chamber cavity and the squish area, and the upper part of the intake valve (refer to images in Figs. 5 & 6 and photo in Fig. 3). The liquid layers formed over the surfaces would be mobile according to gas motions, gravitational force and other effects. Referring to images indicated by 2-a through 2-h in Fig. 5 such as those obtained in 3rd cycle, the layer appear to move downward. In order to gain some insight into what is happening there, let us picture a flame spreading over a flowing fuel on an inclined surface. Since the gas motions are not small and insignificant in the chamber, the reaction front many be blown off if the reactions are improperly anchored where the vapor is formed. Such a stability issue may partially explain the cyclic variations of local reaction centers during the over-rich operation (Fig. 5) as well as some in the stoichiometric combustion (Fig. 6).

Regarding the relationship of deposit layers to the local diffusion reaction centers, although it is difficult to state a conclusive remark at present, some tangible effects have been realized during the course of the present work: The deposit layer formed over the cylinder head surface was inadvertently cleaned in the middle of experiment, and the clean head surface seems to reveal a lower frequency of having local reaction centers than before. A permeable deposit layer holding a liquid fuel would produce different modes and reaction periods of diffusion reaction from those formed over a clean smooth surface covered with the same fuel.

Summary

An optical SI engine equipped with a conventional port injection fuel system was investigated during the cold start period as operated by unleaded regular gasoline. The results obtained in the study include high-speed multiple spectral infrared images from: (1) the very first firing cycle and next seven consecutive cycles; and (2) successive sets of the same images as the former as repeated during the warm-up period in an about two-minute interval. In the former imaging, the engine was operated with a over-rich fuel injection rate (equivalent to 3.5 the stoichiometric), and for the latter, it was run by a near stoichiometric fuel injection. The matching cylinder pressure-time data and temperatures at the intake port and others were obtained at the same time. The spark time was 6CA BTDC. Some of more significant findings are listed in the following.

(1) The very first firing cycle exhibited weak and low-rate flame propagations and insignificant reactions around the liquid fuel layers.

(2) The second, third and fourth cycles exhibited similar p-t results although their radiation intensities vastly varied both in times and locations with some remarkable cyclic variations.

(3) The flame propagations do not appear to be benefited by the over-rich fuel injection after a few or several cycles from the start.

(4) The reaction centers produce radiation after about 15ATDC, indicating the need of a time period for liquid fuel to be heated to support the local diffusion reactions.

(5) Some considerable amount of liquid fuel was flowing down along the cylinder during the over-rich start.

(6) The flame propagations and local diffusion reaction centers change quite dramatically within a few minutes from the start, producing more predictable flame and smaller amount of liquid fuel reactions over the surface.

(7) The reaction centers around the liquid fuel layer continued even (long) after the flame fronts disappeared.

(8) The liquid layers causing the local reactions did not appear in every cycle, and likewise the opposite was found after the engine was well warmed. Some cyclic variations in liquid layer formation is pronounced to exist in the engine cylinder.

(9) The deposit formation over the chamber surface appears to affect the formation-occurrence of local reaction centers around the liquid layers, which needs further study.

(10) The liquid fuel formed over the combustion chamber surface is considered to cause unburned hydrocarbon emissions from both cold and warm engines.

Acknowledgement

The present work has been performed under the sponsorship of the U.S. Army Research Office (Contract No. DAAH04-95-1-0430) and AASERT (DAAH04-94-G-0201), the U.S. Department of Energy (Contract No. ACC-4-14361-01, through National Renewable Energy Laboratory), Ethyl Corporation and Ford Motor Company.

References

1. Song, K., Clasen, E., Chang, C., Campbell, S., Rhee, K.T., "Post-flame Oxidation and Unburned Hydrocarbon in a Spark-ignition Engine," SAE Paper-952543, 1995.

2. Clasen, E., Campbell, S., and Rhee, K.T., "Spectral IR Images of Direct Injection Diesel Engine Combustion with High Pressure Fuel Injection," SAE Paper-950605, 1995

3. Jiang, H., Qian, Y. and Rhee, K.T., "High-Speed Dual-Spectra Infrared Imaging," Optical Engineering, 32 (6), pp. 1281-1289, 1993.

CHAPTER 4

CHARACTERIZATION OF COLD ENGINE PROCESSES

Choongsik Bae
Korea Advanced Institute of Science and Technology

4.1 OVERVIEW

Hazardous exhaust emissions, especially hydrocarbon (HC) emissions, under cold-start conditions in port-fuel-injection (PFI) SI engines are determined by a chain of engine processes such as the spray characteristics of fuel injections, the interaction of the spray with the port wall and valve, the in-cylinder droplet field and associated mixture distribution, the flow/mixture state around the spark plug at ignition timing and the flame/wall interaction. Refined understanding of the combustion and pollutant formation process in SI engines is needed to evolve the engines. This could be achieved by diagnostic techniques, especially the laser-based in-situ diagnostics, on the fuel spray, mixture preparation, flame propagation and pollutant formation.

The diagnostics could be divided into classical ones of physical probing and remote sensing with non-intrusive optical diagnostics. Some optical diagnostics could still be considered as classical ones, though lasers have been introduced as more powerful sources of illumination in these techniques. Laser diagnostic techniques are usually incorporated in optical engines to observe and measure the in-cylinder activity.

Fuel Injection Characteristics and Fuel Delivery into Engine Cylinder— The fuel transport process plays an important role in engine cold start operations, and is initially governed by fuel injection in the intake manifold, leading to the mixture formation.

PFI systems have evolved into an electronic, pulse-width-modulated system utilizing sequentially-timed individual injections into each port, which offers significant advantage in engine transient response and HC emissions. The degree of the preparation of a fuel-air mixture is characterized by fuel atomization and evaporation performances. Liquid phase spray is characterized by the droplet size distribution and droplet number density distribution. The portion of evaporated fuel is also affecting the mixture preparation process. The measurement of spray geometry and penetration depth is enabled by the direct imaging of sprays with front/back illumination with the help of high-speed cinematography and holography, which could also be improved by the introduction of lasers.

Classical methods for determining droplet size distribution, such as collection techniques, fractional separation techniques and electrical methods, have evolved to optical methods where either images of droplets or optical/physical properties are used for determining droplet sizes. Holographic recordings can extend the photographic techniques in three-dimensional measurements. Advanced optical measurements using light scattering of fuel droplets can provide the temporal and spatial information on the microscopic spray fields. Light scattering measurements could be divided into integral methods and particle counting. Fraunhofer diffraction is a typical integral method, measuring large numbers of droplets simultaneously. Laser phase Doppler anemometry (PDA) as a particle counting method views one droplet at a time.

During cold start, the temperature of engine walls is too low to effectively vaporize the liquid fuel. A portion of the fuel droplets is deposited in the form of a fuel film along the manifold walls or on the inlet valve back-surface, whereas another portion of the droplets either vaporizes in the air stream or enters the combustion chamber still in droplet form. The wall wetting of the combustion chamber was found to result in a significant increase in HC emissions. Most wall film measurements had been conducted indirectly by measuring the air/fuel ratio, in-cylinder concentrations or exhaust emissions of HC. UEGO (Universal Exhaust Gas Oxygen) sensor could be used to detect air-fuel ratio (A/F) deviations from stoichiometry. The optical probing technique has been developed to interrogate the behavior of the port film. Laser-based optical techniques such as LIF (Laser Induced Fluorescence) made a breakthrough to the fuel film measurement on the back surface of intake port and cylinder liner. The fuel behavior at the intake port region was investigated by high-speed direct imaging technique. High-speed spectral infrared imaging was also employed to identify some liquid fuel layers formed in the cylinder stemming from intake-port liquid fuel layers.

Measurements of the droplet velocity and size in the optical engine cylinder were obtained as a function of injection timing corresponding to both closed- and open-valve injection strategies. Liquid fuel inflow into the cylinder and the in-cylinder transport of unburned fuel were visualized by planar LIF (PLIF). It was also found that most of the fuel under open-valve injection conditions entered the cylinder as droplet mist, in contrast to closed-valve injection conditions. It was found that the combustion during the early cycles of cold start started with insufficiently vaporized fuel, followed by visible weak flame fronts developed to overall-rich combustion at a higher rate induced by late vaporized fuel portions.

Mixture Distribution and Its Interaction with Flow--Mixture distribution is associated with the interaction of the fuel spray with the port wall and valve, the in-cylinder droplet field, and flow field, which consequently determines flame propagation process and hydrocarbon emissions especially under cold-start conditions.

The detection of the liquid portion of fuel and fuel vapor is needed to identify the degree of fuel evaporation, and could be achieved by laser induced exiplex fluorescence (LIEF) method. Laser extinction and absorption (LEA) is another technique for simultaneous detection of fuel vapor and liquid droplets, generating information of droplet size in the spray.

The combustion stability in SI engines basically depends on A/F ratio. The realistic in-cylinder A/F ratio could be significantly different from the nominal value in the intake mixture during cold-start due to the lag in the fuel transport and the loss of fuel to the crankcase. The global A/F ratio in the engine process could be measured by an oxygen sensor, i.e., UEGO sensor, in the exhaust gas stream, while the realistic in-cylinder air/fuel ratio could be monitored by high speed sampling with a fast-response gas analyzer such as FRFID or FFID (Fast-response Flame Ionization Detector) and laser-based spectroscopic diagnostics. Considering that even the heated exhaust gas oxygen (HEGO) sensors cannot measure the exact A/F ratio of exhaust gas during cold start condition, a new estimator for A/F of the exhaust gas was developed utilizing the exhaust gas temperature measured with a fast-response fine-wire thermocouple, considering the fact that the exhaust gas temperature is a function of the A/F ratio.

Another sampling and analysis technique, a diode laser based spectroscopic technique, allowed the observation that combustion usually begins in the first cycle in which the fuel vapor-air equivalence ratio of the mixture inside the cylinder exceeds the lean flammability limit of the fuel at the desired temperature. The fuel vapor-air equivalence ratio for the start of combustion generally increased at lower temperatures. After the first cycle with combustion, dilution with residual gas can contribute to misfire in succeeding cycles. Hence, the fuel vapor-air equivalence ratio has to be richer than the lean flammability limit for combustion to be sustained in succeeding cycles.

To better understand the mixture formation in port and engine cylinder a number of non-intrusive laser-based techniques have been utilized. Three main laser scattering methods are Rayleigh scattering, Raman scattering and LIF. These measurements provided another evidence of non-uniform fuel distribution owing to liquid wall impingement with a high potential for producing increased HC emissions.

Flow characteristics interact with fuel in the port and cylinder, affecting the fuel delivery into the cylinder, mixture formation/distribution and combustion process. Measurement of air flow has been obtained using laser-Doppler velocimetry (LDV) and particle image velocimetry (PIV) techniques.

Combustion Processes and Pollutant Formations--Laser diagnostics of flame characteristics have been improved a lot by high-speed cameras and powerful lasers in terms of spatial and temporal resolution for analyzing turbulent flow, ignition phenomena and flame growth. LIF has been proved suitable to measure the species concentration of the transient free-radicals in flames that are intermediates in combustion chemistry. FFID could be used to measure the in-cylinder HC concentrations on the post-flame period, mainly to give a better understanding of the mechanism by which HC emissions form from crevices in SI engines. FFID was also used to confirm the beneficial performance of the recent advances in variable valve actuation on the cold-start period. A high-speed spectral infrared imaging method was developed to investigate flame development, particularly for transient processes such as during cold-start, and is a promising tool to improve the understanding of in-cylinder processes. FFID engine out measurements directly after the exhaust valve have shown the instantaneous HC characterizations, proving the potential for simultaneous start-up HC emissions reduction by the optimized cam phasing.

4.2 SELECTED SAE TECHNICAL PAPERS

2002-01-2751

LIF Characterization of Intake Valve Fuel Films During Cold Start in a PFI Engine

Bradford A. Bruno
Union College

D. A. Santavicca and James V. Zello
The Pennsylvania State University

Copyright © 2002 Society of Automotive Engineers, Inc.

ABSTRACT

A Laser Induced Fluorescence (LIF) based technique has been employed to examine the transient fuel film behavior on an intake valve during cold start of a PFI engine. Fluorescence from a tracer in the fuel was collected through a Borescope and imaged onto a CCD camera, providing a 2-D image of the fuel film on the valve. The average intensity of the fluorescence, over a Region of Interest (ROI), was taken to be proportional to the total amount of fuel present in the film.

Images were collected (at a fixed crank angle) on every second engine cycle, resolving changes in the fuel film during the cold start transient. Changes in the fuel film were resolved within a cycle by collecting images at varying crank angles during successive experiments.

Results from four fuel mixtures are reported. Two simulated "single component fuels," one with a higher volatility (HV) and one with a lower volatility (LV), were examined. The remaining two mixtures were both designed to simulate a single two component fuel (2CF); in one case the HV component was traced, while in the other the LV component was traced.

The measurements showed marked differences between the behaviors of the fuels. The films of the LV fuel and the LV component of the 2CF were both found to initially grow for the first five to fifteen cycles and then slowly decline until the end of the experiment. The LV films were found to persist even after the intake valve had opened and the intake charge had been swept into the combustion chamber. The HV components demonstrate a similar initial increase in fuel film followed by a more rapid decline. Very little of the HV fuel or of the HV component in the 2CF remained on the valve after the inlet valve opened.

INTRODUCTION

The vast majority of spark ignition automobile engines produced today use Port Fuel Injection (PFI) as a charge preparation strategy.[1,2,3] Of course engines face ever increasing challenges in the realms of emission control, fuel economy, and performance.

The performance of PFI engines is adversely affected by the formation of fuel films on the port surfaces, caused by fuel spray impinging on these surfaces. In steady state warmed up operation, the fuel mass stored in these films reaches a steady state, and excellent control of charge metering to the cylinder can be obtained.[4] However, during engine transients the port fuel films can act as either a source or a sink for fuel, so it is the imperfectly understood dynamic behavior of these films which actually controls fuel metering to the cylinder. [1,5] The problem of port fuel films is especially critical upon cold start since the port surfaces are cool and retard vaporization, resulting in larger films and larger associated A/F excursions.[7] The magnitude of this issue is illustrated by the fact that PFI engines generate up to 90% of their total unburned hydrocarbon (UHC) emissions during the first 90 sec of the US FTP emission test. [8] Improvement of PFI engine transient performance is dependent on an improved understanding of this fuel film behavior. This paper presents preliminary results of an LIF experiment designed to interrogate the fuel film in one critical region of the port, the back of the intake valve.

The back of the intake valve is a critically important region to study for several reasons. First, it is the primary point of aim for many injection systems, hence direct spray impingement on the valve should be significant. Second, the fuel film in this region is intimately involved in most of the mechanisms that transport fuel into the cylinder. The physical interactions which actually transport fuel into the cylinder are quite complex (e.g. [4]). The transport mechanisms involved include: deposition and drop impact splashing of the fuel films [4,6], preferential vaporization of high volatility components from the fuel films[4,6], liquid fuel being stripped from the films by aerodynamic forces during the induction event (much of this liquid film is redeposited as new liquid films on the cylinder wall and piston surface [9,10]), and fuel splashing (caused by cylinder pressure induced backflow upon valve opening and splashing due to closing of the intake valve) redistributing the port film to areas of the port uncoated by the initial spray film [4, 11, 12]. Again, the fuel film which forms *on the back of the intake valve* is intimately involved in *all* of these mechanisms.

Due to the importance of port fuel films, they have been the focus of many research studies. A greater understanding of the temporal, spatial, and species dynamics of the port fuel film would allow better modeling and control strategies to be implemented. Several research groups have developed strategies to interrogate various aspects of these fuel films. However, due to experimental difficulties, to our knowledge no one has conducted experiments capable of obtaining data which is simultaneously spatially and temporally resolved and species specific, from a fuel film on the back of the intake valve in a running engine. A brief synopsis of some of the relevant previous work in this field is given below.

Senda et al [6] used laser induced fluorescence (LIF) to investigate the fuel film formed when a fuel spray impinges on a plate of glass at ambient conditions. For this simplified condition they were able to show that the LIF technique is capable of making quantitative fuel film thickness measurements with good spatial and temporal resolution. Using this technique they investigated the effect of fuel impingement direction and distance on the formation of the adhered film. They incorporated their results into an enhanced sub-model of port fuel film behavior for the KIVA-II code. Johnen, and Haug [13] used a fiber optic based LIF measurement technique to make fuel film thickness point measurements at several locations on the flat surface of a simplified model intake port. They developed a calibration method for their sensor which allowed them to examine films that are not "optically thin." From these point measurements, they attempted to quantify the total mass of fuel in the film formed in the simplified geometry port. Almkvist et al. [5] made an LIF film thickness measurement along a line of laser light (formed from a laser sheet) focused into the inlet port of a running PFI engine. Other than slight modifications to allow for optical access, the port geometry was unmodified. They used 3-Pentanone as a tracer for iso-octane because the vaporization properties of these two compounds are well matched. [5, 14, 15, 16] They examined the effect of cold versus warm engine operation, and the effect of (intake valve open) IVO versus (intake valve closed) IVC injection timing. They were not able to examine the fuel film in the region of the port near the inlet valves. Finally, Felton et al. [17] used an optical fiberscope to extract LIF signal from a fuel film in the region of the valve guide and septum of a running engine. They interpreted the total normalized florescence intensity signal from the image to be an indicator of the total quantity of film stored in this region. A similar method is used in this study. Using ordinary gasoline which contained a fluorescing marker dye (of unknown volatility), they studied the growth and subsequent disappearance of the fuel film during cold start. They found that the fuel films produced by both of the injection systems that they studied persisted for up to 10 minutes after cold start.

Non LIF methods which have been used to interrogate the behavior of the port film include the following. Evers et al. [18, 19, 20] have developed a film thickness probe which works on the principle of internally reflected light from the film surface. The device is capable of making point measurements in locations where the probe can gain optical access to the bottom of the fuel film. However, this method is not suited to making measurements of the fuel film on the intake valve of a firing engine. Saito et al. [21, 22] used a PFI research engine with hydraulically actuated, electronically controllable valves to isolate the fuel in the cylinder and the fuel in the port at any desired instant in time. The fuel thus isolated was vaporized with hot air, and sent to an FID for quantitative analysis. By stopping the engine and completing this analysis on subsequent cycles during several cold start tests, they could determine the total mass of fuel carried over from cycle to cycle by the port fuel film. However, the exact spatial distribution of the fuel film in the cylinder was not measurable with this technique. Russ et al. [23] used a UEGO sensor to detect A/F deviations from stoichiometric caused by fuels of differing volatility and different injector targeting. Although they did not directly measure the mass of fuel stored in the fuel film with this technique they were able to estimate the mass of fuel in the port film by comparison with the X-τ model [26]. Their results were compared to predictions made in a related modeling study [24] which makes use of the Four Puddle Model (FPM). The FPM attempts to model the physics of the fuel behavior in four locations (puddles) in order to determine the mixing and A/F ratio excursions encountered with PFI injection systems. The four puddles modeled are 1) the port film, 2) the valve film, 3) the upstream film (caused by backflow and valve splashing) and 4) the in cylinder film (caused by liquid fuel stripped from the other films during induction).

Empirical evidence for port fuel film behavior which is temporally, spatially, and species resolved is still sparse, especially for the critical region on the back of the intake valve. This paper presents initial results from an experiment which simultaneously examines the temporal and spatial behavior of a selected fuel component on the back of the intake valve of a running engine. As a case study, we specifically investigated fuel film behavior during a cold start transient. This initial data set illustrates that this experiment can: 1) examine the fuel film in the region directly targeted by the injector spray (typically the back of the inlet valve), 2) produce data with sufficient time resolution to examine the film behavior during cold start transients, 3) produce a sufficiently high spatial resolution, and a sufficiently large area of interrogation to examine the behavior of an appreciable portion of the fuel film during an actual cold start 4) produce data which is fuel component specific and 5) relate the observed fuel film behavior to measured engine cold start performance. This experiment demonstrates strong potential to produce data valuable for developing, verifying and calibrating theoretical models of port fuel film behavior (e.g. the FPM).

EXPERIMENTAL METHODS

TEST ENGINE AND INSTRUMENTATION

The engine used for this study is a single cylinder, extended piston design modified from a production DaimlerChrysler 3.5 L, V-6 engine. The engine is designed to accept both a quartz cylinder and a Bowditch type piston to allow optical access to the cylinder. However, for this study an un-cooled steel cylinder and a aluminum piston were used because optical access to the cylinder was not needed.

The head used was a DaimlerChrysler 3 valve (2 inlet, 1 exhaust) DI head based on a production 4 valve overhead cam design with a single centrally located spark plug. The head was modified to permit optical access to the intake port and measurement instrumentation (as described in detail below). The head was not liquid cooled during these tests. The test cylinder intake was modified from the standard manifold by removing the longer of two runners for that cylinder from the original manifold and using only this to feed air to the intake port of the firing cylinder. The inlet port geometry is not greatly modified from the production engine, but, it is isolated from flow effects caused by the other cylinders. Fuel was supplied to the research cylinder from a single Siemens 27 degree split stream Deka IV fuel injector fed by a ¼ inch line. The distance from the injector to the back of the intake valve was 77 mm. Fuel pressure was supplied by pressurizing a small steel fuel tank with compressed nitrogen regulated to the desired fuel feed pressure. Engine characteristics and operating conditions used for this study are given in Table 1.

Engine Characteristics and Operating Conditions	
Compression ratio	10.9
Bore x Stroke	96mm x 81mm
Clearance Volume	58.6 cc
Inlet Valve Open/ Inlet Valve Close	717°/ 227°
Exhaust Valve Open/ Exhaust Valve Close	469°/ 18°
Engine Speed	889 RPM
Avg. MAP/ fluctuation during cycle	7 psia /2 psia
Equivalence Ratio	1.0
Ignition Timing	35° BTDC Compression
Start of Injection	470° ATDC Intake
Initial Inlet Valve Temp.	24-26° C
Initial Head Temp.	24-26° C
Initial Piston Temp.	25-29° C
Intake Air Temp.	25° C

Table 1: Engine characteristics and operating conditions

The engine was instrumented with thermocouples to monitor the temperature of the firing cylinder's piston surface, head, and port side surface of one of the intake valves. The intake valve thermocouple is particularly important to this study, so some additional comment about it is warranted. No significant difference between the temperatures of the two intake valves is expected with this head design. Therefore, in order to avoid any potential blockage of the optical observation system, it was decided to monitor the temperature of the non imaged thermocouple. In the subsequent data analysis it is assumed that the temperature of the two intake valves is nearly identical. The thermocouple used is a 0.020 in diameter, type K thermocouple located 12.8 mm radially from the center of the valve (maximum valve diameter = 36.45 mm). It is held in firm contact with the valve surface by a thin stainless steel patch spot welded to the valve. The thermocouple is located near the point of aim of the injector and so is exposed to the fuel spray and resultant fuel film. Thermocouples also monitored the ambient air temperature and controlled an intake air heater which maintained a uniform intake air temperature.

The head was also modified to allow in cylinder pressure measurements using a Kistler Model 6123 pressure transducer. Crank angle position and speed were monitored with a BEI shaft encoder which provided both a 1° and a TDC intake reference pulse. Steady state intake air mass flow was measured with a Teledyne Hastings mass flow meter. The exhaust gas was sampled approximately 1.5 ft (45 cm) away from the exhaust valve, and was then sent to an FID (J.U.M. Engineering model 3-200) for analysis of the total engine out UHCs. All of these instruments were tied in to the central DAQ and control computer and were recorded during the experiment.

Control software running on the data acquisition (DAQ) computer operated the entire cold start engine test. The operator set the desired Start of Injection (SOI) timing, duration of injection pulse, spark timing, number of fired cycles in the test, and cycle number(s), and crank angle at which imaging would take place. The test started when the operator started the AC motor which drove the engine. The control computer then began monitoring engine speed. When the engine speed was 800 RPM the control software counted 10 cycles and then initiated fuel injection, spark, and imaging. The engine speed by the 10th cycle reached a steady value of approximately 890. Once the desired number of cycles had been run, the control computer cut off fuel supply and spark, and the operator stopped the electric motor. The engine was allowed to cool at room temperature until the desired initial piston, intake valve, and head temperatures were obtained and the next test cycle was begun.

Each cold start test consisted of 196 sparked / fuel injected cycles. LIF Images were collected on every second cycle, with the first image coinciding with the first sparked and fuel injected cycle. The crank angle at which the images were collected was fixed for a given engine test, but was varied from test to test to resolve transients within the cycle.

OPTICAL ARRANGEMENT AND LIF MEASUREMENTS

The intake port of the firing cylinder was modified to permit the optical access needed for LIF imaging. Two ports were machined into the head, one to accept an 8mm insertion diameter borescope and the other to house a small lens (1cm diameter) and a sealed window to permit incoming laser light. Both of these ports were designed and positioned to cause minimal disturbance to the inlet air flow. Figure 1 shows the relative arrangement of the borescope, laser illumination port, fuel injector, and the intake valves. Both the laser port and the borescope center their field of view on the back of one intake valve, on the side nearer to the septum separating the two intake valves. This field of view was centered on the targeting location for the fuel injector stream associated with that valve. Figure 2 shows a white light (false color) image of the field of view through the borescope imaging system. This figure also shows the "Region of Interest" (ROI) used in subsequent analysis of the images.

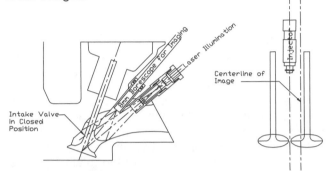

Figure 1: Optical access to Intake Port and Valve

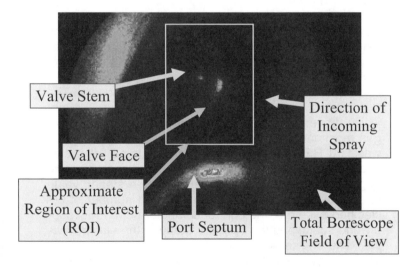

Figure 2: White light, gray scale image of overall system field of view, showing the approximate Region of Interest (ROI) used for subsequent average intensity calculations.

Data transfer rate constraints necessitated a trade off between the spatial resolution and framing rate that could be achieved with the imaging system. Since we were primarily interested in studying the transient response of the fuel film on the back of the inlet valve, we maximized the framing rate. This was accomplished by recording only a selected area of the CCD chip binning. The binning decreases spatial resolution,

improves the framing rate. Using these two techniques the final image was effectively 250 pixels by 340 pixels, and imaged an area of approximately 4 cm by 5 cm. With this arrangement, we were able to collect images on every second firing cycle of the engine while maintaining an acceptable level of spatial resolution. A white light, false color image taken at this camera resolution is given in Figure 3.

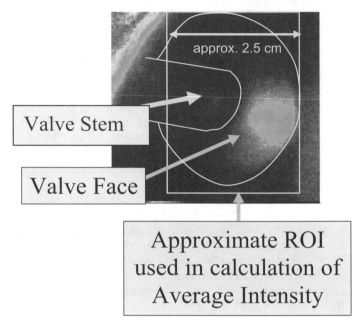

Figure 3: White light, gray scale image showing the region of the borescope image which was recorded, at the resolution used for data collection (2x2 pixel binning)

The remainder of the LIF system is shown in Figure 4. A Continuum Powerlite 8000 Nd:YAG laser with a fourth harmonic generator was used to produce 266 nm light. The fundamental and second harmonic components of the light were separated by a Pellin-Broca prism and sent to a beam dump. The 266 nm beam passes through a plane quartz beam splitter so that the power of each pulse could be measured and recorded by the DAQ system for use in subsequent image corrections. Each laser pulse's power was measured by a Molectron EM500 laser energy meter with a J-50 head and a beam diffuser. The average laser pulse energy was approximately 70 mJ. The beam then passed through a final turning prism and was directed into the laser inlet port in the head. The inlet port window system incorporated a lens to defocus the laser beam so that illumination could be supplied over the desired region of the valve. The final beam diameter is approximately 3 cm at the valve, so the UV power density of a defocused laser pulse in this experiment is about 10^6 W/cm^2.

The fluorescence collection system consisted of an 8mm borescope (AVL / Karl Storz, Hopkins type) coupled to a standard 50mm/f 1.8 camera lens and a blue (BG-12) schott glass filter. Since the borescope and the lens were both standard optical glass, they had a very low transmittance at 266 nm. Thus, they acted as an effective filter to eliminate elastically scattered 266 nm light. The blue schott glass filter removed some remaining scattered 532 nm light which was not removed by the Pellin-Broca prism. It also transmitted the desired fluorescence signal to the camera (the fluorescence signal was centered at about 430 nm).[15] The collected fluorescence light was imaged using a Princeton Instruments intensified array CCD camera. The images were stored on a separate image acquisition computer. The DAQ and control computer synchronized the laser pulse, camera control pulses, and engine position so that images could be collected at the desired crank angle during the cycle.

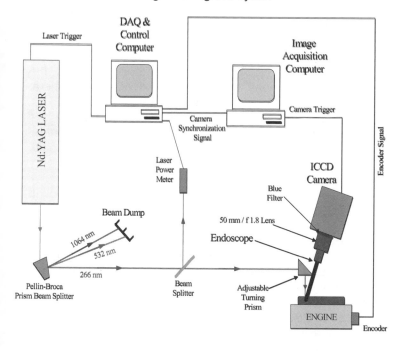

Figure 4: Schematic of LIF intake valve fuel film imaging experiment.

FUEL / TRACER SELECTIONS

For this initial study we tested the experiment's ability to detect differences in both fuel film behavior and the engine's cold start performance with different volatility fuels. The fuel mixtures used were chosen to simulate simple one or two component fuels. Iso-Octane was chosen as the Lower Volatility (LV) fuel component and 2-3 Dimethylbutane was chosen as the Higher Volatility (HV) component.

The LIF technique required the use of a tracer, which would absorb and fluoresce strongly at desirable wavelengths. Various ketones have been successfully used in many previous LIF experiments [15, 16, 17, 27] and their spectral characteristics are well documented. Ketones absorb well at 266nm, and emit strongly with a peak fluorescence at about 430nm.[15] Furthermore, they are relatively insensitive to temperature and pressure variations, oxygen quenching, and dissociation [16]. Because the temperature range in this application is relatively small (maximum range of approximately 50° no attempt is made to correct the fluorescence for tem-

perature dependence. Data for acetone excited at 266 nm from reference 28 indicates that the maximum error due to neglecting temperature effects will be approximately 10-15%.

Because the laser beam was defocused in this experiment to illuminate a larger area, the power density of the illumination was lower than is typically used in LIF or PLIF measurements. Therefore, higher concentrations of the tracer species were needed to achieve acceptable signal strength. Our fuels were a 50:50 mixture by volume of a fuel component and a tracer component. Values of optical properties for acetone published in reference 27 (typical absorption cross section of 4.7×10^{-20} cm^2 and fluorescence lifetime of 4 ns) indicate that the power density of 1.4×10^{6} W/cm^2 used is 3 orders of magnitude below the estimated value required to achieve saturated fluorescence ($\approx 4 \times 10^{9}$ W/cm^2), so the experiment is operated in the linear fluorescence regime. These same values indicate that the representative "optical path length" of our test fuels (in the liquid phase at STP) is on the order of 50 μm. Thus fuel films of several tens of microns of thickness will not be "optically thin". Implications of this fact will be discussed in the results section.

The high tracer concentration required for this work made it especially critical to match the tracer component behavior to the fuel component behavior. It is possible to match the vaporization behavior of a fuel component by choosing a tracer with a similar boiling point. [15,16]. 3-Pentanone was selected as the Lower Volatility Tracer (LVT) and Acetone was selected as the Higher Volatility Tracer (HVT). Additionally, according to correlations in references 14 and 24 the vapor pressures of the tracers remain within 10% of the vapor pressure of their respective fuel components for the temperature range 20°C to 75°C. For this study, it was also important to match the viscosity of the tracer to the fuel component, since viscosity affects the dynamics of the fuel film being stripped off of the valve and into the cylinder during the intake event. Table 2 shows the boiling point, heat of vaporization (h_{fg}), Viscosity (μ), and the binary diffusion coefficient (D_{AB}). These are expected to be among the most important thermo-physical properties of the fuel and tracer components used, with regard to the transport and mixing processes under investigation. Within each volatility category, the tracer was well matched to the fuel component with regard to these parameters. Therefore, the tracers should accurately mimic the behavior of the desired fuel component.

Using these four components it was possible to simulate three different fuels: an HV fuel composed of a 50:50 mix of 2-3dimethylbutane (fuel) and acetone (tracer), an LV fuel composed of iso-octane (fuel) and 3-pentanone (tracer), and a two component (2C) fuel which was a 50:50 mix of an HV and an LV category component. The 2C fuel can be achieved in two ways, either by tracing the high volatility component (50% LV: Iso-octane, 50% HV-T: acetone) or by tracing the lower volatility component (50% HV: 2-3dimethylbutane, 50% LV-T: 3-pentanone). Since the fuel components and tracers are very similar with regard to their thermophysical properties, both of these 2C mixtures should behave similarly in terms of engine performance, and in that sense represent a single 2C fuel. However, by using the two different mixtures we gained the ability to independently trace either the HV or the LV component. The performance similarity of the two 2C fuel mixtures is presented in the results section.

Fuel Component	Boiling Point (C)	h_{fg} @ 30° C (kJ/kgK)	μ @ 30° C (cent-poise)	D_{AB} @ 30° C (cm^2/s)
Iso-Octane (LV)	99.25	34	$4.4*10^{-1}$	$6.5*10^{-2}$
3-Pentanone (LV-T)	102.5	38	$4.2*10^{-1}$	$8.5*10^{-2}$
2-3 dimethylbu-tane (HV)	57.95	29	$3.0*10^{-1}$	$7.8*10^{-2}$
Acetone (HV-T)	56.15	32	$3.0*10^{-1}$	$1.1*10^{-1}$

Table 2: Thermo-Physical properties of fuel components and tracers used in this study. Values from: [14, 25]

Although the tracers and the fuel components were well matched in their thermo-physical properties, they were chemically and structurally different. Therefore, a slightly different quantity of each fuel was injected (for a fixed loading condition with a given amount of air) to maintain an equivalence ratio of 1. Using a fixed injection pressure and start of injection, each fuel required a slightly different injection duration, and hence had a slightly different EOI timing. Before a fuel was run in the engine, the fuel injection duration was manually calibrated to ensure that the proper (stoichiometric) amount of each fuel was being injected. This was accomplished by measuring the average volume of fuel injected during 500 injections into a graduated cylinder. The resulting injection duration and EOI timing for each fuel mixture used is shown in Table 3. The slight change in EOI timing was not expected to have significant effects on the engine performance, but it may be important for the interpretation of the imaging results.

Fuel Name	Fuel Compo-nents (tracer / other)	Vol. per Injec-tion (ml)	Injection Dura-tion, Δt (ms)	Approximate E.O.I. (CAD) EOI = 470 +5.33*Δt
LV	3-Pentanone/ Iso-Octane	0.0156	4.73	495.2
2CF-LVT	3-Pentanone/ 2-3 Dimethylbu-tane	0.0160	4.83	495.75
2CF-HVT	Acetone/ Iso-Octane	0.0171	5.11	497.25
HV	Acetone/ 2-3 Dimethylbu-tane	0.0174	5.40	498.8

Table 3: Fuel mixture injection properties [14, 24]

162

DATA COLLECTION & PROCESSING

Imaging and engine performance tests were conducted on each of the four fuel mixtures, for the ignition and SOI timing listed in Table 1. Images were collected at one selected crank angle during each cold start run. Imaging for each fuel at each selected crank angle was repeated for 3 separate cold start runs to ensure repeatability of the imaging results. The first image was collected on the cycle of the first fuel injection and spark. 98 images were collected during each 196 firing cycle cold start test.

Several crank angle intervals were excluded from imaging for the following reasons:

1. The camera was focused on the back of the intake valve in its closed position, therefore when the intake valve was open (717-227 CAD) the camera was out of focus and imaging was not possible.
2. The DAQ and control computer use the interval between 580 and 720 CAD for computation of control signals and data transfer, hence the camera could not be triggered during this interval.
3. Initial experimentation showed that the (out of focus) fuel spray blocks the image of the valve surface over the crank angle interval 470 to 540 CAD.

Given these constraints, images were first collected at crank angles of 540°, 560°, and 580° crank angle degrees; these images span the interval between when the out of focus fuel spray first clears the field of view until the DAQ blackout period begins. Images were then collected at 250°, 340°, 430°, 450° and 470°; these images span the interval from just after the end of the intake event, when the intake valve is re-closed, until just before the beginning of the next cycle's injection. The regions of the engine cycle excluded from imaging, and the crank angles at which imaging was conducted are shown in Figure 5.

Performance data such as Gross Indicated Mean Effective Pressure (GIMEP), Coefficient of Variation of GIMEP (COV), and total engine out Unburned Hydrocarbons (UHCs) were recorded by the DAQ computer simultaneously with the imaging during each cold start experiment. Performance results were obtained over several days to check for repeatability of the fuel injector calibration and cold start test procedures. No significant day to day variation was found in any of the performance parameters investigated. Between runs with different fuel mixtures the entire injection system was flushed first with iso-octane and then with nitrogen to ensure no residual fuel or tracer remained to affect the next test.

The raw images were processed and corrected using a procedure modified from Lee et al. [15]. In the current procedure images had pixel dark current and background noise subtracted from them, and were normalized with respect to laser pulse power, camera gain. Finally, an average intensity over the ROI was calculated and used to make quantitative comparisons.

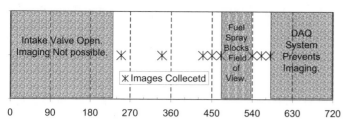

Figure 5: Crank angle positions at which images were collected.

In addition to the desired LIF signal, all of the images also contain camera "dark counts" and background noise (from room light, and imperfectly filtered 532nm and 1064nm laser light scattered from port surfaces). To correct for these effects, first, all of the images have camera dark counts (counts detected by the camera with the lens cap on) subtracted from them.

To correct for background noise, 98 background images of the valve were collected immediately before each engine run. To ensure that the port and valve were free of any fuel residue during the collection of the background images, the engine was motored with no spark or injection for several seconds immediately before background image collection. The engine was then hand cranked to a position where the inlet valves were closed and the set of 98 background images were collected with the engine stationary. These background images were corrected for camera dark counts and normalized for laser power, and then averaged. The corrected, normalized, and averaged image thus produced was subtracted from each data image.

To normalize for laser power fluctuations the intensity of each image was scaled based on the power of the laser pulse used to produce the image. The scaling is done with respect to an arbitrary "standard" laser pulse of 80 mJ, and the resultant normalized image is scaled in arbitrary units.

Typically in spatially resolved LIF measurements (PLIF for example) an additional "uniform field" correction is applied to the images. This correction assumes that an image of a "uniform field" (i.e. an image of a condition where a uniform quantity of tracer is coating all surfaces in the field of view) is available. Due to the geometric and access restrictions associated with imaging the back of the intake valve, we have not obtained such an image for this experiment. A "uniform field" correction would be expected to adjust the image for 1) spatial non-uniformities of the laser power profile over the field of view, 2) camera pixel sensitivity non-uniformities over the imaging chip, and 3) geometric effects caused by the film's presence on the complex surfaces of the intake port (i.e. reflections, highlights, and lensing caused by the curved film surface). None of these effects have been corrected in the data. However, these effects remain essentially constant over each of the conditions examined. The laser profile and the transmitting optics do not change from shot to shot, the sensitivity of each camera pixel does not vary with time, and the geometry of the intake valve does not vary with time. Any errors remaining should be systematic, and can be ignored for the purposes of relative comparisons.

After the above corrections were completed, an average intensity over the selected ROI was calculated. The region of interest was selected to cover the portion of the back of the intake valve which was most directly viewed by the borescope, as shown in Figure 3. Since the LIF signal is proportional to the number density of the tracer species (assuming an optically thin film as discussed below), the average ROI intensity was taken to be an indicator of the total quantity of fuel film on the intake valve as in reference 17. For the processing techniques used the value of the average intensity over the ROI (indicating fuel quantity, with the above qualifications) could change in one of several ways: the fuel film could be present over a greater or lesser region of the ROI, or the intensity of the LIF signal could be greater or lesser at any point in the ROI due to either a higher laser intensity at a given location in the region, or to a thicker fuel film in a region. We have not yet quantitatively calibrated the average ROI intensity to the total mass of fuel in the ROI.

Since the optical path length of the high tracer concentration fuels used is of the order of 50 μm this assumption of an optically thin film could be problematic. For fuel films much thicker than the optical path length the laser will not fully penetrate the film, and the technique will detect only the top layer of the fuel film. This in turn would cause the ROI intensity to reach a peak value, and then remain constant at that value until the (thick) fuel film vaporizes sufficiently so that it becomes optically thin at which time the ROI signal will be able to fall again. If this behavior is observed in the data then the assumption of an optically thin film will have to be reexamined.

A final adjustment was made to the intensity data to facilitate comparison between images using 3-pentanone versus acetone as a tracer. The fluorescence spectrum, duration, and the response of the imaging system to the fluorescence were all expected to differ between tracers. It was also important to rule out effects on intensity due to the mixing of the tracer with the other (non-fluorescing) component. Both of these factors were examined by imaging the fluorescence from each fuel in a 1cm x 1cm x 5cm cuvette at room temperature and pressure and were all collected with the same collection optics used for data collection. These images were corrected for dark counts, camera gain, background noise, and laser power as discussed above. The average intensity of the fluorescence over a ROI corresponding to the illuminated area of the cuvette was calculated for each fuel. The results are given in Table 4. The results indicate that the tracer fluorescence is insensitive to the choice of non – fluorescing fuel component.

Fuel (tracer / other)	Avg ROI Intensity	Avg. for Tracer	3-Pentanone /Acetone Normalization Ratio
3-Pentanone / Iso-Octane	1.672	1.687	
3-Pentanone / 2-3 Dimethylbutane	1.703		1.447
Acetone / Iso-Octane	1.145	1.166	
Acetone / 2-3 Dimethylbutane	1.187		

Table 4: Measured average intensities over the illuminated area for liquid fuel in a cuvette.

The measurements from each tracer differ by less that 4% when the non-fluorescing component is changed. However, the LVT (3-pentanone) yields stronger measured intensities than the HVT (acetone). To facilitate comparison all of the acetone image intensities have been multiplied by 1.447. Thus, all results and comparisons are presented on a "3-pentanone basis."

RESULTS AND DISCUSSION

FUEL PERFORMANCE

The engine performance results (as characterized by GIMEP, COV of GIMEP, total engine out UHCs and intake valve temperature) for the four different fuel mixtures are shown in Figures 6-9. Each plot shows the cycle by cycle averaged results from 5 typical engine cold starts. The error bars represent the standard deviation of the results of the five runs for that engine cycle.

Figure 6 shows the GIMEP results for each fuel. The LV fuel shows the slowest starting behavior. The GIMEP data shows that on average the LV fuel does not start to fire at all until cycle 6. After cycle 6 the average GIMEP for the LV fuel increases over the next 5 cycles until it reaches its "steady" GIMEP value at cycle 11. The HV fuel on the other hand shows a very rapid start. On average the HV fuel experiences only 3 non firing cycles, followed by one partial burn cycle before it achieves its "steady" value of GIMEP in the 5th cycle. The behavior of both 2C fuel mixtures lies between these two extremes, typically beginning to fire by cycle 4, and firing at nearly a steady GIMEP by cycle 10. All of the fuels tested reach a pseudo-steady state within the first 40 cycles. A temperature transient is still present, as indicated in Figure 9, but the other measured performance parameters have achieved nearly steady values. The steady value of the GIMEP is similar for all fuels at approximately 31-34 psi. The 2-C fuels show a slightly lower value than this, of 31-32 psi. However, both 2C fuels behave similarly in this regard. The most marked difference in this pseudo steady regime is the cycle to cycle variation in GIMEP, discussed below.

Figure 7 shows the COV of the GIMEP which is

GIMEP data. In this figure, the difference in the number of cycles required to "start" the engine and the pseudo-steady state "runability" (or roughness) of each of the mixtures becomes more apparent. If we set an arbitrary threshold of 10% COV as the indication of when the engine has started and is running smoothly, we see that the LV fuel requires approximately 28 cycles to achieve this threshold, while the HV fuel requires only 16 cycles. The 2C-LVT achieves sub 10% COV in approximately 24 cycles, while the 2C-HVT does not initially dip below 10% until approximately cycle 40. The 2C-HVT requires more than 55 cycles before it drops below 10% COV permanently. This is the most marked difference between the behavior of the two 2C fuel mixtures. It seems that this long delay before the COV of the 2CF-HVT mix drops below 10% is a function of the poorer steady state "runability," as indicated by its higher "steady" COV. All of the other mixtures achieve pseudo steady COVs which are below 5% rather rapidly, but the 2CF-HVT mixture does not repeatably achieve a sub 5% COV until after cycle 170. The underlying cause of the rough running of the 2CF-HVT is not apparent, but it does not seem to affect the similarity between the two 2C fuel mixtures in other regards, such as total engine UHC output.

Figure 8 shows the total engine UHC output versus cycle number for the four mixtures. The UHC measurement is not cycle resolved because of mixing which takes place in the exhaust pipe prior to sampling. There is also an inherent delay in the UHC results associated with the transit time of the gas sample to the FID. Thus the UHC readings are offset and broadened by 10-20 cycles from the cycles which contribute to the reading. Experience with this system has shown that the delay and the mixing are repeatable from run to run. Thus the results can be meaningfully compared between cases, keeping the above caveats in mind.

The HV fuel has the lowest peak of only 14 Parts Per Thousand (PPT) due to its rapid start characteristics. The HV fuel also has significantly lower pseudo-steady UHC output than the other fuels. The LV fuel produces the largest initial UHC peak of about 48 PPT, associated with its large number of misfire and partial burn cycles. It also shows the highest pseudo-steady UHC output, achieving a minimum value of only about 22 PPT by the final cycles. The two 2C fuel mixtures show very similar behavior to one another with regard to their pseudo steady UCH outputs, both reaching a value of approximately 10 PPT by the end of the test. The peak UHC outputs of the 2C fuel mixtures differ somewhat (25 PPT for the HVT case and 32 PPT for the LVT case) but the error bars indicate that these differences are not statistically significant. Further, they are both well bracketed by the LV and HV results. Thus we can say that the two 2C fuel mixtures behave similarly.

Figure 9 shows the temperature response of the back of the inlet valve for each of the fuel mixtures. This is a parameter of special importance to this study, since it will govern the vaporization response of the fuel film. These curves are characterized by an initial transient reduction in the valve temperature, caused by evaporative cooling of the inlet valve, followed by a steadily

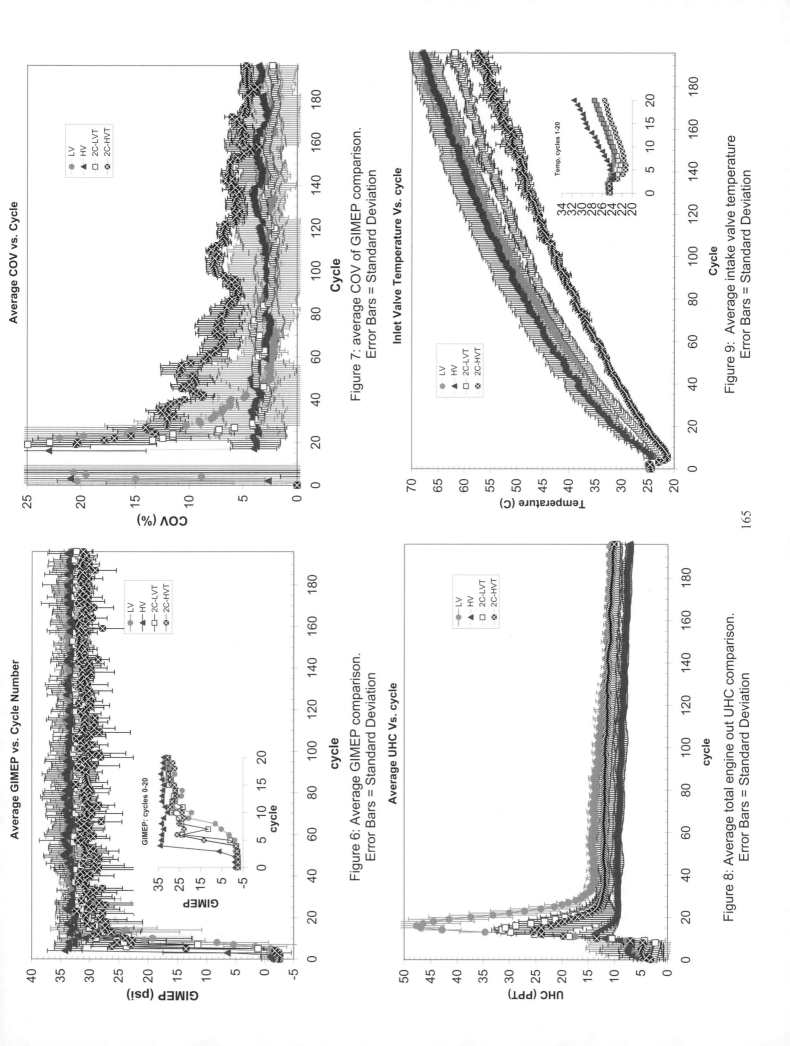

Average GIMEP vs. Cycle Number

Figure 6: Average GIMEP comparison.
Error Bars = Standard Deviation

Average COV vs. Cycle

Figure 7: average COV of GIMEP comparison.
Error Bars = Standard Deviation

Average UHC Vs. cycle

Figure 8: Average total engine out UHC comparison.
Error Bars = Standard Deviation

Inlet Valve Temperature Vs. cycle

Figure 9: Average intake valve temperature
Error Bars = Standard Deviation

165

increasing temperature. However, the valve does not reach steady state temperature during the cold start run.

The temperature results for the two "single component" fuels have very similar responses after the initial transient. Both of these fuels achieve a final valve temperature of about 68° C by the last cycle. Both of the 2C fuels also behave similarly to each other, especially in their initial transient temperature response. However, the final temperature of both 2C fuel mixtures is significantly lower (57°C for the HVT and 62°C for the LVT) than the single component fuels. The HV fuel had the shortest and shallowest initial valve temperature decrease, while the LV component had a longer initial temperature decrease transient. This is consistent with the longer lifetime of the LV fuel film, as discussed below, and with its longer delay before initial firing. The 2C fuel initial temperature transients are both steeper than for either 1C fuel, and the duration of the transient is bracketed by the LV and the HV fuels, as expected.

In general, these performance results show clear differences between the cold start behavior of the 3 fuels used, with increasing average volatility leading to improved cold start performance. Both mixtures used to model the 2C fuel showed similar behavior to each other and their behavior was bracketed by the LV and HV fuels. The one exception to this general trend was the high COV (poor runability) associated with the 2C-HVT fuel. The reason for this exception is not fully understood at this time. However, we can conclude that these three (LV, HV 2C) fuels caused distinguishable differences in cold start characteristics which are attributable to their differences in volatility, and hence form a useful initial test set for the LIF experiment. .

IMAGING RESULTS

Figure 10 shows a typical sequence of corrected, normalized, false color LIF images produced during one cold start run. Every 5th image (every 10th cycle) is shown to conserve space. These particular images are from an experiment using the LV fuel, and an SOI timing of 270°. The images were collected at a constant crank angle of 270° which corresponds to a time just before the fuel spray enters the field of view. Thus the images show the remnants of the fuel film left over after the prior engine cycle induction. The first image in the sequence (first cycle of injection) shows no fluorescence since no fuel has yet reached the valve. The nearly uniform blackness (low count) of this image verifies that the background subtraction and filtering are successfully removing almost all light which does not come directly from LIF.

The next two images (cycles 10 and 20 respectively) show that there is a large fuel film present on the back of the inlet valve. If it is assumed that the integrated intensity of the LIF signal over a region of interest is proportional to the amount of fuel in the film (as discussed previously) then, these images provide an indication that the amount of fuel in this film is growing at least until cycle 20. Subsequent images show that the signal from the fuel film slowly decreases, both in

The fuel film generally retreats from the center towards the edge of the valve. The ability to detect the geometry and extent of coverage of the film is an important capability of this technique. For example, the fact that the film slowly and consistently retreats from the center of the valve to the crevice between the head and valve (a region where it seems likely that liquid would build up) indicates that it is in fact a liquid film that we are observing rather than an out of focus spray, or a cloud of vapor. This ability is important because the LIF technique makes no distinction between fluorescence from the liquid and the vapor phases. By cycle 150 the image shows almost no signal. There is a weak and diffuse residual fluorescence in these later images. This weak diffuse fluorescence may be due to vaporized fuel in the port rather than a film on the valve.

Another way to examine the transient behavior of the fuel film during the cold start test is to compare the average intensity over a fixed region of interest in the image. As discussed previously, the average intensity over this ROI was taken as an indication of the total quantity of fuel in the film. However, since we have not yet calibrated this system, the results of the averaging are in arbitrary units (intensity counts). This fact however does not preclude making quantitative *comparisons* between various cases. Figures 11a-h show plots of average ROI intensity versus cycle number for images collected at various crank angles after start of injection SOI = 470.° (For all of the data shown in Figure 11, recall that the EOI timing varies slightly with fuel type as shown in Table 3). Each data point consists of the results from three engine cold start tests averaged on a cycle by cycle basis. The error bars shown are the standard deviations of the data.

Figures 11a-c show results from imaging at crank angles of 540°, 560°, and 580°, respectively. These images correspond to times in the engine cycle just after the fuel spray has cleared the field of view, but before the intake valve opens (IVO). Figure 11a provides the earliest look at the fuel film, just after the spray impacts the valve. The key feature of this plot is that the average ROI intensities for all of the fuels tested have very similar and relatively high values compared to intensities recorded at later times in the engine cycle. There is a general trend over the first 5-15 cycles, which indicates that the fuel film gradually builds during these initial cycles. The ROI intensity then levels off for the approximate period of cycles 15-40, and begins to slowly decline from cycles 40 – 196. This trend is strongest for the LVT fuels.

Figures 11b and 11c show the same general trend of increasing fuel film for the first 5-20 cycles, followed by a slow decline as the port warms up. In Figure 11b, the 2C-HVT has significantly lower intensities than the other fuels, perhaps caused by preferential vaporization of the HVT. In Figure 11c, slightly later in the engine cycle (at 580°) the two cases with LV tracer are beginning to show behavior which is distinct from the two cases with HV tracer. Both LV tracer cases have a longer period of fuel film build up (at least of the tracer species), and show higher maximum ROI intensities.

This result indicates that the HVT is already being preferentially vaporized prior to imaging, even at this relatively short time after spray impingement.

The next plot in the series, Figure 11d, shows data collected at 250° *after* an induction event, the intake valve has opened and closed again before the image is taken. For early cycles in the cold start there is a marked difference in observed behavior between the two fuels using the LVT and the two with the HVT. This is a strong indication that the HV tracer is being preferentially removed early in the cold start cycle, while the LV tracer is building up in the fuel film. This effect seems to be largely independent of which non fluorescing component is used. Since the viscosity of the components used do not differ significantly, the most likely cause of this preferential removal is preferential vaporization of the HV components, rather than differences in stripping of the fuel film. The HVT in both the HV and 2C fuels, and the LVT in both the LV and 2C fuels, behave similarly in terms of vaporization. In later cycles, as the engine warms up, the difference in behavior between the HV and the LV components slowly disappears which is consistent with the converging engine performance observed for all of the fuels at high cycle numbers.

Figures 11e-h show the behavior for increasing times after the end of induction (after IVC) until just before the next cycle's injection occurs. The same general trend as in Figure 11d is observed in each of these later figures. For higher temperatures (higher cycle numbers), the LVT in the 2C fuel vaporizes slightly more quickly than it does in the LV fuel. This is indicated by the more rapid drop of the 2C-LVT curve. The HVT in the 2C fuel also vaporizes slightly more slowly than the HVT in the HV fuel again, as indicated by the fact that the HV curve generally lies below the 2C-HVT curve in these figures. However, caution is needed when attempting to draw conclusions from these last relatively weak trends in the data. The correction and normalization procedure used are not perfect. This can be seen clearly for the 2C-LVT fuel; the data in Figures 11e and 11g, are "overcorrected" on average by the background subtraction. At high cycle numbers for these cases the average ROI intensity becomes negative, which is clearly physically meaningless. The reason for the overcorrection of these two cases is unknown, but it could be as simple as a slight leak from the fuel injector, or a stray "runt" injection pulse being sent to the injector during the collection of background images for these cases. Improved statistics from larger data sets may enable the technique to distinguish these "finer" differences between the behavior of the LVT in the 2CF fuel and in the LV fuel.

Another important implication of the data in figure 11 is that the fuel film being imaged does not seem to be optically thick, at least not over an appreciable fraction of the ROI. As discussed previously, if the film were optically thick we would expect the integrated intensity over the ROI to "saturate". That is we would expect the integrated ROI intensity to reach a constant and large value and remain there until the film vaporizes sufficiently to again become optically thin, when the value of the signal would again begin to decrease. The data

Figure 11 does not indicate this behavior which seems to vindicate our assumption of an optically thin film.

If we assume that the fluid film on the valve is very thin, and that it quickly comes into thermal equilibrium with the valve surface, then we can use the inlet valve temperature data to calculate the vapor pressure of the tracer species. Since it is this vapor pressure that provides the impetus for vaporization (under the assumption that the film is continuously in thermal equilibrium with the valve) this provides a somewhat more physical (but still greatly oversimplified) means of examining the data trends. Since this study is primarily a "proof of concept" test for this imaging technique, and because the technique has not yet been calibrated for quantitative measurements, a more complex model of the vaporization process has not yet been undertaken.

Figures 12a-h show the same data as Figure 11, however the average ROI intensity was plotted against the calculated vapor pressure of the tracer species. The vapor pressures were calculated based on correlations from reference 14 based on the tracer species only, with no adjustments made for the presence of the other component in the fuel. The range of absolute pressure measured in the intake port during the cycle (from separate experiments with this engine) is also shown on these plots. Of course, at any temperatures higher than this boiling point the assumption that the fuel film is in thermal equilibrium with the valve is violated, and heat transfer rates from the valve to the film will govern the rate of vaporization.

The data taken from crank angles after the intake valve has opened (12d-h) indicate that higher vapor pressures (increased rates of vaporization) correlate with lower measured intensities, as would be expected. In fact, this simple parameter alone, the vapor pressure of the tracer species, is a fairly good predictor of the fluorescence intensity recorded by the system regardless of the fuel used. This fact in turn indicates that under these circumstances the vapor pressure is the primary parameter governing the fuel film removal. The data from the two single component fuels seem to form a single continuous curve when intensity is plotted versus vapor pressure in this manner. The ROI intensity for 2C-LVT fuel decreases more rapidly with respect to tracer species vapor pressure than the LV fuel, indicating that the presence of the HV component is enhancing the vaporization of the LV tracer. Likewise, The 2C-HVT fuel ROI intensity decreases more slowly than for the HV fuel, indicating that the LV component is retarding the vaporization of the HVT. Of course, the data from the initial few cycles, while the film is building up on the valve, do not follow these trends because no account is taken of fuel carried over from previous cycles. As an aside, the correlation of the fuel film behavior with vapor pressure implies that it might be possible to study the nature of a cold start film in a (simpler) steady state, warm engine experiment by matching a research fuel's vapor pressure at the steady valve temperature to the vapor pressure of a more typical fuel at a given point during a warm up transient.

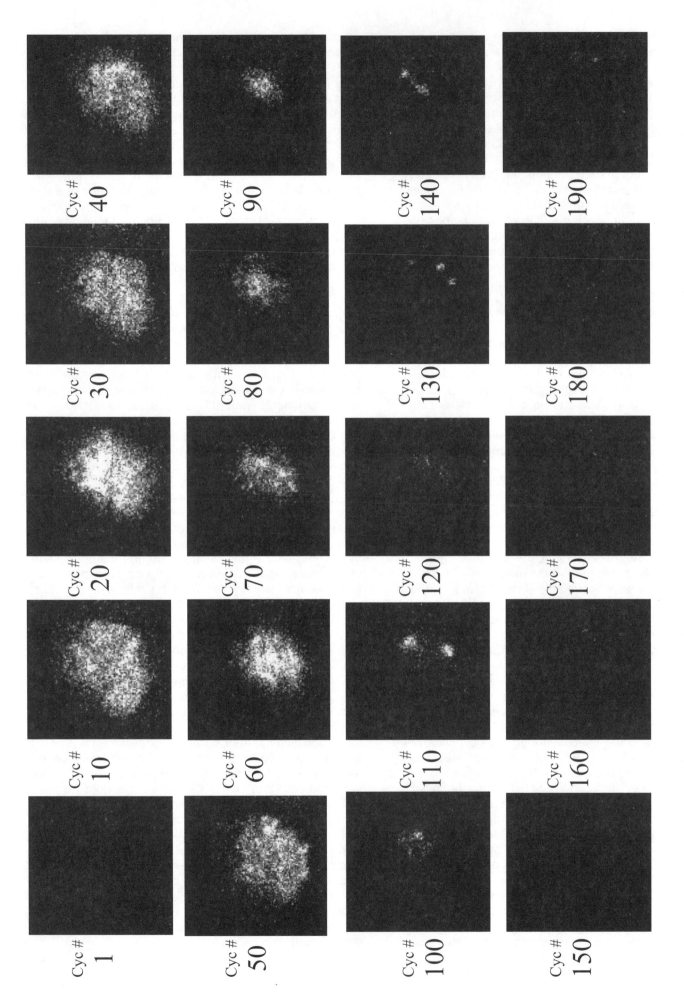

Figure 10: LIF images of the intake valve fuel film for various cycles after cold start.

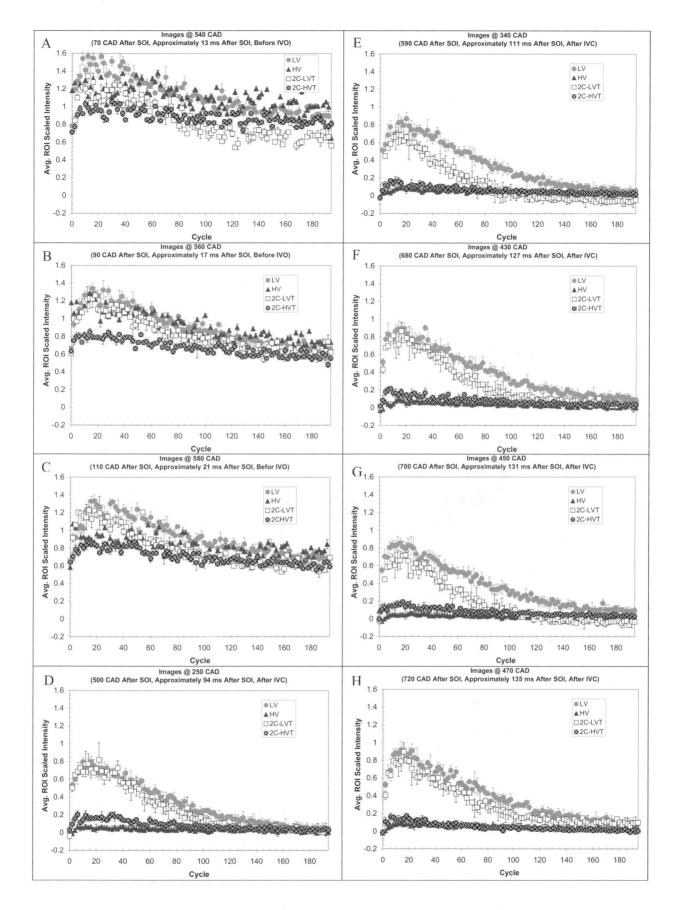

Figure 11: Average ROI intensity versus cycle number at fixed crank angles after SOI (SOI = 470° ATDC Intake). Plots A-C show results for images collected at timings after the fuel spray has impacted on the valve (A), until just before the intake valve opens (C). Plots D-H show results for images collected starting after the intake valve has just closed (D) Error Bars = Standard Deviation.

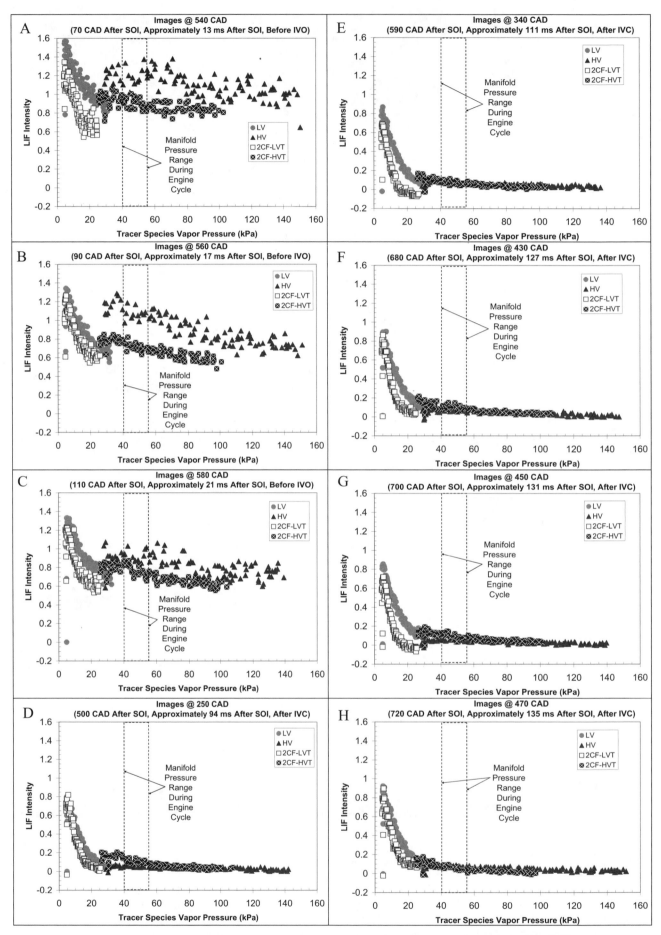

pressure at various crank angles after SOI.

170

The vapor pressure is not as good of a predictor of measured LIF intensity for images collected at early crank angles just after the film is deposited but before the intake valve is open (Figures 12a-c). This implies two possibilities: 1) the vapor pressure predicted is wrong for these cases, which in turn implies that T_{film} is not well approximated by T_{valve} for these cases. In other words, the film is not yet in thermal equilibrium with the valve. Or 2) the predicted vapor pressure is correct (the film is in thermal equilibrium with the valve), but insufficient time has elapsed for any significant vaporization to occur at the time of image collection. In either case, with the passage of time (i.e. examination of images at later crank angles) the data will become better correlated with the vapor pressure. This is exactly what is observed as we proceed from figure 12a to 12c.

Arranging the same data in yet another way, we can more easily examine changes in the fuel film which occur within an engine cycle. By combining data from several cold start runs, each with images collected at a different crank angle, it is possible to examine changes in the fuel film over crank angle space for a fixed cycle number (or range of cycle numbers) during the cold start transient. Figure 13 shows data for the average values of the peak intensities achieved by each fuel versus crank angle. The average "peak intensity" value was calculated by averaging the results of engine cycles from 10 to 30, which roughly corresponds to the region of peak ROI intensity (see figure 11) during the cold start run. Averaging over several cycles rather than selecting one arbitrary cycle has the effect of "smoothing" the variations in the intensity versus cycle plots so that more meaningful comparisons between different crank angles are obtained. The "smoothing" is not perfect however, and run to run variations at a fixed engine condition, imaging angle, and cycle number are significant. Therefore, differences between data points in Figure 13 should not be considered statistically different unless their error bars (shown as standard deviation for the 3 x 20 = 60 cycles composing each point) are well clear of

each other. The greatest (and in each case statistically significant) change in ROI intensity for each fuel occurs across the opening of the intake valve. The LV fuel has the largest ROI intensity remaining (largest quantity of fuel remaining) after the induction event, followed by the LV component of the 2C fuel, the HV component of the 2C fuel and then the HV fuel, as would be expected. Other than this change in level across the induction event, most of the changes in fuel film quantity cannot be viewed as statistically significant. There may be some possibly statistically significant decrease in fuel film quantity from just after the fuel spray clears the field (540°) of view until the last image before the valve opens (580°) for all of the fuels except for the 2C-LVT fuel. However, a larger data set is required to confirm this. After IVC, the fuel film does not change appreciably (in a statistical sense) for the remainder of the engine cycle.

Figure 14 shows the percentage of the deposited fuel film which is removed from the valve during the induction event versus cycle number. This is calculated by taking the difference between the ROI intensity at our last image before before IVO (580°) and just after IVC (250°) and normalizing it with respect to the ROI intensity at 580°. Again, it should be noted that the measurement technique has not yet been calibrated, so we cannot say with confidence if the percent change in ROI intensity is proportional to the mass of fuel removed. However, as discussed below, the qualitative trends observed support this interpretation of the data. During roughly the initial 40 cycles the percentage drop in ROI intensity of the LV components across an induction event is 30-50% implying that an appreciable quantity (50-70%) of the LV fuel and LVT component of the 2C fuel present on the valve before induction remain on the valve after induction. For the same 40 cycles the ROI intensity for the HV fuel decreases by 90-97% across an induction event. This implies that almost all of this fuel is removed even for the initial cycles of the cold start. Also for the same period the ROI intensity decreases 75-80%

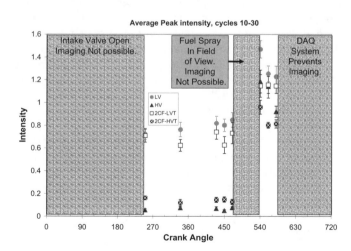

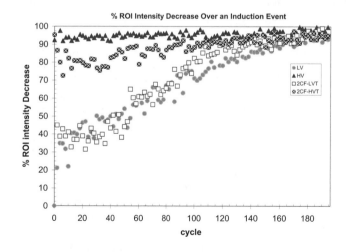

Figure 13: Average Peak ROI intensity (cycles 10-30) versus crank angle. Error Bars = Standard Deviation

Figure 14: Average percentage decrease in ROI intensity over an induction event versus cycle number

for the HVT in the 2C fuel, implying that only 20-25% of the HVT remains on the valve after induction during the initial period of the cold start. This finding lends additional support to models of the intake port fuel film [4,24] which predict that the port fuel film becomes significantly enriched in low volatility species during warm up. As the engine approaches its fully warm condition (cycles 180-196) nearly all of the film (90-99%) is removed during induction. Data of this type could be very helpful in validating and calibrating computational models for port fuel film behavior (e.g. the FPM [24]).

Presumably, the relatively large quantity of LV fuel remaining in the film after induction during the initial cycles implies large effective metering errors for this fuel. This agrees with the poor cold start performance (large number of cycles needed to achieve full GIMEP, and high peak UHC emissions) observed for this fuel. Although the LV component in the 2C fuel is shown to remain after induction, the HV component is mostly removed during induction, even for early cycles. Therefore, it is not necessary to vaporize as much of the LV component with the 2CF fuel in order to achieve a flammable mixture in the cylinder, and so the 2C fuel shows significantly better cold start performance than the "pure" LV fuel. The HV fuel is almost all inducted from the very first cycle, which implies very small metering errors on the initial cycles, and a flammable mixture in the cylinder right away, in agreement with the rapid cold starts observed for this fuel. Near the end of the cold start transient, the difference between fuel film behavior of the fuels nearly vanishes, which is in good agreement with the fact that all of the fuels show similar performance results for later cycle numbers.

CONCLUSION

The experiment discussed in this paper has demonstrated the ability to produce information about the fuel film which forms in a critically important region of the intake port, the back of the intake valve. The information produced is temporally and spatially resolved, and component specific. Further, the experiment has been shown to be able to relate observed differences in the behavior of the fuel film from different fuels to the cold start performance changes brought about by these fuels. The detailed nature of the information provided by this experiment makes it well suited for validation and calibration of computational models. Although only a preliminary, and uncalibrated, data set has been collected to date several conclusions can be drawn from the data.

Observations of the film before an induction event show that initially similar amounts of each of the four test fuels studied are deposited on the valve. Prior to IVO, and especially after IVC, the changes in each of these fuel films relative to each other are small. There is however some indication (possibly statistically significant) that there is preferential vaporization of the HV components prior to IVO.

The amount of each of the fuels which remains on the valve *after* an induction event varies greatly fuel volatility. The HV fuel is

even at early cycles in the cold start. On the other hand a substantially larger (statistically significant) amount of the LV fuel remains on the valve after induction for the first approximately 40 cycles during a cold start. This behavior leads to a much larger film of the LV fuel building up during the initial cycles of the cold start than with the HV fuel. The larger film associated with the LV fuel implies larger effective fuel metering errors during the initial engine cycles, which agrees well with the poorer cold start performance observed with this fuel. Contrary to this, the HV film does not build to very large values, when observed at crank angles after IVC, even very early in the cold start. Again, this agrees with the rapid cold start and low peak UHC output associated with this fuel. When the post IVC data is examined with respect to the vapor pressure of the tracer species, a clear relationship between vapor pressure and the amount of fuel film measured is found. Higher vapor pressures indicate much smaller observed fuel films. The strong relationship between vapor pressure and the amount of fuel remaining in the film after induction implies that it is differences in vaporization which govern the removal process during induction, rather than fluid dynamic differences in the stripping of the fuel from the valves.

The observations of the two 2C fuel mixes show that high volatility components are strongly preferentially vaporized in these mixtures during cold start. When the LV component of this fuel is traced, it shows behavior which is very similar to the "pure" LV fuel in terms of the trends just discussed. Likewise, when the HV component is traced, it shows behavior which is similar to the HV fuel. This conclusion is further strengthened by the result that the 2C fuel (in either mixture) shows cold start behavior which lies between the HV and the LV fuels in terms of both peak UHC output and number of cycles prior to first firing. This implies that the A/F excursions caused by the fuel film associated with this fuel lies between those caused by the HV and the LV fuels. There is some evidence that the presence of an HV component slightly enhances the vaporization of the LV component, and that the presence of an LV component slightly retards the vaporization of the HV component, however these observations are only marginally statistically significant based on the small initial data set.

As the engine warms up the differences in the amount of tracer remaining after an induction event decrease between the fuels, and eventually disappear. Likewise, the performance characteristics of these fuels converge as the cold engine warms up.

An secondary but valuable result of this work is the characterization of the test fuels used. These fuels form a well characterized and useful test set for examining volatility related performance issues in future LIF based experiments.

ACKNOWLEDGMENTS

This work was supported by DaimlerChrysler through a Challenge Fund Grant. We especially want to acknowledge the support and encouragment of Drs. Phil Keller and Fu-Quan Zhao from DaimlerChrysler.

REFERENCES

1. Zhao, Fu-Quan et al., " The Spray Characteristics of Automotive Port Fuel Injection – A Critical Review." *SAE Paper 950506*, 1995.

2. Ferguson, Colin R. and Kirkpatrick, Allan, T., "Internal Combustion Engines, Applied Thermosciences 2nd Ed." John Wiley and Sons, Inc. 2001.

3. Heywood, John B., "Internal Combustion Engine Fundamentals." McGraw- Hill Inc. 1988.

4. Skippon, Stephen M., and Norton, Daniel., "The Effects of Gasoline Volatility on Mass and Composition of the Inlet Port Wall Film in Port Injected SI Engines." *SAE Paper 982517*, 1998.

5. Almkvist, Goran et al., "Measurements of Fuel Film Thickness in the Inlet Port of an S.I. Engine by Laser Induced Fluorescence." *SAE Paper 952483*, 1995.

6. Senda, Jiro et al., "Measurement and Modeling on Wall Wetted Fuel Film Profile and Mixture Preparation in Intake Port of SI Engine." *SAE Paper 1999-01-0789*, 1999.

7. Kelly-Zion, P. L., et al. "Liquid Fuel Behavior in Port-Injected SI Engines." *Automotive Engineering International*, January 1998.

8. Cheng, Wai K., et Al., "An Overview of Hydrocarbon Emissions Mechanisms in Spark-Ignition Engines." *SAE Paper 932708*, 1993.

9. Dawson, Mark and Hochgreb, Simone., "Liquid Fuel Visualization Using Laser-Induced Fluorescence During Cold Start." *SAE Paper 982466*, 1998.

10. Oliverira, Ivan, B. and Hochgreb, Simone., "Detailed Calculation of Heating, Evaporation and Reaction Processes of a Thin Liquid Layer of Hydrocarbon Fuel." *SAE Paper 2000-01-0959*, 2000.

11. Whelan, D. E., et al., "Mixture Atomization From Intake-Port Back-Flow." *Automotive Engineering International*, February 1998.

12. Meyer, Robert., and Heywood, John B., "Liquid Fuel Transport Mechanisms into the Cylinder of a Firing Port-Injected SI Engine During Start Up." *SAE Paper 970865*, 1997.

13. Johnen, T. and Haug, M., "Spray Formation Observation and Fuel Film Development Measurements in the Intake of a Spark Ignition Engine." *SAE Paper 950511*, 1995.

14. Yaws, Carl, L., et al., "Chemical properties handbook: physical, thermodynamic, environmental, transport, safety, and health related properties for organic and inorganic chemicals." McGraw Hill, 1999.

15. Lee, Sihun et al., "A Comparison of Fuel Distribution and Combustion During Engine Cold Start for Direct and Port Fuel Injection Systems." *SAE Paper 1999-01-1490*, 1999.

16. Tong, Kun et al., "Fuel Volatility Effects on Mixture Preparation and Performance in a GDI Engine During Cold Start." *SAE Paper 2001-01-3650*, 2001.

17. Felton, P. G. et al., " LIF Visualization of Liquid Fuel in the Intake Manifold During Cold Start." *SAE Paper 952464*, 1995.

18. Comben-Bourke, Mary and Evers, Lawrence, W., "Fuel Film Dynamics in the Intake Port of a Fuel Injected Engine." *SAE Paper 940446*, 1994.

19. Evers, Lawrence W. and Jackson, Jackson, Kenneth, J., "Liquid Film Thickness Measurements by Means of Internally Reflected Light." *SAE Paper 950002*, 1995.

20. Coste, Timothy, L. and Evers, Lawrence W., "An Optical Sensor for Measuring Fuel Film Dynamics of a Port-Injected Engine." *SAE Paper 970869*, 1997.

21. Saito, Kimitaka et al., "A New Method to Analyze Fuel Behavior in a Spark Ignition Engine." *SAE Paper 950044*, 1995.

22. Takeda, Keiso et al., "Mixture Preparation and HC Emissions of a 4-Valve Engine with Port Fuel Injection During Cold Starting and Warm-up." *SAE Paper 950074*, 1995.

23. Russ, S. et al., "The Effects of Injector Targeting and Fuel Volatility on Fuel Dynamics in a PFI Engine During Engine Warm-up: Part 1 – Experimental Results." *SAE Paper 982518*, 1998.

24. Curtis, Eric et al., "The Effects of Injector Targeting and Fuel Volatility on Fuel Dynamics in a PFI Engine During Warm-Up: Part II – Modeling Results." *SAE Paper 982519*, 1998.

25. Yaws, Carl, L., et al., "Handbook of transport property data: viscosity, thermal conductivity, and diffusion coefficients of liquids and gases." McGraw Hill, 1995.

26. Aquino, C. F., "Transient A/F Control Characteristics of the 5 liter Center Fuel Injection Engine." *SAE paper 810494*, 1981.

27. Lozano, A., Yip, B., and Hanson, R. K., "Acetone: a tracer for concentration measurements in gaseous flows by planar laser-induced fluorescence." Experiments in Fluids. Vol. 13. p 369-376. 1992.

28. Thurber, M. C., et al., "Instantaneous Temperature Imaging with Single-Wavelength Acetone PLIF." AIAA proceedings of the 35th Aerospace Sciences Meeting and Exhibit. Reno, NV. Jan. 6-10, 1997.

1999-01-1107

Cycle-By-Cycle Mixture Strength and Residual-Gas Measurements During Cold Starting

Ather. A. Quader and Richard. F. Majkowski

General Motors Corp. Research and Development Center

Copyright © 1999 Society of Automotive Engineers, Inc.

ABSTRACT

To gain a better understanding of mixture requirements during starting, a diode laser based spectroscopic technique was developed to simultaneously measure the cycle-by-cycle fuel vapor-air equivalence ratio and residual gas CO_2 concentration inside the cylinder of an operating engine. Cranking to startup conditions were simulated in a single-cylinder CFR engine installed in a cold test facility. In separate tests using propane, isopentane, and gasoline as fuel it was found that combustion began in the first cycle in which the fuel vapor-air equivalence ratio exceeded the lean flammability limit of the fuel. In the range of temperatures 22°C to -12°C, richer mixtures were required to start the engine and keep it firing consistently at lower temperatures. Intake charge dilution caused by the residual burned gas left over from the combustion in a previous cycle was found to contribute to misfires in some of the succeeding cycles. Cycle-by-cycle variations occurred in mixture strength as well as residual gas CO_2 concentration in all the tests.

INTRODUCTION

To start gasoline fueled spark ignition engines at low temperatures, a very rich mixture of fuel and air has to be supplied to the engine. Fuel vaporization and the formation of a combustible fuel vapor-air mixture inside the cylinder of an engine are believed to be key elements controlling the starting of spark ignition engines. The need for rich mixtures is speculated to arise because only a small fraction of the gasoline (the low boiling point components) can exist in the vapor state at low temperatures. In a previous study (1) an equivalence ratio (ϕ) of 5.5 (liquid gasoline to air) was found to be the leanest mixture that could start the engine in 10 to 15 cycles at -29°C. Calculations (2), based on equilibrium between the fuel vapor and liquid at the temperature and pressure in the inlet manifold, showed the fuel vapor-air ϕ to be about 0.85. With the port fuel injection system used in the single-cylinder test engine, the actual ϕ in the vapor phase might be leaner than 0.85 because the residence time available is too short to reach equilibrium fuel vaporization. To study and verify some of these speculations,

direct measurements of the mixture strength inside the engine cylinder were needed.

The two main objectives of this study were:

1. To develop a technique to measure the fuel vapor-air equivalence ratio and residual gas concentration of the mixture inside the cylinder of an engine on a cycle-by-cycle basis during cold starting.

2. To apply this technique to measure the fuel vapor-air equivalence ratio during simulated cranking to startup of a single-cylinder engine.

In previous studies using a flame ionization detector (FID) (3,4), the mixture strength and residual gas concentration were measured in spark ignition engines separately, but not simultaneously. Since the residual gas dilutes the fresh fuel air mixture entering the cylinder, it could render the mixture non-combustible and cause misfires during the cold start process. In the present paper we describe the apparatus assembled to determine simultaneously, the equivalence ratio and residual gas concentration in the fuel vapor-air mixture inside the engine cylinder on a cycle-by-cycle basis, while starting the engine at temperatures of 22°C to -12°C. Results with propane (a single component gaseous fuel), isopentane (a single component liquid fuel with low boiling point), and an emission certification gasoline --(Howell EEE), are also presented. The experiments were conducted in a specially designed facility for cold starting tests in a single-cylinder engine (1).

TEST ENGINE AND ACCESSORIES

The single-cylinder split head CFR engine was part of a unique temperature controlled single-cylinder engine cold test facility (1) shown schematically in Figure 1. This facility has a very short cool-down time. The test facility utilizes separate cooling systems for engine intake air, fuel, coolant, and oil, each capable of temperatures down to -30°C. Moreover, the engine is housed inside a temperature-controlled enclosure so that the inside air temperature could be precisely controlled to any value from -30°C to 60°C.

Single-Cylinder Engine Cold Test Facility

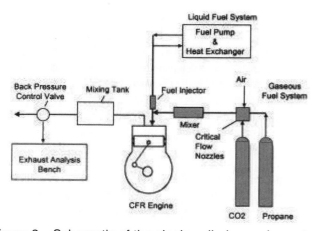

Figure 1. Schematic of single-cylinder engine cold test facility.

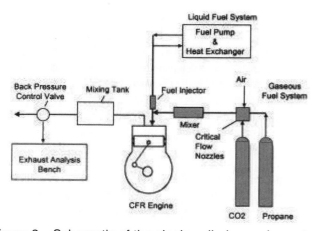

Figure 2. Schematic of the single-cylinder engine setup.

The compression ratio was set at 8:1, and a non-shrouded inlet valve was used. Engine speed was controlled with a dynamometer which was used to motor the engine and absorb the mechanical power output when the engine was firing. The engine was equipped with a programmable high-energy (PHE) ignition system (5). The spark timing was controlled by means of an engine setpoint controller (ESC) interfaced with the ignition system. The ESC used the top dead center (TDC) signal and 360 crank angle signals per revolution supplied by a photoelectric encoder. The output signal from the ESC controlled both the spark timing and spark duration.

The engine was set up to run on gaseous as well as liquid fuels as described (1). Further modifications to the induction and exhaust system used are described here. Figure 2 shows a schematic of the engine induction and exhaust systems. Propane was metered into the intake air stream approximately four feet upstream of the engine and forced through mixing vanes to ensure a uniform mixture in the combustion chamber (6). The ϕ of the mixture delivered to the engine could be controlled by adjusting the propane flowrate. Both propane and air were supplied through calibrated critical flow nozzles so that flowrates could be calculated based on pressures upstream of the nozzles. The ϕ inside the cylinder during

the actual starting test was determined by a special technique described below. This technique required CO_2 to be injected in the inlet air for calibration purposes. Pure CO_2 was injected into the intake air of the engine at the same point as the propane using a separate critical flow nozzle setup also shown in Figure 2.

Liquid fuel was supplied using a Bosch No 280-150-044 port fuel injector (7). The engine setpoint controller controlled the start of injection and the injector pulse duration. The injector was targeted so that the centerline of the fuel spray intersected the centerline of the inlet valve near the valve head. This target location of the injected fuel spray gave low smoke and exhaust hydrocarbon emissions (7). A Max Model 284 512 flow-transmitter was used to monitor fuelflow.

PERIPHERAL EQUIPMENT – The engine was instrumented using thermocouples and pressure transducers so that virtually any temperature or pressure of interest could be monitored. Emission analyzers were used to measure CO, CO_2, O_2, NO, NOx and unburned HC in the exhaust gas. Output from the various transducers was monitored using the Computer Aided Testing System (CATS). The CATS measured and updated all quantities at intervals ranging from 0.02 to 5 seconds. The system also simultaneously calculated other variables such as air and fuel flowrates, which depended on measured values. CATS was further used for data acquisition and processing.

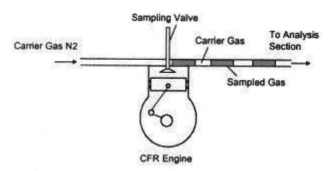

Figure 3. Schematic of the sampling system.

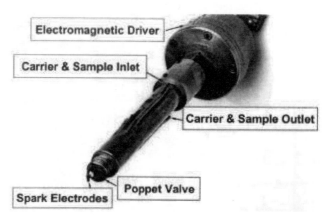

Figure 4. Sampling valve with spark plug used in this study.

Apart from the CATS, cylinder pressure was measured and displayed on an oscilloscope screen. This was particularly useful in determining the onset of combustion. Cycles in which there was no combustion had pressure traces identical to those of a motored engine. The pressure transducer was installed in a specially modified Champion D-16 spark plug used to ignite the mixture.

CYLINDER GAS SAMPLING AND ANALYSIS

The experimental technique involved extracting a small sample of the gas inside the cylinder by means of a specially designed sampling valve. The samples were extracted from the engine cylinder once from each compression stroke, just prior to spark ignition. The sampled gas from each engine cycle was inserted in a continuous stream of inert carrier gas (nitrogen) flowing through the sampling valve as shown in Figure 3. The pressure difference between the higher pressure compressed gas sampled from the cylinder and the nitrogen carrier gas caused a momentary interruption in the flow of nitrogen until the sampled gas displaced the nitrogen in the sampling tube. Thus the samples of the gas from each engine cycle were kept separated by the nitrogen carrier gas and moved to the analysis section by the flow of the nitrogen. This principle of sampling cycle-by-cycle has been used (8), and appears to be simpler than another technique (9). Nitrogen, helium, and argon were tested as carrier gases, but nitrogen was used in most cases unless noted otherwise.

A laser-based CO_2 absorption technique was used to analyze the samples. In brief this technique involved the measurement of the CO_2 content of the sampled gas twice, first as the sample emerged from the cylinder, and second after it was oxidized by passing it over a heated catalyst. From the first measurement the CO_2 concentration in the residual gas in the unburned mixture was determined. The second CO_2 measurement in the oxidized sample was used in arriving at the ϕ of each sample. Details of the various components of this sampling, analysis, and calibration system are discussed in the following sections.

SAMPLING VALVE – A modified version of the electro-magnetically actuated poppet sampling valve designed by Wentworth (10) was used. Figure 4 shows a picture of the modified sampling valve. The modifications included a spark electrode installed in the tip, and an inlet passage for the carrier gas in addition to the single outlet passage in the original design. The tip was designed to fit a 14 mm spark plug hole. The spark gap in the sampling valve was not operational during the tests described in this paper. The electromagnetic driver section was attached to the sampling section of the valve by means of a steel sleeve shown in Figure 4. This arrangement permitted the use of the same electromagnetic driver for various designs of the sampling section.

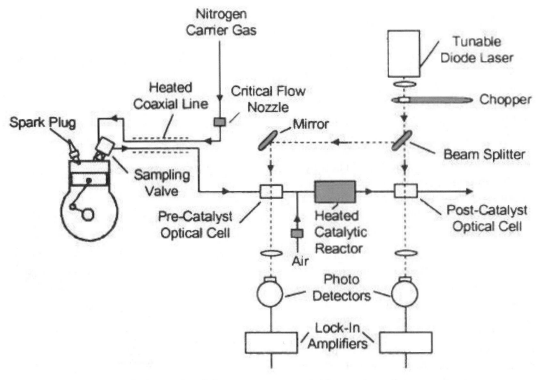

Figure 5. Schematic of the apparatus used to analyze the samples.

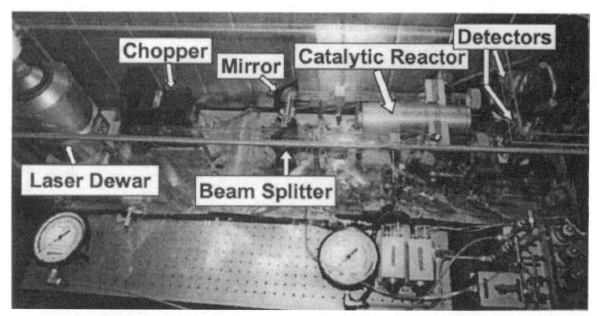

Figure 6. Photograph of optical system used to analyze the sample.

The volume of gas sampled varied with the duration the sampling valve stayed open and the pressure inside the cylinder at the time of sampling. The flow characteristics of the valve are shown in appendix A. Under the conditions tested the sample extracted per cycle was about 7.5 cc at ambient conditions, which corresponds to less than one percent of the compressed cylinder contents at top dead center. The duration of sampling was 1.5 milliseconds, which corresponds to 1.26 degrees of crankshaft rotation at the 140 rpm test speed.

The possibility of fuel droplets getting entrained in the sample and causing problems in the interpretation of the signals was of concern. To prevent fuel droplets from reaching the sampling valve, the tip of the valve was installed in a recessed hole in the cylinder head. Based on the results with gasoline reported here, it appears that fuel droplets were not entrained with the in-cylinder gas samples extracted during the first 20 cycles.

SAMPLE ANALYSIS – The schematic diagram of the sample analysis setup is shown in Figure 5. An insulated coaxial line with carrier gas nitrogen flowing through the outer annulus and the sample and carrier gas flowing out through the inner tube, was used to deliver the sample from the engine cylinder to the optical bench located outside the cold box. The stream of sample and carrier gas as received from the engine was first passed through an optical cell to measure the CO_2 level in the unburned sampled gas. Next the sample was passed over a heated catalyst to burn it. The catalyst was platinum on a ferric alloy honeycomb support and was maintained at a constant temperature of 310°C by means of an electrical heater. The burned sample was passed through a second optical cell to measure the CO_2 level again. The two optical cells were named pre-catalyst cell and post-catalyst cell respectively. The relative concentration of CO_2 in

each sample was determined from the optical attenuation of a diode laser beam transmitted through the sample and tuned to a strong absorption line (at 2223 cm-1) within the 4.4 micron vibration-rotation band of CO_2. Details of a similar technique using an infrared diode laser to measure CO concentration in the engine exhaust have been reported (11,12). The attenuation of the laser beam due to absorption by the CO_2 in the unburned sample passing through the pre-catalyst cell was used to estimate the residual gas fraction. Similarly, the attenuation of the laser beam due to absorption by the CO_2 in the burned sample passing through the post-catalyst cell was used to determine the ϕ of the sampled gas. Brief descriptions of the diode laser, and the optical setup are given below.

Tunable Diode Laser – The laser diode was a double hetero-junction type with a PbTe active layer and PbEuSeTe confinement layers, and at the desired wavelength operated typically at a heat sink temperature around 120°K and an injection current of 0.3 amps (13). The laser was housed in a liquid nitrogen temperature stabilization dewar (14). This dewar was free of the mechanical vibrations which adversely effect the optical quality of the output beam for the closed cycle helium refrigerators which are commonly used with lead salt lasers. This dewar is also substantially cheaper, quieter, and is almost maintenance free compared to the closed cycle refrigerator. During the initial setup, a half meter grating monochromator was used to provide approximate wavelength identification and to determine the conditions (heat sink temperature and injection current) for single mode operation in the 4.4 micron wavelength region of CO_2 absorption. Abrupt shifts (mode hopping) of a laser diode's output optical frequency can occur at the high current end of this operating mode. Also, competing modes can occur within a desired principal mode (multi-mode).

These types of laser operations would be very undesirable. However, the laser used in this study had an excellent single mode at the optical frequency of strong CO2 absorption. Moreover, this single mode was stable over a large current range. This produced a very stable source for the measurements reported in this study. Thus, once the desired conditions of laser operation were established, the monochromator was eliminated from the setup.

Optical Setup – The schematic of the optical setup is shown in Figure 5 and a photograph of the optical bench is shown in Figure 6. A 75 mm focal length lens collimated the laser beam. The collimated beam was passed through a variable speed light chopper (EG&G Model 192/93) and split into two parallel paths. The parallel laser beams passed through two optical cells, each with an aperture of 4 mm diameter and 10 mm length. Barium fluoride (BaF2) windows (2 mm thick) were used at the two ends of each optical cell. The emerging laser beam from each cell was focused on to a HgCdTe(PC) detector (SBRC model No. 40742). The sensing IR element was 1mmx1mm and was housed in a side view metal dewar maintained at liquid nitrogen temperature. To prevent absorption by the atmospheric CO2 in the room, the optical setup was isolated from the ambient atmosphere. All major legs of the optical path were first enclosed inside plastic pipes. These plastic pipes were purged with a continuous supply of nitrogen. The entire setup was then enclosed inside a polyethylene tent that was also purged with nitrogen to further isolate the optical setup from the ambient atmosphere.

The signal from each detector was fed to lock-in amplifiers (EG&G model 128A) set to operate at the chopper frequency. Signal preamplifiers were used ahead of each lock-in amplifier. The difficult and time-consuming job of aligning the optical system was accomplished with the help of an infrared camera and an infrared-sensitive target.

SYSTEM CALIBRATION

Two methods were used to calibrate the absorption signals in the pre- and post-catalyst optical cells.

In the first method, known concentrations of CO2 were mixed with the inlet air supplied to the engine. The engine was motored at the desired speed. All fluid temperatures (coolant, oil, and intake air) were stabilized at the test conditions. The sampling valve was turned on and samples of the gas inside the engine cylinder were extracted at 5° before TDC (top dead center) during the compression stroke. The gas sampled from the engine cylinder was transported to the optical cells by the nitrogen carrier gas flowing at a constant rate of 140 cc/sec. This flow rate provided adequate separation between the samples from successive engine cycles at the test speed of 140 rpm. The samples were exposed to the laser beam in the two optical cells and the attenuations of the incident intensity due to CO2 in the sampled gas were recorded. Typical plots of the recorded intensities of the laser beams for 4.75 percent CO2 in the sampled gas are shown in Figure 7. The attenuation of the laser intensity is related to the concentration of the absorbing species by the Beer Lambert Law which in its simple form for monochromatic radiation can be expressed as:

$$I/Io = \exp(-abc)$$

where

Io = Intensity of the laser beam entering the cell
I = Intensity of laser beam leaving the cell
a = absorption coefficient
b = length of optical path through absorbing medium
c = concentration of the absorbing species

The exponent (abc) is called absorbance and can be expressed as

$$Absorbance = \ln\{Io/I\}$$

The value of the absorption coefficient is nearly constant at any given temperature, and the optical path length through the optical cell is also constant. Hence the absorbance is directly proportional to the concentration of the absorbing species. The value of 'I' the intensity of laser beam transmitted through the cell varies as a function of time because the sample mixes with the nitrogen carrier gas as it flows in the sampling line. We will assume for simplicity that the point at which the value of 'I' is a minimum, called 'Imin' in Figure 7 will give a representative value of absorbance for calibration purposes. The failure of this and other assumptions will result in nonlinear plots of absorbance versus CO2 concentration and can be verified.

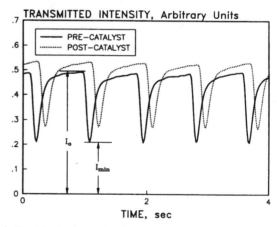

Figure 7. Typical transmitted intensity signals obtained by motoring tge engine, with 4.75% CO2 in the inlet air.

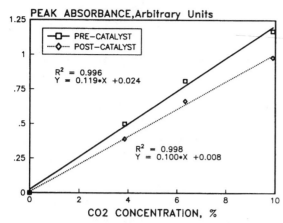

Figure 8. Calibration of the peak absorbance with known CO2 concentrations.

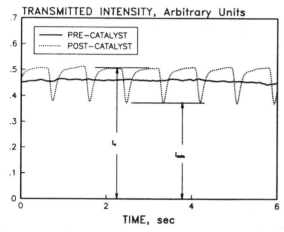

Figure 9. Typical transmitted intensity signals obtained by motoring the engine with propane air mixture. (ϕ =0.6)

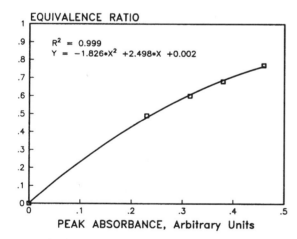

Figure 10. Calibration of peak absorbance in the post-catalyst optical cell with known propane-air equivalence ratios.

Figure 8 shows a typical plot of absorbance for various concentrations of CO2 in the inlet charge of the engine. The absorbance varied linearly with concentration in both the optical cells, meaning that our above assumptions are valid.

In the second method, known concentrations of fuel (propane) were mixed with the air supplied to the engine. The engine was again motored without spark discharge and samples of the gas inside the cylinder were extracted as in the test above. Typical plots of the recorded signals for equivalence ratio ϕ =0.6 are shown in Figure 9. The intensity of the laser beam shows no attenuation in the pre-catalyst optical cell because there is no CO2 in the unburned sample. However, after the sample is burned over the heated catalyst, the laser beam showed absorption due to CO2 in the burned products in the post-catalyst optical cell. Similar data records were obtained for various values of ϕ ranging from 0.5 to 0.8. The calculated peak absorbance in the post-catalyst optical cell is plotted against the equivalence ratio in Figure 10. The calibration thus obtained was used to determine the ϕ of unknown samples during the actual starting tests. Note that this curve is nonlinear reflecting the known nonlinear relation between ϕ and CO2 concentration in burned products (15). During the actual starting tests, after the first firing event there was residual gas in all succeeding cycles. To determine the ϕ of these succeeding samples, the CO2 absorbance due to residual gas had to be subtracted from the total absorbance before the above calibration was applied.

The calibration procedures described above were repeated before each starting test to compensate for shifts in laser intensity, small drifts in laser frequency, and changes in absorption coefficient caused by different test temperatures. The calibration procedure described here is capable of estimating the ϕ values with an error less than 0.02 in the range of interest (ϕ =0.5 to 0.8).

OPERATING CONDITIONS

The operating conditions that were common to all the tests are listed in Table 1, while the fuels, the temperatures, and equivalence ratio values tested are given in Table 2.

Table 1. Operating Conditions Held Constant.

Engine Speed	140 rpm
Airflow	0.625 g/sec or 75% volumetric efficiency
Spark Timing	5° before TDC
Spark Duration	2.5 milliseconds
Spark Gap Width	0.9 mm
Fuel Injection Timing	80° after TDC intake stroke (start of injection)
Compression Ratio	8:1
Non-shrouded Inlet Valve	No Swirl

Table 2. Variables and their Values Tested

Fuel	Temperature, °C	Supplied ϕ
Propane	22°C	0.67
Propane	-12°C	0.67
Isopentane	22°C	0.72
Isopentane	0°C	0.73
Howell EEE	22°C	1.13
Howell EEE	0°C	2.07
Howell EEE	-12°C	2.4

Starting tests were conducted using propane, isopentane, and Howell EEE gasoline (an emission certification test fuel). For each combination of fuel and temperature in Table 2, the ϕ value represents the leanest mixture for starting. In a previous study (1), the leanest mixture for starting was defined as the ϕ supplied to the engine with which the first combustion event occurred in 6 to 10 cycles from the start of propane flow to the engine, and 10 to 20 cycles from the start of gasoline injection. These criteria could not be applied to isopentane injection. With this volatile fuel, the engine either started in the first four cycles after the start of injection or failed to fire consistently if the combustion started after the fourth cycle. Hence the staring ϕ values with isopentane as with propane were also the leanest mixtures that sustained combustion so that the engine ran consistently. At mixtures slightly leaner than those shown in Table 2 for propane and isopentane the engine failed to fire consistently. While with slightly richer mixtures the engine started firing as soon as the fuel air mixture entered the cylinder. This appears to be a characteristic of fuels that are either gaseous or evaporate quickly.

RESULTS OF STARTING TESTS

In this section the typical results of one starting test with propane are presented. The remaining test results with propane are presented in Appendix B. Starting tests similar to those for propane were also conducted with isopentane. We selected isopentane because like gasoline it is a liquid at normal ambient temperature and pressure but unlike gasoline it is a volatile, single component fuel. All the test results with isopentane are presented in Appendix C. Detailed observations of starting tests with a full boiling range gasoline (Howell EEE emission certification fuel) with an RVP of 63.4 kPa are shown in Appendix D. Only comparisons of mixture strength are highlighted in the main body of the text. For details of each test and the residual gas CO_2 concentrations measured the reader should refer to the appropriate appendix.

STARTING TESTS WITH PROPANE AT 22°C – Figure 11 shows plots of engine speed and exhaust gas temperature for a starting test conducted at 22°C with propane as fuel. During this test the temperatures of all the fluids (air, coolant, oil, and fuel) were controlled at 22°C. The engine was motored at 140 rpm and the pro-

pane flow was turned on after the data logging had started. The flow rate of propane was preset to deliver a mixture of ϕ=0.67 to the engine. The top (solid) curve for engine speed shows an average engine speed of 140 rpm with decreases of 5 to 10 rpm during each compression stroke for the first eight cycles. From the recorded propane pressure upstream of the critical flow nozzle it was found that the propane flow was turned on after the second compression stroke. Therefore the fuel will start moving downstream in the inlet pipe of the engine beginning with the third intake stroke labeled as cycle 1 in Figure 11. The engine started to fire after cycle number 6 as evident from the larger excursions of engine speed ranging from 160 to 120 rpm in subsequent cycles. The volume of the inlet system from the point of propane injection to the inlet port of the engine was about 3.5 cylinder displacements. Assuming plug flow, it should take about 4 cylinder displacements for the fuel to reach the cylinder. The lower (dotted) curve for exhaust gas temperature shows a progressive increase in exhaust gas temperature from the 6th cycle onward. The 12th cycle was a misfire as evident from the smaller change in engine speed and a dip in the exhaust gas temperature in cycle 12.

Figure 12 shows traces of the laser intensity transmitted through the pre-catalyst and the post-catalyst optical cells at the peak of a CO_2 absorption line. Note that the transmitted intensity of the laser beam through the optical cell is attenuated whenever CO_2 is present in the sample. Between each attenuation spike, the intensity of the transmitted radiation recovers close to the original unattenuated laser intensity for both the pre and post-catalyst optical cell. This could be inferred to mean that only pure nitrogen is separating the samples. Based on similar behavior of the laser intensity during the starting experiments with isopentane and gasoline, we infer that fuel droplets are not entering the sampling stream and that the samples represent fuel vapor-air mixtures. The peak absorbance values corresponding to the attenuated intensity of the laser beam for each cycle in Figure 12 is shown in Figure 13. The post-catalyst cell absorbance signal showed that the fuel air mixture first started to enter the cylinder in the 4th cycle. In subsequent cycles the absorbance increased and stabilized after the 8th cycle. Combustion started on the 6th cycle after the start of fueling as indicated by the engine speed and exhaust gas temperature traces in Figure 11. This is further confirmed from the appearance of residual gas measured by the pre-catalyst cell absorbance signal. The pre-catalyst cell absorbance signal showed that the first evidence of CO_2 in the sample occurred with the sample extracted from the 7th cycle. This is because combustion started on the 6th cycle and the very first cycle that could have any residual burned gas from the previous cycle was indeed the 7th cycle. The subsequent cycles showed some variation in the absorbance implying variations in the concentration of CO_2 in the residual gas from cycle-to-cycle. Note that cycle 12 misfired, hence the residual gas decrease in the 13th cycle is indicated by a decrease in pre-catalyst CO_2 absorbance signal.

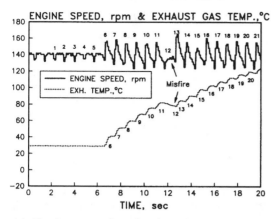

Figure 11. Engine speed and exhaust gas temperature during starting test with propane at 22°C.

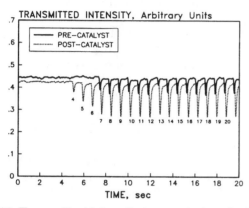

Figure 12. Transmitted intensity traces during starting test with propane at 22°C.

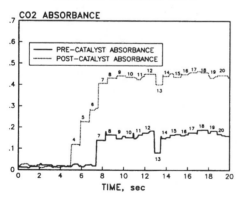

Figure 13. Peak absorbance traces during test with propane at 22°C.

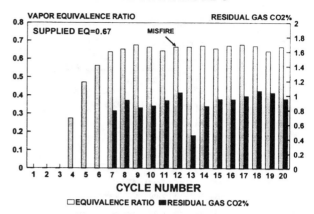

Figure 14. Cycle-by-cycle equivalence and residual gas CO2 for the starting test with propane at 22°C.

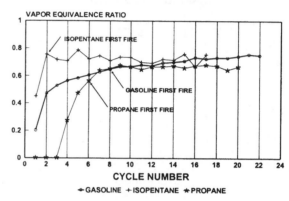

Figure 15. Comparison of cycle-by-cycle equivalence ratios for starting test with propane, isopentane, and gasoline at 22°C.

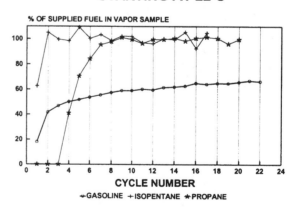

Figure 16. Comparison of fuel supplied per cycle in sampled gas for starting tests at 22°C.

STARTING AT 0°C

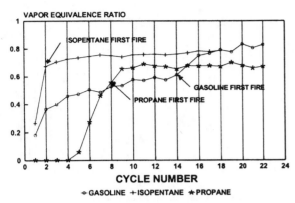

Figure 17. Comparison of cycle-to-cycle equivalence
ratios for starting test with propane,
isopentane, and gasoline at 0°C.

STARTING AT 0°C

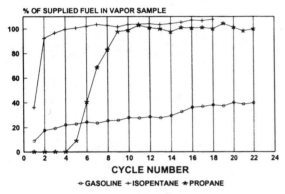

Figure 18. Comparison of fuel supplied per cycle in
sampled gas for starting tests at 0°C.

STARTING AT -12°C

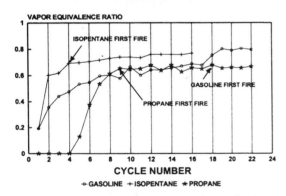

Figure 19. Comparison of cycle-to-cycle equivalence
ratios for starting test with propane,
isopentane, and gasoline at -12°C.

STARTING AT -12°C

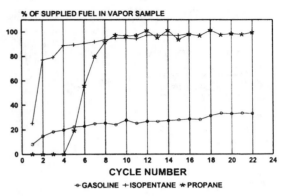

Figure 20. Comparison of fuel supplied per cycle in
sampled gas for starting tests at -12°C.

The absorbance values in Figure 13 can be converted to CO_2 concentration from the pre-catalyst signal and to ϕ of the fuel vapor-air mixture from the post-catalyst signal, using the calibration plots of Figure 8 and Figure 10 respectively. The resulting plots of cycle-by-cycle ϕ and residual gas CO_2 concentration are shown in Figure 14. Note that the absorbance in the post-catalyst optical cell after cycle 7 is due to the combined contribution of the residual gas CO_2 in the unburned gas and the CO_2 formed by the oxidation of the fuel air mixture over the catalyst. The absorbance due to CO_2 in the residual gas was subtracted from the total absorbance in the post catalyst cell. Hence the calculated values of ϕ for cycle 7 onwards are based on the contribution to the absorbance due to CO_2 formed by the oxidation of the fuel air mixture alone.

In Figure 14 the ϕ in cycle number 4 was 0.27, increased to 0.47 in cycle number 5 and reached 0.56 in cycle number 6, when the first combustion occurred. In cycle number 7 the first evidence of residual gas (shown by the solid bars) appeared in the sampled gas. The residual gas CO_2 concentration was calculated to be 0.79% in cycle number 7. It fluctuates from cycle-to-cycle in the succeeding cycles and reaches a concentration of 1.04% CO_2 in cycle number 12 which misfired. The next cycle (number 13) shows a sudden drop in the residual gas CO_2 concentration to 0.46% CO_2 due to the misfire in the preceding cycle. In succeeding cycles the combustion is sustained without any more misfires until the end of the recording. After the start of combustion in cycle 6 the ϕ of the mixture increases in subsequent cycles and reaches 0.67, which is the ϕ of the mixture supplied to the engine based on the measured amounts of air and propane flows.

Why did the engine misfire in cycle 12? The increase in residual gas diluting the inlet charge, may be one of the many factors involved in causing a misfire (16). The residual gas CO_2 concentration reaches its highest value up to that point in cycle 12. This would suggest that flame initiation may have been hindered by the residual gas, thereby causing the engine to misfire. However, in subsequent cycles the CO_2 concentration in the residual gas exceeds the value in cycle 12 but misfire did not occur. Thus factors other than CO_2 concentration in the residual gas are apparently involved in the onset of misfire. The temperature of the inlet charge probably increases progressively with each firing cycle until some steady state is reached. The higher temperature of the inlet charge would probably make ignition easier as the engine warms up. Hence the engine may be able to tolerate higher dilution of the charge without misfire after the engine has warmed up.

COMPARISON OF STARTING WITH DIFFERENT FUELS – To compare the starting performance with different fuels, the cycle-by-cycle equivalence ratio values of the three test fuels were plotted together for each temperature setting. Figure 15 shows plots at 22°C. The plot with asterisks is for propane, the plot with pluses is for isopentane, and the plot with circles is for gasoline. Data for these tree plots are from Figure 14, Figure C-3, and Figure D-3 respectively. In Figure 15 the first fired cycles are also identified. Starting was quickest with isopentane, followed by propane and gasoline.

The quick starting with isopentane may be in part due to port fuel injection as opposed to the upstream injection used with propane. The fuel has a short distance to travel (about 100 mm) from the injector tip to the inlet port before it enters the cylinder. The low boiling point of isopentane at ambient pressure 28°C (82°F), may also contribute to quick fuel vaporization after injection. The quick startup with isopentane suggests that the mixture may have been too rich to be considered the leanest mixture for starting. However, tests using leaner mixtures of isopentane resulted in intermittent misfires and unstable combustion. Hence the demarcation of ϕ for stable and unstable combustion was quite sharp with isopentane.

Moreover, with isopentane the fuel vapor-air ϕ was 0.76 for the first fired cycle, which exceeds the average supplied ϕ of 0.72 during the test. The sampled vapor ϕ also varied significantly from cycle to cycle. The variation in cycle-by-cycle ϕ, and why the vapor-air ϕ exceeded the average supplied ϕ may be due to several factors chief among which are:

1. differences in the amount of fuel injected in each cycle,

2. differences in the amount of fuel passing from the intake port into the cylinder,

3. spatial variations in ϕ causing stratification of the mixture inside the cylinder during each cycle.

It is not possible from the data to identify a definite cause of this cycle-by-cycle variation or to ascribe a specific reason why the ϕ of the sampled gas is greater than the supplied ϕ. An accurate measure of the amount of fuel injected during each cycle is needed to answer some of these questions. The fuelflow meter averaged the fuelflow over a duration of 12 seconds. This was the minimum duration to obtain reasonably repeatable average of fuelflow readings.

It appears from Figure 15 that the ϕ of the sampled mixture has to exceed a minimum value for combustion to start with each fuel. With isopentane this value is between ϕ of 0.45 and 0.76 - the ϕ of the cycle just preceding the first fired cycle and the ϕ of the first fired cycle. For propane it is between 0.47 and 0.56, while for gasoline it is between 0.61 and 0.63. From a fundamental standpoint, for stable combustion, the mixture must be ignitable and flammable. The ϕ values for first fire are probably very close to the limit of combustion for the engine. The lean flammability of propane and isopentane determined in non-engine experiments are 0.51 and 0.50 respectively (17). The interactions of ignition, flame propagation, and lean limits in engines and their relation to flammability limits in non-engine experiments have been discussed (6, 16).

FUEL VAPORIZATION A LIMITING FACTOR

Incomplete vaporization of the liquid fuel could limit the amount of fuel in the fuel vapor-air mixture sampled from the cylinder. The measured values of fuel vapor-air ϕ permits us to estimate the percentage of the supplied liquid fuel in the sampled vapor-air mixture. Figure 16 shows plots of fuel vapor in the sample expressed as a percentage of the fuel supplied per cycle at the 22°C test temperature. Note that with liquid fuels, the fuel vapor sampled in the second cycle onward will have contribution due to fuel supplied in previous cycles. This is because some of the fuel supplied in previous cycles will probably remain on the combustion chamber surfaces and in the inlet port and may continue to vaporize in subsequent cycles. Therefore the fuel vapor percentages shown in Figure 16 represent the total amount of fuel vapor from all these sources expressed as a fraction of the fuel injected per cycle. With isopentane the sample from the first cycle represents 63% of the supplied fuel and reaches complete fuel vaporization from cycle 2 onward. With propane, which should be gaseous under all the conditions tested, it takes roughly 10 cycles for the fuel to reach full strength inside the cylinder. This is because propane is injected upstream in the inlet pipe and mixes with the air as it flows to the engine. With gasoline only 18% of the supplied fuel ends up as vapor in cycle 1, increases to 42% in cycle 2, and increases steadily to 57% in cycle 8 when the first combustion occurs.

From the above observations with gasoline, only about half the supplied fuel per injection appears to be accounted in the vaporized samples. What happens to the remaining fuel that did not vaporize from all the injections? We do not have an answer to this question. We speculate that part of this liquid fuel remains in the fuel films in the inlet port and combustion chamber walls, some of it may get past the rings and end up in the crankcase, while some of it will get exhausted with the exhaust gases as unburned hydrocarbons. After the start of flame propagation some of the fuel in liquid droplets and wall films ahead of the flame, may also vaporize and burn due to the combined effect of compressive heating and heat transfer from the burned gases. Thus the flame probably propagates in a progressively richer mixture ahead of it. More experiments with samples extracted from the end gas at different times during the same cycle would be needed to confirm this hypothesis. Further work is also needed to identify the destination of the liquid fuel since it can have a significant impact on exhaust emissions and engine oil dilution.

Plots similar to Figure 15 and 16 were generated at the other temperatures and are shown in Figure 17 to Figure 20. Comments similar to those made above for the 22°C test temperature are also applicable at the lower temperatures. Some of the observations from these plots are summarized in Table 3. The following general observations can be made about the data in Table 3. With each fuel, first fire took more cycles at lower temperatures. This was in spite of richer liquid fuel air mixtures supplied with isopentane and gasoline at lower temperatures. The fuel vapor-air ϕ of the first fired cycle was greater at lower temperature; implying richer fuel vapor-air mixtures are needed at lower temperatures. Isopentane and gasoline at 22°C were exceptions.

PREDICTION OF FIRST FIRE EQUIVALENCE RATIO – Since fuel vaporization appears to be a limiting step in getting a combustible fuel vapor-air mixture with gasoline, it would be of interest to see if we can predict the amount of fuel needed to start the engine sooner. Suppose we want the engine to fire on the very first cycle after the start of fuel injection. To calculate the fuelflow needed to accomplish this, we know that the percent fuel vaporized is 18% in cycle 1 (see Figure 16) and the fuel vapor-air ϕ needed is 0.63 at 22°C. Thus the liquid fuel to air ϕ needed is 0.63/0.18=3.5. Similarly the liquid fuel to air ϕ needed at 0°C was calculated to be 7.0, and at -12°C the liquid fuel to air ϕ was calculated to be 8.5. To verify these estimates, the engine was run to determine the liquid gasoline to air ϕ at these temperatures. The ϕ values needed to start the engine on the first cycle after fuel injection were found to be 2.5 at 22°C, 6.2 at 0°C, and 9.1 at -12°C. The calculated values are reasonably close to the experimentally measured values.

COMMENTS AND CLOSURE

The results of this study show the feasibility of cycle-by-cycle measurements of fuel vapor-air ϕ and residual gas CO_2 concentrations in an operating engine. Since the completion of this study, we have learned of ongoing research at Fraunhofer-Institute for Physical Measurement Techniques on automotive exhaust emission monitoring by tunable diode lasers (18, 19). They have shown the feasibility of using multiple diode lasers in a single cold head to monitor up to four species in the exhaust gas simultaneously. The possibilities of such a tool are exciting, though the problem of cycle-by-cycle data acquisition remain quite difficult. As pointed out in the introduction, the Cambustion fast FID has also been used to determine cycle based fuel concentrations (3) and other techniques may be applicable as well (20, 21, 22, 23).

The results also show quite clearly that the presence of a flammable fuel vapor-air mixture inside the cylinder determines the onset of combustion. Previous studies (3, 24) have also shown that an equivalence ratio ϕ=0.6 is needed to start combustion. To obtain such a flammable mixture quickly, requires large amounts of gasoline to be delivered in the first few injection events, because only a small fraction of the fuel vaporizes. This is also potentially the source of highest unburned HC emissions. An important observation in the gasoline starting tests in Appendix D is that even after consistent combustion was sustained the ϕ of the sampled mixture was much lower than the ϕ supplied to the engine. This suggests that a significant part of the liquid fuel does not vaporize even after the engine starts firing consistently during and after a cold start. The survival of liquid fuel films on the combustion chamber walls after cold starts has also been visually confirmed (25, 26). Composition of the fuel, spray properties such as drop size of the injected fuel, pressure and temperature of the gas in the inlet port and port walls, and the design of the fuel delivery system can impact the formation of a flammable fuel vapor-air mixture. Further research should focus on methods of generating the desired mixture quickly without the accompanying problems caused by left over unevaporated liquid fuel.

Estimates of fuel vapor-air ϕ values based on equilibrium considerations for the amount of gasoline and air supplied to the engine were shown to vary between 0.85 and 1.1 in the temperature range of -29°C to 20°C (1). The measured values of fuel vapor-air ϕ in the present study were found to be 0.62 to 0.68. Thus the measured values are lower than the equilibrium predictions. This is not surprising, because equilibrium vaporization is probably not reached in the very short time the fuel is exposed to air in the engine.

Table 3. Comparison of Starting Test Results

Fuel	Temp.	Supplied φ	First Fire Cycle #	First Fire Sampled φ	Supplied Fuel in Vapor at First Fire	Supplied Fuel in Vapor after First Injection
Propane	22°C	0.67	6	0.56	84%	-
Propane	0°C	0.67	8	0.56	84%	-
Propane	-12°C	0.67	9	0.65	97%	-
Isopentane	22°C	0.72	2	0.76	105%	`62.6%
Isopentane	0°C	0.73	2	0.67	92%	36.2%
Isopentane	-12°C	0.78	4	0.69	89%	25%
Howell EEE	22°C	1.13	8	0.63	57%	18%
Howell EEE	0°C	2.07	14	0.61	30%	8.7%
Howell EEE	-12°C	2.40	17	0.68	28%	8%

SUMMARY

A diode laser based spectroscopic technique was developed to measure cycle-by-cycle fuel vapor-air equivalence ratio and residual gas CO_2 concentration in the unburned mixture of an operating single-cylinder CFR engine. The technique works as follows:

1. Samples of the mixture inside the cylinder were extracted by means of a specially designed sampling valve mounted in the cylinder head. One sample was extracted during each cycle, just prior to spark ignition.

2. The samples were kept separate and conveyed to an optical bench (for analysis), by a stream of nitrogen carrier gas flowing through the sampling valve.

3. The sample was exposed to a laser beam tuned and calibrated at a strong CO_2 absorption line in the 4.3 micron absorption band. The attenuation of the laser beam was used to determine the residual gas CO_2 concentration in each unburned sample.

4. Each sample was then catalytically oxidized and exposed to a second identical laser beam. The attenuation of this laser beam was used to determine the equivalence ratio of each sample.

This analysis technique was applied to simulated starting tests, conducted with a single-cylinder CFR engine installed inside the cold test facility (1). In separate tests, propane, isopentane, and gasoline, were used as fuel at 22°C, 0°C, and -12°C. For brevity, only the gasoline test results are highlighted below.

1. Fuel vaporization was the major factor controlling the formation of a combustible mixture. The fuel vapor-air mixture sampled from the cylinder at the onset of combustion, expressed as a percentage of the fuel supplied per cycle was 57% at 22°C, 30% at 0°C,

and 28% at -12°C. Even after sustained combustion, a substantial part of the injected gasoline escaped combustion.

2. The equivalence ratios of the fuel vapor-air mixture in the first fired cycle was 0.63 at 22°C, 0.61 at 0°C, and 0.68 at -12°C. These values of equivalence ratio are probably close to the lean flammability limits of the lighter hydrocarbons in the gasoline at the engine operating conditions tested. These equivalence ratios are leaner than the calculated equilibrium fuel vapor-air equivalence ratios based on the amounts of air and fuel supplied.

3. Residual gas CO_2 was detected inside the cylinder on all cycles following the first combustion event. Misfires occurred in cycles that usually had high residual gas CO_2 relative to other preceding cycles. In the cycle following a misfired cycle, the residual gas CO_2 concentration dropped to about half the value in the misfired cycle.

CONCLUSIONS

Based on the information presented we conclude the following:

1. Cycle-by-cycle mixture strength and residual gas concentration can be measured during starting of a single-cylinder engine.

2. Combustion usually begins in the first cycle in which the fuel vapor-air equivalence ratio of the mixture inside the cylinder exceeds the lean flammability limit of the fuel at the desired temperature. In the range of temperatures 22°C to -12°C the fuel vapor-air equivalence ratio for the start of combustion generally increased at lower temperature, i.e. richer mixtures were required at lower temperatures (except isopentane). Increased difficulty of starting may be inferred

from the greater number of cycles to reach a combustible mixture as the temperature was lowered with each fuel.

3. Calculated equilibrium-based fuel vapor-air equivalence ratios were greater than the actual fuel vapor-air equivalence ratios inside the cylinder.

4. After the first cycle with combustion, dilution with residual gas can contribute to misfire in succeeding cycles. Hence, the fuel vapor-air equivalence ratio has to be richer than the lean flammability limit (without residual gas dilution), for combustion to be sustained in succeeding cycles.

5. Cycle-to-cycle variations occur in mixture strength as well as residual gas CO_2 concentration with pre-mixed, gaseous (propane) and port fuel injected liquid fuel delivery systems. It is not clear if these variations are caused by differences in the amount of fuel injected in each cycle, differences in the amount of fuel passing from the intake port into the cylinder, and/or due to spatial variations in ϕ inside the cylinder during each cycle.

ACKNOWLEDGMENTS

We are grateful to Dr. B. D. Peters and Dr. J. A. Sell for many fruitful discussions during the inception of the measuring technique. M. J. Bartch (retired) and D. W. Poma helped in the design of the sampling valve. R. F. Milz assembled parts of the optical bench and assisted in the calibration of the flow system for sample analysis. T. A. Talsma did an outstanding job of keeping the engine and cold test facility operational and helped in the diagnosis and remedy of problems with the sampling valve. J. L. Howes and G. A. Long assisted in setting up the computer aided test system, data logging procedures, and processing cycle-by-cycle laser absorption signals. R. A. Tidrow is cited for his continued support of the engine setpoint controller.

REFERENCES

1. A. A. Quader, "Single-Cylinder Engine Facility to study Cold Starting - Results with Propane and Gasoline," SAE paper 9200001 presented at the Subzero Engineering Conditions Conference, Helsinki, February 1992.

2. C. J. Dasch, N. D. Brinkman, and D. H. Hopper, "Cold Starts Using M-85 (85% Methanol): Coping with Low Fuel Volatility and Spark Plug Wetting," SAE Paper 910865, 1991.

3. J. W. Fox, K. D. Min, W. K. Cheng, J. B. Heywood, "Mixture Preparation in a SI Engine with Port Fuel Injection During Starting and Warm-Up," SAE paper 922170, 1992

4. F. Galliot, W. K. Cheng, C. O. Cheng, M. Sctenderowicz, J. B. Heywood, and N. Collings, "In-Cylinder Measurements of Residual Gas Concentration in a Spark Ignition Engine," SAE paper 900485, 1990.

5. R. W. Johnston, J. J. Newman, and P. D. Agarwal, "Programmable Energy Ignition System for Engine Optimization," Paper 750348, SAE Transactions, Vol. 84, 1975.

6. A. A. Quader, "What Limits Lean Operation in Spark Ignition Engines - Flame Initiation or Propagation?" Paper 760760, SAE Transactions, Vol. 85, 1976.

7. A. A. Quader, "How Injector, Engine, and Fuel Variables Impact Smoke and Hydrocarbon Emissions with Port Fuel Injection," Paper 890623 SAE, 1989.

8. K. Matsui, T. Tanaka, and S. Ohigashi, "Measurement of Local Mixture Strength at the Spark Gap of SI Engines," Paper 790483, SAE Transactions, Vol. 88, 1979.

9. T. Nomura, I. Higashino, and Y. Iwamoto, "A New Method of Cylinder Gas Sampling Cycle by Cycle Using Microprocessor," Bulletin of JSME, p 1420, Vol. 23, No 182, August 1980.

10. J. T. Wentworth, "A New Exhaust Pipe Sampling Valve," Research Publication GMR-766, May 1968.

11. J. A. Sell, R. K. Herz, and D. R. Monroe, "Dynamic Measurement of Carbon Monoxide Concentrations in Automotive Exhaust Using Infrared Diode Laser Spectroscopy," Paper 800463, SAE Transactions, Vol. 89, 1980.

12. J. A. Sell, and M. F. Chang, "Closed Loop Control of an Engine's Carbon Monoxide Emissions Using an Infrared Diode Laser," Paper 820388, SAE Transactions, Vol. 91, 1982.

13. D. L. Partin, " Lead Europium Selenide Telluride Grown by Molecular Beam Epitaxy," Journal of Electronic Materials, Vol. 13, p 493, 1984.

14. P. S. Lee, R. F. Majkowski, and T. A. Perry, "Tunable Diode Laser Spectroscopy for Isotope Analysis - Detection of Isotopic Carbon Monoxide in Exhaled Breath," IEEE Transactions, Vol. 38, p 966, 1991.

15. B. A. D'Alleva, "Procedures and Charts for Estimating Exhaust Gas Quantities and Composition," Research Publication, GMR, 1960.

16. A. A. Quader, "Lean Combustion and the Misfire Limit in Spark Ignition Engines," SAE Transactions, Vol. 83, Paper 741055, 1974.

17. H. C. Barnett, and R. R. Hibbard (editors), "Basic Considerations in the Combustion of Hydrocarbon Fuels with Air," NACA Report 1300, 1959.

18. R. Grisar, "Automotive Exhaust Emission Monitoring by Tunable Diode Lasers," Research Report, Fraunhofer-Institute for Physical Measurement Techniques, 1989.

19. W. Thiel, W. Hubner, R. Grisar, H. Wolf, and W. J. Riedel, "Dynamic Laser Analysis of Exhaust Gas," SAE paper 940825, March 1994.

20. R. Frey, H. Nagel, J. Franzen, H. Betzold, W. Ulke, and U. Boesl, "A Fast Multicomponent Exhaust Gas Analyzer for Dynamic Exhaust Measurements," SAE paper 940826, March 1994.

21. K. C. Chen, W. K. Cheng, J. M. Vandoren, J. P. Murphy III, M. D. Hargus and S. A. McSweeney, "Time Resolved Speciated Emissions from an SI Engine During Starting and Warm-Up," SAE paper 960195, 1996.

22. G. Gruefeld, M. Knapp, V. Beushausen, P. Andresen, W. Hentschel, and P. Manz, "In-Cylinder Measurements and Analysis on Fundamental Cold Start and Warm-Up Phenomena of SI Engines," SAE paper 952394, October 1995.

23. M. Koenig, and M. J. Hall, "Cycle-Resolved Measurements of Pre Combustion Fuel Concentration Near the Spark Plug in a Gasoline SI Engine," SAE paper 981053, February 1998.

24. Y. Shin, "Measurement of In-Cylinder Equivalence Ratio During Starting Using a Fast FID," KSME International Journal, Vol. 11, No. 6, pp. 726-736, 1997.

25. K. Song, E. Clasen, C. Chang, S. Campbell, and K. T. Rhee, "Post-Flame Oxidation and Unburned Hydrocarbons in a Spark-Ignition Engine," SAE paper 962543, October 1995.

26. P. O. Witze, and R. M. Green, "LIF Visualization of Liquid Fuel in the Cylinder of a Spark Ignition Engine," Proceedings of the Eighth International Symposium on Applications of Laser Techniques to Fluid Mechanics, Lisbon, July 1996.

APPENDIX A

SAMPLING VALVE FLOW CHARACTERISTICS

The sampling valve control unit permitted variations in the duration for which the poppet valve could be held open. With this we could control the amount of gas sampled from the engine cylinder by varying the sampling duration. Figure A-1 shows the volume of gas sampled per cycle for various sampling durations at 140 rpm engine speed. The valve was triggered to open at 5°C before TDC of the compression stroke during all the tests. Over the range of sampling durations tested the volume of gas sampled per cycle varied almost linearly with sampling duration. All the tests reported in this study were run with a sampling duration of 1.5 milliseconds. This corresponded to 1.26 degrees of crankshaft rotation, which is a very short interval during the 180° of crank rotation during the compression stroke. During this interval about 7.5 cc of gas were extracted during one engine cycle. The 7.5 cc of sample at atmospheric pressure and temperature corresponded to less than 1 percent of the combustion chamber gas at 8:1 compression ratio. The leakage of gas past the closed valve was less than 30 cc per minute at the operating conditions of the test. This corresponded to less than 6 percent of the sampled gas and was considered negligible.

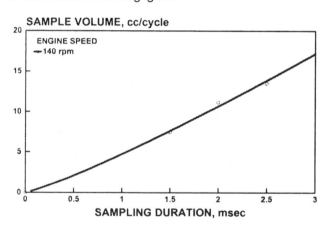

Figure A-1. Sampling valve flow calibration.

APPENDIX B

STARTING TEST WITH PROPANE AT 0°C

Figure B-1 shows plots of engine speed and exhaust gas temperature for a starting test conducted at 0°C. The cycles are identified by numbers. The first cycle after the propane flow was turned on is labeled 1. As in the test at 22°C, the engine was supplied with a propane air mixture of equivalence ratio ϕ=0.67. In this experiment the engine started in cycle number 8 as indicated by the increase in magnitude of the engine speed fluctuations and the accompanying rise in exhaust gas temperature. Cycle number 19 was a misfire as evident from the drop in exhaust gas temperature and engine speed fluctuation.

The ϕ of the mixture corresponding to the values of absorbance in the post-catalyst cell were calculated from a calibration curve similar to that shown in Figure 10. The calibration was generated on the day of the test at the desired operating conditions of the engine just prior to the starting test.

The cycle based ϕ values and residual gas CO_2 concentrations based on the absorbance signals are shown in Figure B-2. The ϕ was 0.06 in cycle 5, 0.27 in cycle 6, 0.46 in cycle 7, and 0.56 in cycle 8 when the first combustion occurred. It should be noted that in the test at 22°C the first combustion had occurred when the ϕ of the sampled gas reached 0.56 also. The implication is that for temperatures at or above 0°C, with propane as fuel, combustion can begin when the ϕ of the mixture inside the combustion chamber reaches or exceeds ϕ=0.56.

In Figure B-2 the residual gas CO_2 concentration was 0.75 percent in cycle 9, the first cycle after combustion started. The residual gas CO_2 concentration reached 0.93 percent in cycle 19 when a misfire occurred. As mentioned earlier there may be other factors involved in the occurrence of a misfire, but the increase in residual gas may be a contributing factor.

STARTING TEST WITH PROPANE AT -12°C

The plots of engine speed and exhaust gas temperarure for a starting test at -12°C with propane ϕ=0.67 supplied to the engine are shown in Figure B-3. In these we see that combustion started in cycle 9. We also see from these plots that there were misfires in cycle 10, cycle 16, and cycle 19. The longer duration to first fire and the increased frequency of misfires are indicative of the increased difficulty for combustion to start and sustain at lower ambient temperatures.

The ϕ and residual gas CO_2 concentrations are shown in Figure B-4. The ϕ was 0.13 in cycle 5, 0.37 in cycle 6, 0.53 in cycle 7, 0.61 in cycle 8, and 0.65 in cycle 9 when the first combustion occurred. The fact that at -12°C temperature, the combustion did not start until the ϕ reached 0.65, while at higher temperatures of 0°C and 22°C the first combustion occurred when the ϕ reached 0.56, suggests that a richer mixture is required to start combustion at lower temperatures.

Figure B-4 shows that cycle 10 had a ϕ of 0.65 and the residual gas CO_2 was 0.86 percent, and as mentioned earlier, cycle 10 misfired. Even though the ϕ of the charge was 0.65, the presence of the burned residual gas with 0.85 percent CO_2 probably prevented the mixture from burning. In cycle 11 the ϕ was again 0.65 but the residual gas was partly scavenged leaving a residual gas containing 0.44 percent CO_2. Combustion occurred in this cycle. The data of cycles 9, 10, and 11 each had a ϕ of 0.65 but only cycle 9 without any residual gas and cycle 11 with low content of residual gas had combustion while cycle 10 with higher residual gas misfired. These three cycles demonstrate the importance of residual gas in rendering a combustible mixture noncombustible.

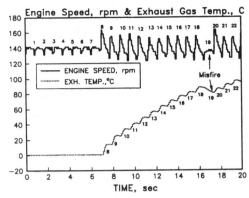

Figure B-1. Engine speed and exhaust gas temperature during starting test with propane at 0°C.

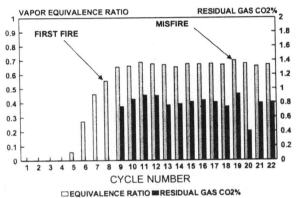

Figure B-2. Cycle-by-cycle equivalence ration and residual gas CO2 for the starting test with propane at 0°C.

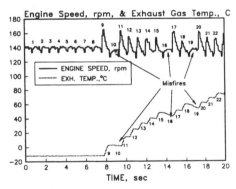

Figure B-3. Engine speed and exhaust gas temperature during starting test with propane at 0°C.

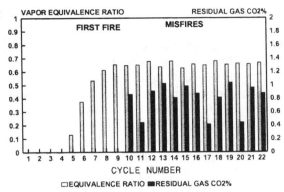

Supplied Equiv Ratio = 0.67

Figure B-4. Cycle-by-cycle equivalence ration and residual gas CO2 for the starting test with propane at 12°C.

APPENDIX C

STARTING TESTS WITH ISOPENTANE AT 22°C

Figure C-1 shows plots of engine speed and exhaust gas temperature for a starting test with isopentane at 22°C. The ϕ based on the measured amount of isopentane and air supplied to the engine was 0.72. Combustion started in the second cycle after the start of fuel injection and there were no misfires in this run

The corresponding cycle-by-cycle ϕ and residual gas CO2 concentrations are shown in Figure C-2. The calibration curve for ϕ versus absorbance was generated with propane. Since the CO2 produced by the combustion of a given ϕ of isopentane is greater than that for propane, an appropriate correction was applied to the measured peak absorbance values in arriving at the correct ϕ for isopentane.

In Figure C-2 the ϕ of the mixture sampled from the first cycle after fuel injection was 0.45, and increased to 0.76 in cycle 2 when the first combustion occurred. In subsequent cycles the ϕ varied from cycle-to-cycle over a range of 0.66 to 0.79 with an average value of 0.73.

The residual gas CO2 concentrations in Figure C-2 also display cycle-to-cycle variations. The mean value of the residual gas CO2 was 1.14 percent with a variation of ±10 percent. It is possible that the cycle-to-cycle variations in the ϕ are the cause of the variations in residual gas CO2 concentrations, but again it cannot be said with certainty.

STARTING TESTS WITH ISOPENTANE AT 0°C

Figures C-3 show plots of engine speed and exhaust gas temperature for a starting test at 0°C with ϕ=0.73 isopentane air mixture supplied to the engine. These plots show that misfire occurred in cycle 7 and cycle 13.

The ϕ and residual gas CO_2 concentrations are shown in Figure C-4. The ϕ of the mixture sampled from cycle 1 was 0.26. It increased to 0.67 in cycle 2 when combustion occurred. In subsequent cycles the ϕ increased further and stabilized at an average value of about 0.76, which is richer than the ϕ supplied to the engine. The reason for this behavior is not clear. The leaner values of ϕ in cycle 1 and 2 relative to the same cycles in Figure C-2 (for the 22°C starting test) are due to less fuel vaporization at lower ambient temperature. However, the stabilization of the ϕ after cycle 6 suggests that isopentane vaporization is not a major factor after combustion has occurred in a few cycles. The fact that misfires were observed in cycle 7 and 13 even after the ϕ had stabilized, indicates increased difficulty of flame initiation due to the lower ambient temperature. The engine is probably operating close to the leanest mixture it can tolerate at this condition.

STARTING TESTS WITH ISOPENTANE AT -12°C

Figure C-5 shows plots of engine speed and exhaust gas temperature for a starting test at -12°C with ϕ=0.78 isopentane air mixture supplied to the engine. These plots show that combustion started in cycle 4 and a misfire occurred in cycle 5 after which the engine ran without any more misfires to the end of the test.

The ϕ and CO_2 concentration in the residual gas are shown in Figure C-6. The ϕ of the gas sampled from cycle 1 was 0.2, increased to 0.6 in cycle 2, further increased to 0.62 in cycle 3 and reached 0.69 in cycle 4 when the first combustion occurred. Several factors are noteworthy here. First, it is taking more cycles for the engine to start at lower temperature. Second, the fuel is not vaporizing as quickly as it did in previous tests with isopentane at higher temperatures. Third, as a consequence of the slower fuel vaporization, it is taking longer for a combustible mixture to form inside the combustion chamber. Hence it is taking longer for the engine to start.

In cycle 5 the ϕ was 0.7 but the presence of residual gas with 0.95 percent CO_2 left over from the previous cycle apparently prevented flame initiation and a misfire occurred. In subsequent cycles the ϕ continued to increase and reached a value of 0.77 in cycle 16. The gradual increase in ϕ of the mixture in each cycle relative to the previous cycle suggests that more fuel is vaporizing in each successive cycle as the engine warms up. This behavior was not as pronounced in the previous tests at higher temperatures where isopentane vaporization was less of a factor.

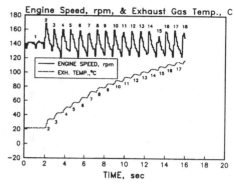

Figure C-1. Engine speed and exhaust gas temperature during starting test with isopentane at 22°C.

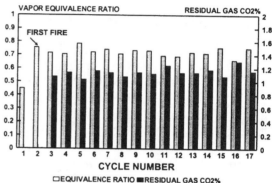

Figure C-2. Cycle-by-cycle equivalence ration and residual gas CO_2 for the starting test with isopentane at 22°C.

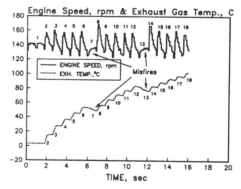

Figure C-3. Engine speed and exhaust gas temperature during starting test with isopentane at 0°C.

ISOPENTANE STARTING 0°C

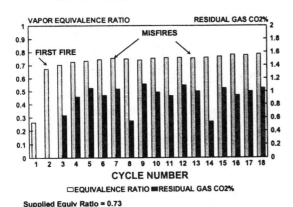

Supplied Equiv Ratio = 0.73

Figure C-4. Cycle-by-cycle equivalence ration and residual gas CO2 for the starting test with isopentane at 0°C.

ISOPENTANE STARTING −12°C

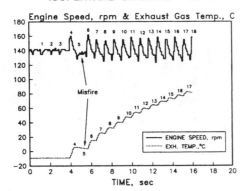

Figure C-5. Engine speed and exhaust gas temperature during starting test with isopentane at -12°C.

ISOPENTANE STARTING -12°C

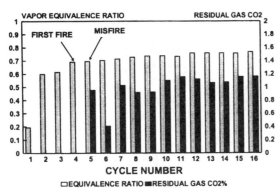

Supplied Equiv Ratio = 0.78

Figure C-6. Cycle-by-cycle equivalence ration and residual gas CO2 for the starting test with isopentane at -12°C.

APPENDIX D

STARTING TEST WITH GASOLINE AT 22°C

The engine speed and exhaust gas temperature for a starting test at 22°C with ϕ =1.13 gasoline air mixture supplied to the engine are shown in Figure D-1. Combustion started in cycle 8 and misfires occurred in cycle 9 and cycle 18. Note that even though a much richer mixture was supplied to the engine, it still took eight cycles for the combustion to start. This is probably because only the more volatile components of the gasoline vaporized under the test conditions.

The ϕ and CO2 concentration of the gas sampled from each cycle are shown in Figure D-2. In the absence of precise knowledge of the gas phase composition of the mixture it was assumed that the more volatile fraction of the fuel would be the C4 to C5 carbon containing hydrocarbons. Hence the calibration based on propane (C3H8) was corrected assuming that the fuel vapor was primarily butane (C4H10). Based on this assumption the ϕ of the mixture was 0.21 in cycle 1, 0.47 in cycle 2, and increased progressively to 0.63 in cycle 8 when the first combustion occurred. The ϕ of the sampled gas continued to increase after combustion started and reached a value of about 0.73 in cycle 22 when the test was terminated. The CO2 in the residual gas was 0.7 percent in cycle 9 and increased to 1.18 percent in cycle 22 at the end of the test. The increase in CO2 appears to be progressive and is different from the trends displayed in the experiments with propane and isopentane. Perhaps the gradual warming of the engine with each combustion event results in the burning of more of the higher carbon containing components of the gasoline which would vaporize at higher temperatures. This may be the cause of the progressive increase in residual gas CO2 concentration in successive cycles.

STARTING TEST WITH GASOLINE AT 0°C

The engine speed and exhaust gas temperature for a starting test at 0°C with ϕ =2.07 gasoline air mixture supplied to the engine are shown in Figure D-3. In this test even though the ϕ of the mixture supplied to the engine was 2.07 it took 14 cycles after the start of fuel injection for combustion to occur. Apparently the first combustion was quite weak as indicated by a small fluctuation in engine speed. The exhaust gas temperature showed quite a large increase of over 30°C for cycle 14, as opposed to the 10°C observed for the first combustion in other experiments shown above. The first combustion event also caused a distinctly audible noise in the exhaust pipe and was accompanied by a sharp pressure spike in the exhaust pressure. Perhaps the unburned fuel vapors accumulated in the exhaust system and exploded when the hot exhaust from cycle 14 ignited them.

The cycle-by-cycle φ and residual gas CO2 concentration for this run are shown in Figure D-4. The φ increased from 0.18 in cycle 1 to 0.50 in cycle 6 quite rapidly after which it showed a gradual increase with cycle-to-cycle fluctuations in the next eight cycles reaching an φ of 0.61 in cycle 14 when the first combustion occurred. After combustion started the φ showed a more rapid increase in the next four cycles until it reached 0.80 in cycle 18. Thereafter the φ fluctuated from 0.8 to 0.83 until the end of the recording. The residual gas CO2 increased rapidly from 0.58% CO2 in cycle 15 to 1.06% CO2 in cycle 18 and fluctuated from cycle-to-cycle thereafter.

The points to note in the above description of events are that with gasoline the vapor phase φ is apparently constrained by the temperature and the amount of fuel supplied. Before combustion starts the fuel vapor-air φ apparently stabilizes around 0.6 with all the fuel supplied in the first 14 cycles. After combustion starts the hot residual gases mix with the incoming charge to vaporize more of the fuel and raise the fuel vapor-air φ to 0.8. The presence of the residual gas tends to dilute the mixture but the combustion is apparently sustained by φ=0.8. What is the fate of the fuel that did not vaporize? Some speculations were discussed in the closure section.

STARTING TEST WITH GASOLINE AT -12°C

The engine speed and exhaust gas temperature for a starting test at -12°C with φ =2.4 gasoline air mixture supplied to the engine are shown in Figure D-5. The engine fired on cycle 17 after the start of fuel injection and showed no misfires in the next 5 cycles after which data acquisition was terminated. As in the previous test at 0°C the first combustion event (cycle 17) was quite weak as indicated by the small fluctuation in engine speed and may have resulted in a slow burn. The accumulation of unburned fuel air mixture in the exhaust system from the first 16 cycles resulted in an explosion in the exhaust pipe when the hot gases from the first combustion event were exhausted. This is also evident from the sudden rise of nearly 30°C in the exhaust gas temperature following the first combustion event.

The cycle-by-cycle φ and residual gas CO2 for this test are shown in Figure D-6. Most of the remarks made on the previous test at 22°C and 0°C are also applicable here, except that the φ of the first combustion cycle was 0.68 in this test which is richer than the 0.63 and 0.61 values observed earlier for 22°C and 0°C respectively. Apparently a richer mixture is required in the gas phase to start combustion at lower temperatures.

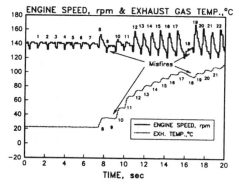

Figure D-1. Engine speed and exhaust gas temperature during starting test with gasoline at 22°C.

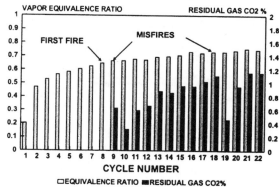

Figure D-2. Cycle-by-cycle equivalence ration and residual gas CO2 for the starting test with gasoline at 22°C.

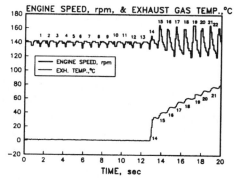

Figure D-3. Engine speed and exhaust gas temperature during starting test with gasoline at 0°C.

191

GASOLINE STARTING 0°C

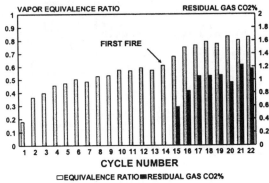

Figure D-4. Cycle-by-cycle equivalence ration and residual gas CO2 for the starting test with gasoline at 0°C.

GASOLINE STARTING −12°C

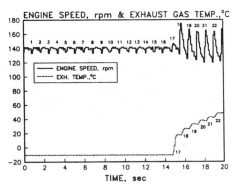

Figure D-5. Engine speed and exhaust gas temperature during starting test with gasoline at -12°C.

GASOLINE STARTING -12°C

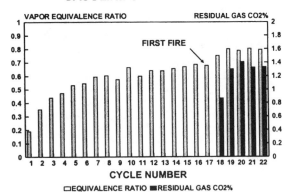

Figure D-6. Cycle-by-cycle equivalence ration and residual gas CO2 for the starting test with gasoline at -12°C.

192

2002-01-1667

Estimation of Air Fuel Ratio of a SI Engine from Exhaust Gas Temperature at Cold Start Condition

Tongwon Lee
Daecheon College

Choongsik Bae
Korea Advanced Institute of Science and Technology

Stanislav V. Bohac and Dennis N. Assanis
University of Michigan

Copyright © 2002 Society of Automotive Engineers, Inc.

ABSTRACT

Wall wetting of injected fuel onto the intake manifold and cylinder wall causes unpredictable transient behavior of air-fuel mixing which results in a significant emission of unburned hydrocarbon (HC) emission during cold start operation. Heated exhaust gas oxygen (HEGO) sensors cannot measure the air-fuel ratio (A/F) of exhaust gas during cold start condition. Precise and fast estimation of air/fuel ratio of the exhaust gas is required to elucidate the wall wetting phenomena and subsequent HC formation. Refined A/F estimation can enable the control of fuel injection minimizing HC emissions during cold start conditions so that HC emissions can be minimized.

A new estimator for A/F of the exhaust gas has been developed. The A/F estimator described in this study utilizes measured exhaust gas temperature and general engine parameters such as engine speed, airflow, coolant temperature, etc. A fast response, fine-wire thermocouple was used to measure exhaust gas temperatures and a fast response flame ionization detector was used to measure HC emissions during the cold start period. A Generalized Regression Neural Network Function Approximation (GRNN) was used to estimate the A/F of exhaust gas. The A/F traces generated by the GRNN algorithm agree very well with measurements.

INTRODUCTION

'Cold Start' is defined as the engine operation period before the coolant temperature of the engine reaches about 80°C. During the early stage of cold start, due to the low temperature of the intake system, the injected fuel does not evaporate easily and is partially in the state of fuel film. The fuel film in the intake manifold and cylinder wall is emitted as a combustion product or unburned hydrocarbon. This is why the A/F of the exhaust gas is lean at the beginning, and it gradually approachs stoichiometric. The cold start duration is divided into two modes. The first mode is the cranking and the second is the warming-up. During the cranking mode, the operation stresses engine startability. Very rich air/fuel mixture is required during this period to compensate for the liquid fuel fraction, which is not easily vaporized for mixture preparation. During the warming-up mode, the commercial lambda sensor (HEGO; Heated Exhaust Gas Oxygen Sensor) is inactive yet due to its own thermal inertia in the cold environment. This forces production control strategies to run in the open-loop mode and to use look-up tables for A/F control during cold start.

The fuel injectors deliver a spray of atomized fuel into the intake port and, under cold running conditions, a significant part of the spray is deposited on the intake port surfaces and valve backs forming a thin film. Fuel is subsequently evaporated from fuel film in the intake system. Fuel is emitted as a form of a unburned fuel or a fuel-originated product, and strongly influence emissions of HC [1-3]. Very rich fuel injection is required because of a discrepancy between fuel injected and fuel induced into the cylinder. This fuel is becoming the main source of automobile hydrocarbon emissions.

In the early stage of a cold start, the HEGO sensor is inactive due to the low temperature of exhaust gas. Even when the HEGO sensors are active, they still only yield information of the A/F mixture being rich or lean of stoichiometry. This switch-like behavior does not enable closed-loop operation at conditions other than stoichiometry. Thus, commercial engines adopt open loop control to calculate fuel injection quantity. Control of A/F is an important factor in reducing exhaust emissions, especially HC. Catalytic efficiency could be secured within a very narrow range of A/F during cold start, where the commercial oxygen sensors are inactive [4]. This requires a new control strategy to replace the open-loop mode using look-up tables for A/F control during cold start. These open-loop systems are unable to compensate for disturbances such as fuel quality, aging over time, and manufacturing tolerances. All of these disturbances can degrade the performance of open-loop strategies [5]. The A/F estimation model during cold start by measuring the torque fluctuation of the engine was attempted by Asik et al. [6].

The mixture is usually set very rich at low coolant temperature. A large amount of hydrocarbon emission is produced under cold start condition, where startability and stability are more important than exhaust emissions. The exhaust gas temperature reaches a maximum value near stoichiometry. This can be used to develop a tool to estimate the A/F in cold start condition utilizing fast response thermocouples.

In this research, by measuring exhaust gas temperature and other engine parameters, A/F is estimated during a cold start. The A/F estimation from exhaust gas temperature measured by fast response thermocouple, was performed to replace the function of the HEGO sensor during a cold start. A mechanism for A/F estimation was established using Generalized Regression Neural Network (GRNN) algorithm which utilizes exhaust gas temperature, coolant temperature, oil temperature and HC concentration as well as wide range lambda sensor (UEGO) data.

ESTIMATION OF AIR FUEL RATIO USING A NEURAL NETWORK

Exhaust gas temperature rises gradually with engine coolant temperature and eventually reaches stationary state after warming up period. Exhaust gas temperature is determined by each engine driving condition such as engine rpm and load. Exhaust gas temperature reaches a maximum at a slightly rich A/F ratio [7]. By measuring the exhaust gas temperature, HC concentration, and coolant temperature simultaneously, an attempt has been made to find estimation model could be utilized for exact engine

control, consequently to reduce HC in cold start condition. A/F was estimated by GRNN (Generalized Regression Neural Network), which is available in the Matlab Neural Network Toolbox. GRNN is a feedforward neural network for system modeling and estimation. GRNN is a kind of radial basis network that is often used for function approximation. It is composed of 4 layers, where the 1st layer is the input layer. The 2nd layer is the pattern layer having one neuron for each input pattern. The 3rd layer is composed of 2 kinds of neurons, S-Neurons, and D-Neurons. S-Neurons calculate the sum of the weighted output, while D-Neurons calculate the sum of the output not weighted. Each output neuron has one S-Neuron and one D-Neuron. The last layer of the GRNN is the output layer, dividing each S-Neuron output with the D-Neuron output. A neural network algorithm is generally used when the system modeling is exceedingly difficult. Input and output parameters of the system are used to identify the system as a form of neural network.

Figure 1 shows the architecture of GRNN used in this study for A/F estimation. As shown in Fig. 1, input parameters are engine rpm (N), air flow rate, exhaust HC concentration, exhaust gas temperature, coolant temperature, oil temperature, and atmospheric temperature. The real exhaust A/F was measured by a wide range oxygen sensor.

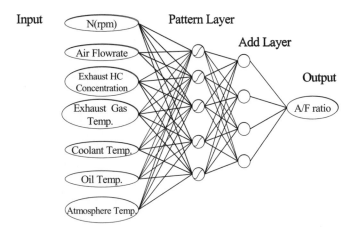

Figure 1. GRNN (Generalized Regression Neural Network) architecture for A/F estimation.

EXPERIMENTAL APPARATUS

ENGINE CONFIGURATION

All the engine experiments were performed at the Walter E. Lay Automotive Laboratory in the University of Michigan, with a 2.5L DOHC engine. It is a 60° V6 engine with aluminum head and block with DOHC and four valves per cylinder. It uses sequential port fuel injection and has two unequally-long intake runners per cylinder. At speeds below 3500 rpm (which accounts for all tests performed in this study) the short runner is closed at the intake manifold, which greatly increases in-cylinder turbulence and burning rate. Two fast light-off catalysts integrated into the double-wall stainless steel manifolds are positioned very close to the engine. Table 1 describes detailed specifications of the engine, and Fig. 2 shows the engine in the test cell.

Table 1: Specifications of 2.5L DOHC engine.

Bore	82.4 mm
Stroke	79.5 mm
Connecting rod length	138 mm
Piston pin offset	1.4 mm
Compression ratio	9.7
Combustion chamber	pent roof
Piston	flat
Intake valve IVS dia. (each)	27.9 mm
Exhaust valve IVS dia. (each)	22.2 mm
EVO @ 1 mm valve lift	43° BBDC
EVC @ 1 mm valve lift	14° BTDC
IVO @ 1mm valve lift	6° ATDC
IVC @ 1mm valve lift	21° ABDC primary 28° ABDC secondary
Maximum torque @ speed	165 lb-ft @ 4250 rpm (advertised)
Maximum power @ speed	170 bhp @ 6250 rpm (advertised)
Minimum BSFC @ speed, load	253 g/kWhr @ 2000 rpm, WOT

Figure 2. Research engine on the dynamometer.

The block assembly and catalysts were new and the heads came from a vehicle with 30,000 miles. Testing began after a break-in procedure for the engine and an ageing procedure for the catalysts. The engine was mounted to a DC dynamometer (Westinghouse) controlled by a Dyn-Loc IV controller (Dyne Systems). Fuel injection and spark timing were set with a Cosworth/Intelligent Controls IC 5460. The IC 5460 modulated injector pulse width so that the desired exhaust lambda value was achieved. Ignition timing at zero load was set to 30° BTDC.

SENSOR POSITIONS

The locations of the sensors in the exhaust manifold are depicted in Fig. 3. Data such as coolant temperature have been acquired via Labtech data acquisition software. Fine wire thermocouples were used for the measurement of exhaust gas temperature while a fast FID and a wide range oxygen sensor were used for the measurement of exhaust gas concentration. Figure 4 shows the exhaust gas temperature sensor and the Fast FID analyzer probe.

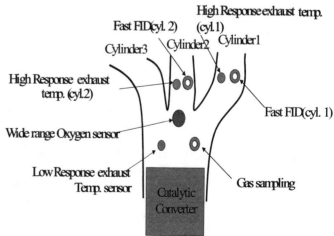

Figure 3. Position of sensors at the exhaust manifold.

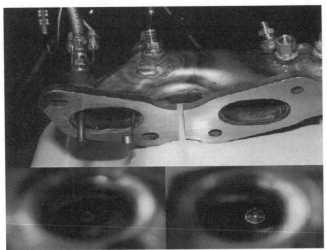

Figure 4. Installation of the fast FID probe and fast temperature sensor located 40mm from the exhaust valve seat.

CYCLE-RESOLVED EXHAUST GAS TEMPERATURES MEASURED BY A FAST-RESPONSE THERMO-COUPLE

A type K fast response thermocouple (Medtherm) was used to measure the exhaust gas temperature (Fig. 5), which has a diameter of $25\,\mu m$ and is surrounded by a triple radiation shield [9]. During the transient cold start condition, it can measure the instantaneous exhaust gas temperature. It cannot withstand the thermal load at high engine speeds or loads but lasts long enough to acquire temperature at low engine speed and load. The fast response thermocouple signals are amplified with 10kHz isolated amplifiers. In these experiments, the signals are recorded with the time based data acquisition system. Response time of the thermocouple was found to be less than 10ms at typical exhaust gas velocities [9].

A/F MEASUREMENT WITH WIDE RANGE LAMBDA SENSOR (UEGO)

During a cold start, A/F of the exhaust gas was measured using a Bosch wide range lambda sensor (UEGO) at several engine speed and load conditions. A/F could be measured during cold start condition since the lambda sensor was preheated well before the start of engine. As shown in Fig. 3, the UEGO sensor is located at the junction of the exhaust manifold.

THC MEASUREMENTS WITH FAST FID ANALYSER

The steady-state emission analyzers provide time-averaged emissions data for locations farther away from the cylinder where the flow is more steady. A fast response analyzer is needed to monitor the realistic fluctuation of the exhaust gas motion, temperature and HC concentration, which may vary significantly through an engine cycle. For this reason, two fast FID (Flame Ionization Detector; Cambustion HFR 500 system) probes are mounted in the exhaust system. Exhaust gas is taken from a point in the exhaust system and passes through a 360 mm heated sample probe to a miniature flame ionization detector. The analyzed data were acquired by a data acquisition system (Labtech). The attached position is at the outlet of the exhaust valve.

The sample probe opens into a constant pressure chamber and also feeds a connecting tube to the FID flame. The constant pressure chamber ensures that a constant flow of exhaust is brought to the flame. A thorough review of the fast FID was given by Cheng et al. [8]. The exhaust gas spends 5.2 ms in the sample probe (transit time) and the FID has a response time of 1.4 ms. The output signal is shifted by the transit time. The response time causes a slight smearing of the signal (17 crank angle degrees at 2000 rpm), which is considered small enough to be neglected.

Figure 5. Fast FID and exhaust gas temperature sensor.

AIR AND FUEL FLOW MEASUREMENT

Airflow into the engine was measured with a laminar flow element (400 cfm, Meriam) and a differential pressure transducer (Sentra). The signal passes through an RC filter and is recorded by a time based data acquisition system installed in a PC (CIO-EXP16 & CIO SAS08 supported by NI Labtech). Airflow was also measured by a production mass airflow sensor (Hitachi). Airflow measurements were corrected for temperature, pressure and humidity. Fuel flow is measured by a positive displacement flow meter (Pierburg 103A-150). The fuel is temperature conditioned and supplied to the engine at 42 psi. The time based data acquisition system converts volume to mass flow and records mass flow versus time.

TEST RESULTS AND DISCUSSIONS

The engine was motored by a DC dynamometer until the coolant temperature reached 25°C. After the coolant temperature reached 25°C, fuel injection was started. Data acquisition takes place at the same time.

AIR-FUEL RATIO

A/F of the exhaust gas was measured in several conditions using the UEGO sensor. The measured exhaust A/F trends are illustrated in Fig. 6 during a cold start at 2000 rpm, 12% throttle opening with supplied air/fuel ratio 13 based on the measured air flow and injected fuel amount. It can be noticed that A/F is very lean at the beginning and gradually converges to A/F 13. The area between the A/F 13 line and the measured A/F line is considered to be the contribution of the fuel film (wall wetting) in the intake manifold and the cylinder and unburned hydrocarbons above and beyond what are generated at steady-state operation. As the engine warms-up gradually, the "wall wetting" amount in intake system and combustion chamber wall is decreased. The amount of fuel supplied to engine is not changed. The A/F of the exhaust gas has been changed gradually from lean to stoichiometric operation.

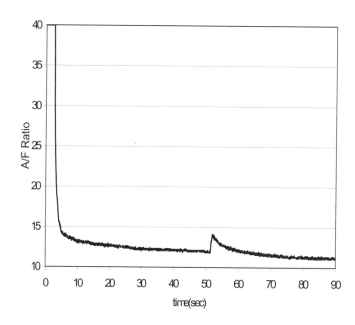

Figure 7. Measured exhaust A/F (UEGO)(1000 rpm, throttle from 6% to 10%(at 50 sec), A/F=11).

EXHAUST GAS TEMPERATURE MEASUREMENT

Exhaust gas temperatures were measured by the K-type fast response (fine wire type) thermocouples and the slow response thermocouple together. The results are given in Fig. 8. The throttle position was also suddenly changed from 6% to 10% at 50 second as shown in Fig.7.

Exhaust gas temperature was measured according to this procedure, and the neural network algorithm estimates the A/F using the measured temperature. There is about 3 ~ 4 seconds of time delay between the slow response thermocouple and fast response thermocouples. The slow response thermocouple data show lower temperature, because it is equipped on farther downstream position in the exhaust manifold and does not have a radiation shield.

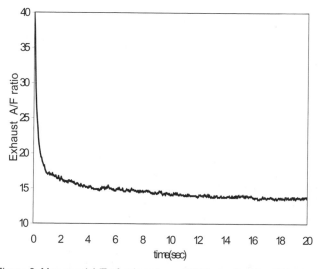

Figure 6. Measured A/F of exhaust gas (2000rpm, throttle 12% open, Supplied A/F=13)

Figure 7 shows measured exhaust A/F at 1000 engine rpm with the supplied A/F=11. Sudden change was made in throttle opening position from 6% to 10% at 50 seconds. Again A/F is very lean at the beginning, and gradually reaches to supplied A/F. At 50 seconds, due to the abrupt change of the throttle opening position, the A/F changes sharply, and gradually reaches normal condition (A/F 11) again.

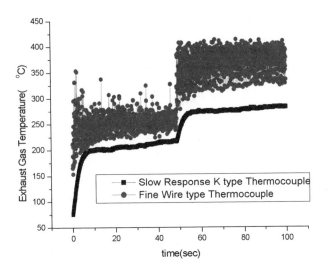

Figure 8. Results of exhaust gas temperature measurement with two different thermocouples corresponding to Fig. 7. (1000rpm, throttle opening from 6% to 10% at 50sec, supplied A/F=11, starting at 25°C coolant temperature).

Figure 9 shows the exhaust gas temperature measured by the fast response thermocouple at 1500rpm, throttle opening 10%, A/F=13 corresponding the conditions of Fig.6.

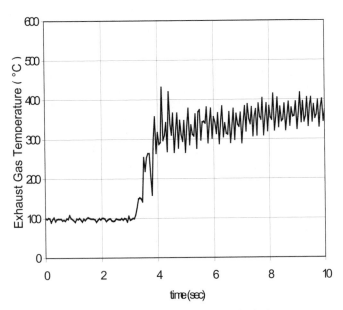

Figure 9. Exhaust gas temperature measured by the fast response thermocouple (1500rpm, throttle 10%, A/F=13).

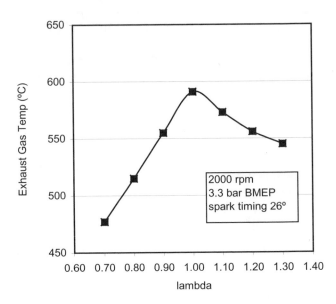

Figure 10. Exhaust gas temperatures before the catalyst brick as a function of lambda (measured by slow response K type thermocouple).

Figure 10 shows how that exhaust gas temperature before the catalyst changes with lambda. This proves that A/F could be estimated from the exhaust gas temperature. Rich mixtures cool down the exhaust more than lean mixtures because a rich mixture has a higher specific heat and also burns faster which allows more expansion work to be extracted.

HYDROCARBON EMISSIONS

The total hydrocarbon (HC) concentration of the exhaust gas, detected by the fast FID, is shown in Fig. 11. During the initial part of a cold start, large amounts of unburned hydrocarbons are produced. The HC concentration briefly decreases, and then gradually increases again. The high HC concentration at the beginning is due to the unburned fuel.

During the period when HC level is gradually increasing, it seems that the fuel film in intake system and combustion chamber is evaporating and being emitted. After about 50 seconds engine operation, the throttle position was suddenly changed from 6% to 10% and the hydrocarbon concentration drops considerably. After that, the hydrocarbon concentration gradually increases and combustion becomes steady.

Figure 12 shows the unburned hydrocarbon concentration at 1500 rpm, throttle opening 10%, A/F=13, when the measurement starts at 25° C coolant temperature using a fast FID analyzer. At the beginning, unburned hydrocarbon concentration is very high, the peak value exceeds 12,000 ppm. After about 2~3 seconds, the concentration becomes stable at about

198

2000 ppm. This graph shows that the hydrocarbon emission is significant at the early stages of cold start.

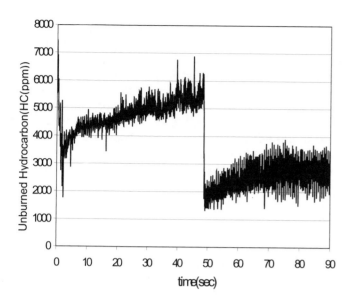

Figure 11. Unburned hydrocarbon concentration (1000 rpm, throttle opening changed from 6% to 10% at 50 sec, supplied A/F=11, measurement starting at 25° C coolant temperature).

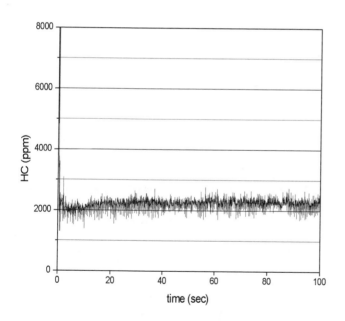

Figure 12. Unburned hydrocarbon concentration (1500 rpm, throttle 10%, A/F=13, measurement starting at 25° C coolant temperature).

ESTIMATION OF A/F BY NEURAL NETWORK ALGORITHM

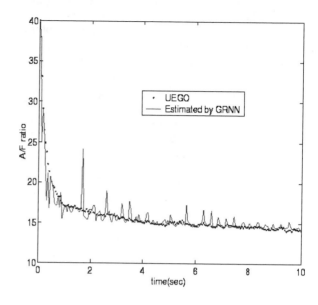

Figure 13 Measured and estimated A/F (estimated by single input GRNN, 2000rpm, 12% throttle opening, supplied A/F 13, GRNN training with a single input (Exhaust Temp. only)).

Figure 13 shows A/F estimated by the Generalized Regression Neural Network (GRNN) algorithm with a single input, compared to the experimental data. The exhaust gas temperature is the only input, and the output is measured A/F at the wide range lambda sensor position.

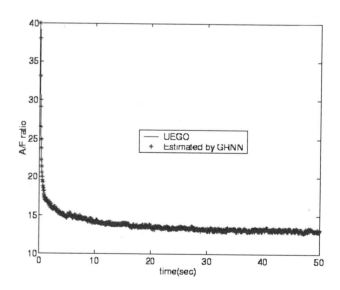

Figure 14. Measured and estimated A/F (estimated by multiple input GRNN, 2000rpm, 12% throttle opening, supplied A/F 13, GRNN training with multiple inputs).

Figure 14 shows A/F estimation by GRNN algorithm with multiple input parameters. A/F was estimated using the Generalized Regression Neural Network (GRNN) algorithm. The input parameters are exhaust gas temperature, engine rpm, air flow rate, coolant outlet temperature, oil outlet temperature, HC concentration in the exhaust gas and air temperature. The output is exhaust A/F ratio. The A/F ratio measured from the UEGO sensor and estimation are in good agreement. Figure 13 and Fig. 14 seem to show that more input parameters cause the model to work better. Three different conditions have been tested and their comparisons are shown in Fig. 15. Each trace was acquired during 10sec in cold start condition. The calculated values using the GRNN algorithm showed good agreement to the experimental results.

For more robust results, experimental data should be acquired for a wider range of engine operating conditions.

Figures 15 verifies that the A/F estimation method proposed here could be used as a tool for A/F control during cold start conditions (See also [10-14]). The fast response thermocouple is very sensitive and useful to measure the exhaust gas temperature in the medium range of engine speed and load conditions in spite of its fine wire size. However, the durability problem and the price of the sensor make it un-practical. Further studies of durability and cost are necessary in the future. The thermocouple measuring the metal surface temperature could be a practical option.

CONCLUSIONS

A new estimator was developed to predict the A/F from the exhaust gas during cold start. In this research, many parameters such as exhaust gas temperature, coolant temperature are used to estimate the A/F of the exhaust gas. Important conclusions are as follows:

1. The concentration of the hydrocarbon emissions in the exhaust gas is very high at the early stage, decreases shortly after, and gradually increases again. It implies that the "wall wetting" phenomenon occurs in the intake manifold, cylinder wall and piston crevice.

2. During cold start, the algorithm to estimate A/F has been developed coupled with exhaust gas temperature measurement. GRNN (Generalized Regression Neural Network) algorithm was used for estimation. From the experiment at the condition of 2000 rpm, 12% throttle opening with supplied A/F=13, it was found that A/F was very lean at the beginning and converged to A/F 13 gradually. The discrepancy between the supplied A/F of 13 and the measured A/F is supposed due to the contribution of fuel film (wall wetting) in the intake manifold and the cylinder.

3. As engine temperature rises gradually, the amount of fuel film due to "wall wetting" decreases, while the amount of fuel supplied to engine is fixed. Fast response thermocouple with a fine wire was used to measure the exhaust gas temperature precisely.

4. A method to estimate the A/F of the exhaust gas is developed using many engine parameters including the exhaust gas temperature. The GRNN algorithm can trace the A/F very well with neglected error.

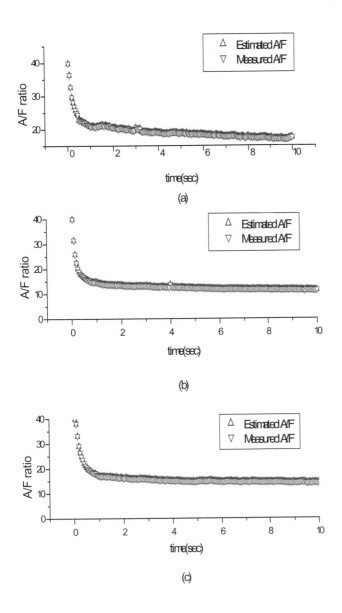

Figure 15. Estimated A/F by GRNN and measured A/F from engine (calculated with combination of 3 different conditions) (a) 2000rpm, 12% throttle, supplied A/F = 13 (b) 2000rpm, 15% throttle, supplied A/F = 11 (c) 2000rpm, 15% throttle, supplied A/F = 16.

ACKNOWLEDGMENTS

The authors would like to express their thanks to the support of Ford Motor Company, the University of Michigan, and the National Research Laboratory (NRL) in Korea.

REFERENCES

1. Keiso Takeda and Takehisa Yaegashi, Kiy onoro Sekiguchi, Kimitaka Saito, and Noburo Imatake, "Mixture Preparation and HC Emissions of a 4-Valve Engine with Port Fuel Injection During Cold Starting and Warm-up", SAE technical paper No. 950074, 1995
2. Juhani Laurikko, Lennart Erlandsson and Reino Abrahamsson, "Exhaust Emissions in Cold Ambient Conditions: Considerations for a European Test Procedure", SAE technical paper No. 950929, 1995
3. Shayler, P. J., Davies, M. T. and Scarisbrick, A., "Audit of Fuel Utilisation during the Warm-Up of SI Engines", SAE technical paper No. 971656, 1997
4. M.J. Anderson, "A Feedback A/F Control System for Low Emission Vehicles", SAE Paper No. 930388, 1993
5. G. W. Pestana. "Engine Control Methods Using Combustion Pressure Feedback", SAE Paper No. 890758, 1989
6. Joseph R. Asik, Jennifer M. Peters+, Garth M. Meyer, and Donald X. Tang, "Transient A/F Estimation and Control Using a Neural Network", SAE technical paper No. 970619, 1997
7. John. B. Heywood, "Internal Combustion Engine", McGRAW-HILL, p770, 1988
8. Cheng, W., Summers, T., Collins, N., "The Fast Response Flame Ionization Detector", Prog. Energy Combust. Sci. Vol. 24, pp. 89-124, 1998
9. M. W. Carbon, H. J. Kutsch, and G. A. Hawkins, "The Response of Thermocouples to Rapid Gas-Temperature Changes", Transaction of the ASME, July, 1950
10. Ulrich Lenz and Dierk Schroeder, "Artificial Intelligence for Combustion Engine Control," SAE technical paper No. 960328, 1996
11. Josehp R. Asik, Garth M. Meyer, and Donald X. Tang, "A/F Estimation and Control Based on Induced Engine Roughness", Proceeding of the 1996 IEEE International Conference on Control Applications, Sep.15-18, 1996
12. Chen-Fang Chang, Nicholas P. Fekete, Alois Amstutz, J. David Powell, 1995," Air-Fuel Ratio Control in Spark-Ignition Engines Using Estimation Theory", IEEE TRANSACTION ON CONTROL SYSTEMS, VOL.3 NO.1, March 1995
13. Kenneth Leistering, Bahman Samimy, Giogio Rizzoni, 1996,"IC Engine Air/Fuel Ratio Feedback Control During Cold Start", SAE technical paper No.961022, 1996
14. Ulrich Lenz and Dierk Schroede, "Transient Air-Fuel Ratio Control Using Artificial Intelligence", SAE technical paper No. 970618, 1997

DEFINITIONS, ACRONYMS, ABBREVIATIONS

ABDC	After Bottom Dead Center
A/F	Air Fuel Ratio
ATDC	After Top Dead Center
BMEP	Brake Mean Effective Pressure
BSFC	Brake Specific Fuel Consumption Rate
BTDC	Before Top Dead Center
ECU	Engine Control Unit
EVC	Exhaust Valve Close
EVO	Exhaust Valve Open
Fast FID	Fast Flame Ionization Detector
FTP75	Federal Test Procedure for testing emissions and fuel economy Performed on chassis dynamometer
GRNN	Generalized Regression Neural Network
HC	Hydrocarbon
HEGO	Heated Exhaust Gas Oxygen sensor
IVC	Intake Valve Close
IVO	Intake Valve Open
Lambda	Air excess ratio
THC	Total Hydrocarbon
UEGO	Universal Exhaust Gas Oxygen sensor
WOT	Wide Open Throttle
λ	Air excess ratio

CHAPTER 5

SPARK RETARDATION FOR IMPROVING CATALYST LIGHT-OFF PERFORMANCE

Stephen Russ
Ford Motor Company

5.1 OVERVIEW

Meeting stringent emissions standards requires the catalyst system to achieve operating temperature and high conversion efficiency (light-off) as soon as possible following a cold start. To meet this goal, sophisticated catalyst heating devices have been proposed including electrically heated substrates, exhaust system combustion devices and secondary air injection into the exhaust. With the exception of secondary air injection, these exhaust heating devices have not found widespread application due to the added cost and complexity accompanying these systems. More rapid catalyst light-off is typically achieved by packaging the catalyst much closer to the engine to minimize heat losses from the exhaust gases and by optimizing the cold start calibration. Cold start calibrations for rapid catalyst light-off typically include three actions:

- **Increased Idle Engine Speed**: Increased cold idle speed increases both the exhaust gas temperature and the mass flow rate to provide more heat to the catalyst at cold start. HC oxidation increases at higher engine speeds presumably due to increased exhaust gas temperatures and improved mixing.
- **Lean Air/Fuel (A/F) ratio**: A slightly lean A/F results in a small amount of excess oxygen to improve HC oxidation. This lowers the feedgas HC emissions during the portion of the cycle when the catalyst is inactive. Additionally a slightly lean A/F ratio also reduces the temperature at which the catalyst becomes active (light-off temperature). This is due to the excess oxygen as well as an increased percentage of reactive olefins in the exhaust.
- **Retarded Ignition Timing**: Retarding ignition timing results in increased exhaust gas temperature as well as increased exhaust mass flow due to the decrease in engine thermal efficiency. This dramatically increases the warm-up rate of the catalyst as well as increases the oxidation of the feedgas HC emissions after leaving the cylinder.

The use of spark retard increases exhaust gas temperatures because the burned gas is not ideally expanded and does not perform as much work on the piston as the MBT case. In fact, with sufficient spark retard, combustion may not even be complete by the time of exhaust valve opening. Subsequent reactions in the exhaust port and manifold release heat to increase the exhaust gas temperature. The use of aggressive spark retard has two main drawbacks. The first is the obvious loss in efficiency that results in a small fuel economy penalty. This penalty is not large as aggressive spark retard is typically used for less than one minute during engine warm-up. The second drawback is that spark retard increases cycle-to-cycle IMEP fluctuations, increasing engine roughness (usually expressed in terms standard deviation of IMEP). The primary cause of cyclic variations in IMEP for retarded spark operation is variations in the combustion phasing (location of 50% mass fraction burn). The expansion ratio decreases rapidly during combustion for retarded spark timing and therefore the combustion phasing of an individual cycle determines the cycle thermal efficiency and the

IMEP produced. Solutions to enable additional spark retard at cold start therefore must focus on making combustion repeatable (similar burn rate cycle-to-cycle) to decrease engine roughness. In addition to operation with retarded ignition, a slightly lean A/F ratio is typically calibrated as outlined previously. The engine must be capable of stable operation at this slightly lean calibrated A/F in addition to having acceptable drivability with less volatile (heavier, higher drivability index) fuels that produce leaner A/F excursions due to additional fuel held up in liquid films on the intake port and valve surfaces.

Typically base engine design changes are made for low emissions applications to improve the engine combustion for lean, retarded spark operation. Two examples of this are:

- **Enhanced Charge Motion:** A higher level of in-cylinder swirl or tumble is frequently used to stabilize combustion. This can be achieved with either a tumble or swirl producing valve in the lower intake manifold or by a cam-switching mechanism that can deactivate one of the intake valves on a multi-valve engine. The higher level of in-cylinder charge motion produces a faster burn rate that is also more repeatable cycle-to-cycle producing lower variations in IMEP. In addition to the improvement in combustion stability, the higher air velocities and turbulence levels in the intake port can also improve the fuel wetting and mixture formation. This results in a reduction in the required fuel injection for start-up with accompanying improvements in HC emissions and drivability.

- **Dual Ignition:** Multiple spark plugs produce a faster burn rate and more stable combustion due to the additional flame area produced by igniting in several locations. The combustion stability or consistency is improved since any non-uniformities of A/F, residual exhaust or turbulence intensity in the chamber are "averaged out" by the two spark locations.

7.2 SELECTED SAE TECHNICAL PAPERS

2000-01-0551

A Quick Warm-Up System During Engine Start-Up Period Using Adaptive Control of Intake Air and Ignition Timing

Masaki Ueno, Shusuke Akazaki, Yuji Yasui and Yoshihisa Iwaki
Honda R&D Co., Ltd.

Copyright © 2000 Society of Automotive Engineers, Inc.

ABSTRACT

Early activation of catalyst by quickly raising the temperature of the catalyst is effective in reducing exhaust gas during cold starts. One such technique of early activation of the catalyst by raising the exhaust temperature through substantial retardation of the ignition timing is well known. The present research focuses on the realization of quick warm-up of the catalyst by using a method in which the engine is fed with a large volume of air by feedforward control and the engine speed is controlled by retarding the ignition timing. In addition, an intake air flow control method that comprises a flow rate correction using an adaptive sliding mode controller and learning of flow rate correction coefficient has been devised to prevent control degradation because of variation in the flow rate or aging of the air device. The paper describes the methods and techniques involed in the implementation of a quick warm-up system with improved adaptability.

INTRODUCTION

While emission regulations for automobiles are being strengthened in various countries in recent years, Honda started mass production of the ACCORD-EX, a passenger car that meets the ULEV (Ultra Low Emission Vehicle) category of the California State LEV regulations in the USA. This is the first vehicle in the world complying with these regulations. The ACCORD-EX makes use of emission reduction technologies suitable for the period during engine start-up, the period immediately after engine start-up, and after the period after engine warm-up. During the period from start-up to immediately after engine start-up, a stable lean burn condition is enabled by the swirl obtained by closing the inlet valve on one side of the VTEC (Variable Valve Timing and Lift Electronic Control) System, as a result of which HC is reduced before activation of the catalyst. After engine warm-up, high precision control of air-fuel ratio[1] is performed by the STR (Self-Tuning Regulator) using adaptive control for utilizing the conversion ability of the three-way catalyst to the maximum extent, and an extremely high catalytic conversion ratio is obtained. In

this way, the amount of hazardous gases after warm-up is already reduced to a very small amount. However, to meet with stricter emission regulations anticipated in the future, the technology for reducing emissions further during engine start needs to be established.

During the period after cold start of the engine until the temperature of the catalyst reaches the catalyst activation temperature, a large quantity of the hazardous emissions are discharged into the atmosphere without being adequately converted by the catalyst. Research on methods for reducing the emission of hazardous gases during engine start-up is being extensively carried out using electric heaters or burners[2]-[4], which raise the temperature of the catalyst to the activation temperature quickly. Moreover, raising the exhaust temperature by increasing the quantity of intake air or by retarding the ignition timing results in the heating up of the catalyst by the exhaust gas. This method, which is similar to the effect obtained by an electric heater or a burner, is well known.

In this research, the focus of attention is on the latter method, which does not entail the use of any special hardware. The development and the effects of a quick warm-up system during the engine start-up period that enables the catalyst to be heated up adequately as required, are reported here.

PRINCIPLE

Individual tests on a 2.2L SOHC-VTEC engine (compression ratio of 8.8) were carried out at constant speeds (1000, 1500, 2000 rpm) in the no-load condition after warm-up for studying the effects of exhaust temperature due to ignition timing. Fig. 1 shows the intake air flow after changing the ignition timing. Fig. 2 shows the exhaust gas temperature at a position 20-mm in front of the close-coupled catalyst. From the results of these figures, it is observed that retarding the ignition timing necessitates a large intake air flow to maintain the engine speed at a constant level. This is because the IMEP (Indicated Mean Effective Pressure) reduces if the ignition timing is retarded while the intake air flow is maintained at a constant rate. It is also observed that the exhaust temperature increases as the ignition timing is retarded. This occurs because of the combination of the decreasing effect of the temperature drop of the combustion gas due to the reduction of adiabatic expansion period during the cycle, and the effect of the increase in the amount of heat because of the increase in the intake air flow.

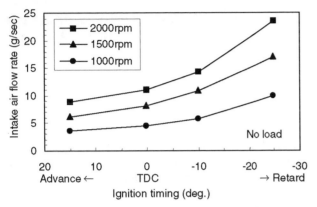

Figure 1.   Effect of intake air flow rate on ignition timing

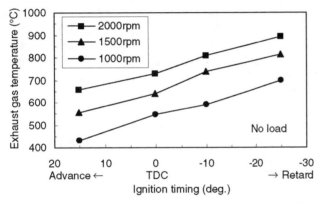

Figure 2.   Effect of exhaust gas temperature on ignition timing

The fundamental principle of the quick warm-up system during engine start-up period is that the temperature rise in the catalyst is promoted by increasing the intake air flow as mentioned above and by properly using retardation of the ignition timing during the cold start.

CONTROL METHODS

BASIC CONTROLS – The air intake control method and ignition timing control method required for realizing the basic principle mentioned earlier, are described here. One of the control methods for quickly raising the exhaust temperature is the method of retarding the ignition timing preferentially and controlling the engine speed by intake air. However, if this method is adopted, the fluctuations in engine speed cannot be fully tracked because of the actuator delay of the intake air system or because of the delay of factors related to the volume of the intake air system. Accordingly, we focused our attention on the fact that there is no delay in manipulated variable for command values in the ignition timing. We concluded that by preferentially supplying a large quantity of air using an air device by feedforward control, and by

controlling the ignition timing so that the engine speed reaches the target value, the accuracy and stability of the quick warm-up system during the start-up period can be improved.

Fig. 3 shows the configuration of the quick warm-up system during the start-up period, and Fig. 4 shows the basic controls. RACV (Rotary Air Control Valve) is an air device that makes use of a rotary valve, which is arranged in parallel with the throttle valve. Firstly, an amount of intake air that is greater than that required during a normal start-up is supplied by feedforward control for the period from the start to the end of warm-up control according to valve opening for the air device that has been set from the time of start-up. If the engine speed exceeds the preset engine speed, the ignition timing control is started by means of PI control so that the engine speed becomes the same as the preset engine speed. Since the quantity of intake air supplied has already been increased, the ignition timing moves to the retardation side, and by and by it reaches a state of equilibrium at a point where the ignition timing has been retarded to balance with the increase in the amount of intake air.

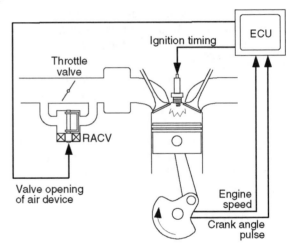

Figure 3.   Configuration of Quick warm-up system during start-up period

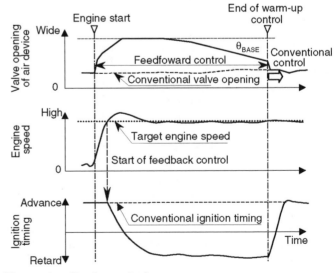

Figure 4.   Basic control

210

Quick warm-up by stable operation is possible in this way, by supplying a large quantity of intake air and by controlling the engine speed by retarding the ignition timing.

CORRECTING THE FLOW RATE OF AIR DEVICE – Air devices are constructed such that variation in air flow rate occurs. Based on an example of the RACV, which is an air device used in mass-produced vehicles, the tolerance in flow rate under design conditions is 10% or greater, as shown in Table 1. Such a large allowance is unavoidable and needs to be set because of cost factors and other factors related to manufacturing technology. Moreover, changes in flow rate characteristics may also be assumed to occur because of aging. If the command value and the actual value of valve opening of the air device do not coincide, the required quantity of heat cannot be supplied to the catalyst because of the change in the flow rate. This may lead to the problem of variation occurring in the achievable temperature of the catalyst when the quick warm-up system during start-up stops. In this way, to compensate for the variation in flow rate of the air device so that the quick warm-up system during start-up can be realized using the method of preferentially supplying intake air through feed-forward control, it is important to correct the intake air flow.

Table 1. Example of RACV Flow rate tolerance

Flow rate (m^3/h)	Tolerance (m^3/h)
1.8 (Minimum)	-0.7, +0.6 (39%)
22.0	±2.3 (10.5%)
64.0 (Maximum)	±6.8 (10.6%)

The items required for correcting the flow rate of the air device are as given below.

1. Avoidance of mutual interference between two-input and two-output systems for ignition timing and intake air

2. Flow rate compensation according to the variation due to factors given below
   - Variation in the actual value of valve opening of air device compared to the command value
   - Steady-state error in the valve opening of air device due to manufacturing error or aging

The description of correction method for flow rate to realize the requirements mentioned above is given below. This method consists of two functions: "Real-time Flow Rate Correction Function" that corrects the flow rate simultaneously with the implementation of the quick warm-up system during the start-up period; and the "Learning Function for Valve Opening of Air Device" that eliminates the steady state error in the valve opening of air device.

Real-time Flow Rate Correction Function – Firstly, the flow rate variation is detected by measuring the actual intake air flow, comparing it with the flow rate of the reference model decided beforehand and finding the error. Next, a controller is designed that outputs the correction value to the valve opening of the air device whenever required, so that the error mentioned above is eliminated. Fig. 5 shows the schematic diagram of the real-time flow rate correction function. The components of the flow correction method are described below. NE represents the engine speed.

Firstly, for the calculations of the reference model, the command value of valve opening of the air device (θ_{OBJ}) is considered to be the representative value of the intake air flow. By multiplying this value with the value of intake air flow of the reference model (G_{OBJ}), the cumulative heat value of the reference model (Q_{OBJ}) is calculated.

For estimation of the intake air flow of engine, the estimated value of the intake air flow (G_{ACT}) per cycle is calculated using the intake air pressure (P_{BA}). By multiplying the above value in a similar manner, the estimated cumulative heat value (Q_{ACT}) is calculated. Here k_1 and k_2 are unit conversion coefficients.

From the above, the error (E_Q) between the cumulative heat value of the reference model and the estimated value can be found. To correct this error (E_Q) in a stable manner quickly, we have made use of sliding mode control[5] in this research as shown in Fig. 6. In this figure, KRCH is the reaching gain, and K_{ADP} is the adaptive gain. In sliding mode control, the switching function ($\sigma(n)$) is defined on the phase plane in which the current value of error ($E_Q(n)$) and the previous value of error ($E_Q(n-1)$) are taken on the axis.

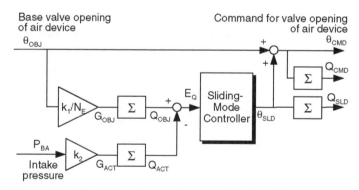

Figure 5. Real-time flow rate correction function

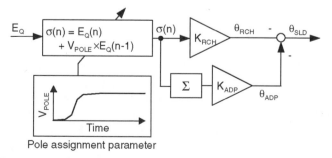

Figure 6. Sliding-mode controller

The main feature here is that the dynamic characteristics of the system can be specified univocally by the parameter (V_{POLE}) that expresses the slope of the switching function ($\sigma(n)$) on the phase plane. Moreover, the converging rate of the error can also be prescribed by V_{POLE}. Accordingly, by setting the flow rate correction feedback at a lower gain compared to the ignition timing feedback, the interference between the two feedback systems can be prevented at all times.

In view of the above, by employing adaptive sliding-mode control for the flow rate correction of the air device, adequate intake air as required can be supplied even if variation occurs in the flow rate of the air device.

Learning Function for Valve Opening of Air Device – If a steady-state error occurs in the valve opening because of aging of the air device or other reasons, there is a possibility that a flow rate correction that includes the same error will be implemented at each cold start of the engine.

To solve this problem, we have introduced a learning function by which the base valve opening of the air device is made to learn the correction amount from the real-time flow rate correction function, then the next time the quick warm-up system during start-up period is implemented, the correction amount is reduced further. Fig. 7 shows the calculation method. Q_{CMD} is the cumulative value of the reference heat when the quick warm-up during start-up period is terminated. Q_{SLD} is the cumulative value of the heat correction in the sliding-mode controller. C_{LRN} is the weighting coefficient used during learning value calculations. K_{LRN} is the learning value of the flow rate correction coefficient, which is multiplied with the base valve opening of the quick warm-up system during start-up period the next time. This learning value is calculated from the correction percentage actually used in real-time flow rate correction. In this way, compensation of steady-state error in the valve opening is possible.

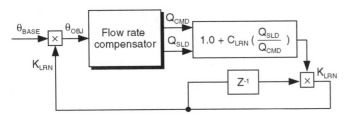

Figure 7.   Learning function for valve opening of air device

TEST RESULTS

EMISSION REDUCTION EFFECTS – The effects of the quick warm-up system during start-up period were verified by using a prototype vehicle based on the '98 model ACCORD (which meets the ULEV category of the LEV regulations) being marketed in California State in the USA. Fig. 8 shows the engine used in the test vehicle, exhaust system and the temperature measurement points. Table 2 shows the specifications of the test vehicle.

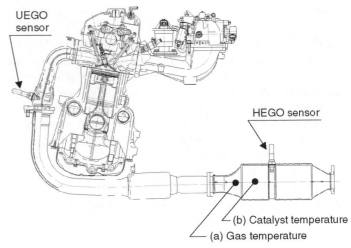

Figure 8.   Cross section of test engine and temperature measurement positions

Table 2.   Specifications of test vehicle

Vehicle	: 98 Model year ACCORD EX-ULEV
Engine	: L-4 , 2.3litters SOHC-VTEC (Prototype)
ECU	: Prototype
Catalyst	: Tri-metal type (Prototype)
	600cell/in^2, 1.7L,100,000mile aging

Fig. 9 shows the effect of temperature rise of the catalyst due to the introduction of the quick warm-up system during start-up period. During the acceleration from standstill 20 seconds after start, it is observed that the temperature of the under-floor catalyst reaches approximately 300°C.

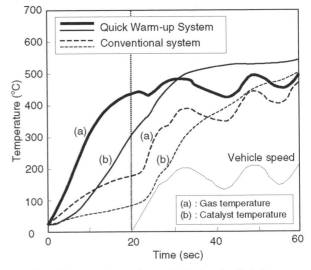

Figure 9.   Temperature rise effect due to Quick warm-up system during start-up period

Fig. 10 shows a comparison of HC and NOx emissions in the first hill for the FTP mode with and without employing the quick warm-up system during start-up period. When the quick warm-up system is employed, the volume of burnt gas immediately after start-up and before activation of the catalyst increases because of which the exhaust emission increases. However, it is noted that since the temperature of the catalyst increases abruptly later, the amount of emission after activation of the catalyst decreases significantly compared to the emission in the conventional method.

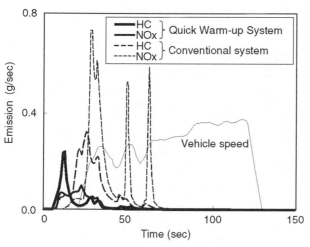

Figure 10. Emission reduction effect due to Quick warm-up system during start-up period

Table 3. Effect of reduction in emission due to quick warm-up system during start-up period (FTP Mode)

CO	NMOG	NOx
50%	45%	60%

Table 3 shows the percentage reduction in emission in the FTP mode when the quick warm-up system during start-up period is introduced. It is observed that the emission has practically halved because of the introduction of the quick warm-up system during the start-up period.

EFFECT OF FLOW RATE CORRECTION OF AIR DEVICE – The effect of flow rate correction air device verified from tests on the actual vehicle is reported here.

The test method used was as follows. During the start-up test with water temperature at 25°C and using a standard air device, the command value of the valve opening of the air device was reduced by 10%. This enabled arbitrary error in flow rate to be generated for the air device. Then the flow rate correction method for the air device was implemented. Fig. 11 shows the transition in the reference heat quantity and estimated heat quantity with the throttle opening in the fully-closed condition after initiation of the quick warm-up mode and until its

termination by a preset timer. Fig. 12 shows the base valve opening and the corrected valve opening of the air device during the period mentioned above. The actual valve opening of the air device was set 10% smaller than the valve opening of the standard air device, therefore the estimated heat quantity was smaller compared to the reference heat quantity. Accordingly, the flow rate correction due to the sliding mode controller (θ_{SLD}) works to increase the valve opening of the standard air device. As a result, the estimated heat quantity gradually approaches the reference heat quantity, and when the quick warm-up mode during the start-up period terminates, the estimated heat quantity reaches the reference heat quantity.

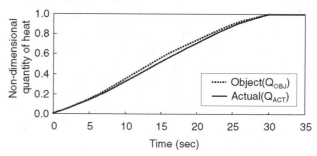

Figure 11. Transition of heat quantity in real-time flow rate correction function

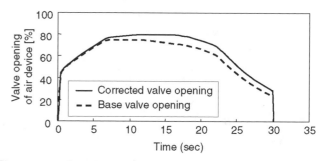

Figure 12. Transition of valve opening in real-time flow rate correction function

By repeatedly performing the start test, it was found that the estimated heat quantity coincided with the reference heat quantity after learning was adequately superimposed several times. It was verified that the repeated performance of the start test led to the elimination of the real-time flow rate correction of the valve opening of the air device, and variation due to steady-state aging and other factors was also eliminated.

EFFECTS OF DISTURBANCE – The quick warm-up system after start-up period makes use of the region of large retardation of ignition timing. This region is not used normally by conventional start-up systems. Generally, if the ignition timing is retarded considerably up to the top dead center, combustion is likely to become unstable, leading to the possibility of large fluctuation in the engine rotation because of the effect of disturbances, such as changes in fuel properties or the environment.

To verify the effects of various disturbances on combustion stability, we decided to confirm the adaptability of the quick warm-up system during start-up period with respect to two factors, namely fuel properties and atmospheric pressure.

Fuel properties – The properties of gasoline generally sold in the market vary considerably depending on the country or region. In the USA for instance, the gasoline properties are classified into five categories (A-E gasoline) according to the ASTM standards, and the use of a specific type of gasoline depends on the region or the season. In summer especially, gasoline with low volatility is used so that vapor locks that are likely to occur in environments in which the outside air temperature is high and degradation in the re-starting ability that frequently occurs in summer, would be prevented. Use of such gasoline leads to the tendency of degradation of cold startability. For the tests, we used re-blended gasoline shown in Table 4 to simulate the low-limit products of A-type gasoline with the lowest volatility that were available in the market, and performed the tests. Fig. 13 shows the results of comparison of the simulated gasoline with normal gasoline.

Table 4.   Properties of low volatile gasoline

Density :	0.750-0.762
RON :	90-93
MON :	82-86
Characters of distillation	
T10%(°C):	57-67
T50%(°C):	112-122
T90%(°C):	165-180

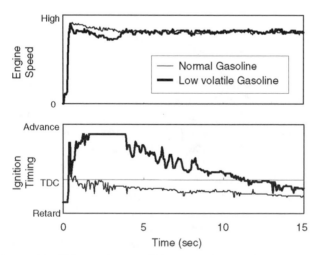

Figure 13. Effects due to differences in gasoline properties

When gasoline of low volatility is used, the engine rotation immediately after start is often impeded. However, as long as the target engine speed is not reached, the retardation control of ignition timing does not start. Therefore, startability equivalent to that

obtained by the conventional starting method can be maintained. After the target engine speed(RPM) is reached, the ignition timing is advanced from value for normal gasoline so that the engine speed does not decrease. Therefore, it is possible to perform quick warm-up while maintaining the steady state of the engine rotation without misfiring.

As mentioned above, it was possible to make the quick warm-up system adapt promptly to differences in gasoline properties. It was also verified that the quick warm-up effects could be utilized fully and in a stable manner for the gasoline properties mentioned above.

Atmospheric pressure – At high altitude, the valve opening of the air device used in intake air controls needs to be corrected because the atmospheric pressure reduces at high altitude. In the quick warm-up system during the start-up period, the correction of the valve opening of the air device is calculated based on the atmospheric pressure at start-up. By increasing the valve opening of the air device so that it becomes greater than the basic valve opening, intake air flow similar to that for low altitude locations is maintained.

For the test, an air pressure of 626 mmHg (83.4 kPa), equivalent to that at an altitude of 1600 m was set in the low-pressure chassis dynamometer laboratory and the test was performed in the LA-4 mode.

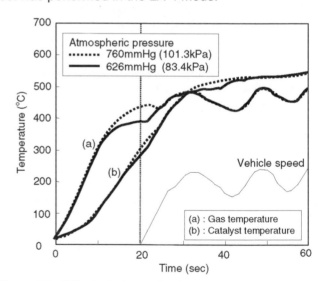

Figure 14. Effect of atmospheric pressure on gas temperature and catalyst temperature

As shown in Fig. 14, the temperature of the under-floor catalyst dropped by about 40°C, 20 seconds after start-up. This is because the command value of the valve opening of the air device exceeded the maximum allowable value of the valve opening, restriction was imposed on the actual valve opening, and full correction could not be applied. If an air device or DBW of higher capacity can be used and the shortage in intake air flow compensated, then a catalytic temperature equivalent to that at low altitudes can be maintained.

CONCLUSIONS

A quick warm-up system during start-up period was developed for reducing emissions during cold start of the engine. For this quick warm-up system, feed-forward control and ignition timing retardation control capable of preferentially supplying intake air were used. The variation in flow rate of the air device and errors due to aging were corrected enabling adequate heat quantity as required, to be supplied to the catalyst.

1. The exhaust emission in the FTP mode was nearly halved by introducing the quick warm-up system during start-up period.

2. By employing the real-time flow rate correction function using adaptive sliding-mode controller, it was confirmed that approximately ±10% variation in flow rate during the start period could be corrected.

3. By using the learning function for the flow rate correction coefficient, it was possible to eliminate the steady-state error that occurred in valve opening of the air device because of aging or other reasons.

4. Since differences in the properties of gasoline could be adapted to promptly, quick warm-up effects suiting the gasoline properties could be attained in a stable manner.

REFERENCES

1. Maki, H., et al., "Real Time Engine Control Using STR in Feedback System" SAE Paper 950007

2. Shimasaki, Y., et al., "Study on Conformity Technology with ULEV Using EHC System" SAE Paper 960342

3. Kollmann, K., et al., "Concepts for Ultra Low Emission Vehicles" SAE Paper 940468

4. Langen, P., et al., "Heated Catalytic Converter Competing Technologies to Meet LEV Emission Standards" SAE Paper 940470

5. Nonami, K. and Den, K., "Sliding Mode Control", Corona Publishing Co.,Ltd. 1994

940963

Time-Resolved Measurement of Speciated Hydrocarbon Emissions During Cold Start of a Spark-Ignited Engine

Edward W. Kaiser, Walter O. Siegl, Lanise M. Baidas, Gerald P. Lawson,
Carl F. Cramer, Kelvin L. Dobbins, Paul W. Roth, and Michael Smokovitz
Ford Motor Co.

ABSTRACT

Speciated HC emissions from the exhaust system of a production engine without an active catalyst have been obtained with 3 sec time resolution during a 70°F cold start using two control strategies. For the conventional cold start, the emissions were initially enriched in light fuel alkanes and depleted in heavy aromatic species. The light alkanes fell rapidly while the lower vapor pressure aromatics increased over a period of 50 sec. These results indicate early retention of low vapor pressure fuel components in the intake manifold and exhaust system. Loss of higher molecular weight HC species does occur in the exhaust system as shown by experiments in which the exhaust system was preheated to 100° C. The atmospheric reactivity of the exhaust HC emissions for photochemical smog formation increases as the engine warms.

INTRODUCTION

The control of hydrocarbon (HC) emissions has become an increasingly important feature of the design of motor vehicles. Recently, modest decreases in the permitted HC emissions have been instituted by both the Federal Government (1990 Clean Air Act Amendments) and the State of California [1]. The new regulations in California will require an additional major reduction of a factor of four in fleet-averaged new automobile HC emissions over the next decade, decreasing from 0.25 gm/mile NMHC in 1993 to 0.062 gm/mile NMOG in 2003. This fleet averaged emission standard represents a HC emissions level that is only 0.5% of the uncontrolled emissions from vehicles prior to the institution of regulation in the late 1960's. In addition, California will require that the reactivity of the emissions for photochemical smog formation, rather than solely the total HC mass, form the basis of the regulations. These new regulations in the state of California provide a demanding challenge to the automotive industry.

Currently, a catalytic converter provides the majority of the engine-out emissions control from spark-ignition (SI) automobiles. The bulk (>50%) of the tailpipe emissions from these vehicles exit during cold start, prior to the time when the catalyst reaches its minimum operating temperature (light off). Thus, the cold start emissions are essentially engine-out emissions, and can be controlled by a combination of a reduction in the engine-out emissions and/or a reduction in the time for catalyst light off. Because of the requirement that the photochemical reactivity of the HC emissions be included in the regulation process, it is crucial to examine both the total emissions and the photochemical reactivity of these emissions as a function of time during the cold-start process. Such experiments are necessary to evaluate the effects of reduced light-off time and changes in engine control strategies on the ability of vehicles to meet the new reactivity-based HC regulations.

Total hydrocarbon emissions during cold start have been discussed in a recent publication by Boam et al [2]. Brown and Woods [3] have examined time-resolved emissions during the warm-up period in a propane-fueled engine with gas chromatographic analysis of species. However, to our knowledge, only one study of the time dependence of the exhaust gas HC composition during cold start has been carried out using gasoline [4]. While reactivity factors as a function of time after cold start were calculated in this paper, detailed information on the distribution of HC species in the exhaust were provided. Studies of HC emissions from gasoline fuel during cold start are of considerable importance because gasoline fuel consists of a multi-component mixture whose individual components have a wide range of boiling points. Thus, the composition of the fuel entering the combustion chamber early in the cold start should be skewed toward the lower boiling point components. This may result in initial exhaust gas emissions with substantially different photochemical reactivity in the atmosphere than those that exit later in the cold start process when the engine is warmer. The time dependence of the reactivity could influence the effectiveness of earlier catalyst light off as a technique to meet HC emissions standards which are based on atmospheric reactivity.

In addition, recent experiments have shown that changing engine operating conditions can affect not only the total mass of engine-out HC emissions but also the distribution of the HC species [5,6,7] and, therefore, the reactivity of these emissions [8]. Thus, studies of the effect of different operating conditions on both the total and individual species emissions are required to assess the effectiveness of possible HC control strategies during cold starts designed to meet reactivity-based standards. In this paper, we measure both the total HC emissions and the individual species emitted from the exhaust system as a function of time after the start of cranking during the 70° F cold start of a production engine with a time resolution of 3 sec.

EXPERIMENTAL

DESCRIPTION OF THE ENGINE - Dynamometer laboratory experiments have been used to generate the data. For these experiments, a production engine (4.6L 2V 8 cylinder engine connected to a dynamometer through an AOD transmission) was used with the normal engine controls and starter motor. During the cold start, a preprogrammed test cycle including engine speed, spark timing, and fuel/air ratio was used. Extreme care was exercised in these experiments to minimize sources of test variability such as: crankshaft/camshaft initial positions (prepositioned to the same locations for each test); electronic engine control (EEC-IV) adaptive fuel tables (reinitialized to the same values for each test); cranking battery state of charge (fully charged for each test); engine system precondition (engine preconditioned identically at the end of each test); and fuel injector leakage into intake manifold during pretest soaking (intake manifold dry at start of cranking). With these precautions, repeatable test results have been achieved, enabling single-test comparisons between different sets of strategy and calibration parameters. Without these precautions, small changes in system behavior caused by strategy and calibration changes could be lost in test repeatability fluctuation. A "bare brick" catalyst support without wash coat or precious metal loading was included in each leg of the dual exhaust system to maintain as close similarity to the performance of a catalyst-equipped engine as possible. Each support was connected to the exhaust manifold by a 5 cm diameter 10 cm long exhaust pipe. Beyond the catalyst supports, the two exhaust pipes merged into a 400 cm long tailpipe which carried the exhaust gases to the CVS sampling system.

Exhaust samples were diluted using a CVS dilution and bagging system. This bagging system was modified by installation of 5 solenoid valves to which were attached Tedlar bags, each having an approximate volume of 10 L. During the cold start process, each solenoid valve could be programmed to open at any predetermined time. The sampling duration was also variable but for these experiments was normally kept at 3 sec. This duration provided sufficient sample for gc analyses of HC species. In addition, exhaust samples diluted by the constant volume air flow were withdrawn from the CVS dilution system and directed through an emissions console for analysis of CO, CO_2, total HC, and NO_x.

Two different cold-start strategies were tested in these experiments. The A/F ratio as a function of time after start is shown in Figure 1A for the baseline strategy.

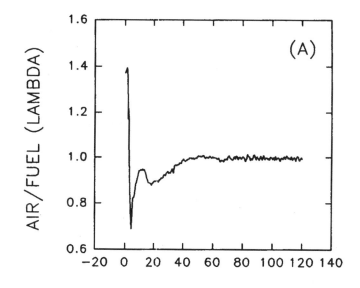

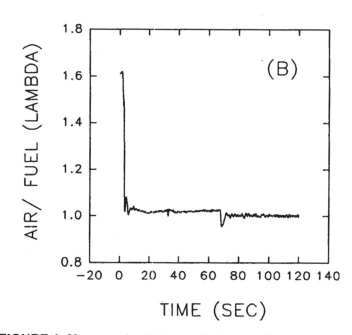

FIGURE 1. Measured relative air/fuel ratio (lambda= {A/F}/{A/F}$_{stoic}$) as a function of time after start of cranking: (A) - Base strategy; (B) - Strategy II.

In this strategy, a predetermined amount of fuel was injected sequentially to each port for the first second to initiate engine firing. This predetermined amount is approximately twice the amount required to achieve an air/fuel ratio of 11/1 in order to achieve firing in each cylinder during the second cycle. The fuel amount was then reduced to achieve the desired fuel-air ratio based on the measured air flow. Ibis is similar to a cold start in a vehicle. A rich spike which lasts 2-3 sec occurs immediately after the engine begins to fire followed by more moderate rich operation during the next 30 seconds and then continuous running at stoichiometric air/fuel ratio. The engine speed is 800 rpm for the first 10 seconds, 20 sec of operation at 1000 rpm, and then a return to 800 rpm for the remainder of the test. The spark timing is retarded approximately 5-10 degrees from MBT for the entire cold start.

A second strategy was developed which reduced the total HC emissions. This was achieved by increasing the idle speed to 1200 rpm for the first 30 sec and to 1000 rpm for the remainder of the test, retarding the spark timing by 20-25 degrees from MBT, and leaning out the mixture to lambda = 1.05 for the first 60 sec and 1.0 for the remainder of the test (see Figure 1B). The intake air flow was 1.5 to 2 times larger than that for baseline because the engine speed was higher and a more open idle-bypass-valve setting was required to run the engine with the extreme spark retardation that was used.

The solid lines in Figures 2A and 2B present the time-resolved, diluted total HC emissions measured as a function of time by the continuous sampling emissions console for the two cold start strategies. Because of the transport delay time between the CVS diluter and the emissions console, the appearance of the HC emissions shows a delay of approximately 10 seconds relative to that of the start of engine cranking (t=O). In fact, the engine begins to fire within 1 second after the beginning of the baseline cold start test as shown by monitoring the in-cylinder pressure. That the 10 sec delay is caused by the sampling system was verified by connecting a MEXA 1300 FKI fast response NDIR CO_2 analyzer to the CVS sampling point. In this test, the CO_2 emissions began to rise rapidly approximately 2 seconds after start of engine cranking. Based on these tests, we estimate that the HC emissions reach their peak at the CVS bag sampling point approximately 2-3 seconds after the beginning of the test. Therefore, the total HC measurements made by the continuous sampling emissions console will show a delay approximately 7.5 seconds longer than that of the mean of the bag samples.

FUEL ANALYSIS - The fuel used in this study was Howell EEE, which meets the US EPA specifications for vehicle certification. The fuel was speciated by a method similar to that used for the emissions samples which will be described in the next section. The fuel speciation data is presented in Appendix B.

SPECIATION OF HC EMISSIONS - Quantitative analyses of individual hydrocarbon species were performed on a Hewlett Packard model 5890 Series II gas chromatograph (GC) outfitted with a split/splitless injector, a flame ionization detector, an automated gas sampling

valve, pressure-controlled flow, and cryogenic cooling. The

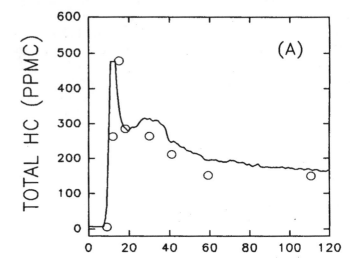

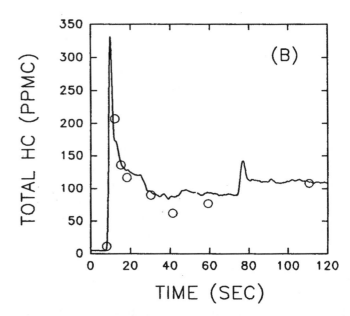

FIGURE 2. Total, diluted HC (ppmC$_1$) for (A) Base and (B) Strategy II. Dilution factor: A = 20-25; B = 10- 17. Time for GC data is (t_{meas}+7.5) [see text].

GC is interfaced to an HP Vectra (386/25) computer. The GC has had its carrier gas "pencil" traps removed; earlier observations have suggested that bleed from these traps contributed to problems with column deterioration. The GC is equipped with a 6-port Valco gas sampling valve (outfitted with a high temperature rotor) and with a 5-mL sample loop. The valve and sample loop are maintained at 110°C and the injection port is maintained at 120°C. The GC uses a 2-meter pre-column of uncoated 0.32 mm i.d. fused silica followed by a 60 meter x 0.32 mm i.d. DB-1 capillary column of 1 μm film thickness (J&W Scientific Co. Folsom, CA). The pre- column is joined to the analytical column using press-fit connectors (J&W Scientific Co. or SUPELCO, Inc.).

Helium is used as the carrier gas and the flow was set to give propane a retention time of 5.4 min, which requires a column head pressure of approximately 21-23 psi. Nitrogen is used as the detector make-up gas. The split vent flow is set to ca. 50 mL/min, measured at 75°C.

The lines from the sampling valve are attached to a vacuum manifold at one end and to a quick-connect fitting (Swagelok) at the other, such that the sample line can be evacuated and then filled from the Tedlar sample bag. The sample loop is taken through 4 flush-fill cycles before a sample is injected.

The temperature program used here is the same as used for Phase I of the Auto/Oil AQIRP program [9]; the column was programmed to hold at -80°C for .01 min, then heat at 20/min to -50°C, hold for 2.5 min, then heat at 6°C/min to 220°C. Between analyses the column may be heated to ca. 250°C to remove possible traces of high boiling materials. This program allowed the effective separation of Cl - C12 HCs in a single analysis of less than 1 hr. Three known co-elutions of significance occur under these conditions. They are the co-elutions of benzene with 1-methylcyclopentene, toluene with 2,3,3-trimethylpentane, and meta- with para-xylene.

1-Methylcyclopentene and 2,3,3-trimethylpentaneare only present in the exhaust when they are present in the fuel (i.e., there is no evidence that they are formed during combustion). Even when present in the fuel, GC/MS (gas chromatograph coupled to a mass spectrometric detector) analysis indicated that 1-methylcyclopentene represented less than 2% of the benzene peak in the engine-out exhaust from a warmed-up engine. In this study, 1- methylcyclopentene (MCP) would represent a serious co- eluter only if it was present in high amount relative to benzene <u>and</u> if the combined MCP-benzene represented a significant fraction of the fuel. For the fuel used in this study, Howell EEE fuel, MCP represents 13% of the combined MCP-benzene peak, and the combined peak represents only 0.38% of the total fuel carbon. Therefore, MCP does not cause a significant interference.

2,3,3-Trimethylpentane (2,3,3-TMP), for the Howell EEE fuel, represents 8.4% of the combined 2,3,3-TMP-toluene peak. However, the combined peak represents 25% of the fuel (2,3,3-TMP represents 2.1% of the fuel). It has been suggested that the concentration of 2,3,3-TMP can be estimated from the concentration of the isomeric 2,3,4-trimethylpentane (which is resolved in the chromatogram) [10,11]. Taking into account the difference in reactivity of 2,3,3-TMP and toluene, we estimate that the co-elution of 2,3,3-TMP with toluene contributes less than a 1% overestimation of the calculated specific reactivities. No correction for the co-elution has been made to either the benzene or the toluene data.

The third co-elution under these GC conditions is that of m-xylene with p-xylene. Earlier work has shown that the meta/para ratio in a wide range of gasolines varies only within a narrow range around 70/30. Furthermore, the meta/para ratio in tailpipe emissions from vehicle tests was, within experimental error, the same as it was in the fuel [12]. On this basis, the ratio of meta-/para-xylene in the exhaust from the experiments with gasoline fuel is estimated to he 70/30, and this ratio is used in the calculation of specific reactivities.

Data collection and peak integration were carried out with Hewlett Packard ChemStation software (Version B.01.01). To facilitate report printing, a printer utility (LaserMaster Corp., Eden Prairie, MN) was used with the Hewlett Packard Laser Jet Series II or III printer.

Hydrocarbons were identified by comparison of observed retention time with retention times calculated from calibration and stored retention index values. The GC retention time scale was updated daily by running a calibration gas mixture at the beginning of each day. The calibration gas is a 23-component mixture containing all C1-C13 n-alkanes, plus ethene, ethyne, 2-methylpropene, 1,3-butadiene, benzene, 2,2,4-trimethylpentane (isooctane), toluene, p-xylene, o-xylene, and 1,2,4-trimethylbenzene, and was prepared by Scott Specialty Gases, Troy, MI; this standard contains 2% NIST-traceable methane, propane and benzene and was prepared for use in the Auto/oil AQIRP program. The response factor for all non-methane hydrocarbons was assumed to be directly proportional to the number of carbon atoms and was calibrated on a propane standard. The methane response factor is determined from methane in the calibration mixture. For individual HCs, the index value is calculated as shown below:

$$\text{retention index} = 100n + 100[(t_{r(x)} - t_{r(n)})/(t_{r(n+1)} - t_{r(n)})]$$

$t_r(x)$ = retention time of unknown species X

$t_r(n)$ = retention time of n-alkane eluting prior to X

$t_r(n+l)$ = retention time of n-alkane eluting immediately after X

n = carbon number of n-alkane with retention time $t_{r(n)}$.

Where doubt regarding the identity of a species existed, the identification was verified by GC/MS. Primarily, hydrocarbons were measured; with the exception of methyl- t-butylether (MTBE) and benzaldehyde, oxygenated organics were not included in the analysis.

The dilute exhaust samples were collected and transported in sample bags made of 4 mil Tedlar film with Kynar hubs (13" x 30") which were obtained from Plastic Film Enterprises, 1921 Bellaire, Royal Oak, MI (Part No. 1330-4-MBK-C). The bags were outfitted with quick-connect fittings (Swagelok Co., Solon, OH) and were covered with black plastic to protect against photo-initiated oxidation. Emission samples were analyzed within 8 hr of collection.

RESULTS

BASE CONDITION - As presented in Figure 2A, the total HC emissions from the on line emissions console show: (1) an initial spike shortly after the engine begins to crank which is over range on the HC instrument; (2) a second, much smaller maximum approximately 18 seconds after this spike during the second rich excursion exhibited in Figure 1A; and (3) a gradual decrease to a steady-state value. Bag samples were withdrawn for the 0-3, 3-6, 6-9, 9- 12, 21-24, 32-35, 50-53 and 100-106 second intervals. The total, diluted HC mole fractions measured by the GC in each bag are included as individual points in Figure 2A and are similar both in trend and magnitude to the mole fractions determined by the emissions console. For the purpose of this plot, we have added 7.5 sec to the mean time of each bag sample to correct for the additional 7.5 sec sample transfer delay in the emissions console measurement (see Engine section).

The HC mole fraction in the 0-3 sec (1.5 sec mean delay time after start of cranking) bag was always very small (<2% of that observed in the 4.5 sec sample), and a GC analysis of the CO_2 in this bag revealed a concentration equal to that of ambient air. The 3-6 sec (4.5 sec mean sample time) bag contained appreciable HC and a CO_2 concentration approximately equal to one half of that observed at longer times. Although CO was not measured in the GC analysis, the on-line sampling measurements indicated that, during the first 3 seconds of engine firing, the CO constitutes approximately 33% of the carbon oxide emissions. This indicates that the engine is firing for a considerable portion (at least 65%) of the time represented by the 4.5 second bag. These results show that the delay time for arrival of the exhaust at the sample bag is of the order of 3 seconds confirming the estimate (2-2.5 sec) in the experimental section.

Approximately 45 species are present in the GC analyses at concentrations greater than 0.2% of the total HC emissions. For clarity, only data from the three most abundant members of each of the three classes of hydrocarbons (fuel paraffins, fuel aromatics, and product olefins) will be presented plus the mole fractions of methane and acetylene. These data illustrate trends also exhibited by the less abundant species. Figures 3A, 3B, and 3C show the percentage contributions to the total HC emissions represented by these eleven species for the base cold start condition. The points represent the average of all data taken for this set of conditions. Three to five separate cold starts were run to obtain the data at 3-6, 6-9, 9-12, and 103-106 seconds in order to assess the experimental reproducibility. Two runs were averaged for the other data points. Two standard deviations of the mean for the data points ranged from ±15-20% for the data at 3-6 sec to <3% at 100-106 sec. The data for short sample times are less reproducible, probably because the mole fractions are changing rapidly and small changes in the time of start can produce significant variation in species concentration.

Figures 3A and 3C present the percentage contributions for fuel-derived paraffinic and aromatic hydrocarbon species as a function of the mean sample time. Figure 3B illustrates the trends for the combustion

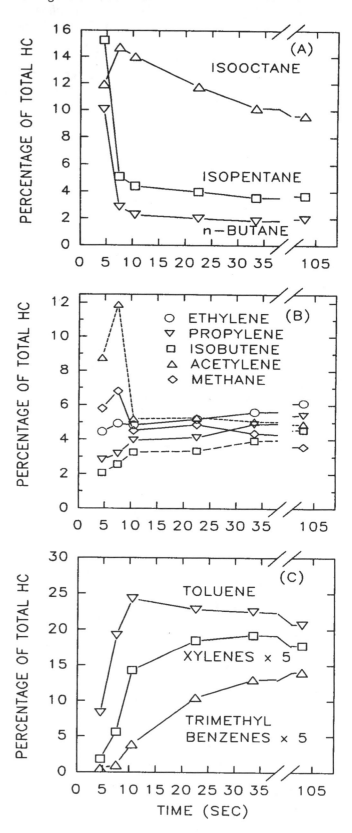

FIGURE 3. Percentage contribution of selected HC species to the total HC emissions for base operation as a function of time after start of cranking.

products. Dramatic changes in the distribution of the unburned fuel species are apparent as the delay time of the sample window increases. For data obtained during the 4.5 sec window, the low-molecular-weight fuel paraffins, butane and isopentane, comprise 25% of the total HC emission. This sample period includes the sharp spike in total HC emissions shown in Figure 2A. For this sample time, the contribution of the three higher-boiling-point aromatic fuel components is relatively small (8%) and is dominated by toluene, the highest vapor pressure member of this class. As the sample time is increased to 7.5 sec, the light paraffinic fuel components decrease very sharply while the aromatic components increase. The percentage contribution of toluene rises by a factor of three between 4.5 and 10.5 sec and then remains nearly constant. The lower volatility component, xylene, rises somewhat more slowly than does toluene, while the sum of the trimethyl benzenes, the fuel components with the lowest volatility in these figures, rises very slowly over a period of approximately 50 sec. The trimethyl benzenes do not rise between the 4.5 and 7.5 sec samples within experimental error, and at 10.5 sec have reached only 25% of their final level. In contrast, toluene reaches 80% of its maximum by the 7.5 sec sample point and xylene shows an intermediate rate of rise. Intermediate boiling point paraffins, illustrated by isooctane, change much less with sample timing than either the low or high boiling point species.

The emissions trends for fuel components with different volatility are those that might be anticipated for a cold start since low vapor pressure components in the fuel may condense temporarily in both the cold intake manifold and in the exhaust system. During the overfueling in the first second of cranking, the lower vapor pressure fuel components collect as a liquid in the intake port and manifold because the surfaces are cool and the engine is not firing. The higher vapor pressure components such as butane and isopentane vaporize quickly even under these cool conditions, accumulate in the intake manifold, and cause the engine to begin to fire. Thus, immediately after cold start, the engine will be running primarily on the lighter ends of the fuel. This is verified by the observation that butane and isopentane are major exhaust constituents in the first bag sample (larger than their mole fraction relative to other components in the fuel), while the aromatic fuel components are either smaller than expected or essentially absent depending upon their boiling points. As the intake manifold is warmed, primarily by exhaust gas backflow, the lighter aromatics begin to evaporate and contribute a greater portion to the cylinder combustion. In contrast, the high-vapor-pressure light alkanes that had accumulated during the initial injection period are purged from the manifold. Thus, the contribution of the light fuel species to the inlet charge (and to the exhaust emissions) drops very rapidly to levels more representative of their proportions in the liquid fuel (see Appendix B). The contribution of the aromatic fuel components, which have higher boiling points than most of the paraffinic species, gradually rises in the exhaust as the engine becomes

sufficiently warm to vaporize them rapidly in the intake manifold. The rate of the rise with sample time becomes slower as the boiling point of the fuel component increases (toluene 111° C; xylenes 138-144° C; trimethylbenzenes = 165-176° C) because higher temperatures are required. Temporary reduction in the exhaust concentration of higher molecular weight species also occurs within the exhaust system early in the cold start process. This was ascertained from experiments presented in Appendix A in which the exhaust system was preheated to approximately 100°C prior to the cold start.

The contribution of selected combustion products to the exhaust HC emissions is presented in Figure 3B. The fractions of the product olefins (ethylene, propylene, and isobutene) in the exhaust. do not change dramatically over the time range investigated, rising slowly by approximately a factor of 1.5. The methane and acetylene peak in the 7.5 sec bag and then drop in the 10.5 sec bag by factors of 1.6 and 2.5, respectively. Subtracting the sampling delay time of 3 sec for the bag samples as discussed earlier, the corrected mean time after start of cranking for this bag is actually 4.5 sec. Thus, the peak in methane and acetylene is associated with the sharp rich excursion in A/F ratio (see Figure 1A) which occurs 4.5 sec after start of engine cranking. A large increase in methane and acetylene in the exhaust relative to lean operation is observed while running an engine fuel rich for all fuels that have been tested during steady-state operation of spark-ignited engines [5,6,7,13]. These earlier experiments with pure fuels and gasoline have also demonstrated that the percentage contribution of olefins to the total HC emissions decreases during rich operation. Thus, the modest reduction in olefin contribution in the first two bags is also consistent with these previous experiments.

STRATEGY II (HIGH-SPEED, LEAN, RETARDED-SPARK OPERATION) - The data in Figure 2B show that the total, diluted HC mass emissions are a factor of 2-3 smaller than the base condition for this second set of operating conditions. Again, the total HC mole fractions measured in the bag samples (individual points) are similar both in trend and magnitude to the emissions console results. Figure 4 presents the contributions of the 11 representative hydrocarbon species to the total HC emissions for this cold start strategy. The light alkanes, butane and isopentane, are high in the 4.5 see bag and drop sharply in the 7.5 sec and succeeding bag samples. Over this time range the aromatic species also increase in importance. However, the rate of rise of the low vapor pressure aromatic species is much faster than was observed in the case of the base strategy, and the absolute difference in the mole fractions between the 4.5 sec bag and the later bags is smaller. For example, the trimethyl benzenes increase by a factor of three between the 4.5 and the 7.5 sec bags and then remain constant. In contrast, the trimethyl benzenes do not increase at all within experimental error between the 4.5 and 7.5 sec bag samples for the base condition, while an increase of a factor of approximately 15-30 occurs gradually over the next 30 seconds. For

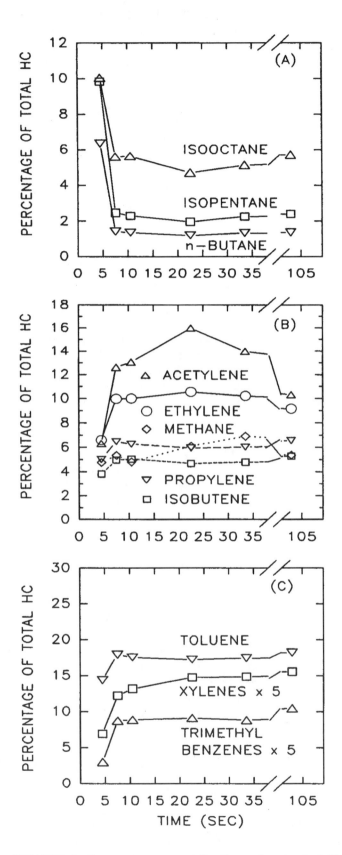

FIGURE 4. Percentage contribution of selected HC species to total HC emissions for strategy II operation as a function of time after start of cranking.

toluene and xylene, the rate of rise is also faster, and the difference between the mole fraction in the 4.5 sec bag and that observed at long times after cold start is much smaller than that observed for the base condition (see Figure 3).

Possible explanations for the faster appearance of the low vapor pressure fuel species may involve several areas in which base and strategy I1 differ. Higher speed operation increases the heat load into the engine, which will both increase the rate at which the temperature rises and the maximum temperature that the engine reaches. Retarded spark operation results in higher temperatures within the cylinder late in the engine cycle because of the reduced efficiency of the engine.Thus, the cylinder gases that flow back into the intake manifold prior to the start of the intake stroke will be hotter, promoting evaporation of the liquid fuel. This will also increase the rate at which the intake port temperature rises. Both of these effects will decrease the sensitivity of the intake charge composition in the cylinder to the vapor pressure of the fuel components and will decrease the time required for all fuel species to be completely vaporized both in the intake and exhaust systems. Finally, the intake air flow is 1.5 to 2 times larger for strategy II. This might also promote more rapid vaporization of the liquid fuel that condenses in the intake manifold during the early stage of cold start.

Figure 5 presents the ratio of the percentage contribution to the total HC emissions for each species for the two cold start strategies (strategy II/base). For an sampling delays, the contribution of the olefin combustion products (ethylene, propylene, and isobutene) is higher while that of the alkane fuel components (butane, isopentane, and isooctane) is lower during strategy II cold starts. This is consistent with steady-state emissions measurements [5,6,7], in which an increase either in engine speed or spark retardation resulted in reduced total HC emissions and an increase in the combustion-derived olefins relative to paraffinic species. This is caused by the increased bum up of fuel species which are stored in crevices or oil films within the combustion chamber. The increased temperatures both within the cylinder late in the cycle and in the exhaust port during strategy II operation increases the secondary burn up of these stored hydrocarbons.

As shown in Figure 5C, the percentage contribution of the aromatics is much higher in the 4.5 sec bag for all three species during strategy II cold starts than for a base start. Xylenes are higher in the 4.5 sec and 7.5 sec bags, while the trimethyl benzenes are larger in the 4.5, 7.5, and 10.5 sec bags. However, for all three aromatic species, the ratio falls below one in the later bags, and, therefore, their percentage contribution to the total HC actually becomes <u>smaller</u> during strategy II operation later in the cold start. Increased burn up with more product formation in the exhaust may account for a portion of this decrease. In addition, less of the heavy aromatic species are stored in the intake manifold and exhaust system early in cold start during strategy II operation because of the faster heating discussed above. During baseline operation, more of these

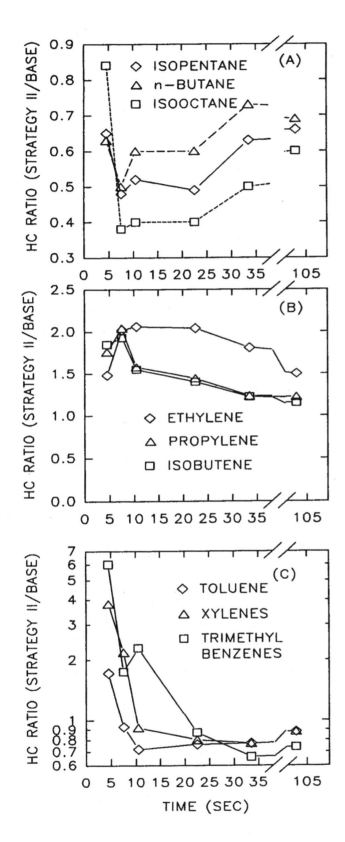

FIGURE 5. Ratio of percentage contribution of individual species to total HC emissions for the two operating conditions.

aromatic species are stored early, resulting in a larger reduction in heavy aromatic species. Later in the cold start these fuel species evaporate and contribute more heavy aromatics than during strategy II operation.

SPECIFIC REACTIVITY OF THE EXHAUST - The large changes in the distribution of the exhaust gas HC species as a function of time during the cold start process might be expected to change the reactivity of these emissions for formation of photochemical smog. According to the California reactivity standards, each HC species is assigned a reactivity factor based on the grams of ozone that the emission of one gram of that hydrocarbon would form in the atmosphere [14]. The specific reactivity (gm O_3/gm HC) of the exhaust gas then can be calculated by summing (over all exhaust species) the product of the fractional contribution of each of the species to the total exhaust HC mass emission and the reactivity factor for that species [8].

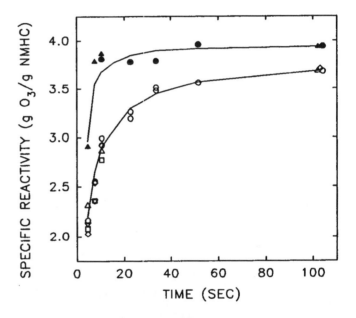

FIGURE 6. Specific reactivity of the emissions. Open symbols = Base; Closed symbols = Strategy II.

Figure 6 presents the specific reactivity for the exhaust gas HC emissions as a function of time after start of cranking for both strategies. For the base strategy (open symbols), several repeat measurements were carried out for each sampling time, and each set of measurements carried out during a single cold start is represented by a different symbol. The exhaust gas reactivity increases by a factor of 1.7 over the first 40 sec. Note that an increase in specific reactivity as time after cold start increases was also observed by Kubo et al [4]. This time dependence of the specific reactivity results from the changes in the HC species distribution that have been discussed earlier. At short times after cold start, the mass contribution of the light alkanes, which have low reactivity in the atmosphere, is large. The mass contribution of the aromatic species, which have higher atmospheric reactivity, is low. The contribution of methane and acetylene is high early, particularly in the 7.5 sec sample, and these species have essentially no photochemical reactivity in the atmosphere. Finally, the light product olefins increase in importance with time after cold start, and these species are very reactive in the atmosphere. Thus, the unreactive methane, acetylene, and light fuel alkanes decrease with time while the contributions of the reactive heavy aromatics and olefins increase. This results in a rise in the atmospheric reactivity of the hydrocarbon emissions as the time after start of cranking increases. Figure 6 shows reduced reactivity in the 4.5 sec bag for strategy I1 operation. However, the reduction is smaller (25%) than for base operation and by 7.5 sec the reactivity has reached its final value. This is consistent with the observation that the distribution of HC species in the exhaust reaches its final value more quickly in the case of strategy II operation.

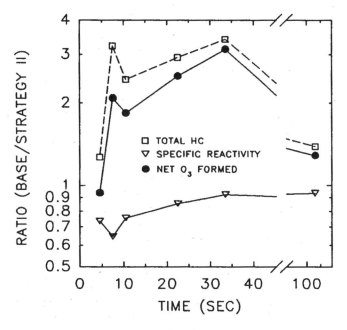

FIGURE 7. Ratio of the results for Base to Strategy II operation.

Figure 7 compares the effectiveness of base relative to strategy II operation for controlling photochemical smog formation in the atmosphere. Three comparisons are included in this figure as functions of sample time: (1) the ratio of total HC mass emissions (open squares) for the bag samples; (2) the ratio of the specific reactivities (open triangles); and (3) the ratios (filled circles) of the net O_3 forming potentials in the atmosphere (=total HC x specific reactivity). As discussed earlier, the total HC emissions for strategy I1 operation are 2-3 times smaller than base operation. However, the specific reactivity of the strategy II emissions is larger. During the first 10 sec after cold start, the major reason for the increased reactivity of strategy II emissions is the reduction in contribution of

alkanes and the increase in the olefinic and heavier aromatic species (particularly the trimethyl benzenes). For longer sample times (>20-30 sec), the difference in reactivity between the two strategies decreases because the contribution of the heavy aromatics actually becomes smaller during strategy II cold start, partially offsetting the increase in olefinic combustion products. As shown in Figure 7, the net benefit in reduced O_3 formation is always less for strategy II operation than would be predicted based on measurements of total HC mass emissions alone. As an example, in the 7.5 sec bag, the total HC emissions in the bag decrease by 70% during strategy II cold start, while the net ozone forming potential of the exhaust gas decreases by 50%.

CONCLUSIONS

In these experiments, diluted HC emissions from the exhaust of a production engine with a catalytically inactive "bare brick" catalyst support have been collected in a Tedlar bag as a function of time (0-100 sec) after the start of engine cranking during a 70°F cold start using two control strategies. For the base strategy, which is similar to a conventional cold start, gc analyses showed that the HC emissions were initially enriched in light fuel alkanes and depleted in heavy aromatic species relative to the fuel composition. Later in the cold start, the light alkanes fell rapidly with increasing sample time while the lower vapor pressure aromatics increased over a period of 50 sec. These results are consistent with the retention of low vapor pressure fuel components in the intake port, manifold, and exhaust system early in the cold start process. The light alkane components such as butane and isopentane vaporize rapidly and provide the initial fuel to start the engine.

The second cold start strategy employed higher speed, a larger spark retardation, and leaner fuel-air ratio. This strategy produced lower total HC mass emissions and a more rapid increase in the contribution of the lower vapor pressure aromatic species in the fuel to the exhaust gases. Thus, steady-state exhaust concentrations of all HC species were attained earlier in the cold start process. However, the distribution of HC species in the exhaust was considerably different from that observed with the base strategy. In particular, the contribution of fuel alkanes to the total HC emissions was reduced by a factor of 1.5-2, while the contribution of product olefins increased by 1.2-2 depending upon species and the sampling time after cold start.

Based on the measured distribution of the HC species in the exhaust, the specific reactivity of the HC emissions for formation of photochemical smog in the atmosphere was calculated using reactivity factors defined by the state of California. For the base start strategy, the reactivity of the HC emissions rose slowly, reaching a final plateau 40 sec after beginning of engine cranking that was 1.7 times higher than at the beginning of engine firing. For the strategy I1 cold start, the reactivity of the emissions was higher than for the base strategy by a factor of 1.5 early in the cold start and 1.1 after 40 sec of running.

The results of these experiments have important implications for meeting future HC emissions standards. They demonstrate:

(1) The species distribution and the atmospheric reactivity of the exhaust HC emissions varies substantially with time after start of cranking. Thus, if the time to catalyst light off is decreased to reduce total HC emissions, the reactivity of the HC emissions may also change significantly,

(2) Changing the engine operating conditions can alter the species distribution, the reactivity, and the time dependence of the HC emissions;

(3) Changing the operating conditions to reduce total HC emissions may increase the reactivity of these emissions for photochemical smog formation, partially counterbalancing the advantage gained from the lower total emissions.

It is critical to recognize that a reduction of total emissions during cold start either by earlier catalyst light off or engine strategy changes may be accompanied by changes in atmospheric reactivity of the exhaust gas. These changes may increase or decrease the apparent effectiveness of the total HC emissions reduction if the emissions standard is based on the reactivity of the exhaust HC emissions for photochemical smog formation. Finally, the exhaust HC species distribution and its reactivity may be affected by the composition of the fuel because of changes in the proportions of the low and high vapor pressure constituents. Therefore, full exhaust gas analyses of HC emissions is required to accurately assess the effectiveness of new HC control methods.

ACKNOWLEDGEMENT

We thank D. Podsiadlik and J. F. O. Richert for assistance with the GC analyses during these experiments. All experiments were carried out at: Research Laboratory, Ford Motor Co., MD 3083/SRL, Dearborn, MI 48121-2053.

REFERENCES

1. Chang, T. Y., Chock, D. P., Hammerle, R. H., Japar, S. M., and Salmeen, I. T. "Urban and Regional Ozone Air Quality: Issues Relevant to the Automobile Industry," Critical reviews in Environmental Control 22, 27 (1992).

2. Boam, D. L, Finlay, I. C., Biddulph, T. W., Ma, T., Lee, R., Richardson, S. H., Bloomfield, J., Green, J. A., Wallace, M., Wallace, S., and Woods, W. A. "The Sources of Unburnt Hydrocarbon Emissions from Spark-Ignited Engines during Cold Starts and Warm-up." I. Mech. E., Combustion in Engines-Technology, Applications and the Environment, paper C448/064 (1992).

3. Brown, P. G., and Woods, W. A. "Measurements of Unburnt Hydrocarbons in a Spark-Ignited Engine during the Warm-Up Period," SAE Technical Paper No. 922233 (1992).

4. Kubo, S., Yamamoto, M., Kizaki, Y., Yamazaki, S., Tanaka, T., and Nakanishi K. "Speciated Hydrocarbon Emissions of SI Engine during Cold Start and Warm up" SAE Technical Paper No. 932706 (1993).

5. Kaiser, E. W., Siegl, W. O., Henig, Y. I., Anderson, R.W., and Trinker, F. H. "Effect of Fuel Structure on Emissions from a Spark-Ignited Engine," Env. Sci. Technol. 25, 2005 (1991); corr. ibid. 26, 2618 (1992).

6. Kaiser, E. W., Siegl, W. O., Cotton, D. E, Anderson, R. W. "Effect of Fuel Structure on Emissions from a Spark-Ignited Engine. 2. Naphthene and Aromatic Fuels," Env. Sci Technol. 26,1581(1992). , 1581 (1992).

7. Kaiser, E. W., Siegl, W. O., Cotton, D. F., and Anderson R. W. "Effect of Fuel Structure on Emissions from a Spark-Ignited Engine. 3. Olefinic Fuels," Env. Sci Technol. 27, 1440 (1993).

8. Kaiser, E. W., and Siegl, W. O. "Atmospheric Reactivity of Engine-Out Hydrocarbon Emissions from a Spark-Ignited Engine Determined by High Resolution Chromatography." Journal of High Resolution Chromatography, in press.

9. Jensen, T. E., Siegl, W. O., Richert, J. F. O., Lipari F., Loo, J. F., Prostak, A., and Sigsby, J. "Advanced Emission Speciation Methodologies for the Auto/Oil Air Quality Improvement Research Program - 1. Hydrocarbons and Ethers," SAE Technical Paper No. 920320 (1992).

10. Siegl, W. O., Richert, J. F. O., Jensen, T. E., Schuetzle, D., Swarin, F. J., Loo, J. F., Prostak, A., Nagy, D., and Schlenker, A. M. "Improved Emissions Speciation Methodology for Phase II of the Auto/Oil Air Quality Improvement Research Program. Hydrocarbons and Oxygenates.," SAE Technical Paper No. 930142 (1993).

11. Hoekman, S. K." improved Gas Chromatography Procedure for Speciated Hydrocarbon Measurement Of Vehicle Emission," J. Chromatog. 639, 239 (1993).

12. Richert, J. F. O., Siegl, W. O., and Chladek, E. presented at the 50th Anniversary Anachem Symposium, November 7, 1991, Farmington, MI.

13. Kaiser, E. W., Rothschild, W. G., and Lavoie, G. A. "The Effect of Fuel and Operating Variables on Hydrocarbon Species Distributions in the Exhaust from a Multicylinder Engine," Combust. Sci and Technol. 32, 245 (1983).

14. a. Carter, W. P. L, and Atkinson, R. "Computer Modeling Study of Incremental Hydrocarbon Reactivity," Environ. Sci. Technol. 23, 864 (1989).

   b. Updated MIR factors from CARB presented in: Tsuchida, H., Ishihara, K., Iwakiri, and Matsumoto, M. SAE Technical Paper No. 932718 (1993).

APPENDIX A. Effect of Exhaust System Preheating.

To assess the importance of loss of HCs via adsorption in the exhaust system early in the cold start, one experiment was performed in which the exhaust system was preheated to approximately 100°C by a heating blanket prior to starting the engine. This was compared to a similar cold start with the blanket heater turned off. During these experiments, the engine was cranked by the dynamometer rather than by the starter motor and was connected directly to the dynamometer instead of through a transmission. In addition, the magnitude of the rich spike in the fuel-air ratio during the first few seconds of firing was reduced. Thus, the conditions, although similar to the baseline strategy, are not identical to those in the main body of the paper. The arrival time of the hydrocarbons at the CVS sampling point (as measured by the total HC analyzer) was 1.4 sec earlier during the preheated experiment than when no preheat was used. Part of the change in the arrival time occurred because of a difference in the onset of engine firing (which is a random effect unrelated to the exhaust system preheating) as shown by the fact that the CO and CO_2 arrival times were 0.4 sec earlier during the preheat experiment. This still leaves a 1 sec difference in arrival times for the hydrocarbons at the sampling point which cannot be explained by a difference in onset of engine firing. This may result from the preheating and is certainly in the right direction since, in the preheated experiments, the total HC peak arrived sooner. However, such an explanation is difficult to accept because a large fraction of the early hydrocarbon emissions are light fuel components and combustion products, which we would not expect to adsorb strongly in the exhaust system at ambient temperature. Thus, it is possible that the difference may be related to the experimental reproducibility. Additional experiments would be required to answer this question.

Figure A1 presents mole fraction vs time profiles for the fuel species presented in Figure 3. The trends in the data are the same as those observed during baseline operation for both experiments: the light

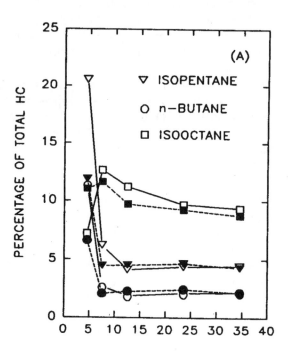

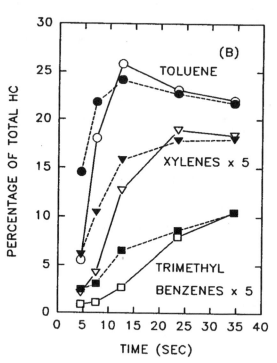

FIGURE A1. Percentage contribution of selected HC species to total HC emissions. Filled symbols= preheated exhaust system; open = no preheating.

components, butane and isopentane, are high at the start and fall rapidly while the mole fractions of the heavier components are reduced at the beginning and rise with time after cold start. The data obtained without preheating of the exhaust system (open symbols) are quantitatively very similar to those presented in Figure 3. Preheating the exhaust system causes the concentrations of the heavier fuel species to increase at short times after cold start while for times longer than 20 sec the results from both experiments (e.g. with and without exhaust system preheating) are identical. The difference is largest for the trimethyl benzenes which are the lowest vapor pressure species presented in the figure. Some of the difference at the shortest times results from the observation discussed above that the hydrocarbon arrival time at the sampling point was 1.4 sec earlier during the preheated experiment. Because of this, the time scale in Figure A1 should be slightly different for the two runs. The delay time of the bag sample relative to the arrival of the initial HC signal will in fact be 1.4 sec longer for the preheated experiment. Increasing the time for the data taken with preheating (closed symbols) by 1.4 sec would produce significantly better agreement between the two data sets at shorter times particularly for the butane, isopentane, isooctane, and toluene. However, there is evidence from these data that adsorption of xylene and trimethyl benzene does occur in the exhaust system for delay times up to approximately 20 sec. This certainly contributes to the very low levels of trimethyl benzenes observed for times less than 10 sec in the baseline experiments presented in Figure 3.

The monolithic "bare brick" catalyst support is the likely location where most of this adsorption takes place because the surface area of the support represents approximately 90% of the total surface area in the exhaust system. In addition, the small, square-cross-section channels through this monolith (0.11 cm per side) provide enhanced opportunity for the gases to contact the surface. Kubo et al [4] observed large decreases in the low vapor pressure components in the exhaust early in the cold start because of absorption within an active catalyst. These results show that an inactive monolithic catalyst support may also adsorb these components.

APPENDIX B. GC analysis of Howell - EEE fuel.

TABLE B1. Composition of fuel.

Species	% of total ppmC
Methane	0
Ethene	0
Ethyne	0
Ethane	0
Propene	0
Propane	0.02
Propadiene	0
Propyne	0
2-Methylpropane	0.39
2-Methylpropene	0
1-Butene	0
1,3-Butadiene	0
Butane	5.04
t-2-Butene	0.02
2,2-Dimethylpropane	0.02
1-Butyne	0
c-2-Butene	0.02
3-Methyl-l-butene	0.02
Ethanol	0
2-Methylbutane	4.66
2-Butyne	0
1-Pentene	0.06
2-Methyl-l-butene	0.1
Pentane	2.18
2-Methyl-1,3-butadiene	0
t-2-Pentene	0.13
3,3-Dimethyl-l-butene	0
c-2-Pentene	0.07
2-Methyl-2-butene	0.2
Cyclopentadiene	0
2,2-Dimethylbutane	0.31
Cyclopentane	0.03
4-Methyl-l-pentene	0
3-Methyl-l-pentene	0.02
Cyclopentane	0.31
2,3-Dimethylbutane~	1.17
MTBE	0
4-Methyl-c-2-pentene	0
2-Methylpentane	2.21
4-Methyl-t-2-pentene	0.03
3-Methylpentane	1.46
2-Methyl-l-pentene	0.04
1-Hexene	0.02
Hexane	1.5
t-3-Hexene	0.04
c-3-Hexene	0
t-2-Hexene	0.06
3-Methyl-t-2-pentene	0.08
2-Methyl-2-pentene	0.05
3-Methylcyclopentene	0
c-2-Hexene	0.03
3-Methyl-c-2-pentene	0.07
2,2-Dimethylpentane	0.1
Methylcyclopentane	0.46
2,4-Dimethylpentane	2.55
2,2,3-Trimethylbutane	0.04
3,4-Dimethyl-l-pentene	0
Methylcyclopentene	0.05
1-Benzene	0.33
3-Methyl-1-hexene	0
3,3-Dimethylpentane	0.12
Cyclohexane	0.09
2-Methylhexane	1.09
2,3-Dimethylpentane	4.75
Cyclohexene	0
3-Methylhexane	1.29
c-1,3-Dimethylcyclopentane	0.07
3-Ethylpentane	0.14
t-1,2-Dimethylcyclopentane	0
1-Heptene	0
2,2,4-Trimethylpertane	12.38
t-3-Heptene	0.02

Heptane	0.91
2-Methyl-2-hexene	0.06
3-Methyl-t-3-hexene	0
t-2-Heptene	0.02
3-Ethyl-c-2-pentene	0.01
2,4,4-Trimethyl-l-pentene	0.04
2,3-Dimethyl-2-pentene	0
c-2-Heptene	0.03
Methylcyclohexane	0.15
2,2-Dimethylhexane	0.01
2,4,4-Trimethyl-2-pentene	0
2,5-Dimethylhexane	0.86
2,4-Dimethylhexane	1.26
3,3-Dimethylhexane	0.06
2,3,4-Trimethylpentane	2.91
2,3,3-Trimethylpentane	2.08
Toluene	22.77
2,3-Dimethylhexane	0.7
1 2-Methylheptane	0.32
4-Methylheptane	0.14
3-Methylheptane	0.44
1-c-2,3-Trimethylcylcopentane	0
c-1,3-Dimethylcyclohexane	0.03
t-1,4-Dimethylcyclohexane	0
2,2,5-Trimethylhexane	0.48
1-Octene	0.03
t-4-Octene	0.02
Octane	0.27
t-2-Octene	0.01
t-1,3-Dimethylcyclohexane	0.02
c-2-Octene	0.01
2,3,5-Trimethylhexane	0.08
2,4-Dimethylheptane	0.05
c-1,2-Dimethylcyclohexane	0.05
Ethylcyclohexane	0.02
3,5-Dimethylheptane	0.02
Ethylbenzene	0.51
2,3-Dimethylheptane	0.04
m&p-Xylene	1.94
2-Methyloctane	0.13
3-Methyloctane	0.1
Styrene	0.05
o-Xylene	0.99
1-Nonene	0.32
Nonane	0.07
i-Propylbenzene	0.26
2,2-Dimethyloctane	0.15
Benzaldehyde	0
2,4-Dimethyloctane	0.12
n-Propylbenzene	0.67
1-Methyl-3-ethylbenzene	2.25
1-Methyl4-ethylbenzene	1.08
1,3,5-Trimethylbenzene	1.21
1-Methyl-2-ethylbenzene	0.82
1,2,4-Trimethylbenzene	2.93
Decane	0.16
i-Butylbenzene	0.04
s-Butylbenzene	0.09

1-Methyl-3-i-propylbenzene	0.09
1,2,3-Trimethylbenzene	0.56
1-Methyl-4-i-propylbenzene	0.03
Indan	0.17
1-Methyl-2-i-propylbenzene	0
1,3-Diethylbenzene	0.07
1,4-Diethylbenzene	0.2
1-Methyl-3-n-propylbenzene	0.34
1-Methyl-4-n-propylbenzene	0.49
1,2-Diethylbenzene	0.05
1-Methyl-2-n-propylbenzene	0.23
1,4-Dimethyl-2-ethylbenzene	0.17
1,3-Dimethyl-4-ethylbenzene	0.16
1,2-Dimethyl-4-ethylbenzene	0.45
1,3-Dimethyl-2-ethylbenzene	0.02
Undecane	0.05
1,2-Dimethyl-3-ethylbenzene	0.07
1,2,4,5-Tetramethyl-benzene	0.15
2-Methylbutylbenzene	0
1,2,3,5-Tetramethylbenzene	0.22
tert-l-Butyl-2-methylbenzene	0.05
1,2,3,4-Tetramethylbenzene	0.11
n-Pentylbenzene	0.06
t-l-Butyl-3,5-dimethylbenzene	0.06
Naphthalene	0.21
Dodecane	0.06
Uncalibrated	4.3
% Speciated	95.7

[**NOTE ADDED IN PROOF** - After completion of this paper, another measurement has been published of the time variation of the specific HC reactivity after cold start in 30 sec steps using gasoline fuel {R. G. Nitschke, SAE Technical Paper No. 932704 (1993)}. The results of this paper also show reduced reactivity of the HC exhaust emissions early in the start.]

Engine-Out and Tail-Pipe Emission Reduction Technologies of V-6 LEVs

**Hideaki Takahashi, Yasuji Ishizuka,
Masayuki Tomita and Kimiyoshi Nishizawa**
Nissan Motor Co., Ltd.

Copyright © 1998 Society of Automotive Engineers, Inc.

ABSTRACT

Compared with in-line 4-cylinder engines, V-6 engines show a slower rise in exhaust gas temperature, requiring a longer time for catalysts to become active, and they also emit higher levels of engine-out emissions. In this study, The combination of a new type of catalyst, and optimized ignition timing and air-fuel ratio control achieved quicker catalyst light-off. Additionally, engine-out emissions were substantially reduced by using a swirl control valve to strengthen in-cylinder gas flow, adopting electronically controlled exhaust gas recirculation (EGR), and reducing the crevice volume by decreasing the top land height of the pistons. A vehicle incorporating these emission reduction technologies reduced the emission level through the first phase of the Federal Test Procedure (FTP) by 60-70% compared with the Tier 1 vehicle. The application of these technologies should enable V-6 engines to comply with Low Emission Vehicle (LEV) standards without adding expensive devices such as a secondary air pump or an electrically heated catalyst.

INTRODUCTION

There have been growing demands for cleaner vehicle emissions in recent years amid the heightened concern over environmental issues. Stricter emission regulations are being adopted worldwide, typical of which are California's Low Emission Vehicle (LEV) standards [1]. The LEV standards mandate that automobile manufacturers progressively reduce the average emission values of their vehicles each year. The LEV and ULEV (Ultra Low Emission Vehicles) standards are initially being directed towards the classes of vehicles for which compliance with these standards is easy, and they will then be directed towards vehicles for which compliance is more difficult. The degree of difficulty in complying with the requirements of the LEV category can vary greatly from one type of engine to another. For example, in-line 4-cylinder (I4) engines can meet the LEV standards by using currently available technologies. One typical approach taken here is to reduce cold-start emissions by positioning a large volume catalyst close to the engine so as to shorten the catalyst light-off time. Another approach is to improve the conversion efficiency of the catalyst following engine warm-up by reducing air/fuel ratio fluctuations through the use of an improved control procedure [2,3]. As a result of using these technologies, several automobile manufacturers, including Nissan, have begun selling 14 engines that meet the LEV standards.

The situation for V-6 engines, however, is more complicated. The vehicle layout makes it difficult to position large-volume catalysts close to the engine, as can be done with 14 engines, especially in the case of front-wheel -drive vehicles. Because of the larger number of cylinders, the exhaust temperature rises more slowly, particularly for double overhead camshaft (DOHC) engines. A further problem is that the engine-out hydrocarbon (HC) level is greater. Moreover, since V-6 engines are typically used on larger vehicles, the heavier vehicle weight is another complicating factor. All of these issues must be addressed through the use of new technologies.

This paper describes the technologies for reducing the emission levels, especially HCs, of a front-wheel-drive vehicle fitted with a V-6 DOHC engine, for which compliance with the LEV standards is seen as being particularly difficult.

OVERVIEW OF EMISSION REDUCTION TECHNOLOGIES

Figure 1 shows an example of the non-methane organic gas (NMOG) emissions measured in the Federal Test Procedure (FTP) for a production vehicle fitted with a 3.0-liter V-6 engine. Cold-start HC emissions produced during the first 1-2 minutes, from engine start to around the catalyst light-off point, account for more than one-half of the HC emissions. Reducing this cold-start portion is therefore essential. Additionally, it is also necessary to reduce the HC emitted after the catalyst becomes operational (i.e., the warmed-up HC portion) in order to comply with the LEV standards.

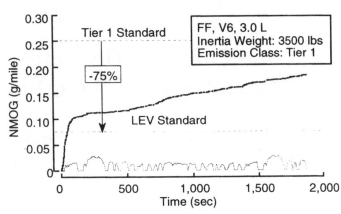

Figure 1. Production Vehicle NMOG Emission in the FTP

The technologies that were used in this study to reduce the HC emissions of a V-6 engine to the LEV level are outlined in Figure 2. Cold-start HC emissions were substantially reduced mainly by using a swirl control valve to obtain a leaner air/fuel ratio and a lower HC level, and by quickening the light-off time of the catalyst because of optimizing control of both the air/fuel ratio and the ignition timing, according to each engine operating parameter. The light-off time was also quickened by adopting a newly developed Pd/Rh catalyst, with excellent low-temperature activity and heat resistance, in the close-coupled location in place of a conventional Pt/Rh catalyst. The warmed-up engine-out HC emissions was reduced by improving the response of the EGR valve through electronic control, and by improving the piston design. These measures, together with the adoption of the new Pd/Rh catalyst with excellent NMHC conversion efficiency and heat resistance, also substantially reduced the warmed-up HC emission as well.

REDUCTION OF COLD-START HC EMISSIONS

USE OF A SWIRL CONTROL VALVE TO OBTAIN A LEANER MIXTURE AND HC REDUCTION – Using a leaner air/fuel ratio after a cold start not only reduces engine-out HC emissions [4], it also has the effect of lowering the catalyst light-off temperature because of the increased oxygen concentration in the exhaust gas. This effect is especially significant for catalysts with a large Pd content. However, simply making the mixture leaner could cause engine stability to deteriorate [5]. The volatility of gasoline on the market has a large variation. A leaner mixture would not pose any problem for gasoline

with average properties. However, for gasoline having heavier properties, a lean air/fuel ratio could result in unstable combustion.

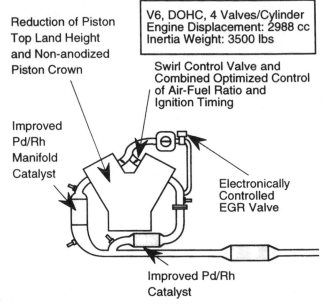

Figure 2. Overview of V6 LEV Emission Technologies

Figures 3 and 4 show that satisfactory engine stability could not be obtained at ordinary air/fuel ratios, $\lambda = 1.0$ at cold idle, $\lambda = 1.1$ in the first quarter of the first phase of FTP, in tests using the heaviest gasoline among the commercial varieties available. When a swirl control valve was used to increase the burning velocity, satisfactory engine stability was obtained. Moreover, by optimizing the geometry of the notch in the swirl control valve, combustion stability was improved further, making it possible to expand the crank angle limit for retarding the ignition timing. As a result, the exhaust temperature was raised and engine-out HC emissions were reduced. Additionally, residual in-cylinder gas flow after the completion of the main combustion period promoted diffusion and oxidation of unburned HCs, which worked to reduce engine-out HC emissions. The results are plotted in Figure 5. It was found that the use of a slanted notch geometry in the swirl control valve achieved both a higher burning velocity and lower HC emissions. Compared with a simple horizontal opening or an upward opening, the slanted notch geometry resulted in lower engine-out HC emissions and a higher exhaust temperature.

230

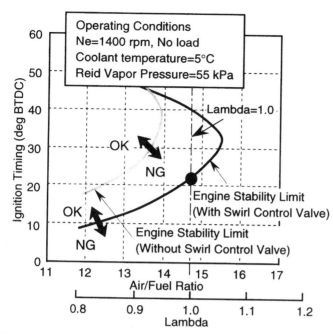

Figure 3. Lean Combustion Limit With Swirl Control Valve at Cold Idle

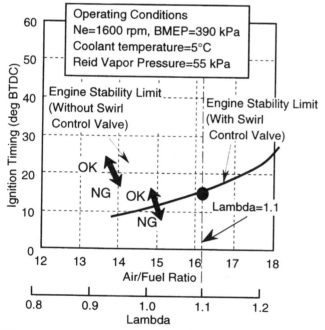

Figure 4. Lean Combustion Limit With Swirl Control Valve in the First Quarter of First Phase of FTP

COMBINED OPTIMIZED CONTROL OF AIR-FUEL RATIO AND IGNITION TIMING FOR EACH OPERATING PARAMETER – As indicated in the foregoing discussion, trying to secure satisfactory engine stability by taking into account the variation in gasoline properties requires a trade-off between the air/fuel ratio and ignition timing. When a lean air/fuel ratio is used, the ignition timing must be advanced, which results in a lower exhaust tempera-

ture. Conversely, the use of a somewhat richer mixture makes it possible to retard the ignition timing and thereby raise the exhaust temperature [5]. Accordingly, an attempt was made to shorten the catalyst light-off time by using the optimum combination of air/fuel ratio and ignition timing. The question of which should be given priority with respect to reducing emissions was determined on the basis of the operating parameters.

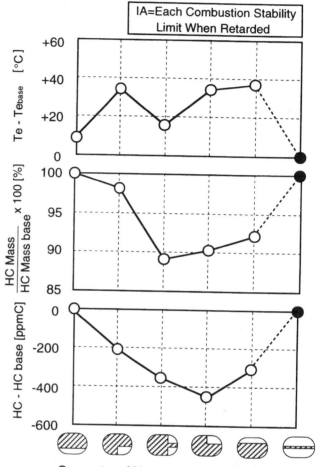

subscript "base": without swirl control valve
Te: Exhaust Temperature

Operating conditions:
N=1600 rpm, BMEP=390 kPa
Ignition Angle=each combustion stability
limit when retarded
Coolant Temperature=40°C
A/F=stoichiometric
Notch area ratio=18.7 %

Figure 5. HC Emissions and Exhaust Temperature versus Notch Geometry in Swirl Control Valve

Air/Fuel Ratio Calibration for Cold Idle – An investigation was made of the optimum air/fuel ratio calibration for engine idle after a cold start. Figure 6 shows the residual HC ratio in the first phase (first 125 s) of the FTP as a function of the excess air ratio at cold idle. The exhaust temperature at the catalyst inlet is also shown in relation to the excess air ratio. Under these conditions, a retarded ignition timing resulted in a higher exhaust temperature and lower HC emissions. Therefore, the ignition timing was retarded to the extent that engine stability could still be assured, taking into account the variation in gasoline properties in the market and the range of variation in the air/fuel ratio. At lean mixture ratios up to $\lambda = 1$, the exhaust temperature at the catalyst inlet showed only a small decrease and the higher oxygen concentration resulting from the leaner mixture had a greater effect on the low-temperature activity of the catalyst. As a result, the residual HC ratio in the first phase of the FTP was reduced. However, at lean mixture ratios greater than $\lambda = 1$, the ignition timing had to be more advanced than that at lean mixture ratios up to $\lambda = 1$, to ensure engine stability. This caused the exhaust temperature at the catalyst inlet to drop substantially, resulting in a larger residual HC ratio. Based on these results, the excess air ratio for cold idle was set near $\lambda = 1$.

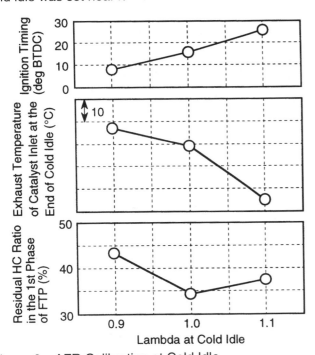

Figure 6. AFR Calibration at Cold Idle

Air/Fuel Ratio Calibration for the 1st Quarter of 1st Phase – A study was then made of the air/fuel ratio calibration for the first quarter (to about 30 s) of the first phase of the FTP to near the catalyst light-off point. Figure 7 shows the residual HC ratio in the first phase of the FTP as a function of the excess air ratio in the first quarter of the first phase of the FTP. The exhaust temperature at the catalyst inlet in approximately the first quarter of the first phase is also plotted relative to the excess air ratio. Figure 8 shows the residual HC ratio in the first half of the first phase for various excess air ratios in the first quarter of the first phase of the FTP. In both cases, the ignition timing was retarded to the limit of engine stability, as mentioned earlier. Similar to the result seen for cold

idle, the exhaust temperature at the catalyst inlet dropped with air/fuel ratios above $\lambda = 1$. Unlike the result seen for cold idle, the residual HC ratio continued to decrease until $\lambda = 1.1$.

Figure 9 shows the change in the light-off temperature of the catalyst at 50% activity as a function of the excess air ratio. The 50% light-off temperature greatly decreased until a lean air/fuel ratio of $\lambda = 1.1$, after which it slightly decreased even though the mixture was made increasingly leaner. In the first quarter of the first phase of the FTP, the exhaust temperature is naturally higher than at cold idle and has reached a sufficiently high level for catalyst light-off. The results presented here indicate that setting the air/fuel ratio so as to obtain the lowest 50% light-off temperature of the catalyst is more effective in reducing HC emissions than setting it so as to raise the exhaust temperature at the catalyst inlet.

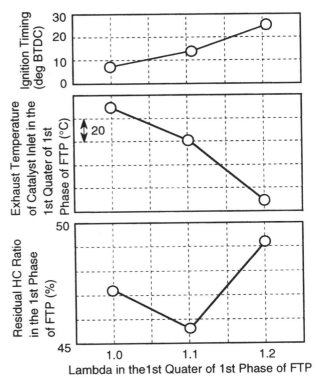

Figure 7. AFR Calibration in 1st Quarter of the 1st Phase of the FTP

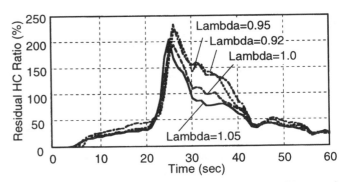

Figure 8. Residual HC Ratio in 1st Harf of 1st Phase of the FTP

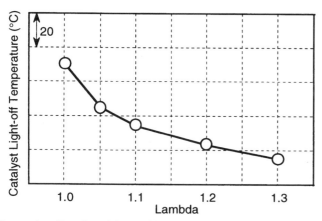

Figure 9. Catalyst Light-off Temperature versus Air/Fuel Ratio

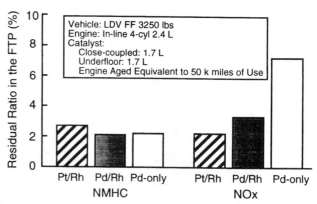

Figure 11. Comparison of Residual NMHC and NOx Ratios

Based on these investigations, the excess air ratio should be set in the vicinity of λ = 1 for cold idle and the ignition timing should be greatly retarded. In the first quarter of the first phase of the FTP, the excess air ratio should be set in the vicinity of λ = 1.1 and the ignition timing should be relatively advanced. These measures will quicken catalyst light-off while assuring stable engine operation.

ADOPTION OF THE NEWLY DEVELOPED PD/RH CATALYST – A newly developed type Pd/Rh catalyst was adopted with the aim of improving the low-temperature catalyst activity and heat resistance. This new catalyst has an optimum arrangement and loading of precious metals on each layer and an optimized third component in terms of the location and loading. Figure 10 compares the residual HC ratio measured in bench tests for aged reference catalysts, the current Pt/Rh catalyst and the new Pd/Rh catalyst. The HC conversion efficiency of the new catalyst exceeded that of the current Pt/Rh catalyst and the reference Pt/Rh, Pd/Rh, and Pt/Pd/Rh catalysts. The Pd-only catalyst displayed an equivalent or a slightly higher level of HC conversion efficiency than the new catalyst. However, as seen in Figure 11, the Pd-only catalyst is noticeably inferior in terms of its NOx conversion efficiency, so the new Pd/Rh catalyst was selected.

Catalysts that use Pd as the precious metal generally tend to be more susceptible to sulfur poisoning [6,7]. However, this problem can be overcome by using a Pd/Rh catalyst in the close-coupled location where exhaust temperatures are higher and a Pt/Rh catalyst in the underfloor location where temperatures are lower. Test results show that catalysts which were initially poisoned by the use of gasoline with a high sulfur content that is found in some markets soon recovered from the poisoning when gasoline with a low sulfur content was used (Figure 12).

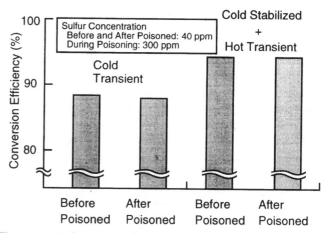

Figure 12. Influence of Sulfur Poisoning on Emissions

The effect of the new Pd/Rh catalyst on enhancing low-temperature activity is shown in Figure 13. Compared with the current Pt/Rh catalyst, its light-off temperature for HC conversion has been reduced. It is well known that the catalyst light-off temperature is lowered when the oxygen concentration of the exhaust gas is increased, such as by making the air/fuel ratio leaner or as a result of large air/fuel ratio perturbations. It is seen in the figure that this effect has been further enhanced with the new catalyst compared with the performance of the catalyst in current production.

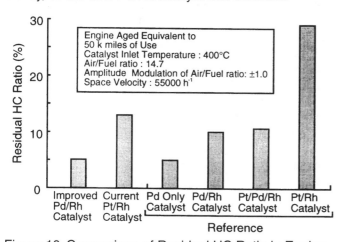

Figure 10. Comparison of Residual HC Ratio in Engine Bench Tests

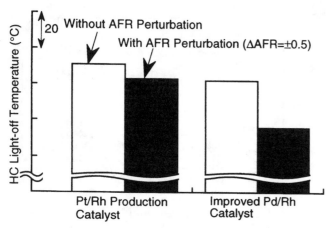

Figure 13. Effect of Improved Pd/Rh Catalyst

REDUCTION OF WARMED-UP HC EMISSIONS

EFFECT OF AN ELECTRONICALLY CONTROLLED EGR VALVE ACTUATED BY A STEPPER MOTOR – An electronically controlled EGR valve actuated by a high-response stepper motor was adopted to improve valve responsiveness and to optimize EGR calibration, making it possible to reduce HC and NOx emissions simultaneously. In order to confirm the effect of this valve on transient emissions, a test was conducted under conditions that replicated gradual vehicle acceleration. The EGR valve lift and engine-out NOx level are shown in Figure 14. The electronically controlled EGR valve substantially reduced NOx emissions during acceleration, although the NOx level was the same during the steady-state portion since an identical EGR rate was used. One presumed factor for this difference is that the mechanical control system using a back pressure transducer (BPT) has a first-order delay with a long time constant, so the actual EGR rate is smaller than the calibrated EGR rate. Another reason, resulting from the previous factor, is that the MBT (minimum advance for best torque) point shifts toward the retarded side, and the relative ignition timing with the mechanical control system occurs at a more advanced crank angle than this point.

Figure 15 shows the reduction in engine-out HC and NOx emissions obtained with the electronically controlled EGR valve and a reduced EGR rate. At the same EGR rate as that of the mechanical control system, the electronically controlled valve reduced NOx emissions without sacrificing the HC level. Reducing the EGR rate to within the allowable margin made it possible to obtain simultaneous reductions in NOx and HC emissions.

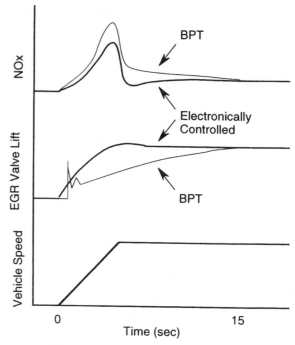

Figure 14. Effect of Electronically Controlled EGR

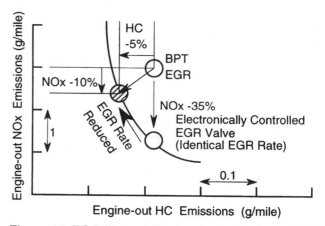

Figure 15. EGR Rate of Electronically Controlled EGR Valve

HC REDUCTION OBTAINED WITH REDESIGNED PISTON – Lowering the top land height of the piston results in a smaller crevice volume, which is effective in reducing HC emissions [8,9]. However, with a lower top land height, the temperature of the piston ring grooves rises, resulting in greater ring groove wear. To prevent wear, an anodized coating is commonly applied to the piston. The surface of this coating, however, contains micro pores which trap unburned HC that is then evacuated in the exhaust stroke, resulting in higher engine-out HC emissions. By limiting the anodized coating to the grooves alone, a substantial reduction in HC emissions was obtained, as indicated in Figure 16.

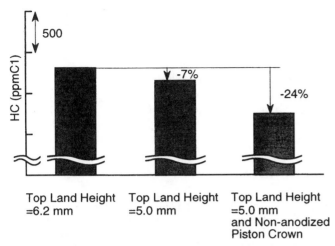

Figure 16. Effect of Reduced Top-land Height and
Non-anodized Piston Crown on HC Emissions

INCREASED CONVERSION EFFICIENCY OBTAINED
WITH THE NEW PD/RH CATALYST – The new Pd/Rh
catalyst achieves substantial improvements with respect
to warmed-up conversion efficiency and high-temperature durability. The effect of this new catalyst on reducing
the residual HC ratio is shown in Figure 17, and its
high-temperature durability is shown in Figure 18. Since
the new catalyst is especially effective in converting olefins and aromatics, it has a large effect on reducing
NMHC emissions.

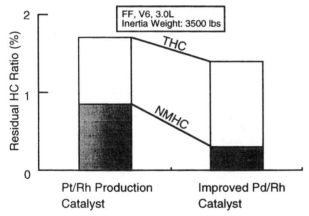

Figure 17. Effect of Improved Pd/Rh Catalyst on Residual
HC Ratio at the 2nd-23rd Phase of FTP

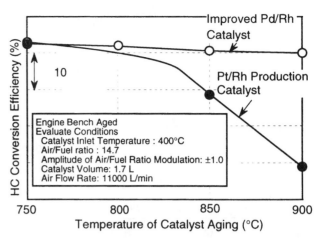

Figure 18. Heat Resistance Improvement of Improved
Pd/Rh Catalyst

TOTAL EMISSION REDUCTION

Figure 19 compares the HC emissions measured during
the first 125 seconds of the FTP for a vehicle complying
with the Tier 1 standards and a vehicle fitted with the
above-mentioned emission reduction technologies. The
total NMOG emissions measured for the two vehicles in
the FTP are shown in Figure 20. The vehicle incorporating these emission reduction technologies reduced the
emission level through the first phase of the FTP by
60-70% compared with the Tier 1 vehicle. This result indicates that the application of these technologies should
make it possible to meet the LEV standards, even taking
into account the variation that occurs under real-world
driving conditions. The emission of the vehicle almost
reached the ULEV levels.

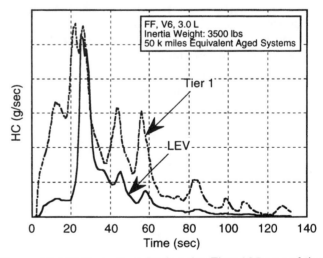

Figure 19. HC Emissions During the First 125 sec of the
FTP

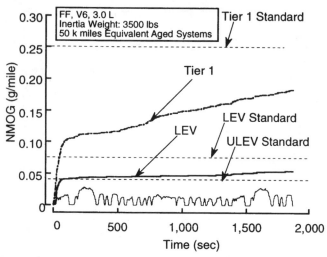

Figure 20. NMOG Emissions in the FTP

CONCLUSION

It has been shown that the application of the following technologies should achieve compliance with the LEV standards for vehicles fitted with V-6 engines, without sacrificing engine performance, such as engine stability.

1. The use of a swirl control valve together with combined optimum control of the air/fuel ratio and ignition timing shortens the catalyst light-off time for cleaner emissions while providing stable engine operation. Moreover, optimizing the notch geometry in the swirl control valve improves engine stability, and simultaneously reduces engine-out HC emissions and increases the exhaust temperature.

2. The use of a newly developed Pd/Rh catalyst in the close-coupled location further quickens catalyst light-off and provides improved heat resistance, making it possible to improve the conversion efficiency of NMHC in the warmed-up phase.

3. The use of an electronically controlled EGR valve actuated by a stepper motor eliminates nearly all response lag even under transient operation and achieves the calibrated EGR rate. This results in a simultaneous reduction of engine-out HC and NOx emissions.

4. Moreover, HC emissions can be substantially reduced by lowering the piston top land height to decrease the crevice volume, and by limiting the application of the anodizing coating to the ring grooves.

REFERENCES

1. State of California, Air Resources Board, "Proposed California Exhaust Emission Standards and Test Procedures for 1998 and Subsequent Model Passenger Cars, Light-Duty Trucks, and Medium-Duty Vehicles," HEV Workshop Draft, August 1995.
2. Y. Hasegawa, S. Akazaki, et al., "Individual Cylinder Air-Fuel Ratio Control, Using an Observer," SAE Paper 940376, 1994.
3. M. Nasu, A. Ohata, and S. Abe, "Model-Based Fuel Injection Control System for SI Engines," SAE Paper 961188, 1996.
4. Y. Shin, W. K. Cheng, and J. B. Heywood, "Liquid Gasoline Behavior in the Engine Cylinder of SI Engine," SAE Paper 941872, 1994
5. Y. Nakayama, T. Maruya, T. Oikawa, M. Kawamata, and M. Fujiwara, "Reduction on HC Emission from VTEC Engine During Cold-Start Condition," SAE Paper 940481, 1994
6. J. C. Summers, J. F. Skowron, W. B. Williamson, and K. I. Mitchell, "Fuel Sulfur Effects on Automotive Catalyst Performance," SAE Paper 920558, 1992.
7. P. Beckwith, P. J. Bennett, C. L. Goodfelloe, R. J. Brisley and A. Wilkins, "The Effects of Three-Way Catalyst Formulation on Sulfur Tolerance and Emissions from Gasoline Fuelled Vehicles," SAE Paper 940310, 1994
8. J. Sterlepper, H.-J. NeuBer, and H. Ruhland "HC-Emissions of SI Engines - Optical Investigation of Flame Propagations in Piston Top Land Crevice," SAE Paper 941994, 1994.
9. Kyoungdoug Min, Wai K. Cheng, and John B. Heywood, "The Effects of Crevices on the EngineOut Hydrocarbon Emissions in SI Engines," SAE Paper 940306, 1994.

1999-01-3506

SI Engine Operation with Retarded Ignition: Part 1 - Cyclic Variations

S. Russ, G. Lavoie and W. Dai
Ford Research Laboratory

Copyright © 1999 Society of Automotive Engineers, Inc.

ABSTRACT

Engine operation with spark ignition retarded from MBT timing is used at cold start to reduce HC emissions and increase exhaust gas temperature; however it also results in increased cyclic variations. Steady-state cold fluids testing was performed to better understand the causes of the cycle-to-cycle variations. Detailed analysis of individual cycles was performed to help gain an understanding of the causes of cyclic variations. The important results were:

- The primary cause of cyclic variations in IMEP is variations in the combustion phasing (location of 50% mass fraction burned). The expansion ratio decreases rapidly during combustion for retarded spark timing and therefore the phasing determines individual cycle thermal efficiency and IMEP. Variations in the late burn have little impact on the IMEP as this combustion occurs close to EVO and does little expansion work.

- A combustion diagram analysis indicates that flame quench and misfire at ignition do not contribute to the increase in cyclic IMEP fluctuations as is the case at the lean limit with standard spark timing. Prior cycle effects which occur at the lean limit with standard spark timing also were not found at cold, retarded spark conditions.

- Improved in-cylinder motion (swirl and tumble) decreased cyclic variations which enabled additional spark retard (relative to MBT timing) and produced lower HC emissions, presumably due to improved mixing.

- Dual spark plugs further improved stability. This was thought to be due to more repeatable combustion phasing as inhomogeneities in the chamber were "averaged out" by the two spark locations.

INTRODUCTION

The use of spark timing retarded from MBT is well known to produce increased gas temperatures due to the fact that the burned gas is not ideally expanded and does not perform as much work on the piston. Hydrocarbon emissions decrease at retarded spark timing due to reduced crevice volume at the end of burn, and increased burned gas temperatures at the end of the expansion stroke and during exhaust which increases oxidation. For these reasons aggressive spark retard is desirable for decreasing cold start hydrocarbon emissions. A cold start strategy which utilized high speed, lean, retarded operation is described in Kaiser et al. [1] and produced approximately 50% lower total HC emissions. Similarly, work done by Mandokoro et al. [2] illustrates that enleanment (10% lean) and the use of spark retard can reduce engine out emissions by 50% with further benefits from improved in-cylinder swirl motion.

The use of spark retard has two main drawbacks, however. The first is the obvious loss in efficiency which results in a small fuel economy penalty. This penalty is not large as spark retard is typically used for less than 1 minute during a cold start. The second is that spark retard increases cycle-to-cycle IMEP fluctuations, increasing engine roughness. The focus of this work is to better understand the causes of the cycle-to-cycle variations. This knowledge will assist in predicting the spark retard capability of future engine designs or hardware modifications.

EXPERIMENT

The engines used in these experiments were a typical 2.0L I4 4V engine and a single cylinder research 2V engine. The 2.0L I4 has a bore and stroke of 84.8 and 88 mm with a compression ratio of 9.5:1. The intake port flow generated in-cylinder tumble motion with a tumble ratio TR~0.5 to increase burn rate for acceptable idle quality and dilution tolerance. The single cylinder engine has a bore and stroke of 90.2 and 90 mm and incorporated a research cylinder head which incorporated 2 spark plugs. This engine also had a compression ratio of 9.5:1. Both engines were controlled by an AC electric dynamometer; dry air was supplied and measured with a critical flow orifice measurement system and a Pierburg system measured the fuel flow. A Horiba emissions console measured the concentration of O_2, CO, CO_2 and NO in dried exhaust and HC using a wet analyzer. Air/fuel ratios were calculated from the exhaust chemistry and measured using a UEGO sensor. The fuel used was 97 RON fully blended gasoline (stoichiometry = 14.6 A/F). Exhaust gas temperature measurements were taken in the empty catalyst can for the 2.0L engine and approximately 2.5 cm from the head face on the single-cylinder engine using 3 mm type K thermocouples.

A steady-state test procedure was used for these experiments. The engine was operated at a steady speed and load of 1200 rpm, 1.0 bar BMEP (2.0L engine) or 2.5 bar IMEP (single-cylinder engine) at a coolant and oil temperature of 20 °C to simulate cold conditions and 90 °C for warmed-up operation. The A/F and spark advance were swept from 14.6 - 16.0 and 15° to -25° until a stability level of 0.3 bar SDIMEP (standard deviation of IMEP) was reached.

EXPERIMENTAL RESULTS

The results for the 2.0L engine are shown below in plots of exhaust gas temperature, HC emissions and SDIMEP as a function of spark advance and A/F (Fig. 1).

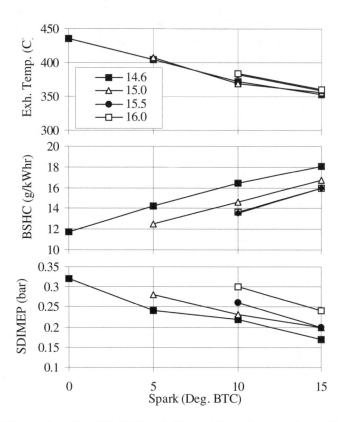

Figure 1. Cold Fluid Spark Retard Results at various A/F (2.0L 4V engine, 1200 rpm, 1.0 bar BMEP, 20 °C Fluids)

These results are similar to those found by Mandokoro et al. [2]). The exhaust temperature increases by about 80 °C as the spark is retarded from 15° BTDC to TDC (14.6 A/F). The HC emissions fall while the cyclic variability as measured by the standard deviation of IMEP (SDIMEP) increases.

SOURCES OF CYCLIC VARIABILITY

The purpose of this section is to examine in more detail the root cause of the increased cycle-to-cycle IMEP fluctuations as the spark is retarded.

FLAME PROPAGATION – It is well established that for lean or dilute engine operation (low loads, high EGR) cyclic variations in IMEP increase. This is due to the fact that dilution slows the laminar flame speed and conditions for flame propagation become unfavorable. This leads to partial burn cycles and eventually to complete misfiring cycles. To compare the engine and combustion stability of retarded ignition to the more well studied phenomena with lean operation a sweep of the air/fuel ratio was acquired at a typical part-load condition for the 2.0L engine. The results as the mixture was leaned out from 14.6 to 20:1 and 22:1 A/F is shown in Fig. 2(a) for fully warmed-up operation (1500 rpm, 2.62 bar BMEP). At 22:1 A/F the engine was just beginning to misfire (1 misfire recorded out of 1200 cycles). The HC emissions were high during this sweep which was taken following the cold fluids tests which probably increased the deposit levels in the engine. At the dilute limit (22:1 A/F) the HC emissions further increase due to the increase in incomplete combustion cycles and the low exhaust gas temperatures (less oxidation). In contrast, Fig. 2(b) illustrates the engine behavior as the spark is retarded at the cold conditions (15:1 A/F). There were no misfiring cycles recorded for this data set or for any of the retarded spark data acquired for this report. As shown, the HC emissions decrease even as the cyclic variability increases. Thus although the combustion may be slow later in the expansion stroke and even incomplete before exhaust valve opening, the combustion appears to continue during the exhaust stroke and in the exhaust port due to the high temperatures.

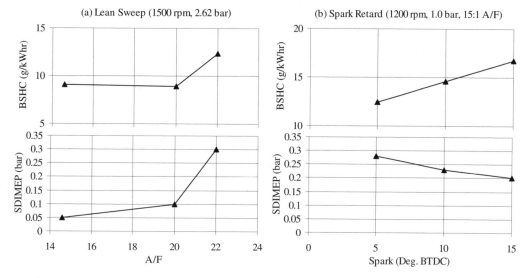

Figure 2. Comparison of HC emission and cyclic variations with enleanment (fully warm, road-load operation, MBT spark) and with spark retard (cold fluid operation)

To examine the stability of the flame under these conditions, the combustion regimes under different engine operating conditions were compared on a Leeds Diagram, which is a combustion regime diagram composed of non-dimensional turbulence intensity (u'/S_L) and integral length scales (L/δ_L) and used to determine the nature of the premixed combustion [3]. The Leeds Diagram can be used to quantify how close the premixed combustion is to the quenching limit and, therefore, examine how stable the combustion is. The farther away from the quenching limit, the more stable the combustion. With the help of a cycle simulation model, parameters for the Leed's Diagram were estimated [3] for both a lean sweep at part-load, MBT spark operation (14.6, 20 and 23 A/F) and for a spark sweep at 1200 rpm, 1 bar BMEP, 15:1 A/F (30°, 15°, 5° BTDC spark timings). The combustion phase diagrams are shown in Figures 3(a) and (b), respectively. Each small line defines the trajectory of unburned gas conditions from the time of spark (left end of line) to the end of combustion (right end of line). The long straight lines define the combustion regimes.

The calculations for the lean sweep illustrate that the flame is becoming less stable and is approaching the quench region as the mixture is leaned out to 23:1, as shown in Figure 3(a). The leanest operating point (23:1 A/F) correlates fairly well with the actual lean-limit of 22:1 AF (Fig. 2). In contrast the calculations for the spark retard sweep illustrate that the flame actually becomes more stable as the spark is retarded, as shown in Figure 3(b).

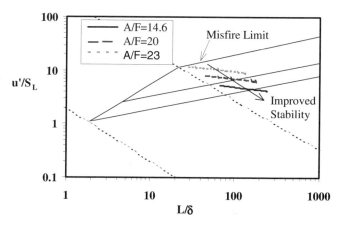

Figure 3a. Flame stability with lean air/fuel (1500 rpm, 2.62 bar BMEP, MBT Spark Timing and 90 °C Fluids)

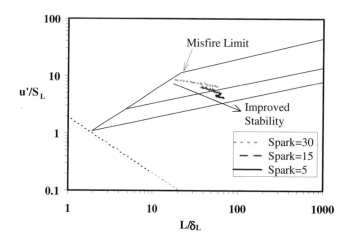

Figure 3b. Flame stability with retarded spark timings (1200 rpm, 1.0 bar BMEP, 14.6 A/F and 20 °C Fluids)

Figures 4(a), (b), (c) and (d) show the individual parameter histories for different spark timings and indicate that, while the turbulence intensity and integral length scales are similar for all of the spark timings, the laminar flame speed increases and laminar flame thickness decreases as the spark is retarded. Both the increase in laminar flame speed and the decrease in laminar flame thickness improve the stability of the combustion. The increase in gas temperature and decrease in residual gas fractions (as a result of the higher intake manifold pressure required to maintain 1.0 bar BMEP) were the causes for the changes in the laminar flame characteristics.

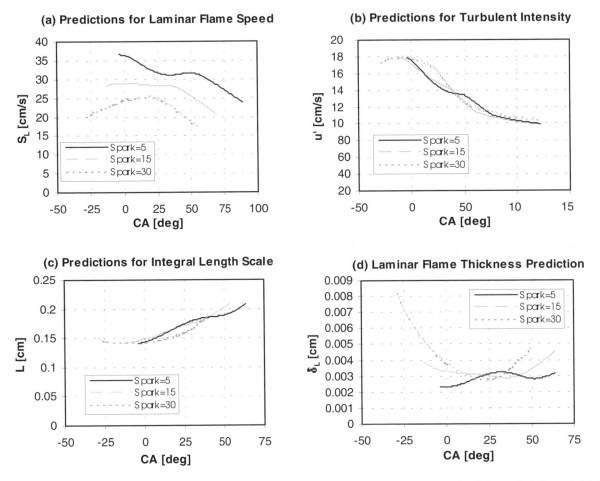

Figure 4. Flame characteristics with different spark timings (1200 rpm, 1.0 bar BMEP, 14.6 A/F and 20 °C Fluids)

The results of this analysis illustrate that for this engine under cold retarded spark conditions, flame quench and misfire are not likely to contribute to the increase in cycle to cycle IMEP fluctuations as is the case at the lean limit with standard spark timing.

PRIOR CYCLE EFFECTS – Combustion stability under very lean or dilute conditions with standard spark timing has been found to be affected by prior cycle interactions [4,5]. This instability arises due to the fact that a bad or slow burn cycle produces higher exhaust gas temperatures and potentially high levels of HC emissions (if the fuel was not completely burned); this results in lower levels of exhaust gas residual and therefore a faster, more complete burn in the following cycle. This fast burn cycle in turn can produce lower exhaust gas temperatures (higher density) and therefore increased exhaust gas residual in the following cycle which can lead to a bad/ slow burn cycle. Return maps which plot the IMEP for each cycle vs. the IMEP for the previous cycle can be used to illustrate this behavior. Fig. 5 presents return maps for lean part-load operation at MBT spark (Fig. 5a) and for cold, retarded spark operation (Fig. 5b) for 1000 consecutive cycles. For warm, part-load operation at air/ fuel ratios leaner than 22:1 the return map illustrates the emergence of some structure and the possibility of prior cycle effects. A more detailed analysis which more strongly proves the evidence of prior cycle effects is documented in a separate report [6]. For the retarded spark conditions, the return maps for both spark advances show no evidence of the high IMEP (fastest burn) cycles following the low IMEP (slowest burn) cycles or visa versa.

Although prior cycle effects may be important at the lean or dilute combustion limit with standard spark timing where misfires and/or partial burn cycles occur, they do not play a significant role in producing cycle to cycle IMEP fluctuations under these less dilute, retarded spark operating conditions.

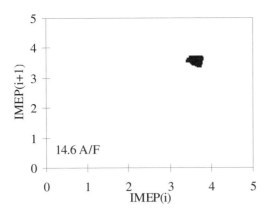
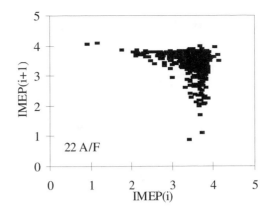

Figure 5a. Return maps for 2.0L engine as A/F ratio is leaned out to limit at MBT spark
(1500 rpm, 2.62 bar BMEP, MBT spark, 90 °C Fluids)

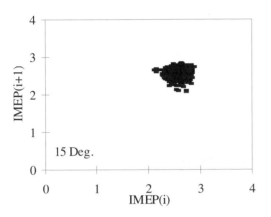
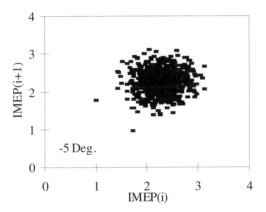

Figure 5b. Return maps for 2.0L engine as spark is retarded (1200 rpm, 1.0 bar BMEP, 14.6 A/F, 20 °C Fluids)

CYCLIC BURN RATE EFFECTS – Using a standard heat release analysis the details of the burn rate and IMEP produced by individual cycles were studied in more detail for the very retarded spark condition (cylinder 1, 2.0L engine, 14.6 A/F, -5° spark) and compared to fully warm MBT spark operation at the lean limit (22.1 A/F, MBT Spark). Fig. 6 presents the correlation of the individual cycle IMEP with 0-2% burn duration, 10-90% burn duration and location of 50% burn. These plots clearly show that faster burning cycles produce increased IMEP (more advanced heat release) with slower burning cycles producing lower IMEP (more retarded heat release) as expected. The correlations are much stronger for the retarded spark condition than observed at the lean limit with MBT spark timing. The plot of IMEP vs. loc. 50% burn location clearly shows that the cyclic variations in IMEP at retarded spark timing can be almost completely explained by the cyclic variations in the burn rate which determines the phasing of the heat release.

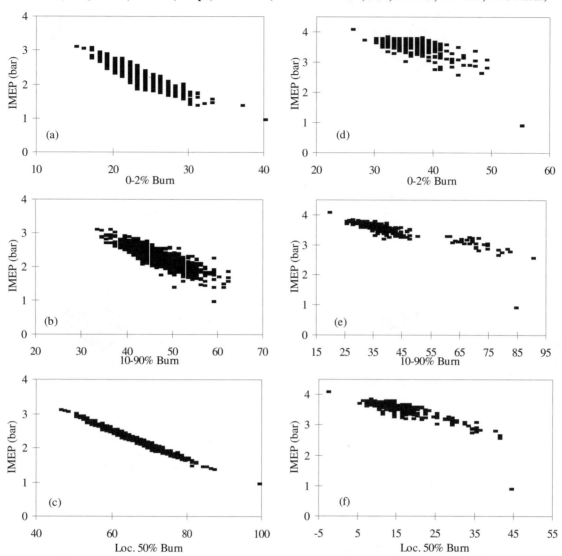

CSS RE (1200, 1.0 bar, 14.6 A/F, -5 Spk, 20 C Fluids)

WWMP (1500, 2.62 bar, 22:1 A/F, 90 C Fluids)

Figure 6. Correlation of individual cycle burn rates with IMEP (a,b,c - 1200 rpm, 1.0 bar BMEP, -5° Spark, 20 °C Fluids; d,e,f - 1500 rpm, 2.62 bar BMEP, 22:1 A/F, MBT spark, 90 °C Fluids)

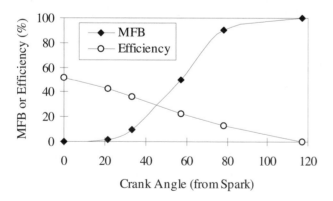

Figure 7. Burn rate and ideal efficiency profile (1200 rpm, 1.0 bar BMEP, -5° spark, 14.6 A/F, 20 °C Fluids)

A plot of the normalized burn rate profile for this condition is shown in Fig. 7. Also shown on the plot is the expansion ratio from the various burn locations to EVO and the corresponding ideal Otto cycle thermal efficiency (assuming $\gamma=1.35$).

Fig. 7 illustrates the fact that at these retarded spark conditions the expansion ratio and therefore the thermal efficiency rapidly decreases during the combustion event. The heat release from the first gas to burn is accompanied by a volume change and therefore produces useful work. The energy release from the end of the combustion event, occurring close to exhaust valve opening, has little accompanying change in volume and therefore does little expansion work and does not contribute to the IMEP of the cycle. For this engine the majority of the heat is released near the 50% mfb location and therefore the 50% mfb location does a good job of representing the average location of the heat release and therefore the expansion ratio for each cycle. Fig. 8 illustrates the combustion phasing, expansion ratio and thermal efficiency for both a slow and fast burn cycle at this operating condition. Note the strong correlation between the IMEP and theoretical efficiency for the fast burn cycle compared to the IMEP and theoretical efficiency for the slow burn cycle.

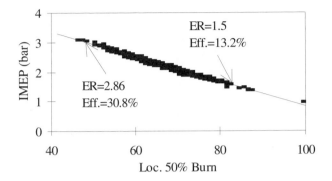

Figure 8. Effect of combustion phasing on cyclic IMEP (1200 rpm, 1.0 bar BMEP, 14.6 A/F, -5° Spark, 20 °C Fluids)

This analysis of individual cycles illustrates that the primary cause of cyclic variations in IMEP is cyclic variations in the burn rate (location of 50% of the mass fraction burned). Differences in the late combustion are not likely to have an effect on cyclic variations in IMEP as this combustion occurs close to EVO and therefore does not contribute to the work produced by the cycle.

EFFECT OF DUAL PLUGS AND CHARGE MOTION – Multiple spark plugs produce faster burn rate due to the additional flame area produced by igniting in several locations. The use of two sparks has also been shown to improve cyclic variations in IMEP at idle and light loads.

For this study a single cylinder research engine was used to investigate the effect of dual spark plugs and compare to the effect of increased charge motion. Fig. 9 is a sketch of the combustion chamber showing the two spark plug locations along with the intake and exhaust valves. The base engine has a high flow port which produces moderate levels of tumble and swirl (TR=0.6, SR=0.4) and a slow burn rate with single plug operation (1500 rpm, 3.8 bar IMEP, MBT spark ~ 32°). The burn rate improves with either the use of a charge motion control valve installed on a shaft approximately 1 cm upstream of the head face (Fig. 9) or igniting with both spark plugs (MBT spark ~26° for both cases). The combination of dual spark plugs with the CMCV closed produces a very fast burn (19° MBT spark) with good mixing. Unfortunately the swirl and tumble rates for the case of CMCV closed were not measured prior to the publication of this report.

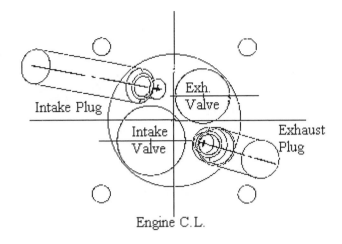

Two-Valve Research Combustion Chamber

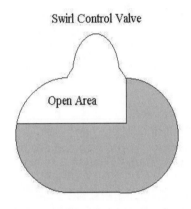

Charge Motion Control Valve (CMCV)

Figure 9. Schematic of dual plug combustion chamber and geometry of charge motion control valve

Results for the four different engine configurations are shown in Fig. 10 for an A/F of 15.5:1. For the base engine (with only the exhaust side plug firing), the SDIMEP is above 0.25 bar at 15° spark and can only retard to 10° to avoid exceeding the limit of 0.3 bar SDIMEP. The CMCV closed improves both the burn rate and the SDIMEP substantially and allows an additional 10° - 15° of spark retard before the 0.3 bar SDIMEP limit is exceeded. This results in improved heat flux and decreased HC emissions relative to the base engine. The dual plug with CMCV open produces almost identical exhaust gas temperatures as the single plug with CMCV closed due to the similar burn rates. Note however that the dual plug produces lower SDIMEP at a fixed spark timing. In fact the 0.3 bar SDIMEP limit is not exceeded even with 20° after TDC spark timing for the dual plug case. This results in increased exhaust temperatures at given level of SDIMEP. The case with the dual spark plugs firing and the CMCV closed produces the fastest burn and lowest SDIMEP at a given spark timing.

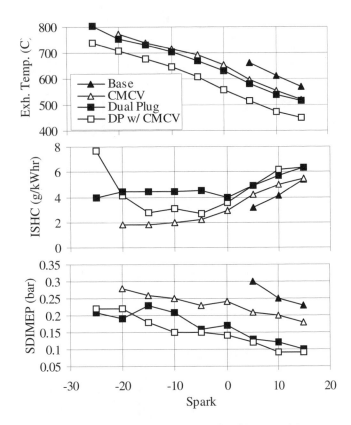

Figure 10. Effect of dual plug and CMCV on cold operation (1200 rpm, 2.5 bar IMEP, 15.5:1 A/ F and 20 °C Fluids)

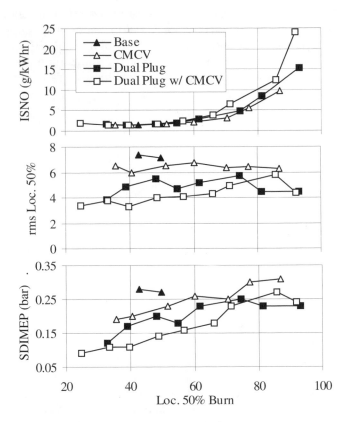

Figure 11. Effect of dual plug and CMCV on cold operation (1200 rpm, 2.5 bar MEP, 15.5:1 A/ F and 20 °C Fluids)

CYCLIC VARIATION DIFFERENCES – The ISNO, rms location of 50% burn and SDIMEP are plotted for the above four cases as a function of the location of 50% burn in Fig. 11. Closing the CMCV did not have a measurable effect on the NOx emissions indicating that the internal exhaust residual level was not affected. The cyclic variability in the combustion phasing (rms location of 50% burn) mirrors the SDIMEP quite closely indicating again that it is the variability in the burn rates which causes the variations in the MEP. At a given location of 50% burn, closing the CMCV produces lower levels of SDIMEP with dual plugs (CMCV open) producing lower levels than the use of a CMCV plate. The case of dual plugs with CMCV closed produces the best results and is slightly better than with dual plugs alone due to less variations in the burn rate (location of 50% mfb).

The cause for the decreased variations in the burn rate and therefore the IMEP with dual spark plugs can be studied by comparing the case of CMCV closed with a single spark to the case of dual spark plugs with the CMCV open. These two cases not only produce similar overall burn rates but the shape of the average burn curves is nearly identical as shown in Fig. 12 below. Because the burn rates for these cases are nearly identical the improvement in the cyclic variability is probably not due to a change in the shape of the burn curve or due to a change in the early burn rate.

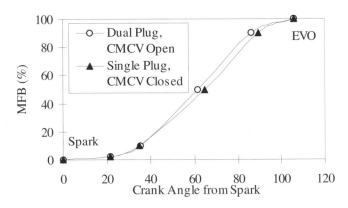

Figure 12. Average burn rate comparison (1200 rpm, 2.5 bar IMEP, -10° Spark, 15.5 A/F and 20 °C Fluids)

A comparison of the variability in the 2%, 10%, 50% and 90% mass fraction burned is presented in Fig. 13. Igniting in two locations produces less variations in the burn event since any inhomogeneities (in turbulence, A/F, residual, etc.) are averaged over two locations. It is much less likely to have very slow (or very fast) flame kernel formation at both of the ignition sites relative to a single site. The continued decreased variability in 10%, 50% and 90% mfb for the dual spark plug case is probably due to both the fact that the very early burn is more repeatable and also due to the increased flame area later in the cycle which again serves to average out local inhomogeneities.

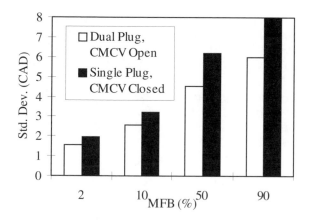

Figure 13. Burn rate variability comparison (1200 rpm, 2.5 bar IMEP, -10° Spark, 15.5 A/F and 20 °C Fluids)

HC EMISSIONS – The HC emissions and exhaust gas temperatures for these cases are re-plotted in Fig. 14 at equivalent levels of combustion phasing (location of 50% burn). The CMCV closed cases have slightly lower exhaust gas temperatures at equivalent combustion phasing due to increased heat transfer caused by the higher in-cylinder turbulence levels. The hydrocarbon emissions are similar for all of the cases with slightly lower levels measured for the cases with the CMCV closed. This is also similar to the results found by Mandokoro et al. [2] with the increased turbulence and in-cylinder motion created with the CMCV closed resulting in improved mixing and HC oxidation during the exhaust stroke.

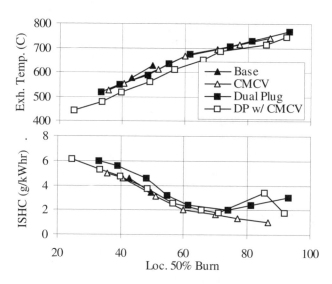

Figure 14. Exhaust gas temperature and HC emissions (1200 rpm, 2.5 bar MEP, 15.5:1 A/F and 20 °C Fluids)

CONCLUSIONS

These studies clearly illustrate that the primary cause of cyclic variations in IMEP for cold, retarded spark operation is variations in the combustion phasing (location of 50% mass fraction burned). The expansion ratio decreases rapidly during combustion for retarded spark timing and therefore the combustion phasing of an individual cycle determines the cycle thermal efficiency and the IMEP produced. Differences in the late burn have little impact on IMEP as this combustion occurs close to EVO and does little expansion work. A Leeds' diagram analysis indicates that flame quench and misfire are unlikely to contribute to the increase in cycle to cycle IMEP fluctuations as is the case at the lean limit with standard spark timing. In fact the calculations of the flame regime indicate that the flame moves further away from the quenching limit with increasing spark retard. Prior cycle effects which occur at the lean limit with standard spark timing also were not found at the stability limit for these cold, retarded spark operating conditions.

Studies on hardware modifications revealed:

- Charge motion control valve (CMCV) improved cyclic variations due to more repeatable combustion phasing (lower rms 50% burn) which enabled additional spark retard (relative to MBT timing). At equivalent combustion phasing, CMCVs produced lower HC emissions, presumably due to improved mixing.

- Dual spark plugs were found to produce even more repeatable combustion events and enable further spark retard when compared with the use of aCMCV. The improved stability was presumably due to the fact that any inhomogeneities in the chamber were "averaged out" by the two spark locations.

ACKNOWLEDGMENTS

The authors would like to acknowledge Randy Bergquist for acquiring this test data and Dave Scholl for assisting in analyzing the prior cycle effects.

REFERENCES

1. Kaiser, E.W., Siegl,W.O., Baidas, L.M., Lawson, G.P., Cramer, C.F., Dobbins, K.L., Roth,P.W. and Smoko-vitz, M., "Time-Resolved Measurement of Speciated Hydrocarbon Emissions During Cold Start of a Spark-Ignited Engine", SAE Technical Paper Series 940963, 1994

2. Mandokoro,Y., S. Kubo, M.Taki and H. Ban, "Reduction in Cold Hydrocarbon Mass Emissions by Combustion Control of Gasoline Engine: Part 1 - The Effect of In-Cylinder Gas Flow on Reactivity Promotion of Hydrocarbons", 14th SAEJ/JSME Internal Combustion Engine Symposium, Paper No. 9737419, September 1997, In: Engine Technology Progress in Japan: Spark Ignition Engine Technology, January 1998.

3. Dai, W., Russ, S.G., Trigui, N. and Tallio, K.V., "Regimes of Premixed Turbulent Combustion and Misfire Modeling in SI Engines", SAE Technical Paper Series 982611, 1998

4. Martin, J.K., Plee, S.L., and Remboski, D.J. Jr., "Burn modes and prior-cycle effects on cyclic variation in lean-burn park-ignition engine combustion", SAE Technical Paper Series 880201, 1988

5. Grunefeld, G., Beushausen, V., Andresen, P. and Hentschel, W., "A major origin of cyclic energy conversion variations in SI engines: cycle-by-cycle variation of the equivalence ratio and residual gas of the initial charge", SAE Technical Paper Series 941880, 1994

6. Scholl, D., and Russ, S., "Air-Fuel Ratio Dependence of Random and Deterministic Cyclic Variability in a Spark-Ignited Engine", SAE Technical Paper No. 1999-3513, 1999

CHAPTER 6

SECONDARY AIR INJECTION
FOR IMPROVING
COLD-START PERFORMANCE

Fuquan (Frank) Zhao and Mark Borland
DaimlerChrysler Corporation

6.1 OVERVIEW

Improving catalyst light-off characteristics during cold start and reducing engine-out (more accurately converter-in) emissions prior to catalyst light-off have been regarded as the keys to meeting future stringent emissions regulations. Many technologies and control strategies have been proposed, and some of them have already been incorporated into production to address these issues. Among these, secondary air injection received a lot of attention.

Engine-out exhaust gas composition varies substantially with engine-fueled air/fuel ratio. When the engine is fueled with a richer-than-stoichiometric mixture, H_2 and CO will be produced and exhausted into the exhaust stream. The concentrations of CO and H_2 rise steadily as the mixture gets richer. A similar trend is also true with engine-out HC emissions. Injecting secondary air into the exhaust port, which allows the secondary air to mix and react with these reactants (H_2, CO, HC) produced during engine-rich operation, leads to an exothermic reaction in the exhaust system. This reaction can effectively reduce HC emissions inside the exhaust manifold and simultaneously accelerate the heating process of the converter following a cold start of gasoline engines. The engine-out NOx can also be reduced due to the lack of oxygen for NOx formation when operating the engine richer, in addition to the evaporative charge cooling effect.

The physical and chemical processes associated with the secondary air injection are, however, quite complicated, and its benefit strongly depends upon the system optimization, which usually varies with the specific application. All the design and operating parameters that determine engine-out emissions, exhaust gas temperature, mixing process of reactants with the secondary air and the mixture residence time inside the exhaust port and manifold will affect the level of success of the secondary air injection strategy.

In general, chemical reactions occurring in the exhaust system with secondary air injection can be categorized into two types of oxidation of reactants. These two types of oxidation are both exothermic, but the temperatures required for these to occur rapidly are dramatically different. One is the oxidation process that happens inside the exhaust port and manifold prior to the catalytic converter, which is referred to as thermal oxidation. The other is the oxidation of HC and CO that occurs inside the catalytic converter, which is significantly different from that of the thermal oxidation due to the presence of a catalyst and is referred to as catalytic oxidation. The thermal oxidation usually requires a temperature of 750°C or higher to become significant, depending on the exhaust species and residence time. It is believed that the production of H_2 in engine-rich combustion can significantly enhance the thermal oxidation process inside the exhaust port and manifold, and CO and HC can well sustain the process once it is established. In contrast, the conversion of HC and CO can start at a similar rate with a much lower temperature (200°C) on the surface of certain types of catalysts. Both thermal oxidation and catalytic oxidation contribute to heating the catalyst for shortening the light-off time. In comparison, catalytic oxidation is more effective in improving the catalyst light-off by heating the catalyst directly when it reaches a certain temperature threshold. This must be provided by cylinder-out exhaust heat plus upstream thermal oxidation. Part of the heat produced via thermal oxidation will be dissipated to manifold walls and as a result, its impact on catalyst light-off will be slightly degraded. However, thermal oxidation can be very effective in removing engine-out HC emissions, if adequate temperatures are sustained. This is extremely important prior to catalyst light-off. Rich engine operation generally produces higher engine-out HC emissions as compared to that of lean operation, which demands a further reduction in converter-in emissions. With the addition of secondary air, the converter-in emissions can be reduced significantly due to the enhanced oxidation in the manifold. A certain level of temperature is, however, necessary to activate the chemistry and sustain the reactions. Adding air too far downstream from the exhaust valve cannot benefit from this emission reduction due to the lack of thermal oxidation, even though it may produce an improvement in catalyst light-off performance.

The effect of exhaust chemistry on the catalyst tends to be ignored in most of the analysis or data interpretation when dealing with secondary air injection. It should be pointed out that the catalyst is not only exposed to the hot exhaust gases, but also to high chemical loads from the leftover HC and CO emissions. The introduction of these chemical constituents at certain concentrations is very important to realize self-sustaining, chain reactions. Therefore, it is strongly believed that the combination of this chain reaction and thermal loading together allows the catalyst to ultimately sustain the high conversion efficiencies following the cold start. This chemical impact should not be ignored at all when locating the catalyst, designing the exhaust system to maximize the benefit of secondary air injection or calibrating the secondary air injection system. A good compromise between the exhaust manifold thermal oxidation and catalytic oxidation of CO and HC emissions is necessary. The former provides the heat necessary to initiate the exothermic reaction of CO and HC on the catalyst surface and reduce the HC and CO loading on the catalyst to minimize any emissions breakthrough; the latter has a direct heating effect on the catalyst surface and requires a certain level of HC and CO emissions to achieve this. If the thermal oxidation consumes too much engine-out reactants, which is important in minimizing HC emissions prior to catalyst light-off during cold start, the catalyst light-off process may be slowed due to a reduced energy release directly inside the catalytic converter. In contrast, if the converter is overdosed with the reactants due to the poor thermal oxidation, there may be more emissions breakthrough even though the catalyst may be lit-off slightly faster. Therefore, the system must be optimized to balance these two oxidation processes in order to maximize the benefits.

It is easy to understand that engine-out reactants can only be oxidized in the exhaust manifold when oxygen is present at a sufficiently high temperature. To maximize the thermal oxidation rate in the manifold, the pathway from the exhaust valve to the converter inlet must provide an environment with a high temperature and a high oxygen concentration over a sufficient period of time. The reaction rate increases exponentially with the temperature. A small increase in the initial temperature can result in the oxidation time of the mixture being cut significantly. Any design and operating parameters such as spark retardation and early exhaust valve opening that can increase engine-out gas temperature will dramatically enhance the thermal oxidation process. Therefore, the early initiation of thermal oxidation around the periphery of the exhaust valve is crucial to the success of secondary air injection. Air injection has to be as close as possible to the exhaust valve where the exhaust gases are hottest, as will be discussed in Section 4, for air injection location optimization. A minimal amount of oxygen is needed to react with the exhaust gases and a minimal amount of reactants (HC, CO, H_2, etc.) and radicals (OH, O, H) must be available to start the reaction. When the secondary air is in an extreme excess ratio or the mixing of exhaust gases with secondary air occurs too rapidly, the thermal reaction will be quenched. Once the reaction is quenched, it will be very difficult to initiate again. On the other hand, if this mixing occurs too slowly, the heat loss to the wall may eventually lower the temperature below that necessary for retaining the thermal oxidation. This leads to the belief that promoting the mixing inside the port during the engine exhaust blowdown process is critical.

The temperature at the exhaust port varies significantly through the cycle. Right after the exhaust valve opens, the temperature initially drops rapidly due to the fact that a large amount of air delivered before the exhaust valve opens is accumulated there to mix with the exhaust gases. Then temperature increases quickly because of the exhaust gas blowdown. During blowdown, exhaust gases exit the engine and flow through the exhaust system at a higher pressure, which prevents the exhaust gases from mixing with the secondary air. At the exhaust port, exhaust blowdown can start thermal oxidation as long as a small amount of secondary air is available. The capability of being able to deliver some secondary air and get it entrained into the exhaust gases is important in utilizing the heat available during the blowdown phase. The mixing between secondary air and exhaust gases is also occurring during this blowdown phase, which prepares the thermal condition for the oxidation once the secondary air and exhaust gases get mixed. After the blowdown, the exhaust gases have lower temperatures but a better mixing rate and longer residence time, which can certainly initiate thermal oxidation if a minimum temperature is maintained. The last part of the mixture from the cylinder flowing through the exhaust port has an even lower temperature, which may prevent any thermal oxidation. After the exhaust valve closes, only secondary air is supplied into the residual gas in the port at a relatively lower temperature until the next engine cycle.

In the pathway of the exhaust gases from the exhaust valve to the catalytic converter, the exhaust gases will be heated by the reaction occurring, lose heat to the walls and either be heated or cooled by the surrounding gas. At the manifold collector, the temperature of the mixture decreases due to the heat losses to the wall and the mixing of the exhaust gases with secondary air coming from other cylinders. The worst case for mixing occurs when the secondary air is forced out of the exhaust port during the exhaust blowdown. However, the exhaust gas blowdown also provides relatively hotter exhaust gases. Any design and operating parameters that can prevent heat losses to the walls and improve mixing with a longer residence time in the manifold are effective in enhancing the completion of the thermal oxidation process. The mixing of exhaust gases between cylinders is as critical as that within each cylinder event. The detailed design of the exhaust manifold such as runner length, runner orientation and manifold volume are all important parameters for optimization. It is worth noting that the effort to improve mixing around this region is crucial both from the point of thermal oxidation and also of catalytic oxidation since the catalytic oxidation requires a homogeneous mixture inside the catalytic converter where the mixing ceases. When considering the residence time for enhancing thermal oxidation in the manifold, the position for a close-coupled catalyst may no longer be so critical as that for the non-secondary-air system. Instead, placing the catalyst slightly distant from the manifold collector will provide more time for oxidation to occur before reaching the catalyst. This will also alleviate the concern of fast thermal deterioration with the closed-coupled location, extend the life of the catalyst and relax the packaging constraints.

Engine enrichment level and secondary air injection quantity are also two determining parameters that must be carefully examined to maximize the secondary air benefit. The level of engine fueling enrichment determines the compositions of engine-out emissions (reactants) and their concentrations. It was reported that mass emissions in Bag I of the FTP test were consistently higher with less engine enrichment. It must be noted that over-enriching the engine may jeopardize combustion stability, which may lead to an engine misfire and subsequently a sharp increase in HC emissions. Regarding the quantity of secondary air, there are two countervailing effects. On one hand, a certain quantity of secondary air is necessary to maximize the reactions in front of the catalyst, which will result in increased temperature. On the other hand, an excess amount of secondary air will cool the exhaust, which may slow down the reactions or in the worst case quench the thermal oxidation inside the exhaust system.

Some concerns such as air pump noise level, its volume for packaging, weight, system complexity, and system durability must be carefully examined before implementing this technology. Since the engine is fueled with an ultra-rich mixture and a large amount of liquid fuel may enter the cylinder directly, particularly during cold start, some of the rich-combustion-related issues must be thoroughly evaluated. These include soot loading on combustion chamber and catalyst, likely oil dilution via repeated and extended fuel-rich operation, and spark plug fouling concerns. In addition, thermal loading in the exhaust system and catalyst must be examined to determine the optimal design of exhaust system and catalyst location. Cold start fuel consumption should also be minimized. Nevertheless, secondary air injection into the exhaust port combined with rich engine operation is a powerful technology to meet the development goal due to its robust and consistent performance and the relative ease of implementation with today's engine system without requiring a major design change.

6.2 SELECTED SAE TECHNICAL PAPERS

The Importance of Secondary Air Mixing in Exhaust Thermal Reactor Systems

Ronald J. Herrin
General Motors Research Labs.

VERY RICH (10-12:1 ENGINE AIR-FUEL RATIO) automotive thermal reactor systems have severe shortcomings which have eliminated them from serious consideration. Most important of these shortcomings is the associated fuel economy penalty. The very rich A/F and retarded spark timing needed for adequate reactor performance decrease fuel economy by 30-40%. In addition, the very high reactor temperatures result in material and durability problems. High underhood and tailpipe outlet temperatures are other areas of concern. If high conversion efficiency were achieved at less rich engine A/F (13-14:1) and cooler exhaust temperatures, the air-injected reactor might become viable.

The effect of mixing on reactor efficiency is explored in this study. Mixing has long been recognized as an important parameter in chemical reactor design. Automotive reactors present a non-classical case of mixing, however, because severe low and high-frequency transients in flow rate, gas composition, and temperature complicate the design of systems already constrained by size, weight, cost, and durability considerations.

Previous research in automotive reactor mixing has dealt primarily with altering the reactor internal configuration to direct gas flow and generate turbulence. In this study the mixing problem is approached at its source—the engine exhaust port, where the secondary air is commonly injected.

THEORY

The mass flow rate of exhaust from one cyl of a piston engine varies cyclically with time, as shown in Fig. 1. Efficient conversion of exhaust combustibles in a rich thermal reactor requires secondary air addition in an amount which will supply sufficient oxygen without excessively cooling the exhaust gas. Thus, an optimum mass flow of secondary air would provide a reactor-inlet mixture having a constant, lean A/F throughout the exhaust event

$$\text{Reactor mass air-fuel ratio} = \frac{\Delta\text{Engine air} + \text{Secondary air}}{\text{Engine fuel}} \quad (1)$$

This condition can be fulfilled if the airflow is synchronized with and proportional to the exhaust gas outflow from the engine cylinder (see dotted line, Fig. 1). However, synchronization and proportioning of the air and exhaust provides no guarantee that the flows leaving the port are unstratified.

ABSTRACT

Automotive thermal reactors have obtained high conversion efficiencies on engines with very rich carburetion, but fuel economy and reactor durability have suffered. Improved mixing of exhaust gas and secondary air in the engine exhaust port was examined as a means of improving reactor efficiency at less rich engine air-fuel ratios. Three air-injection systems which span a broad range of mixing capabilities were examined. Mixing characteristics were deduced from anemometry measurements of instantaneous secondary airflow, and emission performance of each system was generalized by a test program employing four steady-state conditions.

High-pressure, timed air injection provides the best mixing and the best reactor performance. Sparger (radial discharge) air injection tubes provide fair mixing and better performance than conventional open-ended air injection tubes, which exhibit poor mixing characteristics. Performance with sparger tubes is significantly poorer than with timed injection, but sparger tubes are more practical in terms of cost, complexity, and durability.

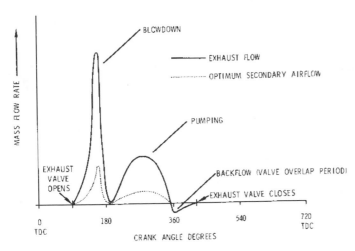

Fig. 1 – Exhaust gas flow and optimum secondary airflow for one cylinder

Therefore, the air and exhaust entering the reactor should be thoroughly mixed to gain full benefit from optimum secondary airflow.

SECONDARY AIR INJECTION SYSTEMS

An experimental technique employing hot-wire anemometry was developed for measuring instantaneous secondary airflow rate. This technique, described in Appendix A, was used to evaluate the airflow characteristics of several air injection systems, of which three were selected for this study.

HARDWARE–The air injection tubes used with two low-pressure secondary air systems are shown in Fig. 2. With both

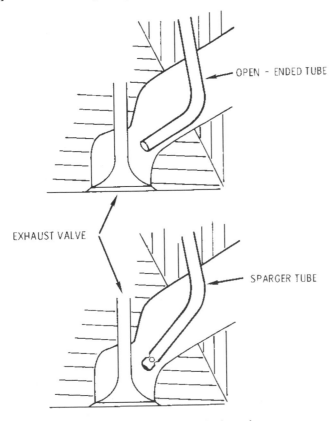

Fig. 2 – Low pressure air injection tubes

systems, an injection tube is located in each exhaust port to discharge air in the vicinity of the exhaust valve. The open-ended injection tube is typical of production A.I.R. systems, and discharges air directly at the valve. The sparger tube is closed at its end, and discharges air through radial holes in the tube body.

The third secondary air system tested is a high-pressure, timed injection scheme designed specifically for this study. High pressure air is injected through several small holes in each exhaust valve seat (Fig. 3). Airflow is terminated when the exhaust valve closes over the injection holes. Details of the timed injection system design are given in Appendix B.

Fig. 3 – Cylinder head modified for high pressure, timed air injection

AIRFLOW CHARACTERISTICS–Anemometry measurements of the airflow characteristics of each system are shown in Fig. 4 and explained in the following sections. The theoretically optimum secondary airflow is included in Fig. 4 for comparison. A detailed interpretation of the anemometer traces can be found in Appendix A.

Open Tube–Airflow from the open-ended tube is terminated during the exhaust event, as reported by Kadlec, et al. (1)*. The exhaust and air flows are unsynchronized, resulting in alternate slugs of air and exhaust entering the reactor. Therefore, mixing occurs only at the interface between slugs.

Sparger Tube–The closed end of the sparger tube shields the secondary air system from the dynamic pressure of blowdown, so airflow is maintained throughout most of the exhaust event. Flow interruption does occur at the height of blowdown, which implies that a significant amount of exhaust

---

*Numbers in parentheses designate References at end of paper.

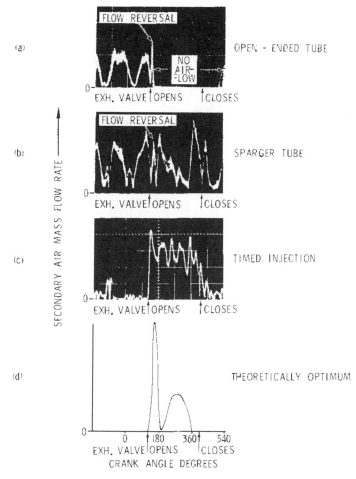

Fig. 4 – Experimentally measured and theoretically optimum secondary airflows

tions in flow with changing crank angle during the portion of the cycle when the exhaust valve is closed. These variations are caused by static pressure pulsations in the exhaust system which arise especially when the intake manifold exhaust cross-over is blocked. The variations in sparger tube flow that occur during the exhaust event are caused by the changing pressures of exhaust blowdown, pumping, and backflow.

Timed Injection– The timed air injection system is designed to overcome the sparger tube mixing deficiencies. As can be seen from the anemometer airflow data (Fig. 4c), secondary air does not flow during the portion of the engine cycle when the exhaust valve is closed. This alleviates the problem of un-mixed cool air entering the reactor, and allows an overall airflow rate which achieves a desirable mixture air-fuel ratio in the port during most of the exhaust event. In addition to solving the problem of flow synchronization, the timed air injection system ensures thorough mixing of air and exhaust. The nine equally spaced air injection holes shown in Fig. 3 provide good dispersion of the air throughout the exhaust port. High velocity air discharging normal to the exhaust flow further enhances mixing.

The cause of rapid fluctuations in the timed injection air-flow profile was investigated by changing the geometry of the secondary air supply system and by changing the exhaust-valve spring preload. From these two experiments, it was concluded that most of the fluctuations in the anemometry output were caused by organ pipe pressure waves, and that the true airflow profile was much closer to a square wave than indicated in Fig. 4c.

Although the airflow profile obtained with timed injection is synchronized with the exhaust event, it does not closely match the theoretically optimum profile (Fig. 4d). This mis-match results in too little secondary air at the peak of blow-down and too much air between the blowdown and pumping portions of the exhaust event. However, since the exhaust pro-file (Fig. 1) varies markedly with engine speed and load, the air delivery characteristics of the experimental timed injection system are a reasonable compromise. A system to proportion air to exhaust flow exactly would be inordinately complex.

The mixing capabilities of the reactor affect conversion even when timed injection is used. The above mentioned mismatch of air and exhaust flows causes a variation in reactor inlet A/F with crank angle. Also, exhaust gas temperature decreases with time during the exhaust event, resulting in cyclic variation in the temperature of the reactor feed. Good backmixing in the reactor itself is therefore helpful in eliminating gradients of concentration and temperature.

TEST PROGRAM

An experiment was designed to evaluate steady-state reactor performance with each of the three air-injection systems. Reactor inlet CO concentration and gas temperature were maintained constant to allow independent evaluation of the effects of both engine speed and flow rate, as well as to generalize the influence of different air injection systems.

gas leaves the port without mixing with air. However, part of the blowdown gas and all of the gas expelled by the rising piston mix with some air, providing a substantial improvement over the open-ended tube.

In spite of the gains in synchronization offered by the sparger tube, it suffers from two deficiencies. The most serious of these is the continuous flow of air during the portion of the engine cycle when the exhaust valve is closed. Thus, the air-flow during the exhaust event cannot be optimized because the excess airflow during the remainder of the cycle cools the reactor and quenches the oxidation reactions. The optimum average airflow rate determined experimentally is a com-promise between providing adequate airflow during the ex-haust event, and obtaining an overall airflow into the reactor which is not so high that its cooling effect retards the reactions.

The second sparger tube deficiency concerns the amount of air-exhaust mixing obtained. As mentioned earlier, synchroni-zation does not ensure thorough mixing. Low-pressure air flowing through large radial holes in the injection tube could result in stratification of the two gas streams. Segregation of the air and exhaust in this manner is undesirable, but is ex-pected to result in better overall mixing than is obtained with open-ended injection tubes.

The airflow profile for the open-ended injection tubes (Fig. 4a), as well as for the sparger tubes (Fig. 4b), has large varia-

An inlet CO concentration of 2.9% was selected as a compromise: low enough to preclude serious fuel economy and reactor temperature problems associated with traditional very rich reactors, but high enough to provide measurable differences in reactor outlet CO concentration. Resolution of the differences in outlet CO was further increased by selecting a small, relatively inefficient thermal reactor. The engine, reactor, and test installation are described in Appendix C.

Emissions are reported as corrected concentrations (see Appendix D). A corrected CO concentration of 2.9% corresponds to an A/F of 13.3:1.

Four reactor operating conditions, produced by the combinations of two engine speeds and two exhaust flow rates as shown in Table 1, were selected to generalize the results. The speeds and flow rates were chosen such that a reactor inlet temperature (as indicated by reactor core gas temperature without air injection) or 760°C (1400°F) could be obtained at each condition by means of ignition timing adjustments.

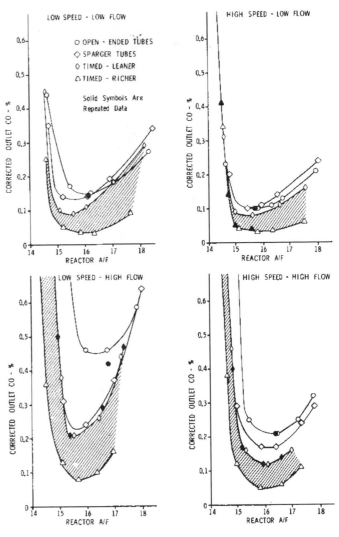

Fig. 5 – Reactor emission performance

Table 1 – Four Reactor Operating Conditions

Exhaust Flow kg/h(lb/h)	Speed	
	1600 rpm	2400 rpm
95(210)	Low Speed Low Flow	High Speed Low Flow
186(410)	Low Speed High Flow	High Speed High Flow

Reactor inlet CO (corrected) = 2.9%
Reactor inlet temp. = 760°C (1400°F)

Each test consisted of measuring reactor outlet emissions at various air injection rates to identify the optimum reactor air-fuel ratio. CO conversion was chosen as the reactor performance indicator because CO is slower to react in a thermal reactor than either hydrogen or hydrocarbons (2). Also, reactor inlet CO concentration is easily controlled via curburetor adjustments because CO is a direct function of A/F when richer than stoichiometric. Details of the experimental procedure are contained in Appendix D.

EXPERIMENTAL RESULTS

The effects of the three air injection systems on reactor outlet CO corrected concentration are shown for each of the four conditions in Fig. 5. The band for timed injection is discussed subsequently. Several test points and three test series were rerun to evaluate data repeatability. A single point at each condition was repeated for the open-ended tubes, and three complete conditions were repeated for the timed injection. All repeated points are shown in Fig. 5 as solid symbols.

SPARGER AND OPEN-ENDED TUBES—The minimum CO concentration obtained with sparger tubes was less than obtained with open tubes at all conditions except high speed–low flow, where the minima were equal. Sparger tubes resulted in the most improvement at the low speed–high flow condition, reducing outlet CO by 50%.

TIMED INJECTION—A difficulty was encountered in the evaluation of the timed air injection system. Because of the location of the injection holes and the high airflow rates, a substantial portion of the injected air was drawn into the engine cyl during the valve overlap period (see Fig. 1). This air leaned the A/F within the cyl, resulting in a reactor inlet CO concentration less than the 2.9% selected for the study. No straightforward technique was discovered to accurately determine the reactor inlet CO with timed air injection. However, two test schemes were devised to ensure that the inlet CO was less than 2.9% in one case and greater than 2.9% in the other. Thus, the desired condition is contained within the band formed by these two tests. A discussion of cyl A/F dilution by timed air injection, and the development of the bracketing test procedures, is given in Appendix B.

The timed air injection data presented in Fig. 5 is labeled Timed-Leaner for the leaner cyl A/F tests and Timed-Richer for the richer cyl A/F tests. The band formed by these two curves demonstrates that, for all conditions tested, timed injection provides considerably lower outlet CO than either open-ended or sparger tubes. Timed injection is consistently

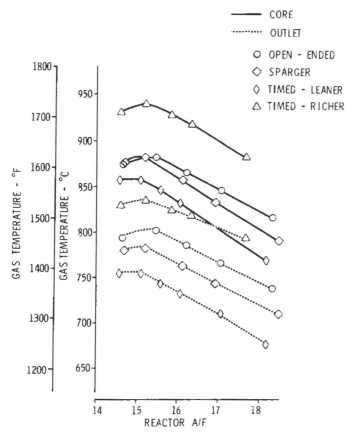

Fig. 6 – Reactor core and outlet gas temperatures for low speed-low flow condition

the relationship of the four conditions remains constant: high engine speed yields better conversion than low speed, and low flow rate yields better conversion than high flow rate. The combination of these two effects results in best conversion at high speed-low flow and poorest conversion at low speed-high flow.

Speed Effect—Improvement in reactor performance with increasing engine speed is believed to be primarily a mixing phenomenon. An increase in speed at the same exhaust mass flow rate results in more individual exhaust events per unit mass of exhaust gas. Consequently, the exhaust is segmented into smaller discrete slugs. Air injection with open-ended tubes is also divided into slugs which, like the exhaust gas slugs, decrease in size with increasing speed.

Thus, the overall effect of increasing speed is to sub-divide the reactor feed into smaller slugs of air and exhaust. The number of slug interfaces increases accordingly, so overall bulk mixing is improved. Also, higher engine speeds increase the frequency of cyl blowdown and the turbulence of the gas expelled during the exhaust stroke. Both of these effects tend to increase small-scale mixing in the reactor. Note that the effects of the increased speed on CO conversion decrease as successively better air injection systems are considered (Fig. 7). With

better even if the upper limit of the band is used for the comparison. If the center of the band is assumed to be representative, timed injection results in 50% less outlet CO (at the optimum A/F) than do sparger tubes at both the low speed-low flow and high speed-high flow conditions.

Reactor core and outlet gas temperatures shown in Fig. 6 correspond with the emission data at the low speed-low flow condition (Fig. 5). These temperature data further substantiate the conclusion that the two timed air-injection test procedures bracket the desired reactor inlet conditions. Note in Fig. 6 that Timed-Richer core and outlet temperatures are higher, and Timed-Leaner temperatures are lower, than the corresponding open-ended and sparger tube temperatures. Differences in exothermic heat release attributable to different outlet CO concentrations cannot account for the large temperature differences between the timed air and the injection tube data, especially since the Timed-Leaner temperature shift is in the wrong direction to explain the improved conversion. Therefore, the gas temperature shifts must be a result of changes in inlet combustible concentrations.

SPEED AND FLOW RATE EFFECTS—The four operating conditions used in this study were found to strongly affect the outlet CO obtained with each system. The effects of speed and exhaust flow rate on the performance of each air injection system are shown in Fig. 7. Although the relative outlet concentrations change significantly from system to system,

259

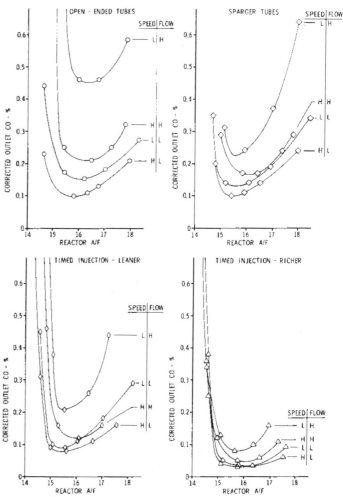

Fig. 7 – Effects of speed and exhaust flow rate on reactor performance

all systems, however, increased engine speed improves reactor performance.

Flow Rate Effect—Low exhaust flow conditions result in significantly lower outlet CO than do high flow conditions, expecially with the open-ended tubes. The data indicate that the small reactor volume suppresses CO conversion at the high exhaust flow rate because residence time is relatively short. Even with the good mixing of timed injection, outlet concentrations for the low speed-high flow condition are more than twice as great as for the low speed-low flow case.

Exhaust flow rate affects gas velocities within the reactor as well as residence time. Internal gas velocities influence both mixing and heat loss, which in turn influence reactor performance. However, it is felt that the velocity effects are small, and that the change in conversion with flow rate is primarily a result of the change in residence time.

HYDROCARBON EMISSIONS—For reasons previously mentioned, CO oxidation was of prime interest in this study. Hydrocarbon emissions were also monitored, however, and were reasonably low for all three air injection systems. The highest outlet HC concentration at optimum reactor A/F was less than 5 ppm with timed injection and less than 15 ppm with the other two systems.

INTERPRETATION OF RESULTS

MIXING EFFECTS—The results of this study demonstrate the importance of mixing in automotive thermal reactor systems. Premixing of the reactants in the exhaust ports effectively increases the reactor volume available for oxidation. Conventional, low-pressure air injection through open-ended tubes results in a segregated feed which the reactor must both mix and react. If mixing is accomplished in the engine ports, the reactor is relieved of the bulk of the mixing chore, and conversion efficiency improves.

The volume utilization effect is clearly illustrated by exhaust gas temperature data for high flow conditions, when the importance of mixing is more pronounced. In Fig. 8 the reactor core and outlet gas temperatures for open-ended tubes are nearly equal. At the richer reactor A/F, the reactor outlet becomes hotter than the core. These characteristics imply that the reaction zone is spread throughout the reactor, with oxidation limited spatially by the amount of mixing which has occurred. As the availability of oxygen decreases attendant with decreasing reactor A/F, the mixing limitation becomes more crucial, and the reaction zone moves downstream in the reactor. Ultimately, the lack of mixed oxygen retards the reaction until stable combustion cannot be maintained in the reactor core. Gas temperatures then drop as outlet CO increases abruptly.

By comparison, the sparger tube temperature data in Fig. 8 show that the core gas is hotter, and the outlet gas cooler, than with open tubes. At an A/F of 16:1, improved CO conversion with sparger tubes could account for a 6°C (10°F) max increase in core temperature, assuming the limiting case in which all additional conversion occurs in the core with no heat loss. The measured increase, however, is approximately 25°C

(45°F). Therefore, the shift in core and outlet gas temperatures between open and sparger tube data must indicate an upstream movement of the reaction zone. This movement is the result of improved mixing.

The timed injection temperature data in Fig. 8 indicate a further upstream shift of the reaction zone. The high core temperature is especially indicative of an early reaction, because Timed-Leaner test inlet combustible concentration and resulting heat of reaction are lower than for the open and sparger tube tests.

Another characteristic of improved mixing is a lower optimum reactor A/F. Note in Fig. 5 that the minimum CO point shifts to the left as injection systems which offer improved mixing are considered. The shift is less pronounced for high speed conditions. It is not apparent in the Timed-Richer data because the inlet combustible concentration is increasing with reactor A/F.

PORT REACTIONS WITH TIMED INJECTION—A special test was performed in which the open-ended injection tubes were used to draw gas samples from the ports during operation with timed air injection. This test was performed to assess port-to-port distribution of the secondary air. However, examination of the data taken at low exhaust flows uncovered an unexpected result: corrected CO concentrations in the port

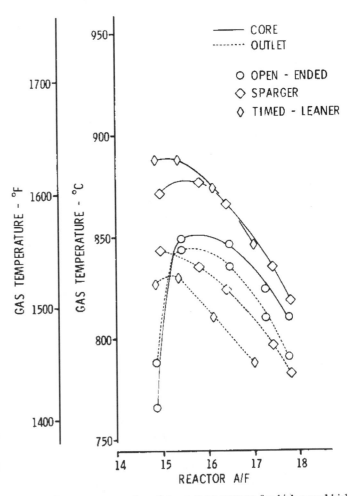

Fig. 8 – Reactor core and outlet gas temperatures for high speed-high flow condition

samples were substantially less than the corrected concentration leaving each cyl.

It was recognized that continuous sampling of cyclic gas flows (such as cyl exhaust events) can often result in species concentrations which are not representative of the bulk gas. To calculate a sampling error, a comparison was made between the averaged corrected NO concentrations measured in the ports and the corrected NO measured at the reactor outlet. The average port-measured concentration of CO was assumed to be in error by the same percentage (36% low). Correcting for this sampling error, average port-measured CO was still 25% lower than the 2.9% CO assumed to be emitted from the engine cyl. (The engine outlet CO was actually some value greater than 2.9%, because these data were taken during a Timed-Richer test series.)

Thus, very rapid oxidation of exhaust gas combustibles in the engine exhaust port can occur if thorough mixing of air and exhaust is achieved at the port inlet. This is not too surprising, since gas temperature in the combustion chamber at the time of exhaust valve opening may be well in excess of 1100°C (2000°F). Mixing secondary air with exhaust gas at the interface between the cyl and exhaust port allows the oxidation reactions to proceed very rapidly before significant heat is lost to the port walls.

STABLE COMBUSTION—Comparison of timed injection data with open-ended tube data at the high speed-high flow condition reveals another characteristic which indicates significant port oxidation. The open-ended tube CO data in Fig. 9 show that, as mentioned earlier, combustion in the reactor becomes unstable and eventually ceases as the reactor A/F is decreased. In constrast, tests with the timed injection system at reactor A/F richer than stoichiometric show that combustion is maintained at all conditions with no abrupt decrease in conversion efficiency. At overall rich A/F, the reactor outlet CO asymptotically approaches the equilibrium concentration, or theoretical minimum, shown as a dashed line in Fig. 9 (3).

A qualitative explanation for the markedly different behavior of the reactor with the two different air injection systems can be obtained from fundamental reactor theory. Levenspiel (4) describes the permissible operating states of a reactor as the conditions obtained from the simultaneous solution of the energy and material balances, shown graphically as the intersection of the two curves in Fig. 10. On a plot of fraction converted versus temperature, the energy balance line predicts the gas temperature which would be obtained at each level of conversion. The intercept of the energy balance line with the abscissa defines the reactor feed temperature. Chemical heat release decreases the slope, while reactor heat loss increases the slope. The material balance curve depicts the disappearance of the particular reactant throughout the temperature range for a given residence time. At low temperatures, the rate of reaction is slow, and very little conversion occurs. As temperature is increased, the reaction rate grows exponentially and the material balance curve rises steeply. When high levels of conversion are reached, the reaction rate once again slows down due to reactant depletion, and the curve asymptotically approaches complete conversion.

261

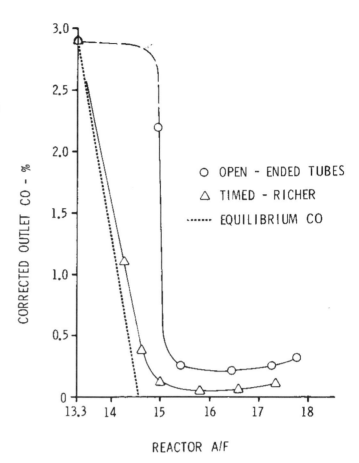

Fig. 9 – Outlet CO for open-ended tubes and timed injection at high speed-high flow condition

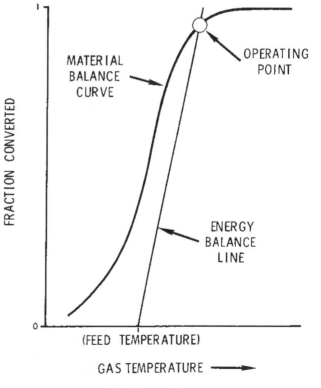

Fig. 10 – Reactor theory

CO oxidation, with at least a stoichiometric amount of oxygen and perfect mixing, is qualitatively described by the dotted material balance curve in Fig. 11. If a mixing limitation is imposed, the material balance curve departs from the perfect mixing case, as shown by the dashed curve. At this level of imperfect mixing, reductions in available oxygen shift the curve further down and to the right. Therefore, the solid material-balance curves describe the effect of decreasing the reactor A/F (and available oxygen), as was done in the previously mentioned experiments. Although the mixing limitation is significantly less with the timed injection system than with open tubes, the major influence on the shift of the material balance curves for conditions richer than stoichiometric is oxygen shortage. Thus, the curves shown in Fig. 11 are representative for both injection systems.

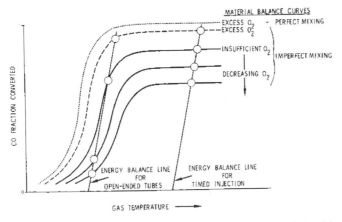

Fig. 11 – Reactor operating points for open-ended tubes and timed injection at conditions of decreasing oxygen

The slope of the energy balance line to depict a particular experimental condition should be identical for all injection systems because inlet combustible concentration and reactor heat loss are nearly constant. The only variable left which can explain the different performance characteristics between open-tube and timed injection systems is the reactor feed temperature, which fixes the position of the energy balance line.

The reactor performance with open-ended tubes can be explained by the energy balance line to the left in Fig. 11. As the reactor A/F is decreased, the material balance curve shifts down and to the right until there is no longer a high conversion intercept of the two lines. At this condition, the fraction converted drops abruptly to a low value.

The energy balance line to the right in Fig. 11 can be used to explain the reactor performance with timed air injection. Reductions in reactor A/F cause the CO conversion to decrease in a manner characteristic of a shortage of oxygen and a lack of perfect mixing. At no condition, however, does the reaction become unstable and drop to a low level of conversion.

The conclusion drawn from this simplified argument is that the reactor feed temperature must be considerably higher with timed injection than with open-ended tubes. This seems inconsistent with the experiments because the engine outlet exhaust gas temperature was maintained nearly constant for all tests. The inconsistency can be explained, however, by the fact that the mixing of air and exhaust occurs at considerably different locations for the two systems. With open-ended tubes, significant mixing is not obtained until the air and exhaust have entered the reactor. The exhaust gas loses a considerable amount of heat to the engine port, resulting in a relatively low reactor feed temperature.

In contrast, with timed injection, good mixing of the secondary air and exhaust is achieved at the inlet to the engine port. As a result, the port becomes part of the reactor, and the effective reactor inlet temperature approaches the very high cylinder gas temperature. This explanation satisfactorily describes the performance of the timed injection system, and supports the experimental indication of oxidation in the port as well.

PRACTICAL CONSIDERATIONS

Improved mixing provides several benefits for an automotive thermal reactor system. An increase in exhaust-secondary air mixing allows low outlet CO and HC concentrations to be achieved with either a leaner engine A/F, a smaller reactor, less insulation, or a combination of these factors. Furthermore, the combined effects of improved mixing and leaner engine A/F reduce the secondary airflow requirements considerably. An increase in fuel economy and a reduction in reactor system size and complexity are possible benefits of better mixing.

Unfortunately, greatly improved mixing bears the penalty of increased expense and complexity of the secondary air system. The timed air injection system used in this study resulted in outlet CO as much as 67% lower than obtained with open-ended tubes. However, this timed injection system was designed as a research expedient. For production vehicle application it would be expensive, difficult to control, and probably lacking in durability. The system would require an engine-driven, high-pressure air pump which provides outlet pressure proportional to speed. In addition, a controller would be required to proportion secondary airflow to engine exhaust flow by throttling the pump inlet or diverting part of the outlet air (5). The possibility of eventual injection hole plugging and reduced valve seat life further decreases the practicality of the system.

Less complex schemes of timing secondary air injection can undoubtedly be developed. However, any such scheme would likely lose the benefit of air injection via high velocity jets at the inlet to the port, with a resulting decrease in reactor efficiency. Ultimately, then, a timed air injection system is not viable unless the need for a more efficient thermal reactor can override substantial penalties in cost and complexity.

A more realistic and immediate benefit lies in the use of sparger tubes. The overall improvement offered by sparger tubes is considerably lower than can be obtained with timed injection. However, the improvement appears large enough to warrant further consideration. Development work on such sparger tube design variables as hole size, orientation, and tube location may further improve performance. In the early stages of this study, hot-wire anemometry measurements of secondary airflow were made on a different engine equipped with sparger tubes. These measurements indicate that airflow was not shut off during the peak of cyl blowdown. A significantly different orientation of the injection tube in the exhaust port is the probable cause of the difference in air delivery characteristics. Whatever the cause, a further improvement in conversion over open-ended tubes can be expected when air injection is continuous throughout the exhaust event.

CONCLUSIONS

The following conclusions are based on the CO conversion performance of a fully warmed-up thermal reactor operating as part of an engine exhaust system.

1. Improved mixing of secondary air and exhaust gas in the engine port can substantially increase thermal reactor conversion efficiency.

2. The secondary airflow required to obtain optimum conversion decreases as the air-exhaust mixing is improved.

3. Reactor conversion efficiency improves with increasing engine speed and decreases with increasing exhaust flow rate.

4. Moderate improvements in mixing and as much as 50% lower outlet CO can be obtained by replacing conventional open-ended air injection tubes with sparger (radial discharge) tubes.

5. Large improvements in mixing and as much as 67% lower outlet CO can be obtained by replacing low-pressure injection tubes with a high-pressure, timed air injection system.

6. Significant CO oxidation in the engine port is obtained with timed air injection, which markedly improves reactor performance at oxygen-deficient conditions.

7. Although less effective than timed injection, sparger tubes are more practical in terms of cost, complexity, and durability.

ACKNOWLEDGMENTS

The author wishes to thank Mr. D. J. Pozniak for providing the air tube hardware, and Mr. J. R. Abel for considerable help in obtaining the first anemometer data.

REFERENCES

1. R. H. Kadlec, E. A. Sondreal, D. J. Patterson, and M. W. Graves, Jr., "Limiting Factors on Steady-State Thermal Reactor Performance." Paper 730202 presented at SAE January Meeting, Detroit, January 1973.

2. H. A. Lord, E. A. Sondreal, R. H. Kadlec, and D. J. Patterson, "Reactor Studies for Exhaust Oxidation Rates." Paper 730203 presented at SAE January Meeting, Detroit, January 1973.

3. B. A. D'Alleva, "Procedure and Charts for Estimating Exhaust Gas Quantities and Compositions." GM Research Laboratories Research Publication, GMR-372, May 1960.

4. Octave Levenspiel, "Chemical Reaction Engineering." New York: John Wiley and Sons, Inc., 1962, pp. 225-229.

5. D. J. Pozniak and R. M. Siewert, "Continuous Secondary Air Modulation--Its Effect on Thermal Manifold Reactor Performance." Paper 730493 presented at SAE Mid-Year Meeting, Detroit, May 1973.

6. "Hot Film and Hot Wire Anemometry--Theory and Application." Thermo-Systems Inc., Bulletin TB5, St. Paul, Minnesota.

7. D. L. Stivender, "Development of a Fuel-Based Mass Emission Measurement Procedure." Paper 710604 presented at SAE Mid-Year Meeting, Montreal, Quebec, Canada, June 1971.

APPENDIX A

INSTANTANEOUS SECONDARY AIRFLOW MEASUREMENT

ANEMOMETER AND PROBES—The previous use of a constant-current hot-wire anemometer in obtaining a single air injection flow measurement verified the potential of such instrumentation for the thermal reactor application (1). In developing a measurement technique for this study, however, a constant temperature rather than a constant current anemometer was chosen because of its superior high frequency response. The anemometry equipment used was a DISA 55M system which incorporated an electronic signal linearizer to simplify interpretation of results. Cylindrical hot-film probes were used to record airflow with the air tubes, and hot-wire (10μ dia, platinum—10% rhodium) probes were used with the timed injection system.

The constant temperature anemometer/hot-film probe combination demonstrated completely acceptable response characteristics for detailed secondary airflow measurements. At an engine speed of 1200 rpm, for example, the anemometer could resolve (3 time constants) flow rate changes within one-half crank angle degree at high airflow rates and within two crank angle degrees at zero flow. Using hot-wire probes, the resolution was even greater.

PROBE MOUNTING—Instrumentation extensions were fabricated to allow installing an anemometer probe in the flow path between the secondary air manifold and the point of air injection. These extensions provided a straight, constant ID tube of the same size as the individual runners on the air manifold, and had a tapped hole for inserting the probe perpendicular to the flow (Fig. A1). In addition, two other tapped holes were used for thermocouple and pressure transducer ports. Conax fittings for the anemometer probes were modified with Teflon sleeves to provide electrical insulation for the probe body. Probes were mounted with the sensor intersecting the centerline of the air supply tube and perpendicular to the direction of flow. The distance from the probe to the point of air injection was approximately 150 mm (6 in).

IN-PLACE CALIBRATION—The most useful information to be gained from the anemometry was the cyclic variation in airflow, not the absolute value of the flow rate. In light of this, an in-place calibration was most useful in adjusting the anemometer linearizer such that the output voltage varied directly with mass flow rate. Proper calibration of the

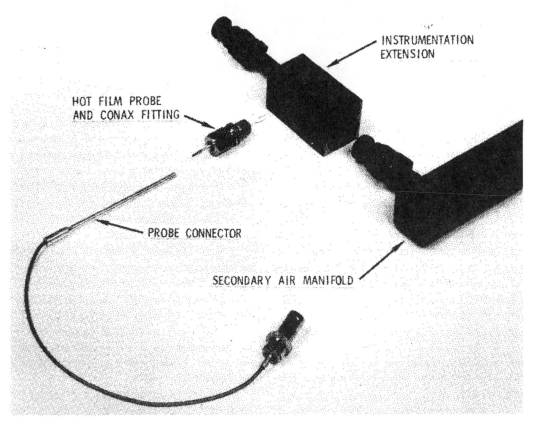

Fig. A1 – Anemometer probe installation

linearizer insured that the time-varying airflow profiles would be accurate.

The test cell engine/thermal reactor system used regulated auxiliary high-pressure air as the secondary air supply. Two in-series pressure regulators and a sharp-edged orifice meter in the supply line provided accurate control and measurement of secondary airflow rates. The anemometer probes were installed in the instrumentation extensions, and anemometer readings were obtained at several flow rates with the engine stopped. This technique allowed quick linearization and calibration of each anemometer probe used.

INTERPRETATION OF MEASUREMENTS—The measurements of airflow through open-ended injection tubes were initially puzzling because the linearized anemometer output went to zero volts during the exhaust event (Fig. 4a). An output reading of zero indicates no bulk or turbulent air motion in the air injection tube, which is most unlikely. In hope of clarifying the results, nonlinearized anemometer output was examined and also found to be zero during the same period. This result was unexpected because the nonlinearized anemometer output at zero flow is a positive voltage (6).

A hot-film anemometer operates by losing heat to the gas, so the only explanation for a zero anemometer output is that the probe is in an environment hotter than the operating temperature of the sensor. Since the hot film probes were operating between 200-250°C, it appeared that the airflow was reversing and hot exhaust gas was flowing back into the air injection system. A 1.59 mm (1/16 in) sheathed thermo-couple was installed in the instrumentation extension near the anemometer probe, and subsequent measurements confirmed that hot exhaust gas was indeed flowing into the air injection tube. With this knowledge, a zero anemometer output becomes a convenient means of assessing backflow of exhaust into the air tube.

The occurrence of flow reversal is indicated by an abrupt decrease in flow to near-zero when the exhaust valve opens, followed by a sharp positive spike. Air trapped downstream of the anemometer probe at the time of reversal is forced back into the air manifold by the exhaust gas entering the air tube. Since an anemometer cannot detect direction of flow, the reverse flow of air appears as a positive spike on the oscilloscope. At the peak of the spike, the hot exhaust gas flowing into the tube reaches the sensor, which terminates its heat loss and causes the anemometer voltage to go to zero.

During the period of zero anemometer output, nothing can be deduced about the flow magnitude or direction. Since the air tube is filled with exhaust gas, however, it follows that no secondary air can be flowing into the port. In Fig. 4a, the period of airflow shutoff actually exceeds the exhaust valve opening duration. Evidently the dynamic pressure head of the exhaust blowdown forces hot gas into the air tube, instrumentation extension, and part of the air manifold. When the pressure in the exhaust port decreases after the exhaust event, the hot gas flows back out of the air tube into the port. The first portion of the flow is the accumulated exhaust gas, which explains the zero anemometer voltage for a period of time after the exhaust valve closes.

264

TIMED AIR INJECTION SYSTEM

HARDWARE DETAILS—High-pressure air injection through holes in the valve seat was accomplished by modifying the cyl head casting and pressing in machined Stellite valve seats. A modified 350 engine cyl head was made from casting cores which had been shaved to fill in water jacket areas for air supply passages. These passages were drilled from the outboard side of the cyl head below the exhaust ports. The two center exhaust valve seats were fed by a single air passage located between them, while the two end seats were fed by individual passages.

The Stellite valve seat inserts were designed with a semi-circular annulus on the OD. This annulus aligned with the air supply passage when the seat was inserted, and distributed air to the injection holes which were drilled through the seat face into the annulus (Fig. B1).

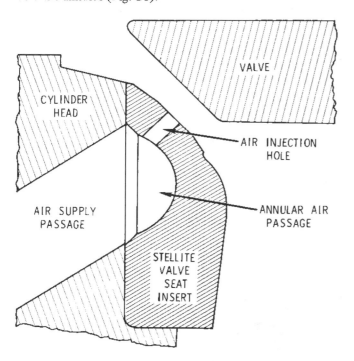

Fig. B1 – Cross-section of timed air injection valve seat

The air injection holes were sized such that the highest required airflow would be obtained at the maximum available supply pressure. This design constraint allowed the use of the smallest possible total injection hole area, which resulted in high air-jet velocity for good mixing and insured sonic flow at low injection rates. Nine 0.76 mm (0.03 in) holes were used in each seat, which resulted in measured flow rates of 19.5 kg/h (43 lb/h) per seat at a supply pressure of 620 kPa (90 psig). Note that relatively high flow rates were required because injection occurred only when the exhaust valve was open.

Positive sealing of the air injection holes by the closed exhaust valve is crucial to the performance of the timed injection system. To obtain good sealing, a seat width of approximately 2 mm (0.08 in) was chosen. Furthermore, each valve was hand-lapped prior to assembly.

PROBLEMS IN EMISSION PERFORMANCE EVALUATION—Examination of preliminary emission data from the timed air injection system revealed that corrected NO concentration increased almost linearly with air injection rate. This trend was not observed with open-ended and sparger injection tubes, as shown in Fig. B2. Furthermore, thermal reactor temperatures were much too low to result in the formation of additional NO. It was therefore assumed that the engine operating condition was changing due to secondary air being swept into the cyl during the valve overlap period and leaning the A/F. This assumption was substantiated by engine torque measurements, which increased with air injection rate. Also, a separate test was performed in which valve overlap was reduced by introducing valve lash. This change in valve overlap resulted in smaller increases in NO and torque with increasing air injection, further verifying the suspicion that secondary air was entering the cyl.

The major consequence of secondary air leaning the cylinder A/F is a reduction in the reactor inlet CO concentration. Since reactor conversion efficiency depends strongly on combustible concentration because of reaction heat release, timed injection performance cannot be directly compared with injection tube data unless reactor inlet CO is the same.

A test procedure was devised to compensate for the leaning effect of secondary air entering the cyl. At each air injection rate tested, the engine A/F was decreased until the corrected

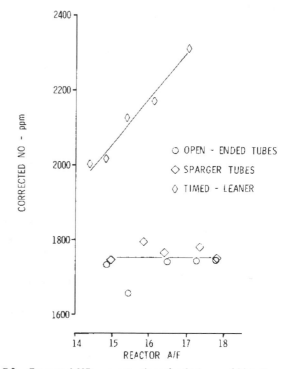

Fig. B2 – Corrected NO concentrations for high speed-high flow condition

NO equaled the value obtained without air injection. It was hoped that this procedure would result in the same cyl A/F (and therefore the same reactor inlet CO) at each test point.

Very low outlet CO was obtained with this procedure. However, examination of reactor temperature data for this test series revealed that the increase in core temperature (above the open and sparger tube data) was much greater than could be explained by improved conversion or an upstream shift in the reaction zone. At this point, it was realized that the secondary air which entered the cyl not only increased the A/F, but also displaced some exhaust gas in the residual fraction. The result was a reduction in the internal exhaust gas recirculation, and an increase in the NO formation at any given A/F.

Therefore, cyl NO production with increasing air injection was a function of two additive effects (increasing A/F and de-creasing residual fraction), and adjusting the carburetor A/F to maintain a constant corrected NO concentration did *not* result in a constant cylinder A/F. In actuality, the cylinder A/F was always less than the desired 13.3:1 for this test procedure.

No straightforward means of determining the cyl A/F when using timed secondary air injection was found. However, in the initial tests when the carbureted A/F was maintained constant as air injection was varied, the reactor inlet CO was always less than the desired concentration. Likewise, when the carburetor A/F was decreased with increasing air injection to maintain constant corrected NO, the reactor inlet CO was always greater than the desired concentration. Therefore, the band formed by the data obtained with these two procedures must bracket the condition needed to compare timed injection performance with injection tube data.

APPENDIX C

EXPERIMENTAL APPARATUS

ENGINE—A 5.74 l (350 CID) V-8 coupled to an electric dynamometer was used for this study. The following table itemizes non-standard engine details:

Compression Ratio—11.25:1
Carburetor—Modified four barrel (primaries only)
Cam—High lift, 114 deg overlap, hydraulic lifters
Cyl. to Cyl. A/F distribution—±0.5 A/F
Exhaust Crossover—Blocked
Fuel—Indolene Clear

The modified carburetor had a seal float chamber that could be pressurized to increase flow in the main metering circuit and decrease A/F.

THERMAL REACTOR—A small volume (2.46 l, 150 in^3) cast iron reactor was used for all tests (see Fig. C1). This reactor consisted of a cast iron housing, a steel cover, and stain-less steel core and radiation shield sheet metal. No external insulation was used. Core gas temperature was obtained from a radiation-shielded Cr-Al thermocouple inserted through the end of the cover and located in the center of the core. Outlet gas temperature was measured with a triple-shielded Cr-Al thermocouple centrally located in the cast iron outlet. Outlet gas samples were drawn from scarfed tubes centered in the exhaust pipes 80 mm (3.15 in) from the reactor outlet flanges. Regulated and filtered shop air was used as the secondary air supply.

INSTRUMENTATION—The anemometer system is described in Appendix A.

Air Meter—Total engine airflow was measured with a viscous-type air meter.

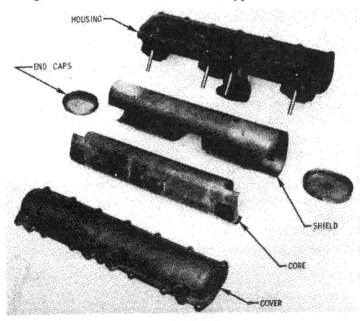

Fig. C1 – Reactor

Fuel Burrette–Total engine fuel flow was monitored by means of an electrolyte-displacement fuel burrette system.

Crank Angle Indicator–Crank angle markers for intra-cycle positioning of the anemometer signal were obtained from a magnetic pickup placed over a slotted wheel on the crankshaft pulley. Narrow slots 10 deg apart produced small pulses, with wider slots at 90 deg, 180 deg, and 270 deg and a very wide slot at TDC producing larger pulses.

Oscilloscope–A Type 564 dual beam Tektronix storage oscilloscope with camera was used to record anemometry output and crank angle markers.

Emissions Instrumentation–The gas species sampled, and the instruments used appear below:

Hydrocarbons–Beckman Model 109A Flame Ionization Detector

Carbon Monoxide–Beckman Model 315B NDIR with 3 ranges and 2 sample cells
Beckman Model 215A NDIR with 3 ranges

Carbon Dioxide–Beckman Model 315 NDIR
Nitric Oxide–TECO Series 10 Chemiluminescent Analyzer
Oxygen–Beckman Process Oxygen Analyzer
Bailey Heat Prover

INSTALLATION–A schematic of the test setup is included as Fig. C2. The reactor on the right bank of the engine was used to evaluate emission performance. No secondary air was added to the reactor on the left bank, which served only as a manifold.

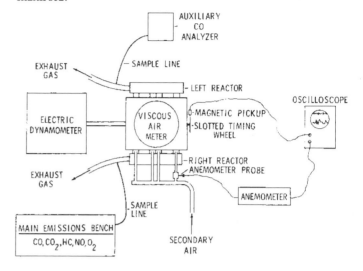

Fig. C2 – Test installation

APPENDIX D

EXPERIMENTAL PROCEDURE

The engine was warmed up at the test speed with no secondary air addition. Engine airflow (measured by the viscous air meter) was set at the desired level by adjusting the throttle. The right reactor outlet was sampled with the main emission bench, and carburetor fuel flow was adjusted until the desired exhaust CO concentration was obtained. Since the engine A/F was richer than stoichiometric and there was no secondary air added, the reactor outlet CO was assumed to equal the reactor inlet CO.

While maintaining the desired CO concentration at the right reactor outlet, the engine ignition timing was adjusted until the shielded thermocouple in the reactor core indicated the desired gas temperature. This measurement was assumed to represent the temperature of the gas entering the reactor core. At this point, the engine was operating at the condition desired for the test series. Engine speed, airflow, fuel flow, and left reactor CO concentration and temperatures were recorded.

The test series was initiated by supplying enough secondary air to the right bank of the engine to achieve a reactor A/F of approximately 18:1. Right reactor temperatures and outlet emissions were allowed to stabilize (requiring 10-20 min), and then all engine and emission data were taken. Airflow, fuel flow, and left reactor temperatures and outlet CO were monitored continuously to detect any shift in engine condition. An auxiliary CO analyzer was used to sample the left reactor outlet, thus allowing continual adjustment of the carburetor fuel metering to maintain a constant reactor inlet CO concentration. The use of the auxiliary CO analyzer freed the main bench of intermittent sampling of the left bank, and prevented hydrocarbon hangup in the main sample line.

Successive data points were taken by reducing secondary air injection in steps of approximately one reactor A/F until the stoichiometric condition was approached. Test points were arranged in the order of decreasing air injection rate to avoid the problem of hysteresis in reactor performance with varying amounts of secondary air (1).

At the end of each test series, all secondary air was shut off, the system was allowed to stabilize, and the right reactor temperatures and outlet CO were rechecked to ensure that the reactor inlet conditions had remained constant.

Reactor A/F was calculated from exhaust gas specie measurements using both a carbon and an oxygen balance (7). These two calculations agreed within 0.5 ratios. Their average was used in reporting emission data.

Gas concentrations were corrected for dilution according to the equation:

$$[X] \text{ corrected} = \frac{15[X]}{6[HC] + [CO] + [CO_2]} \quad \text{(D-1)}$$

2002-01-0744

Investigation of Post Oxidation and Its Dependency on Engine Combustion and Exhaust Manifold Design

Christoph Koehlen, Eberhard Holder and Guido Vent
DaimlerChrysler AG

Copyright © 2002 Society of Automotive Engineers, Inc.

ABSTRACT

In response to ever more stringent emission limits (EURO IV, SULEV), engine developers are increasingly turning their attention to engine start-up and warm-up phases. Since in this phase the catalytic converter has not yet reached its operating temperature, problems occur especially with regard to hydrocarbon emissions (HC) which are emitted untreated. Secondary air injection represents one option for heating up the catalytic converter more quickly. The engine is operated during the heating up cycle with retarded ignition angles and a rich mixture. Ambient air (secondary air) is injected close to the exhaust valve seat. During the spontaneously occurring post oxidation phase, the reactive exhaust components ignite and heat up the catalytic converter while simultaneously reducing HC.

The various processes which affect the post oxidation, are not well known up to now. In order to achieve concrete improvements, detailed knowledge of its influences are necessary. This paper will give an overview of these complex relationships. For the investigations fast response measuring techniques are installed to record the reaction of the secondary air and exhaust gas so the local formation of the reaction zone can be detected. A high-speed camera is used to display the reaction fields on an exhaust manifold made of quartz glass. The internal dimensions of this glass manifold correspond to the series production manifold. A comparison with the appropriate image sequences from CFD calculations aids in understanding and interpreting the processes. We can see that not only the quality of the mixture formation and combustion in the combustion chamber but most especially the distribution of flow, temperature and concentration of air and exhaust gas in the exhaust manifold are decisive for optimum post oxidation.

INTRODUCTION

A secondary air system can be used to assist modern three-way catalytic converters in the phase in which the light-off temperature of the catalytic converter (approx. of 300°C) has not yet been reached. Below this temperature, the catalytic converter is mostly ineffective. As a consequence, more than 80% of the hydrocarbon emissions measured over the entire FTP-75 drive cycle are emitted within the first 20 seconds after the engine has been started,. Furthermore, thermal ageing of the catalytic converter over its service life results in delayed activation [1].

Post oxidation occurs during rich-mixture engine operations and the injection of secondary air close to the exhaust valve seat. The exhaust gas contains reactive components which may have participated in the reaction in the combustion chamber but have not yet been completely burnt off. These components are predominantly carbon monoxide (CO), hydrogen (H_2) and hydrocarbon (HC). If appropriately high temperatures occur, an immediate reaction takes place [7,10]. During this phase, the engine is operated with retarded ignition angles [3,4], and a rich fuel-air mixture. Mixture formation plays a decisive role here. If with rich fuel-air operations, this deteriorates to such an extent that a majority of the fuel enters the combustion chamber in droplet form, these parts cannot participate in the combustion process. When the engine is cold, these parts mostly deposit themselves in gaps, on the piston base, on the cylinder wall and close to the exhaust valves [5,6]. As a result of flame quenching, the flame does not reach this fuel. It enters the exhaust port in an uncracked form. Concentrations of CO and H as high as possible are however required to produce a flame in the exhaust port. The hydrocarbons only react when a flame is already burning because they have a longer ignition delay and a higher threshold temperature at the start of the reaction. Higher exhaust gas temperatures and a heat loss as low as possible at the exhaust port and manifold walls therefore accelerate the reaction [7].

EXPERIMENTAL

EXPERIMENTAL SETUP

The experiments are conducted on a Mercedes Benz V6 gasoline engine with a capacity of 3.2 liters, which is fitted with secondary air injection as standard. The engine, which fulfills the ULEV standard, has a power of 165 KW, and a maximum torque of 315 Nm at 3000 rpm. Special characteristics of the engine are 3-valve-technology and double ignition. The double ignition affects combustion and exhaust gas emission favorably. Through double ignition, the flame front captures more than 50 % of the piston top land, at the time the flame from a centric spark plug would just reach it [2].

The engine is fitted onto a stand together with the automatic transmission. Since only the first 20 seconds of the FTP-75 cycle are to be reproduced, the engine is operated at idling speed for a time period of 15 seconds after start-up. The gear is then engaged. Since the transmission output is blocked, this corresponds to the status of a vehicle with its foot-operated brake depressed. Once the test is completed, the engine is operated until a coolant temperature of 70°C is reached. This is required to ensure a sufficient ability to reproduce the measurements. The engine is then subjected to enforced cooling using a coolant temperature of 20°C. The next test can be conducted after 1.5 hrs. Fig. 1 shows the arrangement of the measurement probes. All measurement points mentioned in this report relate to two measurement levels. Measurement level 1 is directly at the transition between the exhaust port of cylinder 2 and the manifold and measurement level 2 is upstream of the catalytic converter.

The secondary air injection systems currently used in series production consist of an electrically driven radial pump and of various pipes and valves. The pump is attached to either the bodywork or the engine block and connected via pipes with the air filter or to the exhaust side of the engine. A diagram of such a system is shown in Fig. 2. The secondary air pump starts immediately after engine start and delivers a constant air mass flow of 25 kg/h. This air volume is distributed symmetrically on the 6 exhaust ports, and injected direct at the exhaust valve seat.

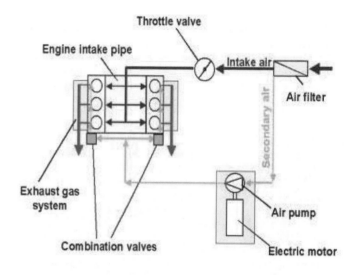

Fig. 2 Principle of the secondary air system

MEASUREMENT TECHNIQUES

Cylinder pressure recording

The cylinder pressure is measured with the assistance of a measurement spark plug with integrated cylinder pressure sensor. The measurements indicated here are only measurements taken in cylinder no. 2 of the engine. The pressure sensor is a piezoelectric recorder whose signal is prepared by a boost amplifier.

Fast Response FID (FFID)

A fast flame ionization detector (HFR 400), produced by Cambustion Ltd., is used to measure the concentration of hydrocarbons. The principle behind the function of this 2-channel device has already been described sufficiently in the bibliography [8,9]. Only the response time of the FFID is to be referred to here briefly. The device sucks a partial flow out from the exhaust gas flow by a small capillary. This is conducted into the hydrogen flame where the

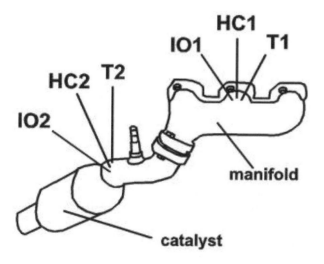

Fig. 1 Arrangement and designation of measurement probes

ionization of the hydrocarbons takes place. For the transport of this partial flow a certain time is required. With the available FFID the time to reach around 90% of the final indicator value is four ms. Additional the exhaust gas needs a certain time interval to run from the exhaust valve to the measuring point. This time can be measured and is at the available arrangement three ms. Both time delays are corrected in all cycle-resolved diagrams shown in this paper.

Rapid thermocouple

The temperature measurements are taken using rapid thermal elements. These are NiCr-Ni thermal elements whose temperature-sensitive transition points consist of 0.08 mm thick wires (Fig. 3). Temperature fluctuations can be recorded very quickly due to the low heat capacity of these thin wires. Temperature differences of 40°Kelvin can therefore be measured over a time difference of 1 ms. This is sufficient to allow us to observe temperature

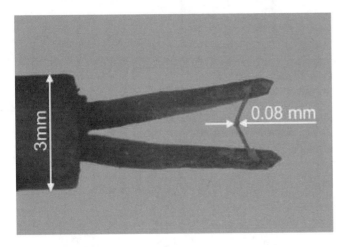

Fig. 3 Rapid Thermocouple

fluctuations at speeds of 1200 rpm (triggered by cycles). A 1mm-thermocouple shows a temperature difference of only 0.3°Kelvin in the same time interval with a comparable change of the gas temperature. Of course the temperature measurements does not fully correspond to the actual gas temperature but allows for a relative comparison to be made between different variants. Surprisingly the thermocouples showed up as expressed stable. During the investigations a few thermocouples got broken due to the stress by the exhaust gases, but particularly during installation and removal. Even at the higher engine speeds when conditioning the engine, the thermocouples survived without problems.

Ion flow probe

An ion flow probe can be used to detect flame phenomena. An electric field exists between two electrodes subject to a DC of 250 volts. If a flame passes

through this field, ions and free electrons migrate and can be recorded in the form of a low current. This current is amplified and recorded. It represents a measurement for the local intensity of the flame as well as a measurement for the time for which it is present at the measurement point.

TIME HISTORY MEASUREMENT OF POST OXIDATION

After the start the engine is driven in the idle with a speed of 1250 rpm. Exactly 15 seconds after the start the gear is engaged. Thus speed sinks on 850 rpm. Fig. 4 shows the speed profile during the test. The direct comparison of the operation with and without secondary air shows the effect of the post-reaction. In both cases the engine runs under the same conditions, i.e. the ignition angle lies at 1°TDC and the air/fuel ratio lies at about 12.

In Fig. 4 the temperatures and HC-concentrations before the catalyst are represented. The temperature difference between the operation with and without secondary air is clearly recognizable. By injected air the exhaust gas is first cooled down. Despite this cooling substantially higher temperatures are achieved. The exothermic reaction causes a nearly constant temperature rise of 100°C during the first 15 seconds. A break-down of the reaction is recognizable clearly after the gear is engaged. The reaction can compensate only the cooling by injected air, causes however no additional temperature rise.

With the engine running without secondary air the rich mixture is recognizable from the high HC-concentrations. The HC-emission does not drop after the actual starting

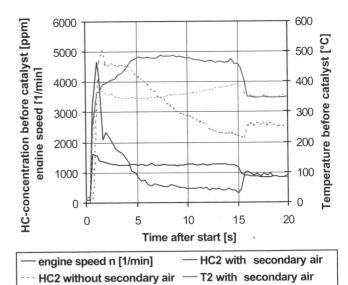

Fig. 4 Measurement with and without injection of secondary air

270

peak, but remains on a high level. In order to compare the measurements with secondary air directly to those without, the dilution caused by the injected secondary air was corrected. The difference of the two HC-concentrations shows the rate of HC oxidation by the post-reaction. The beginning of secondary air injecting is recognizable clearly one second after the start. Six seconds after the start a HC-level is reached and remains approximately constant until 15 seconds after the start. This indicates that between seconds 1 and 6 after start the air/fuel mixture improves due to heating effects, then an optimal condition is obtained. When the gear is engaged the HC-concentration rises by a factor of 2, since post oxidation partly breaks down as showed already by the temperature profile.

VISUAL INVESTIGATIONS ON MANIFOLD WITH QUARTZ GLASS WINDOW

The visible part of the post oxidation flame is investigated to allow us better to evaluate the formation and development of the post oxidation reaction, as well as to optimize the design of manifold geometries. A quartz window, through which the exhaust valves can be seen, is fitted to a manifold for this purpose (Fig. 5). A highly sensitive high-speed camera (Kodak HSV 4540) is used to record the flame development of several operating cycles 10s after the start. The speed of the recording is 4500 images/second.

Fig. 7 shows an image sequence of 20 images together with the flame phenomena visible in the exhaust port

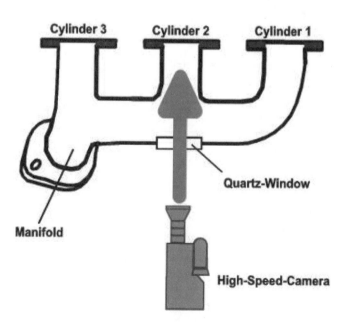

Fig. 5 View through manifold into exhaust port

(viewed direction indicated in Fig. 5). The exhaust valve of cylinder No. 2 starts to open 85°aTDC. As soon as there is a small gap between exhaust valve and valve seat, a bright yellow flame can be seen. As the gap widens, the yellow flame becomes brighter. The yellow color of the flame indicates very rich fuel-air. As a result of the retarded ignition angle and the very delayed combustion process, the reactions in the combustion chamber have not yet been completed at this time. When the exhaust valves open, the hot reactive gases flow out at great speed. When these make contact with the oxygen present outside the valve, this results in the creation of this yellow luminous flame phenomena [4].

60°CA after the exhaust valve is opened (152°aTDC), this flame disappears again. In the next phase, no flame can be seen in the exhaust port until 202°aTDC. Then a blue flame is visible, which moves out of the combustion chamber into the exhaust port. This flame is present for 50°CA before it disappears from the range of vision of the camera (261°aTDC). This flame is a premixed form of combustion and has already been observed by the Ford Research Laboratory [4] simply during operations with retarded ignition angles. Unlike the first flame, this flame moves through the exhaust port and manifold area and is therefore dependent on the local flow parameters and temperature gradients.

VISUAL INVESTIGATIONS ON QUARTZ GLASS MANIFOLD

In order to be able to observe the post oxidation flame as it passes through the manifold, a quartz glass manifold is produced. The interior dimensions of this manifold fully correspond to the inner shell of the series production LSI manifold (Fig 6).

Fig. 6 Glass manifold on a bank of the V6 engine

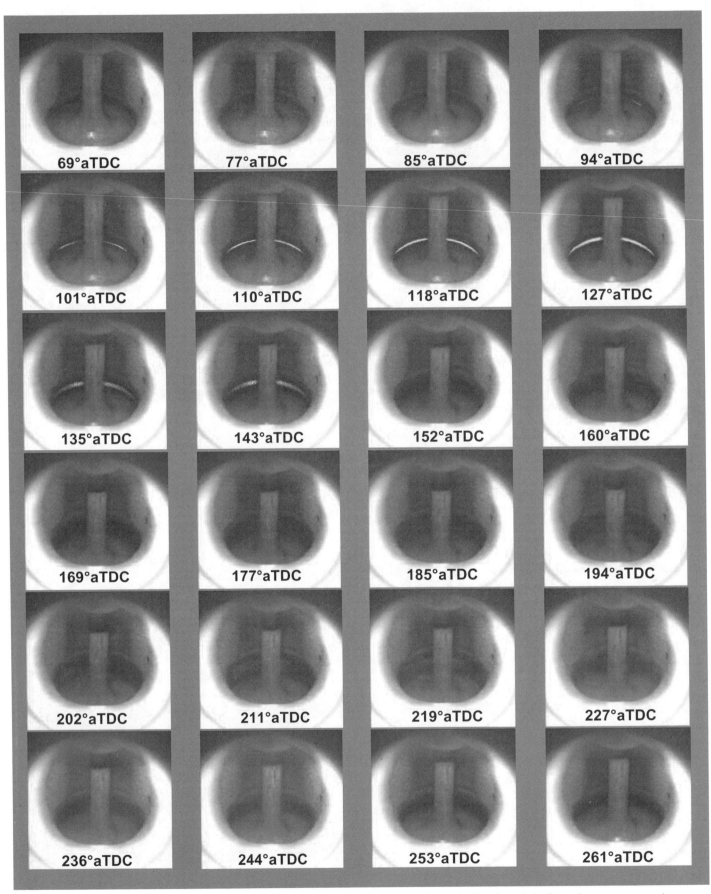

Fig. 7 View of the cylinder 2 exhaust valve through the manifold's quartz glass window (crankshaft angle measurements relate to cylinder 2)

272

During quartz glass manifold operations, the post oxidation flame can be seen as a very rapidly pulsating blue luminous phenomenon during the engine start-up and warm-up. Fig. 9 shows the development of the flame over the time of one operating cycle. The outline of the glass manifold is shown as a narrow line. The image sequences only show the time periods within which a flame is visible. For cylinder 1, this applies as of 488°aTDC when a flame first appears in the glass manifold from the exhaust port. After 110°CA, the flame has passed the manifold and disappears heading towards the face end catalytic converter. At this time, the flame appears like a burning exhaust gas droplet. Swirls and recirculating areas caused by mechanical flow influences such as pipe junctions or pipe bends can be clearly seen. One can also see that the entire pipe cross-section is not always reached by the flame.

Cylinder 3 follows as the next part of the ignition sequence. The flame from this exhaust port only passes over a small visible distance before disappearing into the exhaust pipe. It is however clear that partial burning of the flame occurs in the direction of the other two cylinders. This shows that the flame is connected to the flow and follows the flow mechanical influences. A similar pattern can be seen in the exhaust gas of cylinder 2. A flame can be seen between 240°aTDC and 350°aTDC.

Flame observations clearly show that a flame is only present in the manifold at the most during only half of the time period of the cycle. All flames come from the opened exhaust valves whereby the blue flame, unlike the yellow flame, moves downstream with the exhaust gas. There is no contact between the flames from different cylinders. One can observes however that some flames caused by pulsations and swirling motions in the exhaust gas system move towards the closed exhaust valves but not to such an extent that we can speak of the burning away of the reactive exhaust gas components present at this point.

COMPARISON BETWEEN VISUAL INVESTIGATION AND NUMERICAL SIMULATION

To calculate the flow field in the exhaust port and the manifold, a complete grid of one cylinder bank was prepared with the complete exhaust gas system up to the catalytic converter. The flow field simulation was carried out with the three-dimensional CFD-code STAR CD [14]. As boundary conditions, measurements of the cylinder pressure and the exhaust port system pressure were taken. In the calculations no reaction kinetics model was implemented. This means, that in the direct comparison between pictures of the simulation and pictures of the visual investigations the effect of the post oxidation is recognizable as a difference. Fig. 8 shows a section of the exhaust port of cylinder no.2 with the gas velocities at 210°aTDC. In the gap between valve and valve seat high

velocities are visible. The position of the secondary air bore through that secondary air is injected is also visible. Velocities reached in this hole are rather low in comparison to velocities in the valve gap. So during emission of exhaust gas, not enough secondary air and oxygen is injected to let the reaction continue uninterrupted. Fig. 7 indicates that after 250°aTDC no further reaction takes place.

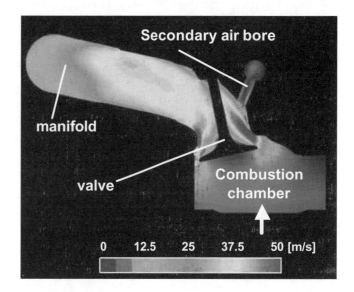

Fig. 8 Gas velocity in the exhaust port of cylinder 2 at 210°aTDC

Fig. 10 shows a picture sequence of the visual investigation. Every image is compared with a calculated picture, which shows the temperature field at the same time. Thus it is possible to realize the current temperature situation when the flame is appearing. At 225°aTDC a flame is entering the glass manifold. At the same time CFD-calculation indicates high temperatures at this position. Until 265°aTDC the flame is moving in the same direction as the hot gases. Subsequently the flame is moving further even though CFD-calculation shows no more movement. This indicates the effect of the post oxidation.

Through altering the geometry of the manifold it should be possible to optimize the conditions for the post oxidation. It is important that the reaction zone does not arrive too early in cold areas, where the ignition delay is too long. So CFD-calculation could be a good instrument to improve manifold design. Interesting would be a coupling between a simple reaction kinetics model with the three-dimensional CFD-calculation. Such calculations could then be adjusted by comparing the CFD-results with images from a glass manifold.

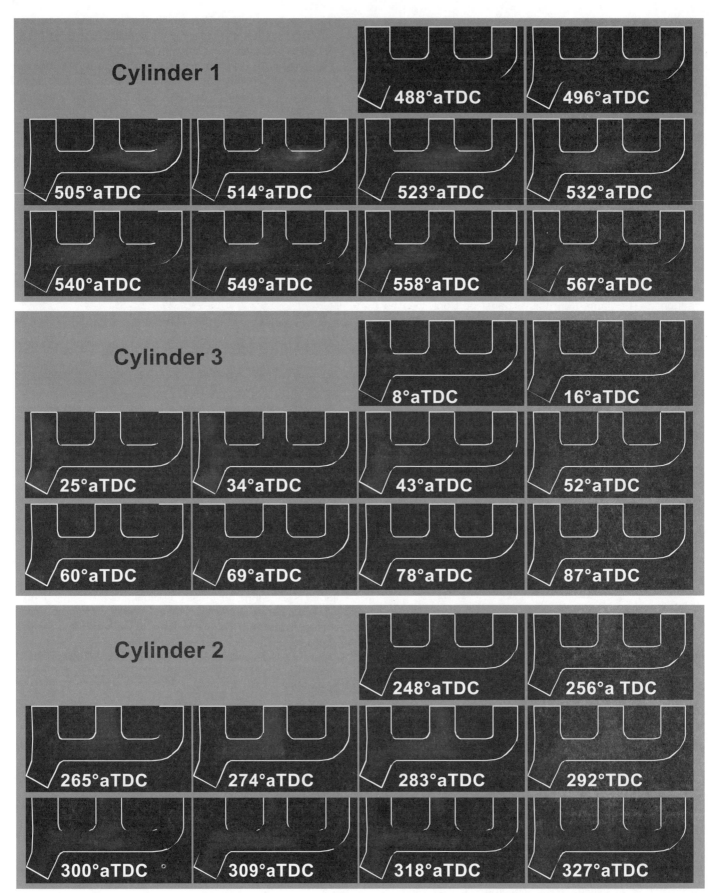

Fig. 9 View on the quartz glass manifold and the post oxidation flame (crankshaft angle measurements relate to cylinder 2)

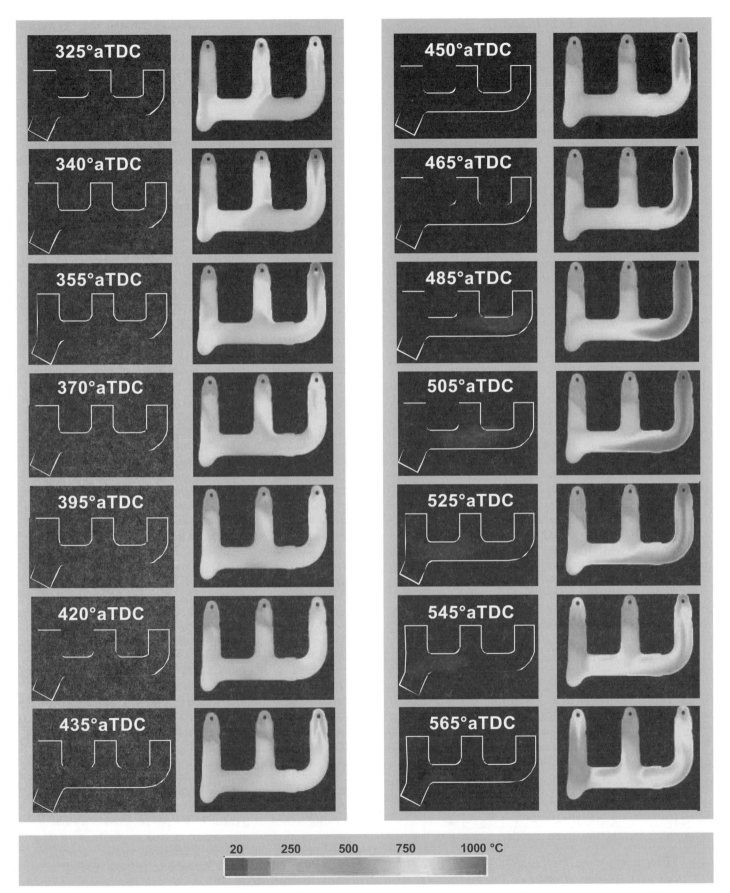

Fig. 10 Comparison of the post oxidation flame with pictures of the calculated temperature field in the exhaust manifold (crankshaft angle measurements relate to cylinder 2)

CYCLIC-TRIGGERED MEASUREMENTS

A pressure recording was conducted on the test body. By way of example, the measurements on cylinder no. 2 are to be explained here. A rapid temperature probe, an ion flow probe and a FFID probe are fitted in measurement position 1 (Fig 1). A relationship can therefore be established between combustion in the combustion chamber and post oxidation in the exhaust port. Fig. 12 shows the mean value of a measurement of 25 operating cycles, which were recorded 10 sec after the start. The retarded ignition angle of 1°aTDC and therefore very delayed combustion can be clearly seen in the cylinder pressure characteristics. In this example, the secondary air pump is switched off so that a subsequent reaction does not occur in the exhaust gas. The exhaust gas oxygen sensor therefore corresponds to the combustion chamber oxygen sensor and is set to λ=0.85.

Measurement position 1 is approx. 80 mm downstream of the exhaust valve. According to [8], mixture processes as well as chemical reactions which affect the characteristics of the non-combusted hydrocarbons occur over this section. However the measured HC concentration (Fig. 11) displays characteristics typical of gasoline engines as described in the bibliography [8,9]. Once, the exhaust valve has opened, the HC concentration firstly increases slightly. This is caused by non-combusted or fluid hydrocarbons which have become deposited in the valve gap. When the exhaust valve is opened, these are entrained by the exhaust gas flow. The HC concentration then falls to lower values because during this time period, exhaust gas is emitted from the preceding combustion process. Just before the exhaust valve closes, the HC concentration increases greatly. This is caused by the "roll up vortex" which comes about as a result of the peeling of the wall film during the upwards movement of the piston. Fluctuations in the concentration can also be seen when the exhaust valve is closed. These are caused by pulsation movements of the exhaust gas column whenever the other exhaust valves open.

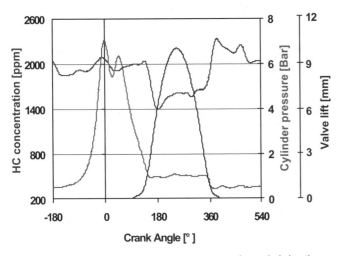

Fig. 11 HC concentration without secondary air injection

During operations with secondary air, the exhaust gas oxygen sensor changes to a value of λ=1.25. Fig. 12 shows the characteristics of the parameters measured as a mean value of 25 operating cycles. Once the exhaust valve has been opened, the HC concentration can be seen to fall. It remains at a low level and only increases just before the valve again closes. Until the exhaust valve is opened again, the HC concentration stabilizes at a level of 1000 ppm, whereby the small fluctuations in concentration are caused by pressure fluctuations in the exhaust gas system.

When the exhaust valve opens, the ion flow probe displays a strong signal. This occurs at the same time at which the diffusion flame appears in the valve gap. However, no similar movement was observed with the yellow flame. We are therefore assuming that this signal is created by particles of the soot created during the diffusion combustion and transported from exhaust gas flow to the ion flow probe. According to [11,12], the soot may be electrically charged as well as electrically conductive and is considered as a cause of the strong signal. During the time period in which a flame phenomenon cannot be established during the visual investigations, an ion flow signal was not recorded either. At 200°aTDC, another ion flow signal occurs, although it is not as distinct as the first signal. It occurs at the same time as the blue flame established during the visual investigations. We can therefore assume that we are dealing with ionization by the post oxidation flame which moves through the manifold. Just after the exhaust valve is opened, the temperature probe displays a rapid increase of more than 100°C. The temperature stabilizes at a plateau before it again increases by more than 100°C at the same time as the second ion flow signal occurs.

The exhaust gas mass flow calculated using a 1-dimensional gas exchange calculation program (PROMO, [13]) provides clarification for the occurrence and the temporal process of the two flames in the exhaust port. Firstly, the yellow flame occurs as the hot exhaust gas flows out and comes into contact with the oxygen already present in the exhaust port. As soon as sufficient oxygen ceases to be available and/or the exhaust port is filled with exhaust gas in the area around the valve gap, the flame is extinguished. It does however disappear at the latest, when the exhaust gas-air mixture returns to the combustion chamber. If exhaust gas then again flows out of the combustion chamber, the blue flame can be seen. If this has also caused the combustible exhaust gas components and/or the remaining oxygen to react, a flame can no longer be seen directly in the exhaust port. It moves further into the manifold with the exhaust gas until it extinguishes as a result of the temperature continuing to fall as it moves downstream or due to the poor mixture and therefore the lack of a reaction partner. During the entire remaining period for which the exhaust valve is open, hydrocarbons which do not participate in post oxidation flow out.

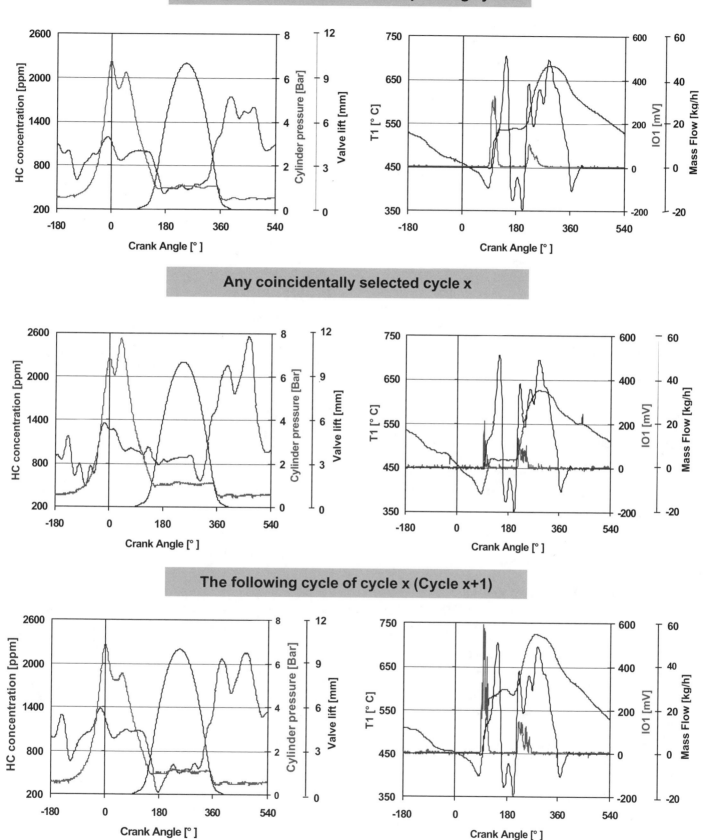

Fig.12 Cycle-triggered measurements recorded 10 seconds after the start

Fig. 12 shows two coincidentally selected examples of the 25 operating cycles recorded to clarify the cyclic fluctuations of post oxidation and its causes. These two operating cycles were recorded directly after one another. The first (cycle x) is an example of rapid combustion in the combustion chamber. One can see that this is due to the early and very high increase in pressure. Parallel to this, when the exhaust valve is opened, only a small ion flow signal can be seen as well as a slight increase in temperature. This indicates that the post reaction is only weak. The HC concentration stabilizes at a relatively high level during the extension phase. The characteristics displayed in the subsequent operating cycle are very different. The maximum pressure measured during combustion occurs much later. This is significantly delayed combustion. The post oxidation is much more powerful and is expressed in a very clear ion flow signal. Even the increase in temperature of 200°C is twice as high as in the first operating cycle. As a result of the strong post oxidation flame, the HC concentration briefly falls to a very low level. The link between post oxidation and combustion in the combustion chamber is therefore clear. Delayed combustion creates favorable conditions for post oxidation because, as a result of the long combustion process, little time remains between the end of combustion and the exhaust valves opening for complete completion of all reactions. A larger number of radicals and reactive components come into contact with the oxygen present in the exhaust port. Post oxidation occurs in an appropriately more powerful manner. Theoretically, this relationship can be recorded via the 90% energy conversion point. While the 90% energy conversion point is at an average of 67°aTDC, a value of 62°aTDC is measured in operating cycle x and a value of 72°aTDC is measured in cycle x+1.

Post oxidation while gear is engaged

A measurement taken while the gear is engaged underlines the significance of the 90% energy conversion point (Fig. 13). When the gear is engaged, the speed is reduced from 1250 rpm to 850 rpm. In an applicative manner, all relevant parameters such as combustion chamber oxygen sensor, ignition angle and exhaust gas oxygen sensor are controlled so that post oxidation has to continue to function. An interruption to the reaction should however be noted and can be seen from the temperature signal and the HC concentration. As a result of the increased load, combustion in the combustion chamber is conducted faster and more completely. The 90% conversion point occurs earlier. Post oxidation almost comes to a complete standstill as a result of the lower reactivity of the exhaust gases.

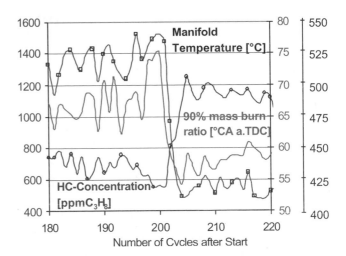

Fig. 13 Gear selection in automatic transmissions

CONCLUSION

The quality of post oxidation of an engine is heavily dependent on the quality of carburation. If there is a little fuel in the combustion chamber during the cold start and warm-up and if all hydrocarbons which enter the exhaust port participated in combustion, these can be included in post oxidation and incinerated. The design of the exhaust manifold is decisive to the post reaction. Long single-flow pipes result in a large amount of heat loss per unit distance covered by the exhaust gas. The exhaust gas temperature therefore falls rapidly, what increases the ignition delay of the reactive components. This allows post oxidation to come to a standstill. Exhaust manifolds with a rapid joining of the exhaust gases of different cylinders (a bar-type manifold, similar to Fig. 5) have the advantage that hot gases flow through pipes which have already been heated by other cylinders. Less heat is therefore lost to the walls. However when using what is commonly referred to as a bar-type manifold, the post reaction flame does not "jump" from one cylinder to the next. The flames are only ever briefly present in the manifold and are closely linked to the cycle fluctuations of the appropriate combustion chamber combustion process. Delayed combustion processes with retarded ignition timing establish favorable conditions for the post reaction whereas perfect conditions are not established during rapid combustion processes. The ability of an engine to be operated when cold with retarded ignition timing and low cycle fluctuations is therefore decisive.

In order to improve post oxidation in future engines, the manifold geometry must be optimized through the use of three-dimensional CFD calculations. The objective must be to create a geometry which allows the post oxidation flame to burn for as long as possible as well as for the non-combusted exhaust gas components also to be captured by other cylinders. Consideration should also be given to introducing a recirculation area which stabilizes the flames in the manifold and retards the movement downstream to cold areas. Co-ordinating the individual cylinders to one another is just as important as optimizing the distance covered by the exhaust gases before reaching the catalytic converter.

ACRONYMS

λ : Oxygen sensor

°aTDC : degree Crank angle after top dead center

HC1 /2 : Hydrocarbons measurement position 1 and/or 2

IO1 /2 : Ion flow signal measurement position 1 and/or 2

ppm   : parts per million C_3H_8 (Propane)

T1 /2   : Temperature measurement position 1 and/or 2

FFID   : Fast flame ionization detector

FTP 75 : Federal Test Procedure

TDC   :Top Dead Center (always related to cyl. 2)

REFERENCES

1. Kollmann K., Abthoff J., Zahn W., "Secondary Air Injection with a New Developed Electrical Blower for Reduced Exhaust Emissions", SAE 940472

2. Niefer H., Weining H., Bargende M., Waltner A., "Verbrennung, Ladungswechsel und Abgasreinigung der neuen Mercedes-Benz V-Motoren mit Dreiventiltechnik und Doppelzündung", MTZ 58, 1997

3. Umehara K., Tateishi T., Nishimura H., Mismi M., "HC-Reduction System for Cold Start and Warm-Up Phases - Improvement of Catalyst Warm-Up by Retarded Ignition", JSAE Review 18(1997) 57-82

4. Russ S., Thiel M., Lavoie G., "SI Engine Operation with Retarded Ignition: Part 2-HC Emissions and Oxidation", SAE 1999-01-3507

5. Meyer R., Heywood J., "Liquid Fuel Transport Mechanisms into the Cylinder of a Firing Port-Injected SI Engine During Start Up", SAE 970865

6. Takeda K., Yaegashi T., Sekiguchi K., Saito K., Imatake N., "Mixture Preparation and HC Emissions of a 4-Valve Engine with Port Fuel Injection During Cold Start and Warm-up", SAE 950074

7. Crane M., Thring R., Podnar D., Dodge L., "Reduced Cold-Start Emissions Using Rapid Exhaust Port Oxidation (REPO) in a Spark-Ignition Engine", SAE 970264

8. Finaly I., Boam D., Bingham J., Clark T., "Fast Response FID Measurement of Unburned Hydrocarbons in the Exhaust Port of a Firing Gasoline Engine", SAE 902165

9. Henein N., Tagomori M., Yassine M., Asmus T., "Cycle-by-Cycle Analysis of HC Emissions during Cold Start of Gasoline Engines", SAE 952402

10. Ma T., Collings N., Hands T., "Reduction of Exhaust Emissions during Engine Cold Start by means of an Afterburner Operating in the Hydrogen Ignitable Regime", 4th International EAEC Conference on Vehicle and Traffic Systems Technology, Strasbourg, Volume 2, Pages (365-372), 1993

11. Kessler M., "Ionenstromsensorik im Dieselmotor" Dissertation, University Karlsruhe, 2002

12. Gerhardt P., Homann K., Löffler S., Wolf H., "Large Ionic Species in Sooting Acetylene and Benzene Flames", 20th Symposium (International) on Combustion, Pages (1121-1128), 1988

13. Seifert H., "Twenty Years successful Development of Gas Exchange Program PROMO", Motortechnische Zeitschrift, Pages (478-488), 1996

14. Nebel M.,"Transient simulation of the flow in an exhaust manifold", Konferenz-Einzelbericht: Berechnung u. Simulation im Fahrzeugbau, VDI-Berichte, Volume 1283, Pages (401-409), 1996

1999-01-1540

A Study on the Practicability of a Secondary Air Injection for Emission Reduction

Geon Seog Son, Dae Jung Kim and Kwi Young Lee
Institute for Advanced Engineering

Eui Rak Choi and Young Woong Kim
Daewoo Motor Co.

Copyright © 1999 Society of Automotive Engineers, Inc.

ABSTRACT

In this study, feasibility tests of secondary air injection technology and lean A/F control technology were performed for LEV program using the FTP75 test on a 2.0 DOHC A/T vehicle. Second-by-second emissions and temperatures were evaluated. The temperatures of exhaust gas were measured at exhaust manifold, front of warm up, and the center of warm up converter. At first, amount of secondary air injection was determined with a bench aged warm up converter and a fresh UCC. And then, the performances of secondary air injection and lean A/F control strategy were compared with 80,000km vehicle aged converters(warm up converter, UCC). Both secondary air injection and lean A/F control technologies satisfied the ULEV regulation. This study shows that the lean A/F control strategy can be one of the potential technologies to meet the LEV/ULEV regulations without an active system that need a cost up.

INTRODUCTION

As one of the promising technologies for LEV program, secondary air injection system has been studied widely[1]. It supplies fresh oxygen to exhaust gas for post reaction of unburned hydrocarbon. This post reaction is an exothermic reaction, which increases the temperature of the exhaust gas[2]. It also gives sufficient air to the catalyst during the cold start fuel enrichment that is necessary to prevent driveability problems. The post reaction and lean A/F of exhaust gas promote the HC light-off of a Pd converter[3]. Because a use of Pd is effective to improve HC conversion and enhance a HC light-off performance in lean exhaust gas. Even though it shows good performance in reduction of emission, the secondary air injection technology is an active system that needs a pump, injection lines, a controller, and etc.

There were some discussions about its economical efficiency compared with a passive system. Because lean A/F control strategy during the initial dozens of seconds can produce similar effects to air injection. The initial lean A/F control makes low HC engine out emissions. Also a Pd catalyst itself has an advantage of lean A/F, because of a better light–off, which occurs sooner under lean conditions[1].

In this study the performances of secondary air injection and lean A/F control strategy were compared. The tests were done on the same vehicle with FTP75 mode test. The engine out and tailpipe emissions were recorded with temperatures during FTP75 test, which were measured at exhaust manifold, front of warm up converter and the center of warm up converter.

QUANTITY OF SECONDARY AIR INJECTION

EXPERIMENTS – To determine a quantity of air injection, 2.0 DOHC A/T vehicle(97MY) was used. The vehicle was equipped with EGR. The original ECU(electronic control unit) level was to meet TLEV regulation. After a base emission test the ECU was modified to get rich A/F control during the initial 80 seconds through a injection control of fuel. The aftertreatment system was composed with a warm up converter and a underbody catalyst converter(UCC). The specifications of the converters were listed in Table 1. The warm up converter was aged using the exhaust gas from a 3.0 liters gasoline engine for 50hrs. The aging condition contains a 40 sec cruise mode with inlet temperature of 850?. The 40 sec cruise mode was followed by 20 sec power enrichment(3%). During the power enrichment air was injected by 3%. But UCC was fresh. The Phase II fuel was used for the aging and FTP tests.

Air flow was controlled with a MFC and a solenoid valve operated by a timer. Factory air was used instead that of air pump for a convenience. Secondary air at 50, 100, 130 and 150L/min were injected into a exhaust manifold port for 40sec(for 100, 130, 150L/min) or 80sec(for 50L/min) after the engine began to run. Fig. 1 shows the locations of air injection and thermocouple at exhaust manifold. According to Kollmann's report, this is not an optimum injection point[1]. Air was supplied from right side of the engine through 6mm tube. The temperatures were measured at exhaust manifold, front of the warm up converter and the center of the warm up converter with skinless K-type thermocouples.

Table 1. Specification of Warm Up and Underbody Catalyst Converter

Converter	Loading, g/L	PM Ratio (Pt/Pd/Rh)	Substrate
Warm Up	5.3	0/1/0	400cpsi, ceramic, 0.67L
UCC	1.8	5/0/1	400cpsi, ceramic, 0.9L x 2

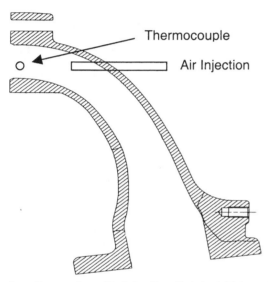

Figure 1. Secondary Air Injection Point at Exhaust Manifold

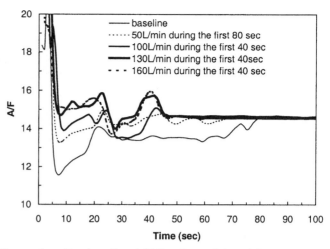

Figure 2. Engine Out A/F Profiles of the Vehicle According to Secondary Air Injection

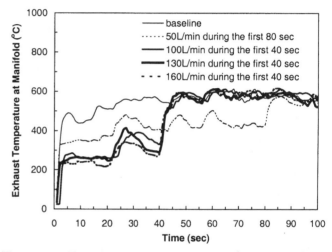

Figure 3. Temperature Profiles Measured at Exhaust Manifold According to Secondary Air Injection

RESULTS AND DISCUSSIONS – Fig. 2 shows A/F trends according to secondary air injections. Result of base emission test(0L/min) shows that the A/F of modified ECU was on the rich side after cold start for 80 sec. The rich A/F gradually moved to lean side according to increasing of air injection. It can satisfy the condition of a post reaction at exhaust gas, because there are sufficient unburned HC and fresh oxygen.

Fig. 3 shows temperature profiles measured at exhaust manifold. As the injection amounts are increasing the temperatures at exhaust manifold are decreasing, because the temperature of ambient air is much lower than that of exhaust gas. After the air injection the temperatures are returned to the level of base line test(0L/min).

Fig. 4 shows the temperature profiles of exhaust gas measured at inlet of the warm up converter. Even though air injection increases, there is no difference in the temperatures of exhaust gas. It may be due to an exothermal energy coming from the post reaction that occurs at from exhaust manifold to inlet of warm up converter

Fig. 5 shows the temperature profiles of the center of the warm up converter. Air injection promotes increasing of bed temperature and reducing of light-off time compared to base line(0L/min). The highest temperature is shown at 130L/min injection. The light-off times of the warm up converter are shown in Table 2. The light-off temperature(T_{50}, 50% HC conversion) was assumed at 350°C. The case of 130L/min injection shows the fastest time, 15 sec, to reach at 350?. Air injection reduces light-off time by 13sec compared to base line test(0L/min). These data mean that there is an optimum air injection amount for an aftertreatment system and in this case 130L/min injection is the optimum condition. The less air injected than 130L/min, the less the exhaust temperature will increase. The more cold air injected than 130L/min, the more the catalyst light-off will be delayed. In case of 160L/min injection, the amount of air injection doesn't seem cool down the exhaust gas to below the start point of post reaction. The temperature of 160L/min injection at manifold was lower than that of 130L/min injection but at inlet of warm up converter the temperature of 160L/min injection was similar to that of 130L/min injection. This means that there was sufficient post reaction to compensate the temperature loss. The low temperature of 160L/min injection at the center of warm up converter may come from shortage of oxidants like as HC and CO in exhaust gas. Most oxidants may be consumed before warm up converter by post reaction already.

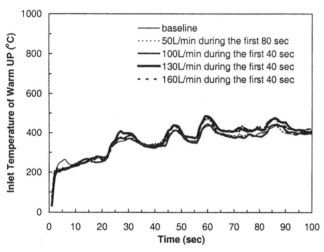

Figure 4.   Temperature Profiles Measured at Inlet of Warm Up Converter According to Air Injection

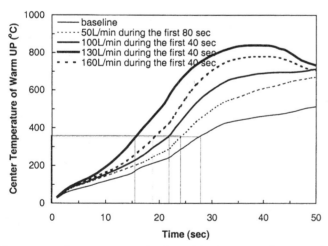

Figure 5.   Temperature Profiles Measured at Center of Warm Up Converter According to Air Injection

Table 2.   Light Off Time According to Secondary Air Supply Condition

Secondary Air	0 L/min	50 L/min	100 L/min	130 L/min	160 L/min
Light Off	28sec	24sec	22sec	15sec	18sec

Fig. 6 and 7 show THC and CO emissions measured at tailpipe during FTP test. HC and CO emissions are reduced drastically by air injection. The emission improvements in Fig. 6 and 7 are mainly due to the faster light-off of the catalyst. 130L/min injection showed the best result. This result concurs with the results of Fig. 5 and Table 2.

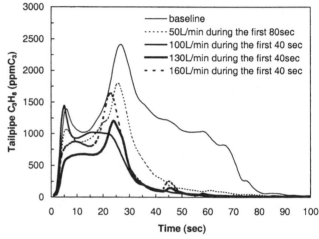

Figure 6.   Tailpipe THC Emission According to Air Injection

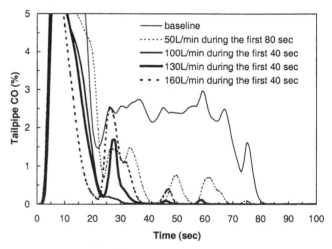

Figure 7.   Tailpipe CO Emission According to Air Injection

Fig.8 shows NOx emission measured at tailpipe during FTP test. Compared to the base line test, the NOx emission is increased rapidly at 100L/min. Also a breakthrough of NOx emission can be seen at around 20sec in all cases. Generally a three-way catalyst has a shortage in NOx conversion at the lean side. Because of the air injection exhaust gas is lean as shown at Fig. 2 and it is out of A/F-window of the converters.

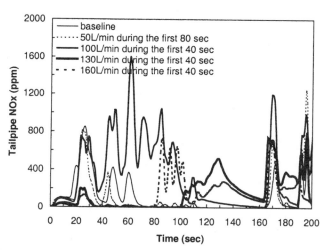

Figure 8.　Tailpipe NOx Emission According to Air Injection

Table 3.　FTP75 Emission Test Results According to Secondary Air Injection

Air Injection L/min	ECU Level	Total (g/km)		
		THC	CO	NOx
0	TLEV	0.051	0.846	0.056
50	"	0.035	0.496	0.113
100	"	0.036	0.741	0.079
130	"	0.026	0.392	0.081
160	"	0.025	0.440	0.092

Table 3 shows the results of FTP75 tests according to secondary air injection. The test results are the average of 5 tests. Compared to base line test, NOx was increased in all injection cases. And the increase was proportional to air injection except 50L/min injection. In case of 50L/min injection, NOx abruptly increased, because air injection kept for initial 80sec. In case of THC, it was decreased in all injection cases. And the decrease was proportional to air injection. These mean that THC and NOx are in the relation of trade-off. In a comparison of 130L/min injection and 160L/min injection, THC was on the same level but NOx and CO were high at 160L/min. When THC, CO and NOx emissions and the amount of air injection were considered, 130L/min injection was the best condition for this study.

SECONDARY AIR INJECTION VS. LEAN A/F CONTROL

EXPERIMENTS – The same vehicle was used in the comparison of secondary air injection and lean A/F control. The TLEV ECU was changed for a LEV ECU, because the vehicle was available as a TLEV and a LEV package. And then the ECU was modified to get rich A/F control and lean A/F control during the initial 40 seconds through a control of duration of fuel injection. The lean A/F was limited to 15.5 to get smooth running and to avoid buckling or stalling of the car. The aftertreatment system was composed with a warm up converter and a underbody catalyst converter(UCC). The specifications of the converters are listed in Table 4. Both converters were aged on a chassis-dynamometer to 80,000km with AMA(automotive manufacture association) cycle as a system in the same locations as this test. Secondary air was supplied with factory air instead of air pump for 40sec with 130L/min. Other experimental conditions were the same with air quantity experiments.

Table 4.　Specification of Warm Up and Underbody Catalyst Converter

Converter	Loading	PM	Substrate	Aging
Warm Up	5.29g/L	0/1/0	400cpsi, ceramic, 0.67L	50k MAD*
UCC	8.83g/L	10/0/1	400cpsi, ceramic, 0.9L x 2	50k MAD

* MAD : mileage accumulation driving system

RESULTS AND DISCUSSIONS – Fig. 9 shows the A/F profiles of base line(before modification of ECU), lean, rich A/F control and (rich A/F control + 130L/min secondary air injection). The modification of lean and rich A/F control was only for initial 40sec and the rest of test had the same ECU control. Base line A/F shows middle A/F between rich A/F control and lean A/F control. (rich A/F control + 130L/min secondary air injection) shows the leanest A/F.

Fig. 10 shows the temperature profiles at exhaust manifold. Lean A/F control shows the highest temperature and (rich A/F control + 130L/min secondary air injection) shows the lowest temperature during the air injection. The temperature drop is mainly due to the injection of cold ambient air.

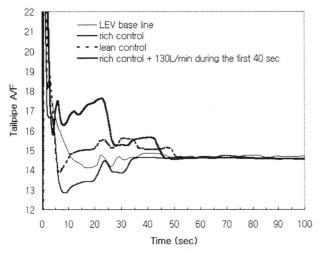

Figure 9. A/F Profiles of According to A/F Control and Air Injection

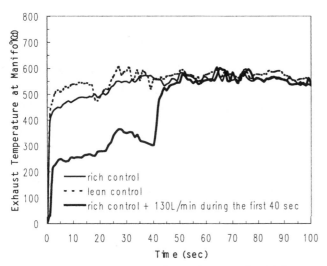

Figure 10. Temperature Profiles Measured at Exhaust Manifold

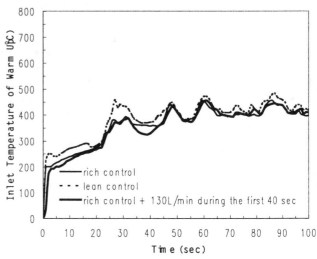

Figure 11. Temperature Profiles Measured at Inlet of Warm Up Converter

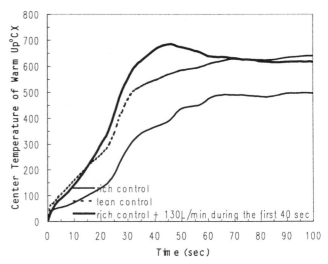

Figure 12. Temperature Profiles Measured at the Center of Warm Up Converter

Fig. 11 shows the temperature profiles at inlet of the warm up converter. Lean A/F control shows still the highest temperature and rest two cases show similar temperatures. In case of air injection, post reaction seems to compensate the temperature for the loss of air injection.

Fig. 12 shows the temperature profiles at the center of the warm up converter. Air injection shows the highest temperature for the first 60sec. Rich A/F control shows the lowest temperature till 100sec. We can confirm the results of **QUANTITY OF SECONDARY AIR INJECTION**.

Fig. 13 shows the THC emissions measured at tailpipe. All cases show perfect reduction after 80sec. Lean A/F control shows the lowest emission and rich A/F control shows the highest emission.

Table 5 shows the results of FTP emission tests. The results are the average of 5 tests. In case of THC and CO, lean A/F control shows very similar results with air injection. In case of NOx, rich A/F control shows the lowest result and air injection shows the highest results. Lean A/F control shows increased NOx emission compared to rich A/F control but it still meets ULEV regulation safely. When we consider the cost and the performance, the lean A/F control strategy can be a potential method to meet ULEV regulation without any active system. Also if loading of warm up converter will be increased, we can get better results with lean A/F control strategy.

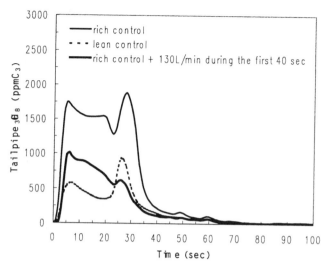

Figure 13. THC Profiles Measured at Tailpipe

Table 5. FTP75 Emission Results with the 80.000 km vehicle aged converters (warm-up converter and UCC)

Test	ECU Level	HC (g/km)	CO (g/km)	NOx (g/km)
ULEV Standard 80k km	-	0.02485	1.0563	0.1242
Non-2ndary Air (Rich Control)	LEV	0.0348	0.318	0.0702
Non-2ndary Air (Lean Control)	"	0.0198	0.164	0.0801
130L/min, 40sec 2ndary Air	"	0.0192	0.161	0.1124

CONCLUSIONS

The comparison of performances between secondary air injection technology and lean A/F control technology was conducted with a 2.0 DOHC A/T vehicle. At first, the air injection quantity was determined with a TLEV ECU and then the practicability of air injection was studied with a LEV ECU.

The initial test with the TLEV ECU had a bench aged warm up converter and a fresh UCC.

1. The optimum air injection amount in the experimental condition was 130L/min. With a TLEV level ECU, secondary air injection reduced HC emission by 51% but it increased NOx emission by 64%.

   The aftertreatment system(warm up converter, UCC) used for the practicability study was aged on a chassis-dynamometer with AMA cycle to 80,000km.

2. With a LEV level ECU that was modified for initial 40sec, lean A/F control reduced HC emission and it met ULEV regulation safely(79% of ULEV regulation). In case of NOx, it increased emission but it still met ULEV regulation safely(64% of ULEV regulation).

3. With a LEV level ECU that was modified for initial 40sec, rich A/F control + 130L/min reduced HC emission and it met ULEV regulation safely(79% of ULEV regulation). In case of NOx, it greatly increased emission and it barely met ULEV regulation(90% of ULEV regulation). Also the NOx emission was higher than that of lean A/F control.

4. Compared to the secondary air injection technology, the lean A/F control technology has an advantage in both cost and emission. And this study shows that lean A/F control can be a potential method to meet ULEV regulation without any active system.

REEFERNCES

1. K. Kollmann, J. Abthoff, W. Zahn, H. Bischof, and J. Gohre, "Secondary Air Injection with a NEW Developed Electrical Blower for Reduced Exhaust Emissions", SAE Paper 940472, 1994.

2. H. Katahiba, R. Nishiyama, Y. Nishimura, Y. Hosoya, H. Arai, and S. Washino, "Development of an Effective Air-Injection System with Heated Air for LEV/ULEV", SAE Paper 950411, 1995.

3. N. Kishi, H. Hashimoto, K. Fujimori, K. Ishii, and T. Komatsuda, "Development of the Ultra Low Heat Capacity and Highly Insulating(ULOC) Exhaust Manifold for ULEV", SAE Paper 980937, 1998.

2002-01-2803

Application of Secondary Air Injection for Simultaneously Reducing Converter-In Emissions and Improving Catalyst Light-Off Performance

Mark Borland and Fuquan Zhao
DaimlerChrysler Corp.

Copyright © 2002 SAE International

ABSTRACT

Improving catalyst light-off characteristics during cold start and reducing engine-out (more accurately converter-in) emissions prior to catalyst light-off have been regarded as the keys to meeting future stringent emissions regulations. Many technologies and control strategies have been proposed, and some of them have already been incorporated into production, to address these issues. Among these, secondary air injection received a lot of attention. This study was initiated to investigate the thermal and chemical processes associated with secondary air injection inside the exhaust system in order to maximize the simultaneous benefit of improving catalyst light-off performance and reducing converter-in emissions. The effects of several design and operating parameters such as secondary air injection location, exhaust manifold design, spark timing, engine enrichment level, and secondary air flow rate were carefully examined. It was found that proper design and optimization of secondary air injection can significantly improve catalyst light-off characteristics and reduce converter-in HC emissions.

1. INTRODUCTION

Improving catalyst light-off characteristics during cold start and reducing engine-out (more accurately converter-in) emissions prior to catalyst light-off have been regarded as the keys to meeting future stringent emissions regulations. As a result, a majority of the development efforts have been focused on the cold start period. Accelerating the heating process of the catalytic converter is commonly known as the most effective approach to drastically reduce cold-start HC emissions. Since it generally takes some time to light-off the catalyst (the time varies depending on the emissions goal), reducing converter-in HC emissions prior to catalyst light-off is as important as lighting-off the catalyst faster, both of which must be simultaneously obtained. Many technologies and control strategies have been proposed and some of them have already been incorporated into production. The following is a summary of these technologies:

- Catalytic converter design (high precious metal loading, thin substrate wall combined with high cell density) [2, 22]
- Passive [13,15] or active [25,26] HC traps
- Close-coupled catalyst
- Electrically-heated catalyst [5,17]
- Energy-storage type catalyst such as vacuum-isolated converter [1]
- Exhaust manifold improvement for enhancing mixing and thus oxidation [33]
- Exhaust manifold design (light weight, air gap wall)
- Secondary air injection [7,12,19,21,31]
- Exhaust gas ignition [4]
- System control optimization [30,34]
- Fuel reforming [18].
- Light-component fuel through fractionation [27]
- On-board H_2 generation [6]
- Spark retardation
- Higher idle speed.

Some of those technologies are applied independently and some of them have to be combined with other approaches to maximize the benefit. Among these, secondary air injection into the exhaust port in combination with rich engine operation received a lot of attention due to its robust and consistent performance to meet the development goal and its relative ease of implementation [3,7,16,19,20,21,28,29,32]. Coupling the ultra-fuel-rich engine combustion with the injection of precisely-metered air into the exhaust port results in an exothermic reaction in the port and exhaust manifold before the exhaust gases reach the converter. This can effectively reduce HC emissions inside the exhaust manifold and simultaneously accelerate the heating process of the converter following a cold start of gasoline engines. When compared with other approaches, secondary air injection can be implemented relatively easily with today's engine system without requiring a major design change.

This study was initiated to investigate the thermal and chemical processes associated with secondary air injection inside the exhaust system in order to maximize the simultaneous benefit of improving catalyst light-off performance and reducing converter-in emissions. Several design and operating parameters are examined carefully. These parameters include secondary air injection location, exhaust manifold design, spark retardation, engine-enrichment level and secondary air flow rate. To our knowledge, the effect of exhaust manifold design on secondary air injection has never been addressed in the literature. This paper is presented in five major sections. Section 2 is prepared to present a general description of the physical and chemical processes associated with the secondary air injection. All the key design and operating parameters that will affect the success for implementing the secondary air injection strategy are outlined first, followed by a comparison of thermal oxidation and catalytic oxidation. The role of mixing and temperature in enhancing the thermal and catalytic oxidation and the effects of engine enrichment and secondary air injection quantity are presented in order to sort out the key issues that must and will be addressed in the current investigation. Some other considerations associated with practical application are also briefly mentioned in this section. In Section 3, a detailed description of the experimental setup and the test procedures are provided. The key observations from this secondary air injection study are detailed in Section 4. Finally the principal conclusions derived from this investigation are summarized in Section 5.

2. PHYSICAL AND CHEMICAL PROCESSES ASSOCIATED WITH SECONDARY AIR INJECTION

2.1 KEY DESIGN AND OPERATING PARAMETERS

It is well known that engine-out exhaust gas composition varies substantially with engine-fueled air/fuel ratio. When the engine is fueled with a richer-than-stoichiometric mixture, H_2 and CO will be produced and exhausted into the exhaust stream. The concentrations of CO and H_2 rise steadily as the mixture gets richer [23,24]. A similar trend is also true with engine-out HC emissions. Injecting secondary air into the exhaust port, which allows the secondary air to mix and react with these reactants (H_2, CO, HC) produced during engine-rich operation, leads to an exothermic reaction in the exhaust system. This reaction can effectively reduce HC emissions inside the exhaust manifold and simultaneously accelerate the heating process of the converter following a cold start of gasoline engines. The engine-out NOx can also be reduced due to the lack of

oxygen for NOx formation when operating the engine richer, in addition to the evaporative charge cooling effect.

The physical and chemical processes associated with the secondary air injection are, however, quite complicated, and its benefit strongly depends upon the system optimization, which usually varies with the specific application. All the design and operating parameters that determine engine-out emissions, exhaust gas temperature, mixing process of reactants with the secondary air and the mixture residence time inside the exhaust port and manifold will affect the level of success of the secondary air injection strategy. Some of the key design and operating parameters are summarized as follows:

- Engine-out emissions (reactants) concentrations and temperature
  - Engine air/fuel ratio (enrichment level)
  - Spark timing (level of spark retardation)
  - Idle speed
  - Exhaust cam timing
- Mixing of secondary air with the reactants and mixture residence time
  - Air injection location
  - Air injection orientation
  - Air flow rate
  - Capability for delivering secondary air during exhaust blowdown phase
  - Exhaust port design
  - Exhaust valve opening timing (blowdown flow vs. displacement flow)
  - Exhaust manifold design for enhanced mixing (manifold runner length, runner orientation of each cylinder, plenum volume, collector location, cross-talk between cylinders)
- Thermal conditions of the exhaust system
  - Exhaust port geometry (affecting heat rejection to the port wall)
  - Exhaust manifold design and material (affecting heat loss to the manifold wall)
- Reaction inside the converter
  - Mixing quality of the mixture prior to entering the converter
  - Composition of the mixture.

The design and operating parameters that are investigated in this study are: engine enrichment level, spark retardation, secondary air injection location and orientation, secondary air flow rate, and exhaust manifold shape. Significant efforts have been directed towards optimizing the exhaust manifold design in order to maximize the benefit for simultaneously improving catalyst light-off performance and reducing the catalyst-in HC levels.

2.2 THERMAL OXIDATION VS. CATALYTIC OXIDATION

In general, the chemical reactions occurring in the exhaust system with secondary air injection can be categorized into two types of oxidation of reactants. These two types of oxidation are both exothermic, but the temperatures required for these to occur rapidly are dramatically different. One is the oxidation process that happens inside the exhaust port and manifold prior to the catalytic converter, which is referred to as thermal oxidation in this paper. The other is the oxidation of HC and CO that occurs inside the catalytic converter, which is significantly different from that of the thermal oxidation due to the presence of a catalyst and is referred to as catalytic oxidation. The thermal oxidation usually requires a temperature of 750°C or higher to become significant, depending on the exhaust species and residence time. It is believed that the production of H_2 in engine combustion can significantly enhance the thermal oxidation process inside the exhaust port and manifold, and CO and HC can well sustain the process once it is established [3]. In contrast, the conversion of HC and CO can start at a similar rate with a much lower temperature (200°C) on the surface of certain types of catalysts [11]. Both thermal oxidation and catalytic oxidation contribute to heating the catalyst for shortening the light-off time. In comparison, catalytic oxidation is more effective in improving the catalyst light-off by heating the catalyst directly when it reaches a certain temperature threshold. This must be provided by cylinder-out exhaust heat plus upstream thermal oxidation. Part of the heat produced via thermal oxidation will be dissipated to manifold walls and as a result, its impact on catalyst light-off will be slightly degraded. However, thermal oxidation can be very effective in removing engine-out HC emissions, if adequate temperatures are sustained. This is extremely important prior to catalyst light-off. Rich engine operation generally produces higher engine-out HC emissions as compared to that of lean operation, which demands a further reduction in converter-in emissions. With the addition of secondary air, the converter-in emissions can be reduced significantly due to the enhanced oxidation in the manifold. As will be discussed in the next section, a certain level of temperature is necessary to activate the chemistry and sustain the reactions. The results that will be presented in Section 4 demonstrated that adding air too far downstream from the exhaust valve cannot benefit from this emission reduction due to the lack of thermal oxidation, even though it may produce an improvement in catalyst light-off performance.

The effect of exhaust chemistry on the catalyst tends to be ignored in most of the analysis or data interpretation when dealing with secondary air injection. It should be pointed out that the catalyst is not only exposed to the hot exhaust gases, but also to high chemical loads from the left-over HC and CO emissions. The introduction of these chemical constituents at certain concentrations is very important to realize self-sustaining, chain reactions. Therefore, it is strongly believed that the combination of this chain reaction and thermal loading together allows the catalyst to ultimately sustain the high conversion efficiencies following the cold start. This chemical impact should not be ignored at all when locating the catalyst, designing the exhaust system to maximize the benefit of secondary air injection or calibrating the secondary air injection system. A good compromise between the exhaust manifold thermal oxidation and catalytic oxidation of CO and HC emissions is necessary. The former provides the heat necessary to initiate the exothermic reaction of CO and HC on the catalyst surface and reduce the HC and CO loading on the catalyst to minimize any emissions breakthrough. While the latter has a direct heating effect on the catalyst surface and requires a certain level of HC and CO emissions to achieve this. If the thermal oxidation consumes too much engine-out reactants, which is important in minimizing HC emissions prior to catalyst light-off during cold start, the catalyst light-off process may be slowed down due to a reduced energy release directly inside the catalytic converter. In contrast, if the converter is overdosed with the reactants due to the poor thermal oxidation, there may be more emissions breakthrough even though the catalyst may be lit-off slightly faster. As will be discussed in Section 4, the system must be optimized to balance these two oxidation processes in order to maximize the benefits.

2.3 ROLE OF TEMPERATURE AND MIXING IN ENHANCING THE THERMAL OXIDATION PROCESS

It is easy to understand that engine-out reactants can only be oxidized in the exhaust manifold when oxygen is present at a sufficiently high temperature. To maximize the thermal oxidation rate in the manifold, the pathway from the exhaust valve to the converter inlet must provide an environment with a high temperature and a high oxygen concentration over a sufficient period of time. The reaction rate increases exponentially with the temperature. A small increase in the initial temperature on the order of 10 to 20°K can result in the oxidation time of the mixture being cut in half [7]. Any design and operating parameters such as spark retardation and early exhaust valve opening that can increase engine-out gas temperature will dramatically enhance the thermal oxidation process. Therefore, the early initiation of thermal oxidation around the periphery of the exhaust valve is crucial to the success of secondary air injection. Air injection has to be as close as possible to the exhaust valve where the exhaust gases are hottest, as will be discussed in Section 4 for air injection location optimization. A minimal amount of oxygen is needed to react with the exhaust gases and a minimal amount of reactants (HC, CO, H_2 etc.) and radicals (OH, O, H) must be available to start the reaction. When the secondary air is in an extreme excess ratio or the mixing of exhaust gases with secondary air occurs too rapidly, the thermal reaction will be quenched [14,29]. Once the reaction is quenched, it will be very difficult to initiate again. On the other hand, if this mixing occurs too slowly, the heat loss to the wall may eventually lower the temperature below that necessary for retaining the thermal oxidation. This leads to the belief that promoting the mixing inside the port during the blowdown process is critical [8~10].

The temperature at the exhaust port varies significantly through the cycle. Right after the exhaust valve opens, the temperature initially drops rapidly due to the fact that a large amount of air delivered before the exhaust valve opens is accumulated there to mix with the exhaust gases. Then temperature increases quickly due to the exhaust gas blowdown [7]. During blowdown, exhaust gases exit the engine and flow through the exhaust system at a higher pressure, which prevents the exhaust gases from mixing with the secondary air. At the exhaust port, exhaust blowdown can start thermal oxidation as long as a small amount of secondary air is available. The capability of being able to deliver some secondary air and get it entrained into the exhaust gases is important in utilizing the heat available during the blowdown phase. The mixing between secondary air and exhaust gases is also occurring during this blowdown phase, which prepares the thermal condition for the oxidation once the secondary air and exhaust gases get mixed. After the blowdown, the exhaust gases have lower temperatures but a better mixing rate and longer residence time, which can certainly initiate thermal oxidation if a minimum temperature is maintained. Visualization of the exhaust port revealed that after the blowdown phase, a blue luminous zone is observed. This blue zone moves downstream with the exhaust stream and can reach the exhaust manifold collector area[20]. The last part of the mixture from the cylinder flowing through the exhaust port has an even lower temperature, which may prevent any thermal oxidation. After the exhaust valve closes, only secondary air is supplied into the residual gas in the port at a relatively lower temperature until the next engine cycle.

In the pathway of the exhaust gases from the exhaust valve to the catalytic converter, the exhaust gases will be heated by the reaction occurring, lose heat to the walls and either be heated or cooled by the surrounding gas. At the manifold collector, the temperature of the mixture decreases due to the heat losses to the wall and the mixing of the exhaust gases with secondary air coming from other cylinders. The worst case for mixing occurs when the secondary air is forced out of the exhaust port during the exhaust blowdown. However, the exhaust gas blowdown also provides relatively hotter exhaust gases. Any design and operating parameters that can prevent heat losses to the walls and improve mixing with a longer residence time in the manifold are effective in enhancing the completion of the thermal oxidation process. The mixing of exhaust gases between cylinders is as critical as that within each cylinder event [33]. As will be presented in Section 4, the detailed design of the exhaust manifold such as runner length, runner orientation and manifold volume are all important parameters for optimization. It is worth noting that the effort to improve mixing around this region is crucial both from the point of thermal oxidation and also of catalytic oxidation since the catalytic oxidation requires a homogeneous mixture inside the catalytic converter where the mixing ceases. When considering the residence time for enhancing thermal oxidation in the manifold, the position for a close-coupled catalyst may no longer be so critical as that for the non-secondary-air system. Instead, placing the catalyst slightly distant from the manifold collector will provide more time for oxidation to occur before reaching the catalyst. This will also alleviate the concern of fast thermal deterioration with the closed-coupled location, extend the life of the catalyst and relax the packaging constraints.

2.4 REQUIREMENTS ON ENGINE ENRICHMENT AND SECONDARY AIR INJECTION QUANTITY

Engine enrichment level and secondary air injection quantity are also two determining parameters that must be carefully examined to maximize the secondary air benefit. The level of engine fueling enrichment determines the compositions of engine-out emissions (reactants) and their concentrations. It was reported that mass emissions in Bag I of the FTP test were consistently higher with less engine enrichment [3]. For a leaner mixture, a relatively warmer exhaust gas can be obtained but it contains much less reactants. As a result, the energy conversion is low even though the oxidation is thermally favorable. Therefore, a relatively leaner in-cylinder mixture is not the favored strategy for secondary air injection [7]. It must be noted that over-enriching the engine may jeopardize combustion stability, which may lead to an engine misfire and subsequently a sharp increase in HC emissions. Regarding the quantity of secondary air, there are two countervailing effects. On one hand, a certain quantity of secondary air is necessary to maximize the reactions in front of the catalyst, which will result in increased temperature. On the other hand, an excess amount of secondary air will cool the exhaust, which may slow down the reactions or in the worst case quench the thermal oxidation inside the exhaust system.

2.5 OTHER ISSUES

In addition to the issues listed in Sections 2.2 ~ 2.4, some other concerns such as air pump noise level, its volume for packaging, weight, system cost, and system durability must also be carefully examined before implementing this technology. Since the engine is fueled with an ultra-rich mixture and a large amount of liquid fuel may enter the cylinder directly, particularly during cold start, some of the rich-combustion-related issues

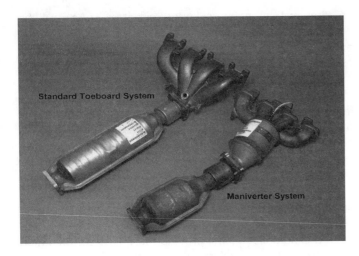

Figure 1   Production and Prototype Exhaust Systems for the 2.4L Dodge Stratus

must be thoroughly evaluated. Those include soot loading on combustion chamber and catalyst, likely oil dilution via repeated and extended fuel-rich operation, and spark plug fouling concern. In addition, thermal loading in the exhaust system and catalyst must be examined to determine the optimal design of exhaust system and catalyst location. Cold start fuel consumption should be minimized.

3. EXPERIMENTAL SETUP AND METHODOLOGY

All of the experiments for this study were conducted on a DaimlerChrysler Corporation 2.4L four-cylinder engine. Some of the 2.4L engine specifications are given in Table I. The engine was installed in a 2001 model year Dodge Stratus with an automatic transmission. The vehicle had a conventional exhaust system downstream of the exhaust manifold. All of the experimental exhaust manifolds had their outlets at the same position as the production manifold. The catalytic converter face is about 450 mm from the cylinder head. The layout of the production exhaust manifold and catalyst are shown in Fig. 1, labeled with "Standard Toeboard System". The system labeled with "Maniverter System" in Fig. 1 is a prototype system that was developed. The cast exhaust manifold part of the "maniverter" system was also tested in this study.

Unless otherwise noted, the temperature and emissions data in this study were collected immediately following a cold start, with the vehicle idling in neutral. While most of the test data of interest were collected within 30 seconds, it was an important part of the test procedure to be sure to fully warm up the engine after every start. If the engine were not fully warmed up, the lube oil would quickly become contaminated. Also, within 30 seconds, one cannot be sure that the liquid fuel in the intake ports

is completely cleared out, which may lead to a residual impact on the subsequent test data by the liquid fuel film.

Table I   Engine Specifications

Displacement	2429 cc
Bore	87.5mm
Stroke	101mm
Compression Ratio	9.4:1
Intake Valve	35.5mm
Exhaust Valve	30.5mm
Firing Order	$1-3-4-2$
Other Features	4 valve/cyl Dual Overhead Cam Cast iron block Aluminum cylinder head

The cold start soak conditions for these tests are the same as those for the FTP-75 test. An accelerated soak period of 8 hours with the hood open and a cooling fan operating was found to produce results consistent with the 12-hour soak required by the FTP-75. This shortened soak time allowed us to speed up testing substantially. All results are averaged over a number of tests with both full and shortened soak times.

K-type thermocouples with a diameter of 1.6 mm were used to monitor the temperatures in the collector of the exhaust manifold, the inlet cone of the catalyst, and 38 mm behind the front face of the first catalyst brick. The latter is the standard location for the measurement of catalyst temperatures within the Chrysler Group. The thermocouple at the collector of the exhaust manifold was placed near the production O2 sensor location. This position, shown in Fig. 2, was chosen since it is designed to equally monitor the exhaust gases from all cylinders.

The HC concentration measurements were conducted with a Cambustion fast FID analyzer. The exhaust gas sample was taken from the collector of the exhaust manifold very near the location of the thermocouple as is schematically illustrated in Fig. 2. The HC concentrations in this paper have been normalized to the HC concentration of the same engine running fully warm

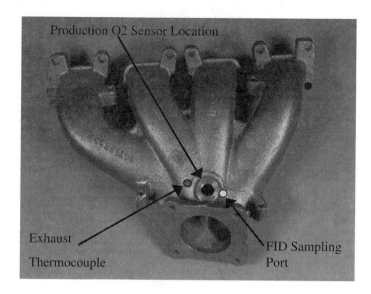

Figure 2   Production Exhaust Manifold with Instrumentation Locations

at 1600 rpm and 2.4 bar BMEP with 30 deg BTDC spark timing.

Throughout this paper two lambda values will be referred to. The first is the engine lambda. This is a calculated value intended to represent the lambda value inside the combustion chamber based upon the amount of air entering through the intake manifold and the amount of fuel injected. The second is the exhaust lambda. This value is measured near the catalyst inlet using a wide-range O_2 sensor. This exhaust lambda is measured in a location where the exhaust gas and the injected secondary air are assumed to be well mixed.

The idle speed of this vehicle has been raised to 1300 rpm while the transmission remains in park or neutral. The nominal spark advance while idling in this condition is about 30 deg BTDC. The spark timing at idle is algorithmically adjusted within a calibratable band in order to stabilize idle speed.

During the test program, a number of parameters were investigated over a wide range in a search for the optimum combination. Table II summarizes these various parameters that were explored.

Table II   Investigated Paramters

Parameters	Ranges
Exhaust Manifold Configuration	5 variations tested: Versions V1 and V2 were designed with a priority for increased residence time through higher volume. Versions M1 and M2 were designed with a priority for exhaust gas mixing through the application of short, opposing runners. Version H1 is a hybrid of the volume and mixng designs. H1 represents a production-viable compromise of the design features found in V1, V2, M1, and M2.
Secondary Air Injection Location	Air injection locations in the head and outside the head in the exhaust manifold are examined.
Secondary Air Flow Rate	Four values of air flow rate between 16 and 38 kg/hr were tested with a constant exhaust lambda of 1.25.
Exhaust Lambda	Six values of exhaust lambda between 1.15 and 1.75 were tested with a constant secondary air mass.
Spark Retardation	Three values of spark retardation from the nominal idle value of 30 deg BTDC were tested. They varied from 0 to 43 degrees.

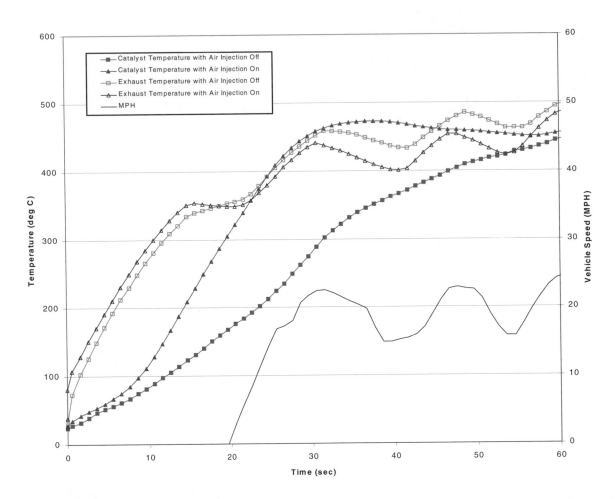

Figure 3 Comparison of Exhaust and Catalyst Temperatures With and Without Secondary Air Injection System Operating

4. RESULTS AND DISCUSSION

4.1 EFFECT OF SECONDARY AIR INJECTION LOCATION

As discussed in Section 2.2, the thermal oxidation reaction is characterized by a dramatic increase in exhaust gas temperature and a corresponding reduction in the HC and CO concentrations of the exhaust gases prior to the catalyst. The catalytic oxidation reaction is characterized by an increase in the rate of catalyst temperature rise when compared with a standard exhaust system. Generally, catalytic oxidation will happen, with air injected into a rich mixture, once the catalyst reaches the light-off temperature. Thermal oxidation only happens when secondary air injection conditions are optimized. Catalytic oxidation will always accompany thermal oxidation. To produce thermal oxidation, the secondary air must be injected as close to the exhaust valve as possible. Since the exhaust gases

cool rapidly after leaving the cylinder, mixing the secondary air into the exhaust gases as soon as possible is crucial. To produce the catalytic oxidation reaction only, the secondary air can be injected further downstream as long as sufficient mixing of the secondary air and exhaust gases can take place prior to catalyst entry. Both types of applications have been used successfully in production to meet various manufacturers' specific goals.

The strategy of injecting air downstream from the exhaust ports, was used in 2000 model year production vehicles by several manufacturers. Some brief testing of one of those vehicles was conducted to compare the effect of the secondary air injection system on exhaust system temperatures. The results of the exhaust and catalyst temperature measurements are shown in Fig. 3. The data show that injecting air downstream of the exhaust port has almost no effect on the exhaust temperatures around the collector of the exhaust manifold. This strategy does have a dramatic effect on catalyst temperature. The catalyst temperature rises significantly after 20 seconds of the FTP-75 test as

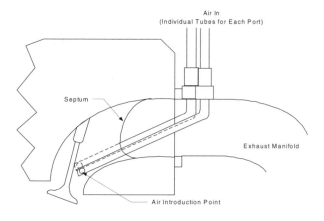

Figure 4 Configuration of Early Secondary Air Injection Experiments

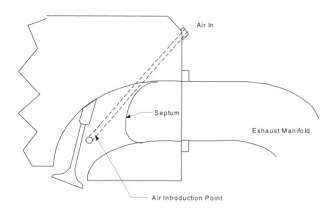

Figure 5 Design of Prototype Cylinder Head

compared to the test with the secondary air disabled. The testing clearly confirmed that injecting secondary air downstream from the hot areas around the exhaust port does still have the capability of improving catalyst light-off performance via catalytic oxidation. This injection strategy misses the opportunity to enhance thermal oxidation of engine-out emittants which can further speed up catalyst light-off and drastically reduce the emissions entering the converter.

To maximize catalyst light-off performance and minimize the converter-in HC emissions, significant efforts have been directed toward the many parameters associated with thermal oxidation. The first task in this study was to determine the optimal location for secondary air injection. The initial versions of air injection were tubes directed straight towards the valve, similar to the sketch in Fig. 4. It was found that the oxidation reaction in the exhaust gases was not achieved with this design. Subsequently, a sparger-type design was constructed. This design had

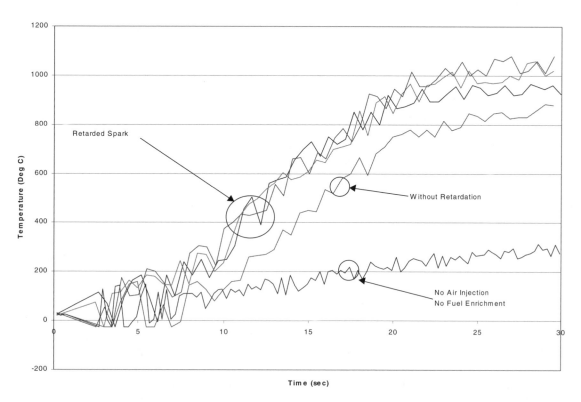

Figure 6 Time Histories of Catalyst Temperature at Varying Levels of Spark Retardation

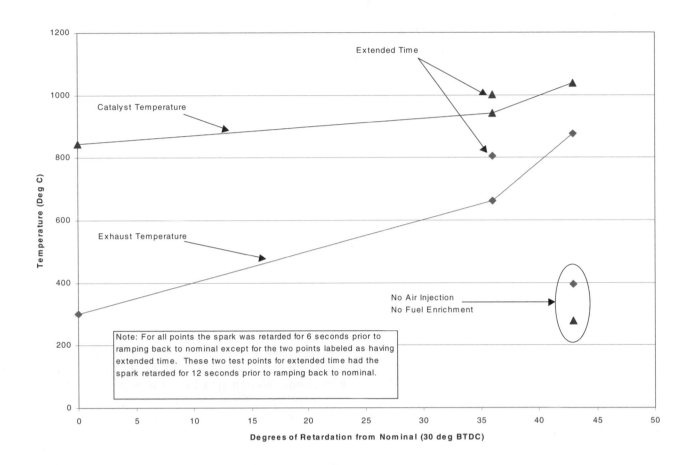

Figure 7   Effect of Spark Retardation on Exhaust Gas and Catalyst Temperatures

a closed end of the tube and a hole drilled on the side of the tube such that the secondary air flows out of the tube perpendicular to the exhaust flow. This design resulted in successful thermal oxidation. It was considered that the pressure from the exhaust gas during the blowdown phase of the exhaust process was enough to shut off the flow of air out of the tube and leave the exhaust gases largely unmixed with secondary air[8]. Changing the air outlet to the side allows the secondary air and exhaust gases to mix better during the blowdown phase. This study reinforced the importance of secondary air mixing with the exhaust gases. This is particularly important during the blowdown phase where the exhaust gases are hottest and most likely to initiate the reaction.

It is concluded that the air injection location is one of the keys to determining whether the thermal oxidation reaction will take place in the exhaust gases or not. For the reaction to be initiated, the secondary air must meet the reactants (H_2, CO, and HC) in an environment above 750°C. Since the exhaust gases cool rapidly as they pass through the exhaust port, it is important to mix air at a point where this high temperature is available. Once the oxidation reaction is started, it produces a tremendous amount of heat and is self-sustaining. From this study, a prototype cylinder head with a secondary air passage design, as illustrated in Fig. 5, was developed. The rest of the study was based on this prototype cylinder head.

4.2 EFFECT OF SPARK RETARDATION

Spark retardation is a widely used technique for raising exhaust gas temperatures. Higher exhaust gas temperature enhances the initiation of the thermal oxidation reaction. The time histories of the catalyst temperatures with different spark retardations are shown in Fig. 6. Clearly, the advantage with the spark retardation is that hotter exhaust gases, resulting from thermal oxidation, heat the catalyst more quickly to the catalyst light-off temperature. This figure also shows that once the catalytic oxidation starts, the catalyst can attain the same heating rate as the other cases even without the thermal oxidation reaction, but its curve is shifted in time. This indicates that, for a given exhaust lambda, the catalytic oxidation reaction will provide a fixed rate of energy release once the light-off temperature has been reached. This heating rate is probably related to the catalyst cell density, the precious metal loading, and the converter-in mixture quality, which is affected by many other operating variables.

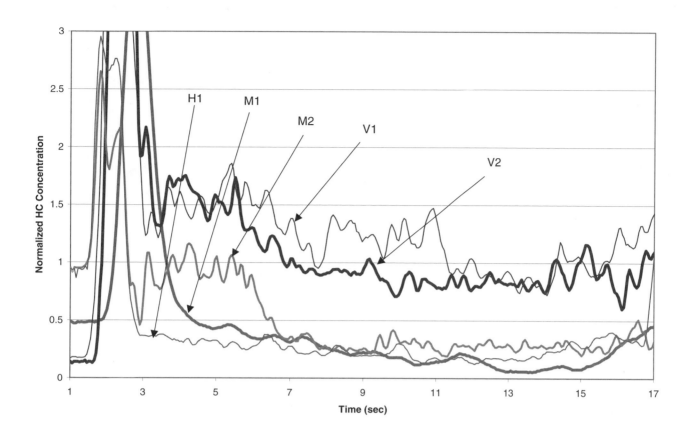

Figure 8 Time Histories of HC Concentration with Various Exhaust Manifolds

Figure 7 plots the exhaust gas and catalyst temperatures measured using various levels of spark retardation. The points in Fig. 7 are the average of the temperatures once they reach their steady state values around 25 seconds in Fig. 6. The points in Fig. 7 with 0 deg of retardation show the exhaust temperatures with air injection and fuel enrichment, but without spark retardation. In this case, the thermal oxidation reaction does not seem to be initiated in the exhaust gases. The data show that the catalytic oxidation reaction was initiated, and the air injection still has a dramatic effect on the catalyst temperature as already shown in Fig. 3.

In order to study the benefit of the period of spark retardation on catalyst light-off, two cases of spark retardation for 36 deg were examined. Both cases are shown in Fig. 7, with each case holding the spark retardation for 6 and 12 seconds, respectively, after which the spark is ramped back toward the normal idle value around 30 deg BTDC. Using the exhaust gas temperature as an indicator of the thermal oxidation reaction, it can be seen that the additional thermal energy added to the exhaust by holding the retarded spark longer does enhance the rate of the oxidation reaction. It was also found that extended use of spark

retardation has a rather smaller benefit on the catalyst temperature increase.

At the spark retardation of 43 deg, the retardation was held for 6 seconds before being ramped back. The extra spark retardation raises the peak exhaust gas temperature and, to a much lesser degree, the catalyst temperature. Unfortunately, HC concentration data accompanying these temperatures are not available. It is believed that the HC concentrations would be lower for 43 deg of retardation than for 36 deg. This lower HC concentration is a result of the higher oxidation rate caused by the higher temperature in the exhaust gases. The higher oxidation rate will consume more of the reactants, leaving less available to react in the catalyst. The higher exhaust gas temperature will also result in greater heat losses along the path to the catalyst. These two causes together explain the greater effect of spark retardation on exhaust gas temperature than on catalyst temperature.

It is concluded that some spark retardation needs to be used in this engine, with this secondary air injection configuration, in order to initiate the thermal oxidation reaction in the manifold. The benefit in improving the reaction rate must be examined carefully against the

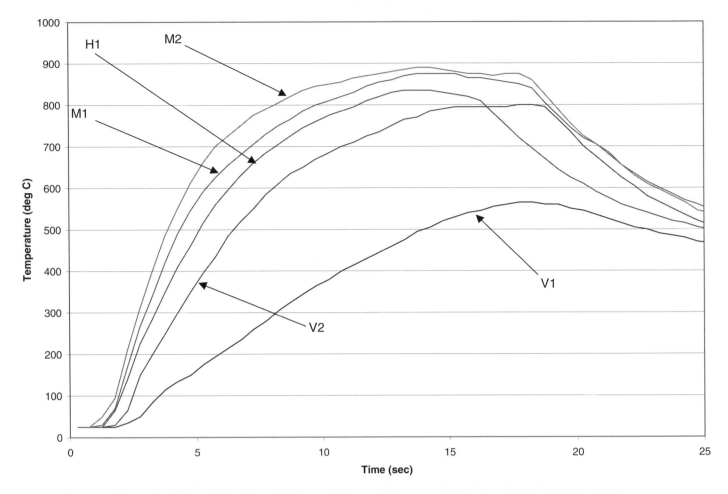

Figure 9   Time Histories of Exhaust Temperatures with Various Exhaust Manifolds

concerns with combustion quality with extreme spark retardation. For further investigations in this study, 36 deg of retardation from nominal held for 6 seconds is used.

4.3 EXHAUST MANIFOLD DESIGN

In addition to air injection location and exhaust temperature, exhaust manifold design is another key factor in the successful implementation of secondary air injection. As discussed in Section 2.3, the exhaust manifold design plays an extremely important role in enhancing the mixing of secondary air and the reactants from the engine. This mixing is crucial in sustaining the reaction. All of the modified versions of the exhaust manifold are based on either the 2001 production manifold or the "maniverter" manifold, as shown in Fig. 2. Figures 8 – 10 show the time histories of HC emissions, measured temperature at the exhaust manifold collector, and measured temperature inside the converter for various versions of exhaust manifold design. The measurements were conducted with the same air injection strategy to obtain the same exhaust lambda for various manifold designs. Figure 11 shows the

comparison of temperature and normalized HC emissions for various manifold designs. The temperature and HC emissions shown in this figure are the values from the time history averaged over the steady-state portion of the curves in Figs. 8 – 10.

Two important factors in reducing catalyst-in emissions would be residence time and exhaust gas mixing, both of which can be influenced significantly by exhaust manifold design. The first modifications (V1 and V2) were focused on increasing the volume of the manifold in order to increase the residence time. This will allow more opportunity for the gases to mix and the oxidation reaction to take place. Also, the larger volume will reduce the amount of quenching of the reaction when the exhaust gases contact the manifold wall. Another goal of these designs was to promote mixing by shortening the runners to improve cross-talk between cylinders. Exhaust manifold V1 was created by shortening the runners of the production exhaust manifold to the extent possible while still maintaining manufacturing tool clearance. V1 had the two middle runners connected together and the length of the two outer runners shortened. For the V2 design, the V1 manifold was used

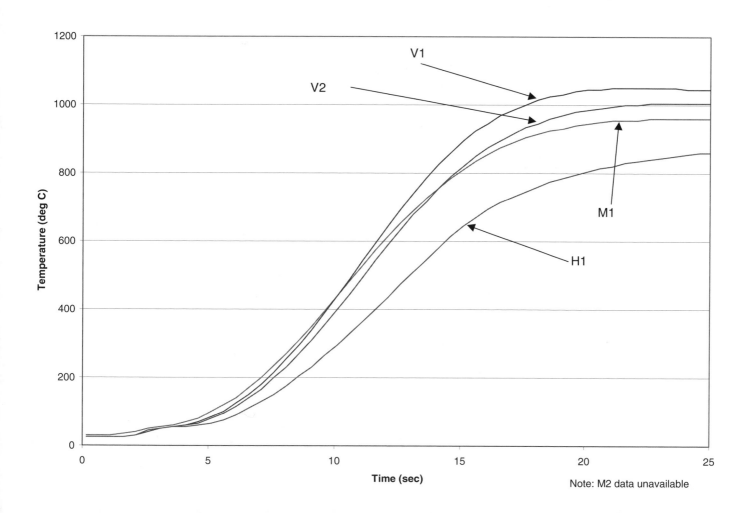

Figure 10   Time Histories of Catalyst Temperatures with Various Exhaust Manifolds

and a large volume was added to the concave side of the manifold to increase the volume. When comparing the performance between manifolds V1 and V2, it is clear that some reduction in the HC concentration was achieved with the V2 manifold. In addition, the V2 manifold produced a dramatic increase in the exhaust gas temperature along with a decrease in the catalyst temperature. The exhaust gas temperature increase can be attributed to the longer residence time, which allows for more thorough mixing of the exhaust gases and more time for the thermal oxidation reaction to complete. The drop in catalyst temperature is a result of more reactants being consumed in the manifold and less being available to the catalyst for the exothermic reaction there. This confirms the trade-off between thermal oxidation in the manifold and the catalytic oxidation in the catalytic converter, as discussed early in Section 2.2. An enhanced thermal oxidation in the manifold will consume more reactants that would otherwise be available to heat the catalyst directly through the catalytic oxidation inside the converter.

The other design path to be explored was the effect of mixing. An exhaust manifold with an integral catalyst, shown in Fig. 1 as part of the "maniverter" system, was tested. With the outer two exhaust runners directly opposed to each other, the opportunity for good mixing seems to be high. Therefore, a fabricated version of this "maniverter" was made. This manifold maintained the same basic shape, but the exhaust runners were increased in diameter to increase the overall volume. This manifold is labeled as M1. M1 also had the integral catalyst replaced by a short length of empty catalyst can. As shown in Figs. 8 – 11, this manifold showed a tremendous reduction in the HC concentrations while maintaining good exhaust and catalyst temperatures. Though it has less volume than that of the V2 manifold, it is believed that it produced better exhaust gas mixing than either of the increased volume configurations (V1 and V2).

Another investigation was conducted on a modified version of the cast iron "maniverter" as shown in Fig. 1. This version has less volume than that of the M1 manifold, but essentially the same geometry. This

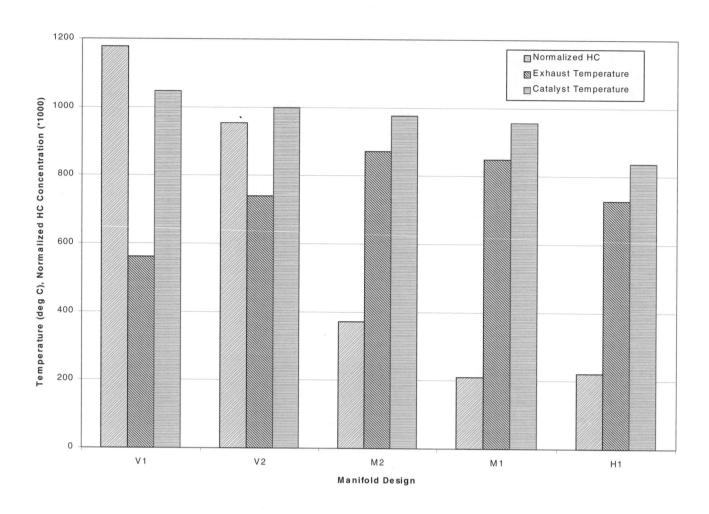

Figure 11 Effect of Exhaust Manifold Design on HC Concentrations and Temperatures

manifold is labeled by M2. The M2 manifold was the cast iron part of the "maniverter" exhaust system with the two center runners connected together and some additional volume added to the concave side of the manifold. It shows higher HC concentrations than the M1 design with similar exhaust temperatures. Unfortunately, catalyst temperature data is not available for this exhaust manifold. It is expected that higher HC concentrations would have lead to higher catalyst temperatures as compared to the M1 manifold. The last manifold tested is a hybrid design. This new design is made of cast iron. It incorporates a high volume and a design feature to promote mixing without a substantial penalty in power. The results for this manifold are labeled by H1. This manifold shows much-reduced HC emissions, but neither the exhaust nor the catalyst temperatures are as high as that for the M1 manifold.

As discussed previously, there is a trade-off between low HC concentrations in the exhaust gases and the rate at which the catalyst heats up. With less reactants entering the catalyst, there is less energy conversion in the converter in its exothermic reaction. From this section it

is very clear that the design of the manifold can have a major effect on the degree of HC concentration reduction as well as on the exhaust temperature.

4.4 EFFECT OF EXHAUST LAMBDA

To quantify the effect of the exhaust lambda, a number of tests were conducted with the H1 exhaust manifold. The injected air mass and injection locations were held constant for these tests. Changing the engine lambda varies the exhaust lambda. All data was collected in idle neutral with the soak conditions described earlier. The exhaust and catalyst temperatures were allowed to reach their steady state values during the test. Likewise, the HC concentrations would reach a steady state value during the test. The time histories of temperatures and HC concentrations are shown in Fig. 12 for the test condition of exhaust lambda = 1.25. The time history clearly shows the events that occur during the cold start. First, the engine is started just after 5 seconds. Immediately, there is a spike in the HC concentration associated with the fuel enrichment required for good startability. Then, about 8 seconds, the thermal oxidation

298

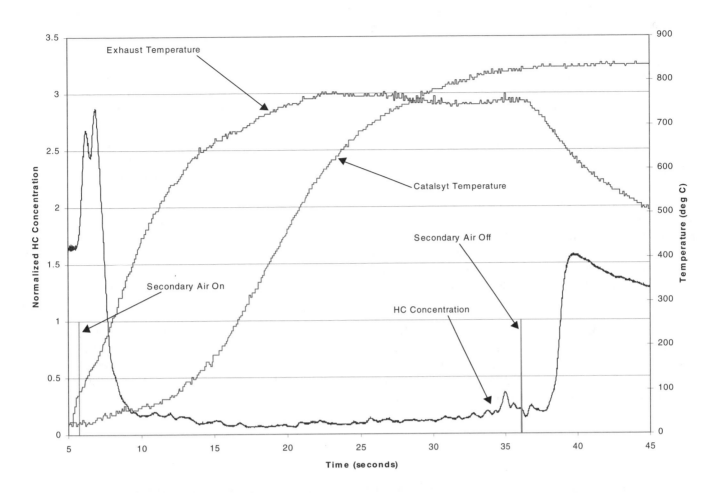

Figure 12 Time Histories of Temperatures and HC Concentrations at the Exhaust Lambda of 1.25

begins to dominate and the HC concentration drops dramatically. This low value is held throughout the idle until approximately 38 seconds when the air and excess fuel enrichment are cut off.

The steady state values of HC concentration and temperature were time-averaged and are shown in Fig 13. The exhaust and catalyst temperatures increase linearly as the exhaust lambda decreases. Since the air mass injected is constant, the engine lambda changes proportionately with the exhaust lambda. As the engine lambda decreases, more reactants are available for enhancing temperature increases. The effect of exhaust lambda on HC concentration is definitely non-linear, showing a minimum around the lambda of 1.25. For a mixture leaner than the optimum, there are not enough reactants produced to match the injected air. For a mixture richer than the optimum, the engine-out HC dominates the trend.

4.5 EFFECT OF SECONDARY AIR FLOW RATE

Finally, an evaluation was conducted on the effect of air flow rate. Using the same setup as for the study of

exhaust lambda, a valve was installed in the secondary air injection plumbing in order to restrict the air flow. Using this set up, measurements were conducted at the optimum exhaust lambda (1.25) observed previously, but at various air flow rates. Constant exhaust lambda with various air flow rates required the engine lambda to change accordingly. The time histories for these tests look very similar to those shown in Fig. 12 and are not plotted here. The average steady state values of exhaust temperature, catalyst temperature, and HC concentration are plotted in Fig. 14. The HC concentration was averaged over the steady-state period of the test and the results are normalized. It can be seen that the HC concentration steadily increases as air flow rate decreases. It is speculated that the trend toward a lower HC concentration with increasing air flow will continue to a point where the engine lambda required to maintain the exhaust lambda is so low as to cause rich misfire.

The temperatures shown in Fig. 14 do not show the same linear response to the air flow increase. As the air flow increases, the temperatures increase as expected up to a point, and then begin to decrease. This decrease is likely due to the fact that the extra air mass injected to maintain the constant exhaust lambda cools the mixture and slows down the reaction. It would be interesting to evaluate the rate of decrease as the air flow rate is further increased. Eventually, the HC concentrations are expected to start increasing as well, which was not observed in the air flow range tested.

5. CONCLUSION

This study was initiated to investigate the thermal and chemical processes associated with secondary air injection inside the exhaust system in order to maximize the simultaneous benefit of improving catalyst light-off performance and reducing converter-in emissions. Some of the conclusions derived from this study are summarized as follows:

1. Both thermal and catalytic oxidation reactions of CO, HC and H_2 with secondary air injection affect the catalytic converter light-off performance. Catalytic oxidation alone can be used to improve catalyst light-off performance, but this strategy gives up many of

the benefits that come with thermal oxidation.

2. Promoting thermal oxidation in the manifold is effective in raising exhaust gas temperature and is also effective in lowering the converter-in HC concentration. An improvement in the thermal oxidation reaction may degrade the catalytic oxidation reaction due to the increased consumption of reactants upstream of the catalyst. There is a trade-off between the performance of thermal oxidation and catalytic oxidation reactions.

3. Mixing is the key to success of the secondary air injection strategy. Proper mixing immediately downstream of the air injection location is critical. The optimal secondary air quantity is highly dependent on the mixing quality.

4. Exhaust manifold design has a major influence on the performance of the thermal oxidation reaction and the subsequent catalytic oxidation. Enhancing mixing and increasing residence time are the design goals.

5. The exhaust lambda must be optimized in order to minimize the emissions entering the catalyst. The amount of air injected influences the catalyst-in emissions and temperatures by increasing engine enrichment (and therefore, increased production of

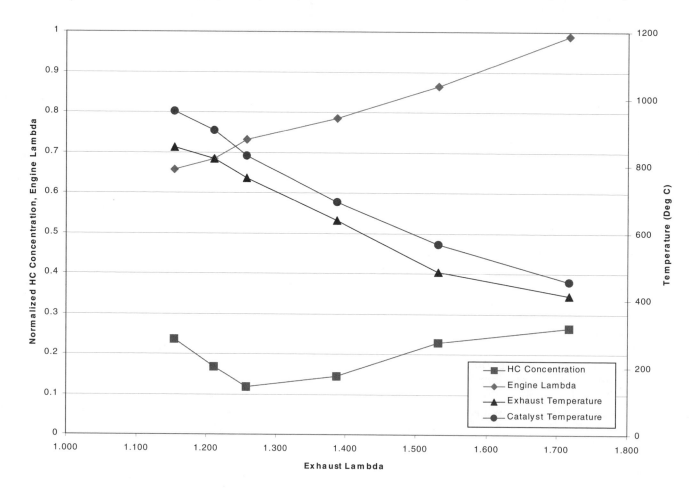

Figure 13   Effect of Exhaust Lambda on Exhaust Temperature, Catalyst Temperature, and HC Concentrations

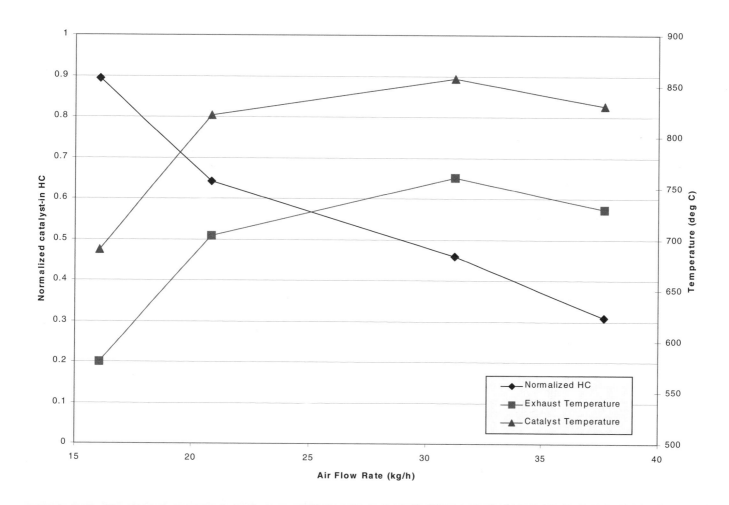

Figure 14 Effect of Secondary Air Flow Rate on Exhaust Temperature, Catalyst Temperature, and HC Concentrations

reactants) while maintaining the optimum exhaust lambda.

6. Spark retardation may be required to initiate the thermal oxidation reaction. Holding retarded spark for an extended period of engine operation will add more energy into the exhaust gases and enhance thermal oxidation in the exhaust manifold.

REFERENCE

1. Adamczyk A. *et al.*, Experimental and modeling evaluations of a vacuum-insulated catalytic converter, *SAE Technical Paper*, No. 1999-01-3678 (1999).
2. Ball D. *et al.*, Ultrathin wall catalyst solutions at similar restriction and precious metal loading, *SAE Technical Paper*, No. 2000-01-1844 (2000).
3. Crane M. *et al.*, Reduced cold-start emissions using rapid exhaust port oxidation (REPO) in a spark ignition engine, *SAE Technical Paper*, No. 970264 (1997).
4. Eade D. *et al.*, Fast light-off of underbody catalysts using exhaust gas ignition (EGI), *SAE Technical Paper*, No. 952417 (1995).
5. Heimrich M., Air injection to an electrically-heated catalyst for reducing cold-start Benzene emissions from gasoline vehicles, *SAE Technical Paper*, No. 902115 (1990).
6. Heimrich. *et al.*, On-board hydrogen generation for rapid catalyst light-off, *SAE Technical Paper*, No. 2000-01-1841 (2000).
7. Hernandez J., A study of the thermochemical conditions in the exhaust manifold using secondary air in a 2.0 L engine, *SAE Technical Paper*, No. 2002-01-1676 (2002).
8. Herrin R., The importance of secondary air mixing in exhaust thermal reactor systems, *SAE Technical Paper*, No. 750174 (1975).
9. Herrin R., Lean thermal reactor performance characteristics - a screening study, *SAE Technical Paper*, No. 760319 (1976).

10. Herrin R., Emissions performance of lean thermal reactors - effects of volume, configuration, and heat loss, *SAE Technical Paper*, No. 780008 (1978).

11. Heywood J., Internal Combustion Engine Fundamentals, McGraw-Hill, 1988.

12. Hummel K.-E. *et al.*, Secondary air charger – an innovative system to reduce cold start emissions, *Proceedings of the 10th Aachen Colloquium – Vehicle and Engine Technology*, 2001.

13. Inoue T. *et al.*, Challenge for the cleanest car – part 2: tailpipe emission reduction, *JSAE Technical Paper* (in Japanese), No. 20005188 (2000).

14. Kadlec R. *et al.*, Limiting factors on steady-state thermal reactor performance, *SAE Technical Paper*, No. 730202 (1972).

15. Kanazawa T. *et al.*, Development of the automotive exhaust hydrocarbon adsorbent, *SAE Technical Paper*, No. 2001-01-0660 (2001).

16. Katashiba H. *et al.*, Development of an effective air-injection system with air for LEV/ULEV, *SAE Technical Paper*, No. 950411 (1995).

17. Kemmler R. *et al.*, Current status and prospects for gasoline engine emission control technology – paving the way for minimal emissions, *SAE Technical Paper*, No. 2000-01-0856 (2000).

18. Kirwan J. *et al.*, Advanced engine management using on-board gasoline partial oxidation reforming for meeting super-ULEV (SULEV) emissions standards, *SAE Technical Paper*, No. 1999-01-2927 (1999).

19. Kochs M. *et al.*, Innovative secondary air injection systems, *SAE Technical Paper*, No. 2001-01-0658 (2001).

20. Koehlen C. *et al.*, Optimization of catalyst light off by investigating motion of the post oxidation flame and its dependence from combustion and manifold geometry, *SAE Technical Paper*, No. 2002-01-0744 (2002).

21. Kollmann K. *et al.*, Secondary air injection with a new developed electrical blower for reduced exhaust emissions, *SAE Technical Paper*, No. 940472 (1994).

22. Lafyatis D. *et al.*, Ambient temperature light-off aftertreatment system for meeting ULEV emissions standards, *SAE Technical Paper*, No. 980421 (1998).

23. Leonard L., Fuel distribution by exhaust gas analysis, *SAE Technical Paper*, No. 376A (1961).

24. Lord H. *et al.*, Reactor studies for exhaust oxidation rates, *SAE Technical Paper*, No. 730203 (1973).

25. Nishizawa K. *et al.*, Development of new technologies targeting zero emissions for gasoline engines, *SAE Technical Paper*, No. 2000-01-0890 (2000).

26. Nishizawa K. *et al.*, Development of second generation of gasoline PZEV technology, *SAE Technical Paper*, No. 2001-01-1310 (2001).

27. Oakley A. *et al.*, Feasibility study of an online gasoline fractionating system for use in spark-ignition engines, *SAE Technical Paper*, No. 2001-01-1193 (2001).

28. Sim H. *et al.*, Effect of synchronized secondary air injection on exhaust hydrocarbon emission in a spark ignition engine, *Proc Instn Mech Engrs*, Vol. 215, Part D, pp. 557-566 (2001).

29. Son G. *et al.*, A study on the practicability of a secondary air injection for emission reduction, *SAE Technical Paper*, No. 1999-01-1504 (1999).

30. Ueno M. *et al.*, A quick warm-up system during engine start-up period using adaptive control of intake air and ignition timing, *SAE Technical Paper*, No. 2000-01-0890 (2000).

31. Yamada T. *et al.*, Technologies for reducing cold-start emissions of V6 engine, *Proceedings of the JSAE Spring Convention* (in Japanese), No. 1997-5, pp. 309-312 (1997).

32. Yamamoto T. *et al.*, Reduction of cold-start emissions by combination of new-type catalytic converter and secondary air injection system, *Proceedings of SAE-Japan Convention* (in Japanese), No. 54-00, Paper No. 20005121, pp. 17-20 (2000).

33. Yamamoto S. *et al.*, Useful combustion in cylinder during exhaust stroke and in exhaust port with gasoline direct injection, *Proceedings of COMODIA*, pp. 187-192 (2001).

34. Yasui Y. *et al.*, Secondary O_2 feedback using prediction and identification type sliding mode control, *SAE Technical Paper*, No. 2000-01-0936 (2000).

CHAPTER 7

EFFECT OF FUELS AND FUEL REFORMING ON COLD-START PERFORMANCE

James A Eng
General Motors Research

7.1 OVERVIEW

The main reason for the increase in engine-out HC emissions during a cold-start is that only a fraction of the fuel vaporizes at low temperature. Equilibrium calculations indicate that only 10-20% of the fuel vaporizes during the first few cycles of a cold-start. As a result, the engine must be significantly over fueled in order to create a combustible mixture within the cylinder. This over fueling results in large amounts of liquid fuel entering the cylinder that are a major source of HC emissions. Significant reductions in HC emissions could be achieved if the engine could be operated closer to stoichiometry during a cold-start.

Mixture preparation during a cold–start is affected by both the fuel injection system and the vaporization properties of the fuel. While improved fuel atomization does not eliminate the problem of fuel vaporization, better atomizing injectors do dramatically reduce the amount of fuel that is deposited on the walls of the intake port. By maintaining more of the fuel in the air, rather than on the intake port wall, the engine can be operated with less over-fueling. Engine warm-up tests have indicated that the effect of the fuel on HC emissions is similar in magnitude to the effect of the fuel injector. The effects of fuels on cold-start HC emissions and potential means of reforming the fuel for lower HC emissions are discussed in the following.

Gasoline Properties - A fully-blended gasoline is composed of hundreds of hydrocarbon compounds with different molecular structures and boiling points. The composition of gasoline varies widely around the world and both regionally and seasonally within the U.S. In the U.S. a typical non-oxygenated gasoline is composed of roughly 60% paraffins (single C–C bond), 30% aromatics (benzene ring), and 10% olefins (double C=C bond). Oxygenates have been blended into fuels for many years in order to improve the octane rating of gasoline. The addition of oxygenates also causes an emissions reduction in older vehicles without closed-loop emissions control and exhaust gas aftertreatment due to the lean-shift in the air/fuel ratio. While the largest emissions reduction obtained from the addition of oxygenates is in CO emissions, there is also an HC emissions benefit. Federal regulations mandate that reformulated gasolines contain oxygen levels between a minimum of 1.5% to a maximum of 2.7% by weight.

The volatility of a fuel is characterized by its vapor pressure and boiling range, or distillation curve. The volatility of a fuel is commonly characterized by the Reid Vapor Pressure (RVP) and the ASTM distillation curve. The RVP is the pressure of an equilibrium fuel/air mixture in a test apparatus at a temperature of $37.8^\circ C$ ($100^\circ F$). EPA regulations established in 1992 restrict the maximum summer RVP in northern states to 62 kPa (9 psi) and in the southern states to 54 kPa (7.8 psi). The distillation curve for a fuel is determined using ASTM Procedure D86. A 100 mL sample is distilled under specified conditions at atmospheric pressure using a controlled temperature history for the fuel. Important parameters determined from the distillation curve are the initial and final boiling points, and the temperatures for 10%, 50% and 90% fuel evaporation. Typical ranges for T_{10} are 40 to $60^\circ C$, T_{50} from 90 to $110^\circ C$, and T_{90} from 160 to $180^\circ C$.

The cold weather driveability of a gasoline is defined in terms of whether it will start easily, idle smoothly, and have a good acceleration with no hesitations under cold ambient conditions. Good volatility of gasoline under cold ambient conditions is crucial for avoiding driveability problems. However, too much vaporization under hot ambient conditions can result in hot weather driveability problems. The driveability of a fuel is the result of a complex interaction between the vehicle, engine, fuel handling system, and volatility characteristics of the fuel. The driveability of a fuel under cold operating conditions has a direct influence on cold-start HC emissions under real world operating conditions. The fuel parameters that influence driveability are not simple and can vary widely from vehicle to vehicle. In the U.S. one of the most commonly used indexes is the driveability index (DI). The DI is calculated from the ASTM distillation curve by the equation $DI = 1.5\,T_{10} + 3\,T_{50} + T_{90}$. The DI for a typical U.S. gasoline ranges from 1000 to 1300°F (540 to 700°C).

Reformulated gasoline (RFG) was introduced into the market in order to reduce both tailpipe and evaporative emissions from vehicles. Reformulated gasoline was first introduced by ARCO in September 1989 in southern California. It was intended to reduce the emissions from older vehicles that were not equipped with catalytic converters. The emission reductions were obtained by lowering the RVP to reduce evaporative emissions, and reducing aromatics and olefins to reduce the ozone forming potential and generate lower levels of toxins in the exhaust gas, such as benzene. Oxygenates were added to the fuel to reduce CO and HC emissions.

Effects of Fuels on HC Emissions - Vehicle tailpipe emissions are a complex interaction between the engine design, emissions control system, ambient temperature, and fuel properties. The overall impact of the fuel on emissions is small relative to the effect of the vehicle design and ambient temperature. Test results of cold-start emissions obtained from a range of vehicles and fuels indicate that the vehicle design has the single largest effect on HC emissions. At the LEV and TLEV emissions level, the effect of the vehicle design on emissions is roughly 7-10 times larger than that of the fuel. However, from a clean air standpoint the effect of fuels is significant since it affects all vehicles that are currently on the road.

The fuel DI has a direct impact on tailpipe HC emissions. While HC emissions are essentially independent of DI for levels below 1200°F, for DI increases above this amount the HC emissions increase more or less linearly. It has been proposed that the EPA should develop legislation to limit gasoline DI values to below 1200°F as an effective means to reduce emissions from the entire vehicle fleet. As the HC emission standards are reduced to ULEV and SULEV levels the effect of fuels on emissions will increase. In FTP tests performed with a ULEV-certified vehicle, the first cycle HC emissions increased 45% by changing the fuel from California Phase II gasoline with a DI of 1150°F to a 1280°F DI fuel.

On-Board Fuel Reformers – The poor volatility of gasoline at low temperatures represents a significant barrier to reducing cold-start HC emissions. All of the control strategies and improved fuel injection systems have been developed as means of dealing with the problem of obtaining a combustible mixture during the first few engine cycles of the cold-start. As an alternative, if the fuel could be prevaporized before entering the engine it is expected that mixture preparation would improve, resulting in significant reductions in engine-out HC emissions. One potential means to prevaporize the fuel is through the use of on-board fuel reformers.

Fuel reformers partially oxidize gasoline into reformate containing high concentrations of CO and H_2, with only small concentrations of HC. The reformate is then used to cold-start the engine rather than liquid gasoline. After the engine is started and the catalyst reaches its light-off temperature the engine is transitioned into operation on liquid fuel. A distinct advantage of starting the engine on reformate is that the engine will be more robust to changes in fuel volatility. Recently there has been an increased interest in fuel reformers as a means to produce hydrogen for fuel cell vehicles. In the absence of a hydrogen infrastructure, fuel reformers represent a potential near-term solution to generate hydrogen from gasoline on board the vehicle. The side benefit of this to SI engine development is that fuel reformer technologies resulting from fuel cell development work can be directly applied to gasoline-fueled SI engines.

Fuel reformers use one of the following three major processes: 1) steam reforming, 2) partial oxidation (POx) reforming, and 3) autothermal reforming. In a steam reformer the fuel is reacted with water into CO and H_2. The energy required to steam reform a typical hydrocarbon fuel is roughly 1305 kJ per mole of fuel. In a POx reformer the fuel is reacted with air in a fuel-rich environment. POx reforming is exothermic and releases 630 kJ per mole of fuel. Autothermal reforming uses a combination of steam reforming and POx reforming such that the overall reaction is energy neutral. The autothermal reforming process is the most attractive one for vehicles since it minimizes energy losses.

The benefits of cold-starting an engine with reformate has been extensively evaluated using both synthetic reformate and reformate generated from a prototype POx fuel reformer. All of the experiments reported to date have been performed on a single-cylinder engine under a simulated cold-start transient. The control strategy is to start the engine on POx gas during cranking and the initial first idle, and then after catalyst light-off the engine is transitioned to operate on gasoline. Engine-out HC emissions are reduced by 75% relative to those obtained with starting on gasoline. While starting the engine with POx has been shown to result in significant reductions in HC emissions, the challenge now becomes one of "cold-starting" the fuel reformer. The reformer must be brought up to its operating temperature before it will efficiently reform the fuel. This is one of the most significant development issues for fuel reformers that needs to be addressed since there can be significant HC breakthrough before it is at its operational temperature. In order to rapidly heat the reformer catalyst a strategy was developed where a stoichiometric air/fuel mixture was initially supplied to the reformer to preheat the catalyst. After the reformer temperature was increased, the equivalence ratio was adjusted to fuel-rich operation ($\Phi = 2.5$) for reformer gas production. The performance of the reformer is very sensitive to the preheat time, and an important development issue is to shorten the preheat time requirements for the reformer.

An additional potential use of the reformate fuel is to add it to the exhaust gas during the cold-start to reduce the catalyst light-off time. Hydrogen reacts with air over a standard three-way catalyst at room temperature. In addition, it is known that CO reacts at lower temperatures than hydrocarbons over a catalyst when sufficient O_2 is available. Thus, it is expected that adding reformate to the exhaust will increase the rate of catalyst heating. In simulated cold-start tests with POx gas added to the catalyst the time to obtain 50% CO conversion in the catalyst is reduced from 50 seconds with gasoline to 22 seconds with POx gas.

On-Board Fuel Distillation - One of the distinguishing characteristics of gasoline is that it vaporizes over a wide range of temperatures. A typical gasoline vaporizes over a range of temperatures from a low of 30-40°C to a high of 190-200°C. If the low boiling point compounds in gasoline could be separated from the rest of the fuel it would be possible to start an engine at low ambient temperatures with these compounds and avoid the problem of over-fueling the engine to obtain a combustible mixture.

An on-board fractioning system has been developed to separate gasoline into two separate streams with different octane ratings. The higher boiling point compounds in gasoline are largely aromatic species (toluene and other alkylated aromatics) that have higher knock resistance than the lower boiling point straight-chain paraffinic species. The objective of the fuel fractioning system was to obtain improved engine performance by providing a fuel with increased octane rating during high-load operation. The system works by flowing fuel into a temperature controlled vaporization chamber. The fuel is separated into two streams consisting of a vapor fraction with lower boiling point compounds and a heavy fraction with relatively higher boiling point compounds. A copper coil immersed in the fuel was used to heat the liquid in the vaporization chamber. The system was designed to work at steady-state conditions to produce a steady flow of distilled fuel. It was difficult to produce a stable vapor stream from the system at transient conditions, and the system temperatures had to be adjusted to maintain a stable flow. As expected, the light fraction is composed primarily of paraffinic compounds while the heavy end is higher in aromatics. While the primary objective of the system was to improve full-load performance, a system like this could also be used to reduce cold-start HC emissions. For cold-start operation it will be necessary to develop a storage system to store the distilled fuel on-board so that it is readily available for a cold start.

An On-Board Distillation System (OBDS) was developed to aid in the cold-starting of an E85 fueled vehicle. The distillation unit was designed to address both the cold-starting issues on the FTP as well as starting at low ambient temperatures and improving cold weather driveability. In contrast to the vaporization system discussed above, the OBDS unit was developed specifically for improving vehicle cold-starts and included an additional fuel tank on-board the vehicle to store the distilled fuel for starting the engine. The distillation unit consisted of a heat exchanger, distillation column, vapor condenser and a cold-start fuel tank to store the distilled fuel. A heat exchanger is used to heat the fuel and vaporize the lower boiling point compounds out of the mixture. Engine coolant is used to heat the fuel to a temperature of 85 to 90°C. The distillation column was a vertical tube packed with steel wool to obtain a large surface area for vaporization. The vapor temperature was controlled with a cooling fan. The heated fuel entered the middle section of the column and flows down to the bottom of the column. The important operating parameters are the temperature gradient along the column and the differences in the boiling point temperatures of the different components in the mixture. The fuel flow into the OBDS unit is the return fuel from the engine. A sensor is used to detect the fuel level in the cold-start tank, and a control unit activates the distillation process when the fuel level in the cold-start tank drops below a specified level. A third generation fuel distillation system has been developed and tested using a fully-blended gasoline. A distillation curve of the OBDS processed fuel showed that it is composed primarily of C_5 hydrocarbons (pentanes) that vaporize at temperatures near 30°C.

Cold-start tests have been performed using gasoline and the OBDS system. The cold-start calibration for the vehicle was modified to take full advantage of the improved distillation characteristics of the OBDS processed fuel. In particular, the fueling levels during the crank and the initial speed flare were reduced so that the engine operated closer to stoichiometry during the cold start. A full FTP test was performed with the vehicle to determine the potential of the system to reduce HC emissions over the entire FTP cycle. The calibration was modified so that the catalyst light-off times were reduced from 40 sec to 20-30 sec. Tailpipe hydrocarbon emissions over the FTP were reduced by 40 to 45% using the OBDS system, and CO emissions were reduced by roughly 75%. A breakdown of the emission over the FTP cycle showed that the same percent reductions in engine-out emissions were measured from the cold-start and the first test cycle. This is the same overall level of change in HC emissions as obtained by reducing the fuel DI by 130°F. The OBDS system appears to have a significant potential to obtain large reductions in cold-start HC emissions.

7.2 SELECTED SAE TECHNICAL PAPERS

2003-01-3239

An On-Board Distillation System to Reduce Cold-Start Hydrocarbon Emissions

Marcus Ashford, Ron Matthews, Matt Hall and Tom Kiehne
The University of Texas

Wen Dai, Eric Curtis and George Davis
Ford Research Laboratory, Ford Motor Company

Copyright ©2003 SAE International

ABSTRACT

An On-Board Distillation System (OBDS) was developed to extract, from gasoline, a highly volatile crank fuel that allows the reduction of startup fuel enrichment and significant spark retard during cold starts and warm-up. This OBDS was installed on a 2001 Lincoln Navigator to explore the emissions reductions possible on a large vehicle with a large-displacement engine. The fuel and spark calibration of the PCM were modified to exploit the benefits of the OBDS startup fuel.

Three series of tests were performed: (1) measurement of the OBDS fuel composition and distillation curve per ASTM D86, (2) measurement of real-time cold start (20 °C) tailpipe hydrocarbon emissions for the first 20 seconds of engine operation, and (3) FTP drive cycles at 20 °C with engine-out and tailpipe emissions of gas-phase species measured each second. Baseline tests were performed using stock PCM calibrations and certification gasoline. The OBDS fuel used throughout the test program was derived from certification gasoline.

The key benefits provided by the OBDS fuel were (1) emissions reductions over the FTP drive cycle of >50% for hydrocarbons and ≈73% for CO, (2) decrease in catalyst light-off time >50%.

INTRODUCTION

Estimates vary, but it is generally agreed that in contemporary automobiles 60-80% of all HC emissions occur during the first 2 minutes of engine operation. In PZEV (SULEV warranted to 150K miles + zero fuel-based evaporative emissions) vehicles, the portion of HC emissions due to cold start most likely exceeds 90%. Toyota

their 2003 MY PZEV vehicles occur during the cold start period [1].

The reasons for high cold-start emissions are numerous, but can be traced to two main factors: low fuel volatility and inactive catalysts. The relatively low volatility of gasoline means that more liquid fuel must be injected in order to supply enough vapor for an ignitable mixture [2, 3]. A significant portion of the remainder enters the exhaust, probably due to liquid gasoline that is deposited on in-cylinder surfaces and evaporates slowly [2,4]. The higher engine-out emissions, left unchecked by the catalysts, become high tailpipe emissions; conventional three-way catalysts may not reach "light-off" (50% conversion efficiency) temperature for 30 seconds or more [5].

It follows that significant cold-start HC emissions reductions can be achieved through the use of a high-volatility startup fuel. Ideally, this fuel would be volatile enough to allow stoichiometric or lean starts, thus eliminating the need for startup enrichment. This concept has been proven in gaseous-fueled engines, where no startup enrichment is necessary.

ON-BOARD DISTILLATION SYSTEM (OBDS)

There are numerous technologies being developed to solve the cold-start problem. Most focus on aftertreatment (HC storage and improved catalyst performance); some address engine-out emissions through improved components and sophisticated engine control [1, 6 - 8]; few – notably Delphi's On-Board Reformer [9] – attack the root cause of the problem, the fuel itself. The OBDS is also focused on the problems caused by the low volatility of gasoline during cold start and warm-up.

THEORY OF OPERATION

The On-Board Distillation System (OBDS) attacks the fuel issue by generating a highly volatile fuel for use during starting and warm-up. OBDS was developed to address the cold-start problem in E85-fueled vehicles. The idea was to use distillation to extract the lighter fraction of the gasoline portion of E85 for use as a start/warm-up fuel [3, 10]. OBDS was very effective, generating top scores for emissions and cold-starting in the 1999 and 2000 Ethanol Vehicle Challenge contests.

The OBDS presented here differs from its predecessors only in execution and main fuel; the concept is identical. Figure 1 shows the OBDS process and flow diagram. As before, engine coolant is the heat source driving the distillation process. For cold starts, OBDS supplies the starting fuel to the engine, and switches fuel to gasoline at some point during the warm-up. Detailed descriptions of the modes of operation can be found in References 3 and 10.

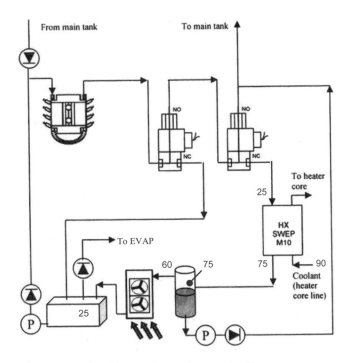

Figure 1. OBDS Flow Diagram (temperatures in °C)

The major difference between this OBDS and earlier versions is in how it perceives its environment. This iteration of OBDS has the ability to communicate with the engine control computer (PCM). Ambient environmental conditions and the state of vehicle operation can be determined directly, without need of additional sensors. This is a sensible development; should OBDS become production hardware, surely it would be controlled by the PCM.

One benefit of this communication with the PCM is in the determination of the time to switch from OBDS fuel to gasoline during a cold-start. OBDS now performs this fuel control.

confident that the wetted engine components (particularly the backs of the intake valves) are sufficiently warmed so that stoichiometric gasoline scheduling is possible without emission penalty.

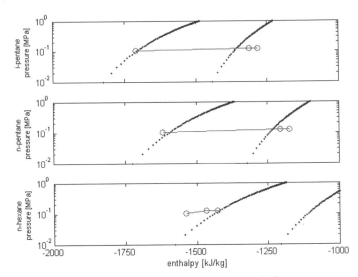

Figure 2. Saturation curves with OBDS process data [13].

Figure 1 shows typical fluid temperatures [°C] within the OBDS for fully-warmed, idling engine operation. Figure 2 depicts the saturation curves of i-pentane, n-pentane and n-hexane, respectively, along with three fluid states along the distillation process. The states are listed in Table 1. Note the large temperature drop from flash tank inlet to flash tank overhead. The layout of the vapor overhead line is such that it ascends a significant vertical run before descending to the condenser. The 60 °C temperature given was measured at the highest point in the vertical run. This is essential, for any vapors that condense above this temperature will flow back into the flash tank. The data presented in Figure 2 offer insight into the composition of the distillate without performing full vapor-liquid equilibrium calculations. Our expectations are that the distillate should be rich in hydrocarbons with boiling points lower than n-hexane.

Table 1. Fluid State Data for Figure 2		
State	Temperature	Pressure
Flash Tank Vapor Space	75 °C	125 kPa
Flash Tank Overhead	60 °C	125 kPa
OBDS Fuel Tank	25 °C	108 kPa

INSTALLATION

Most new emissions reduction technologies have targeted smaller vehicles with smaller displacement engines. Yet more than half of US new car sales are light-duty trucks and SUVs. With comparatively little emissions research being conducted on large-displacement gasoline engines, it was decided that OBDS should be installed and tested on a full-size truck or SUV.

This OBDS was installed on a 2001 Lincoln Navigator donated new from Ford Motor Company and equipped with Ford's 5.4L InTech 32-valve V-8 engine. The engine has bore/stroke of 90.2 x 105.8 mm and compression ratio of 9.5. Exhaust gases from each bank are treated by one close-coupled (CC) catalyst and one dual-brick underfloor catalyst (4 catalyst canisters total). The Navigator was originally calibrated to meet federal Tier I emissions standards. Ford supplied a tool, the Research Console (RCON), that is capable of making PCM calibration changes and performing data acquisition. The RCON has I/O features that are used to communicate with the OBDS. Also supplied by Ford was a break-out box (BOB) that allowed access to all 104 PCM I/O pins. At the time testing began, the vehicle had accumulated approximately 8000 miles.

Two 0.51 mm (0.020 inch) diameter Type-K thermocouples were used to measure left-hand bank exhaust temperatures. One thermocouple was located in the inlet to the CC catalyst; the other was inserted into the CC catalyst brick approximately 25 mm from the face.

Wide-range lambda sensors were mounted in the exhaust pipe leading from each bank, near the locations of the stock oxygen sensors, immediately downstream from the exhaust manifolds.

EXPERIMENTS PERFORMED

Testing proceeded in three phases. The first phase consisted of fuel analyses during OBDS development and again during FTP testing. In Phase II (Calibration Development), the fuel and ignition timing limits of OBDS fuel were explored. FTP testing verified the effectiveness of promising Phase II calibrations.

PHASE I: FUEL CHARACTERIZATION

During development of this iteration of OBDS, the quality of the fuel generated was quantified by determination of its distillation curves. The parent fuel used for initial development was commercially available 93-Octane pump gasoline, a summertime Central Texas blend. Distillation curves were generated per ASTM D86 in University of Texas (UT) laboratories.

The parent fuel used for Phases II and III of this development program was EEE certification gasoline, supplied by Haltermann Products. This same certification fuel was used for all baseline testing. Detailed distillation and compositional analyses were performed of the EEE parent fuel, the OBDS daughter fuel, and the residual fuel. Residual fuel is the bulk fuel remaining in the main fuel tank. As OBDS removes light fractions, the volatility of the bulk parent fuel decreases. These tests, which provided distillation curves (ASTM D86), detailed composition (ASTM D5134) and research/motor octane numbers (ASTM D2699M, D2700M), were performed Southwest Research Institute (SwRI).

PHASE II: CALIBRATION DEVELOPMENT

Cold start tests were performed after overnight soaks. These tests were intended to mimic the initial seconds of the FTP drive cycle. No chassis dynamometer was available for driving, so data were taken for only 20 seconds, with a shift into drive at 15 seconds, as per the FTP. The vehicle was maintained at 20 °C for the overnight soaks, which were approximately 18 hours in length. The cold starts were performed at 20 °C.

Engine and vehicle data were recorded at 100 Hz by the RCON. Real-time tailpipe hydrocarbon emissions were measured via the UT-developed Fast-Spec. The Fast-Spec uses the principle of IR absorption to measure the mass density of total hydrocarbons (THC) in its chamber [11, 12]. Measurements were made at the tailpipe exit, with great care taken to prevent the entry of condensed water into the test chamber. Fast-Spec data, after processing, had a time resolution of 100 Hz. Certain parameters were recorded by both the Fast-Spec and the RCON to ease time synchronization between data sets.

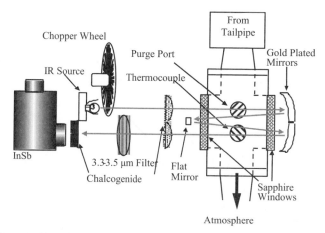

Figure 3. Fast-Spec Layout

The goals of these tests were to discover the benefits and limitations of OBDS fuel. "Fuel-only" calibrations explored the lean-limits of cranking and warm-up, with no changes to stock spark timing. "Retarded-spark" calibrations explored the extent to which warm-up spark retard could be applied to bring about faster catalyst light-off when using the volatile OBDS fuel. All tests were performed with regard to idle quality and hydrocarbon emissions levels.

These tests were performed at the General Motors Foundation Automotive Research Laboratory at The University of Texas.

PHASE III: CALIBRATION VALIDATION

FTP testing was conducted primarily to verify initial findings with industry-standard testing methods. FTP testing also allowed observation of OBDS characteristics during a controlled drive. Any changes in parameters such as fuel economy or emissions would be noteworthy. Each test was performed at 20 °C after an EPA UDDS (FTP Bags 1 & 2) pre-conditioning cycle and 20 °C overnight soak. Engine-out and tailpipe emissions of gas phase species were measured continuously (second-by-second) and also from bag samples. A CVS system was in use for all testing. Engine and vehicle data were recorded by the RCON at 10 Hz.

The OBDS fuel and calibration were in use for the first 20 seconds (approximately) of operation after engine start. The switch from OBDS fuel to gasoline was commanded when the vehicle transitioned into closed-loop fuel control. To ease the changeover to gasoline operation and to eliminate the chance of fuel pressure loss, the main fuel pump (gasoline) was activated for 2 seconds prior to deactivation of the OBDS fuel pump.

Calibration validation, via FTP and Cold 505 (FTP Bag 1) testing, was conducted at SwRI, overseen by investigators from UT.

PHASE I. FUEL CHARACTERIZATION

Fuel analyses are presented in two sections, the first section covering distillation tests performed during OBDS development. The second section summarizes more comprehensive test results from samples collected at the conclusion of Calibration Validation.

OBDS DEVELOPMENT

Distillation curves of OBDS fuel and its parent gasoline (93 Octane [(R+M)/2] pump gas) are shown in Figure 4. Shown on the same graph are the boiling points for i-pentane, n-pentane and n-hexane. In Figure 5 the compositions of distillate and parent are characterized by boiling point range. The data indicate that the OBDS fuel is rich in C_5's, as expected. Informal analysis not presented here detected a strong presence of pentane in the OBDS fuel.

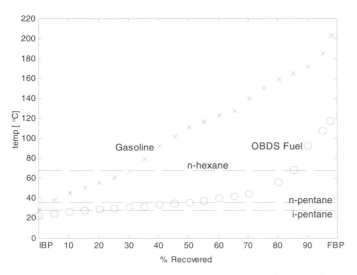

Figure 4. Distillation Curves. OBDS Distillate with Pump Gasoline Parent

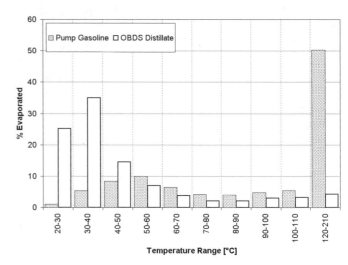

Figure 5. Fuel Evaporation by Temperature Range. OBDS Distillate with Pump Gasoline Parent

DETAILED ANALYSES

Figure 6 shows distillation curves for a sample of OBDS Distillate (×) and its Federal Certification Fuel parent (○). For comparison, the OBDS Distillate (◇) and Pump Gas Parent (□) distillation data from Figure 4 are included. Immediately apparent is the fact that the pump gasoline-based OBDS distillate contained significantly more light-ends than the certification fuel-based OBDS distillate. This can be attributed to differences in parent fuel composition (more light ends in pump gasoline than certification fuel) and distillation conditions. The pump-gas OBDS fuel was distilled at 72 °C with ambient temperature of 28 °C., whereas the certification-fuel OBDS distillation temperature was 80 °C with 40 °C ambient.

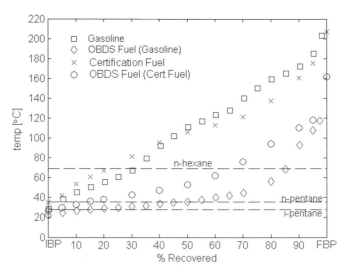

Figure 6. Distillation Curves. OBDS Distillates with Federal Certification Fuel and Pump Gasoline Parents

Distillation curves for the certification fuel parent, the OBDS distillate and residual fuels are shown in Figure 7, with boiling point ranges given in Figure 8. Composition by carbon number is shown for the certification fuel parent and the OBDS distillate in Figure 9. The distillation curve for the residual fuel in the region below T_{50} reflects the transfer of light ends to the distillate.

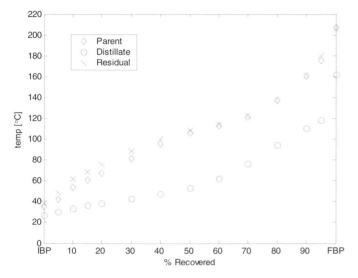

Figure 7. Distillation Curves. OBDS Distillate and Residual Fuel with Federal Certification Fuel Parent

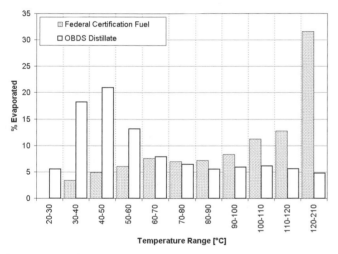

Figure 8. Fuel Evaporation by Temperature Range. OBDS Distillate with Federal Certification Fuel Parent

Composition by carbon number of distillate and parent fuels are shown in Figure 9. Detailed compositions (top 90% by volume) of the parent, distillate and residual fuels are listed in the Appendices. Notable is the anticipated increase of light-ends and decrease of heavy-ends in the distillate. Equally notable, however, are the levels of heavier compounds (toluene, iso-octane, iso-nonane) present in the distillate. The indication is that the distillation temperature was too high. We are currently designing an OBDS model that will help predict the optimum distillation temperature and could assist in system control. This model will be the subject of a future publication.

We expected the reduction of the light ends (or concentration of heavy-ends) in the residual fuel would give rise to an increased octane rating. Table 2 indeed confirms slight RON and MON increases in the residual fuel. A more selective separation, one that would exclude heavier, octane-boosting compounds like toluene from the distillate, would yield higher octane ratings for the residual fuel.

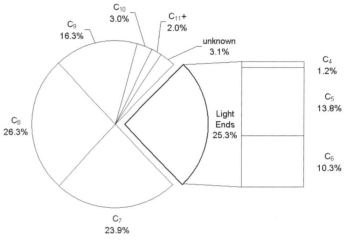

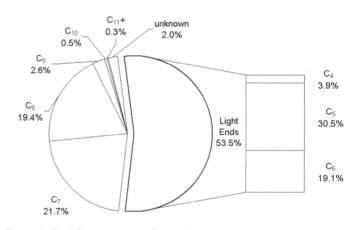

Figure 9. Fuel Composition by Carbon Number. OBDS Distillate with Federal Certification Fuel Parent

Table 2. Distillation Effect on Octane Number		
	Certification Fuel Parent	Residual Fuel
RON	96.4	97.1
MON	88.1	88.4

PHASE II. CALIBRATION DEVELOPMENT

The goal of this research was to quantify an emissions reduction attributable to the use of OBDS fuel for FTP cold-starts. We anticipated this reduction to be realized through lower engine-out emissions coupled with faster catalyst light-off. We began the calibration task by dividing engine operation into four distinct regimes, shown graphically in Figure 10: Crank, Flare, Warm-up and Normal Operation. Our basic calibration approach was to crank the engine with a stoichiometric or lean mixture and apply generous spark retard during warm-up.

Expectations were that a lean crank and warm-up mixture would provide lower engine-out HCs firstly by eliminating startup enrichment. Secondly, lean combustion has a slower burn rate, delivering less energy to the piston and resulting in higher energy exhaust gases that could accelerate catalyst light-off.

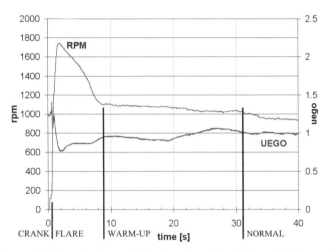

Figure 10. Four Regimes of Engine Operation During Starting and Warm-up

We also anticipated that a spark timing significantly retarded from MBT would reduce engine-out HC emissions. (With gasoline, this engine is intolerant of spark retard during starting and warm-up.) Decreased cylinder pressures associated with retarded spark timing would lessen crevice packing, a key source of engine-out HCs [2]. Further, considerably less fuel energy would be delivered to the piston as ignition timing is retarded, resulting in high (thermal) energy exhaust gases which should accelerate catalyst light-off.

Two OBDS calibration strategies were developed, one where the fuel strategy alone was changed, and another where both fuel and ignition timing strategies were altered from stock. Throughout development, considerable effort was expended to maintain consistent operating conditions to isolate the effects of the OBDS fuel. For example, the idle air bypass calibration was manipulated to compensate for decreasing engine speeds caused by leaner air/fuel mixtures and retarded ignition timing.

The calibrations presented here are relatively immature. For this research, the goal was simply to demonstrate OBDS benefits, not develop a production-ready calibration. In the fuel-only calibration, for example, stock ignition timing was used, rather than timing optimized for

contend with the switch from OBDS fuel to

making it necessary that all modifications from stock expire before the fuel switch occurred.

FUEL-ONLY CALIBRATION

Engine speed, excess air ratio (λ) and ignition timing for a 20 °C (68 °F) cold start with stock calibration are shown in Figure 11. Excess air ratio at crank is $\lambda \cong 0.65$. Through flare and warm-up, λ increases nearly linearly to $\lambda \cong 0.9$ at ≈ 20 seconds after engine start, when the transition to closed-loop fuel control occurs ($\lambda = 1$).

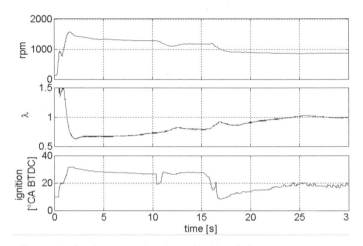

Figure 11. Stock excess air ratio and ignition timing. (Neutral → Drive shift at 15 seconds.)

The objective for the OBDS fuel-only calibration was a stoichiometric or lean crank, lean warm-up, and transition to closed-loop fuel control at the stock time. Our development tactic was to focus on crank and flare, then warm-up, and finally the transition to normal operation. Shown in Figure 12 are data from four 20 °C cold-starts. Data include engine speed, excess air ratio (λ), tailpipe HC emissions and exhaust/catalyst temperatures. The amount of crank fuel injected for each start is expressed as a percentage of the stock amount. For each warm-up period $\lambda = 1.1$, and the transmission was shifted into drive at 15 seconds after start. Stock ignition timing was employed for all four starts.

Immediately apparent from the engine speed trace in Figure 12 is that 13% of the stock crank amount is not enough fuel for robust starting. This is echoed in the emissions trace where the "13%" emissions were higher than all but stock, evidence of unstable combustion. Stock emissions are significantly higher than all, including the poor-combustion 13% start, demonstrating a definite effect of startup enrichment on HC emissions. The lowest catalyst temperatures are those of the stock calibration; up to 15% more time is required to reach the same temperatures as in the lean warm-up starts. Testing revealed the lowest reliable crank fuel amount was around 20% of stock; this amount yielded a mixture slightly lean of stoichiometric.

318

Attention then turned to the warm-up phase. We expected the lean limit for stable combustion to increase as the intake valves warmed. The data shown in Figure 13 support this assertion. Excess air ratio, engine speed and tailpipe emissions are shown for three cold starts. In the excess air ratio plots, both the intended λ (dotted line) and measured λ (solid line) are shown. For each of these starts the crank was strong, with a start time better than or equal to stock. The Figure 13 data demonstrate an inability to schedule aggressively lean mixtures ($\lambda > 1.2$) until ≈10 seconds after start. Unstable combustion, seen as "hunting" in λ and rpm plots; was corroborated by the emissions data. We determined that the best warm-up fuel schedule (for emissions and smoothness) was stoichiometric or slightly lean until the engine was running stably. From then on λ could be increased, so long as $\lambda < 1.2$ until 10s after start, and $\lambda \leq 1.3$ always.

RETARDED IGNITION CALIBRATION

Cold start spark retard is an increasingly popular technique for decreasing catalyst light-off time. Early in the development of the retarded-ignition OBDS calibration, it became evident that OBDS fuel can tolerate a much greater degree of retard than can gasoline. As ignition timing was retarded about 20° from stock, the air flow required to maintain stock engine speeds could not be sustained by the (now fully open) idle-air bypass valve. As seen in Figure 14, air flow in calibrations "ret 1" and "ret 2" is nearly identical, though engine speed decreased appreciably with the level of ignition retard.

The lower engine speeds had an effect on catalyst temperatures. The "ret 1" calibration is 8-10° retarded from "ret 2" yet catalyst temperatures are not much higher. Consider "ret 3," where airflow is increased enough to maintain near-stock engine speeds. Catalyst temperatures are drastically higher. In Phase III testing we found that catalyst temperature at light-off (50% efficiency) is consistently 370 °C [700 °F], and that the stock light-off time is ≈39 seconds. The "ret 1" calibration, with additional air, would achieve light-off in 15 seconds or less.

THC emissions are reported in Figure 14 by both mass rate and concentration. With 20-30% variation in exhaust flow rates between the three retarded-timing calibrations and stock, concentration is not as reliable an indicator of emissions levels as mass rate. Figure 14 reveals a marked improvement in THC emissions over the stock calibration. "Ret 3" emissions did increase near 12 seconds after start, most probably due to excessively retarded ignition timing (-18° BTDC when the increase began). Cumulative tailpipe THC emissions, shown in Figure 15, compared favorably over the first 15 seconds with those of the best "fuel-only" calibration, "fuel 1." Total Bag 1 emissions should be lower due to faster catalyst light-off.

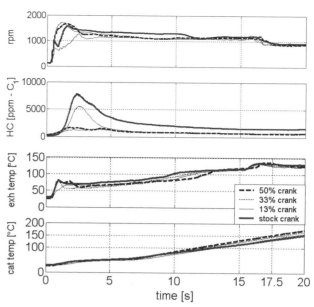

Figure 12. OBDS at various crank fuel amounts as percentage of stock, followed by warm-up at $\lambda = 1.1$.

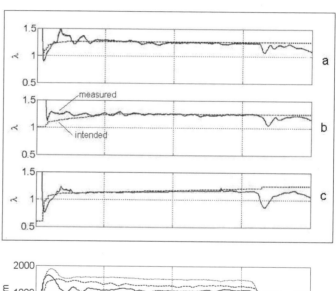

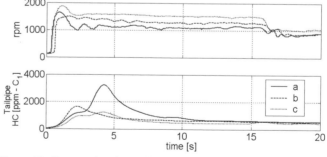

Figure 13. Excess air ratio, engine speed and HC emissions for three fuel-only calibrations with ramped warm-up λ

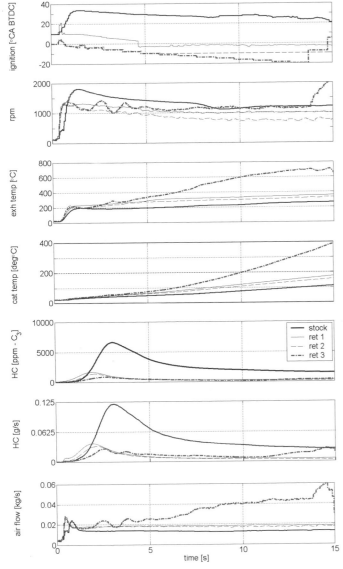

Figure 14. Three calibrations with retarded ignition timing compared to stock. Note that the ignition timing for "ret 2" is the same as that of "ret 3", but limited to −15° CA BTDC.

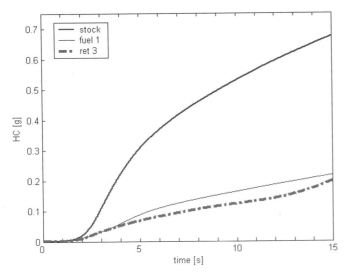

Figure 15. Cumulative THC emissions: stock *vs.* best fuel-only calibration (fuel 1) *vs.* best retarded-ignition calibration (ret 3).

LEAN FUEL CALIBRATION

Fuel scheduling of the Cold 505 "lean fuel" calibration is shown in Figure 16. The emissions data, summarized in Table 3, show substantial HC and CO benefits from the use of OBDS fuel, even with the stock ignition timing. Key is the elimination of startup enrichment and the use of warm-up enleanment. Hydrocarbon emissions were reduced by 43-46% over stock, CO reduced by over 75%. Any NO_x emission benefit or penalty was inconclusive: the measurements were within the confidence interval of the baseline. Light-off time was also reduced by 15-18%. The importance of HC catalyst conversion is well illustrated by these tests. Over the Cold 505 cycle, tailpipe HC's for "fuel 1" were 6% lower than "fuel 2," although engine-out HC's were 7% *higher*.

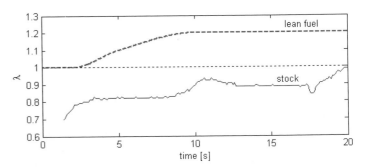

Figure 16. Lean-Fuel Cold 505 Fuel Scheduling vs. Stock A/F Ratio

OBDS distillate alone was supplied for the start and approximately the first 20 seconds thereafter; both OBDS and gasoline were supplied for the next 2 seconds, then gasoline alone for the duration of the test. Second-by-second engine-out HC emissions are plotted in Figure 18 for stock and OBDS calibrations. There is a clear benefit for OBDS use for the first 20-40 seconds, during its expected "sphere of influence." (The OBDS fuel remains in the injectors and fuel rail after the switch to gasoline.) A notable benefit can be seen over the 120 seconds shown in Figure 18. This benefit actually extends approximately 300 seconds into the test. We believe that the higher emissions of the stock tests are due to the re-emergence of "lost fuel," excess start-up gasoline that remained in the cylinder as a liquid or had escaped into the crankcase past the piston rings. With OBDS fuel, almost all of the start-up fuel vaporizes, generating little or no lost fuel. Research by Shayler *et al*. [4] indicates the process of returning lost fuel to the bulk mixture can last for 300 seconds or more. Shayler also points out that lost fuel returned to the bulk mixture is poorly mixed and enters the exhaust as unburned hydrocarbons. Thus, the lost fuel phenomenon is a plausible explanation for the high HC emissions of the stock gasoline calibration versus the OBDS fuel calibrations for a duration that extends well beyond both the switch to gasoline and the time required to consume/purge OBDS fuel from the injectors and fuel rail.

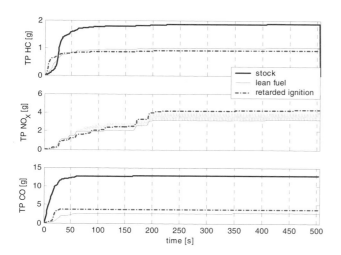

Figure 17. Cold 505 cumulative tailpipe emissions

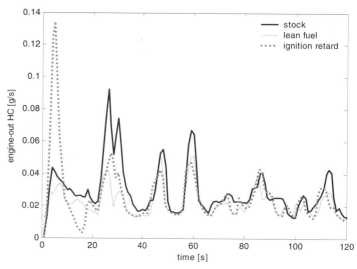

Figure 18. Second-by-second engine-out HC emissions, Cold 505

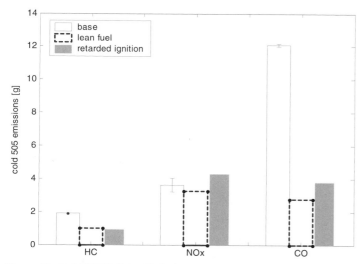

Figure 19. Cold 505 Tailpipe Emissions

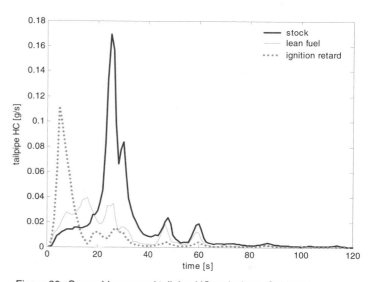

Figure 20. Second-by-second tailpipe HC emissions, Cold 505

Table 3. Cold 505 Tailpipe Emissions and Percent Reduction Compared to Stock

	Stock	Lean Fuel	Retarded Ignition
Light-off [s]	40.4 ± 1.1	31	20
		-22.5%	-50.5%
HC [g]	1.90 ± 0.01	1.02	0.92
		-46.3%	-51.6%
NO$_x$ [g]	3.61 ± 0.41	3.25	4.28
		NSD*	+18.6%
CO [g]	12.10 ± 0.71	2.75	3.79
		-77.3%	-68.7%

* NSD = not significantly different from baseline

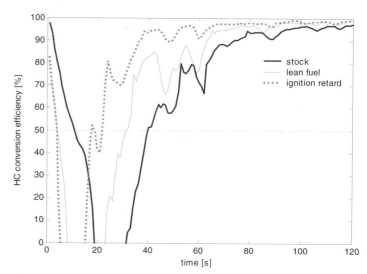

Figure 21. Second-by-second HC conversion, Cold 505

RETARDED IGNITION CALIBRATION

During heavy ignition timing retard, the engine required airflow beyond the capability of its throttle bypass to generate enough torque to maintain a desired idle speed. An effective means of supplying this air was via the cruise-control actuator. Microcontroller-based electronics replaced the stock controller. Fuel mapping, and distillate/gasoline scheduling were the same as those for the lean fuel calibration. Ignition timing for this calibration is given in Table 4. The throttle was opened 10-15% from 1 to 15 seconds after start.

Table 4. Ignition Timing, Retarded Ignition Calibration	
Time After Start [s]	Ignition Retard from Stock [°CA]
crank	0
0-1	0
5-15	≈ 40
15+	0

A prominent peak in start-up engine-out HC emissions is apparent in Figure 18. Preliminary results from a mixture-preparation model under development indicate that after throttle opening, a significant portion of injected fuel remained liquid rather than join the combustible air/fuel mixture. The higher MAP removed a significant driving force for fuel evaporation. Throttle opening should have been delayed somewhat to allow more heat transfer into the intake valves. Nevertheless, tailpipe HC emissions over the entire Cold 505 were 52% lower than the stock baseline. We attribute this to very low engine-out HC during the retard period and fast catalyst light-off, which occurred in about 20 seconds, or half the time required by the base configuration.

Curiously, retarded ignition calibrations showed tailpipe HC improvement in (FTP) Bag 2 of about 25% relative to stock, despite averaging 7% higher engine-out HC. This behavior was not seen with the lean-fuel calibrations, where the influence was confined to Bag 1. We are currently exploring the effects warm-up period ignition timing on catalyst efficiency.

FTP WEIGHTED EMISSIONS

Full FTP testing with the stock gasoline calibration was conducted for reference. Full FTP testing with the OBDS calibrations are scheduled for the near future. However, the Cold 505 tests discussed above included data acquisition not only for Phase 1, but Phase 2 also. We cannot assume that the influence of OBDS does not extend beyond Bags 2 and 3 of the FTP cycle, but results from Bag 3 of the reference FTP can be used to provide a reasonable estimate of FTP-weighted emissions from the OBDS Cold 505 tests. These results are shown in Figure 22. Applying NMOG = THC x 0.943 [14], the OBDS conversion

LEV-compliant for hydrocarbons (LEV NMOG max = 0.09 g/mi), with a factor of safety of 32% for the lean fuel calibration, and 38% for the retarded ignition calibration.

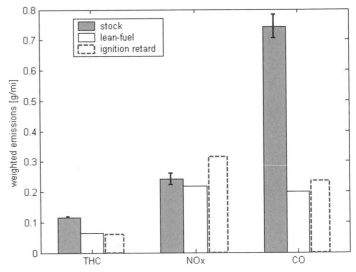

Figure 22. FTP weighted emissions

Table 5. FTP Weighted Emissions and Percent Reduction Compared to Stock			
	Stock	Lean Fuel	Retarded Ignition
HC [g/mi]	0.116 ± 0.001	0.065	0.059
		−44.0%	-49.2%
NOₓ [g/mi]	0.242 ± 0.018	0.220	0.315
		NSD	+30.2%
CO [g/mi]	0.744 ± 0.038	0.200	0.236
		−73.1%	-68.3%

* NSD = not significantly different from baseline

SUMMARY AND CONCLUSIONS

The fuel generated by the On-Board Distillation System can eliminate the need for startup enrichment. Furthermore, OBDS distillate fuel allows warm-up enleanment and accelerates catalyst light-off. OBDS fuel is much more tolerant of spark retard than is gasoline. Warm-up ignition timing retarded 40° or more from MBT is possible, without significant detriment to idle stability.

Catalyst light-off time was reduced by >50%. Hydrocarbon and CO reductions of >50% and >65%, respectively have been demonstrated over Bag 1 of the FTP drive cycle, using rough calibration techniques. Optimized calibration of fuel, ignition and airflow (MAP), derived from knowledge gained through modeling, should yield ULEV-capable results.

ACKNOWLEDGMENTS

This research was funded by Ford Motor Company, the U.S. Department of Energy (under Grant No. DE-FG04-99AL66262) and the Texas Advanced Technology Program (Grant 003658-0810-1999). We wish to express our sincere gratitude to the personnel of Southwest Research Institute without whose help the data could not have been acquired. We would also like to thank Scott Bohr, Joe Grahor and Ray Willey of Ford for their invaluable contributions. The opinions and findings presented herein do not necessarily reflect the views of the U.S. DOE or any other Federal or State agency.

REFERENCES

1. Kidokoro, T, K Hoshi, K Hiraku, K Satoya, T Watanabe, T Fujiwara and H Suzuki (2003), "Development of PZEV Exhaust Emission Control System," SAE Technical Paper 2003-01-0817.

2. Matthews, RD (2003) Internal Combustion Engines and Automotive Engineering, draft textbook.

3. Kane, E, D Mehta and C Frey (2001), "Refinement of a Dedicated E85 1999 Silverado With Emphasis on Cold Start and Cold Driveability," SAE Technical Paper 2001-01-0679.

4. Shayler, PJ, MT Davies and A Scarisbrick (1997), "Audit of Fuel Utilisation During the Warm-Up of SI Engines," SAE Technical Paper 971656.

5. Stovell, CH, RD Matthews, BE Johnson, HK Ng and B Larsen (1999) "Emissions and Fuel Economy of a 1998 Toyota With a Direct-Injection, Spark-Ignition Engine," SAE Technical Paper 1999-01-1527.

6. Johnson, TV (1999), "Gasoline Vehicle Emissions – SAE 1999 In Review," SAE Technical Paper 2000-01-0855.

7. Kishi, N, S Kikuchi, N Suzuki and T Hayashi (1999), "Technology for Reducing Exhaust Gas Emissions in Zero Level Emission Vehicles (ZLEV)," SAE Technical Paper 1999-01-0772.

8. Oguma, H, M Koga, K Nishizawa, S Momoshima and S Yamamoto (2003), "Development of Third Generation of Gasoline P-ZEV Technology," SAE Technical Paper 2003-01-0816.

9. Quader, A, JE Kirwan and MJ Grieve (2003), "Engine Performance and Emissions Near the Dilute Limit With Hydrogen Enrichment Using An On-Board Reforming Strategy," SAE Technical Paper 2003-01-1356.

10. Ku, J, Y Huang, RD Matthews and MJ Hall (2000), "Conversion of a 1999 Silverado to Dedicated E85 With Emphasis on Cold Start and Cold Driveability" SAE Technical Paper 2000-01-0590.

11. Mizaikoff, B, P Fuss and MJ Hall (1998), "Fast-Spec: An Infrared Spectroscopic Diagnostic to Measure Time-Resolved Exhaust Hydrocarbon Emissions from S.I. Engines," 27th Symposium (International) on Combustion, Combustion Institute.

12. Alger TF, MJ Hall and RD Matthews, "The Effects of In-Cylinder Flow Fields and Injection Timing on Time-Resolved Hydrocarbon Emissions in a 4-Valve, DISI Engine," SAE Technical Paper 2000-01-1905

13. ALLPROPS Property Package v.6/4/96, The Center For Applied Thermodynamic Studies, The University of Idaho (1996).

14. Lindhjem, CE, (1997) "Conversion Factors for Hydrocarbon Emission Components," U.S. EPA Office of Mobile Sources Report No. NR-002.

Appendix I: Federal Certification Fuel Parent Composition.

COMPONENT	%VOL	COMPONENT	%VOL
Toluene	17.935	n-Octane	0.193
2,2,4-Trimethylpentane	13.746	2-Methylheptane	0.187
Isopentane	7.863	Isopropylbenzene	0.177
1,2,4-Trimethylbenzene	5.3	14Dimethyl2ethylbenzene	0.175
n-Pentane	5.014	4-Methylheptane	0.174
2,3,4-Trimethylpentane	4.468	3-Methylheptane	0.171
1-Ethyl-3-Methylbenzene	3.064	n-Nonane	0.16
2-Methylpentane	2.272	1,3-Diethylbenzene	0.159
n-Hexane	1.801	trans-1,2-DimethylcycloC5	0.155
2-Methylhexene+C7-Olefin	1.772	C9-Isoparaffin	0.152
2,3-Dimethylbutane	1.687	P-Xylene	0.146
2,5-DimethylC6+c8-Olefin	1.564	2Mnaphthalene	0.127
2,4-Dimethylhexane	1.552	C11-Isoparaffin	0.123
1,3,5-Trimethylbenzene	1.544	2,2-Dimethylpropane	0.123
1-Ethyl-4-methylbenzene	1.423	trans-1,3-Dimethylcyclope	0.12
cis-1-Ethyl -3methylcyC5	1.326	C8-Olefin+C7-Diolefin	0.118
1-Ethyl-2-methylbenzene	1.321	2-Methyl-2-butene	0.116
2,4-Dimethylpentane	1.309	Naphthalene	0.113
3-Methylpentane	1.259	2,5&3,5-Dimethylheptane	0.108
n-Butane	1.215	cis-1,3-DimethylcycloC5	0.106
2,3-Dimethylhexane	1.176	1235-tert-Methylbenzene	0.105
Methylcyclopentane	1.146	cis-1,3-DimethylcycloC6	0.102
1,2,3-Trimethylbenzene	1.106	2,2,3-Trimethylbutane	0.098
Cyclohexane	1.041	2-Methylbutylbenzene	0.084
Methylcyclohexane	1.02	2-Methyldecane	0.083
Propylbenzene	0.926	C10-Aromatic	0.078
n-Heptane	0.641	3-Ethylhexane	0.074
3-Methylhexane	0.523	trans-2-Pentene	0.074
1-Methyl-3-propylbenzene	0.518	1,1,2-Trimethylcyclohexan	0.07
2,2-Dimethylbutane	0.487	C9-Olefins	0.066
Benzene	0.48	1-Decene	0.062
2-Methylnonane	0.417	33DimethylC5+5-M1hexene	0.057
O-Xylene	0.413	1234tetramethylbenzene+C1	0.056
Cyclopentane	0.385	Ethylcyclohexane	0.056
M-Xylene	0.377	Butylbenzene	0.054
3-Ethylnonane	0.371	3-Methyldecane	0.053
nC13	0.329	2-Methyl-1-butene	0.053
3,5-Dimethyl-1-Ethylbenze	0.282	n-Undecane	0.051
Indan	0.27	2-Methyl-4-ethylhexane	0.051
Ethylbenzene	0.236	Butylcyclopentane	0.05
1-Methyl-4-propylbenzene	0.226	C11-Isoparaffin	0.045
12DM4Ebenz+C1indan	0.214	Ethylcyclopentane	0.044
13Dimethyl2ethylbenzene	0.209	3-Methyloctane	0.044
C9-Olefin	0.198	t-1,2-Dimethylcyclohexane	0.043
n-Decane	0.194	1-Methyl-4-Isopropylbenze	0.042
		5-Methyldecane	0.041

Appendix II: OBDS Distillate Fuel Composition.

COMPONENT	%VOL
Isopentane	17.001
Toluene	15.632
2,2,4-Trimethylpentane	12.248
n-Pentane	11.607
2-Methylpentane	4.641
n-Butane	3.849
2,3-Dimethylbutane	3.6
n-Hexane	3.172
2,3,4-Trimethylpentane	2.787
3-Methylpentane	2.391
Methylcyclopentane	1.828
2-Methylhexene+C7-Olefin	1.758
2,4-Dimethylpentane	1.736
Cyclohexane	1.437
2,2-Dimethylbutane	1.214
2,5-DimethylC6+c8-Olefin	1.065
2,4-Dimethylhexane	1.03
Cyclopentane	0.89
Methylcyclohexane	0.878
1-Ethyl-3-Methylbenzene	0.719
2,3-Dimethylhexane	0.697
Benzene	0.668
cis-1-Ethyl -3methylcyC5	0.644
n-Heptane	0.56
3-Methylhexane	0.504
2,2-Dimethylpropane	0.457
1,3,5-Trimethylbenzene	0.339
1-Ethyl-4-methylbenzene	0.336
1-Ethyl-2-methylbenzene	0.289
Propylbenzene	0.225
1,2,3-Trimethylbenzene	0.215
2-Methyl-2-butene	0.17
trans-1,2-DimethylcycloC5	0.16
2,2,3-Trimethylbutane	0.136
trans-1,3-Dimethylcyclope	0.123

Appendix III: OBDS Residual Fuel Composition.

COMPONENT	%VOL	COMPONENT	%VOL
Toluene	18.316	nC13	0.355
2,2,4-Trimethylpentane	13.824	M-Xylene	0.291
Isopentane	6.956	1-Methyl-4-propylbenzene	0.285
1,2,4-Trimethylbenzene	5.617	3,5-Dimethyl-1-Ethylbenze	0.284
C7-Diolefin	4.651	Indan	0.276
n-Pentane	4.606	cis-2-Octene	0.233
1-Ethyl-3-Methylbenzene	3.308	13Dimethyl2ethylbenzene	0.22
2-Methylpentane	2.227	n-Octane	0.206
n-Hexane	1.845	12DM4Ebenz+C1indan	0.203
2-Methylhexene+C7-Olefin	1.718	2-Methylheptane	0.189
2,5-DimethylC6+c8-Olefin	1.661	4-Methylheptane	0.18
1,3,5-Trimethylbenzene	1.657	3-Methylheptane	0.168
2,3-Dimethylbutane	1.654	Isopropylbenzene	0.168
2,4-Dimethylhexane	1.621	trans-1,2-DimethylcycloC5	0.165
1-Ethyl-4-methylbenzene	1.539	1,3-Diethylbenzene	0.164
cis-1-Ethyl -3methylcyC5	1.441	n-Nonane	0.16
1-Ethyl-2-methylbenzene	1.419	C9-Isoparaffin	0.155
2,4-Dimethylpentane	1.304	Ethylbenzene	0.148
2,3-Dimethylhexane	1.251	14Dimethyl2ethylbenzene	0.138
3-Methylpentane	1.235	C11-Isoparaffin	0.133
Methylcyclopentane	1.182	2Mnaphthalene	0.133
1,2,3-Trimethylbenzene	1.161	C8-Olefin+C7-Diolefin	0.123
Methylcyclohexane	1.128	trans-1,3-Dimethylcyclope	0.122
Cyclohexane	1.119	2,5&3,5-Dimethylheptane	0.119
Propylbenzene	0.994	cis-1,3-DimethylcycloC6	0.113
n-Butane	0.837	P-Xylene	0.112
n-Heptane	0.682	cis-1,3-DimethylcycloC5	0.109
2,2,3-Trimethylpentane	0.615	Naphthalene	0.109
1-Methyl-3-propylbenzene	0.545	2,2,3-Trimethylbutane	0.102
3-Methylhexane	0.518	2,2-Dimethylpropane	0.099
Benzene	0.483	2-Methyldecane	0.089
2,2-Dimethylbutane	0.468	1235-tert-Methylbenzene	0.081
2-Methylnonane	0.442	3-Ethylhexane	0.078
3-Ethylnonane	0.387	C10-Aromatic	0.078
Cyclopentane	0.365	1,1,2-Trimethylcyclohexan	0.075
O-Xylene	0.356		

Fast Start-Up On-Board Gasoline Reformer for Near Zero Emissions in Spark-Ignition Engines

John E. Kirwan, Ather A. Quader and M. James Grieve
Delphi Automotive Systems

Copyright © 2002 Society of Automotive Engineers, Inc.

ABSTRACT

This paper describes recent progress in our program to develop a gasoline-fueled vehicle with an on-board reformer to provide near-zero tailpipe emissions. An on-board reformer converts gasoline (or another hydrocarbon-containing fuel) into reformate, containing hydrogen (H_2) and carbon monoxide (CO). Reformate has very wide combustion limits to enable SI engine operation under very dilute conditions (either ultra-lean or with heavy exhaust gas recirculation (EGR) concentrations). In previous publications, we have presented engine dynamometer results showing very low emissions with bottled reformate. This paper shows results from an engine linked to an experimental, fast start-up reformer. We present both performance data for the reformer as well as engine emissions and performance results. Program results continue to show an on-board reforming system to be an attractive option for providing near-zero tailpipe emissions to meet low emission standards.

INTRODUCTION

The Ultra Low Emission Vehicle (ULEV) II standards proposed for 2004 introduction in California include a Super-ULEV (SULEV) standard. Gasoline-fueled vehicles that robustly meet SULEV standards over their useful lives offer a significant step toward eliminating the automobile as a source of regulated pollutants. Developing SULEVs can significantly reduce an OEM's fleet average non-methane organic gas (NMOG) emissions. Further, the standards currently offer partial credit toward the zero emission vehicle fleet requirements (P-ZEV credits) for vehicles meeting SULEV emissions standards. As the California standards are currently written, up to 60% of the zero emission vehicle (ZEV) requirement could be fulfilled with P-ZEV credits from SULEV vehicles that also have zero evaporative emissions. The maximum P-ZEV credits possible from SULEVs would be obtained by an OEM if 30% of the vehicles that it sells in California were certified as SULEV vehicles.

Consequently, significant efforts are under way to develop SULEV vehicle systems. Beginning with the 2000 model year, Honda [1][1] and Nissan [2] each have certified gasoline-fueled vehicles meeting the SULEV standards. In both cases, sophisticated aftertreatment and engine management systems have been developed to meet the SULEV standard. More detailed descriptions of these SULEV systems have been recently published [1-4].

Both the Honda and Nissan SULEV vehicles represent formidable achievements. Their approaches may not be the most appropriate solution under all conditions, however. Aftertreatment systems such as these may have difficulty meeting SULEV with larger, higher emitting engines. Also, poor volatility fuels can be a problem. (This is especially significant for the Northeastern states that have adopted the California emissions standards but not the California fuel requirements.) Both Honda and Nissan rely on running relatively lean with spark retard during warm-up for lower engine-out emissions and faster catalyst warm-up. These strategies can be susceptible to poor warm-up performance with low volatility fuels.

One of Delphi's SULEV strategies under development uses an on-board gasoline reformer. This approach is targeted at a number of advantageous features, including:
- robustness for vehicles with larger engines and heavier vehicles.
- robustness to fuel variation due to low volatility (high Driveability Index).
- robustness under off-cycle conditions (e.g. low ambient temperature).
- much lower loading of precious metals in the catalyst system (especially Pd).
- no compromise in exhaust system backpressure.
- Synergies with automotive fuel cell systems, especially solid-oxide fuel cell (SOFC).

----

[1] Numbers in Brackets designate references listed at the end of the paper

The reformer combines gasoline and air under very fuel rich conditions. Schematically, the partial oxidation (POx) reaction within the reformer can be represented as:

$$\text{Gasoline} + \text{Air} \rightarrow H_2 + CO + N_2 + \text{Heat} \qquad (1)$$

$$+ (\text{trace } CO_2, H_2O, HCs)$$

Reforming gasoline provides an on-board source of H_2. During the 1970's, researchers at the Jet Propulsion Laboratory (JPL) recognized that adding H_2 to gasoline allows an engine to run very lean, due to hydrogen's wide flammability limits [5]. They subsequently developed a method for on-board H_2 generation using POx reforming of gasoline [6], and demonstrated very low NOx on a vehicle [7]. However, a reformer could not compete at that time with a 3-way exhaust catalyst and closed loop fuel control for meeting the NOx standard. A recent review paper describes a number of additional studies that have investigated supplementing gasoline with H_2 to extend lean operation [8].

Hydrogen-rich reformate has a number of attributes that make it an attractive fuel for very low emissions. It offers a fuel source that is very low in hydrocarbons. (Ideally reformate converts all hydrocarbons in gasoline to CO and H_2; with actual reformate roughly 10% - 15% of the gasoline may break through the reformer as hydrocarbons during start-up). Reformate also promotes low temperature catalyst light-off, and it has very wide flammability limits. Our H_2 enrichment strategy with an on-board reformer implements each of the above attributes of reformate to augment (not replace) a 3-way catalytic aftertreatment system. The basic elements of the system are shown schematically in Figure 1. Briefly, the H_2 enrichment strategy consists of:

- fueling with 100% reformate under ultra-lean conditions during cold start for near-zero engine-out HC and NOx;
- supplying reformate to the lean exhaust stream during cold start for rapid exhaust catalyst light-off;
- using gasoline supplemented with a modest fraction of reformate fueling at light and medium loads to permit high EGR dilution for ultra-low engine-out NOx.

The present paper is focused only on the first element. The catalyst light-off and high EGR elements will be the subject of future publications. We are currently developing a demonstration vehicle to implement our H_2 enrichment strategy. The demonstration vehicle is a recent model production vehicle with a 2.4 L 4-cylinder engine and a manual 4-speed transmission (see [9] for additional vehicle details). The vehicle program's objective is to demonstrate our on-board gasoline reformer system as an enabler for meeting SULEV emissions standards. As the program progresses, we

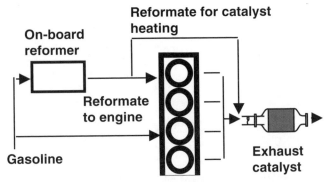

Fig. 1: Schematic of the On-Board Reformer System

are periodically documenting its performance. In previous publications, we have outlined the H_2 enrichment strategy in more detail, and indicated its potential as a SULEV enabler with data from both engine dynamometer and vehicle tests running with ideal (bottled) reformate [9 - 13]. One paper [9] also provided early performance data from a warmed-up reformer fueled with gasoline operating on a flow bench.

As discussed above, the H_2 enrichment SULEV strategy relies on very low cold start HC emissions provided by starting the engine with 100% reformate. Thus, fast start-up of the on-board reformer is essential to enable quick engine starting. The focus of the experiments described in this paper were two-fold:

1. to characterize start-up performance of an experimental reformer using gasoline combustion to preheat the reformer;
2. to evaluate the performance and emissions of a single-cylinder engine fueled with reformate generated by the reformer.

Described below are details of the study performed using an experimental reformer coupled to a single-cylinder engine. This arrangement enabled simultaneous characterization of reformer output and reformate-fueled engine combustion and emissions performance during reformer start-up. In parallel, we are currently fitting a reformer system to our development vehicle and developing the control software and calibrations to fully demonstrate the strategies. Vehicle development work, however, is beyond the scope of this paper.

EXPERIMENTAL

SINGLE-CYLINDER ENGINE AND ACCESSORIES - A 2.4 L displacement, 4-valve per cylinder production engine head was mounted on a single-cylinder CFR (corporate fuel research) engine crankcase [10]. The head for cylinder number 3 was used. The single-cylinder engine was coupled to an electric dynamometer. Overall, the engine represents one cylinder from the

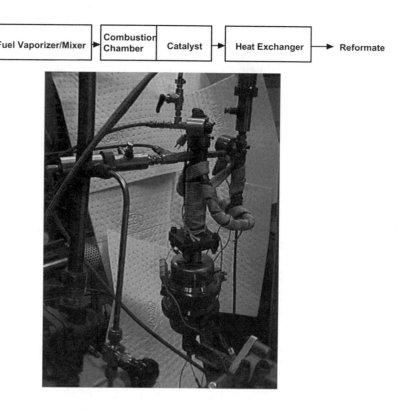

Gasoline ⟶
Air ⟶ | Fuel Vaporizer/Mixer | ⟶ | Combustion Chamber | Catalyst | ⟶ | Heat Exchanger | ⟶ Reformate

Fig. 2: Schematic layout and photograph of experimental reformer.

engine in our development vehicle. A special crankshaft and a cylinder sleeve were fabricated and installed to duplicate the stroke and bore of the production engine. Production piston, rings, connecting rod, piston pin, and bearings were used. The compression ratio was 9.5:1.

The engine was set up to run on gaseous as well as liquid fuels as described in [10]. For the current experiment, the engine generally was run on 100% reformate fuel generated by the experimental reformer described below. Baseline tests were also performed using port fuel injected (PFI) gasoline for comparison with results with reformate.

EXPERIMENTAL REFORMER SETUP - Figure 2 shows a schematic layout and photo of the experimental reformer. It is made up of components which perform the following functions:

- Air/Fuel metering
- Fuel Vaporization and Mixing
- Combustion
- Catalytic Reforming
- Heat Exchange

Further details of the reformer components are proprietary. Two experimental reformers were fabricated and tested. One was a full-scale reformer sized to satisfy the flow requirements for a 2.4 liter 4-cylinder engine. The second reformer tested was a quarter-scale version sized more appropriately for our single-cylinder test engine.

DATA ACQUISITION EQUIPMENT - Emission analyzers were used to measure CO, CO_2, O_2, NOx and unburned HC in the engine exhaust gas. The same emission analyzers were used to alternatively sample from the engine intake pipe to monitor the reformate CO, CO_2, and HC levels blended with the engine combustion air. The engine and reformer were instrumented using thermocouples and pressure transducers so that virtually any temperature or pressure of interest could be monitored. Output from the emissions bench and various transducers was monitored using an HP 1000 computer.

Engine performance characteristics were monitored and recorded by a separate system. An air cooled pressure transducer (Kistler Mod. No 6121) mounted in the cylinder head and a shaft encoder (Hewlett Packard HEDS6310) were used in conjunction with an ACAP (DSP Technologies) combustion analysis system. The ACAP analyzed the cylinder pressure to determine and record engine performance characteristics on a cycle by cycle basis in real time.

TEST PROCEDURE - California Phase 2 (CP2) Certification fuel was used in this study, both for the baseline PFI gasoline tests as well as for fueling the reformer. CP2 represents a low-sulfur gasoline formulated to provide relatively low exhaust emissions in gasoline-fueled engines.

Prior to a test, the single-cylinder engine was motored at a steady speed with the appropriate engine air flow for the test. Engine coolant and oil temperatures were set at 23°C. The appropriate reformer air flow was established and the vaporizer section was heated to 100°C. In order to describe the fueling strategy during testing, it is important to first recognize that two separate quantities of air are required for a reformate-fueled engine. The first quantity of air is combined with gasoline and fed to the reformer under very fuel rich conditions (roughly 2.5 to 3 times richer than stoichiometric). The reformer consumes essentially all of the oxygen provided to it to make reformate for fueling the engine. A separate quantity of air is provided directly to the engine. It burns with the reformate fuel in the cylinders to power the engine.

In this paper, Φ is used to represent fuel air equivalence ratio according to the following equation

$$\Phi = \frac{F/A}{(F/A)_{stoich}} \qquad (2)$$

Where F/A is the mass fuel-air ratio, and $(F/A)_{stoich}$ is the stoichiometric fuel-air ratio for complete combustion. For a stoichiometric mixture, $\Phi = 1$; lean fuel mixtures have $\Phi < 1$, and rich mixtures have $\Phi > 1$. As a consequence of the two separate air quantities, two separate fuel-air equivalence ratios existed during engine testing. $\Phi_{reformer}$, represented the fuel-air mixture provided to the reformer. This was determined based on the gasoline flow to the reformer plus the air flow provided only to the reformer. The second equivalence ratio represented the fuel-air mixture provided to the engine, Φ_{engine}. This parameter included in its calculation the additional air fed to the engine for combustion of the reformate in the cylinders.

Tests for this study were performed with pre-vaporized gasoline and air. (An optimized mixture preparation system for the reformer start-up period is under parallel development, but not described in this paper.) Figure 3 shows schematically the fueling strategy used to enable rapid gasoline-fueled preheat of the reformer-catalyst during start-up. $\Phi_{reformer} = 1.0$ was supplied for a short duration and burned in the combustion chamber of the reformer. The heat released by burning the stoichiometric gasoline mixture elevated the temperature of the catalyst. The magnitude of the temperature rise depended on the duration for which the stoichiometric mixture was supplied, called preheat-time in Figure 3.

Following the catalyst preheat, the fueling rate was enriched to $\Phi_{reformer} = 2.75$ to begin reformate production. The engine would begin firing when the reformer began producing sufficient quality reformate to support combustion.

- Fueling Strategy
  - Stoich heating at $\Phi_{reformer} = 1$
  - Preheat-time $\Delta t_{stoich\ fuel} = 1$ to 10 sec
  - Reformate production at $\Phi_{reformer} = 2.75$

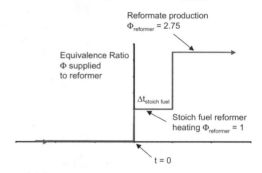

Fig. 3: Fueling strategy for the reformer.

OPERATING CONDITIONS - A series of tests were run in which the reformer preheat-time was varied from 1 second to 10 seconds as listed in Table 1. These tests

Table 1. Engine operating conditions with full-scale reformer
75 kPa Manifold pressure (MAP), 23°C coolant and oil
$\phi_{engine} = 1$, 100% reformate fueling
MBT spark timing (determined during steady operation with 100% reformate)

Preheat-Time seconds	Reformer Airflow g/s	Engine Speed RPM
1	1.6	1300
2	1.6	1300
3	1.6	1300
4	1.6	1300
5	1.6	1300
6	1.6	1300
7	1.6	1300
8	1.6	1300
9	1.6	1300
10	1.6	1300

were done with the full-scale reformer that is sized for a 4 cylinder engine. The reformer-airflow was 1.6 g/s during these tests. This airflow was estimated to correspond to the reformer-airflow required to generate sufficient reformate to start and idle the 2.4 L 4-cylinder engine in our demonstration vehicle on 100% reformate fuel at 23°C. Subsequent to these tests, preliminary operation of our vehicle fueled by a reformer shows that the reformer-airflow required during warm-up is closer to 2.5 g/s. Thus these early vehicle data indicate an error in our estimate. However, reformer bench testing has shown that reformer start-up performance is quite insensitive to reformer flow rate over the range of 1.5 g/s to 2.5 g/sec of reformer-airflow (see Appendix). Therefore, we expect little difference in results with the reformer operating at 2.6 g/s airflow compared to the results presented below at 1.6 g/s reformer airflow. Both engine speed and load experienced by the single-cylinder engine during our experiments with the full-scale reformer were significantly higher than typical idle conditions for the vehicle during engine warm-up. Engine speed for the single-cylinder engine was approximately 30% higher than idle, while the load during the single-cylinder engine tests was approximately 2 times higher than typical warm-up idle engine load. Baseline tests with PFI gasoline fueling were run on the single-cylinder engine at identical speed and load conditions as the reformer tests. Engine results with this full-scale reformer are not intended to be a direct measure of performance and emissions expected from a vehicle. Rather, the engine data provide a measure of reformate combustibility and offer an estimate of engine-out HC emissions during starting with a reformer compared to PFI gasoline operation. In Table 1 $\Phi_{engine} = 1$ was maintained for tests listed with the full-scale reformer. A stoichiometric mixture offers more rapid engine firing than an ultra-lean mixture, especially with weaker reformate mixtures produced early during reformer start-up.

The ability to run an engine very lean is a fundamental characteristic of fueling with H_2-rich fuels [8]. As described in the introduction, lean cold start fueling comprises part of our overall H_2 enrichment vehicle strategy. Ultra-lean operation has been documented with bottled reformate in both our engine and demonstration vehicle [9-11]. To investigate the lean starting capability of the engine under conditions more representative of a single cylinder engine, the quarter-scale reformer was tested. Table 2 shows the operating conditions for these tests. The preheat time was 10 seconds in these tests, while the Φ_{engine} was varied from 1.0 to 0.68.

Optimizing Φ_{engine} for engine performance and emissions was beyond the scope of this paper, but will be the subject of significant effort in our vehicle development program.

RESULTS OF TESTS WITH PREHEAT-TIME VARIATIONS

EFFECT OF PREHEAT-TIME ON REFORMER TEMPERATURES - The effect of preheat-time on catalyst temperature was measured in tests without reformate production at a reformer air flow rate of 1.6 g/s in the full-scale reformer. For these tests, $\Phi_{reformer} = 1$ was maintained during the preheat-time. However, following the preheat, rather than enriching the mixture to begin reformate production, the fuel to the reformer was immediately turned off. In these tests reformer temperature rise was due solely to reformer preheat,

without the additional temperature increase caused by the exothermic reforming reactions.

Table 2. Engine operating conditions with quarter-scale reformer

45 kPa Manifold pressure (MAP), 23°C coolant and oil
Preheat time=10 sec, 100% reformate fueling
MBT spark timing (determined during steady operation with 100% reformate)

Φ_{engine}	Reformer Airflow g/s	Engine Speed RPM
1	0.7	1330
0.93	0.7	1330
0.86	0.7	1330
0.81	0.7	1330
0.75	0.7	1330
0.68	0.7	1330

The data in Figure 4 show the measured combustion chamber and midpoint catalyst temperatures at the end of preheat as a function of preheat-time. Two important points should be noted. First, the measured combustion chamber temperature rises much more rapidly than does the midpoint catalyst temperature. The hot combustion products lose significant energy as they pass through the reformer catalyst. Therefore, temperature at the front of the catalyst during preheat will be greater than the midpoint catalyst temperature shown in Figure 4. Conversely, temperature at the rear of the catalyst will be lower than the midpoint catalyst temperature.

Second, increase in preheat-time significantly increases midpoint catalyst temperature. At the end of preheat, measured midpoint catalyst temperature varies from 150°C to 440°C with preheat-times from 2 to 10 seconds. The higher temperatures at increased preheat-time are a consequence of the increased amount of energy released from burning a larger amount of fuel.

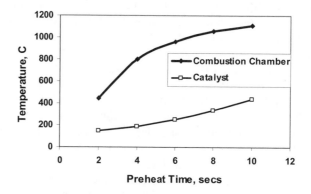

Fig. 4: Combustion chamber temperature and midpoint catalyst temperature at the end of reformer preheat.

Reformer temperatures were measured using 1/16 inch shielded thermocouples that showed a significant (but unquantified) lag in their time response. For example, from reformer bench tests performed with visual access into the reformer, we have observed a blue flame inside the combustion chamber immediately at the beginning of the preheat-time. Thus, combustion gas temperatures immediately rise to above 1000°C. In contrast, the thermocouple inside the combustion chamber requires 7 seconds before its temperature exceeds 1000°C. Therefore the measured combustion chamber temperature is a poor estimate of the combustion gas temperature. Because midpoint catalyst temperature increases much more slowly, measured midpoint catalyst temperature in Figure 4 is a better estimate of its actual temperature. Still, since the thermocouple time response is unknown for these tests, the measured midpoint catalyst temperature should be regarded as a lower bound for the actual catalyst midpoint temperature.

EFFECT OF PREHEAT-TIME ON REFORMATE PRODUCTION - Let us now consider results from tests with reformate production at 1.6 g/s air flow through the full-scale reformer. Figures 5 and 6 show CO selectivity and hydrocarbon (HC) breakthrough profiles during reformer start-up for 2 second, 5 second and 10 second reformer preheat-times.

These two reformer performance metrics, defined below, were determined from emissions bench measurements of HC, CO and CO_2 by sampling the reformate-air mixture in the intake port of the engine.

$$CO \; selectivity = \frac{moles \, CO}{(moles \, CO + moles \, CO2)} \qquad (3)$$

$$HC \; breakthrough = \qquad\qquad\qquad (4)$$
$$\frac{moles \, C \, in \, HCs}{(moles \, C \, in \, HCs + moles \, CO + moles \, CO2)}$$

HC breakthrough indicates the fraction of gasoline that escapes the reforming process and remains as HCs downstream of the reformer. CO selectivity indicates the fraction of carbon in the reformed gasoline that is converted to CO (as desired), rather than being completely oxidized to CO_2. Perfect reformate would have 0% HC breakthrough and 100% CO selectivity. High CO selectivity is important because production of CO_2 degrades engine combustion, both due to lower reformate fuel energy and the addition of diluent to the fuel-air mixture in the engine intake. HC breakthrough is undesirable because it provides a source of HCs in the fuel that lead to higher engine HC emissions. We did not have a means to measure H_2 production for these tests. However, the Appendix provides mass spectrometer data from bench tests using a similarly-configured

reformer. Mass spectrometer data from these bench tests show that the profile of H_2 production by the reformer during start-up is similar to, but slightly lags, the CO production profile. Thus comparing CO selectivity and HC breakthrough profiles between tests provides a good indication of overall reformate quality.

The abscissas for Figures 5 and 6 indicate time after the beginning of reformate production. Negative values of time represent reformer preheat with $\Phi_{reformer} = 1$, and t=0 represents the instant that $\Phi_{reformer}$ is switched to 2.75. Before t=0, the figures provide an indication not of reformer performance, but rather combustion characteristics in the reformer combustor section during reformer preheat. Perfect stoichiometric combustion would result in oxidation of all fuel carbon to CO_2, so that

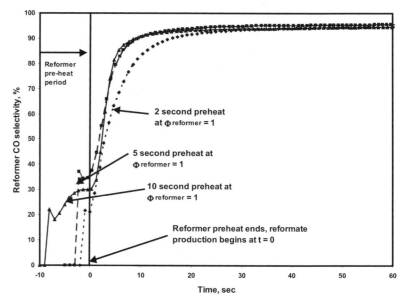

Fig. 5: CO selectivity profiles during reformer start-up.

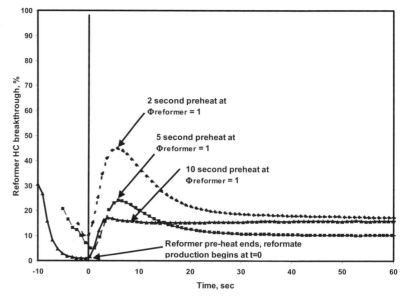

Fig. 6. HC breakthrough profiles during reformer start-up.

both HC breakthrough and CO selectivity would be 0%. Low CO selectivity and HC breakthrough indicate greater energy release for faster reformer preheat. Low HCs are also important during this preheat phase because the engine does not fire during this period so that HC emissions generated by the reformer during this time contribute directly to engine exhaust emissions. (The impact of preheat HCs on total engine-out HC emissions is discussed later).

Of more interest in this section are the characteristics of Figures 5 and 6 after reformate production begins (t ³ 0). The 10 second data shows the best reformer performance during reformer start-up. CO selectivity rises rapidly with a HC breakthrough of approximately 15%. The 5 second reformer preheat data shows performance that is only slightly worse than the 10 second test. CO selectivity is nearly identical to the 10 second data. HC breakthrough is slightly higher, reaching a peak value of 25% approximately 6 seconds after reformate production begins. Reformer performance is severely degraded for the 2 second preheat data. CO selectivity rises significantly slower than for the other tests, indicating substantial quantities of CO2 diluting the intake charge ingested by the engine. Further, HC breakthrough peaks at nearly 50% early during reformer start-up.

SINGLE-CYLINDER ENGINE PERFORMANCE RESULTS - The engine performance as measured by the net mean effective pressure (NMEP) during start-up is shown in Figure 7 for the 2, 5, and 10 second preheat-times. For comparison the NMEP values with PFI gasoline are also plotted in Figure 7. For the PFI data, a rich pulse of fuel (Φ_{engine} = 1.8) had to be supplied initially (for 10 engine cycles) and decayed down to equivalence ratio of 1.0 to get a quick start. (This is akin to cold start enrichment calibrations typically used in production vehicles for rapid engine starting.) NMEP is a measure of net engine work output and is plotted from the start of reformate fueling. Note that several engine cycles go by before the engine starts to fire for all three data sets plotted. The initial delay in NMEP (comprising 5 engine cycles) is due to the transport time required for the reformate to reach the engine from the reformer. With 2 second preheat-time, the engine exhibits unacceptable start-up, with misfires and unstable combustion before the NMEP stabilizes above 400 kPa. With 5 second preheat-time, initial engine combustion is much stronger, and cycle-by-cycle variability is markedly reduced during start-up. With 10 second preheat, the engine starts up as soon as reformate enters the cylinder and NMEP was quite stable almost immediately after start-up.

The engine performance results described in this section correlate well with reformer performance documented in Figures 5 and 6. The superiority of the engine starting performance with the 10 second preheat is clearly evident. PFI gasoline started the engine one cycle quicker than reformate with 10 second preheat. But the engine stability

as indicated by the fluctuation of NMEP was much better with the reformate than with PFI gasoline.

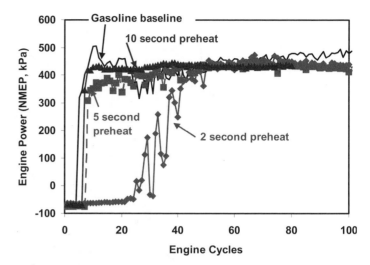

Fig. 7: Effect of preheat-time on engine NMEP from the start of reformate fueling. PFI gasoline baseline included for comparison.

SINGLE-CYLINDER ENGINE HC EMISSIONS RESULTS - Hydrocarbons were observed both during reformer preheat as well as during reformate production. The magnitude of the HC emitted depends on many factors and their composition varies from pure fuel components to partial oxidation species. The simple measurements made do not permit us to identify the sources or the composition of the HC emissions with certainty.

The exhaust HC emissions corresponding to the start-ups with reformate and PFI gasoline discussed above are shown in Figure 8. With the 2 second preheat the misfires and poor combustion during start-up result in extremely high HC emissions. For the 5 second preheat, fairly significant HC emissions occurred during the preheat, but decreased steadily with the onset of reformate delivery to the engine, reaching a steady value of about 150 ppm. With the 10 second preheat a small amount of HC breakthrough occurred during the preheat. This was followed by a gradual increase of HC during start-up until it reached a steady value of 150 ppm. Ideally with the engine running on reformate there should be no HC emissions. However, recall from Figure 6 that HC breakthrough for these tests settled at a steady value of 10-15%. The engine combustion process appears to burn most of these hydrocarbons but a small amount (150 ppm) of unburned HC escapes combustion and is emitted in the exhaust. With PFI gasoline the exhaust HC emissions rise quickly during start-up and reach a steady value of roughly 1500 ppm.

In Figure 9 we have plotted the cumulative mass of HC emissions from the start of reformate fueling up to 20 seconds of engine running. The 2 second preheat-time

333

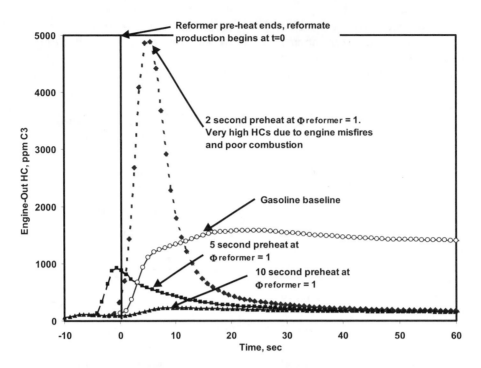

Fig. 8: Effect of preheat-time on engine-out exhaust HC concentrations. PFI gasoline baseline included for comparison.

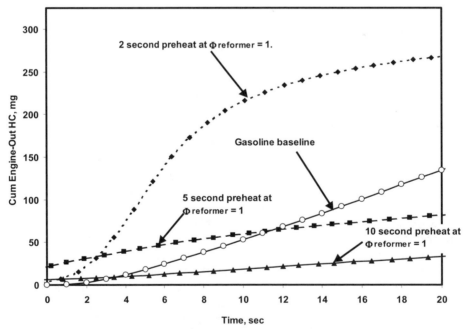

Fig. 9: Effect of preheat-time on cumulative engine-out mass HC emissions. PFI gasoline baseline included for comparison.

In Figure 9 we have plotted the cumulative mass of HC emissions from the start of reformate fueling up to 20 seconds of engine running. The 2 second preheat-time as expected resulted in extremely high cumulative mass HC at the end of 20 seconds because of the misfires and poor combustion during start-up. The cumulative mass HC with 5 seconds preheat-time starts with an offset at time zero because of HC breakthrough during the preheat-time but ends with about a 40 percent reduction in cumulative mass HC emissions compared with PFI gasoline. With the 10 second preheat-time the cumulative mass HC emissions were lowest, providing a 75 percent reduction compared with PFI gasoline after 20 seconds of engine operation.

The HC breakthrough during the preheat-time was subject to significant variations, and depended on the ignition and early flame development in the combustion chamber of the reformer A repeat test with 5 second preheat, gave much lower HC breakthrough during the preheat-time – probably because of quick ignition and fast early flame development. The low HC breakthrough during preheat with the repeat 5 second test shows cumulative mass HC emissions very similar to the 10 second preheat start.

RESULTS OF TESTS WITH Φ_{ENGINE} VARIATIONS

The purpose of these tests was to characterize the ability of the reformate produced by the experimental reformer to burn mixtures leaner than $\Phi_{engine} = 1.0$ during start-up. Test with the quarter scale reformer suited for the single-cylinder engine were run with lean Φ_{engine} supplied to the engine. The reformate flow was less than half the amount produced in the full scale reformer tests described above. For comparison the engine was also fueled with California phase 2 gasoline using port-fuel injection (PFI) during start-up. To achieve a quick start with PFI gasoline a rich pulse of $\Phi_{engine} = 1.8$ was initially supplied and decayed down to $\Phi_{engine} = 1.0$ in 40 engine cycles. The steady speed and load were 1330 rpm and 216 kPa NMEP after the initial transient fluctuations during start-up. Only data for $\Phi_{engine} = 1.0$ and 0.75 are shown for the tests with the scaled down reformer.

START-UP NMEP - Figure 10 shows the NMEP during start-up with gasoline and reformate. The NMEP with gasoline was initially higher than 216 kPa (the steady state target value) because of the rich fueling to get a quick start. With reformate, the NMEP rose quickly to within 85% of the steady value and stabilized thereafter to the steady value. During stoichiometric reformate start-up, the cyclic variations in NMEP were very low.

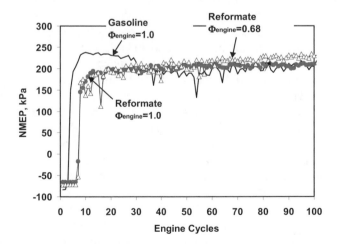

Fig. 10: Effect of Φ_{engine} on engine NMEP from the start of reformate fueling with the quarter-scale reformer. PFI gasoline baseline included for comparison.

With lean reformate fueling $\Phi_{engine} = 0.68$, the cyclic variations in NMEP were comparable if not slightly lower than those for the gasoline test after the initial start-up. As explained above, the transport delay for the reformate to reach the engine cylinder caused an initial delay of 4 to 5 cycles for the rise in NMEP with reformate. Note that the quality of the early reformate generated by the reformer is likely to be somewhat poor with higher amounts of unburned fuel, CO_2 and water vapor in the first few fired cycles. Even so the engine appears to start-up with negligible hesitation.

EMISSIONS - Figure 11 shows plots of NOx, HC, and CO emissions respectively for the gasoline and reformate starts. The stoichiometric and lean reformate starts gave substantially lower NOx and HC emissions when compared with gasoline. In fact the NOx emissions with the lean reformate start were at least an order of magnitude lower than those for PFI gasoline operation. Clearly, the ability to run lean with reformate is a large factor in the lower NOx emissions compared to gasoline. However, there is also a second factor at work. Reformate contains CO_2 and water as described earlier in Eq. (1). These constituents produced by the reformer serve the same function as EGR to dilute the fuel-air mixture to the engine and lower combustion temperatures. This explains why even stoichiometric operation with reformate provides much lower NOx than the gasoline baseline.

The poorer quality reformate likely to be produced during the early part of the reformer start-up did not appear to adversely impact HC emissions. CO emissions were high during the start-up with reformate but after the engine stabilized the CO emissions with lean reformate were also significantly lower than with gasoline. The initial spike in CO could be due to either incomplete combustion during the preheat time of the reformer or the incomplete burning of the leading edge of the initially poor quality reformate air mixture which may be beyond the flammability limit. Further work is needed to address this issue.

DISCUSSION

The results provided above represent a significant step in developing a fast start-up on-board reformer system. Results from our experiments show an excellent correlation between reformate production and engine combustion and HC emissions. Overall, with our current generation of experimental hardware, the data indicate that gasoline-fueled reformer preheat offers a means of rapid reformate production that can enable cold starts with up to a 75% reduction in engine-out HCs compared to baseline PFI gasoline operation. The ability to start the engine quickly with lean reformate mixtures and get at least an order of magnitude reduction in NOx

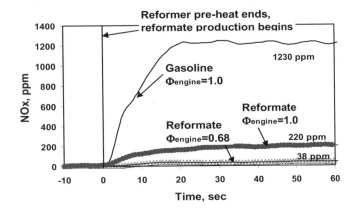

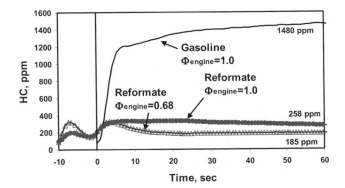

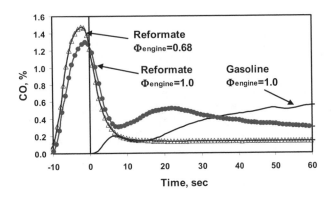

Fig. 11: Effect of Φ_{engine} on engine-out emissions with the quarter-scale reformer. PFI gasoline baseline included for comparison.

emissions compared with PFI gasoline has also been demonstrated. Additional reductions in tailpipe emissions should be realized when reformate heating of the exhaust catalyst is employed for faster light-off. Overall, we find these results to be encouraging and continue to show promise for our on-board reforming system to enable meeting SULEV.

However, significant work remains to be done on several fronts. Currently, reformer preheat-time is 5 to 10 seconds with a pre-vaporized gasoline-air mixture being fed to the reformer. While this is relatively fast, We believe that this ultimately must be reduced to roughly 2 seconds, and include a means for rapid fuel vaporization.

Significant efforts are under way (details are proprietary) to enable rapid fuel vaporization and decrease reformer preheat-time. HC emissions during the reformer preheat period can comprise a significant fraction of total cold start HCs with reformate. The magnitude and start-to-start variability of HC emissions due to reformer preheat must be minimized. Ultimately, the system must exceed our current cold start HC reduction capability. Finally, significant efforts are still required to couple the reformer system to a vehicle. One challenge here is providing appropriate transient response. A second challenge is adapting and calibrating the reformer-vehicle system to optimize its performance with full implementation of our H_2 enrichment strategy.

SUMMARY AND CONCLUSIONS

This paper provides an update of work-in-progress toward our development of a H_2 enrichment system for meeting SULEV emission standards in spark-ignition engines. Results from this paper have focused on tests using gasoline combustion to rapidly warm-up the reformer catalyst to enable reformate-fueled engine cold starts with very low HC emissions. Tests for this study were performed using an experimental reformer system with pre-vaporized gasoline and air. (An optimized mixture preparation system for the reformer start-up period is under parallel development, but not described in this paper.)

The following conclusions are supported by the data presented in this paper.

- Gasoline combustion was able to rapidly preheat our experimental reformer. A preheat period of 5 to 10 seconds (with pre-vaporized gasoline) enabled production of good quality reformate for fast engine starting. Shorter preheat times led to significant engine misfires.
- Reformer temperature was a strong function of preheat-time. Measured reformer midpoint catalyst temperature varied from 150°C to 440°C with preheat-times from 2 to 10 seconds. Due to thermocouple lag, these measured temperatures are regarded as the lower bound for the actual midpoint catalyst temperature at these times.
- HC mass emissions from reformate fueled cold starts in a single-cylinder engine were up to 75% lower than with gasoline fueling. The reformer preheat period showed significant test-to-test variability in HC emissions. Ultimately, cold start HCs must be consistently lowered to enable a viable system for meeting SULEV.
- Starting the engine with lean reformate air mixtures was demonstrated. An order of magnitude reduction in NOx emissions compared with gasoline operation was observed.

Based on these findings, on prior published work, and on the expected additional benefits in catalyst light-off and warmed up emissions, we remain optimistic on the successful practical implementation of the onboard reformer strategy to meet SULEV emissions in gasoline SI engines.

ACKNOWLEDGMENTS

Mark Smigielski is cited for fabricating the experimental reformer for the single-cylinder engine. He also installed the reformer, ran the experiments, maintained the single-cylinder engine test facility, and helped with the data processing. Dave Schumann and Jeff Weissman supplied the reformer catalyst samples for this study. Rick Nashburn, Mike Salemi, Jonathon Bennett, Pete Crawford, and Brian Allston are providing critical support in reformer development, testing and vehicle implementation.

REFERENCES

1. Kitagawa, H.; Mibe, T.; Okamatsu, K.; and Yasui, Y. "Design of L4 Engine for Super Ultra Low Emission Vehicle" SAE Paper 2000-01-0887, Society of Automotive Engineers, March 2000.

2. Nishizawa, K.; Momoshima, S.; Koga, M.; Tsuchida, H.; and Yamamoto, S. "Development of New Technologies Targeting Zero Emissions for Gasoline Engines" SAE Paper 2000-01-0890 Society of Automotive Engineers, March 2000.

3. Masaki, U.; Akazaki, S.; Yasui, Y.; and Iwaki, Y. "A Quick Warm-up System during Engine Start-up Period Using Adaptive Control of Intake Air and Ignition Timing" SAE Paper 2000-01-0551, Society of Automotive Engineers, March 2000.

4. Nishizawa, K.; Mitsuishi, S.; Mori, K.; and Yamamoto, S. "Development of Second Generation Gasoline P-ZEV Technology" SAE Paper 2001-01-1310, Society of Automotive Engineers, March 2001.

5. Breshears, R.; Cotrill, H.; and Rupe, J. "Partial Hydrogen Injection into Internal Combustion Engines – Effect on Emissions and Fuel Economy" First Symposium on Low Pollution Power Systems Development, Ann Arbor, MI, 1973.

6. Houseman, J.; and Cerini, D. J. "On-Board Hydrogen Generator for a Partial Hydrogen Injection Internal Combustion Engine" SAE paper 740600, Society of Automotive Engineers, 1974.

7. Houseman, J.; and Hoehn, F. W. "A Two-Charge Engine Concept: Hydrogen Enrichment" SAE paper 741169, Society of Automotive Engineers, 1974.

8. Jamal, Y.; and Wyszynski, M. L. "On-Board Generation of Hydrogen-Rich Gaseous Fuels – A Review," International Journal of Hydrogen Energy, Vol. 19, No. 7, pp. 557-572, Elsevier Science, 1994.

9. Kirwan, J. E.; Quader, A. A.; and Grieve, M. J. "An On-Board Gasoline Reforming System for Meeting SULEV Emissions Requirements in a Spark-Ignition Engine" Proceedings of the Global Powertrain Congress, Detroit, Michigan, June, 2000.

10. Kirwan, J. E.; Quader, A. A.; and Grieve, M. J. "Advanced Engine Management Using On-Board Gasoline Partial Oxidation Reforming for Meeting Super-ULEV (SULEV) Emissions Standards" SAE Trans. Paper 1999-01-2927, Society of Automotive Engineers, August 1999.

11. Grieve, M. J.; Kirwan, J. E.; and Quader, A. A. "Integration of a Small On-board Reformer to a Conventional Gasoline Internal Combustion Engine System to Enable a Practical and Robust Nearly-zero Emission Vehicle" Proceedings of the Global Powertrain Congress, Stuttgart, Germany, October 1999.

12. Grieve, M. J. "Hydrogen Leveraging for Near Zero-Emission Vehicles with Conventional or Mild Hybrid Powertrains and Gasoline Fuel", Proceedings of the Global Powertrain Congress, Detroit, Michigan, October 1998.

13. Kirwan, J. E.; Quader, A. A.; and Grieve, M. J. "Development of a Fast Start-up On-Board Gasoline Reformer for Near Zero Emissions in Spark-Ignition Engines," Proceedings of the 10th Aachen Colloquium on Automobile and Engine Technology (Aachener Kolloquium Fahrzeug- und Motorentechnik), 8-10 October, 2001.

APPENDIX – BENCH TESTS WITH EXPERIMENTAL REFORMER

In addition to engine testing with the reformer, bench tests are also ongoing to characterize reformer performance during start-up. The reformer used for bench tests is different hardware from that used on the engine, but is essentially the same size as the full-scale reformer and has the same components as described above in the text (vaporizer, combuster and reformer-catalyst).

In these bench tests, a mass spectrometer provides time-resolved measurements of reformate composition. Figure A-1 shows typical reformate composition from a start-up test with 9 second reformer preheat and two different reformer airflow rates. The graph on the left is with 1.5 g/s reformer air flow the graph on the right is with 2.5 g/s reformer air flow. Notice that there is very little difference in the start-up behavior between the two reformer air flow rates for the first 15 seconds. During the preheat phase, stoichiometric gasoline combustion produces mostly CO_2. After 9 seconds, the fueling mixture is enriched and reformate production begins rapidly. CO and H_2 concentration profiles are well-correlated, indicating that our measurement of CO in engine testing is a good indicator of H_2 production as well. Note that within 3 seconds after reformate production begins, CO and H_2 yields are both greater than 60% of their final value. The reformate yield is somewhat higher for the lower flow rate after 30 seconds. At 1.5 g/s air flow, CO and H_2 are about 21% each compared to about 17% each for 2.5 g/s airflow.

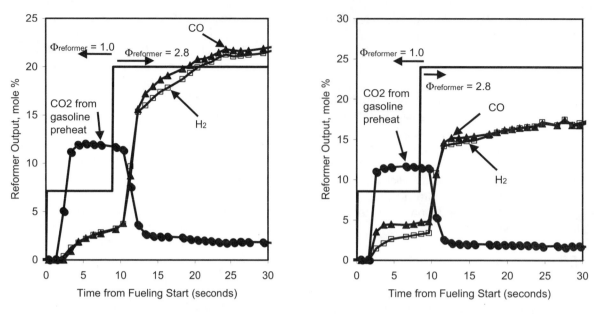

Figure A-1. Reformer output during start-up from bench tests at reformer airflow rates of 1.5 g/s (left figure) and 2.5 g/s (right figure).

2001-01-1193

Feasibility Study of an Online Gasoline Fractionating System for use in Spark-Ignition Engines

Aaron Oakley, Hua Zhao and Nicos Ladommatos
Brunel University, UK

Tom Ma
Ford Motor Co.

Copyright © 2001 Society of Automotive Engineers, Inc.

ABSTRACT

A fuel fractionating system is designed and commissioned to separate standard gasoline fuel into two components by evaporation. The system is installed on a Ricardo E6 single cylinder research engine for testing purposes.

Laboratory tests are carried out to determine the Research Octane Number (RON) and Motoring Octane Number (MON) of both fuel fractions. Further tests are carried out to characterize Spark-Ignition (SI) and Controlled Auto-Ignition (CAI) combustion under borderline knock conditions, and these are related to results from some primary reference fuels.

SI results indicate that an increase in compression ratio of up to 1.0 may be achieved, along with better charge ignitability if this system is used with a stratified charge combustion regime. CAI results show that the two fuels exhibit similar knock-resistances over a range of operating conditions.

INTRODUCTION

The exhaust emissions from internal combustion (IC) engines have been reduced by tenfold during the last two decades and have to be reduced by at least another 50% in the next several years in order to meet the stage IV or ULEV2 emission standard. In recent years, reduction of exhaust emissions from SI engines has been achieved through the use of three-way catalysts. However, SI engines have to operate at stoichiometric air/fuel (A/F) ratio in order to use three-way catalysts, thereby suffering poorer fuel economy at part-load conditions. The fuel economy of SI engines has been lowered further by the reduction of maximum compression ratio when lead alkyl compounds are removed from gasoline fuels. Therefore, there is a strong industrial need for an IC engine with simultaneous improvements in fuel economy and exhaust emissions.

It is well known that the reduction in CO_2, or fuel consumption, from SI engines can be achieved by increasing the compression ratio and by means of lean-burn or exhaust gas re-circulation (EGR) dilution at part-load. Traditionally, efforts have been concentrated on the mechanical and fluid mechanical aspects of engine design in order to extend the lean-burn or dilution limits, and to increase the maximum allowable compression ratio. These include the optimisation of combustion chamber design, in-cylinder flow, and air/fuel mixture distribution.

On the other hand, IC engines rely on the release of fuel chemical energy to produce mechanical power output. The properties of gasoline fuel have significant effects on combustion and pollutant formation in the cylinder. They also have significant influence on the driveability and performance attributes of the engine, such as reliable cold starting, fast warm-up, rapid-response during acceleration, knock-free combustion and durable operation. To meet all these attributes, the gasoline fuel is a carefully formulated blend of many hydrocarbons of different properties to give the best combination for all-round performance, and as such the fuel is treated as one entity based on the averaged properties of its many components.

Here, for the first time, we propose to use the standard gasoline fuel in a more flexible manner, by targeting the superior properties of the separate fuel components within the blend to match each engine attribute more specifically. For example, the lighter fractions of the gasoline are more ignitable and are better suited for cold start and lean burn, whereas the heavier fractions of the gasoline have higher octane number and are more resistant to knock. This has led to an exciting new concept of charge stratification by fuel components, rather than by air/fuel ratio of the averaged components. According to this new concept, the whole gasoline fuel is consumed completely in the engine every cycle just like an ordinary engine, but it can be used more strategically by concentrating the lighter and heavier fractions in different parts of the combustion chamber. In particular, stratifying the high octane fraction at full load will allow increased compression ratio for better performance and efficiency, while stratifying the low octane fraction will extend the lean burn limit and improve the cold start quality of the engine.

Gasoline fuel is composed mainly of paraffins and aromatics. Lovell [1] showed that generally speaking, the aromatics contained in gasoline have higher knock resistance than the paraffins. Coincidentally, most of the aromatics contained in gasoline also have higher boiling points than the paraffins. Thus, for the components present in standard gasoline, octane rating scales approximately with boiling point. If gasoline is subjected to external heating, the fractions obtained from the initial boiling temperature to an intermediate boiling point should comprise mainly saturated paraffins. Correspondingly, the fractions obtained from the intermediate boiling point to the final boiling temperature should be made up mostly of aromatics. Therefore, standard gasoline can be separated into light (mainly paraffinic) and heavy (mainly aromatic) fractions so as to take full advantage of the high-volatility and high-octane of the individual steams respectively, to suit different engine operating conditions.

Owen and Coley [2] discuss the distribution of octane quality that occurs through the boiling range of gasoline. Due to the multi-component nature of gasoline, it is possible to have two fuels that meet the same specification in terms of RON and MON, but have vastly different light and heavy fraction octane quality. For normal SI engine operation, this property is important mainly for transient knock phenomena: under engine acceleration when fuel flow into the inlet manifold is high, light fractions will preferentially vapourise, leading to a temporary octane reduction in the fuel entering the cylinder. In theory, for the present system it would be more desirable to have a large octane distribution over the boiling range to obtain the largest octane difference between light and heavy fractions. However, there are still limiting practical considerations such as Reid Vapour Pressure.

A preliminary investigation to find if paraffins and aromatics could be separated by means of heating has been undertaken. 95 Octane unleaded gasoline was sent to Saybolt UK Ltd. for distillation to an intermediate boiling point of 90 °C, and subsequent detailed analysis. Table 1 shows that the light fraction comprises mainly paraffinic hydrocarbons, while the heavy fraction contains mostly aromatic hydrocarbons. Olefins are distributed roughly equally, with most residing in the light fraction. Despite the octane raising presence of Olefins in the light fraction, octane numbers for the light and heavy fractions are recorded as 90 and 99.4 RON respectively, indicating that the paraffin/aromatic octane difference is dominant.

Table 1 Properties of RON 95 unleaded gasoline fuel and its distilled fractions at atmospheric pressure

	Standard Gasoline	Light fraction (Distillation from IBP to 90 °C)	Heavy fraction (Distillation from 90 °C to FBP)
% of total fuel	100	~49	50
Paraffins (% vol)	58.3	79	24.3
Olefins (% vol)	7.1	9.2	6.6
Aromatics (% vol)	34.1	11.8	69.1
RON	94.7	90	99.4

SCOPE

The work presented here covers the development of a fuel fractionating system to investigate the feasibility of 'online' gasoline fractioning in terms of engine fuel delivery requirements. With an operational system, the knock-limited spark advances (KLSA) for each of the fractions at various compression ratios are mapped. These results are compared to those gained from standard gasoline and some primary reference fuels (PRF) of known octane numbers.

Controlled Auto-Ignition (CAI) combustion is a growing area of research in IC engines. CAI combustion involves the compression ignition of a premixed combustible charge. The in-cylinder charge temperature required to achieve auto-ignition is obtained in practice through the use of hot EGR in combination with ordinary charge compression. EGR also acts as a diluent, limiting the heat release rate, thus preventing engine knock. In much the same way as spark-induced auto-ignition, the type of fuel used in CAI combustion helps determine the maximum knock-limited load, and the maximum and minimum permissible EGR rates to maintain stable and non-knocking combustion respectively. Many studies [3-8] have chronicled how various fuels affect CAI combustion. In particular, Thring [6] investigated the effects of A/F ratio, EGR rate, fuel type, and compression ratio on the attainable CAI combustion region and engine-out emissions. A similar approach is taken in this work, where CAI combustion is achieved using inlet charge heating to simulate the EGR heating effect. The gasoline fractions are tested to ascertain the upper load limit (knock limit) at various EGR rates. Comparisons between these results and those gained from testing standard gasoline and some PRF fuels are presented.

DEVELOPMENT OF THE FUEL FRACTIONATING SYSTEM

A number of systems were proposed and tested during the course of this research work. However, only the final version of the system shall be presented here. The entire system including measurement and control systems is presented in Figure 1.

ENGINE SETUP - The fuel fractionating system converts gasoline into two separate fuel streams. The light and heavy fraction streams are output from the system in vapour and liquid form respectively. Development and commissioning of the system requires that these streams be consumed simultaneously in an engine. For this purpose, a Ricardo E6 single cylinder research engine has been selected. The engine is of the single cylinder type with two overhead poppet valves, and has a bore of 76 mm and a stroke of 111 mm. The combustion chamber is cylindrical in shape. The compression ratio of the engine is continuously variable between 4.5 and 20, and may be changed during engine operation by means of a worm gear that controls the cylinder head height relative to the crankshaft. The engine is coupled to a swinging field AC dynamometer allowing accurate manual speed control. Although this engine does not allow facility for fuel stratification, it is extremely versatile with regard to compression ratio and motored speed control, which are more important for prototype development purposes.

transferred to the fuel from a copper coil, through which water at 90°C is passed. This temperature is chosen for two reasons:

1. Results presented in Table 1 show that gasoline is split into two streams of equal mass flow-rate if the intermediate boiling temperature is set at 90°C.
2. For a practical system, heat energy can be transferred either from hot exhaust gases or from the engine coolant system. Since the engine coolant system is normally maintained at approximately 90°C, this matches the distillation requirements of the fuel.

The temperature and pressure of the fuel inside the vapourisation chamber solely determine the fuel cut that is obtained. That is, if gasoline is separated into two equal streams at a temperature of 90°C and atmospheric pressure, an increase in temperature and/or decrease in pressure within the vapourisation

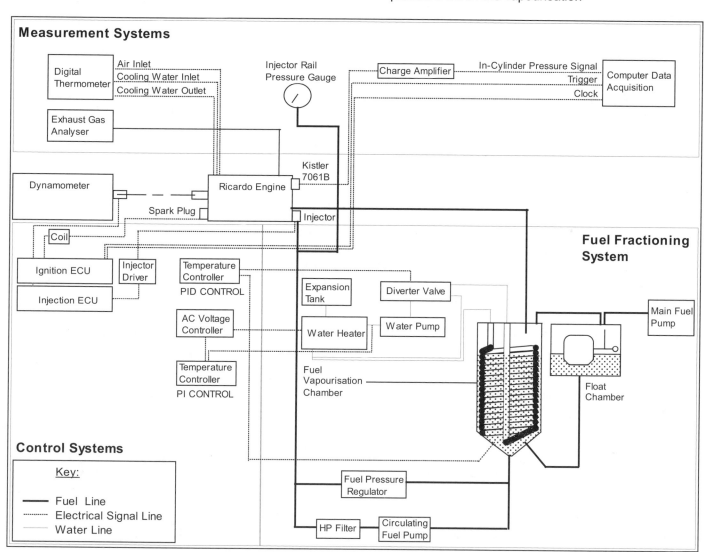

Figure 1: Schematic of the Fuel Fractionating and Associated Control and Measurement Systems

FUEL FRACTIONATING SYSTEM - Fuel fractioning is made to occur in a vapourisation chamber. Heat is

chamber will cause a shift in the vapour/liquid equilibrium, and more vapour will be produced than liquid. Thus, if the temperature and pressure prevalent in the vapourisation chamber can be controlled accurately, then vapour and liquid flow-rates output from the system will be fixed proportional to the flow-rate of gasoline into the system. In this way the hydrocarbon compounds present in the light and heavy fractions remain unchanged, and the fuel properties are constant regardless of the total flow-rate through the system.

The flow-rate of gasoline into the system is controlled by means of a float chamber such that the volume of fuel contained in the vapourisation chamber is fixed. Heavy fraction in liquid form drawn from the bottom of the vapourisation chamber is fed into a standard fuel rail for injection via a Bosch port injector. The pressure maintained in the fuel rail is 2.7 bar, controlled by a fuel pressure regulator. Unused fuel from the rail is circulated back to the inlet of the standard fuel circulation pump. In a practical system, it would not be necessary to circulate the fuel from the rail back to the main fuel tank, as is normal for injector circuits to prevent fuel overheating. This is because the fuel present in the rail is entirely heavy fraction, having had all of the lighter fractions removed to a temperature of 90°C within the vapourisation chamber. Vapour bubbles are unlikely to form in this fuel despite moderate engine bay temperatures.

The flow-rate of heavy fraction through the injector is controlled by a separate electronic system, which is used to set the injector pulse width. The flow-rate of heavy fraction directly determines the flow-rate of gasoline into the fractionating system, since the fuel-cut is fixed by the temperature and pressure conditions of the vapourisation chamber.

Temperature Control of the Vapourisation Chamber - The simulated coolant circuit consists of a 4 kW heater, a water pump, a diverter valve, and a copper coil immersed in the fuel within the vapourisation chamber. The fluid contained in the circuit is a mix of 70% water and 30% ethylene Glycol. The fluid is representative of that used normally in a production engine in terms of viscosity, thermal conductivity, and heat capacity. The heater maintains a coolant temperature of 90 ± 0.2°C by means of separate Proportional + Integral (PI) temperature controller to simulate the (idealised) operation of a standard engine thermostatic valve. The flow-rate of coolant through the coil is controlled by the on/off diverter valve, and depends on the feedback temperature measured at the base of the vapourisation chamber. Coolant is diverted back to the heater inlet when the coil flow is stopped; this prevents excessive pressure build-up at the exit to the water pump.

Due to the poor thermal conductivity of hydrocarbon compounds, thermal gradients exist within the chamber during normal operation. Thus, the Proportional + Integral + Differential (PID) temperature controller that maintains the vapourisation chamber temperature controls the thermocouple 'local' temperature at 65 ± 0.1 °C. Despite this, vapourisation still occurs at the copper/fuel interface at 90°C, maintaining the correct fuel cut.

Pressure Control of the Vapourisation Chamber - Some means is necessary to transport the (vapour) light fraction to the manifold of the engine, while maintaining a constant pressure within the vapourisation chamber. This has been achieved by fitting a constant-depression type carburettor to the engine manifold. Vapour is fed directly to the throat of the venturi. An on/off valve is situated in the vapour line for safety reasons during shutdown of the system.

COMMISSIONING AND PRELIMINARY TESTING - The parameters that define the correct operation of the fuel fractionating system are:

1. A rate of vapourisation of 50% of the total fuel flow
2. Stable (non-oscillating) and continuous vapour production
3. A reasonable octane difference between light and heavy fractions

Various methods of heat input were tried during development of the system. A copper coil immersed in the fuel was found to be a very good method of heat exchange, allowing a large enough heat flow for 50% of the fuel to be vapourised on a continuous basis. Stable vapour production was much harder to achieve. The critical factor in maintaining stable vapour production is that all elements of the system must operate continuously. That is, the fuel flow into the system must approximate continuous despite the on/off operation of the float mechanism. Also, heat input to the system must approximate continuous despite the on/off operation of the diverter valve controlling coolant flow to the coil. Despite these drawbacks, the system is capable of feeding the Ricardo engine with light and heavy fractions (recombined in the inlet manifold) such that variations in lambda measured by a Richard Oliver 650 exhaust gas analyser are smaller than the resolution of the machine (to 2 decimal places) at a stoichiometric A/F ratio.

Table 2 shows the comparison of octane quality between light and heavy fractions produced by the original distillation method (Table 1) and the continuous method used by the fuel fractionating system. As predicted, the light and heavy fractions produced by the fractionating system have lower and higher RON respectively than standard gasoline. However, the system does not appear to separate the paraffins from the aromatics as well as the distillation method, resulting in a smaller octane spread (RON) between light and heavy fractions. Fuel sensitivity (RON-MON) reflects how the knock resistance of a fuel changes as the engine operating conditions (e.g. initial charge temperature, engine speed, ignition timing) become more severe. Heavy fractions clearly have a much higher sensitivity than light fractions, with gasoline roughly in the middle. Thus, as engine-operating conditions are made more severe, the anti-knock quality of the (higher RON) heavy fraction is reduced at a greater rate than that of the (lower RON) light fraction, exemplified by converging values of MON for light and heavy fractions, and standard gasoline.

Table 2 Comparisons of hydrocarbon blends and RON of light and heavy fractions of gasoline using two methods of fuel fractioning

	Gasoline	Light fraction		Heavy fraction	
		Original*	Current**	Original	Current
% of total fuel	100	~49	~48	50	50
Paraffins (% vol)	58.3	79	72.1	24.3	42.6
Olefins (% vol)	7.6	9.2	12.4	6.6	9.9
Aromatics (% vol)	34.1	11.8	15.5	69.1	47.5
RON	94.7	90	93.5	99.4	98
MON	84.5	84	85	87.1	86.5
Fuel Sensitivity	10.2	6	8.5	12.3	11.5

       * Original: by distillation
       ** Current: by evaporation

ENGINE TEST OBJECTIVES AND PROCEDURES

A series of tests have been devised to further ascertain the combustion qualities of the fuel streams produced by the fractionating system. The light and heavy fractions, and standard gasoline are tested under a number of operating conditions, outlined in Table 3. Primary reference fuels consisting of mixtures of Isooctane and n-Heptane are tested under similar conditions to determine what differences exist, if any.

Table 3 Summary of Engine Conditions for Spark-Ignition and Controlled Auto-Ignition (CAI) Tests.

Operating Conditions		Tests A (SI)	Tests B (CAI)
Engine Speed	(rpm)	1500	1500
Throttle		WOT	WOT
lambda		0.95	Variable
Spark Advance	(° CA)	For MBT and Knock	No Spark
Compression Ratio		7, 8, 9, 10, 11, 12, 13	11.5
Inlet Charge Temperature	(° C)	40	320
Coolant Temperature	(° C)	70	80
Oil Temperature	(° C)	55	55
External EGR rate	(% by mass)	0	0-60

The main objectives for both SI and CAI tests are to characterise combustion of the different fuels under borderline knocking conditions, and to ascertain differences in engine operating conditions that cause similar degrees of knock.

A real-time analysis program has been developed at Brunel University based on the Labview® data acquisition system. The system reads inputs from a crankshaft encoder, and an in-cylinder water-cooled pressure transducer (Kistler type 7061B). An array of pressure traces is generated, from which heat-release data (10, 50, and 90% burn CA), and net-indicated mean effective pressure (IMEP) may be calculated. Real-time knock analyses are carried out by setting a digital band-pass filter to single out the characteristic engine knock frequency ($\approx$ 8 kHz), for the Ricardo E6 engine. Measurements of the amplitude of this filtered trace can result in very accurate determination of the knock-limited boundary of operation for each fuel. An amplitude

threshold of 0.5 bar is set to define whether knock has occurred for each individual cycle. Cyclic variations within the combustion chamber caused by variations in gas motion, homogeneity, and mixture composition from cycle to cycle lead to cyclic variations in knock intensity. Consequently, when measuring incipient and non-destructive knock phenomenon within the engine, a sample number of cycles will contain both knocking and non-knocking cycles. The Knock Occurrence Frequency (KOF) is a measure of the percentage of knocking cycles (knocking above the predefined 0.5 bar threshold) out of the total number of cycles recorded.

SI TEST PROCEDURES - Auto-ignition occurs in SI engines when part of the unburned charge (end-gas) is compressed to a high enough pressure and temperature, and for a long enough time period for chemical reactions to occur within the gas that lead to combustion before arrival of the spark-ignited flame front. The temperature and pressure histories of the end-gas are dependant on the rate of charge compression and final pressure due to the piston action (compression ratio), the rate of end-gas compression due to spark-ignited combustion, which are in turn dependant on engine speed. Engine knock occurs if the conditions are such that auto-ignition has taken place and there is sufficient unburned charge to support spontaneous bulk combustion in the end-gas region, involving extremely rapid heat-release and resulting in sonic pressure oscillations. Thus, for an SI engine operating at WOT conditions, the compression ratio, spark-advance, and engine speed are the three major factors in determining whether knock occurs for any particular fuel. Throughout the SI experiments, the inlet charge is kept constant at 40 °C. A comprehensive review of how engine operating conditions effect borderline knock is presented by Russ [9].

The Ricardo E6 engine has a relatively limited speed range of 0-2500 rpm. So, a parametric study of compression ratios and spark advances at a single speed should provide enough data for the comparison of the various fuels under borderline knocking conditions. Each fuel is tested at a range of compression ratios (7,8,9,10,11,12, and 13). At each compression ratio, the spark timing is gradually advanced towards Maximum Spark Advance for Best Torque (MBT) timing, and beyond, where the pressure and temperature rise in the end-gas results in auto-ignition and knock in some cases. The spark advance to cause 10% KOF (at a filtered threshold of 0.5 bar) is designated the Knock Limited Spark Advance (KLSA).

CAI TEST PROCEDURES - Unlike SI engine operation, CAI combustion timing is not controlled directly. Instead, ignition occurs when the in-cylinder temperature, pressure, and chemical conditions are favourable. Thus, ignition timing is controlled through a combination of initial charge temperature, compression ratio, A/F ratio, and EGR dilution for a particular fuel. Furthermore, as with SI operation, heat release rate is dependant on combustion timing, compression ratio, engine speed, A/F ratio, and degree of EGR dilution. The rate of in-cylinder pressure rise is directly determined by the heat release rate; high values can lead to engine knock. So, there are a number of parameters that may be varied to determine how different fuels behave relative to one another. In this study, compression ratio, engine speed, and inlet charge temperature have been fixed so that fuel and dilution effects can be studied independently. For each fuel and a range of external EGR rates, the A/F ratio is gradually reduced at WOT (increasing fuel rate) until 10% KOF is observed. This limit is designated the Knock Limited Lambda (KLL).

An analytical approach for calculating the overall A/F ratio and EGR rate in the cylinder has been developed. A set of equations relating inlet and exhaust gas species including CO and unburned HC allow the simultaneous calculation of A/F ratio and EGR rate without measuring inlet airflow directly (Heywood [10] and Stone [11]). This has advantages over less accurate methods (e.g. UEGO sensors) that do not account for all of the exhaust unburned hydrocarbons, which can be a significant proportion of the injected fuel under some CAI combustion conditions. The external EGR and gas speciation systems are shown schematically in Figure 2.

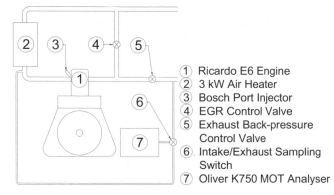

① Ricardo E6 Engine
② 3 kW Air Heater
③ Bosch Port Injector
④ EGR Control Valve
⑤ Exhaust Back-pressure Control Valve
⑥ Intake/Exhaust Sampling Switch
⑦ Oliver K750 MOT Analyser

Figure 2: External EGR and Gas Sampling Systems

RESULTS AND DISCUSSION

SI RESULTS AND DISCUSSION - Figure 3 shows the measured KLSA for six fuels and a range of compression ratios. Although tests were carried out at compression ratios as low as 7, none of the fuels were forced into knocking combustion under those conditions. As would be expected the KLSA is retarded with increasing compression ratio for every fuel. Higher compression ratios cause higher in-cylinder pressures and temperatures, leading to higher heat release rates and the spark advance to cause a similar degree of knock is retarded.

Comparisons between gasoline and heavy fraction results show that if the heavy fraction is used to occupy the end-gas region of the combustion chamber, increases in maximum compression ratio can be made. An engine normally having a high compression ratio would benefit over one of low compression ratio, since the benefits of switching to the heavy fraction are increased at higher compression ratios. For example, an

engine with a maximum compression ratio of 9 and normally operated with standard gasoline fuel could have its compression ratio increased by approximately 0.3 for the same degree of KLSA if the heavy fraction were stratified to occupy the end-gas region of the combustion chamber. However, an engine with maximum compression ratio of 12 could have it increased to approximately 13 for the same KLSA, if heavy fraction fuel stratification were employed.

Differences between gasoline fractions and Primary Reference Fuels (PRFs) are most notable for the higher-octane blends. The heavy fraction (98 RON) clearly has a much flatter response to increasing compression ratio than even the 95 RON PRF. On the other hand, the Light fraction appears to have the same knock resistance as the 90 RON PRF despite having higher RON and MON.

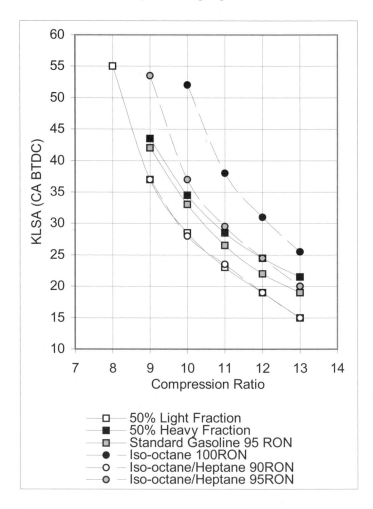

- □ — 50% Light Fraction
- ■ — 50% Heavy Fraction
- ▣ — Standard Gasoline 95 RON
- ● — Iso-octane 100RON
- ○ — Iso-octane/Heptane 90RON
- ◐ — Iso-octane/Heptane 95RON

Figure 3 Spark Advance (BTDC) to Cause 10% KOF Against Compression Ratio for Various Fuels

It is interesting to speculate why the gasoline fractions show different trends to PRFs. One explanation is this: The end-gas compression to cause auto-ignition results more from combustion at lower compression ratios, and more from the piston action at higher compression ratios. This is because as compression ratio is increased, the mass fraction burned at the onset of end-gas auto-ignition becomes progressively less. The combustion event up to the onset of auto-ignition can be divided into two distinct phases: Ignition (flame development), and flame propagation. Figure 4 shows the ignition and heat release data (10, 50, and 90% burn CA) for each of the six fuels at each compression ratio and MBT timing. Generally speaking, the 10-90% burn CA for each fuel is equal at each compression ratio. This assertion is supported by the work of Davis and Law [12], who showed that under laboratory conditions, aromatics and paraffins exhibit similar laminar flame speeds - the defining characteristic of combustion duration if turbulence is discounted (for similar compression ratio).

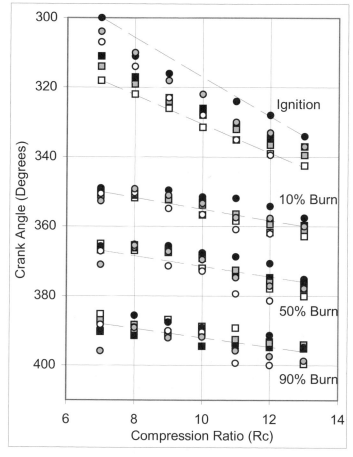

Figure 4   MBT Ignition Timings and Burn History against Compression Ratio for Various Fuels (for legend see figure 5)

However, to maintain MBT the PRFs require more advanced timing than the gasoline fractions, as do the higher-octane fuels compared to the lower-octane fuels of each class. This indicates that the flame development angle (0-10% burn) is the key combustion characteristic in determining the KLSA for a particular fuel. Since development angles for PRFs are significantly larger than for gasolines, this explains spark-timing differences to cause similar degrees of knock. Furthermore, these differences are reduced at higher compression ratios because auto-ignition is less dependent on the spark-ignited combustion event.

Although this explanation does account for the differences between 100 and 95 RON PRFs, and heavy and standard gasoline distillates respectively, it does not explain why the light fraction appears to have the same knock resistance as the 90 RON PRF irrespective of

compression ratios, despite having dissimilar flame development angles. Fuel sensitivities given in Table 2 offer further explanation for these trends. PRFs by definition have zero fuel sensitivity and as such, they are the benchmark for measuring other fuel types. Results in Figure 3 appear to suggest that the severity of engine operating conditions increases with decreasing compression ratio, due to the more advanced ignition timings required to obtain knock. The multi-component fuels containing significant amounts of aromatics (standard gasoline and heavy fraction) have increasingly retarded KLSA values as the compression ratio is decreased, relative to a PRF of the same RON. Since the light fraction comprises mainly paraffinic compounds with low fuel sensitivity, it is expected to behave more like the PRFs over a range of operating conditions.

Comparisons between light fraction and gasoline results in Figure 4 show that if fuel stratification were employed such that the light fraction occupies the region around the spark plug, the ignitability of the charge is increased. A shorter flame development angle for the light fraction leads to lower overall combustion duration, which results in more efficient and stable combustion. However, it is not clear whether the shortened flame development angle for the light fraction is a result of chemical ignition characteristics of the fuel or better charge homogeneity, since only the light fraction enters the engine manifold as a vapour.

CAI RESULTS AND DISCUSSION - Figure 5 shows the 10% KOF knock limited lambda for each of the fuels tested. For each of the PRFs, the minimum lambda obtainable during testing is much higher than 1.0. The presences of CO_2 and H_2O species in the EGR are responsible for retarding combustion timing and increasing combustion duration, resulting in high combustion variability at high EGR rates and eventual misfire. This response is a function of how the fuel's auto-ignition chemistry is affected by the exhaust gas species. Clearly, the gasoline fractions are more tolerant to EGR than the PRFs under these conditions. Although the gasoline fractions are capable of CAI combustion at lambda 1.0 and high EGR rates, combustion is effected in a similar way to the PRF combustion, resulting in instability and misfire at higher EGR rates (>45%).

Among the PRFs, there is clearly a relationship between minimum lambda attainable (KLL) at each EGR rate and fuel octane, with Isooctane being most resistant to knock and 80 RON PRF least resistant. Although this trend is intuitive, the same does not follow for the gasoline fractions. In fact, all three gasoline fuels have considerably lower knock resistance than even the 80 RON PRF over the entire EGR range, despite having higher RON and MON values. Furthermore, the differences in KLL trends of each fraction do not appear to bear any relation to octane quality or fuel blend.

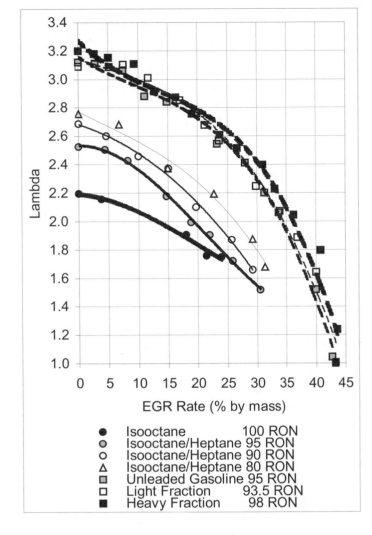

●	Isooctane	100 RON
◉	Isooctane/Heptane	95 RON
○	Isooctane/Heptane	90 RON
△	Isooctane/Heptane	80 RON
▣	Unleaded Gasoline	95 RON
▢	Light Fraction	93.5 RON
■	Heavy Fraction	98 RON

Figure 5 Lambda against EGR rate for various fuels defining knock limited CAI boundary

To understand the different behaviours shown by the gasoline fractions regarding KLSA and KLL, one needs to look at the engine conditions for SI and CAI combustion. CAI combustion is achieved here through the use of elevated intake charge temperatures. The entire charge (air, fuel and EGR) is formed at a temperature of 320°C in the intake manifold. However, the charge temperature during the SI tests carried out previously was maintained at only 40°C. Thus, the conditions are made more severe for the CAI tests. MON tests are usually carried out at an initial charge temperature of 149°C, so the CAI test conditions are even more severe than the MON test. This assertion is supported by the fact that the gasoline (84.5 MON) exhibits a considerably lower knock resistance than the 80 MON PRF.

Results presented in Table 2 show that the octane qualities of the gasoline fractions tend to converge as conditions approach those of the MON test. This occurs because the fuel sensitivity of aromatic compounds (higher RON) is higher than for paraffinic compounds (lower RON). If the sensitivities are extrapolated beyond MON values to simulate even more severe conditions such as in these CAI tests, the light fraction may even

exhibit higher knock resistance than the heavy fraction, as indicated in the trends for light and heavy fractions at higher EGR rates.

The results of the CAI tests are inconclusive. The test conditions are responsible for each of the gasoline fractions exhibiting similar trends of KLL. More tests are required under conditions that better approximate those found in a CAI engine that uses internal EGR to heat the intake charge during the induction and compression strokes.

CONCLUSIONS

A system has been designed and built to separate standard gasoline into two equal fractions by evaporation. Octane tests on the resulting fractions show that they exhibit differences in RON and MON. The specific properties of each fraction may be used to improve various engine attributes such as cold start operation, and faster and more knock-resistant combustion. The system has been commissioned using a research engine, and optimised for steady state operation testing to better understand the characteristics of the fuels produced. The following conclusions can be drawn from these tests:

1. Fuel fractions obtained by laboratory distillation to an intermediate boiling point of 90°C show a large Research Octane spread of 9 ON. However, fractions obtained by the online experimental system show a smaller Research Octane spread of 4.5 ON.
2. Fuel sensitivities of the different fuel types conspire to cause their knock resistances to converge under more severe engine operating conditions, resulting in similar MON values.
3. The light fraction is more easily ignitable than the heavy fraction or standard gasoline under SI operation. This result is observed through a reduction in flame development angle (0-10% burn) of up to 10% compared to that of standard gasoline. However, some or all of this effect may be attributable to better charge homogeneity, since the light fraction enters the inlet manifold of the engine as a vapour. In any case, greater ignitability facilitates faster and more stable combustion.
4. The heavy fraction is more resistant to knock than either the light fraction or standard gasoline. This may allow increases in compression ratio if fuel stratification is employed.
5. Increases in compression ratio resulting from fuel stratification can be larger for higher compression ratio engines (11-12). Results indicate increases as much as 1.0 may be achieved with the current fuel fractioning system.
6. The CAI test operating conditions selected have proven to be more severe than the MON tests. Consequently, Knock Limited Lambda (KLL) trends with EGR rate for the gasoline fuels tend to converge, indicating there is little or no difference in knock resistance between light and heavy fractions.

More tests are required and this is the subject of further work.

In order to take full advantage of the stratified charge fuel fraction concept, a better fuel fractionating system is required for more efficient separation of the paraffinic and aromatic compounds contained in gasoline. In addition, modifications to gasoline fuel to increase the octane spread between light and heavy fractions will help to realise the potential of this technology. Despite strict control of the octane quality of commercial gasolines, composition of gasoline from region to region may vary greatly, which in turn may affect the calibrated operation of a fuel fractionating system. Further work is necessary to determine how sensitive the system is to variation in gasoline fuel blend.

ACKNOWLEDGMENTS

The authors would like to acknowledge the financial support to the work reported here by EPSRC and the Ford Motor Company

REFERENCES

1. Lovell, W. G., "Knocking Characteristics of Hydrocarbons", Ind. Engng. Chem.40, pp. 2388-2438, 1948.

2. Owen, K., and Coley T., 'Automotive Fuels Reference Book', pp. 108-111, SAE publications, ISBN 1-56091-589-7, 1995.

3. Onishi, S., Hong Jo, S., Shoda, K., Do Jo, P., and Kato, S., "Active Thermo-Atmosphere Combustion (ATAC) - A New Combustion Process for Internal Combustion Engines", SAE Paper 790501, 1979.

4. Noguchi, M., Tanaka, Y., Tanaka, T., and Takeuchi, Y., "A Study on Gasoline Engine Combustion by Observation of Intermediate Reactive Products during Combustion", SAE Paper 790840, 1979.

5. Najt, P. M., and Foster, D. E., "Compression-Ignited Homogenous Charge Combustion", SAE Paper 830264, 1983.

6. Thring, R. H., "Homogenous Charge Compression-Ignition (HCCI) Engines", SAE Paper 892068, 1989.

7. Pucher, G. R., Gardiner, D. P., Bardon, M. F., and Battista, V., "Alternative Combustion Systems for Piston Engines Involving Homogenous Charge Compression Ignition Concepts - A review of Studies Using Methanol, Gasoline and Diesel Fuel.", SAE Paper 962063, 1996.

8. Stanglmaier, R. H., and Roberts, C. E., "Homogenous Charge Compression Ignition (HCCI): Benefits, Compromises, and Future Engine Applications", SAE Paper 1999-01-3682, 1999.

9. Russ, R., "A Review of the Effect of Engine Operating Conditions on Borderline Knock", SAE Paper 960497, 1996.

10. Heywood, J. B., "Internal Combustion Engine Fundamentals", pp. 148-152, McGraw-Hill Book Company, ISBN 0-07-100499-8, 1988.

11. Stone, R., "Introduction to Internal Combustion Engines", Third Edition, pp.535-541, Macmillan Press, ISBN 0-333-74013-0, 1999.

12. Davis, S.G., and Law, C.K., "Determination of and Fuel Structure Effects on Laminar Flame Speed of C_1 to C_8 Hydrocarbons", Combust. Sci. and Tech.140, pp.427-449, 1998.

2000-01-1884

The Effects of Driveability on Emissions in European Gasoline Vehicles

Roberto Bazzani
BP Amoco

Klaus Kuck
Aral AG

Yeong Kwon
Esso

Murray Brown and Margret Schmidt
Shell Global Solutions

Copyright © 2000 CEC and SAE International.

ABSTRACT

Fuel volatility and vehicle characteristics have long been recognised as important parameters influencing the exhaust emissions and the driveability of gasoline vehicles. Limits on volatility are specified in a number of world-wide / national fuel specifications and, in addition, many Oil Companies monitor driveability performance to ensure customer satisfaction. However, the relationship between driveability and exhaust emissions is relatively little explored.

A study was carried out to simultaneously measure driveability and exhaust emissions in a fleet of 10 European gasoline vehicles. The vehicles were all equipped with three-way catalysts and single or multi-point fuel injection. The test procedure and driving cycle used were based on the European Cold Weather Driveability test method. Six experimental test fuels of varying volatility were tested in the vehicle fleet, at temperatures of -5°C and +10°C, to provide a range of driveability performance and allow the relationship between driveability and emissions to be investigated.

The main differences in demerits were seen between different vehicles and changes in ambient temperature; fuel volatility and oxygenate content had a relatively minor effect. Analyses of the emissions data showed that a change in the exhaust emissions occurred when driveability malfunctions such as stumble or hesitation were observed. These findings were limited to the data generated at -5°C only as there were too few demerits produced at +10°C. Driveability demerits gave rise to a statistically significant increase in HC and CO_2 emissions, an apparent increase in NOx, but no correlation with CO emissions. The data also showed that the control of air/fuel ratio during engine warm up was a factor influencing driveability and emissions performance.

INTRODUCTION

It is known that under certain conditions a vehicle can generate driveability malfunctions. Malfunctions associated with cold weather driveability (e.g., hesitation, stumble, surge, backfire) can occur at (but are not restricted to) lower ambient temperature or with use of a low volatility fuel or a combination of both, although the severity of malfunction is dependent, to a large extent, on the vehicle technology and the type of fuel system employed. The driveability malfunctions caused by any of these combinations are perceived to adversely influence certain exhaust emissions.

A significant amount of prior work exists [1][2] that links volatility with emissions and volatility with driveability, although the link between driveability demerits and exhaust emissions is relatively little explored. Previous studies which investigated the relationship between exhaust emissions and driveability malfunctions were carried out in the US [3][4] and they indicated that poor driveability is correlated to higher exhaust emissions, especially hydrocarbon emissions. This work, however, was limited to vehicles representative of the US fleet with driveability tests carried out on a US test cycle. Only a single test temperature (ca. 5°C) was used.

To further investigate the effect of emissions and driveability in European vehicle technology, the Intercompany Emissions Group (IEG) conducted a dynamometer based programme with 4 member companies (5 participating laboratories) represented, namely Aral, BP Amoco, Exxon and Shell. Ten three-way catalyst equipped European vehicles were tested on the European driveability test cycle at -5°C & +10°C to determine the extent of any correlation between driveability and exhaust emissions under critical and non-critical conditions; as well as to evaluate possible causes and their relative magnitudes.

METHODOLOGY

Simultaneous measurements of exhaust emissions and driveability were conducted at -5°C and +10°C using 10 European gasoline vehicles, equipped with three-way catalysts, over a slightly modified version of the standard European cold weather driveability (CWD) test cycle [5]. As this programme was focused specifically upon the generation of driveability demerits, a suitable test cycle had to be selected. For the main test programme, the CEC CWD cycle was used in preference to the standard European Emissions "Euro 2" test cycle as the latter cycle is not designed to produce a severe driving regime and thus promote the generation of driveability demerits. The selection of test temperature, in association with the test cycle, was critical in view of the programme objectives.

TEST VEHICLES – The details of the test vehicles used are summarised in Table 1.

Vehicle	Model Year	Capacity (cc)	Power (kW)	Injection System
A	1994	1587	85	MPI
B	1994	1124	33	SPI
C	1995	1799	90	MPI
D	1994	1389	44	SPI
E	1995	1998	100	MPI
F	1996	1396	76	MPI
G	1994	1590	82	MPI
H	1997	1196	33	SPI
I	1996	1796	85	MPI
J	1995	1998	100	MPI

Table 1. Test Vehicle Matrix

The mileages on the above vehicles were all in excess of 5000km to ensure stable engine conditions and catalyst performance.

TEST FUELS – The details of the test fuels and their blended values are summarised in Table 2. The Base fuel was blended to meet the European year 2000 gasoline specification, with the exception of benzene which was blended to a level of 5% max. It was considered that the benzene level would have little impact on driveability and regulated emissions. The qualities of the remaining test fuels were designed specifically for this test programme and were not intended to represent market quality.

The fuel matrix was designed to allow a single step change in fuel property relative to Base fuel except for Fuel 2, which was specifically blended as a high volatility fuel, and was used only when no demerit free operation was achieved on the other test fuels (as a consequence not all member companies tested with this fuel). The test fuels were not blended, and thus did not range sufficiently in volatility properties, to specifically investigate the correlation between fuel quality and driveability demerits.

Fuel	RVP mbar	E70 % v/v	E100 % v/v	E120 % v/v	MTBE % v/v	Remarks
Base	530	23.0	50.5	65.9	0	EU2000 Spec.
Fuel 1	930	25.5	48.5	65.1	0	High RVP
Fuel 2	891	42.5	61.7	71.5	0	High RVP, E70, E100
Fuel 3	505	22.0	53.0	67.5	15.8	High MTBE
Fuel 4	492	18.5	41.9	68.0	0	Low E100
Fuel 5	520	21.5	42.3	53.1	0	Low E100 + E120

Other fuel parameters were kept as constant as was possible, e.g. sulphur = 50ppm m/m max, aromatics = 30% v/v max, E150 = 85 - 90% v/v, and olefins = 3 - 5% v/v.

Table 2. Test Fuel Matrix

TEST CYCLE – The cold weather driveability (CWD) test cycle is based on the CEC CWD Procedure (CEC M-08-T-83) as shown in Figure 1 and was used with the following modifications:

1. The first 15 seconds of idle period were deleted
2. The rating of the vehicle's driveability and regulated emissions were measured over six CWD cycles

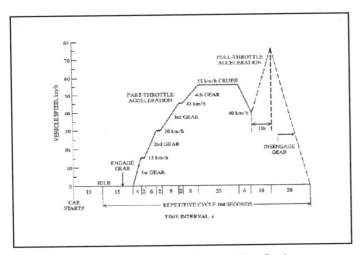

Figure 1.   CEC Cold Weather Driveability Cycle

TEST MEASUREMENTS – Prior to commencing the driveability test programme, each vehicle underwent a single 40 second idle ECE+EUDC regulated emissions test to check the functionality of emissions control equipment.

1. The following procedure was used for measuring emissions from the driveability cycle:

2. The regulated emissions were measured from ignition key-on.

3. The phasing of the CWD emissions bags as shown in Figure 2.

All vehicles were pre-conditioned by driving 3 CWD cycles on the test fuel, then soaked for a minimum of six hours at the test temperature.

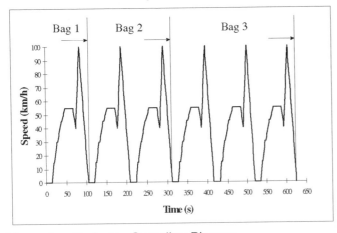

Figure 2.   Emission Sampling Phases.

TEST TEMPERATURES – All vehicles were tested at temperatures of +10°C and -5°C on the modified CEC driveability test cycle. These temperatures were selected on the basis of achieving a sufficient range of demerit and demerit free tests within the fuel matrix in combination with the range of test vehicles shown in Table 1. Careful consideration was given as to the appropriate lower temperature to be used, as a too low a temperature, in combination with a critical fuel, could cause severe demerits or indeed driving stalls where effectively zero emissions would be generated for part of the test cycle. This was to be prevented in order to maximise the number of valid tests.

The temperatures selected were thought to be the best compromise to investigate fully the link between emissions and driveability under critical conditions. Furthermore, it would allow the change in emissions caused by driveability malfunctions and the change in emissions caused by different temperatures to be studied and quantified separately.

DRIVEABILITY CORRELATION BETWEEN TEST LABORATORIES – Prior to the commencement of the test programme a comprehensive exercise was carried out with raters representing each testing Laboratory. Each rater was subjected to a range of different vehicle and fuel combinations representing a wide range of driveability malfunctions as a means of establishing the degree of correlation between the raters.

On the basis of these results, the authors were satisfied that the testing programme could proceed.

RESULTS

INITIAL DATA INSPECTION – The initial statistical analysis was to determine whether for the emissions and driveability demerits the test repeatability showed any consistent trends across all vehicles, fuels and temperatures.

Visual inspection of the standard deviation / mean plots suggested possible outliers in the data. These were confirmed by a statistical analysis using the Hawkins Test [6] assuming that the standard deviation increased linearly with the mean.

The decision to retain or reject results from a test which contained statistical outliers was based on the number of tests conducted on that specific fuel. If statistical outliers were identified in duplicate tests on a fuel, then both tests were retained in the database used for subsequent analysis. This ensured that for any vehicle/fuel combination at least two test results would be included. In those instances where triplicate tests were conducted and statistical outliers were identified, then engineering judgement was used to decide whether that particular test was rejected. This judgement was applied so as to minimise the number of tests actually rejected from the data-set.

The repeatability standard deviations relative to the mean were determined to be:

CO_2	1.6 %
CO	5.1%
HC	10.0%
NOx	14.5%

Despite the tests being carried out both at low temperature and generated between different laboratories, the overall repeatability proved to be well in line with more traditional (25°C) ambient temperature emissions studies [1].

DRIVEABILITY DEMERITS OF INDIVIDUAL VEHICLES AND FUELS – Prior to the start of the main analysis, the driveability demerits generated by each vehicle (averaged over all fuels) at both temperatures were plotted to investigate the influence of vehicle technology against driveability demerits. These data are shown in Figure 3.

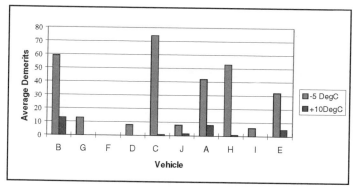

Figure 3.   Driveability Demerits of Individual Vehicles

Although the level of average demerits is relatively small, as expected at +10°C, the demerits generated at -5°C exhibit a much larger variation, which is dependent on the test vehicle. Of the 10 vehicles, 5 appear to be fairly non-sensitive to fuel volatility at the lower ambient temperature with average demerits of 12 or below (indicating few driveability malfunctions), whereas the remainder of the fleet experience, to some varying degree, an increased sensitivity to fuel volatility. This is reflected in the larger demerit values, especially with vehicle C.

In the same fashion, the driveability demerits generated by each individual fuel (averaged over all vehicles) at both temperatures was plotted to investigate the influence of fuel volatility on demerits among the different vehicles. These data are shown in Figure 4.

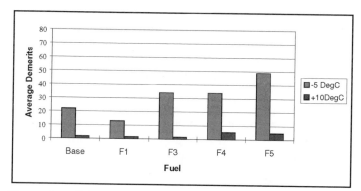

Figure 4.   Driveability Demerits of Individual Fuels

Comparison of Figures 3 and 4 suggest that the influence of varying fuel volatility on driveability demerits appear to have less of an effect than that due to the individual vehicles, although these data do demonstrate that driveability demerits are clearly influenced by both vehicle and fuel technology, in addition to ambient temperature.

TEMPERATURE, VEHICLE AND FUEL QUALITY EFFECTS ON EXHAUST EMISSIONS – In this section, a "confidence level of 95%" (5% of chance of accepting a false hypothesis, 0.05 * p) was the minimum accepted. Use of the word "significant" will mean "statistically significantly different from the null hypothesis at the 95% confidence level". In general the null hypothesis will be that there is no difference between two means or that a model effect has a coefficient of zero.

Although many data are available investigating the effect of ambient temperature on exhaust emissions, there appears to be no available data on a drive cycle designed to generate driveability problems. In order to quantify this, the exhaust emissions were averaged over all fuels and vehicles at both test temperatures, a summary of which is shown in Table 3.

	THC	CO	NOx	CO$_2$
10°C	1.83	16.89	0.34	201.00
-5°C	3.78	25.04	0.33	210.91
% Change	+107	+48	-3	+5
Significant	Yes	Yes	No	Yes

Table 3.   Average Exhaust Emissions (g/km) at Different Temperatures

The percentage difference in average emissions together with the ranges among individual vehicles are also depicted in Figure 5.

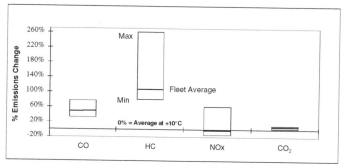

Figure 5.   Emissions Change due to a Reduction in Temperature (+10°C to -5°C)

351

Hydrocarbon Emissions – It can be seen from Table 3 that the fleet hydrocarbon emissions increased significantly with decreasing temperature. Decreasing ambient temperature from +10°C to -5°C caused the HC emissions to increase by over 100%, from an average of ca. 1.8 g/km up to ca. 3.8 g/km. The range of emissions change within individual vehicles varied immensely, with values between ca. 80% and 260%.

This increase in HC emissions is believed to be due mainly to:

- increased mixture enrichment required for cold start combustion and warm-up
- driveability deterioration resulting from lower ambient temperature (caused by poor combustion and misfire)
- increase in catalyst light off time at the lower ambient temperature

Carbon Monoxide Emissions – Fleet CO emissions also increased significantly with decreasing temperature. Reducing ambient temperature from +10°C to -5°C increased CO emissions by ca. 50% from ca. 17 g/km to 25 g/km. Individual vehicles increased in emissions between ca. 33% and 76%.

The causes of CO emissions increase with the reduction in temperature is believed to be similar to that of HC emissions.

Nitrogen Oxides Emissions – A reduction in temperature produced a slight, but non-significant, decrease in fleet NOx emissions of approximately 3%, from 0.34 to 0.33 g/km. However, among individual vehicle types there was a notable spread in NOx emissions ranging from ca. -15% to 60%.

This variation in NOx emissions response with both temperature and individual vehicles is a phenomenon already seen in previous studies [7].

Carbon Dioxide Emissions – A reduction in temperature produced a statistically significant increase in fleet CO_2 emissions of approximately 5%, from ca. 201 to 211g/km. Individual vehicle variations ranged between ca. 2% and 10%.

RELATIVE EFFECTS OF TEMPERATURE, VEHICLE & FUEL QUALITY – The relative contributions of temperature, fuel quality and vehicle differences to exhaust emissions were compared by calculating the percentage differences between the maximum and minimum of each contributor (vehicle, temperature and 3 individual fuel quality parameters). For example, to estimate the vehicle contribution the difference between the highest and lowest emitting vehicle (averaged over all test fuels) was calculated and the difference expressed in percentage terms. Similarly, the fuel parameters were expressed by comparing the highest and lowest emitting

fuel (averaged over all test vehicles), in percentage terms. As each fuel represented a distinct change in gasoline parameter, each set of results showed the largest effect of any single fuel quality parameter. In the case of temperature, the highest -5°C results (averaged over all vehicles and fuels) were compared in percentage terms with the lowest +10°C results.

These differences are illustrated and compared for HC, CO and NOx in Figure 6.

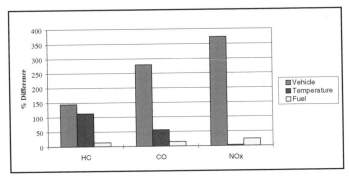

Figure 6.   Relative Effects of Vehicle, Temperature and Fuel Quality Effect on Emissions

For all regulated emissions, vehicle-to-vehicle effects were dominant compared to both changes in temperature and fuel quality.

CORRELATION OF DRIVEABILITY DEMERITS WITH TEMPERATURE AND EMISSIONS IN THE FLEET – In order to establish an initial trend as to the emissions response of the test fleet, the exhaust emissions (HC, CO, NOx and CO_2) of each individual test vehicle were plotted against driveability demerits at both test temperatures.

By examining these data alone it became clear that in the fleet of vehicles, exhaust HC and CO_2 emissions and, to a lesser extent, NOx emissions showed good correlation with driveability demerits, with all vehicles for HC and CO_2 exhibiting positive slopes (i.e., increasing driveability demerits increased these emissions), whereas CO showed little consistent correlation among individual vehicles. These results, in general, support previous reported work albeit on different vehicle technology and drive cycles [3][4].

However, in view of the limited amount of demerits measured at +10°C with the majority of test vehicles, it was decided to concentrate the subsequent analysis only with the data-set measured at -5°C.

HC Emissions – A model was used to determine the correlation between HC emissions and driveability as a function of vehicle at a single test temperature of -5°C. These data are shown in Figure 7.

For this analysis, each point on the graph represents individual vehicles whose results are averaged over all fuels.

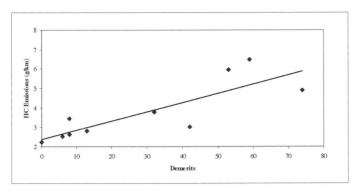

Figure 7.   HC Emissions vs. Driveability (Vehicle Effect)

Figure 7 confirms the large variation in both demerits and HC emissions between the test vehicles. However, a good correlation ($R^2 = 0.69$) is nevertheless shown and confirms the earlier trend that increasing demerits increases HC emissions.

In order to minimize the influence of the vehicle to vehicle variation in terms of absolute emission level, the emissions data were normalised relative to the base fuel at -5°C to determine the correlation between HC emissions and demerits as a function of test fuel. These data, shown at the 95% confidence level, are depicted below in Figure 8.

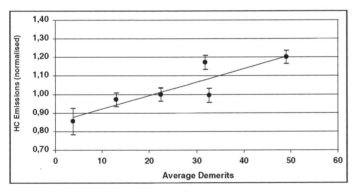

Figure 8.   Normalised HC Emissions vs. Average Driveability Demerits at the 95% Confidence Level (Fuel Effect)

The above data shows a very good linear correlation ($R^2 = 0.79$) between HC emissions and driveability demerits. Applying a best fit regression line to the data indicates that increasing demerits by 10 increases HC emissions by approximately 7% across the test fuels.

Although the above data do indicate that a correlation exists between HC emissions and driveability demerits, these data are averaged across many fuels / vehicles. Therefore it is unclear as to the relative contribution to the HC emissions from either the demerit or demerit free tests and so it is not possible to determine the influence of the driveability demerits alone on exhaust HC emissions.

Therefore, in order to try and de-couple relative contributions from driveability and volatility to HC emissions, all tests at -5°C were again selected. In order to balance the vehicle / fuel groups between demerit and

demerit free tests (NB it is not possible to have the same fuel/car effect both with and without demerits), the demerit free fleet at -5°C included all 6 test fuels and 8 vehicles (58 tests), and the demerit fleet included all 6 test fuels and 9 vehicles (64 tests). An ANOVA (analysis of variance) was carried out against HC emissions whilst the fuels and the vehicles were used as independent variables. Total HC emissions for both these data set are shown in Figure 9, at the 95% confidence level, and summarised in Table 4.

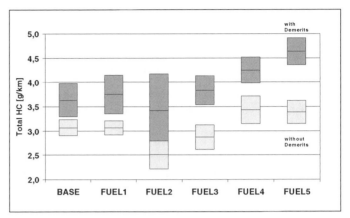

Figure 9.   Comparison of HC Emissions in the Demerit and Demerit-free Fleets, at -5°C at the 95% Confidence Level

Table 4.   Comparison of HC Emissions in the Demerit and Demerit-free Fleets

g/km	Base	Fuel 1	Fuel 2	Fuel 3	Fuel 4	Fuel 5
Zero Demerits	3.07	3.06	2.51	2.87	3.43	3.39
With Demerits	3.61	3.76	3.42	3.83	4.26	4.64
% Increase	22.1	22.9	36.3	33.4	24.2	36.9

The difference between the data set with demerits and the demerit-free data set, averaged over all the fuels, was ca. 29%.

The present study thus indicates that there is a substantial contribution to exhaust HC emissions derived purely from driveability malfunction, rather than from differences in combustion brought about by fuel volatility changes. The individual vehicle regressions also confirmed this trend.

NOx Emissions – In a similar fashion to the data analysis for HC emissions (in order to minimise the influence of the vehicle to vehicle variation in terms of absolute emission level), the emissions data was normalised relative to the base fuel at -5°C to determine the correlation between Nox emissions and demerits as a function of test fuel as shown in Figure 10.

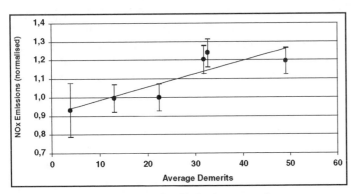

Figure 10. Normalised NOx Emissions vs. Average Driveability Demerits at the 95% Confidence Level

Although it is possible to interpret the above data as 2 groups of 3 data points, at 2 distinct levels, a linear interpolation has nevertheless been applied to investigate the general trend.

These data show, at the 95% confidence level, that increasing demerits increases NOx emissions, with a good correlation as determined by the best fit regression line (R^2 = 0.74). On this basis, these data show that an increase in 10 demerits provides an increase in NOx emissions of approximately 7%.

In view of these trends, a detailed analysis of the NOx emissions data was also carried out in an attempt to quantify further these findings. A breakdown of the second-by-second NOx emissions in relationship with the demerit data failed to highlight areas in the test cycle where NOx increased in parallel with increasing demerits (e.g., as found for HC emissions).

Overall, in view of the limited number of data points and by the nature of the data-set, these results relating NOx emissions and demerits cannot be considered as conclusive.

NB. Due to the very low absolute level of NOx emissions at -5°C this projected increase of 7% is relatively insignificant.

CO Emissions – As briefly described earlier, the response of individual test vehicles with demerits and CO emissions was relatively inconsistent and this is very much reflected in the data presented in Figure 11.

The normalised analysis shows the correlation of CO emissions with increasing demerits, the significance of which is very poor (R^2 = 0.24) as measured at the 95% confidence level.

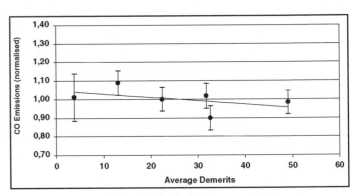

Figure 11. Normalised CO Emissions vs. Average Driveability Demerits at the 95% Confidence Level

CO_2 Emissions – Normalised emissions were determined relative to the base fuel at -5°C to determine the significance of the CO_2 correlation. These data, shown at the 95% confidence level, are depicted below in Figure 12.

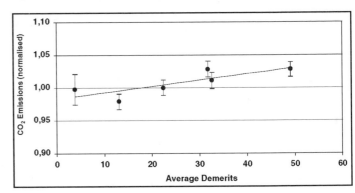

Figure 12. Normalised CO_2 Emissions vs. Average Driveability Demerits at the 95% Confidence Level

The above data shows a good linear correlation (R^2 = 0.65) of the data-set and indicates that increasing demerits by 10 increases CO_2 emissions by approximately 1% across the test fuels.

In summary, the above data indicates that HC emissions possess the most significant correlation with driveability demerits relative to other emissions and it is for this reason that the remainder of the report will focus on furthering the understanding of the mechanisms associated with the HC emissions increase.

CORRELATION OF DRIVEABILITY WITH HC EMISSIONS IN A SINGLE VEHICLE – As already mentioned, individual vehicles showed good correlations of driveability with HC emissions on different fuels.

Figure 13 depicts an example of a second by second HC emissions trace generated on a critical test vehicle (vehicle E), operating on Fuel 1 and Fuel 4 which are representative of high and low volatility respectively, at -5°C. It can be seen that during the initial part of cycle 1 under part throttle acceleration, a very steep rise in HC emissions occurred with Fuel 4 (peaking at over 7000ppm, the upper limit of the emissions equipment). This coincided with the vehicle exhibiting severe hesitation and moderate stumble on this fuel as rated by a trained technician.

Fuel 1 (high volatility), however, generated zero demerits in this initial part of the cycle and experiences a small rise in HC compared to Fuel 4. The cruise portion, upon which the vehicle is driving at constant speed, allowed the instantaneous emissions of the 2 fuels to converge under a malfunction free environment. Under full throttle acceleration, approximately the same amount of HC emissions were generated for both fuels, which can again be explained by the zero demerits achieved by both fuels during this portion of cycle 1.

During the second cycle, as the vehicle begins to warm up, there were no additional demerits generated on both test fuels during the part load portion of the test cycle. This demerit free environment is again reflected by the very stable HC emissions trace between 120 and 180 seconds. Full load acceleration again generated no driveability malfunctions with both test fuels generating approximately the same amount of HC emissions. This evidence clearly shows that the majority of the HC emissions occur where the driveability demerits are at their greatest.

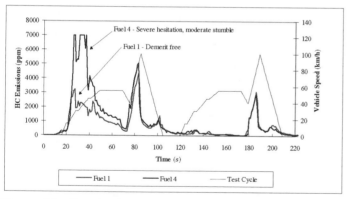

Figure 13. Continuous (1Hz Sample Rate) HC Emissions for a Critical Vehicle (E) over Cycles 1&2

In contrast, Figure 14 shows a similar trace of HC emissions, measured at 1 Hz, on a non-critical vehicle (vehicle F), i.e., a vehicle that did not exhibit any form of driveability malfunctions. The data is shown with the same two test fuels as for the critical vehicle. Although there are minor deviations in HC emissions between the fuels, these are most likely caused by their different volatility characteristics. There are no peaks associated with any driveability malfunctions.

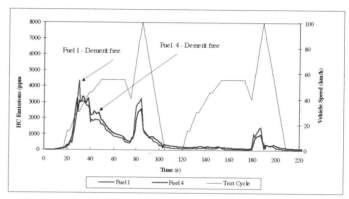

Figure 14. Continuous (1Hz Sample Rate) HC Emissions for a Non-Critical Vehicle (F) over Cycles 1 & 2

This data suggests that on a non-critical vehicle (or a vehicle that possesses good driveability), the influence of fuels of different qualities on exhaust HC emissions is minimal, providing the vehicle remains demerit free. This could be explained by the vehicle enrichment operation at low temperature masking any potential fuel effects.

In terms of cumulative emissions for the critical vehicle E, Figure 15 confirms that the majority of the HC emissions generated for the whole test occurred during cycle 1. The difference between the two fuels in terms of total HC emissions, is a function of the absolute level of driveability demerits experienced by the vehicle during the first part of the cycle. The rate of change of HC emissions for both test fuels after cycle 1 (i.e., demerit free operation) were almost identical.

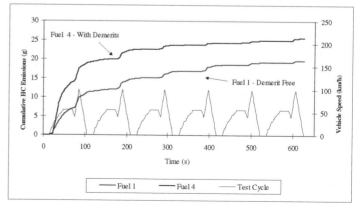

Figure 15. Cumulative HC Emissions for Critical Vehicle (E)

The percentage contributions of individual Phase HC emissions and demerits on vehicle E, relative to the entire test cycle, is shown in Table 5.

Parameter	Fuel 1 HC Emissions	Fuel 4 HC Emissions	Fuel 1 Demerits	Fuel 4 Demerits
Contribution of Phase 1 to Total Cycle (%)	55.7	73.1	zero	92.3
Contribution of Phase 2 to Total Cycle (%)	28.8	18.7	zero	7.7
Contribution of Phase 3 to Total Cycle (%)	15.5	8.1	zero	0

Table 5. Percentage Contribution of individual Phase HC Emissions and Demerits

MECHANISM FOR DRIVEABILITY DEMERITS – A

spark ignition engine will fire if an ignitable air-to-"fuel vapour" ratio can be created. Unvapourised fuel is only burned after ignition. Mixture enrichment at cold start is used to ensure that an ignitable vapour is formed, but this amount is generally limited in today's current vehicle technology in order to avoid increased emissions even though, at the time of writing, low ambient temperature emissions legislation in Europe has yet to come into effect. Decreasing the amount of fuel injected makes it more difficult to form an ignitable mixture because a larger fraction of the higher boiling point components must be vapourised.

There are of course other vehicle dependent factors which can influence cold weather driveability, such as fuel system technology (e.g. carburetted, single or multi-point injection) and thus the quality of fuel distribution, inlet system design or retention of fuel in inlet system deposits.

Some of the vehicles tested in this programme produced little or no driveability demerits on even the most critical fuels (Fuel 4 and Fuel 5) whereas other vehicles exhibited some degree of driveability demerits on fuels expected to be non-critical. Although this was clearly dependent on the test temperature, the variability in driveability demerits seen in the test vehicles was found to be linked to the control of the Air Fuel Ratio (AFR) as measured in different vehicles. An example of an AFR trace is shown below in Figure 16. This shows a second by second account of the AFR across 2 of the 6 driveability cycles for two test vehicles, vehicle A exhibiting good driveability and vehicle C exhibiting poor driveability, on a non-critical fuel.

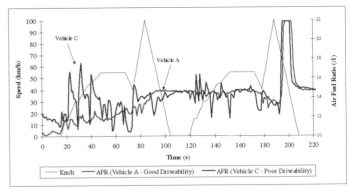

Figure 16. Continuous (1Hz Sample Rate) Air Fuel Ratio

It can be seen that the fuel enrichment experienced during the cold start is shown during the initial idle period (first 15 seconds), by both vehicles, prior to initial drive-away. Upon drive-away and under increasing engine load, vehicle A exhibited good AFR control with few fluctuations gradually rising to AFR ~ 14.7 or stoichiometric at the end of cycle 1. In comparison, vehicle C showed substantially poorer AFR control during the part load operation in cycle 1 - these lean excursions ultimately translating into driveability demerits probably due to partial combustion. The severe lean excursion shown during cycle 2 by both vehicles is explained by fuel cut-off during the deceleration (over-run) which at this stage occurs under closed loop engine management control.

CONCLUSIONS

- The following correlations between driveability demerits and exhaust emissions in a European vehicle fleet were limited to the data generated at -5°C as too few demerits were produced at +10°C.

  ⇒ Driveability demerits gave rise to a statistically significant increase in HC and CO_2 emissions, an apparent increase in NOx emissions, but no correlation with CO emissions at the 95% confidence level. The apparent increase in NOx could not be explained from the detailed breakdown of the driveability and emissions data.

  ⇒ An increase in 10 driveability demerits caused an increase in HC and CO_2 emissions of ca. 7% and 1%, respectively, averaged across all test fuels as tested on the CEC driveability test cycle.

  ⇒ De-coupling demerit and demerit free tests showed, on average, ca. 29% higher HC emissions for the vehicle fleet with driveability demerits averaged over all test fuels. Individual fuels ranged from 22% to 37%.

  ⇒ Control of air/fuel ratio during engine warm up was identified as a factor influencing driveability and emissions performance.

- A reduction in temperature from +10°C to -5°C showed a statistically significant increase in HC, CO and CO_2 emissions and a small, but non-significant, reduction in NOx emissions.

- Overall, the influence of vehicle technology and a 15°C change in ambient temperature were far greater than changes to fuel quality for all regulated emissions.
- This study demonstrated the suitability of the current CEC cold weather driveability test procedure as a means for assessing the driveability performance of modern vehicle technology.

ACKNOWLEDGEMENTS

The authors would like to thank BP Amoco, Sunbury, UK for the blending and supply of the experimental fuels used in this programme.

REFERENCES

1. European Programme on Emissions, Fuels and Engine Technologies, ACEA and EUROPIA, EPEFE Report (1995).

2. AUTO/OIL Air Quality Improvement Research Program, Phase 1 Final Report, May 1993.

3. Jorgensen, SW and Benson, JD, A Correlation between Tailpipe Hydrocarbon Emissions and Driveability, SAE 962023.

4. Jorgensen, SW and Benson, JD, Measurements of Driveability and Emissions at Cool Ambient Temperatures, SAE 941870.

5. CEC M-08-T-83, Cold Weather Driveability Test Procedure, CEC Tentative Code of Practice.

6. ISO 4259 Petroleum Products - Determination and Application of Precision Data in Relation to Methods of Test, 2nd Edition 1992-12-15.

7. CONCAWE, The Effect of Gasoline Volatility on Vehicle Exhaust Emissions at Low Ambient Temperatures, Report No. 93/51.

CHAPTER 8

ADVANCED CATALYST DESIGN

Paul J. Andersen, Todd H. Ballinger and David S. Lafyatis
Johnson Matthey

8.1 OVERVIEW

Regulated emission levels from automobiles continue to be lowered around the world in order to improve air quality. The most stringent regulations for gasoline-powered vehicles are now approaching near-zero levels for tailpipe emissions. Between 2003 and 2009 automotive manufacturers will phase in the production of light-duty vehicles (passenger cars), medium-duty passenger vehicles (minivans, SUVs, and full-size vans with gross vehicle weights up to 10,000 lbs.), and light-duty trucks (gross vehicle weights up to 8,500 lbs.) that meet near-zero emission standards.

The central exhaust component necessary to meet the near-zero tailpipe emissions is the advanced three-way catalyst, which contains individual or a combination of the platinum group metals (PGM) Pd, Pt, and Rh to oxidize the hydrocarbons and carbon monoxide (to carbon dioxide and water) and to reduce the nitrogen oxides (to nitrogen). Modern advanced three-way catalysts can achieve virtually 100% conversion of the three pollutants (hence near-zero emissions) when two important criteria are met: the catalyst must be at operating temperatures (temperature >300°C, ideally temperature >450°C), and the exhaust gas mixture is at the optimum stoichiometric air-to-fuel ratio (AFR). Under most driving conditions, the exhaust temperature is high enough and an on-board computer (using oxygen sensor feedback) maintains engine conditions such that the exhaust gas is near the stoichiometric point, so near-zero emissions can be realized. However, if near-zero emissions are to be achieved under all driving conditions, two important problems must be overcome: (1) brief transients away from the stoichiometric AFR during hard accelerations and decelerations, and (2) rapid warm-up of the catalyst to reach operational temperatures when a vehicle is started with a cold engine (so-called cold starts).

Three-way catalyst (TWC) improvements have had significant impacts on near-zero emissions. Washcoat components such as stabilized alumina and stabilized ceria-zirconias (oxygen storage components) have greatly improved the thermal durability of TWCs. In addition, optimized PGM-support interactions and layered catalyst structures are also used to improve catalyst activity and provide thermal durability. Consideration should also be given to the types of PGM used in near-zero emission systems. Pd-Rh and tri-metal catalysts are used in close-coupled locations for rapid hydrocarbon light-off and warmed-up HC and CO oxidation. Rh has been demonstrated to be the most effective PGM for NOx reduction; hence, near-zero emission systems use Pd-Rh or Pt-Rh underfloor catalysts that have been optimized for high NOx conversion activity.

The factors involved in advanced catalyst design for near-zero emission vehicles will be discussed below. The first part demonstrates the effect of PGM, washcoat components, and catalyst formulations on improving TWC activity. The second part shows how high activity TWCs are incorporated into an exhaust system and how the combination of other exhaust strategies are used to produce near-zero emission systems.

Advanced Three-Way Catalyst Concepts and Design - Three-way catalysts (TWC) for meeting near-zero emissions must have very high hydrocarbon (HC) and NOx conversion activities. Studies have been conducted in which the three platinum group metals (PGMs) employed in TWCs were examined to determine their utility for converting pollutants. HC emissions were clearly a strong function of Pd or Pt load. Additionally, Pd was more effective for converting HC emissions than Pt. For NOx emissions, Rh-containing systems had substantially lower emissions. Thus, catalyst development efforts have focused on the Pd and Rh washcoat functionalities.

For the Pd component, the washcoat consists primarily of alumina, a ceria-zirconia mixed oxide (Oxygen Storage Material or OSC component), and a basic promoter such as La or alkaline earth materials. Modifying the distribution of Pd between the alumina and CeZrOx supports in a Pd-only catalyst can have a strong impact on the activity and the response of the catalyst to sulfur poisoning. The catalyst with the highest amount of Pd/OSC interaction shows the highest activity under zero sulfur conditions, but the catalyst with no Pd/OSC interaction showed the highest activity after sulfur poisoning. Thus, the Pd/OSC interaction can be adjusted to give catalysts with the optimum balance of low sulfur and high sulfur performance for a given application. In addition, careful selection of the basic promoter significantly improves activity under both low and high sulfur conditions.

Pd-based catalysts are usually located in close coupled positions and thus are exposed to high temperatures during operation. Consequently, the thermal durability of the Pd containing catalysts is also critical to performance. Due to the larger sensitivity of CO/NOx activity to thermal degradation, studies have focused primarily on improving activity for converting those components after high temperature aging. Much of the focus has been on improving the stability of the OSC component in the catalyst. In general, it is desired to have a homogeneously mixed CeZrOx phase which does not separate into CeO_2 and ZrO_2 phases after exposure to high temperatures. Studies have shown that improvements in the CeZrOx properties can substantially affect thermal durability. For example, CO and NOx conversions were not affected in a catalyst prepared with an improved OSC when the aging temperature was increased from 1050 to 1100°C.

Significant improvements in catalyst activity have also been possible by improving the TWC activity of the Rh component. The main focus of development work to improve the Rh component was optimizing the composition and the preparation method of the CeZrOx OSC component used to support Rh in the catalyst. As with Pd-only catalysts, the preparation method for the OSC Rh support and the manner in which Rh was deposited on that support was critical to achieving high activity and thermal durability. Finally, Pt and Pd components can be placed in separate washcoat layers to avoid negative interactions that may arise between catalyst components after high temperature exposures.

Catalyst System Design Principles for Ultra-Clean Vehicles - Implementation of these improved washcoats into an optimal system design gives the opportunity to approach near-zero emission with passive means. Strategies for both the close-coupled and underfloor catalyst to greatly reduce both NOx and NMHC emissions over the FTP cycle can be demonstrated.

For the close-coupled (cc) catalyst, the first challenge is light-off. A conventional vehicle will fail the ultra-clean vehicles such as PZEV standard for hydrocarbon in the first 20 seconds of the FTP test, while the vehicle is still idling after the cold start. Improvements in the design of the cc-catalyst can significantly reduce light-off emissions. One well-known method of improving cold-start HC and NOx emissions is to increase the PGM content on the cc-catalyst. Pd is an excellent PGM for light-off purposes following high temperature aging: increasing the Pd loading from 54 g/ft^3 to 200 g/ft^3 leads to a significant decrease in the Bag 1 emissions for NMHC, NOx and CO.

Close-coupled catalyst washcoat formulations can also lead to emission improvements during the cold-start. When the HC light-off performance of two different Pd-Rh catalysts was studied at the identical PGM loading (following an accelerated aging to represent 50,000 miles), an improved formulation leads to significantly improved HC performance during the first 60 seconds of the FTP drive cycle as well as during the transient conditions of Hill 1.

The close-coupled catalyst is also a critical component for controlling NMHC and NOx during the warmed-up sections of the driving cycle. In addition to the improvement in Bag 1 performance that was already noted, increasing the Pd loading in cc-catalysts also leads to improved Bag 2 and Bag 3 emissions. Another PGM strategy for improving FTP performance from the cc-catalyst portion of the exhaust system is the addition of Rh to the catalyst. In advanced washcoats, the addition of just an additional 6 g/ft^3 Rh to a Pd catalyst reduces NOx emissions by roughly 20%, while leaving HC emissions unchanged.

In addition to the close-coupled catalyst, near-zero exhaust systems are very likely to have additional catalysts further downstream in the exhaust system, commonly referred to as underbody catalysts. These catalysts play a complimentary role to the cc-catalysts for exhaust emission control. Although they generally play little role in light-off performance, they are critical for cleaning up emissions that pass through the cc-catalyst during warmed-up operation due to either high-space velocity mass transfer limited operation or air-fuel excursions during transient driving events. The underbody catalyst requirements for thermal durability are reduced compared to cc-catalysts, because of their greater distance from the engine. It is important in system design to recognize that the function of these catalysts is significantly different from cc-catalysts, and to optimize the system configuration to account for this.

An example of the dramatic effect on overall emissions that the design of the underbody catalyst can have is shown in a study conducted on a sport-utility vehicle, using a common set of cc-catalysts but varying the underbody catalysts. Each of the systems was aged to 50,000 miles. Three different underbody catalyst systems were tested: a standard Pd-Rh catalyst (0:9:1/60), an advanced Pd-Rh catalyst (0:9:1/60) and an advanced Pt-Rh catalyst (9:0:1/60). The data indicate that the change of the underbody catalyst formulation from the standard Pd-Rh to the advanced Pt-Rh has reduced NMHC tailpipe emissions by approximately 10%, but more impressively NOx emissions have been reduced by 75%!

All of the above have focused on changes that can be made to catalysts in the exhaust stream to reduce emissions while using a fixed vehicle architecture and calibration strategy. However, it is important to recognize that parallel to the development of better catalyst components and system design strategies, vehicle manufacturers have revolutionized the vehicle control algorithms during both cold start and warmed-up conditions. The results of these changes when combined with catalyst component optimization can be truly remarkable.

The combination of an advanced vehicle calibration with the improved exhaust system design led to an 89% reduction in NOx emissions. Significant improvements in all 3 phases of the FTP cycle were seen for NOx, while the improvements are mainly in Bag 1 for both NMHC and CO emissions. One key to the Bag 1 improvements for all 3 pollutants is the rapid catalyst warm-up demonstrated by the advanced system, which is a combination of both the changes to the vehicle calibration and the close-coupled substrate. The critical time for warming the close-coupled converter up to light-off temperature has been reduced by approximately 50%, resulting in a major benefit in emissions during the critical first 50 seconds of the FTP cycle.

Tailpipe emission levels from mobile sources continue to be dramatically reduced around the world. These new emission limits have required the entire industry to focus on cost-effective emission control technologies. Improved catalyst washcoat materials that have dramatically increased the activity and durability of three-way catalysts containing Pt, Pd and Rh have played an important role in this process. By combining advanced catalyst washcoats with other exhaust system improvements with revolutionary advances in the calibration and control strategies, standards such as PZEV have been achieved at a minimal increase in system cost and complexity.

8.2 SELECTED SAE TECHNICAL PAPERS

2002-01-0349

Utilization of Advanced Three-Way Catalyst Formulations on Ceramic Ultra Thin Wall Substrates for Future Legislation

J. Schmidt, J. Franz and N. Merdes
DaimlerChrysler AG

M. J. Brady
DaimlerChrysler Corp.

W. Mueller, D. Lindner and T. Bog
OMG AG & Co. KG

D. Clark
OMG Corp.

T. Buckel, W. Stoepler, R. Henninger and H. Ermer
Faurecia Abgastechnik GmbH

F. Abe and M. Makino
NGK Insulators Ltd.

A. Kunz and C. D. Vogt
NGK Europe GmbH

Copyright © 2002 Society of Automotive Engineers, Inc.

ABSTRACT

The LEV II and SULEV/PZEV emission standards legislated by the US EPA and the Californian ARB will require continuous reduction in the vehicles' emission over the next several years. Similar requirements are under discussion in the European Union (EU) in the EU Stage V program. These future emission standards will require a more efficient after treatment device that exhibits high activity and excellent durabilty over an extended lifetime. The present study summarizes the findings of a joint development program targeting such demanding future emission challenges, which can only be met by a close and intensive co-operation of the individual expert teams. The use of active systems, e.g. HC-adsorber or electrically heated light-off catalysts, was not considered in this study. The following parameters were investigated in detail:

- The development of a high-tech three-way catalyst technology is described being tailored for applications on ultra thin wall ceramic substrates (UTWS).
- The influence of UTWS type and precious metal loading are tested in model gas atmosphere, on an engine, and on three different vehicles with US as well as EU calibrations.

- Pressure drop controlled canning technologies were investigated to ensure the canning of UTWS like 900 cpsi/2.5 mil or 1200 cpsi/2.5 mil.

The test data indicates that improvements can be made by switching from a 600 cpsi substrate to a 900 cpsi substrate, but further only small advantages can be realized for aged systems by changing from 900 cpsi to 1200 cpsi substrates. From the performance point of view, 900 cpsi substrates seem to be the optimum. For the studies described in this paper the influence of the precious metal (PGM) loading is different for the individual applications and generally more pronounced as compared to the effect of an increased cell density – in most cases a PGM increase beyond 200 g/ft³ does not lead to major changes in the test results.

INTRODUCTION

The global increase in passenger car traffic and the possible negative effects on the environment especially in the centers of population led to more and more stringent emission limits. The U.S. was the first country introducing efficient measures followed by the European Union more than one decade later. While in the U.S the so-called SULEV/PZEV emission standards fixed in the California Phase II program request a close to 99 plus x% conversion level (see Figure 1), in the EU similar actions are being considered and might be introduced with the discussed EU V legislation. In both cases, the additional request for an extended durability leaves no room for a major loss of efficiency for the entire exhaust gas aftertreatment system.

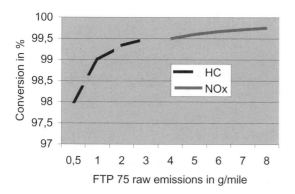

Figure 1: Required conversion levels to meet SULEV.

Certainly, this cry for both, a more active and a more durable catalytic aftertreatment, necessitates a very close co-operation between the development experts of the individual components that can contribute to the overall performance.

Regarding engine design, improved light-off strategies, in combination with an excellent lambda control, are taken into account resulting in decreased light–off times as well as less breakthroughs of pollutants during non-stationary driving conditions. The use of low heat capacity materials for the exhaust pipe canning support the fast light-off strategy even more. Additionally, advanced UTWS offers a significant improvement potential with respect to both, lowering the mass that needs to be heated up during cold-start, and the increased geometric surface area positively influencing the dynamic characteristics of the catalyst. Nevertheless, in order to take advantage of these physical features, a highly sophisticated catalyst technology tailored to the UTWS characteristics is absolutely necessary. Such formulations have been successfully invented and are available for different PGM mixtures that are Pt-rich as well as Pd-rich formulations. Details regarding the development of advanced aftertreatment systems are given in [1-32].

The present study concentrates on the following two design parameters:

- ceramic substrates ranging from 600 to 1200 cpsi
- precious metal loadings in the range of 100 to 300 g/ft³

All catalysts went through serious model gas or engine aging procedures and were tested in model gas atmosphere, on an engine, and on three different vehicles for EU and US legislation, respectively.

ULTRA THIN WALL SUBSTRATES

For this study, the most advanced UTWS have been selected. The characteristics of the UTWS are shown in Table 1 in comparison to the standard substrate 400cpsi/6.5mil.

Table 1: Substrate characteristics

Cell Structure [cpsi/mil]	400/6.5	600/3.5	900/2.5	1200/2.5
Wall Thickness [mm]	0.17	0.09	0.06	0.06
Geometric Surface Area [cm²/cm³]	27.3 100%	35.1 129%	43.7 160%	50.3 184%
Hydraulic Diameter [mm]	1.10	0.95	0.79	0.67
Open Frontal Area [%]	75	83	86	84
Bulk Density [g/cm³]	0.43	0.29	0.24	0.28
Pressure Drop [%]	100	108	133	171

UTWS provide significantly increased surface area which improves catalytic performance [30]. As shown in Table 1, the 1200cpsi/2.5mil UTWS has a 84 % increased geometric surface area compared to the standard 400cpsi/6.5mil substrate. Another key parameter for the light off behavior is the heat capacity related to the geometric surface area, i.e. the specific heat multiplied by the bulk density and divided by the volumetric surface area [20]. The UTWS provides desirable low values for this parameter. The specific heat capacity per cell structure is demonstrated in Figure 2.

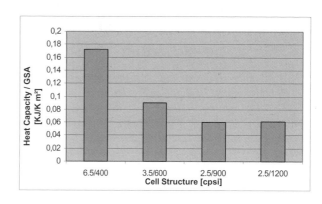

Figure 2: Specific heat capacity per cell structure.

The substrate sizes which have been used in this study are shown in Table 2. The substrate volume has been kept constant, cell geometry has been varied from 600cpsi/3.5mil to 1200cpsi/2.5mil.

Table 2: Evaluated substrate sizes and cell structures (D = diameter)

Substrate size	Substrate Volume [l]	Cell structure [cpsi/mil]	Total GSA [m²]
D 4.33" x 100 mm	0.88	600/3.5	3,11
D 4.66" x 80 mm	0.88	900/2.5	3,85
D 4.66" x 80 mm	0.88	1200/2.5	4,43
D 4.33" x 100 mm	0.95	600/3.5	3,35
D 4.33" x 100 mm	0.95	900/2.5	4,15
D 4.33" x 100 mm	0.95	1200/2.5	4,78

MECHANICAL DURABILITY

As the wall thickness of the substrate is decreased from conventional 6.5 mil (0.17 mm) to 2.5 mil (0.06 mm) for UTWS, there is the concern that aggressive flow of the exhaust gas especially, in close-coupled (CC) position of the converter, may cause erosion of the substrate matrix.

Extensive tests have been conducted to determine the impact of high exhaust flow on CC positioned UTWS.

Table 3 describes the test conditions used for the engine bench evaluation.

Table 3: Test conditions engine bench

Engine	4.5 liter V8
Mode	1 cycle: 1 min. 5000 rpm + 5 sec. Idle
Exhaust gas temperature	Max. 920 °C
Flow rate	3.6 Nm³/min per bank
Duration	112 hours (6100 cycles)
Cell structures evaluated	400/6.5; 600/3.5; 900/2.5; 1200/2.5

For the test, a Japanese V8 engine has been used. The position of the converter is shown in Figure 3.

All substrate types which have been used in the durability test program have shown excellent erosion resistance even under severe exhaust flow conditions in CC positions. In case of foreign objects in front of the substrate are excluded, no substrate erosion could be detected after completion of the 112 hour test cycle.

Figure 3: Engine bench with converter position.

ISOSTATIC STRENGTH

An overview of the isostatic strength of the three substrate types investigated in this study compared to the standard substrate is given in Figure 4. The value for a standard substrate (400cpsi/6.5mil) is set to 100 %. In comparison the values of the UTWS are decreased, therefore, canning conditions need to be adjusted.

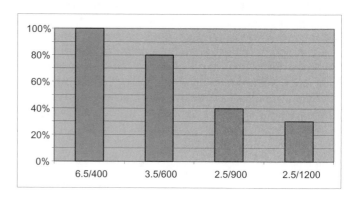

Figure 4: Relative isostatic strength of evaluated substrates.

CANNING

Faurecia has developed the canning designs and has built the different catalytic converters used for the emission tests. Three different DC-cars were equipped.

The aim of the study was to compare the compatibility of two existing technologies (clam shell already in series production for DC Vehicle B for example and calibrated Rohrkat) with UTWS applications. All substrates have round cross sections. Three new applications were checked: the new Vehicle B, Vehicle A and Vehicle C.

Due to higher legal durability requirements and a lower mechanical strength of thinwall- as well as UTWS in comparison to standard ceramic substrates, the following part of this paper describes the new canning requirements.

Vehicle C

Vehicle B

Vehicle A,
left and right stream

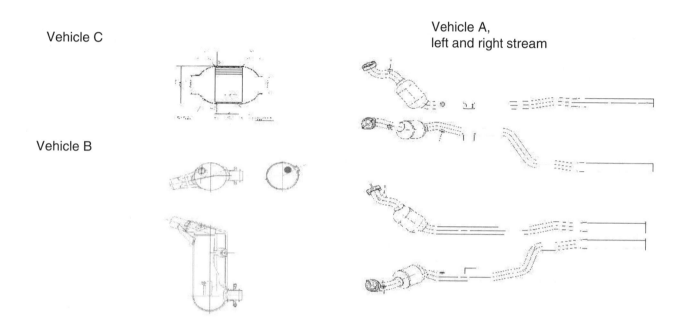

Figure 5: Catalytic converters used for the tests.

PROCESSES

Clam shell

One of the oldest and cheapest technology is the well known clam shell technology. Figure 6 describes the process steps for a clam shell catalytic converter.

Step 1: Edge stabilization of the mat

Step 2: Substrate with mat positioning in the lower half shell

Step 3: Assembling the upper half shell and closing the can

Step 4: Welding the half shells

Step 5: Marking the can by means of dotpining: coating charge, date of fabrication, part-number

Step 6: Calibration of the endcones

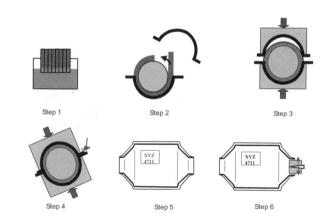

Figure 6 : Clam shell catalytic converter - process steps.

371

Rohrkat

This technology creates a minimum of bending stress and shear stress on the substrate. The "Calibrated Rohrkat" combines the advantages of a "Modular Canning Technology" and compensates the dimensional tolerances of the substrate in size and ovality.

Figure 7 describes the process steps of the Calibrated Rohrkat.

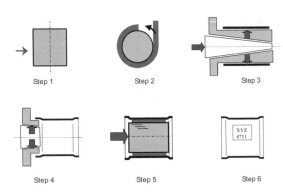

Figure 7: Calibrated Rohrkat - process steps.

Step 1: Each ceramic substrate is measured by a Laser. Size, ovality and conformity with the dimensional specification of the substrate is controlled (first 100% control).

Step 2: Preassembling monolith with mat.

Step 3: Calibration of the tube corresponding to the measured size of the substrate.

Step 4: Calibration of the tube ends.

Step 5: Push-in of the preassembled mat and ceramic into the tube. The push-in-force (PIF) is measured and controlled at 100%. Parts out of tolerance are separated from further process steps.

Step 6: Marking of the can by means of dotpining: push-in-force, coating charge, date of fabrication (up to min/sec), assembly line, time, part-number.

The process of the "Rohrkat" is automated and 100% process controlled. No operator error can occur.

QUALITY AND DOCUMENTATION

- Each substrate is measured and checked according to its substrate specification.

- Each substrate gets its own tube.

- Push-in-force (canning quality) is controlled.

- Permanent actualized control charts with control limits directly in the fabrication.

Documentation (Box and Plant):

Part-no., coating-charge-no., production date (year/day/hour/min/sec), push-in-force (PIF), production line

In addition, the plant memorizes the substrate diameter and the gap.

Process compatibility

Figure 8 contains the experience with ceramic standard thin wall and UTWS in catalytic converters for the two canning technologies Clam Shell and Calibrated Rohrkat. Additionally, the influence of intumescent mat and LCF mat were investigated.

It turns out that canning UTWS requires LCF mats in any cases. Intumescent mats are limited to thin wall applications.

INFLUENCE OF SUBSTRATES

As described in Figure 4 the advanced Ultra-Thin-Wall substrates like 900/2.5 or 1200/2.5 show significantly reduced isostatic strength values in comparison to the standard 400/6.5.

Therefore, the canning technologies have to be adapted to this new generations of substrates in order to ensure safe canning conditions and reliable converter performance in the field.

During converter assembling the mat creates forces on the ceramic. The value of these forces is strongly dependent on the canning technology. The assembling of a Clam Shell catalytic converter as well as of a tourniquet catalytic converter produces high forces on the ceramics. Due to the reduced mechanical strength the risk of breaking Ultra-Thin-Wall substrates is higher. In contrast to those canning technologies the mat in the Rohrkat technology produces very low shear forces. Therefore, it is an excellent and advantageous technology to can catalytic converters.

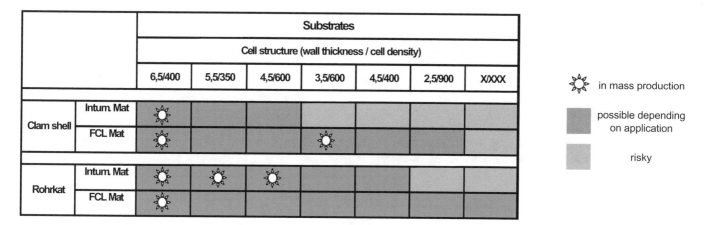

		Substrates						
		Cell structure (wall thickness / cell density)						
		6,5/400	5,5/350	4,5/600	3,5/600	4,5/400	2,5/900	X/XXX
Clam shell	Intum. Mat	☀						
	FCL Mat	☀			☀			
Rohrkat	Intum. Mat	☀	☀	☀				
	FCL Mat	☀						

☀ in mass production

▓ possible depending on application

░ risky

Figure 8: Substrate applications in catalytic converters today and tomorrow.

INFLUENCE OF MATS

To keep the ceramic substrate during the lifetime of the catalytic converter in the correct position in the steel can, an elastic material (mat) is used. The objectives of this mat are:

- to compensate the different thermal expansion of ceramic and steel.

- to hold the substrate in place by exercising a sufficient pressure on the ceramic - despite the vibrations of the exhaust-line and under exhaust gas temperatures up to 980°C.

Figure 9 shows the compression curves of two different mats used for the canning of ceramic substrates. The x-axis represents the gap between substrate and mat. The 0-value corresponds to the nominal gap of the two mats. Left of this value, the gap increases, right of 0 the gap decreases.

The y-axis describes the residual pressure the mat creates at different gaps. The graph compares the influence on mat pressure in the gap range for the canning technologies Calibrated Rohrkat and Clam Shell. The real gap in a catalytic converter is mainly affected by the actual values of:

- size and out of round contour of the substrate - area density of the mat.

- size of the can (clam shell, tube).

The Clam Shell catalytic converter has the largest gap range. Clam Shell converters are noncalibrated and have to take into account the tolerances mentioned above. The Calibrated Rohrkat compensates the size of the substrate and, therefore, has a smaller gap range. The pressure of the mat changes over the possible gap range of the catalytic converters. To keep the substrate under all driving conditions in position a minimum pressure of the mat is required. This minimum pressure is not a constant value. It depends strongly on the kind of mat and the different applications in a car. The maximum pressure of the mat is limited by the specified isostatic strength of the type of substrate.

Additionally, this diagram clearly shows that intumescent mats are not recommended to can UTWS.

CONCLUSION CANNING

This study showed that it is possible to can UTWS with existing technologies (Rohrkat or clam shell + FLC mat) despite the change in properties of the substrates. Rohrkat gives higher safety by reducing the tolerances on the gaps. It is also to notice that all designs passed the validation (dyno test, hot shake test…).

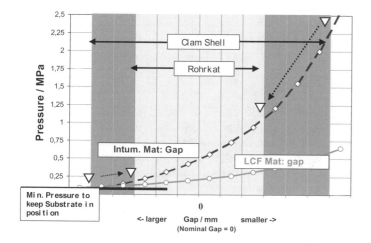

Figure 9: Influence of mat compression on assembling process for Clam Shell catalyst and Rohrkat.

WASHCOAT TECHNOLOGY

In order to take full advantage of the UTWS potentials, a special catalyst design must be realized as it was explained in detail in [32]. Major design criteria for such high-tech products are given below.

First of all, a balance between pressure drop, washcoat loading and durability aspects is required. The results of a basic study were published in [32]. From this data it was concluded to utilize medium washcoat loaded systems offering the best compromise between pressure drop and durability.

In the next step the influence of washcoat thickness, precious metal concentration as well as oxygen storage capacity was determined for Pd/Rh formulations on 900 cpsi/2.5 mil substrates. A description of the converters is given in Table 4, results of vehicle tests are shown in Figure 10 (for details see [32]).

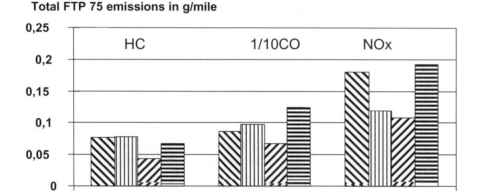

Figure 10: Vehicle test results – washcoat optimization study (LEV certified vehicle, Vcat/Veng = 0.71) [32].

Table 4: Catalyst matrix of washcoat optimization study (reprint from [32])

catalyst	cpsi/mil	amount of WC, %	relative washcoat thickness	relative precious metal concentration	Amount of OSC per converter, %
E	400/6.5	100	high	low	100
F	900/2.5	100	medium	low	100
G	900/2.5	80	low	medium	95
H	900/2.5	52	lowest	high	70

Based on these results the high temperature stable catalyst technology coded "G" was investigated in further tests on an EU IV certified vehicle using 400 cpsi/6.5 mil, 600 cpsi/3.5 mil, 900 cpsi/2.5 mil, and 1200 cpsi/2.5 mil substrates. In this study the 900 cpsi/2.5 mil converter showed the best performance with slight advantages as compared to 1200 cpsi/2.5 mil. It was indicated in [32] that the catalyst mass (substrate plus washcoat) is one important parameter especially, for the light-off phase. The system 900 cpsi/2.5 mil is offering the lowest mass of the converters tested and demonstrating the best heat-up characteristics.

EXPERIMENTAL

CATALYSTS AND TEST PROCEDURES

In the present study, the Pd/Rh technology being tailored for UTWS described in the former paragraph was applied. The PGM-loading ranged between 100 and 300 g/ft³. The washcoats were investigated on different UTWS - 600 cpsi/3.5 mil, 900 cpsi/2.5 mil, and 1200 cpsi/2.5 mil.

Model gas lambda sweep tests were performed at 350°C after calcination in wet air at 985°C. Light-off tests at lambda 0.999 (1Hz +/- 0.5 A/F) and sweep tests under different temperatures and pertubations were performed after engine aging at 870°C inlet temperature in a fuel cut mode for 25 hours on the engine.

For the model gas and engine evaluation the following precious metal loadings were used: 100 g/ft³, 150 g/ft³, and 200 g/ft³ at a Pd/Rh mass ratio of 14/1.

The vehicle study included tests on three different vehicles, details are summarized in Table 5. On all vehicles 600 cpsi/3.5, 900 cpsi/2.5, and 1200 cpsi/2.5 ceramic substrates were evaluated. The PGM loading ranged from 100 to 300 g/ft³ at a Pd/Rh mass ratio of 14/1. Not all possible combinations were investigated. In order to optimise the cost – information – ratio, a DOE approach was taken and all data were analysed by statistical methods. All converters were tested fresh and after aging on an engine in a fuel cut mode at catalyst inlet temperatures of 940°C for 96 hours. This aging cycle correlates well to 100000 km real road aging under German Autobahn driving conditions.

Table 5: Major parameters of the vehicle study.

	Vehicle A	Vehicle B	Vehicle C
Engine displacement in litre	4.3	1.4	2.4
Catalyst volume in litre	1.8	0.9	0.9
Calibration	Improved ULEV	EU IV	ULEV
Heat-up time during cold-start	short	medium	medium
Test cycle	FTP 75, MVEG-B	MVEG-B	FTP 75

TEST RESULTS

MODEL GAS TEST RESULTS

Seven aged catalyst systems were evaluated. Figure 11 summarizes the CO/NO crossover values achieved in the sweep tests. From this data, the following can be concluded:

Comparing different substrates at equal PGM loadings, the 900 cpsi/2.5 mil substrate leads to the most promising test results, no further improvements become visible for 1200 cpsi/2.5 mil. The influence of the PGM content is much more pronounced and determining the conversion level – the 200 g/ft³ sample always provides the best activity under the given test and aging conditions.

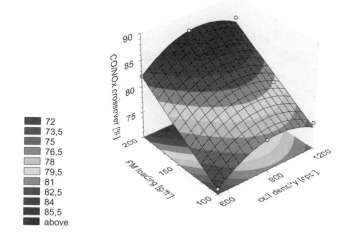

Figure 11: Model gas test results, inlet temperature 350°.

375

ENGINE BENCH TEST RESULTS

The data regarding the light-off tests and the results of the sweep tests are presented in Figure 12 and Figure 13. Again, similar to the findings of the model gas investigation, the activity is mainly ruled by the PGM loading and the effect of different substrates is obvious, but less pronounced. This trend becomes clearer at the lower cell density and slows down for the 900 and 1200 cpsi/2.5 samples.

The largest gain in activity was observed by switching from 100 to 150 g/ft³, whereas, a further increase up to 200 g/ft³ resulted in minor improvements. With respect to the substrates a real distinction is not possible at the highest PGM loading of 200 g/ft³. At the lower PGM loadings, the 900cpsi/2.5 mil converter shows a higher activity as compared to 600 cpsi/3.5 mil, whereas, a further increase of the cell density up to 1200 cpsi/2.5 mil indicates some advantages in the sweep tests, but not in the light-off experiments.

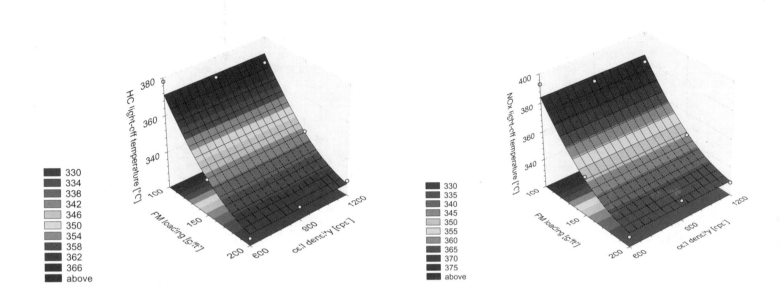

Figure 12. Engine test results; light-off tests at lambda 0.999, 1Hz ± 0.5 A/F.

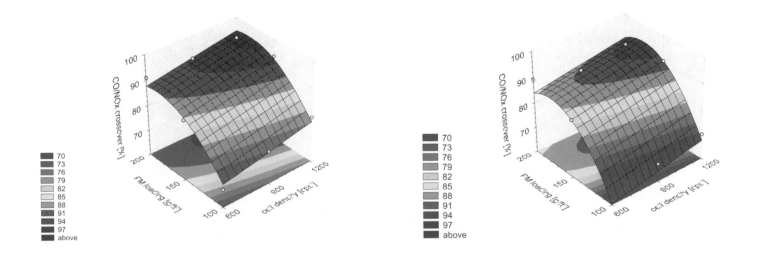

Figure 13: Engine test results; sweep tests at 400°C (+/- 0.5 A/F; right) and 450°C (+/-1A/F;left).

VEHICLE TEST RESULTS

- Vehicle A (Figure 14 and Figure 15)

FTP 75, fresh:

While the CO and NOx emissions are on a very similar level for all converters, the HC values differ by a factor of about 3. Here, both the type of substrate and the PGM loading show a major effect. At the highest PGM-loading the influence of the substrate is minor.

FTP 75, aged:

Again, the CO and NOx emissions are less sensitive to the PGM-loading and/or type of substrate. After aging the NOx emissions were approximately doubled and the CO values increased by a factor of about 1.4. For the HC, the influence of the substrate type is negligible and the amount of PGM determines the test result. Improvements can be achieved by switching from the low to the medium loading, a further increase does not lead to any significant advantages. All data is well below the ULEV 2 emission standards.

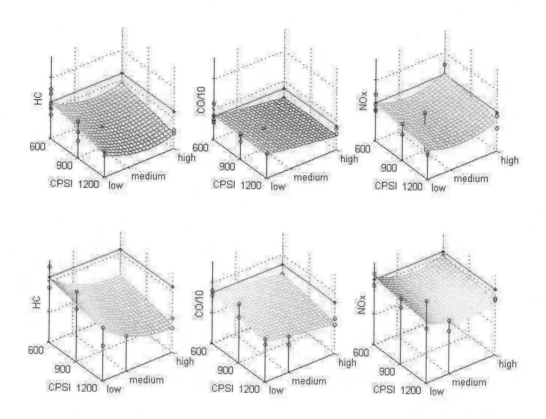

Figure 14: Vehicle A, relative FTP 75 test results (fresh top and aged bottom).

377

MVEG-B, fresh:
Similar to the findings in the US driving cycle, in the European driving cycle the fresh systems hardly showed any differences especially for CO and minor for NOx. For the HC, both PGM-loading as well as substrate type have an impact on the activity.

MVEG-B, aged:
The NOx emissions, being increased by a factor of about 1.2, are close to independent on the substrate type and

PGM-loading.For CO and HC, these two parameters have an impact on the test result. In this case the type of substrate is the more dominant factor. The deterioration during aging can be described by a factor of about 2 for CO and of about 1.4 for HC. The emission level for all three exhaust gas components was well below the EU IV standards.

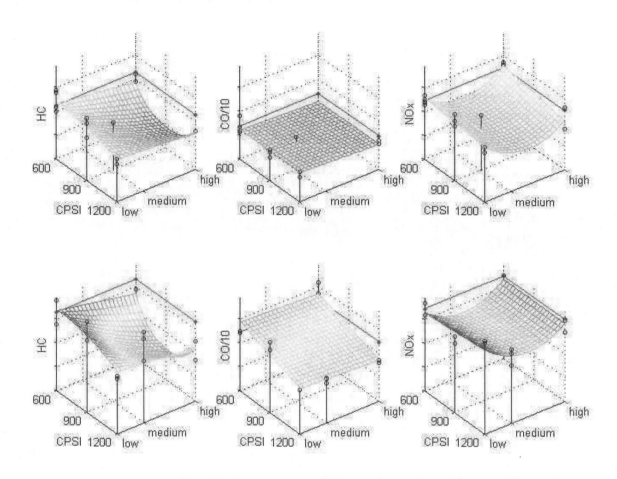

Figure 15: Vehicle A, relative MVEG-B test results (fresh top and aged bottom).

- Vehicle B (Figure 16)

MVEG-B, fresh and aged:
All fresh tests were on a very similar level for CO, HC, and NOx; neither substrate nor PGM loading had a noticeable influence on the results. This situation changed after aging. Increasing the substrate cell density from 600 cpsi to 900 cpsi demonstrates significant improvements for NOx, but for HC and CO

the effects are also obvious. This is true for the lower PGM-loadings and not that pronounced for the highly loaded converters. Further improvements by applying the 1200 cpsi substrate are negligible. The positive impact of the PGM-loadings is especially present in case of CO and NOx. For HC this trend is less, but still exists. The emission levels after ageing were increased by factors of 1.5 to 3.

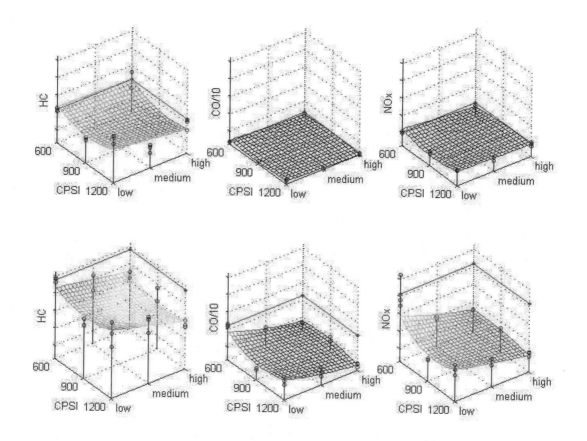

Figure 16: Vehicle B, relative MVEG-B test results (fresh top and aged bottom).

• Vehicle C (Figure 17)

FTP 75, fresh:
While the CO data are relatively stable for the given test matrix, for HC a positive effect of the PGM-loading and increased cell density was observed. The NOx emissions are mainly influenced by the PGM-loading.

FTP75, aged:
Most of the test results are on a very similar level. Especially, this was observed for NOx and partly CO. For HC, a positive trend is indicated for an increase of PGM.

In summary, model gas and engine bench tests demonstrate benefits by changing the substrate from 600 cpsi/3.5 mil to 900 cpsi/2.5 mil and that a further increase of the cell density does not result in a significant additional gain in activity. Generally, the amount of PGM in the system mainly determines the activity levels under the given test conditions.

Most vehicle results of the study confirmed these findings. Precious metal loadings exceeding 200 g/ft^3 and cell densities above 900 cpsi did not significantly improve the system performance.

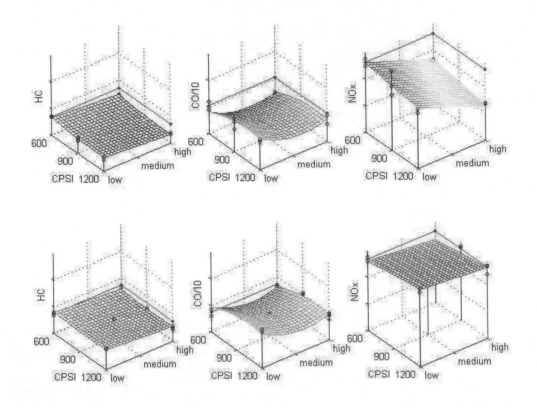

Figure 17: Vehicle C, relative FTP-75 test results (fresh top and aged bottom).

INFLUENCE OF BACKPRESSURE ON POWER AND TORQUE

On the engine bench the 1200 cpsi/2.5 mil and the 600 cpsi/3.5 mil substrate were compared at full load conditions. The resulting 25% increase of back pressure led to a 5% loss of power and 2% loss of torque output. The 900 cpsi/2.5 mil substrate showed only a 10 % increase in back pressure. The influence on the power and torque output was negligible.

CONCLUSIONS, OUTLOOK

- The combination of the Calibrated Rohrkat with the LCF mat makes it possible to can all thin wall and ultra thin wall substrates.

- The geometric surface area and the heat capacity are positively influenced by increasing the cell density at low wall thicknesses for ceramic materials.

- On the contrary, higher cell densities even at low wall thicknesses lead to the undesired effect of an increased pressure drop.

- A balance of these two design parameters can be achieved by the application of advanced three-way catalyst coating technologies which offer excellent aging stability at moderate washcoat loadings.

- Different design features need to be tailored – the washcoat loading, the washcoat thickness, and the precious metal concentration. Furthermore, these features can also be dependent on the degree of aging.

- Model gas and engine bench tests demonstrate benefits by changing the substrate from 600 cpsi/3.5 mil to 900 cpsi/2.5 mil

- Precious metal loadings exceeding 200 g/ft^3 and cell densities above 900 cpsi did not significantly improve the system performance on the vehicle.

- In future programs, the interactions between substrate weight, physical aspects like heat capacity, and catalyst properties like washcoat thickness or precious metal concentration will be investigated in more detail for different ageing levels.

ACKNOWLEDGEMENT

The authors want to thank all colleagues who contributed to this work for the valuable discussions and the high quality work.

LITERATURE

1. K. Kollmann, J. Abthoff, W. Zahn, "Concepts for Ultra Low Emission Vehicles", SAE Paper 940468 (1994).
2. M.E. Crane, R.H. Thring, D.J. Podnar, L.E. Dodge, "Reduced Cold-Start Emissions Using Rapid Exhaust Port Oxidation (REPO) in a Spark-Ignition Engine", SAE Paper 970264 (1997).
3. E. Achleitner, W. Hosp, A. Koch, W. Schürz, "Electronic Engine Control System for Gasoline Engines for LEV and ULEV Standards", SAE Paper 950479 (1995).
4. Y. Shimasake, K. Kato, H. Muramatsu, T. Teshirogi, T. Aoki, A. Saito, G. Rodrigues "Study on Conformity Technology with ULEV using EHC System", SAE Paper 960342 (1996)
5. K. Nishizawa, T. Yamada, Y. Ishizuka, T. Inoue, "Technologies for Reducing Cold-Start Emissions of V6 ULEVs", SAE Paper 971022 (1997).
6. Y. Shimasaki, H. Kato, F. Abe, S. Hashimoto, T. Kaneko, "Development of Extruded Electrically Heated Catalyst System for ULEV Standards", SAE Paper 971031 (1997).
7. P. Ahlvik, L. Erlandsson, A. Laveskog, "The Influence of Block Heaters on the Emissions from Gasoline Fueled Cars with Varying Emission Control Technolgy at Low Ambient Temperatures", SAE Paper 970747 (1997).
8. S. Roychoudhury, G. Muench, J.F. Bianchi, W.C. Pfefferle, "Development of MicrolithTM Light-Off Preconverters for LEV/ULEV", SAE Paper 971023 (1997).
9. J.P. Day, "Substrate Effects on Light-Off - Part II - Cell Shape Contributions", SAE Paper 971024 (1997).
10. K. Kollmann, J. Abthoff, W. Zahn, H. Bischof, J. Gohre, "Secondary Air Injection with a New Developed Electrical Blower for Reduced Emissions", SAE Paper 920400 (1992).
11. B. Pfalzgraf et al., "The System Development of Electrically Heated Catalyst (EHC) for the LEV and EU-III Regulations", SAE Paper 951072 (1995).
12. W. Held, M. Rohlfs, W. Maus, H. Swars, R. Brück, F.W. Kaiser, "Improved Cell Design for Increased Catalytic Conversion Efficiency" SAE Paper 940932 (1994).
13. G. Faltmeier, B. Pfalzgraf, R. Brück, C. Kruse, W. Maus, A. Donnerstag, "Catalyst Concepts for Future Emission Legislation demonstrated at a 1.8 Liter, 5 Valve Engine", Vienna Motorensymposium, Vienna, Austria, p. 398-417 (April 1996).
14. D. Lindner, E.S. Lox, R. van Yperen, K. Ostgathe, T. Kreuzer, "Reduction of Exhaust Gas Emissions by Using Pd-based Three-way Catalysts", SAE Paper 960802 (1996).
15. J. Hepburn, K. Patel, M. Meneghel, H.S. Gandhi, "Development of Pd-only Three-way Catalysts", SAE Paper 941058 (1994).
16. D.J. Ball, "A Warm-up and Underfloor Converter Parametric Study", SAE Paper 930386 (1993).
17. B.H. Engler, E.S. Lox et al., "Recent Trends in the Application of Three Metal Emission Control Catalysts", SAE Paper 940928 (1994).
18. B.H. Engler, D. Lindner, E.S. Lox, K. Ostgathe, A. Schäfer-Sindlinger, W. Müller, "Reduction of Exhaust Gas Emissions by Using Hydrocarbon Adsorber Systems", SAE Paper 930738 (1993).
19. G. Waltner, Loose, A. Hirschmann, L. Mußmann, D. Lindner, W. Müller, "Development of Close-Coupled Catalyst systems for European Driving Conditions", SAE Paper 980663 (1998).
20. J. Schmidt, A. Waltner, G. Loose, A. Hirschmann, A. Wirth, W. Müller, J.A.A. van den Tillaart, L. Mußmann, D. Lindner, J. Gieshoff, K. Umehara, M. Makino, K.P. Biehn, A. Kunz, "The Impact of High Cell Density Ceramic Substrates and washcoat Properties on the Catalytic Activity of Three-Way catalysts", SAE Paper 1999-01-0272 (1999).
21. N. Higuchi and T. Nakamura, "Ceramic Honeycomb", Industrial Material, Vol. 31, (1984).
22. J. Howitt, "Thin Wall Ceramics as Monolithic Catalyst Supports," SAE Paper 800082 (1980).
23. S. Mochida, "Ceramic Honeycomb in the Spotlight," Chemical Engineering MOL, March (1984).
24. H. Yamamoto, H. Horie, J. Kitagawa and M. Machida, "Reduction of Wall Thickness of Ceramic Substrate for Automotive Catalysts", SAE Paper 900614 (1990).
25. H. Yamamoto, F. Kato, J. Kitagawa and M. Machida, "Warm-up Characteristics of Thin Wall Honeycomb Catalysts," SAE Paper 910611 (1991).
26. M. Machida, T. Yamada and M. Makino, "Study of Ceramic Catalyst Optimization for Emission Purification Efficiency," SAE Paper 940784 (1994).
27. T. Hijikata, H. Kurachi, F. Katsube and H. Van Honacker, "Thermal Reliability and Performance Improvement on Close-Coupled Catalytic Converter," SAE Paper 960565 (1996)
28. Sword, G. Morgan, R. O'Sullivan, D. Winerbone and M. Machida, "Development of a High Performance Catalytic Converter for a Turbocharged Gasoline Engine Using Thin Wall Ceramic Technology," SAE Paper 930943 (1993).
29. N. Tamura, S. Matsumoto, M. Kawabata, M. Kojima, M. Machida, "The development of Automotive Catalyst with Thin Wall (4 mil / 400 cpsi) Substrate," SAE Paper 960557 (1996)
30. K. Umehara, T. Hijikata and F. Katsube, "Catalytic Performance Improvement by High Cell Density / Thin Wall Ceramic Substrate," ISATA Paper 96EN044 (1996).
31. K. Umehara, T. Yamada, T. Hijikata, Y. Ichikawa and F. Katsube," Advanced Ceramic Substrate : Catalytic Performance Improvement by High Geometric Surface Area and Low Heat Capacity," SAE Paper 971029 (1997).
32. R. Domesle, D. Lindner, W. Mueller, L. Mussmann, M. Votsmeier, E.S. Lox, T. Kreuzer, M. Makino, C.D. Vogt, "Application of Advanced Three-way Catalyst technologies on High Cell Density Ultra Thin-Wall Ceramic Substrates for Future Emission Legislations", SAE Paper 2001-01-0924 (2001).

CONTACT

Jürgen Schmidt

DaimlerChrysler AG
HPC C 300
D-70546 Stuttgart, Germany
Email:juergen.s.schmidt@daimlerchrysler.com
Phone: 0049 711 17 32902
Fax: 0049 711 17 55650

Wilfried Mueller

OMG AG & Co. KG
Rodenbacher Chaussee 4
D-63403 Hanau, Germany
Email: wilfried.mueller@dmc-2.de
Phone: 0049 6181 59 4018
Fax: 0049 6181 59 4517

Henninger, Richard

Faurecia Abgastechnik GmbH
Herboldshofer Str. 35
D-90765 Fürth, Germany
Email: rhenninger@stadeln.faurecia.com
Phone: 0049 911 7610 175
Fax: 0049 911 7610 378

Kunz, Alexander

NGK Europe GmbH
Westerbachstr. 32
D-61476 Kronberg; Germany
Email   : akunz@ngk-e.de
Phone : 0049 6173 993 148
Fax.   : 0049 6173 993 170

2001-01-3840

Pushing the Envelope to Near-Zero Emissions on Light-duty Gasoline Vehicles

Joseph E. Kubsh
Engelhard Corporation

Copyright © 2001 Society of Automotive Engineers, Inc

ABSTRACT

The integration of advanced emission control technologies including advanced three-way catalysts and advanced, high cell density, ultra-thin wall substrates with advanced gasoline powertrains and advanced engine controls is necessary to achieve near-zero tailpipe emission requirements like California's SULEV or PZEV light-duty certification categories. The first gasoline vehicles meeting these near-zero regulations have been introduced in California in 2001. Advanced three-way catalysts targeted for these near-zero regulations feature layered architectures, thermally stable oxygen storage components, and segregated precious metal impregnation strategies. Engine calibration strategies focused on tight stoichiometric air/fuel control and fast catalyst heat-up immediately after engine start are important enablers to achieve near-zero hydrocarbon and NOx emissions.

INTRODUCTION

Recent light-duty vehicle regulatory actions in the United States and the state of California have created vehicle emission certification categories that approach near-zero levels for tailpipe emissions of regulated pollutants including non-methane hydrocarbons, carbon monoxide, and nitrogen oxides. These certification categories include the ULEV-2 (Ultra-low Emission Vehicle), SULEV (Super Ultra-low Emission Vehicle) and PZEV (Partial Zero Emission Vehicle) categories defined in California's Low Emission Vehicle II program and the so-called Bin 2 and Bin 3 categories associated with the United States EPA's Tier 2 emission regulations. Tailpipe emission standards for these certification categories and mileage requirements associated with intermediate and full useful life durability are provided

in Table 1. Note that these near-zero certification levels are part of larger regulatory programs that include light-duty certification levels at higher emission levels than those shown in Table 1. Auto

Table 1. California LEV II and U.S. EPA Tier 2 Certification Categories with Near-Zero Tailpipe Emission Requirements

Emission Category	Durability Mileage	NMOG* (g/mi)	CO (g/mi)	NOx (g/mi)
California ULEV-2	50K mi	0.040	1.7	0.05
	120K mi	0.055	2.1	0.07
California SULEV	120K mi	0.010	1.0	0.02
California PZEV	150K mi	0.010	1.0	0.02
U.S. EPA Tier 2, Bin 3	120K mi	0.055	2.1	0.03
U.S. EPA Tier 2, Bin 2	120K mi	0.010	2.1	0.02

*NMOG = Non-methane organic gases

manufacturers are expected to phase-in the production of light-duty vehicles (passenger cars, light-duty trucks up to 8500 lbs. gross vehicle weight, and medium-duty passenger vehicles with gross vehicle weights up to 10,000 lb., including minivans, SUVs, and full-size vans) meeting these near-zero emission standards over the 2003-2009 timeframe in the United States. In California two different manufacturers have already introduced gasoline vehicles meeting the SULEV category and one manufacturer has introduced a gasoline vehicle meeting the PZEV emission levels shown in Table 1.

This paper discusses emission control technologies (e.g., advanced three-way catalysts, advanced high cell density, ultra-thin wall substrates) that enable light-duty gasoline passenger cars and trucks to reach the near-zero emission levels associated with the certification categories shown in Table 1. In order for these high performance emission systems to reach the high efficiencies associated with the emission standards shown in Table 1, it is essential that they be coupled with advanced engine hardware and advanced engine control strategies. Examples are given here of the impact of these advanced control strategies on reaching near-zero tailpipe emission levels with advanced emission control technologies.

FIRST SULEV/PZEV INTRODUCTIONS

Before examining in detail the attributes of advanced emission control technologies targeted for the near-zero emission performance requirements shown in Table 1, it is interesting to examine the range of technologies included with the first gasoline vehicles introduced in the last year in California to meet SULEV/PZEV emission regulations. These vehicles include the Honda Accord SULEV, the Toyota Prius SULEV, and the Nissan Sentra-CA PZEV models. Each of these models combine advanced powertrains, advanced engine control strategies, and advanced three-way catalysts to achieve near-zero tailpipe emissions.

HONDA ACCORD SULEV – This gasoline SULEV vehicle is an upgrade to Honda's first ULEV vehicle, the 1998 model year Accord. The Accord SULEV features Honda's 2.3 liter, 4 cylinder engine with Honda's VTEC variable valve control technology. The variable valve technology facilitates a lean engine cold-start strategy. The engine control module features a 32 bit microprocessor that allows Honda to utilize sophisticated control strategies including individual cylinder air/fuel control. The emission system features an underfloor-only catalytic converter (1.7 liters in volume) and a insulated manifold and exhaust pipe. Important characteristics of the underfloor converter include a two layer, tri-metallic (Pt/Pd/Rh) advanced catalyst architecture and use of a 1200 cpsi/2.0 mil wall ceramic substrate. Both the catalyst and the substrate were upgraded for this SULEV application relative to the Accord ULEV application. The insulated exhaust system makes use of a thin metal inner wall, air-gap construction for fast and efficient heat transfer to the underfloor converter.

Additional details on Honda's SULEV system have been provided in a recent SAE reference (1).

TOYOTA PRIUS SULEV – The Prius is Toyota's first hybrid electric vehicle introduced in the United States. The hybrid electric powertrain includes a 1.5 liter gasoline engine and an 33 kW electric drive in a series-parallel hybrid configuration. The gasoline engine, like the Honda Accord includes a variable valve control technology. The hybrid powertrain and its control algorithms allow for the gasoline engine to be stopped during low speed and idle situations and provides for a smart engine restart strategy. For example during the U.S. FTP test cycle the Prius engine shuts down and restarts more than 20 times. This shut-down/smart restart strategy improves fuel consumption and reduces emissions. The emission system on the Prius includes close-coupled + underfloor converters both with advanced three-way catalyst formulations. The close-coupled converter makes use of a 900 cpsi/2 mil wall ceramic substrate (0.9 liter volume) combining low thermal mass and large geometric surface area for fast heat-up after engine start. The underfloor converter on the Prius has a complex design that combines a 600 cpsi/3 mil wall ceramic substrate (1.1 liter volume) surrounded by a metallic substrate matrix (0.7 liter volume). An advanced three-way catalyst is present on the ceramic substrate while a hydrocarbon adsorber coating (zeolite-based) is placed on the metallic matrix. Upstream of this underfloor converter is a large butterfly type valve and valve actuator. During cold-starts this valve is closed and exhaust flow is directed only through the hydrocarbon adsorber function of the underfloor converter in order to trap hydrocarbons in the early stages of the start. After the close-coupled converter has reached its operating temperature, the underfloor converter valve opens and the exhaust flows through the three-way catalyst present on the ceramic underfloor substrate. As this underfloor ceramic catalyst warms-up, this heat is transferred to the adsorber metal matrix causing trapped hydrocarbons to desorb and flow through the underfloor converter where they are oxidized. Details on the Prius' sophisticated emission control system are provided in recent publication (2).

NISSAN SENTRA-CA PZEV – Nissan's Sentra-CA model is the first PZEV certified vehicle introduced in California. The PZEV certification requires SULEV tailpipe emission performance with 150K mile durability and an evaporative emission system that meets California zero evaporative standards. Details on Nissan's zero evaporative system's characteristics have been published recently (3). With respect to tailpipe emissions, the Sentra-CA includes a 1.8 liter, 4 cylinder gasoline engine with a swirl valve design for lean cold-start operation.

Advanced controls also provide for precise control of the exhaust air/fuel ratio near the optimum stoichiometric point. The emission system features three catalytic converters in series: a close-coupled converter, followed by two separate underfloor converters. Each of the underfloor converters uses an advanced multi-layer three-way catalyst architecture that includes a zeolite-based hydrocarbon adsorber layer to minimize cold-start hydrocarbon emissions. The close-coupled converter includes a 900 cpsi/2 mil wall ceramic substrate coated with a multi-layer three-way catalyst. Additional details on Nissan's PZEV emission system can be found in two recent SAE publications (4,5). Additionally this PZEV vehicle includes another advanced emission technology, a radiator coated with a direct ozone decomposition catalyst (4,6). This coating decomposes ground level ozone that flows through the vehicle's radiator during normal driving.

Each of these production vehicles features a unique systems solution to meeting SULEV or PZEV emission standards. There are, however, some common features to each of these emission systems: advanced multi-layer three-way catalysts, advanced high cell density, ultra-thin wall substrates, and calibration strategies that facilitate fast heat-up and operation of the catalytic converter during a cold-start. Each of these system features will be considered in some detail in the following sections.

ADVANCED THREE-WAY CATALYSTS

State-of-the-art three-way catalysts with high conversion efficiencies for all three criteria pollutants (hydrocarbons, CO, NOx) are an important part of systems solutions targeting near-zero tailpipe emissions. These catalysts feature layered architectures with segregated noble metal impregnation strategies to maximize the thermal stability and performance characteristics of each noble metal function. Advanced, thermally stable oxygen storage materials (e.g., stabilized ceria/zirconias) are incorporated to promote noble metal activity in these catalysts. Thermally stable three-way catalysts are especially important for durable performance in close-coupled converter locations. Performance characteristics of an advanced tri-metallic catalyst suitable for close-coupled SULEV or PZEV applications are detailed in the next section.

ADVANCED TRI-METAL CATALYST – The thermal stability characteristics of an advanced tri-metal catalyst targeted for close-coupled applications were demonstrated by a vehicle durability test on German autobahns with substantial vehicle operation at high speeds. A late model European vehicle

equipped with a 3.2 liter V6 engine was used as a test vehicle in this program. The test vehicle has a dual exhaust system with a close-coupled and underfloor converter on each cylinder bank of the V6 engine. Each of the close-coupled converter catalyst volumes was 1.09 liters with each underfloor converter having a catalyst volume of 0.80 liters. Substrates used in all converter locations were ceramic with standard 400 cpsi/6.5 mil wall characteristics. An advanced tri-metal catalyst formulation (100 g/ft^3, Pt/Pd/Rh= 1/13/1) was used in the close-coupled locations and an advanced Pd/Rh formulation (50 g/ft^3, Pd/Rh= 5/1) was used in the underfloor converters. The test vehicle was run for a total of 50,000 km of high speed, autobahn driving with emissions measured at 2000, 15,000, and 50,000 km intervals using the revised Euro 3/Euro 4 European test cycle with emissions measured immediately after engine start. Average tailpipe emissions measured at these aging intervals from multiple emission test cycles are summarized in Figure 1.

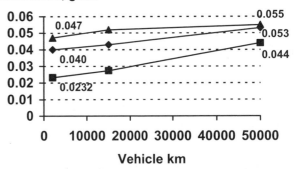

Emissions, g/km

Figure 1. European MVEG test cycle emissions during autobahn durability test with CC+UF converter system on a 3.2 liter test vehicle

This autobahn durability test lasted 5 weeks, with average vehicle speeds over the entire 50,000 km program of 148 km/h with 20.5% of the aging time at vehicle speeds in excess of 200 km/h. During the test program the vehicle was fueled with standard European pump-grade gasoline. Temperatures measured in the exhaust system of the test vehicle indicated that 28-32% of the time on the vehicle resulted in close-coupled catalyst temperatures (measured 25 mm from the inlet substrate face) in excess of 950°C, with 10-12% of the aging history with close-coupled catalyst temperatures in excess of 1000°C. Percentage ranges reflect temperature differences between each cylinder bank of the V6 engine. Maximum close-coupled catalyst temperatures experienced in this vehicle test were in excess of 1050°C. Despite this long term, high temperature vehicle operation, the advanced tri-metal close-coupled catalyst exhibited high efficiencies in the European test cycle with low deterioration rates for all three criteria pollutants. THC emissions increased by 32 % from the 2000 km to 50,000 km test points while NOx emissions increased by 17% over the same time. The largest portion of the observed deterioration occurred between the 2000 km and 15,000 km test points. Comparing data obtained at the 15,000 km test point with the 50,000 km test point, THC emissions increased by 23%, and NOx emissions increased by only 6%.

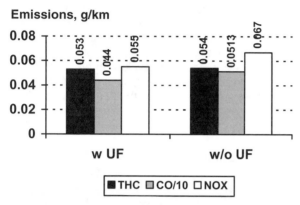

Emissions, g/km

Figure 2. European MVEG test cycle results after 50,000 km autobahn driving – CC only or CC+UF converters

In systems that include close-coupled and underfloor converters, the close-coupled converter generally provides the dominant oxidation requirements for hydrocarbons and CO. The underfloor converter generally provides additional NOx reduction performance. Such was the case with the system evaluated in the autobahn durability test program. Figure 2 shows European test cycle emission tests run after the 50,000 km durability history with and without the underfloor converters installed on the test vehicle. Essentially all of the THC performance is associated with only the close-coupled converters, with the underfloor converters contributing to the NOx conversion efficiency of this system.

ADVANCED PLATINUM CATALYSTS - Increasing demand for palladium-based three-way catalyst formulations and the subsequent steep increases in palladium market prices during the 1990s focused attention on developing high performance, platinum-based technologies using the same advanced catalyst design features (layered architectures, segregated noble metal impregnation strategies, thermally stable oxygen storage materials and supports) first developed for palladium-based catalysts. The result has been the introduction of Pt/Rh catalysts primarily targeted for underfloor converter applications, including near-zero emission

applications meeting ULEV-2 and SULEV performance requirements. These advanced Pt/Rh catalysts exhibit performance on a par with or even surpassing the performance of commercial Pd/Rh formulations. The performance advantages of these advanced Pt/Rh catalysts are especially evident in applications that do not utilize low-sulfur gasoline formulations. Figure 3 compares aged sweep performance of a two-layer advanced Pt/Rh catalyst (40 g/ft^3, Pt/Rh= 5/1) with a commercial Pd/Rh catalyst (60 g/ft^3, Pd/Rh= 9/1). Equal volume converters of both catalyst technologies were aged for 75 h using an aging schedule with inlet converter temperatures of 850°C and maximum catalyst temperatures of approximately 900°C. Sweep tests were run on an engine dynamometer following aging of the converters at a temperature of 450°C with 1.0 Hz, +/- 0.5 air/fuel perturbations and a gas hourly space velocity of 80,000/h. Sweep tests were run with two different gasoline sulfur levels: 40 ppm S and 300 ppm S.

Figure 3. CO/NOx crossover measured in engine

CO/NOx Crossover, %

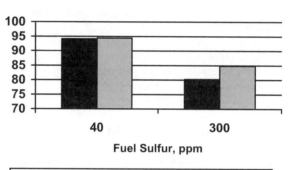

sweep tests after 75 h of engine aging, comparing Pt/Rh vs. Pd/Rh catalyst technologies with different fuel sulfur levels

The sweep results of Figure 3 show equivalent performance for the lower loaded, advanced Pt/Rh technology at the lower fuel sulfur level but superior performance at the higher fuel sulfur level due to platiunum's superior resistance to sulfur poisoning compared to palladium.

ADVANCED SUBSTRATES

Each of the commercial SULEV/PZEV vehicles discussed previously employ the latest generation of high cell density, ultra-thin wall substrates in close-coupled and/or underfloor locations. These new substrate designs were developed to improve gas contacting efficiency with the active catalyst by increasing the geometric area of the substrate through the use of smaller cell dimensions. These high cell density substrates also incorporate ultra-thin walls to reduce the thermal mass and improve the pressure drop characteristics of the substrate. The resulting lower thermal mass substrates heat-up faster during the dynamic light-off process on the vehicle, an important characteristic for near-zero cold-start emissions. The relative cold-start performance of this new class of ceramic substrates was evaluated in a test program.

SUBSTRATE IMPACT ON COLD-START PERFORMANCE - A series of converters were prepared using ceramic substrates with a variety of cell density, wall thickness combinations. Each of these substrates were coated with a Pt/Pd catalyst formulation (250 g/ft^3, Pt/Pd= 1/10) developed for high oxidation performance of hydrocarbons during cold-start operation. Table 2 summarizes important physical characteristics of the ceramic substrates investigated in this study.

Table 2. Ceramic substrate physical properties

Cell Geometry cpsi/wall	OFA* %	GSA* m^2/liter	Bulk Density g/liter	Bulk Density (coated) g/liter
600/4.3 mil	80.0	3.45	324	543
600/3.5 mil	83.6	3.53	238	493
600/2.5 mil	88.1	3.62	196	441
900/2.5 mil	85.6	4.37	238	507
1200/2.5 mil	83.4	4.98	273	541

*OFA=open frontal area, GSA=geometric surface area

The physical properties contained in Table 2 show how important characteristics such as geometric surface area and bulk density vary with cell geometry. Increasing cell density at a constant wall thickness increases geometric area but also increases bulk density. Decreasing wall thickness at a constant cell density increases geometric area but reduces bulk density.

Total catalyst volume for all test parts was held constant at 0.67 liters. Each converter was aged for 100 h using a fuel cut schedule with an inlet converter temperature of 800°C and maximum catalyst temperatures of approximately 900°C. After aging, each converter was evaluated for simulated cold-start hydrocarbon emissions on a 5.7 liter engine dynamometer test unit. Total exhaust flow was held constant at 30 scfm during the test with exhaust

stoichiometry slightly lean of the stoichiometric point. Inlet temperature to the converter was controlled through a heat exchanger system and total system pressure was held constant with the use of an orifice plate downstream of the test converter. A moderate temperature ramp was used to simulate a start situation with the inlet temperature of the converter reaching 350°C in 25 seconds after the start of exhaust flow to the converter.

Results from this cold-start simulation are shown in Figure 4. Accumulated total hydrocarbon emissions for each aged part are reported after 33 s of the simulated cold-start test and after 44 s of the test.

Accumulated THC, g

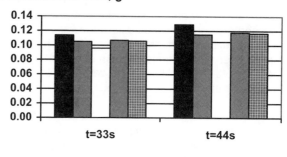

Figure 4. Accumulated total hydrocarbon emissions measured in simulated cold-start tests for catalysts made using the ceramic substrates detailed in Table 2 (A=600 cpsi/4.3 mil, B=600 cpsi/3.5 mil, C=600 cpsi/2.5 mil, D=900 cpsi/2.5 mil, E=1200 cpsi/2.5 mil; all substrates coated with same catalyst technology)

Performance in this dynamic heat-up test are generally in line with the coated bulk densities of each substrate, demonstrating the importance of thermal mass on the converter light-off process. The heaviest sample (600 cpsi/4.3 mil wall) heats up the slowest of these test parts and exhibits the highest hydrocarbon emissions, while the lightest part (600 cpsi/2.5 mil wall) exhibits the smallest hydrocarbon emissions. Samples made with the 600 cpsi/3.5 mil wall and 900 cpsi/2.5 mil wall ceramic substrates have comparable intermediate coated weights and similar emission performance in this simulated cold start test. It appears that the heavy weight associated with the 1200 cpsi/2.5 mil wall substrate is in part compensated for by the high geometric surface area of this monolith.

These high cell density, ultra-thin wall substrates help facilitate fast converter light-off, critical to reaching near-zero emission targets. The large geometric areas offered by these substrates also improve converter efficiencies during warmed-up operation, helping to drive emissions during warmed-up conditions to essentially zero levels. Although the test program reported here focused on ceramic substrates, metallic substrates with high cell densities and ultra-thin walls (foil thickness down to 1 mil) have also been developed recently for near-zero emission applications.

ADVANCED ENGINE CALIBRATIONS

Equally important to advanced catalysts and advanced substrates are the development of advanced engine calibration strategies that reduce engine-out emissions during the critical cold-start operation, accelerate catalyst heat-up during the cold-start , and precisely control air/fuel near the optimum stoichiometric point during warmed-up operation. Each of the three commercial SULEV/PZEV vehicles discussed previously makes use of a variety of advanced calibration/control strategies to achieve their near-zero emission levels. An example of the impact of cold-start calibration on cold-start emissions is provided below.

COLD-START CALIBRATION IMPACTS – U.S. FTP emission test cycles were run on a late model test vehicle (2.4 liters) calibrated for California LEV I emission performance. A prototype close-coupled converter was fabricated for this test vehicle incorporating a cascade design with two substrates in series. The first of these substrates was 0.52 liters in volume, utilized a 900 cpsi/2 mil wall ceramic design, and was coated with a Pd-only catalyst (300 g/ft^3). The second substrate was 1.19 liters in volume, utilized a 600 cpsi/4 mil wall ceramic design, and was coated with the same advanced tri-metal catalyst formulation (250 g/ft^3, Pt/Pd/Rh= 1/22/2) used in the autobahn test program discussed previously. Prior to FTP tests this cascade converter design was stabilized on an engine dynamometer for 10 h using a fuel-cut aging schedule with a maximum catalyst temperature of approximately 950°C. Triplicate FTP cycles were run with this close-coupled converter installed on the test vehicle with the LEV I cold-start calibration strategy, and a modified cold-start calibration strategy targeted for a potential SULEV application. This SULEV cold-start calibration included lean air/fuel operation during the initial idle of the FTP cycle, additional spark retard during the initial seconds after engine start, and higher engine idle speed during the initial idle. All these calibration changes were focused on accelerating catalyst light-off during the cold-start process. Additional calibration modifications included the disablement of all engine fuel-cuts during vehicle decelerations and some rich biasing of the air/fuel during the first acceleration after

engine start. These calibration modifications were aimed at improving catalyst NOx performance.

Figure 5 compares average cold-start total hydrocarbon and NOx performance with the stabilized, close-coupled converter with both the LEV I calibration and the SULEV calibration strategies. Included in this figure are emissions measured after the first 140 s of the FTP test cycle (includes engine crank, first idle and the first "hill" of the drive cycle) and for the first 505 seconds of the test cycle (Bag 1 of the FTP). Dramatic reductions in both hydrocarbons and NOx emissions were obtained in the cold-start portion of the cycle with the SULEV calibration strategy. Over the complete Bag 1 portion of the cycle, the SULEV calibration strategy reduced THC emissions by 65% and NOx emissions by 76% relative to the LEV I calibration. Temperatures measured in the close-coupled converter during these FTP tests, showed faster heat-up of the close-coupled converter with the SULEV start strategy. The lean start strategy also significantly reduced hydrocarbon emissions during the initial vehicle idle of the FTP compared to the LEV I start strategy which includes

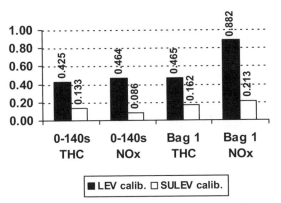

significant cold-start enrichment after engine crank.

Figure 5. Cold-start calibration impacts on hydrocarbon and NOx emissions for two different calibration strategies: LEV I vs. SULEV

The addition of an underfloor converter to this test vehicle utilizing 600 cpsi/4 mil wall ceramic substrates coated with the same tri-metal catalyst formulation used in the autobahn test program (250 g/ft^3, Pt/Pd/Rh= 1/14/1, total converter volume = 1.8 liters) further reduced Bag 1 hydrocarbon and NOx emissions using the SULEV calibration strategy relative to the performance shown in Figure 5 for the close-coupled system alone. Using the Bag 1 non-methane hydrocarbon (NMHC) emissions for the complete close-coupled + underfloor system, total FTP NMHC emissions were calculated assuming Bag 2 and Bag 3 NMHC emissions were zero. This calculation yielded 0.005 g/mi NMHC emissions with the SULEV calibration, emissions in the range of the SULEV NMOG standards (see Table1).

CONCLUSION

Emission technologies are available to reach near-zero emission targets such as California's ULEV-2, SULEV, PZEV on light-duty gasoline passenger cars and trucks. The first of these near-zero emission vehicles have been introduced in California featuring advanced powertrains, advanced engine controls, high performance catalysts, and high cell density, ultra-thin wall substrates.

Advanced three-way catalysts have been developed using multi-layer architectures, segregated noble metal impregnation strategies, and thermally stable oxygen storage materials. These advanced catalysts include tri-metal catalysts suitable for close-coupled applications and advanced Pt/Rh catalysts suitable for underfloor locations. Autobahn durability tests of an advanced tri-metal catalyst showed high emission performance and relatively low deterioration rates after 50,000 km of high speed driving.

New high cell density, ultra-thin wall substrates have been developed featuring high geometric surface areas and low thermal mass. These advanced substrates heat-up quickly to facilitate fast catalyst light-off. Engine tests run with a simulated cold-start procedure with substrates with a range of cell densities and wall thickness rank hydrocarbon emission performance generally in line with coated substrate bulk density. This result indicates that low thermal mass, high cell density substrate designs are important for improved cold-start emission performance.

The cold-start vehicle performance of advanced catalysts coated on high cell density substrates is a strong function vehicle cold-start calibration. SULEV calibration strategies can include lean start strategies, spark retard, and higher idle engine speeds after start to accelerate catalyst heat-up during the critical early phases of the cold-start operation.

ACKNOWLEDGMENTS

The author acknowledges the work of Engelhard's Environmental Technologies Group Research & Development staff for their work in completing the experimental results reported in this paper for the autobahn durability program and tests aimed at understanding cold-start calibration impacts on catalyst performance.

Corning Inc. is acknowledged for their assistance in completing the characterization of the emission performance of catalysts made with high cell density, ultra-thin wall ceramic substrates.

REFERENCES

1. H. Kitigawa, T. Mibe, K. Okamatsu, and Y. Yasui, "L4-Engine Development for a Super Ultra Low Emissions Vehicle," SAE paper no. 2000-01-0887.

2. T. Inoue, M. Kusada, H. Kanai, S. Hino, and Y. Hyodo, "Improvement of a Highly Efficient Hybrid Vehicle and Integrating Super Low Emissions," SAE paper no. 2000-01-2930.

3. H. Matsushima, A. Iwamoto, M. Ogawa, T. Satoh, and K. Ozaki, "Development of a Gasoline-Fueled Vehicle with Zero Evaporative Emissions," SAE paper no. 2000-01-2926.

4. K. Nishizawa, S. Momoshima, M. Koga, and H. Tsuchida, "Development of New Technologies Targeting Zero Emissions for Gasoline Engines," SAE paper no. 2000-01-0890.

5. S. Yamamoto, K. Matsushita, S. Etoh, and M. Takaya, "In-line Hydrocarbon Adsorber System for Reducing Cold-Start Emissions," SAE paper no. 2000-01-0892.

6. J. Hoke, R. Heck, and T. Poles, "Premair Catalyst System – A New Approach to Cleaning the Air," SAE paper no. 1999-01-3677.

UltraThin Wall Catalyst Solutions at Similar Restriction and Precious Metal Loading

Douglas J. Ball, Russell Richmond, Charles Kirby and Glenn Tripp
Delphi Automotive

Burton Williamson
ASEC Manufacturing

Copyright © 2000 CEC and SAE International.

ABSTRACT

FTP and ECE + EUDC emissions are measured from six converters having similar restriction and platinum group metals on two 1999 prototype engines/calibrations. A 2.2L four cylinder prototype vehicle is used to measure FTP emissions and an auto-driver dynamometer with a prototype 2.4L four cylinder engine is used to determine the ECE + EUDC emissions. The catalytic converters use various combinations of 400/3.5 (400cpsi/3.5mil wall), 400/4.5, 400/6.5, 600/3.5, 600/4.5, and 900/2.5 ceramic substrates in order to meet a restriction target and to maximize converter geometric surface area. Total catalyst volume of the converters varies from 1.9 to 0.82 liters. Catalyst frontal area varies from 68 cm^2 to 88 cm^2. Five of the six converters use two catalyst bricks. The front catalyst brick uses either a three-way Pd washcoat technology containing ceria or a non-ceria Pd washcoat technology. Pd loadings are 0.1 troy oz. of Pd. The rear catalyst brick uses a Pt/Rh washcoat at a loading of 0.06 t.o. and a ratio of 5/1. Each converter was aged for 125 hours of RAT^{sm}-820 prior to the emission evaluations. The FTP results show that washcoat technology and catalyst substrate can have a significant impact on HC and NOx emissions. Results suggest that 900 cpsi substrates with three-way Pd catalyst may be a critical part of the ULEV-II emission solution. ECE+EUDC emission results suggest that non-ceria Pd front brick converters may be best for high speed NOx emissions.

INTRODUCTION

It is evident that high cell density and/or ultra-thin wall ceramic substrates are being used to improve the efficiency of catalytic converters. Such converters are becoming an integral part of the overall emission solution for stringent emissions regulations (1-9). Many have attributed the success of these ultra-thinwall substrates to parameters such as low mass (bulk density) and geometric surface area (higher cell density). 1989 Socha et al (10) claimed that the high geometric surface area of 600 cpsi ceramic thinwall catalysts had significantly better light-off and warm-up performance than 400 and 200 cpsi catalysts. Yamamoto (11) et al found that the low mass of the 400/4mil substrates reduced light-off temperatures by a 20°C and reduced converter back pressure by 10% over 400 cpsi 6mil product. Pfalzgraf (1) et al compared the performance of various start-up converters with various ceramic and metallic substrates having different cell densities. They claimed that CO and NOx emissions performance of these converters are more dependent upon geometric surface area than the mass of substrate. Umehara (4) et al compared a variety of ceramic (900 cpsi /2 .0 through 400/ 4.0 mil) close-coupled converters on a four cylinder vehicle running the FTP. They found that increasing geometric surface area reduces emissions during all portions of the FTP. Reduction in bulk density (substrate mass) reduces only cold start emissions. Blanchett et al (12) modeled the physical attributes of various monolithic substrates. Using the NTU (Number of Transfer Units) method they predicted and verified that catalyst volume could be reduced in inverse proportion to cell density. Also they stated with respect to thermal response that heat capacity per catalyst NTU has the greatest impact on thermal response. Ball et al (5) compared 600 cpsi 3.5 mil (600/3.5) substrates to the 400/6.5 and 600/4.3 products. Results suggested that the high geometric surface area substrates helped HC light-off and that the low bulk density of the 600/3.5 substrate was significantly better for NOx emissions. At similar catalyst volumes the 400/6.5 and 600/3.5 catalyst had similar restriction. Kikuchi et al (6) evaluated 600, 900 and 1200 cpsi substrates. They claimed that at "very high" cell densities, such as 1200 cpsi, the precious metal concentrations must be increased with respect to the increase in the surface area of the substrate. Also thinner wall (e.g. 2 mil), lower mass substrates permit improved HC and NOx cold start emissions. Kishi et al (9) used this substrate (1200/2.0) to demonstrate ZLEV emission levels with a hybrid catalyst technology that consists of a three-way catalyst layer atop a HC adsorber layer.

This paper investigates possible solution sets of catalytic converters given specific constraints of precious metal loadings and converter restriction. It is widely known that palladium catalysts have excellent HC performance and high temperature durability (13-17). It is not surprising that the demand for palladium has increased as well as its price. During 1998 the average monthly price of Pd varied from $200 to $350/ t.o. (18). It has become more important to optimize the use of palladium in catalytic converters. In this experiment, total Pd usage is kept within a range between 0.1 and 0.12 t.o. for each converter. Similarly each converter uses 0.01 t.o. Rh. In order to enhance catalytic performance and meet the converter backpressure criteria, various combinations of ultra thin wall substrates were used, from 400/3.5 to 900/2.5. Catalyst frontal area is also varied in an attempt to investigate the tradeoff between catalyst frontal area and catalyst volume. Do smaller converters at higher precious metal concentration perform better than larger converters with the same precious metals?

SCOPE

CATALYTIC CONVERTERS – Table 1 describes the six converters tested in this study. Converters 1 – 5 contain two catalyst bricks. The front converter brick uses a Pd catalyst at a loading of 0.1 troy oz. The lengths of the front catalyst bricks are 63.5mm. The rear catalyst brick uses a 3-Way Pt/Rh catalyst technology at a loading of 0.05 troy oz Pt and 0.01 troy oz. Rh. The Baseline Converter, 1.9L 600/400, has the largest frontal area of the six, 86 cm^2. A Pd light-off catalyst technology is applied on a 600/4.3 ceramic substrate at a Pd concentration of 5.5 g/l. The 400/6.5 rear catalyst brick has a Pt/Rh loading of 1.39 g/l. Converters 2-6 use smaller substrates that have a catalyst frontal area of 68 cm^2. The smaller frontal area substrates permit higher Pd concentrations at constant Pd loading for improved light-off performance. Converters 2-5 have a Pd concentration of 7.22 g/l in the front brick. Converters 1 – 4 use the same front and rear brick catalyst technologies. The front brick uses a non-ceria Pd light-off catalyst technology and the rear brick uses a Pt/Rh 3-Way technology.

More specifically, Converter 2, 1.45L 400/400, uses 400/3.5 substrates in the front and rear. Since this substrate has the lowest restriction per unit volume, 1.03 liters of rear substrate was used to achieve similar restriction. Converter 3, 1.L 900/400, uses 0.42 liters 900/2.5 substrate in the front followed by 0.61 liters of 400/4.5 substrate in the rear in order to achieve similar restriction. Due to the smaller rear brick of Converter 3, the PGM concentration is 3.09 g/l compared to 1.39 and 1.80 g/l respectively for Converters 1 and 2. Converter 4, 1.L 600/600, is identical in catalyst volume and PGM concentration to Converter 3 except 600/3.5 substrates are used for both of the front and rear catalyst bricks. Converter 5, 1.L 3-Way 600/600, is identical to Converter 4, except it uses a Pd 3-Way catalyst in the front brick. Converter 6, 0.82L 3-Way 900, is the only single brick converter. Only 0.82 liters of catalyst could be used to achieve similar restriction with the 900/2.5 substrate. A 3-Way Pd/Rh catalyst technology is used with this substrate at a PGM loading of 0.12 troy oz. Pd and 0.01 troy oz. Rh. This converter contains no Pt and has 20% more Pd than the other converters. Since the total PGMs in Converter 6 are spread evenly across one catalyst brick, the PGM concentration for catalyst light-off is only 4.94 g/l.

AGING – Each converter was dynamometer aged for 125 hours. Four converters were aged at once in parallel using a 7.4L engine. Engine speed is held constant at 3500 rpm. Since this study consists of six converters, two of the converters were aged with similar converters having similar restriction used in another study. Figure 1 describes this four mode dynamometer aging schedule (19). The first 40 seconds of the cycle the converter is exposed to stoichiometric exhaust at a temperature of 820°C. Then at 40 seconds the engine is operated rich to produce 3.0% CO. This step lasts six seconds. Then at 46 second into the cycle the engine continues to operate rich and air is injected in front of the converter to produce 3.0% O$_2$. During this mode the CO and O$_2$ combust in the converter to produce a 290°C exotherm. This third mode lasts 10 seconds. Finally at 56 seconds into the cycle the engine returns to stoichiometric control with the air injection for four seconds.

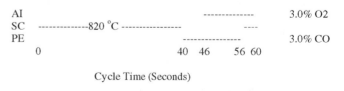

Cycle Time (Seconds)

AI - Air Injection
SC - Stoichiometric Cruise
PE - Power Enrichment

Figure 1.   Dynamometer Aging Schedule

Table 1. Catalytic Converter Descriptions

Converter	Front Catalyst Brick					Rear Catalyst Brick					Total Catalyst		
	Vol. (L)	Subs. (cpsi) (wall)	Cat. Tech.	PGM Pt/Pd/Rh (t.o.)	Load. (g/l)	Vol. (L)	Subs. (cpsi) (wall)	Cat. Tech.	PGM Pt/Pd/Rh (t.o.)	Load. (g/L)	Front. Area (cm2)	Total Vol. (L)	Total PGM Pt/Pd/Rh
1 Baseline 1.9L 600/400	0.56	600 4.3	Pd L-Off	0/.1/0	5.55	1.34	400 6.5	3-Way	.05/0/.01	1.39	86	1.90	.05/.1/.01
2 1.45L 400/400	0.42	400 3.5	Pd L-Off	0/.1/0	7.22	1.03	400 3.5	3-Way	.05/0/.01	1.80	68	1.45	.05/.1/.01
3 1.L 900/400	0.42	900 2.5	Pd L-Off	0/.1/0	7.22	0.61	400 4.5	3-Way	.05/0/.01	3.09	68	1.03	.05/.1/.01
4 1.L 600/600	0.42	600 3.5	Pd L-Off	0/.1/0	7.22	0.61	600 3.5	3-Way	.05/0/.01	3.09	68	1.03	.05/.1/.01
5 1.L 3-Way 600/600	0.42	600 3.5	Pd 3-Way	0/.1/0	7.22	0.61	600 3.5	3-Way	.05/0/.01	3.09	68	1.03	.05/.1/.01
6 0.82L 3-Way 900	0.82	900 2.5	Pd/Rh 3-Way	0/.12/.01	4.94						68	0.82	0/.12/.01

FTP EMISSION TESTING – These six converters were evaluated on a prototype 2.2L four cylinder vehicle using the FTP test. A single 48 inch roll chassis dynamometer was used with Horiba emission benches. Vehicle inertia and HP were 3375 lbs and 6.8 HP, respectively. Bag and modal emissions were measured. California Phase II fuel was used for all emission testing. The FTP results represent an average of at least three FTP tests. Typical engine-out emissions of this vehicle were 1.2, 12.7, and 4.1 gm/mile for nmHC, CO, and NOx, respectively. The converters were located in a close-coupled location approximately 16cm from the exhaust manifold. Converter inlet temperatures were measured about 75 mm upstream of the inlet converter flange with 1/16" K type radiation shielded thermocouple.

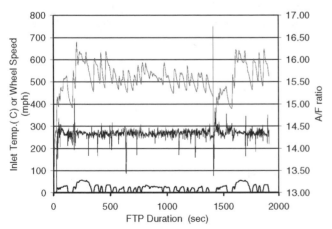

Figure 2. Air/Fuel and Catalyst Inlet Temperature of Prototype 2.2L 4 Cylinder Vehicle.

Figure 2 presents the engine-out A/F ratio and the catalyst inlet temperatures during the FTP. Converter inlet temperature reaches 400°C prior to the 1st cycle acceleration. Peak converter inlet temperatures of about 650°C are achieved during Cycles 2 and 20. Throughout Cycles 6-18 converter inlet temperatures vary around 500°C. The air/fuel trace suggests that good A/F control is achieved during the 1st cycle cruise.

ECE +EUDC TESTING – The ECE + EUDC emissions testing was performed in an auto-driver dynamometer performance cell (19). A 1999 2.4L four cylinder engine and transmission is driven by a computer and electric dynamometer. Two modal benches are used to measure emissions before and after the converter. In between emission evaluations the engine oil and coolant was forced cooled to room temperature. This permitted 2 or 3 ECE + EUDC emission evaluations per 8 hour shift. Typical engine-out emissions were 1.1, 5.4, and 2.6 gm/km for HC, CO, and NOx, respectively. The six converters were evaluated twice in random order. Similar to the FTP testing, the converters were located in a close-coupled location about 23 cm. from the manifold. Indolence fuel was used for all of the ECE +EUDC evaluations. Figure 3 shows the temperature and air/fuel characteristics of this engine. Converter inlet temperatures were measured approximately 75 mm ahead of the converter flange. Figure 3 shows that converter inlet temperatures exceed 400°C after the 2nd cycle acceleration. Once warmed-up the converter inlet temperatures vary between 380°C and 600°C during the ECE portion of the test. During the EUDC portion the converter inlet temperatures nearly reach 800°C. The Air/Fuel trace shows that good closed-loop fuel control is obtained about 35 seconds into the cycle. Also the air/fuel control tends to go richer during the first 6 cycles of ECE portion of the test. Please note that the engine is calibrated for the FTP emission cycle. No attempt was made to change the calibration to improve the ECE + EUDC emissions.

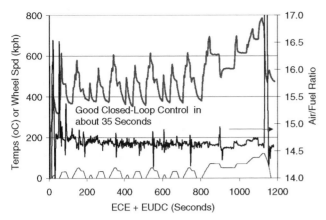

Figure 3.   Air/Fuel and Catalyst Inlet Temperature of 2.4L 4 Cylinder Engine in Auto-Driver Performance Cell.

RESULTS

FLOW TESTING – A laboratory cold flow apparatus was used to measure the flow restriction of each converter. Air at standard temperature and pressure is drawn through each converter at flow rates up to 100 g/sec. Figure 4 presents these results. Flow restriction is plotted versus mass flow rate. The legend lists the converters in order of restriction. Converter 1.45L 400/400, is the least restrictive at 100 g/sec with a converter restriction of 250 mm H_2O. Converters 1.9L 600/400 and 0.82L 3-Way 900 have the highest restriction at 100 g/sec of about 300 mmH_2O. The three 1.0 L converters had similar restriction of 275 mm H_2O at 100 g/sec. In summary, at 100 g/sec all of the converters were within 10% of 275 mm of H_2O.

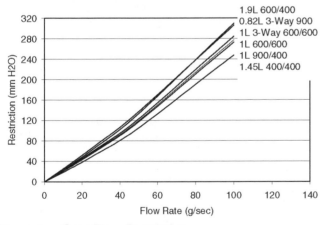

Figure 4.   Cold Flow Restriction

No attempt was made to adjust catalyst length in order to achieve identical flow restriction.

2.2L FTP RESULTS – Figures 5-8 present Bag and Modal emissions results of the six converters. Figure 5 presents the Bag results and Figures 6-8 present the modal emissions results for HC, CO, and NOx, respectively. In addition, the Appendix presents a statistical comparison of the Bag 1 and overall Bag emissions to a 95% confidence interval.

Figure 5 summarizes the FTP Bag emission results of the six converters. The stacked bars present the Bag 1, 2, and 3 emissions for nmHC, CO, and NOx. CO emissions are divided by 10. Bag 1 represents the Cold Start emissions of the FTP, the first five cycles. Bag 2 emissions are from the warmed-up portion of the FTP, Cycles 6-18. Finally, the Bag 3 emissions are from the Warm Start portion of the FTP, Cycles 19-23. The methane tailpipe emissions (wtmg/mile) of each converter are shown above the nmHC emission bars.

With respect to nmHC emissions, the converters have similar emissions. The average emissions of the six converters varied from 57 to 61 mg/mile. Converters 1.L3-Way 600/600 and 0.82L 3-Way 900 may have better Bag 1 nmHC emissions. It is interesting to note that the tailpipe methane emissions of converter 0.82L 3-Way 900 are 26 mg/mile, about 10 mg/mile higher than any other converter. This converter is the only converter that does not contain a Pt/Rh catalyst. With respect to CO emissions, the largest converter, Converter 1, 1.9L 600/400, had the lowest CO emissions at 1.02 g/mile. The other converters had slightly higher CO emissions, ranging from 1.1 g/mile for converter 1L 3-Way 600/600 to 1.31 g/mile for converter 1.45L 400/400. There are considerable differences in the NOx performance of the six converters. Converter 1.45L 400/400 has the highest NOx emissions of 0.17 g/mile. The baseline converter, 1.9L 600/400, has the lowest NOx emissions of 0.087 g/mile. Of LEV-II significance is the Bag 1 NOx emissions of converter 0.82L 3-Way 900. At 0.035g/mile, this converter has the best NOx light-off performance.

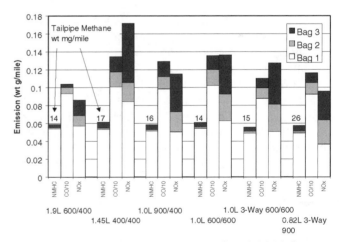

Figure 5.   Average Bag Emissions of 2.2L Vehicle

Figures 6a and 6b present the HC modal emission results (including methane). Figure 6a compares the HC light-off performance of the converters during cold start, whereas, Figure 6b shows the modal HC emissions for all of the 23 cycles. Please note that modal data contains total hydrocarbons, unlike the Bag results in Figure 5 that just present nmHC emissions. Like the Bag results, the modal results are the average of at least three FTP evaluations. The legend in the upper right hand corner of each graph lists each converter in order of cumulative emission performance.

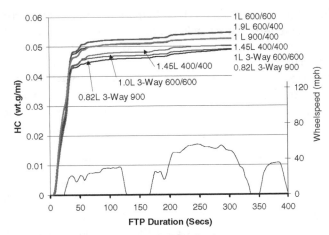

Figure 6a. 2.2L HC Light-Off Emissions

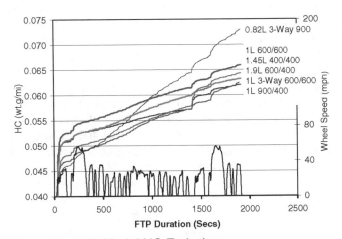

Figure 6b. 2.2L Modal HC Emissions.

Converters 0.82L 3-Way 900, 1.L 3-Way 600/600 and 1.45L 400/400 appear to have better HC light-off emissions than the other three converters. It is difficult to draw any direct conclusions as to why this might be. The 1.45L 400/400 converter has the lowest mass substrates for light-off. The other two converters have a ceria containing front brick. The 0.82L 3-Way 900 converter lights off rather well considering it has a PGM concentration of only 4.94 g/l compared to 7.22 g/l for the front bricks of converters 1.L 3-Way 600/600 and 1.45L 400/400.

Figure 6B presents the modal data for the remainder of the FTP. What was one of the best converters for HC light-off, converter 0.82L 3-Way 900, appears to have the worst total HC emissions. In Figure 5, this converter has comparable nmHC emissions to the other converters, however it has 10 mg/mile higher methane emissions. The results in Figure 6b suggest that the differences in modal HC emissions between converter 0.82L 3-Way 900 and the other converters is methane. The other five converters have HC emissions within 5 mg/mi. Converter 1L 900/400 may be the most efficient for HC emissions after light-off.

Figure 7 presents the modal CO results for the converters. The two converters with the 3-Way catalyst in the front brick , 1L 3-Way 600/600 and 0.82L 3-Way 900,

appear to have the best CO light-off performance. However, once warmed-up, the 1.9L 600/400 converter is the most efficient and has the lowest overall CO emissions.

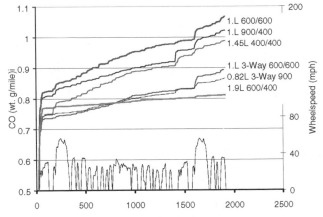

Figure 7. 2.2L CO Modal Emissions

Figure 8a shows the NOx light-off performance of the various converters. This figure shows how catalyst substrate and technology can affect NOx light-off emissions. The single brick converter, 0.82L-3Way 900, has the best NOx light-off emissions. This converter lights-off prior to the 1st Cycle cruise. The other five converters light-off later. There appears to be a synergy of effects when combining the 3-Way Pd/Rh catalyst technology with the 900 cpsi substrate. This can be stated since this converter, 0.82L 3-Way, has lower PGM concentration (4.94 vs. 7.22 g/l) for light-off than converters, 1L 900/400 and 1L 3-Way 600/600. Converters 1.45L 400/400 and 1.9L 600/400 have the poorest NOx light-off performance. Converter 1.45L 400/400 has significant NOx emissions during the 2nd Cycle acceleration.

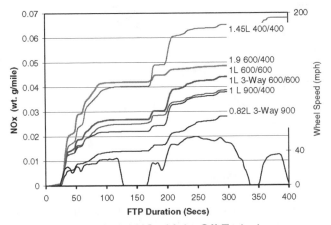

Figure 8a. 2.2L Modal NOx Light-Off Emissions

The overall NOx modal emissions presented in Figure 8b continue to show the poor NOx performance of converter 1.45L 400/400. Conversely, the largest converter, 1.9L 600/400, is the most efficient after Cycle 2. This converter also excels during Cycle 19 and has the lowest total tailpipe NOx emissions of 0.073 g/mile. Converter 0.82L 3-Way 900 is next best at 0.087 g/mile.

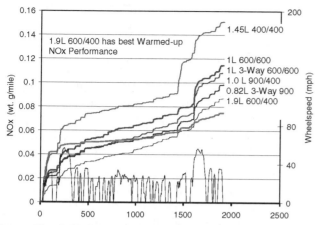

Figure 8b. 2.2L Modal NOx Emissions

ECE +EUDC RESULTS – The ECE + EUDC results of the same six converters are presented in Figures 9 – 11 for HC, CO, and NOx emissions from the 2.4L engine.

Figure 9a shows the HC light-off performance during the first four cycles of the ECE. As in Figure 6a these results also suggest that having a 3-Way catalyst in the front brick may improve HC light-off performance since converters 1 L3-Way 600/600 and 0.82L 3-Way 900 performed best. Converters 1.9L 600/400 and 1.45L 400/400 appear to have the poorest HC light-off performance.

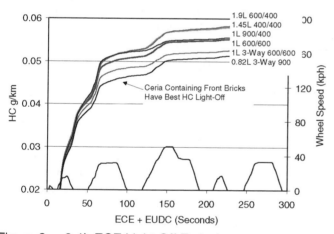

Figure 9a. 2.4L ECE Light-Off Emissions

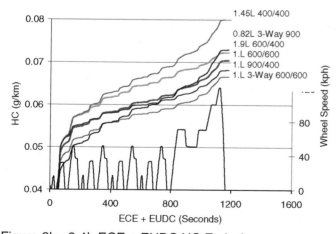

Figure 9b. 2.4L ECE + EUDC HC Emissions

Converter 1.45L 400/400 continues to have the poorest HC emissions once warmed-up (Figure 9b). The largest converter, 1.9L 600/400, does have relatively good HC efficiency after light-off. Converter 0.82L 3-Way 900, which had the best light-off performance, is not the most efficient once it is warmed-up. This behavior was observed earlier in Figures 5 and 6b and is probably due to higher methane emissions. The best two converters for overall ECE + EUDC HC emissions, appear to be 1L 900/400 and 1L 3-Way 600/600.

Converter 0.82L 3-Way 900 has the poorest CO light-off of any of the converters. Conversely, the converter with the lowest mass substrates, 1.45L 400/400, has the best CO emissions (Figure 10). It is interesting to note that after light-off the slopes of all of the CO emissions lines are similar during the remainder of the ECE, indicating that these all of these converters have similar warmed-up CO efficiencies. During the hard accelerations of the EUDC the largest converter, 1.9L 600/400, may be best for CO emissions.

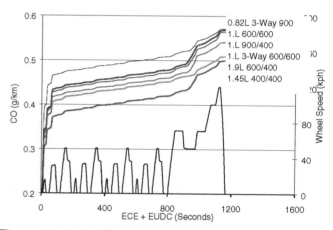

Figure 10. 2.4L ECE + EUDC CO Emissions

Figure 11a shows the converters containing 900 cpsi substrates, 1L 900/400 and 0.82L 3-Way 900, have the best NOx light-off performance about 150 seconds into the ECE. The 3-Way catalyst technology appears to facilitate NOx lightoff since the PGM concentrations of the two converters for light-off are 7.22 and 4.94 g/l, respectively. (See Table 1.)

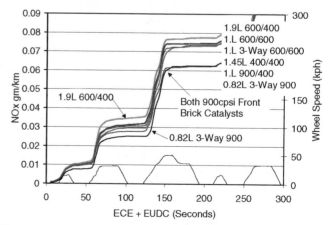

Figure 11a. 2.4L ECE NOx Emissions

Figure 11b presents the NOx emissions for the entire ECE + EUDC. The converters with the 3-Way catalyst in front, 1L 3-Way 600/600 and 0.82L 3-Way-900, have the highest NOx emissions. Meanwhile all of the converters that had the non-ceria Pd light-off catalyst in the front brick had improved NOx emissions during the high speed acceleration near the end of the EUDC. Converters 1L 900/400 and the largest converter, 1.9L 600/400, have the lowest total NOx emissions.

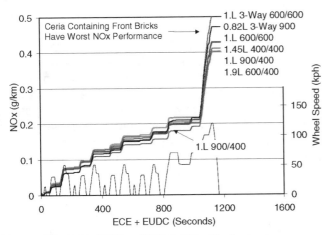

Figure 11b.2.4L ECE + EUDC NOx Emissions

DISCUSSION

The Bag results in Figure 5 suggest that there exists a variety of emissions solutions. On the 2.2L vehicle the nmHC are not statistically different for these six converters. (See Appendix.) Non-methane HC emissions range from 57 to 61 mg/mile. Yet these converters vary significantly in design. Catalyst volume of the six converters varies from 1.9 to 0.82 liters. Cell densities and wall thickness of the ceramic substrates ranged from the low mass 400/3.5 substrate to the high geometric surface area substrate of 900/2.5. By coincidence, combining these substrates to yield similar restriction yields similar nmHC emissions at similar PGM loading. With respect to the HC emissions, additional studies are warranted to investigate the differences in the nmHC and HC emissions of converter 0.82L 3-Way 900. This converter appears to be best for HC light-off emissions but is not the most efficient once warmed-up. This difference appears to be methane emissions. Is this difference an artifact of the Pd/Rh catalyst technology or its' low PGM loading? Are Pt/Rh catalysts best for methane emissions? A similar HC emission trend is exhibited during the ECE and EUDC evaluations (see Figures 9a and 9b).

CO emissions of all of these converters are well below the ULEV and Euro-3 emissions standards. However, these converters did behave differently on the emissions cycles. Converter 0.82L 3-Way 900 lights-off well during the 1st Cycle of the FTP. However, it lights off poorly during the ECE. Is this due to the relatively low PGM concentration of 4.94 g/l? The CO light-off emissions during the ECE +EUDC appear to be dictated by converter substrate mass. Converter 1.45L 400/400 with the 400/3.5 substrate performs best. This result appears to contradict Schmidt's et al (7) modeled CO results. During the FTP evaluation on the 2.2L vehicle, the CO emissions are both a function of light-off and steady state performance. Converters 1L 3-Way 600/600, 0.82L 3-Way 900 and 1.45L 400/400 appear to have the best light-off performance. However, once warmed-up, the converters with the 3-Way catalyst in the front brick and the large 1.9L converter are more efficient for CO.

The various combinations of ultra-thinwall substrates and catalyst technologies have a significant impact on NOx light-off emissions during the 2.2L FTP evaluations. Converter 0.82L 3-Way 900 lights-off during the 1st Cycle acceleration whereas the other converters light-off later during the 1st cycle cruise (see Figure 8a). This type of performance separation is not seen during the ECE (see Figure 11a). Both of the converters containing 900 cpsi substrates appear to have better NOx emissions at about 150 seconds into the ECE. In general, the emission evaluations on the FTP and ECE+EUDC emissions cycles suggest that 900 cpsi substrates can improve NOx light-off. This result is similar to that of Lafyatis et al (20).

It is difficult to draw any conclusions regarding the warmed-up NOx performance of the 2.4L engine during the ECE + EUDC test, since this engine is not specifically calibrated for the schedule. However, the results may imply that converter 1L 900/400 may be best. Additional testing on a vehicle calibrated for this cycle is warranted. Also it would be interesting to further investigate the effects of converters with and without ceria in the front brick. During the last acceleration of the EUDC the converters with the 3-Way catalyst in the front brick have the highest NOx emissions. During the FTP evaluations larger differences in NOx emissions among the converters were observed. The largest converter, 1.9L 600/400, lights-off slowly but has superior warmed-up performance. Converter 1.45L 400/400 has significant NOx emission breakthroughs during the cool, transient cycles of the FTP, Cycles 1, 2, 19, and 20.

SUMMARY AND CONCLUSIONS

The data presented in this paper suggests that there exists a set of catalytic converters (emission solutions) that have similar emission performance at similar restriction and precious metal loading while allowing catalyst volume to vary as much as 100%. Ultra-thinwall substrates and combinations thereof, can offer significant packaging advantages (5,12). These substrates can be used to significantly reduce catalyst volume while maintaining emissions performance, restriction and precious metal loadings. In addition, using combinations of ultra-thinwall substrates can permit the use of Pd catalyst in the front brick for fast light-off and Pt/Rh catalyst in the rear brick for warmed-up three-way activity. A strategy such as this may reduce precious metal dependence.

Within the scope of our investigation:

2.2L Vehicle FTP Evaluations

1. NOx emissions can be significantly affected by choice of catalyst and substrate. A Pd/Rh 3-Way washcoat applied on a 900 cpsi catalyst exhibited excellent NOx light-off performance. This strategy may be part of the LEV-II emission solution.

2. Non-methane HC performance was similar for all of the converters as catalyst volume and catalyst substrates were varied to maintain similar restriction.

3. CO emissions are well below the ULEV emission level. However, catalyst washcoat and substrate type does have an affect on CO light-off emissions and warmed-up efficiency.

Dyno 2.4L ECE + EUDC Evaluations

Since the calibration of the 2.4L engine was originally designed for the FTP emission cycle, the ECE+EUDC results should be interpreted with this in mind. Additional studies are warranted on a European Calibrated vehicle.

1. CO emissions are well below the Euro-3 Standards. However, CO emissions appear to be a strong function of CO light-off performance. Most interesting, the converter with the lowest mass 400/3.5 substrates, 1.45L 400/400, had the best CO light-off, while converter 0.82L 3-Way 900 had the worst light-off.

2. Pd catalysts applied on 900/2.5 cpsi substrate may improve NOx light-off performance. Once warmed-up, the combination of 900/2.5 substrate with a non-ceria Pd light-off catalyst with a rear brick of 400/4.5 substrate with a Pt/Rh 3-Way catalyst may be the most efficient. The data also suggests that the converters with 3-Way catalyst in the front brick may not be best for EUDC NOx emissions.

3. Ceria in the front catalyst brick appears to help HC light-off. The converters, 1L 3-Way 600/600 and the 1L 900/400 appear to have the best combination of light-off and warmed-up performance.

ACKNOWLEDGMENTS

The authors wish to thank Mike Nellet, John Deer, John Boenhke, Jim Forro for the vehicle and converter builds. We would also like to thank the competent drivers and technicians in Delphi's emission laboratory, Dan Trytko, Jeff Webb, Tom Horton, Dave Lewis, Jacque Benoit, Jim Soderlund and Gary Zemeski.

REFERENCES

1. B. Pfalzgraf, M. Reiger and G. Ottowitz, "Close-coupled Catalytic Converters for compliance with LEV/ULEV and EGIII Legislation – Influence of Support Material, Cell Density and Mass on Emission Results", SAE Paper No. 960261, 1996.

2. N. Kishi, S. Kikuchi, Y. Seki, A. Kato and K. Fujimoro, "Development of the High Performance L4 Engine ULEV System", SAE Paper No. 980415, 1998.

3. E. Otto, F. Albrecht and J. Liebl, "The Development of BMW Catalyst Concepts for LEV / ULEV and EU III / IV Legislations 6 Cylinder Engine with Close Coupled Main Catalyst", SAE Paper No. 980418, 1998.

4. K. Umehara, T. Yamada, T. Hijikata, Y. Ichikawa and F. Katsube, "Advanced Ceramic Substrate: Catalytic Performance Improvement by High Geometric Surface Area and Low Heat Capacity", SAE Paper No. 971029, 1997.

5. D.J. Ball, G.E.Tripp, L.S. Socha, A. Hiebel, M.Kulkarni, P.A. Weber and D.G. Linden, "A Comparison of Emissions and Flow Restriction of Thinwall Ceramic Substrates for Low Emission Vehicles", SAE Paper 1999-01-0271

6. S. Kikuchi, S. Hatcho, T. Okayame, S. Inose and K. Ikeshima, "High Cell Density and Thin Wall Substrate for Higher Conversion Ratio Catalyst", SAE Paper 1999-01-0268.

7. J.Schmidt, A.Waltner, G.Loose, A.Hirschmann, A.Wirth, W.Mueller, J.A. van den Tillaart, L.Mussmann, D. Lindner, J.Gieshoff, K.Umehara, M.Makino, K.P.Biehn and A.Kunz, "The Impact of High Cell Density Ceramic Substrates and Washcoat Properties on the Catalytic Activity of Three Way Catalysts", SAE Paper 1999-01-272.

8. P.Ehmann, N.Rippert,K.Umehara and C.D.Vogt, "The Development of a BMW Catalyst Concept for LEV/EU3 Legislation for a 8 Cylinder Engine by Using Thin Wall Ceramic Substrates", SAE Paper 1999-01-0767.

9. N.Kishi,S.Kikuchi,N.Suzuki and T.Hayashi, "Technology for Reducing Exhaust Gas Emissions in Zero Level Emission Vehicles (ZLEV)", SAE Paper 1999-01-0772.

10. L. S. Socha, JR, J.P. Day and E.H. Barnett, "Impact of Catalyst Support Design parameters on FTP Emissions", SAE Paper No. 892041, 1989.

11. H. Yamamoto, H Horie, J Kitagawa and M. Machida, "Reduction of Wall Thickness of Ceramic Substrates for Automotive Catalysts", SAE Paper No. 900614, 1990.

12. S. Blanchett, R. Richmond and G. Vaneman, "Implementation of the Effectiveness-NTU Methodology for Catalytic Converter Design", SAE Paper No. 980673, 1998..

13. J.C. Summers, J.J. White, W.B.Williamson, "Durability of Padallium-Only Three-Way Automotive Emission Control Catalysts", SAE Paper 890794, 1989.

14. J.S. Hepburn, K.S.Patel, M.G. Meneghel, H.S. Gandhi, Engelhard Development team and Johnson Matthey Development Team, "Development of Pd-Only Three Way Catalyst Technology", SAE Paper 941058, 1994.

15. J.G.Nunan, W.B. Williamson, H.J.Robota and M.G. Henk, "Impact of Pt-Rh and Pd-Rh Interactions on Performance of Bimetal Catalysts", SAE Paper 950258, 1995.

16. R.J.Brisley, G.R.Chandler, H.R. Jones, P.J. Anderson and P.J.Shady, "The Use of Palladium in Advanced Catalysts", SAE Paper 950259, 1995.

17. S.Sung, R.F.Ober, R.Casperand G. Miles, "A 45% Engine Size Catalyst System for MDV2 ULEV Applications", SAE Paper 982553, 1998

18. "Platinum 1999", Johnson Matthey, May 1999.

19. D. J. Ball, A.G. Mohammed and W. Schmidt, "Application of Accelerated Rapid Aging test (RAT[sm]) Schedules with Poisons: The Effects of Oil Derived Poisons, Thermal Degradation and Catalyst Volume on FTP Emissions", SAE Paper No. 972846, 1997.

20. D.S. Lafyatis, N.S.Will, A.P.Martin,J.S.Rieck,J.P.Cox and J.M.Evans, "Use of High Cell Density Substrates and High Technology Catalysts to Significantly Reduce Vehicle Emissions", SAE Paper No. 2000-01-0502, 2000

CONTACT

Douglas J. Ball
Delphi Automotive Systems
M/C 485-220-070
1601 N. Averill Avenue
Flint Mi 48556
Email doug.ball@delphiauto.com

APPENDIX

The intervals plotted are based on Fisher's least significant difference (LSD) procedure. They are constructed in such a way that if two means are the same, their intervals will overlap 95.0% of the time. The intervals are calculated by dividing the pooled standard deviation by the square root of the number of observations at each level. The calculated intervals for any of the converters are equal, if their number of observations is the same. Reference Statgraphics Plus, version III, "User Manual", pp 10-61 through 10-63.

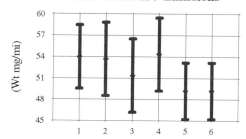

Phase 1 NM HC Emissions

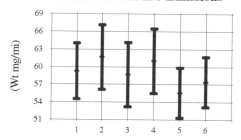

Total FTP NM HC Emissions

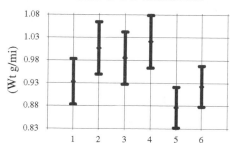

Phase 1 CO Emissions

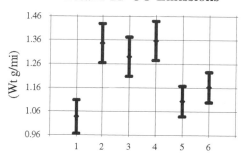

Total FTP CO Emissions

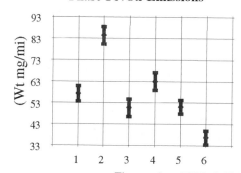

Phase 1 NOx Emissions

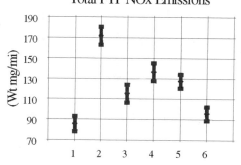

Total FTP NOx Emissions

Figure 1. 1999 2.2L Vehicle FTP Statistical Emission Comparisons
Converter Identification: 1 – 1.9L 600/400; 2 – 1.45L 400/400; 3 – 1.0L 900/400;
4 – 1.0L 600/600; 5 – 1.0L 3 Way 600/600; 6 - 0.82L 3 Way 900

ECE HC Emissions

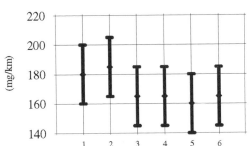

ECE/EUDC HC Emissions

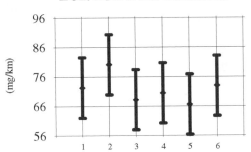

ECE CO Emissions

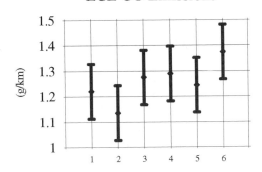

ECE/EUDC CO Emissions

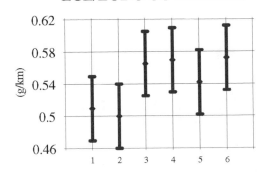

ECE NOx Emissions

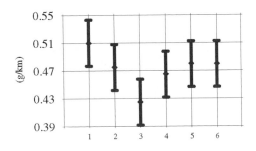

ECE/EUDC NOx Emissions

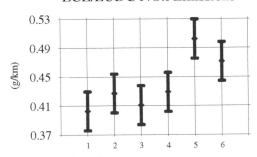

Figure 2.   1996 2.4L Autodriver ECE & EUDC Statistical Emission Comparisons
Converter Identification: 1 – 1.9L 600/400; 2 – 1.45L 400/400; 3 – 1.0L 900/400;
4 – 1.0L 600/600; 5 – 1.0L 3 Way 600/600; 6 - 0.82L 3 Way 900

1999-01-0308

Improvements in Pd:Rh and Pt:Rh Three Way Catalysts

Paul J. Andersen and Todd H. Ballinger
Johnson Matthey, Inc.

Copyright © 1999 Society of Automotive Engineers, Inc.

ABSTRACT

As one method of meeting current and future emission regulations on vehicles, automakers have increased PGM loadings in three-way catalysts. Engine dynamometer and FTP testing after accelerated engine agings were performed to compare current Pd:Rh and Pt:Rh catalysts with new Pd:Rh and Pt:Rh catalysts. This comparison demonstrated that enhanced three way performance can be obtained in the new catalysts with reduced Pd loadings or with the use of Pt:Rh instead of Pd:Rh. These improved catalysts will reduce the demand for high PGM loadings as well as provide flexibility in the PGM combinations used in exhaust systems.

INTRODUCTION

As automotive emission standards have tightened with the LEV, ULEV, and SULEV standards, the amount of platinum group metals (PGM) in catalyst systems has increased(1). This is particularly true for Pd due to the development and wide-spread use of high Pd load three way catalysts (TWC). It has been found that hydrocarbon emissions could be reduced by increasing the amount of Pd in Pd-Only (2-5), Pd:Rh (5-7) and Pt:Pd:Rh (5,6,8) catalysts. This was primarily because increasing Pd load reduced light-off temperatures and improved warmed-up hydrocarbon activity; both necessary to reduce hydrocarbon emissions. Figure 1 shows a air/fuel sweep profile for a standard Pt:Rh TWC (5:1 Pt:Rh with 1.41 g/L total PGM) which can be compared to Figure 2, the sweep profile of a standard Pd:Rh TWC (9:1 Pd:Rh with 2.12 g/L total PGM). These sweeps demonstrate the large activity improvements which have been achieved with Pd based washcoat technology. As a consequence of these improvements, the amount of Pd now exceeds the amount of Pt and Rh used in vehicle catalyst systems (1,9).

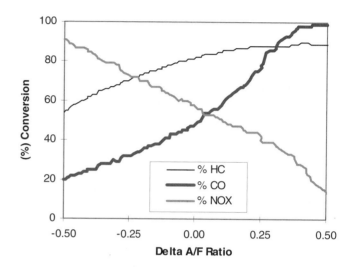

Figure 1: Air/Fuel Sweep - Standard Pt:Rh TWC, 5:1 Pt:Rh - 1.41 g/L total PGM

Currently, the use of high Pd loads in automotive catalysts is attractive both technically and economically. However, catalyst formulations which allow a more flexible balance of precious metal demand will be required in the future. Thus, new catalyst technology that reduces Pd requirements while maintaining or improving overall three-way activity is desirable. This paper examines the activity of current technology Pd:Rh three way catalysts with reduced Pd loads to highlight areas for improvement. Next, washcoat improvements are demonstrated which allow activity to be maintained while reducing Pd content. Finally, Pt will be examined as a substitute for Pd allowing the preparation of high activity, non-Pd containing three way catalysts. These developments should provide a greater flexibility in both the amount and type of PGM used in improved catalyst systems.

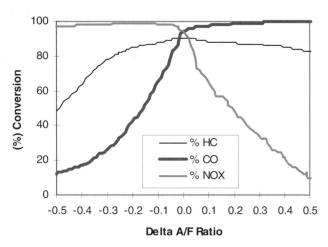

Figure 2: Air/Fuel Sweep - Standard Pd:Rh TWC,
9:1 Pd:Rh - 2.12 g/L total PGM

EXPERIMENTAL

CATALYST PREPARATION – Catalyst formulations were coated on 1.24 L ceramic substrate with 62 cells/cm^2 (400 cells/in^2) and 0.15 mm (6 mil) wall thickness.

DYNAMOMETER AGING CYCLE – Catalysts were aged with a 4-Mode cycle for 120h on a 4.6 L engine fueled with 0 Pb, 75-ppm S gasoline. This aging cycle was equivalent to 100,000 miles of road aging. The 4-mode cycle consisted of rich, stoichiometric, lean, and leaner A/F ratio modes that repeated every minute of the aging. The maximum bed temperature, measured 2.5 cm from the front face of the substrate during the lean mode, was 1000°C. Aging traces were compared to ensure A/F ratio differences between the agings were minimal.

ENGINE DYNAMOMETER EVALUATION – Light off profiles and Air/Fuel sweeps were collected for the aged catalysts on a 4.6 L engine fueled with California Phase-2 fuel.

For the AFR sweeps, HC/CO/NOx conversions were measured as lambda was continuously varied from 1.07 to 0.93 using an A/F ratio perturbation of ±0.5 (1.0 Hz frequency) at a space velocity of 85,000 hr^{-1}. The exhaust temperature at the catalyst inlet was 450°C. Before evaluation, the catalysts were pretreated on the engine at 660°C for 30 minutes at an A/F ratio of 13.5.

Light off profiles were measured with the same 4.6L engine/dynamometer setup. HC/CO/NOx conversions were measured as the inlet exhaust temperature was increased at 30°C/minute. The gas mixture was stoichiometric with an A/F ratio perturbation of ±0.5 (1.0 Hz frequency) at a space velocity of 85,000 hr^{-1}.

VEHICLE EVALUATION – The FTP performance of the aged catalysts was evaluated on a single bank of a 1995 vehicle with a 4.6 L, dual-exhaust engine. This setup resulted in a catalyst volume/engine displacement ratio of 0.54. The vehicle was fueled with 75-ppm S fuel and the catalyst was located 38 cm from the manifold. Reported conversions are the average of multiple tests.

RESULTS AND DISCUSSION

EFFECT OF Pd LOADING IN Pd:Rh CATALYSTS – In order to demonstrate the effect of Pd loading in Pd:Rh catalysts, three standard Pd:Rh catalysts were prepared with Pd loadings which were reduced in decrements of approximately 0.70 g/L. Figures 2, 3, and 4 show the sweep results for these Pd:Rh catalysts with precious metal loads of 2.12 g/L (Pd:Rh=9:1), 1.41 g/L (Pd:Rh=5:1), and 0.64 g/L (Pd:Rh=2:1) respectively. Several features of the sweep and light off are representative of important catalytic properties:

- CO/NOx Crossover - defined as the CO and NOx conversions at the AFR where the two conversions are equal - representative of the overall CO/NOx activity.

- AFR Window Width - defined as the AFR when the NOx conversion equals 50% minus the AFR when the CO conversion equals 50% - representative of the ability of a catalyst to convert CO and NOx during relatively large Air/Fuel transients which often occur under vehicle operating conditions.

- T50 - defined as the inlet exhaust temperature when 50% of a pollutant is converted under a stoichiometric AFR condition - representative of the catalyst cold start activity.

Table 1 summarizes these characteristic features of the sweeps and light offs. For the 2.12 g/L catalyst, the CO/NOx crossover conversion, the HC conversion at the crossover AFR, and the AFR window width were relatively high. The 1.41 g/L catalyst had somewhat lower the performance but this Pd reduction did not result in a drastic reduction in sweep activity or increase in T50. However, removing an additional 0.64 g/L drastically lowered the CO/NOx crossover and HC conversion. Also, the window width was essentially reduced to zero and the T50's increased by more than 50°C.

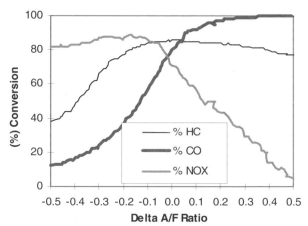

Figure 3: Air/Fuel Sweep - Standard Pd:Rh TWC,
5:1 Pd:Rh - 1.41 g/L total PGM

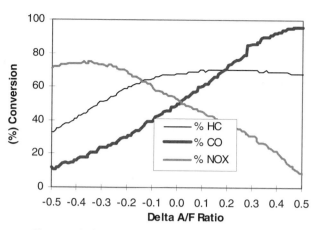

Figure 4: Air/Fuel Sweep - Standard Pd:Rh TWC,
2:1 Pd:Rh - 0.64 g/L total PGM

Table 1.   Sweep and light off results as a function of
precious metal load

Pd:Rh Ratio/ Loading	Crossover Conversions (%)		Window Width	Temperature for 50% Conversion (C)		
	CO/NOx	HC		HC	CO	NOx
9:1 / 2.12 g/L	94	90	0.35	358	376	381
5:1 / 1.41 g/L	75	86	0.26	357	377	395
2:1 / 0.64 g/L	51	68	0.03	418	443	444

The three catalysts were also FTP tested as described in the Experimental section. Table 2 shows the weighted emission for HC, CO, and NOx over the FTP cycle. The HC emissions increase with Pd removal was nearly linear. Inspection of the individual bags showed a steady increase in Bag 1 emissions. In contrast, the NOx emissions showed a much larger increase for the PGM decrease from 1.41 to 0.64 g/L than from 2.12 to 1.41 g/L.

Table 2.   FTP conversions as a function of precious metal load

Pd:Rh Ratio/ Loading	Weighted Total FTP Conversions (%)		
	HC	CO	NOx
9:1 / 2.12 g/L	94.3	89.3	95.4
5:1 / 1.41 g/L	91.7	83.8	91.9
2:1 / 0.64 g/L	90.8	84.9	84.2
0.21 g/L Rh	78.5	84.0	79.2

Thus, removal of a small amount of Pd resulted in a moderate loss of activity. However, further Pd load reductions resulted in a drastic loss of catalytic activity, especially NOx activity. One might propose that, at the 2:1 / 0.64 g/L composition, Pd is at such a low load that it no longer makes a significant contribution to catalyst activity. This is contradicted by the data for the non-Pd containing catalyst (Table 2) which demonstrated the large contribution that Pd makes to NOx and especially HC activity, even at these low loadings.

This study on the impact of Pd load on catalytic activity highlights the key contribution of Pd to the overall catalytic activity of the current high-activity, Pd-based catalyst technology. It is apparent that alternate catalyst technology is required if Pd loadings are to be reduced without having a large negative impact on pollutant conversions. The next sections will describe improvements in catalyst washcoat technology which will allow equivalent pollutant conversions at reduced Pd loads. An alternate strategy which will be examined is to substitute Pt, a catalytically active precious metal with different supply and demand characteristics, for Pd.

IMPROVEMENTS IN LOW Pd-LOADED Pd:Rh CATALYSTS – The Pd load study described in the previous section indicated that there were significant opportunities for improvement in the activity of the 2:1 / 0.64 g/L catalysts under these test conditions. Thus, the effects of washcoat materials, catalyst promoters, and catalyst preparation techniques were examined in an attempt to improve the activity of 2:1 / 0.64 g/L catalysts.

Modification of the washcoat materials and promoter properties had a dramatic effect on the activity of these low Pd catalysts. Figure 5 shows the sweep curves for the improved catalyst (Pd:Rh-A). Substantial increases in the conversion of all three pollutants were observed across the entire Air/Fuel range (compare to Figure 4). Additionally, the window width was increased and the T50's were decreased 30 to 50°C compared to the standard 2:1 / 0.64 g/L catalyst. The increased activity demonstrated in the engine dynamometer testing was also observed under FTP test conditions (Table 3).

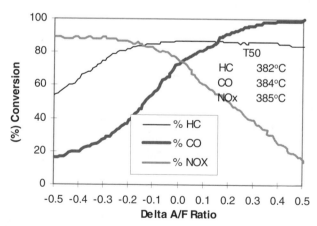

Figure 5: Air/Fuel Sweep and light-off temperatures -
Pd:Rh A, 2:1 Pd:Rh - 0.64 g/L total PGM

Table 3. FTP conversions for improved Pd:Rh catalysts

Catalyst	Weighted Total FTP Conversions (%)		
	HC	CO	NOx
Pd:Rh A - 2:1 / 0.64 g/L	92.0	83.3	88.2
Pd:Rh B - 2:1 / 0.64 g/L	91.3	86.0	91.1
Pd:Rh C - 2:1 / 0.64 g/L	92.1	86.1	93.2
Pd:Rh C - 9:1 / 2.12 g/L	94.7	85.2	94.9

Table 3 also shows the results of further development focused on improvements in the catalyst preparation methods (Pd:Rh B) and the catalyst promoter composition (Pd:Rh C). The main impact of these additional improvements was increased NOx activity. Note that the NOx emissions from Pd:Rh C were 42% lower than those from Pd:Rh A. Comparing these data to the FTP results in Table 2, one can conclude that the improvements represented by catalyst Pd:Rh C allow a 2:1 / 0.64 g/L catalyst to achieve higher FTP conversions than the standard Pd:Rh catalyst with 5:1 / 1.41 g/L Pd:Rh. This represents more than a 50% reduction in Pd usage.

Finally, Pd:Rh C was prepared with the 9:1/2.12 g/L load to determine if catalyst improvements could be combined with high PGM loads to give a very low emission system. The FTP results (Table 3) indicated this formulation had activity which was equivalent to the original 9:1/ 2.12 g/L catalyst. One interpretation of this result is that other aspects of the overall emission system are actually limiting the pollutant conversions and it will not be possible to observe the effects of improved catalyst formulations at higher PGM loads until those limitations are removed. Further work is required to verify this proposal.

Pt:Rh CATALYSTS – As an alternate to Pd removal, an investigation was performed to determine if Pt could be substituted for Pd. Initially, a catalyst was prepared with materials which, based on results described previously, were believed to be an improvement over materials included in the standard catalyst. Unfortunately, the air/fuel sweeps for this modified catalyst (Pt:Rh A shown in Figure 6) showed only moderate improvement over the standard Pt:Rh catalyst (Figure 1). However, since previous reports indicated negative interactions may develop between Pt and Rh catalyst components (3,7,10-12), a second catalyst (Pt:Rh B) was prepared with the same improved materials but which also had negative Pt-Rh interactions minimized. Pt:Rh B did show substantially higher sweep activity than the standard Pt:Rh catalyst (Figure 7 and Table 4).

The improvements in Pt:Rh sweep activity were not accompanied by similar improvements in light off activity. Pt:Rh A had somewhat higher T50's than the Standard Pt:Rh; however, the T50's for Pt:Rh B were approximately equivalent to the T50's for the Standard Pt:Rh catalyst.

Table 4. Sweep and light off results for 5:1 / 1.41 g/L Pt:Rh catalysts

Pd:Rh Ratio/ Loading	Crossover Conversions (%)		Window Width	Temperature for 50% Conversion (C)		
	CO/NOx	HC		HC	CO	NOx
Standard Pt:Rh	54	83	0.09	367	367	346
Pt:Rh A	60	84	0.22	390	389	381
Pt:Rh B	78	86	0.38	384	384	371

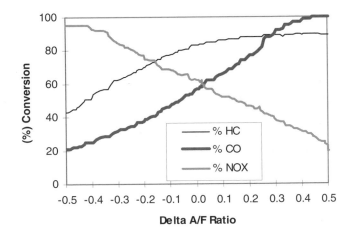

Figure 6: Air/Fuel Sweep - Pt:Rh A, 5:1 Pt:Rh - 1.41 g/L total PGM

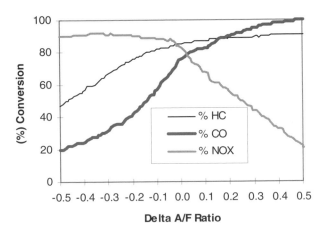

Figure 7: Air/Fuel Sweep - Pt:Rh B, 5:1 Pt:Rh g/L total PGM

FTP test results for the Pt:Rh catalysts are shown in Table 5. The standard Pt:Rh catalyst was not evaluated since there was little difference between the standard and improved catalysts in the sweep test. As found in the sweep testing, Pt:Rh B had higher activity than Pt:Rh A. Pt:Rh B with a 5:1/1.41 g/L PGM load had similar activity to Pd:Rh C with 2:1 / 0.64 g/L. As was mentioned previously, both catalysts had higher activity than the standard Pd:Rh catalyst with 5:1 / 1.41 g/L of Pd:Rh.

Table 5.   FTP conversions for improved Pt:Rh catalysts

Catalyst	Weighted Total FTP Conversions (%)		
	HC	CO	NOx
Pt:Rh A - 5:1 / 1.41 g/L	91.3	79.0	90.2
Pt:Rh B - 5:1 / 1.41 g/L	91.7	85.1	93.2

CONCLUSIONS

For standard Pd:Rh catalysts, the Pd:Rh load could be reduced from 9:1/2.12 g/L to 5:1/1.41 g/L without drastically lowering the three way activity. However, a further decrease in Pd:Rh load to 2:1 / 0.64 g/L dramatically lowered three way activity, especially NOx activity. Improved catalyst technology was developed which increased the activity of a 0.64 g/L Pd:Rh catalyst to higher levels than the 1.41 g/L standard Pd:Rh catalyst. This improved activity should allow substantial Pd load reductions or improved activity at standard Pd:Rh loads. Pt:Rh catalysts were also examined as an alternative to Pd. Improved Pt:Rh formulations were developed which also showed higher activities than the standard 1.41 g/L Pd:Rh catalyst. This improved catalyst technology increases the options available for the management of precious metal demand.

REFERENCES

1. Cowley, A, Platinum 1998, Johnson Matthey, May 1998.

2. J.S. Hepburn, K.S. Patel, M.G. Meneghel, H.S. Gandhi, Engelhard TWC Dev. Team, Johnson Matthey TWC Dev. Team, "Development of Pd-only Three Way Catalyst Technology", SAE 941058 (1994).

3. R.J. Brisley, G.R. Chandler, H.R. Jones, P.J. Andersen, P.J. Shady, "The Use of Palladium in Advanced Catalysts", SAE 950259 (1995).

4. Z. Hu, R.M. Heck, "High Temperature Ultra Stable Close-Coupled Catalysts", SAE 950254 (1995).

5. J.E. Thoss, J.S. Rieck, C.J. Bennett, "The Impact of Fuel Sulfur Level on FTP Emissions - Effect of PGM Catalyst Type", SAE 970737 (1997).

6. N.R. Collins, G.R. Chandler, R.J. Brisley, P.J. Andersen, P.J. Shady, S.A. Roth, "Catalyst Improvements to Meet European Stage III and ULEV Emissions Criteria", SAE 960799 (1996).

7. J.G. Nunan, W.B. Williamson, H.J. Robota, M.G. Henk, "Impact of Pt-Rh and Pd-Rh Interactions on Performance of Bimetal Catalysts", SAE 950258 (1995).

8. A. Punke, U. Dahle, S.J. Tauster, H.N. Rabinowitz, T. Yamada, "Trimetallic Three-Way Catalysts", SAE 950255 (1995).

9. E.R. Becker, R.J. Watson, "Future Trends in Automotive Emission Control", SAE 980413 (1998).

10. G. Zhang, T. Hirota, Y. Hosokawa, M. Muraki, "Thermally Stable Pt/Rh Catalysts, SAE 972909 (1997).

11. R.E. Lakis, Y. Cai, H.G. Stenger, Jr., C.E. Lyman, "Alumina-Supported Pt-Rh Catalysts", J. Catal., v. 154, p. 276 (1995).

12. Z. Hu, F.M. Allen, C.Z. Wan, R.M. Heck, J.J. Steger, R.E. Lakis, C.E. Lyman, "Performance and Structure of Pt-Rh Three-Way Catalysts: Mechanism for Pt/Rh Synergism", J. Catal., v. 174, p. 13 (1998).

CHAPTER 9

HYDROCARBON TRAP

Kimiyoshi Nishizawa
Nissan Motor Co.

9.1 OVERVIEW

The hydrocarbon (HC) trap is the key, and the most recent, technology to reduce cold-start emissions. The 2000 Nissan Sentra CA was the world's first partial zero emission vehicle (P-ZEV). This car used an HC trap system as the main technology to reduce emissions to the level required to meet PZEV standards. An HC trap catalyst consists of an HC trapping material, such as zeolite, and a catalyst coating. Before the catalyst is warmed up the trapping material captures the HC. After the catalyst is warmed up, the trapping material releases the HC and the catalyst coating converts it.

Functions of HC Trap - The HC trap system that is used for the 2000 Nissan Sentra CA consists of a close-coupled three-way catalyst followed by an HC trap catalyst located under the oil pan and a second HC trap catalyst located under the floor. Following a cold-start, the following sequence of events occurs:

- The first HC trap catalyst traps the HC until the close-coupled catalyst is warmed up.
- After some time, the first HC trap catalyst begins releasing trapped HC and converting some of it.
- The HC released from the first trap that is not converted is then trapped on the second HC trap catalyst.
- After some time, the HC is released from the second HC trap catalyst with some conversion.

The material used for trapping HC is a kind of zeolite that is coated on a substrate. Usually a three-way catalyst is over-coated on it. During the cold start, the zeolite traps the HC molecules in the holes in its porous, crystalline structure. The holes in the zeolite structure are 0.5 to 0.8 nm, while the HC molecules range in size from 0.4 to 0.7 nm. After the zeolite temperature reaches 150-250°C the trapped HC is released from the material. The zeolite has an aluminosilicate crystal structure. SiO_4 and AlO_4 tetrahedrons with oxygen in common form a three-dimensional network. The shape of the structure and the size of the pores are affected by the way the network is connected.

Factors Affecting the Efficiency - Even after aging, it is not difficult to make a trap catalyst with good trapping performance, but it is very difficult to maintain good conversion performance. Catalyst aging, especially at high temperature, typically shifts the light-off temperature of the catalyst higher. If shifted high enough, the HC trap will release before the catalyst is lit off. The most important, and the most difficult, aspect in the design of the HC trap system is to maximize the conversion of the trapped HC, minimizing the amount of unconverted HC from HC trap.

One way to improve the conversion efficiency of an HC trap system is to select and develop a trapping material that can release the trapped HC slowly.

For this purpose, selecting the zeolite that has better pore size and distribution for HC trap has been studied.

Besides selecting and developing the trapping material, another way to improve the efficiency of the HC trap catalyst system is to improve the performance of the catalyst coating on the HC trap catalyst. This can be done chiefly through a reduction in the light-off temperature of the catalytic coating. There are two ways to lower the light-off temperature of the three-way catalyst. The first is to increase the precious metal loading, and the other is to apply improved materials in the catalyst coating.

The substrate on which the HC trap catalyst coating is applied affects the desorbed HC conversion performance. It is known that increasing geometric surface area (GSA) is effective in improving the light-off and conversion performance of three-way catalysts. While increasing GSA improves the light-off slightly, the HC desorption rate increases causing lower conversion rates. Reducing GSA has the effect of delaying desorption resulting in higher conversion rates. By seeing the effect of GSA on the above characteristics, optimizing substrate can improve the efficiency of HC trap.

Measures for Improving System Efficiency - Many measures to delay the release of HC and/or to achieve faster catalyst light-off were considered. One of the methods tried was active control of the exhaust gas path. Examples of these measures are discussed below.

The by-pass trap system consists of a conventional three-way catalyst, a by-pass loop containing the HC trap, and a second three-way catalyst downstream. The by-pass loop can be isolated from the exhaust stream by means of valves. The by-pass system functions as follows. At the first, the exhaust gas goes through the HC trap. When the HC trap reaches a certain temperature, the valves are switched as the exhaust gas bypasses the HC trap and goes through the second three-way catalyst. When the second three-way catalyst is warmed up, the valves are switched as the exhaust gas goes through the HC trap.

The system consists of a first catalyst, an adsorber with a hole in the center, and a second catalyst downstream. The hole in the adsorber is intended to allow some fraction of the exhaust gases to pass directly to the second catalyst in order to heat it to its light-off temperature more quickly. This design requires a trade-off between trapping efficiency and the speed with which the second catalyst lights off. A smaller hole will increase the trapping efficiency but slow down the light off of the second catalyst. A larger hole decreases trapping efficiency but speeds up light off of the second catalyst.

A step taken to improve system efficiency without the addition of active components is to implement a two-stage trap system. This system consists of two HC trap catalysts in sequence in the same exhaust flow. In the two-stage trap system, a temperature differential exists between the two HC trap catalysts due to the heat capacity of the first unit. Hydrocarbons are initially trapped and released from the first HC trap catalyst. Even though the first HC trap catalyst is hot enough to release the HC, the second is still cold enough to adsorb the HC. This mechanism improves system performance significantly.

The application of HC trap catalyst systems has just started and the number of auto manufacturers using such systems is still small. Although many improvements still need to be made to HC trap catalyst systems and their application is not so easy to accomplish, they have a large potential for reducing HC emissions. HC trap catalyst systems are the most promising way to eliminate HC emissions ultimately. It is expected that the performance of HC trap catalysts will be continually improved and that their application will be expanded widely in the near future.

9.2 SELECTED SAE TECHNICAL PAPERS

In-line Hydrocarbon (HC) Adsorber System for Reducing Cold-Start Emissions

Shinji Yamamoto, Kenjirou Matsushita, Satomi Etoh and Masahiro Takaya

Nissan Motor Co., Ltd.

Copyright © 2000 Society of Automotive Engineers, Inc.

ABSTRACT

An adsorber system for reducing cold-start hydrocarbon (HC) emissions has been developed combining existing catalyst technologies with a zeolite-based HC adsorber. The series flow in-line concept offers a passive and simplified alternative to other technologies by incorporating one additional adsorber substrate into existing converters without any additional valving, purging lines, secondary air, or special substrates. Major technical issues to be resolved for practical use of this system are 1) the ability to adsorb a wide range of HC molecular sizes in the cold exhaust gas and 2) the temperature difference between HC desorption from the adsorber and activation of the catalyst to convert desorbed HCs. This paper describes the current development status of hydrocarbon adsorber aftertreatment technologies. We report results obtained with a variety of adsorber properties, washcoat structures of adsorber catalyst and start-up and underfloor catalyst system combinations. The system was evaluated in FTP tests using a 2.4-liter L4 vehicle. The system reduced up to 60% of cold-start HC emissions beyond the three-way catalyst-only baseline system. This in-line HC adsorber system could be one of the potential technologies to meet ULEV/SULEV and future regulations, without the need for ancillary electrically heated catalyst (EHC) hardware and its associated costs.

INTRODUCTION

Emission regulations are met today by using three-way catalysts, which reduce exhaust emissions through catalytic reactions [1-3]. However, these systems are inactive during cold start and warm-up of an engine, as they require a temperature level typically around 300°C for sufficient conversion. As a result, the most significant part of tailpipe emissions occurs during the cold-start phase of the FTP test cycle. Compliance with tighter emission regulations in the future will require a further reduction of cold-start hydrocarbon (HC) emissions. The concept of using an HC trap to reduce the amount of cold-start HCs emitted by a vehicle is well known [4-15].

Although activated carbons have also been considered, zeolites are the most common HC trap materials. When the exhaust gas is cold at engine start, the HC trap materials (referred to here as adsorbers) adsorb HCs. As the exhaust gas heats up, the temperature of the adsorber increases, causing the HCs to desorb. After the stored HCs are released from the adsorber, they may be converted by a traditional three-way catalyst located either on the same monolith as the adsorber or downstream. In the first part of this paper, we present the advances made in the development of HC adsorber technologies. In the second part, we describe the results obtained by applying these technologies to a vehicle system. The objective of this study was to demonstrate the potential of HC adsorber technologies in passive flow in-line systems with respect to meeting low emission regulations.

EXPERIMENTAL

ADSORBER SCREENING – Medium-pore zeolite is capable of adsorbing small to medium size hydrocarbons while large-pore zeolite is more efficient at adsorbing large size molecules. Tests were carried out using various adsorbers having different frame structures, Si/2Al ratios (Si/2Al=30-2000) and combinations of medium and large pore size zeolite. For the adsorber evaluation, test samples were prepared by applying a washcoat of a binder material and a zeolite on a cordierite substrate and catalyzing it with precious metals.

ENGINE AGING – The engine aging conditions are shown in table 1. Adsorber aging was carried out for 50 hours on an engine dynamometer. The air/fuel ratio of the 3.0 liter 6-cylinder engine was adjusted so that the nominal adsorber inlet temperature was 750°C. The close-coupled TWC aging temperature was 850°C. All test samples were aged. For in-vehicle evaluation of the HC adsorber system, some catalysts were aged for even longer period to simulate a deteriorated state.

LABORATORY TEST – Hydrocarbon adsorption-desorption properties and conversion performance of the test samples were evaluated utilizing laboratory test apparatus after engine aging. The test apparatus can mix several kinds of source gases simulating engine cold-start exhaust gas, and passes the gas through a sample that monitors HC concentrations at its inlet and outlet. After this operation, the total amount of adsorbed and desorbed HCs was calculated from the difference between the two concentrations. The detailed conditions are shown in table 2-1. The HC adsorption test was carried out with the inlet gas temperature controlled to 110. In every HC adsorption test, the sample was calcinated at 500 for the sake of complete HC desorption. Toluene was used as an adsorbent because it is one of the major HC species in engine exhaust. After a series of adsorption operations, an HC desorption test was performed every time at 110. After the HC supply was stopped, the quantity of HC released with elapsed time was examined. A conversion performance test of the three-way catalyst components was carried out with the inlet gas temperature controlled to 300. Propene or propane was used as an HC species because it is one of the major HC species in engine exhaust. The detailed conditions are shown in table 2-2.

Table 1   Accelerated Engine Aging Conditions

Equipment: 3.0 liter, V6 Engine
Gas Temperature: 750°C at inlet
Aging Cycle: 60 seconds Cruise Mode (A/F ratio=14.6)
                    5 second fuel cut-off and 5 second rich spike
Duration: 50 Hours
Aging Fuel: Japanese domestic fuel
                  (Sulfur≦30 ppm, unLead)

Table 2-1   Laboratory Adsorption-Desorption Test Conditions

Space Velocity: 62,000 per hour

	Adsorption	Desorption
Simulation Gas: HC(Toluene)	1050 ppmC	none
CO	0.6 %	0.6 %
CO_2	10 %	10 %
H_2	0.2 %	0.2 %
O_2	0.51 %	0.51 %
NO	500 ppm	500 ppm
H_2O	10 %	10 %
N_2	Balance	Balance

Adsorption Temperature: 110 °C
Adsorption Period: 600 seconds
Desorption Temperature: 110°C
Desorption Period: 600 seconds

Table 2-2   Laboratory Conversion Test Conditions

Space Velocity: 62,000 per hour

Simulation Gas: HC(C_3H_6 or C_3H_8)	1050 ppmC
CO	0.6 %
CO_2	10 %
H_2	0.2 %
O_2	0.51 %
NO	500 ppm
H_2O	10 %
N_2	Balance

Conversion Temperature: 300 °C

TEST VEHICLE – For the HC adsorber and the HC adsorber system evaluation, an FTP test cycle was carried out using a 1995 U.S. vehicle equipped with a 4.1-liter V-8 engine, with electronically controlled fuel injection and a 4-speed automatic transmission. The car was certified to meet Tier-1 regulations. Phase fuel was used. All FTP tests were performed on a chassis dynamometer. Vehicle emissions were measured using a test vehicle equipped with a 2.4-liter L4 engine that was improved to meet ULEV regulations.

SYSTEM DESIGN – The hydrocarbon adsorber system consisted of a close-coupled three-way catalyst and HC adsorbers that were connected downstream. A schematic diagram of the basic hydrocarbon adsorber system is shown in as (I) in Fig. 1. The 1.7-liter first close-coupled catalyst consisted of a three-way catalyst formulation coated on a cordierite substrate (600cell/inch^2) with a precious metal loading of 10.6 g/L (300 g/ft^3) with a Pd/Rh ratio of 11:1. The second catalyst consisted of a double-layer structure coated on a cordierite substrate (400cell/inch^2). The inside layer was an adsorber formulation with a binder and the outside layer was a three-way catalyst formulation with a precious metal.

Cold-start emission behavior, selection of the washcoat structure, in-vehicle conversion behavior of the HC adsorber system, and the effects of the catalyst volume and substrate structure were investigated using this system. The effect of temperature rise was examined with the pre-underfloor system (II) and close-coupled system (III) in Fig. 1. The effect of additional air was investigated using the basic underfloor system (I). Improvement of the vehicle system was investigated using the two-stage system having two integrated HC adsorbers.

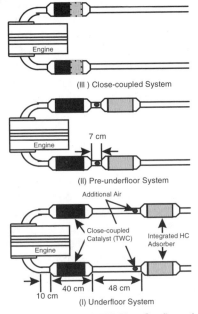

Fig. 1   Illustration of FTP Test Configuration in Vehicle

TEST PROCEDURE – For convenience in analyzing the FTP test results, the A-bag was divided into two bags, A-bag 1 and A-bag 2. A-bag 1 was used in the interval when HCs were adsorbed on the adsorber and contained mostly cold-start emissions. A-bag 2 was used in the interval when the desorbed HCs from the adsorber were converted. The A-bag 1 time and the time for the first and second catalyst to reach light-off were determined from continuous HC and temperature readouts in the baseline runs. For the initial run (A-bag 1 time), HCs were generally monitored before and after the first catalyst (close-coupled TWC). For the remainder of the test, HCs were monitored before and after the second catalyst (integrated HC adsorber).

RESULTS AND DISCUSSION

PART 1 - HC ADSORBER DEVELOPMENT – COLD-START HC EMISSIONSBEHAVIOR An integrated hydrocarbon adsorber must perform two functions. 1) It serves as an appropriate adsorber to trap and store HCs. 2) It provides a catalytic function to convert HCs as they are being released from the adsorber. An appropriate adsorber requires sufficient adsorption-desorption properties to withstand the hot temperature and the flow velocity likely to be encountered in an automotive aftertreatment environment.

Figure 2 shows the HC emission level, vehicle speed and inlet temperature of the close-coupled catalyst during the early part of the FTP test. As this TWC system is inactive during cold start and warm-up of an engine, most of the cold-start HC emissions are emitted from the tailpipe without being converted. As the exhaust gas heats up, the first catalyst inlet temperature rises rapidly to light-off temperature and begins to convert HCs. As a result, 70 to 90% of tailpipe HC emissions occur during the cold-start phase of the FTP test cycle, making it imperative to reduce them.

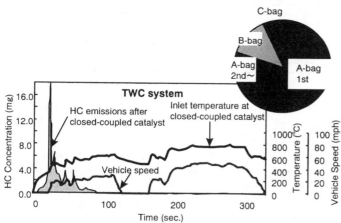

Fig. 2 Behavior of HC Emission Concentration, Vehicle Speed and Exhaust Gas Temperature during FTP Test (A-bag)

Figure 3 shows the composition of the HC species in the exhaust during the cold-start phase as obtained with a flame ionization detector. Although the composition varies depending on the fuel and engine model used and the combustion state, about 100 HC species of different molecular sizes are contained in the exhaust. They consist of about 10% methane, about 30% alkenes such as ethylene or propene, about 30% alkanes such as pentane or hexane, about 20% aromatics such as toluene or xylene, and about 10% other species. Accordingly, adsorber materials must be selected to correspond to these HC species.

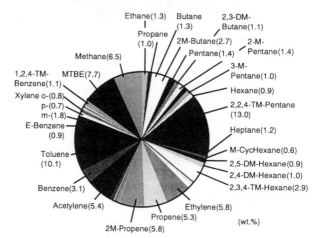

Fig. 3 HC Composition in Engine-out Exhaust Gas (LA-4 A-bag, 0~63sec.)

ADSORBER SCREENING – The HC adsorption capacities of test samples were evaluated utilizing laboratory test apparatus after engine aging. Figure 4 shows a comparison of the adsorption performance of various samples using Toluene as a model gas. Each adsorber had a suitable pore size for the adsorption of toluene. HC adsorption performance varied considerably from 50% to 70% because of differences in zeolite pore size, pore size distribution and frame structure. Among these adsorbers, adsorber J showed the best adsorption performance of 80% or more because of its suitable pore size distribution. Hydrocarbons trapped by adsorbers during the cold-start phase are desorbed as a result of a rise in the exhaust gas temperature.

The following are important as desorption properties: 1) delayed desorption and 2) a slow HC desorption rate as the adsorber is heated. Figure 4 also shows a comparison of the desorption rates of the various samples. Although a large difference was not found between the onset of desorption and temperature, the desorption rates varied. Adsorber G was inferior in adsorption performance but superior in its desorption restraint effect because of its effective frame structure. Compared with the other adsorbers, adsorber J displayed a superior restraint effect because its desorption rate was small. From these results, it is estimated that the adsorption-desorption of cold-start HCs depends on the pore size and frame structure of zeolite. Therefore, optimization of the pore size and distribution is needed to improve adsorption-desorption performance.

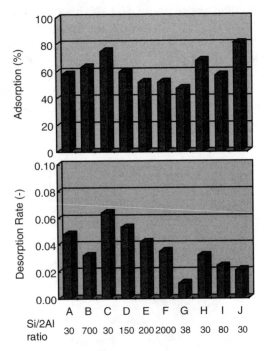

Fig. 4 Adsorber Performance and Desorption Rate of Various Adsorbers using Toluene as Model Gas

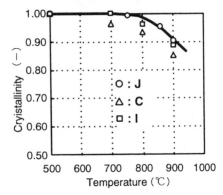

Fig. 5 Thermal Stability of HC Adsorbers in H2O 10% - Air

Figure 5 shows the results obtained for several candidate adsorbers that were investigated for structural integrity after being steamed at temperatures as high as 900°C. Sample J exhibited superior structural stability for meeting high-temperature durability requirements. Based on these test results, adsorber J was selected for its overall high potential.

SELECTION OF WASHCOAT STRUCTURE – (1) Using adsorber J, three different types of hydrocarbon adsorber structure were investigated. In the first trial, the adsorber, binder, and the catalyst ingredients were formed in a homogeneous washcoat, which is referred to as the NH1-type integrated adsorber. In the second trial, the adsorber and binder were applied to the substrate first as an underlayer, followed by the catalytic layer as an overcoat. This structure is referred to as the NH2-type integrated adsorber. In the third trial, the catalyst ingredients were arranged on the substrate first as an underlayer, followed by the adsorber and binder as an overcoat. This structure

is referred to as the NH3-type integrated adsorber. Roughly equivalent loadings of adsorber, binder and catalyst ingredients were applied to all three structures. In each case, the precious metal loading was a total of 2.4 g/L (80g/cf3) with a Pd/Rh ratio of 11:1. The effects of the catalyst structure were evaluated in vehicle tests using the basic underfloor system (), and the results are shown in table 3. For reference purposes, the results obtained with a two-brick converter, having the adsorber and TWC arranged separately, are also shown in table. Although adsorption performance was superior for the NH3-type integrated adsorber, desorption was faster than that of the other structures, with the result that desorbed HCs from the adsorber could not be converted. The NH1 and NH2-type integrated adsorbers were a little inferior in adsorption performance to the NH3-type integrated adsorber, but they were more efficient in converting desorbed HCs. The NH2-type integrated adsorber showed the best conversion performance. The two-brick converter also showed superior adsorption performance, but the light-off time of the downstream TWC was later than that of the other structures, with the result that HCs desorbed from the adsorber could not be converted as efficiently. Accordingly, this type was selected as the integrated adsorber structure.

Table 3 Structure of Integrated HC Adsorber and Performance

Washcoat Structure	Adsorption (%)	Conversion of Desorbed HCs (%)
Separate Bricks	73	~10
Adsorber TWC NH2	70	20 - 30
Adsorber TWC NH3	73	~ 10
Adsorber +TWC NH1	68	15 - 25

PART 2 - HC ADSORBER APPLICATIONS IN AFTERTREATMENT SYSTEMS CONVERSION BEHAVIOR OF HC ADSORBER SYSTEM – In an HC adsorber system, the three processes of adsorption, desorption and conversion proceed in succession. Figure 6 shows an outline of the reduction mechanism of cold-start HC emissions using the basic underfloor system (I).

418

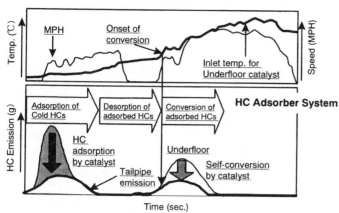

Fig. 6 Emission Reduction Mechanism of HC Adsorber System in Vehicle
Engine displacement = V 4.1 L, Vehicle Inertia Weight = 1925 kg
Evaluation mode = LA4 A-bag
3-way catalyst (Closed-coupled catalyst) = 1.7 L (Aged)
Adsorber substance = 1.3 L (Aged)

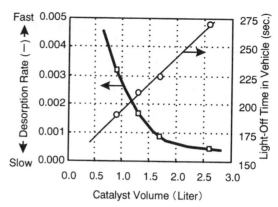

Fig. 7 Relation between Desorption Rate, Light-Off Time
and Underfloor Catalyst Volume in HC Adsorber System
(Fig. 1 (III) Underfloor System)

(I) Adsorption process: The NH2-type integrated adsorber located in the underfloor position adsorbs HC emissions during the cold-start phase before activation of the close-coupled TWC. In this process, 80% or more of cold-start HCs can be adsorbed for a significant reduction of the emission level.

(II) Desorption process: Hydrocarbons trapped by the adsorber are released as a result of a rise in the exhaust gas temperature and an increase in the exhaust gas flow velocity with increasing vehicle speed. In the FTP test cycle, desorption of HCs begins in the second acceleration area of the A-bag because of increasing exhaust gas flow. Desorption takes place under similar temperature and gas flow velocity conditions in European and Japanese test modes.

(III) Conversion process: Desorbed HCs are converted by the TWC layer provided as the upper part of the adsorber layer. Desorbed HCs can be converted efficiently by synchronizing the activation of the TWC layer and the onset of HC desorption. In an HC adsorber system, the third process in this series is the most difficult.

IMPROVEMENT OF DESORBED HC CONVERSION PERFORMANCE – Delaying HC desorption and accelerating TWC layer activation are two important issues in reducing cold-start HC emissions. Accordingly, the catalyst volume, temperature rise and provision of additional air were examined as ways of improving the system technology.

EFFECT OF CATALYST VOLUME – Increasing the catalyst volume or the adsorber quantity is effective in delaying desorption because it increases the force for holding HCs. For this reason, the volume of the underfloor catalyst was increased in this HC adsorber system (basic underfloor system (I) in Fig. 1). Figure 7 shows the relations obtained in an FTP test between the desorption rate, catalyst light-off time and underfloor catalyst volume.

Increasing the catalyst volume slowed down the rise in the catalyst bed temperature. As a result, HC desorption was delayed and the shape of the desorption peak became broader as well. However, the conversion efficiency of desorbed HCs did not improve simultaneously because activation of the catalytic layer was delayed and the timing of desorption and conversion was not synchronized. In each case, the conversion rate was about 25%. Accordingly, increasing the catalyst volume does not enhance conversion performance for desorbed HCs. On the other hand, increasing the adsorber quantity delays HC desorption, resulting in improved conversion efficiency. However, a large conversion improvement can not be obtained because there is a limit to how much the overcoat quantity can be increased.

(2) EFFECT OF TEMPERATURE RISE As the next step, the adsorption, desorption and conversion intervals of the integrated HC adsorber were compared for each catalyst position. The close-coupled system (III) and pre-underfloor system (II) were compared with the basic underfloor system (I). It is seen in figure 8 that the intervals varied depending on the location. With the integrated HC adsorber, the series of processes proceeds independently in each position. The adsorption, desorption and conversion intervals varied on account of differences in the temperature rise characteristic due to the location of the integrated HC absorber. Changing the position of the integrated HC absorber so that the temperature of the catalyst layer rose faster shortened the time to light-off of the three-way catalyst, resulting in improved performance. Figure 9 shows the desorption rate and desorbed HC conversion performance as a function of the TWC light-off time. When the temperature rise of the catalyst is quickened, the desorption rate becomes worse. However, because the effect of shortening the light-off time of the TWC layer is larger, conversion performance can be improved.

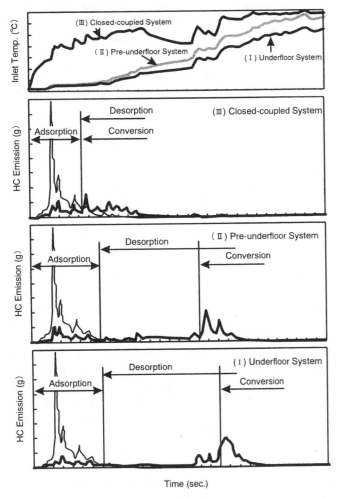

Fig. 8   Emission Reduction Behavior of HC Adsorber System in Vehicle

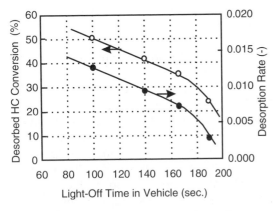

Fig. 9   Relation between HC Conversion and
Light-Off Time in HC Adsorber System

(3) EFFECT OF ADDITIONAL AIR - The HC release behavior of the integrated HC adsorber was analyzed in an FTP test, and the conversion atmosphere was examined. Engine-out HC emissions before TWC activation are trapped by the adsorbers, and adsorbed HCs are subsequently released in the following interval. Desorption occurs due to increased temperature and gas flow quantity. As upstream TWC activation is achieved in this interval, desorption and conversion of adsorbed HCs are accomplished because of the increased temperature

and gas flow quantity of the exhaust gas from the close-coupled TWC. Therefore, the HC and oxygen concentrations were examined at the time HCs were released from the underfloor catalyst. The HC concentration and A/F were measured at the underfloor catalyst outlet. An investigation was made of the atmosphere in the catalyst layer, expressed by the value indicating the ratio between the oxidation constituents and reduction constituents. This value referred to as the Z value, was calculated with the following equation.

$$Z=([O_2 \text{ conc.}]+[NO \text{ conc.}])/([HC \text{ con.}]+[CO \text{ conc.}])$$

The change in the Z value with elapsed time from the onset of desorption was estimated and the results are shown in figure 10. With the onset of desorption in the desorption interval, the Z value shifts greatly toward the insufficient oxygen side and the TWC layer is exposed to a state of insufficient oxygen. Accordingly, mitigating this insufficient oxygen state so as to achieve an atmosphere (Z value) suitable for conversion was presumed to be an important factor in improving desorbed HC conversion efficiency. To accomplish that, the introduction of outside air by means of a pump was examined. The effect of this secondary air on improving conversion is shown in Fig. 11. Air was introduced through the inlet of the catalytic converter after 160 seconds of the FTP test, just before light-off of the catalyst layer of the underfloor integrated HC adsorber. The basic underfloor system (I) in Fig. 1 was used in the evaluation. When the quantity of additional air was varied, conversion of desorbed HCs improved in proportion to the quantity of air up to 100 L/m. However, the catalyst layer temperature fell when the quantity of additional air was increased further, which had the contrary effect of worsening conversion performance. The improvement effect was not sufficient when the air introduction time was short or when the quantity of air added was not adequate. Accordingly, in trying to overcome the insufficient oxygen state through the addition of secondary air, determining a suitable air quantity, introduction timing and duration is effective in improving conversion of desorbed HCs.

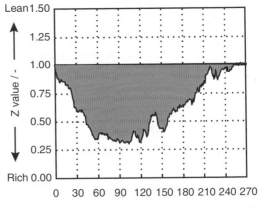

Fig. 10   Relation between Z value and HC Desorption

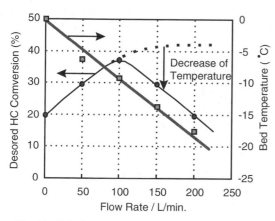

Fig. 11 Relation between Conversion of Desorbed HC and Flow Rate of Additional Air in HC Adsorber System

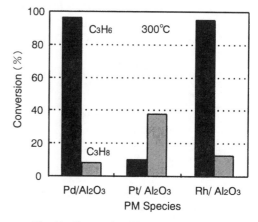

further if cerium oxide can supply a large quantity of oxygen at low temperature.

Fig. 12 Conversion Characteristics of Precious Metals

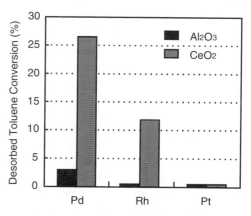

Fig. 13 Conversion Characteristics of Precious Metals

IMPROVEMENT OF TWC OVERCOAT – Hydrocarbon desorption properties depend on the pore size. Hydrocarbon species of a small molecular size are released in the first half of the desorption interval and those of a large size are released in the latter half. An investigation of the desorbed HC species revealed that normal chain HCs of a small carbon number are released in the first half. Aromatic hydrocarbons and those having a side chain are released in the latter half. Accordingly, in the low temperature region, it is advisable to select noble metals that are superior for converting normal chain HCs. Therefore, an investigation was first made of the conversion characteristics of each noble metal for desorbed HCs. Figure 12 shows the conversion characteristics obtained for palladium, platinum and rhodium using a model gas. Test samples were prepared by coating the precious metal catalysts on alumina or cerium oxide substrates. Samples were aged under the same conditions as the HC adsorbers in table 1. Compared with platinum, palladium and rhodium showed superior conversion performance for normal chain HCs. Further, the order of this conversion performance was connected with the light-off temperature (T50). Palladium and rhodium were therefore selected as a combination of noble metals for use with the integrated HC adsorber. Conversion of desorbed HC proceeds in an insufficient oxygen atmosphere. Therefore, it is thought that, in an insufficient oxygen atmosphere, desorbed HC conversion performance can be improved if palladium can make use of oxygen released by cerium oxide.

Figure 13 shows the results of a model gas evaluation of the conversion characteristics of catalysts that combined palladium with alumina or cerium oxide substrates. Better conversion performance was obtained with the cerium oxide substrate than with the alumina one. A correlation was found between the quantity of oxygen released by cerium oxide, as estimated from TPR measurements, and desorbed HC conversion performance. It is assumed that oxygen supplied from cerium oxide is used effectively in the conversion process. Accordingly, it is thought that conversion performance can be improved

EFFECT OF SUBSTRATE STRUCTURE – An attempt was made to improve the substrate on which the integrated HC adsorber washcoats were arranged in order to enhance desorbed HC conversion performance. Generally, GSA is skillfully used as the structural parameter of substrates. It is known that increasing GSA is effective in improving the light-off performance and conversion performance of three-way catalysts. Figure 14 shows the desorption rate and desorbed HC conversion rate as a function of GSA in the basic system (I). NH2-type integrated HC adsorbers having the same washcoat on different GSA substrates (200-600cell/inch^2) were examined. Light-off performance improves only slightly when GSA is increased and the effect is small. Conversely, the desorption rate increases, and conversion deteriorates. On the other hand, reducing GSA has the effect of delaying desorption, resulting in improved conversion. It is supposed that these effects of GSA originate in changes in the gas diffusion state in the overcoat layer. Based on these results, the substrate of the integrated HC absorber was improved and desorbed HC conversion of 45 to 50 % was then obtained.

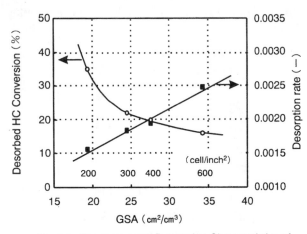

Fig. 14 Conversion and Desorption Characteristics of GSA in HC Adsorber System

IMPROVEMENT OF VEHICLE SYSTEM – Ways of improving the vehicle system were investigated using a system in which the integrated HC adsorber was located in the close-coupled position. As explained earlier, the series of processes in the adsorption-desorption-conversion cycle occurs in each position where the integrated HC adsorber is located. Accordingly, multiple execution of the cycle was thought to be effective in further improving conversion performance. An investigation was made of a two-stage HC adsorber system in which one more catalyst was added downstream to provide two conversion cycles. Figure 15 shows the two-stage HC absorber system arranged in two configurations (IV) and (V). The basic system in Fig. 1 was improved to a two-stage system, which included one TWC and two NH2-type integrated HC adsorbers. On the basis of the result in Fig. 8, combinations of NH2-type HC adsorbers with reduced GSA (200cell/inch^2) were examined.

Table 4 shows the evaluation results for each vehicle system. Desorbed HC conversion performance of 40 to 50% was obtained with the basic system. The two-stage system having integrated HC absorbers at the close-coupled and pre-underfloor, and at the close-coupled and underfloor positions improved conversion performance substantially.

With this system, HC emissions that the front catalyst cannot convert are trapped by the downstream catalyst again. It is assumed that the downstream catalyst carries out the same cycle of processes. As a result, the two-stage system achieved HC conversion of 70 to 75% (cold-start HC reduction of 60 to 65%). Desorbed HC conversion was improved further to 75 to 85% (65 to 70%) when air was added to mitigate the insufficient oxygen state. Based on these results, the improved vehicle system in which several catalysts were arranged in-line reduced cold-start HC emissions by 60% or more (reduction rate of 70% or more for desorbed HC adsorbers can reduce cold-start HC emissions by 60% or more when the catalysts are still in a state of little deterioration. Even after undergoing aging equivalent to

100,000 miles of operation, the two-stage system can reduce cold-start HC emissions by 40% or more.

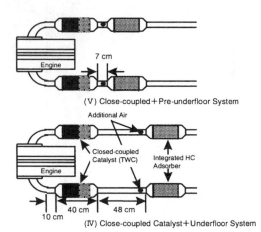

Fig. 15 Illustration of Two-Stage HC Adsorber System in Vehicle

Table 4 Performance of HC Adsorber System for Cold Start Emissions

System	Closed-coupled Catalyst PM Loading (g/cf)	**HC Adsorber First PM Loading (g/cf)	Second PM Loading (g/cf)	Adsorption (%)	Conversion of Desorbed HCs (%)	Conversion of Cold-start HC Emissions (%)
Integrated HC Adsorber TWC	Pd/Rh =300-11/1	Pd/Rh =320-11/1		79	36	29
TWC Integrated HC Adsorber	Pd=480	Pd/Rh =320-11/1		60	66	40
TWC Integrated HC Adsorber	Pd=480	Pd/Rh =320-11/1	Pd/Rh =320-11/1	85	70	60
		*with Additional Air		*85	*75	*64
TWC Integrated HC Adsorber	Pd=480	Pd/Rh =320-11/1	Pd/Rh =320-11/1	83	63	52

**Each substrates having 200 cells per square inch

Table 5 Emission Performance of HC Adsorber System in Test Vehicle*

Aging Condition	Cold-start HC Emission Reduction (%)／A-bag
5,000 miles equivalency	76
50,000 miles equivalency	64
100,000 miles equivalency	45

*Improved to meet ULEV regulations

SUMMARY AND CONCLUSIONS

A durable adsorber was found which can withstand an exhaust environment with high temperature and steam. The newly developed adsorber showed superior adsorption-desorption performance for cold-start HCs.

A catalyst structure and a three-way catalyst layer were developed that are effective for conversion of desorbed HCs. An integrated HC adsorber that incorporates this new type of structure was developed.

The substrate of the integrated HC absorber and the position of the catalytic converter were optimized to improve the conversion efficiency of desorbed HCs.

An in-line HC adsorber system consisting of a close-coupled TWC and an integrated HC adsorber reduces cold-start HC emissions more than a typical TWC system.

This in-line HC adsorber system was devised as a result of analyzing the working conditions of the integrated HC adsorber in each catalyst position. This system markedly reduced cold-start HC emissions after engine aging, demonstrating its substantial HC reduction effect. The results confirmed that this new cold-start HC reduction technology has the potential to comply with even more stringent regulations expected to be adopted after the ULEV standards.

As a result of having investigated various in-line HC adsorption systems, we identified the HC adsorber technology that can significantly reduce cold-start HC emissions. Because this new system can substantially reduce emission levels compared with the performance of a typical TWC system, it is expected to be a key technology for meeting more rigorous requirements after the ULEV regulations. As the technology is now under development, there are still problems to be overcome, including the high cost, insufficient durability and limited usage environment. Nonetheless, because it falls under the category of a typical TWC system that does not use additional equipment such as an EHC, this HC adsorber technology may be superior in cost performance in the future.

ACKNOWLEDGMENTS

The authors would like to thank various individuals for their invaluable cooperation in connection with this research.

REFERENCES

1. Summers, J. C., Skowron, J. F. and Miller, M. J., "Use of Light-Off Catalyst to Meet the California LEV/ULEV Standards", SAE 930386, 1993.

2. Heck, R. M., Hu, Z., Smaling, R., Amundsen, A., and Bourke, M. C., "Close Coupled Catalyst System Design and ULEV Performance after 1050 Aging", SAE 952415, 1995.

3. Hu, Z. and Heck, R. M., "High Temperature Ultra Stable Close Coupled Catalysts", SAE 952554, 1995.

4. Heimrich, M. J., et al., "Cold-Start Hydrocarbon Collection for Advanced Exhaust Emission Control", SAE 920847, 1992.

5. Engler, B. H., Lindner, D., et al., "Reduction of Exhaust Gas Emissions by Using Hydrocarbon Adsorber Systems", SAE 930738, 1993.

6. Kollman, K., Abthoff, J., and Zahn, W., "Concepts for Ultra Low Emission Vehicles", SAE 940468, 1994.

7. Burk, P. I., Presentation at Toptec, "Hydrocarbon Traps for Automotive Pollution Abatement", Jan. 1995.

8. William, J. I., Patil, M. D., and Hertl, W., "By-Pass Hydrocarbon Adsorber System for ULEV", SAE 950410.

9. Hertl, W., Patil, M. D., and Williams J. I., "Hydrocarbon Adsorber System for Cold Start Emissions", SAE 960347, 1996.

10. Patil, M. D., Hertl, W., Williams, J. I., and Nagel, J. N., "In-Line Hydrocarbon Adsorber System for ULEV", SAE 960348, 1996.

11. Noda, N., Takahashi, A., and Mizuno, H., "In-line Hydrocarbon (HC) Adsorber System for Cold Start Application", SAE 970266.

12. Buhrmaster, C. L., Locker, R. J., et al., "Evaluation of In-Line Adsorber Technology", SAE 970267, 1997.

13. Ballinger, T. H., Manning, W. A., and Lafyatis, D. S., "Hydrocarbon Trap Technology for the Reduction of Cold-Start Hydrocarbon Emissions", SAE 970741, 1997.

14. Silver, R. G., Dou, D., Kirby, C. W., Richmond, R. P., Balland, J., and Dunne, S., "A Durable In-Line Hydrocarbon Adsorber for Reduced Cold Start Exhaust Emissions", SAE 972843, 1997.

15. Abthoff, J., Kemmler, R., Klein, H., Matt, M., Robota, H., Wolsing, W., Wiehl, J., and Dunne, S., "Application of In-Line Hydrocarbon Adsorber Systems", SAE 980422, 1998.

By-Pass Hydrocarbon Absorber System for ULEV

J. L. Williams, M. D. Patil, and W. Hertl
Corning, Inc.

Copyright 1996 Society of Automotive Engineers, Inc.

ABSTRACT

A by-pass zeolite adsorber system consisting of a first catalyst, a by-pass loop containing the zeolite adsorbers followed by a downstream second catalyst was FTP tested using a U.S. vehicle equipped with a 3.8 L, V6 engine. The system exhibited ULEV emissions performance with hydrocarbon adsorption and regeneration (desorption and oxidation) within the FTP cycle and required only a single diversion valve within the exhaust line. Adsorption takes place during the initial 70 seconds of the FTP cycle. The adsorbers were regenerated with the exhaust gas plus injected air.

INTRODUCTION

Motor vehicle emissions have been reduced dramatically since the 1970's, and in 1990 the U.S. Congress passed a new set of Clean Air Act Amendments to further reduce tailpipe emissions responsible for photochemical smog formation [1]. CARB (California Air Resources Board) has set in place a program for future more stringent regulatory standards that are designed to beimplemented in stages to low levels. Table 1 lists California's Low Emissions Vehicle (LEV) and Ultra Low Emissions Vehicle (ULEV) standards for 50,000 miles of vehicle use. The European Economic Council is also considering phasing in more s gent standards using a multi-staged approach. In order to achieve such low level emissions, it is important to reduce cold start hydrocarbon emissions which contribute 70-90% of total tailpipe emissions. Several approaches including electrically heated catalysts [2-4], close-coupled catalysts [5], pre-catalysts [6], exhaust gas burners [7,8], and hydrocarbon traps [9-12] have evolved over the last few years.

This paper describes FTP 75 testing of our by-pass zeolite adsorber system that exhibits performance capable of meeting California's ULEV emissions standard, 0.04 g/mi NMHC. Hydrocarbon (HC) adsorption and regeneration (desorption and oxidation) was demonstrated during the FTP cycle. Figure 1 illustrates the by-pass system consisting of a first catalyst three valves, a by-pass loop containing the zeolite adsorbers, and a downstream second catalyst. The basic function of the system is: 1) zeolite adsorption of hydrocarbons by physisorption during cold start; 2) selectively heating the downstream second catalyst above its light-off temperature; and 3) hydrocarbon desorption and oxidation with additional air over the second catalyst.

Adsorption occurs during the initial 70 seconds of the FTP cycle followed by heat up of the second catalyst to its light-off temperature by the exhaust gas. The adsorbers are then purged and regenerated with the hot exhaust gas plus injected air from 250-505 seconds in the FTP cycle when the car is running near maximum speed. All testing was performed at Southwest Research Institute (SwRI) using a U.S. vehicle equipped with a 3.8L, V6 engine.

EXPERIMENTAL DESIGN

ADSORBERS - Tests were carried out using a combination of medium and large pore zeolite adsorbers. The medium pore zeolite is capable of adsorbing small to medium size hydrocarbons, while the large pore zeolite is more efficient at adsorbing larger size molecules. In one case, a zeolite capable of adsorbing both small and large hydrocarbons was used. The zeolites were washcoated and catalyzed with precious metal on cordierite cellular substrates with 62 cells/cm^2 (400 cells/in^2). Generally four substrates each 9.30 cm diameter x 7.62 cm length (3.66" x 3") were used for a total volume of 2.06 liters.

CATALYST - The first catalyst was a standard three-way catalyst formulation coated on a ceramic substrate with precious metal loading of 3.18g/L (90g/ft^3) with Pt/Rh ratio 5:1, 1.7 liter volume. The second catalyst used was two 7.2 cm diameter x 7.2 cm length (3.3" x 3") ceramic substrates coated with a standard TWC, 7.03g/L (199g/ft^3) Pt/Rh=5:1.

ENGINE DYNAMOMETER AGING - The first catalyst and second catalyst were aged at 850°C for 44 hours on a 7.5 L, 8 cylinder engine with throttle body fuel injection. The adsorbers were also aged under these conditions. However during aging the zeolite adsorber bed temperature ranged from 730-800°C. The catalysts and adsorbers were aged separately. After aging the components were installed on the vehicle and FTP tests conducted.

VEHICLE EVALUATION - Tests were conducted using a 1991 U.S. vehicle equipped with a 3.8 L, V6 engine with electronically controlled port fuel injector and 4-speed automatic transmission. The car was certified to meet 1991 California emissions standard. The engine management and electrical system were not modified.

Emissions performance was determined by using the Federal Test Procedure (FTP 75). The system was fitted with an array of thermocouples and gas sampling probes (P1-P6) as shown in Figure 1 to monitor temperature, continuous HC, O_2 and CO emissions. This allows for measurement of the contribution of each component For convenience of analyses, Bag 1 of the FTP cycle was split into two bags. Bag 1A represents 0-220 seconds and Bag 1B, 221-505 seconds. Phase II fuel was used for all testing. In addition to measuring FTP emissions, the converter temperatures, adsorber temperatures and continuous HC, CO and NOx emissions were measured. The continuous hydrocarbon emissions were measured on a carbon basis.

INJECTED AIR QUANTITIES - Air injection was provided by a belt driven air injection automotive pump rated at 212 L/min. (7.5 scfm). For test purposes this pump was belt driven with a 0.37 kW (0.5 hp) electrical motor rated at 1750 rpm. With a 10.16 cm (4") diameter pulley approximately 297 L/min. (10.5 scfm) of air could be delivered. A gate valve was used to regulate the desired injected air flow.

SYSTEM DESIGN - The by-pass adsorber system consists of a first catalyst followed by a by-pass loop containing the zeolite adsorbers, which could be isolated using two vacuum actuated valves. In the main exhaust pipe line a vacuum actuated valve could be set to open, closed or partially open for diversion of hot exhaust gases to desorb the Hcs. The second catalyst is located downstream to oxidize HCs as they are released from the adsorbers. A schematic of the system is shown in Figure 1. In later tests only one diversion valve was used, see Figure 2.

The key functions of the system are (a) HC adsorption during cold start, and (b) HC desorption from the adsorber and oxidation over a downstream second catalyst. Adsorption takes place from 0-70 seconds. The best strategy to desorb the HCs is to use the hottest exhaust gas produced when the vehicle is running at maximum speed, and the second catalyst is hot (i.e. higher than the light-off temperature). Additional air is also essential to convert the desorbed HCs to H_2O and CO_2 over the second catalyst.

DIVERSION VALVE CALIBRATION - The function of the main diversion valve (V1) is to control exhaust flow to the by-pass loop and the downstream second catalyst for maximum adsorption and oxidation efficiency. With the engine and second catalyst at a steady-state running temperature (vehicle speed is 55 mph) the main diversion valve is partially opened to a given setting, allowing some of the hot exhaust gas to flow through the by-pass loop over the cold adsorbers and the remainder to pass directly through the main exhaust pipe to the second catalyst. The adsorbers and second catalyst temperatures were monitored. Air was also injected into the system before the adsorbers, since this was also used during the FTP tests.

When steady-state temperatures were obtained, the main valve (V1) was opened completely and the adsorber cooled with shop air. The procedure was repeated with different valve openings. These measurements were continued until a valve opening was obtained in which the second catalyst temperature did not drop below 250°C (later 270-300°C), an arbitrary estimated light-off temperature, and the adsorbers attained an exit bed temperature of at least 250-300° C within the 285 seconds available during the latter part of Bag 1 in the FTP cycle.

TEST PROCEDURE - During the FTP test exhaust gas is passed over the adsorbers from 0-70

Table 2: FTP test results summary for by-pass zeolite adsorber system tested on a 3.8L, 1991 U.S. vehicle.

	System	Test Condtions	BAG 1A g Total	%HC Adsorbed	BAG 1B g Total	%ADS. HC Oxidized	BAG 2 g Total	BAG 3 g Total
A1	First catalyst aged		1.36	-	0.023	-	0.065	0.098
A2	First catalyst + Second catalyst		1.40	-	0.017	-	0.066	0.092
B1	First catalyst + Second catalyst + 2,06L Adsorber Fresh	56.6 L/m 2nd air, Before-TWC Configuration 1	0.652	64%	0.338	79%	0.197	0.126
B2	First catalyst + Second catalyst + 2.06L Adsorber Fresh	141.6 L/m 2nd air, Before-Adsorber Configuration 1	0.705	61%	0.203	91%	0.037	0.08
B3	First catalyst + Second catalyst + 2.06L Adsorber Fresh	141.6 L/m 2nd air, Before-Adsorber Configuration 1	0.485	65%	0.142	86%	0.086	0.055
C1	First catalyst + Second catalyst + 2.06L Adsorber Aged	141.6 L/m 2nd air, Before-Adsorber Configuration 1	0.633	55%	0.068	93%	0	0.052
D1	First catalyst + Second catalyst + 2.06L Adsorber Aged	141.6 L/m 2nd air, Before-Adsorber, No Insulation Configuration 1	0.618	56%	0.186	78%	0.01	0.068
E1	First catalyst + Second catalyst + 2.06L Adsorber Aged	Two by-pass valves (V2 & V3) left open, No Insulation Configuration 2	0.623	56%	0.189	78%	0.01	0.066
*F1	First catalyst + Second catalyst + 2.06L Adsorber Aged	Two by-pass valves (V2 & V3) left open, No Insulation Configuration 2	0.605	57%	0.072	91%	0.026	0.081

First catalyst: 3.18 g/L, Pt/Rh=5/1; Adsorber: Catalyzed medium pore & large pore zeolites; Second catalyst: 7.03 g/L Pt/Rh=5/1;
*F1: Zeolite capable of adsorbing small and large hydrocarbons.

seconds with valves V2 and V3 open and V1 closed. This is the time determined from the continuous HC monitoring readout (see Figure 3), for the first catalyst to begin functioning and the HC concentration to drop to ca. 100 ppmC.

At 70 seconds V1 is opened while valves V2 and V3 are closed. Complete opening of V1 allows unrestricted flow of the exhaust gases for rapid heat up of the second catalyst. At t=250 seconds when the temperature monitors showed that the downstream second catalyst was at operating temperature, valves V2 and V3 were completely opened and V1 partially opened. Secondary air injection was also started. At the end of Bag 1 (505 seconds) the air injection was turned off, V1 was completely opened and valves V2 and V3 closed. Air injection and adsorber purging (as in Bag 1) was also performed during the time period t=250 to 505 seconds of Bag 3, although the continuous HC readouts measured after the adsorbers indicated that complete desorption and regeneration had occurred during the Bag 1B purge.

For FTP test analyses, three bags of exhaust emissions are collected, Bags 1,2 and 3, and analyzed for HC, CO, NOx and CH$_4$. To evaluate the efficiency of the adsorbers and second catalyst the exhaust emissions for Bag 1 is subdivided into Bags 1A and 1B. For these tests the exhaust emissions from 0-220 seconds are collected in Bag 1A and from 221 to 505 seconds in Bag 1B.

The bags do not correspond exactly with the adsorption and regeneration cycles. As mentioned above, from 0-70 seconds the exhaust gases are passed over the adsorbers and the by-pass loop is closed off. From 71-220 seconds exhaust emissions are collected as part of Bag 1A. However HCs in the exhaust after 70 seconds are very small since the first catalyst is functioning efficiently. Complete speciation analyses were carried out in some tests on Bag 1A exhaust emissions.

RESULTS AND DISCUSSION

GENERAL

The overall goal of this investigation was to demonstrate that our by-pass zeolite adsorber system is capable of exhibiting ULEV emissions performance, 0.04 g/mi NMHC with adsorber regeneration within the FTP test cycle. A second objective was to determine whether some of the by-pass valves could be eliminate& Table 1

shown below lists California's LEV and ULEV emissions standards for 50,000 miles of use.

Table 1: California's LEV and ULEV.
Standards for 50,000 miles

Category	Exhaust Emissions g/mi		
	NMHC	CO	NOx
TLEV	0.0125	3.4	0.40
LEV	0.075	3.4	0.20
ULEV	0.04	1.7	0.20

Figure 3 shows engine out and HC emissions after the zeolite adsorber plotted against time. Emissions were measured at P2 after the first catalyst and at P6 after the adsorber. The plot clearly shows ~60% adsorption of HCs within the first 70 seconds after engine start up.

DIVERSION VALVE

The vacuum actuated main diversion valve (V1) is readily closed completely to pass all the exhaust gases over the adsorbers or opened to allow the exhaust gases to pass directly through the main exhaust pipe line. In order to regenerate the adsorbers with the hot exhaust gas it is necessary to heat the adsorbers while not allowing the second catalyst to drop below its light-off temperature. Should the second catalyst temperature drop too much it would allow the desorbed HCs to be emitted to the atmosphere through the tailpipe. The fractional opening of the vacuum actuated valve was varied and controlled by bleeding air into the vacuum line through a needle valve.

FTP TESTS

Table 2 lists specific adsorber systems, test conditions, and HC analyses of each bag. The table also lists percent of HCs adsorbed and the fraction oxidized by the second catalyst. A complete FTP composite summary including NMHC of all systems tested is given in Table 3.

The base line emissions for system A1 using the first catalyst only and system A2 with the first catalyst and second catalyst were measured and found to be 0.085 and 0.081 g/mi NMHC, respectively. The base line emissions of A1 were used to calculate the adsorber efficiency as shown in equation 1.

$$HC \text{adsorbed} = \frac{HC \text{ base } 1A - HC \text{ test } 1A}{HC \text{ base} 1A} \quad (1)$$

Baseline Bag 1A contained 1.36g of HC (Table 2). The fraction of HCs oxidized by the second catalyst was calculated using system A1, Bag 1B emissions as shown in equation 2 below.

$$Fraction \text{ HC Oxid.} = 1 - \frac{HC \text{ test } 1B - HC \text{ base } 1B}{HC \text{ base } 1A - HC \text{ test } 1A} \quad (2)$$

The FTP tests were conducted using a combination of zeolites washcoated on separate honeycomb substrates with a total volume of 2.06 liters. Both aged and fresh zeolite adsorbers, were tested. The entire by-pass system including the catalysts were initially insulated with fibrous furnace insulation to minimize heat loss. In addition, the effect of secondary air addition was evaluated on systems B1 and B2.

Table 3. Summary of FTP composite results.

FTP COMPOSITE				
System	Total HC	CO	NOx	NMHC g/mi
A1	0.09	0.963	0.111	0.085
A2	0.098	0.937	0.083	0.081
B1	0.093	0.574	0.421	0.062
B2	0.063	0.624	0.125	0.045
B3	0.052	0.724	0.083	0.035
C1	0.044	0.827	0.093	0.033
D1	0.053	0.756	0.101	0.038
E1	0.053	0.734	0.093	0.039
F1	0.049	0.766	0.098	0.032

During the FTP test of system B1, 56.6 L/min. (2 scfm) of secondary air was introduced in front of the first catalyst. The second catalyst oxidation efficiency was 79%. Earlier tests with lower amounts of secondary air, 15.2 L/min (0.5 scfm) resulted in poorer HC oxidation. Therefore the amount of air was increased to 56.6 L/min. The rationale was to introduce sufficient oxygen to react with the CO and HCs, and to minimize cooling of the adsorber during desorption and oxidation by the second catalyst. Excess CO is produced during acceleration within the test cycle. The excess CO preferentially consumes all the available oxygen allowing desorbed HCs to escape. Thus the additional secondary air resulted in a higher fraction of adsorbed HCs being oxidized. The NMHC is 0.062 g/mi for system B1. Also NOx emission was high for B1 (Table 3) due to the addition of secondary air in the exhaust passing over the first catalyst.

The addition of secondary air was increased for system B3 to 141.6 L/min. (5 scfm) and added just in front of the zeolite adsorber in the by-pass loop. The second catalyst oxidation efficiency increased from 79% for B1 to 91%. Bags 2 and 3 emissions also decreased. Adsorption efficiency was similar with NMHC = 0.045 g/mi. In subsequent tests 141.6 L/min (5 scfm) air became the standard.

Systems C1, D1, E1 and F1 tests were run on aged zeolite adsorbers. C1 was tested with 141.6 L/min. of air added in front of the adsorber, the second catalyst efficiency was 93% with NMHC = 0.033 g/mi. The performance of fresh and aged zeolite adsorbers were equivalent. The system as well as the adsorbers maintain Their performance after multiple FTP tests with adsorber regeneration.

The insulation was removed from system D1 before the FTP test The adsorption in Bag 1A remained the same. However the second catalyst efficiency decreased to 78%. The system barely met the ULEV standard with NMHC = 0.093 g/mi Removal of the insulation may have allowed the exhaust gases to cool more thus lowering the inlet temperature to the second catalyst and reducing its activity. Speciation analysis shows (Figure 4) that the aged zeolite adsorber is very efficient at adsorbing C_4 or greater HCs. The fraction of each HC species is calculated from the Bag 1A test emissions divided by the Bag 1A baseline emissions, see equation 2.

To determine whether some of the by-pass valves could be eliminated, configuration 2 was used for testing system E1. During the FTP test the two by-pass valves V2 and V3 were left open. Therefore, the exhaust gas flow through the main pipe and the by-pass was controlled by a single diversion valve (V1) only. The results were essentially the same as in the previous test (system D1), 0.039 g/mi NMHC. This test demonstrates that the by-pass unit functions well with a single control valve and does not require the additional two valves.

Similarly F1 was tested using configuration 2 (no insulation, single control valve) with a different type aged zeolite. The zeolite used was capable of adsorbing small and large hydrocarbons. The adsorption behavior was similar to the zeolite combination used in the previous run, however the second catalyst oxidation efficiency increased to 91%. The NMHC emissions were 0.032 g/mi. Speciation analysis shown in Figure 5 was very similar to that for the combination of adsorbers.

BACK PRESSURE

Figure 6 gives a plot of back pressure measured during the initial 160 seconds of Bag 1 during the FTP test The figure shows back pressure measurements for the first catalyst with and without the second catalyst and adsorbers. There is an increase in back pressure when the second catalyst is in line. Comparison of analyses during the base line tests with and without the second catalyst showed no adverse effect on the emissions. Measurements taken with the first catalyst plus second catalyst and adsorbers in the by-pass loop showed that the adsorbers caused only a minimal increase in back pressure.

SUMMARY

A by-pass zeolite adsorber system was demonstrated that exhibited ULEV emissions performance with adsorption and regeneration (desorption and oxidation) within the FTP cycle and required only a single diversion valve within the exhaust line. NOx and CO emissions standards were also within the ULEV standards. The adsorbers in a by-pass loop effectively removed up to 60% of exhaust hydrocarbons.

ACKNOWLEDGMENT

The authors wish to thank the following individuals from Coming Incorporated Mr. Ken Zaun and Ms. Liz Wheeler for extrusion and sample preparation; Mr. Tom Rosenbusch, Mr. Lou Socha and Dr. I.M. Lachman for helpful discussions during designing and testing experiments. We appreciate the efforts of Southwest Research Institute, Mr. Melvin Ingals and Mr. Phil Weber for guidance and testing of the adsorbers.

REFERENCES

1. Calvert, J.G., et al., "Achieving Acceptable Air Quality: Some Reflections on Controlling Vehicle Emissions," Science, Vol. 261, July (1993).

2. Socha, L.S., Thompson, D.F. and Weber, P.A., "Optimization of Extruded Electrically Heated Catalysts," SAE 940468 (1994).

3. Laing, P.M., "Development of an Alternator-Powered Electrically Heated Catalyst System," SAE 940465 (1994).

4. Heimrich, M.J., et al., "Electrically Heated Catalyst System Conversions on Two Current Technology Vehicles," SAE 910612 (1991).

5. Summers, J.C., et al., "Use of Light-off Catalysts to Meet the California LEV/ULEV Standards," SAE 930386 (1993).

6. Socha, L.S., et al., "Advances in Durability and Performance of Ceramic Preconverter Systems", SAE 95047 (1995).

7. Oser, P., "Novel Emission Technologies with Emphasis On Cold Start Improvements Status Report On VW-Pierburg Burner/Catalyst Systems," SAE 940474 (1994).

8. Ma, T., et al., "Exhaust Gas Ignition - A New Concept For Rapid Light-off Of Automotive Exhaust Catalyst," SAE 920400 (1992).

9. Lachman, I.M., Patil, M.D., and Socha, L.S., "Dual Converter Engine Exhaust System for Reducing Hydrocarbon Emissions," U.S. Patent 5,125,231.

10. Engler, B.H., Lindner, D., Lox, E.S., Ostgathe, K., Schafer-Sindlinger, A. and Muller, W., "Reduction of Exhaust Gas Emissions by Hydrocarbon Adsorber Systems," SAE 930738, (1993).

11. Burke, P.L., Hochmuth, J.K., Anderson, D.R., Sung, S., Tauster, S.J., et al., "Cold Start Hydrocarbon Emissions Control," SAE 950410 (1995).

12. Kollman, K., Abthoff, J., and Zahn, W., "Concepts for Ultra Low Emissions Vehicles," SAE 940469 (1994).

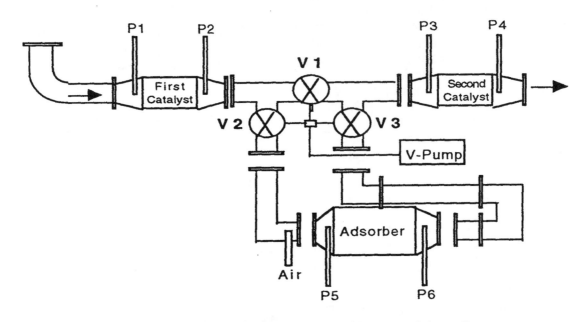

Figure 1: Configuration 1 of By-pass Adsorption System Showing First Catalyst, Zeolite Adsorber, Three Valves, and Second Catalyst.

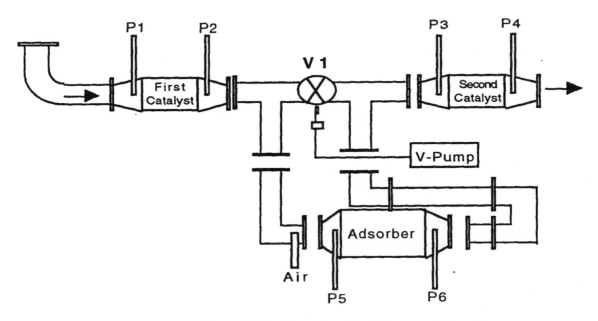

Figure 2: Configuration 2 of By-pass Adsorption System Showing First Catalyst, Zeolite Adsorber, Single Valve, and Second Catalyst.

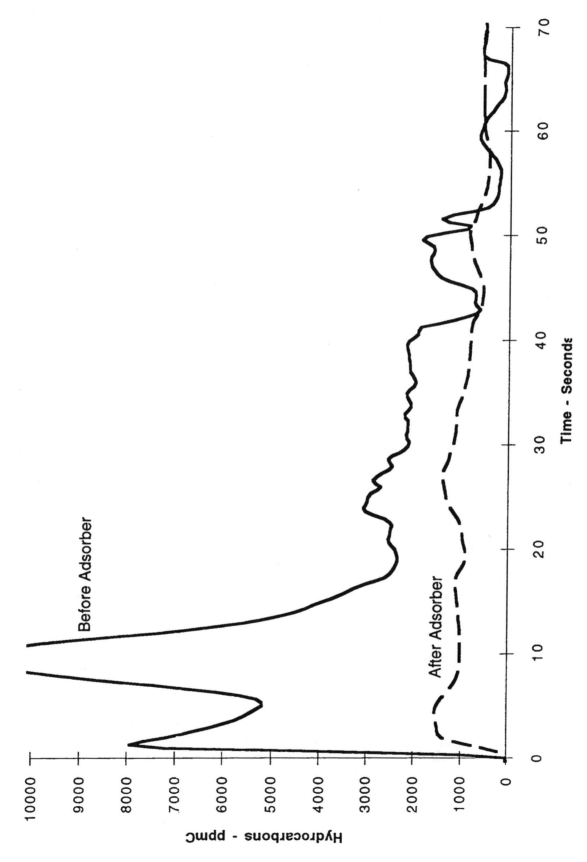

Figure 3: Continuous HC Emissions (ppmC) for 0-70 Seconds Before and After the Zeolite Adsorber, 60% HC Adsorption.

431

Figure 4: Bag 1A speciation analysis as per cent HC adsorption vs. carbon number for aged zeolite adsorber system, D1.

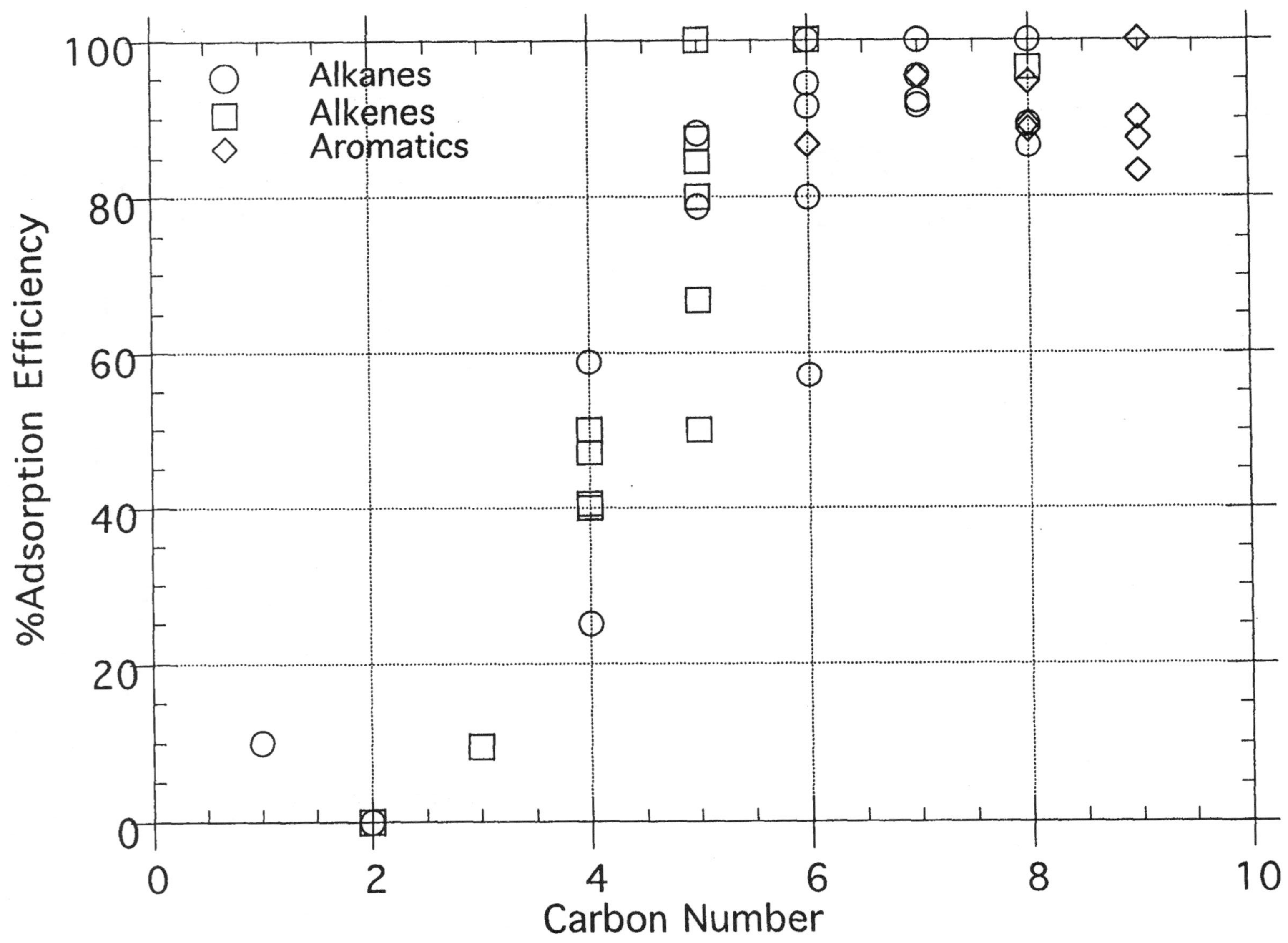

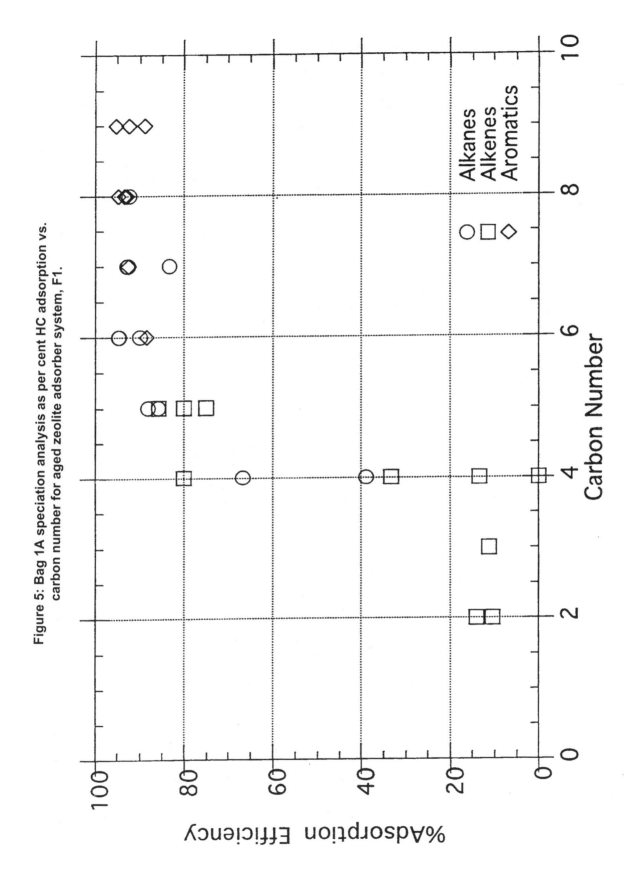

Figure 5: Bag 1A speciation analysis as per cent HC adsorption vs. carbon number for aged zeolite adsorber system, F1.

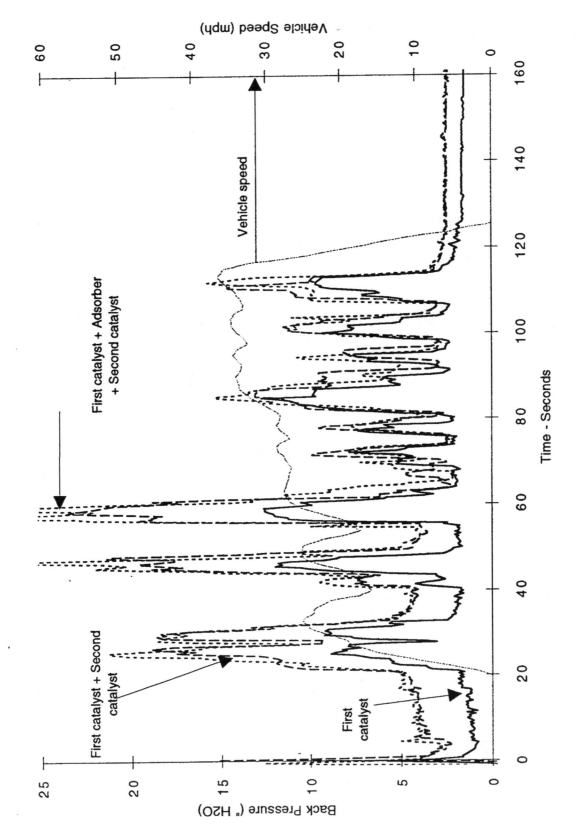

Figure 6: Back pressure measurements for the first catalyst with and without second catalyst and adsorbers in by-pass loop. The back pressure was measured during the initial 160 seconds (bag 1) of the FTP cycle.

Hydrocarbon Adsorber System for Cold Start Emissions

W. Hertl, M.D. Patil, and J.L. Williams
Corning, Inc.

Copyright 1996 Society of Automotive Engineers, Inc.

ABSTRACT

A new adsorber concept has been tested. A zeolite adsorber with a central hole is mounted below the first catalyst, with a second catalyst downstream. During the cold start, when the adsorber is cool and the HC concentration high, HCs are adsorbed from the gas fraction passing through the channels. The small fraction of exhaust gas passing through the hole impinges directly on and heats the second catalyst. The rationale is to design the hole to maximize the second catalyst heating rate, minimize desorption during heat-up and simultaneously keep the HCs which pass through the hole at an acceptably low level. FTP test results on the 3.8 L engine give 0.081 g/mi NMHC with no hole (same as base line) and decrease to 0.056 g/mi NMHC with the hole. This concept exhibits NMHC performance in the LEV range.

INTRODUCTION

A by-pass adsorber system was previously demonstrated [1] which exhibited NMHC emission performance in the California ULEV range. The system included adsorber regeneration within the FTP cycle and required only a single diversion valve within the exhaust line. In the by-pass adsorber system the HC emission was minimized by optimizing the following steps:

a) HC adsorption (trapping) during cold start;

b) selectively heating the second catalyst above its light-off temperature;

c) HC desorption and oxidation with additional air.

Data from the by-pass adsorber system showed that with the particular test vehicle used the %HC adsorption should be above 57% and the oxidation efficiency of the desorbed HCs should be above 85% to give performance within the ULEV range.

For reference, the allowed maximum emission limits for the regulated gases are listed in Table 1.

Table 1: Exhaust Emission Limits (g/mile) at 50,000 miles.

Category	NMHC	CO	NOx
TLEV	0.125	3.4	0.40
LEV	0.075	3.4	0.20
ULEV	0.040	1.7	0.20

The goal was to design and demonstrate a system that did not require any moving mechanical parts within the exhaust system. This report describes the initial testing with this new adsorber concept.

This concept involves using an adsorber with a central hole and a second catalyst downstream. The in-line adsorber consists of wash-coated zeolites on a cordierite honeycomb support. During the cold start, when the adsorber is cool and the hydrocarbon (HC) concentration high, the HCs are adsorbed from the gas fraction passing through the channels. The smaller fraction of exhaust gas passing through the hole impinges directly on and heats the second catalyst. The rationale in the design is to design the hole to maximize the second catalyst heating rate and simultaneously keep the HCs which pass through the hole at an acceptably low level.

EXPERIMENTAL DESIGN

ADSORBERS- Tests were carried out using a combination of medium and large pore size zeolite adsorbers. The medium pore zeolite is capable of adsorbing small to medium size hydrocarbons while the large pore zeolite is more efficient at adsorbing larger size molecules. The zeolites were wash-coated on cordierite cellular 62 cells/cm^2 (400 cells/in^2) substrates and catalyzed with precious metal.

ENGINE AGING- Aging was carried out for 44 hours with an engine dynamometer. The air/fuel ratio of the 7.5 L 8 cylinder engine was adjusted so that the nominal adsorber bed temperature (2.54 cm from front face) was initially 850°C. The catalytic oxidation reaction exotherm was unable to maintain this temperature so that the actual temperature was about 800°C.

An aged 1.7 liter volume first catalyst was used for tests. This consisted of a three way catalyst formulation (TWC) coated on a ceramic substrate with a precious metal loading of 3.2 g/L (90 g/ft^3) with Pt/Rh ratio 5:1. A non-aged 10.6 (300 g/ft^3) Pd second catalyst was used.

TEST VEHICLE- FTP testing was carried out using a 1991 U.S. vehicle with 3.8 liter V-6 engine, electronically controlled port fuel injection and 4-speed automatic transmission. The car was certified to meet 1991 California emission standards. Phase II fuel was used. All FTP tests were performed at Southwest Research Institute (SwRI).

EXHAUST FLOW RATES- Exhaust flow in an automobile varies with vehicle speed, engine size, transmission gear ratio etc. The exhaust flow of the vehicle during the FTP test is shown in Figure 1. This was measured using a laminar flow element fitted to the intake manifold. Also shown in Figure 1 are the vehicle speeds during the early part of the FTP cycle.

SYSTEM DESIGN- A schematic diagram of the adsorption system is shown in Figure 2. Nearest the engine at a distance of 48 cm is the first catalyst. Connected to this via a cone to cone connector is the adsorber with 11.8 cm (4.66") diameter and nominal (no hole) volume = 2.2 liters. Each adsorber had a hole drilled through the center. The cone connected to the adsorber can was equipped with 8 peripheral jets for air injection to provide sufficient oxygen to completely oxidize the desorbed HCs during regeneration. The adsorber can was connected downstream via a cone connector to the second catalyst can. The 0.8 L second catalyst consisted of two 7.62 cm (3") lengths of 7.62 cm diameter honeycombs with 10.6 g/L Pd.

Gas probe positions are also shown in Figure 2. The exhaust gas was sampled for the continuous HC, NOx, CO and O$_2$ concentration monitoring. In the continuous monitoring data differential plots of instantaneous concentrations are given; the concentrations are not linearly related to the gas masses since the volume flows (Figure 1) vary by at least 2X as the engine speed is varied. They do however give a rough qualitative indication of the exhaust gas conditions. Thermocouples were also

mounted at various positions for monitoring the temperatures.

Air injection was provided by a belt driven automotive air injection pump rated at 212 L/min (7.5 scfm). For test purposes this pump was belt driven with a 0.37 kW (0.5 hp) electrical motor rated at 1750 rpm. With a 10 cm diameter pulley approximately 297 L/min (10.5 cfpm) of air could be delivered. A gate valve was used to regulate the desired injected air flow.

INJECTED AIR QUANTITIES- Additional air is required in the exhaust stream during HC desorption (Bag 1B, 66 to 505 seconds) to provide sufficient oxygen to oxidize the desorbed HCs since the exhaust gas composition is nearly stoichiometric. Ibis additional air was injected through jets in the transition cone connected to the adsorber unit. The notion was to also keep the adsorber cool with the secondary air.

Incremental increases of injected air flow, with continuous CO, HC and O$_2$ monitoring in successive runs, showed that 99 L/min (3.5scfm) air was insufficient to completely oxidize the CO spikes during accelerations when the oxygen was completely consumed. CO preferentially reacts with oxygen at less than stoichiometric concentrations so that less HC oxidation occurs. An addition of 127 L/min (4.5scfm) air provided a slight oxygen excess in the desorption phase (Bag 1B).

Figure 3 shows a typical plot of HC concentration exiting the second catalyst when 99 L/min (3.5 scfm) and 127 L/min (4.5 scfm) injected air is used. The HC concentration entering the second catalyst is typically 400-500 ppm. As the plot shows, 200 ppm HC exits the catalyst when 99 L/min (3.5 scfm) air is used but only about 100 ppm HC when 127 L/min (4.5 scfm) air is used. In all the results reported here (Table 2), except the base line runs, 142 L/min (5 scfm) air was used.

TEST PROCEDURE- The Bag 1A time (0-65 sec) and the time for the first catalyst to light off, was determined from the continuous HC and temperature readouts in the base-line runs. The only variable available during the test was the time and volume of injected air, which was 142 L/min in Bag 1B in the later runs.

During the test the HC concentrations in the exhaust are continuously monitored at various points. For the initial 65 seconds (Bag 1A time) HC was generally monitored before and after the adsorber (P2 and P3 in Figure 2). For the remainder of the test the HCs were monitored before and after the second catalyst (P3 and P4 in Figure 2). This is in addition to obtaining the discrete bags for analysis.

FLOW VELOCITY MEASUREMENTS- Flow measurements through the adsorbers were carried out in the laboratory using adsorbers with a range of hole sizes. The adsorber in its can was attached to a 5.08 cm (2") pipe and air was passed through the adsorber using a range of flow rates. Generally 566 to 1132 L/min. (20 to 40 scfm) air is comparable to the exhaust flow in this car during the Bag 1A time. The flow velocities exiting the honeycomb were measured in the central hole and at various points on the exit face. Measurements were made with an Omega Model HHT-610 hot wire anemometer flow meter. These substrates and the ones used in FTP testing had 62 cells/cm^2 (400 cells/in^2) with nominal 70% open frontal area.

RESULTS

The FTP results are summarized in Table 2. Col. 1 = test number; Col. 2 = hole size in the adsorber; col. 3 = Bag 1A HC analysis in g (0-65 sec); col. 4 = % of the emitted HCs which are adsorbed 0-65 seconds; col. 5 = Bag 1B HC analysis in g (66-505 sec); col. 6 = % of the adsorbed HCs which were oxidized; col. 7 and 8 = Bag 2 and Bag 3 HC analyses in grams; col. 9 = FTP test value in g/mi total HC; col. 10 = g/mi NMHC. Table 3 gives the measured FTP values for all the regulated gases.

The values in columns 4 and 6 are calculated using equations (1) and (2) with the Bag 1A and 1B values in columns 3 and 5.

$$\text{Fraction HC Adsorbed} = \frac{HC_{base\ 1\ A} - HC_{test\ 1\ A}}{HC_{base\ 1\ A}} \quad (1)$$

$$\text{Fraction of Adsorbed HC oxidized} = 1 - \frac{HC_{test\ 1\ B} - HC_{base\ 1\ B}}{HC_{base\ 1\ A} - HC_{test\ 1\ A}} \quad (2)$$

The base line values from base-line test 1 were used for these calculations.

FLOW DISTRIBUTION STUDIES- The flow distributions across the exit faces of adsorbers with various hole sizes were measured for a range of air flow rates. Some typical data are plotted in Figure 4 (no hole) and Figure 5 (1.27 cm hole) as the measured air velocity against the position on the exit face.

Comparison of the data in the plots shows that at the higher flow rates with no hole there is only a slight preference for higher flow through the center. With the center perforation there is a high preferential flow rate through the hole due to the much lower back pressure compared with the honeycomb channels. Note that this is raw velocity data. While the velocities in the hole are about 4X to 8X higher than in the peripheral regions, the 1.27 cm hole only represents about 1.7% of the total open area through the substrate.

The rationale for using this concept is to use the relatively low volume/ high velocity exhaust flow passing through the hole with little thermal loss to heat up the second catalyst. Ideally the second catalyst would attain its light off temperature when the remainder of the substrate reaches a temperature to desorb the HCs (ca. 100-200°C).

FTP TESTS- The base-line FTP values (no adsorber) are given in Table 1, tests 1 and 2. The continuous HC and temperature data were used to assist in determining the hole size effects. For this concept to work, it is necessary to:

a. initially adsorb a large fraction of the HCs

b. minimize the second catalyst heating time so that it has attained light off temperature by the time the HCs desorb.

c. maximize retention of adsorbed HC while the second catalyst is heating up.

The rationale for dividing Bag 1 exhaust gas for analysis into Bag 1A (0-65 sec) and Bag 1B (56-505 sec) was to assist in analyzing the data. The HCs in Bag 1A compared to the base line is a measure of the adsorption efficiency; these values can be compared with the continuous HC readouts monitored from probes P2 and P3 (Figure 2). The HCs in Bag 1B, after the first catalyst lights off, when corrected for the base line HC, is a measure of the oxidation efficiency of the desorbed HCs and can be compared with the continuous HC readouts monitored from probes P3 and P4 (Figure 2).

The %HC adsorbed in Bag 1A time (Table 1, col. 4) shows a trend for larger hole sizes to give smaller adsorption efficiencies. With no hole (test 3), adsorption efficiencies were typically near 60%, as with the by pass adsorber system; this drops to 52% with a 2.54 cm hole (test 5) and 33% with a 5.1 cm hole (test 7). The efficiency drop is due to two discrete factors. First, with larger hole sizes, a larger fraction of the exhaust gas passes through the hole (cf. flow distribution plots in Figures 4 and 5) and thus cannot be adsorbed. Second, with increasing hole size the total available adsorption volume of zeolite decreases from 2.18 L (no hole) to 2.08 L with a 2.54 cm hole.

Figure 6 shows some typical temperature traces measured at the second catalyst inlet when using adsorbers with a range of hole sizes. With no adsorber hole the catalyst heating rate is very slow (e.g. 170 sec to reach 250°C); with, an adsorber hole the catalyst heating rate is much faster (ca. 110 sec. to reach 250°C). The catalyst heating rates do not vary much with the hole size, at least in the range used here (1.27, 1.91, 2.54 cm).

The third important factor is to minimize HC desorption during the second catalyst heat-up. Figure 7 gives plots of the continuous HC readouts (measured between the adsorber and second catalyst) against time. While the hole size, as described above, had little effect on the heating rate, Figure 7 clearly shows that the hole size does affect HC desorption. The peak desorption occurs at about 105 seconds with no hole, and increases in time with increasing hole size to 180 sec with a 2.54 cm hole. The longer the desorption can be delayed, the better will be the oxidation.

The salient results of %HC adsorbed the per cent of the adsorbed HC which is catalytically oxidized and the FTP test values are summarized in Table 2 (columns 4, 6 and 10). The %HC adsorbed drops from 61% (no hole) to 33% (5.08 cm, hole). The per cent of this adsorbed HC which is oxidized increases with hole size from 0% with no hole to 35% with a 2.54cm hole and 73% with a 5.08 cm hole. In agreement with these values, the FTP test results give 0.081 g/mi NMHC with no hole (same as the base line value) and decrease to about 0.056 g/mi NMHC with 1.91, 2.54 and 5.08 cm holes.

This in-line adsorber configuration exhibits NMHC emission performance in the LEV range. Testing this configuration ceased at this point and the studies shifted to using fluidic diversion of the exhaust gas away from the hole during the Bag 1A collection time [2].

ACKNOWLEDGEMENT: We gratefully acknowledge the skillful supervision of the FTP testing by Melvin N. Ingalls and Janet Cleary of Southwest Research Institute, and of Elizabeth Wheeler and Thomas F. Rosenbusch of corning, Inc for assistance in preparing materials and carrying out laboratory measurements.

REFERENCES

1. J.L. Williams, M.D. Patil, an W. Hertl, "By-pass Hydrocarbon Adsorber System for ULEV," SAE 960343, (1996).
2. M.D. Patil, W. Hertl, J.L. Williams, and J.N. Nagel, "In-line Hydrocarbon Adsorber System for ULEV," SAE 960348 (1996).

Table 2: Hydrocarbon Adsorber System FTP Summary (1991 U.S. vehicle, 3.8L engine)

COL. 1 TEST NUMBER	COL. 2 ADSORBER HOLE SIZE	COL. 3 BAG 1A	COL. 4 %ADS. Bag 1A	COL. 5 BAG1B	COL. 6 % Oxidation of Adsorbed HC	COL. 7 BAG 2	COL. 8 BAG 3	COL. 9 FTP COMOSTE (TOTAL HC)	COL. 10 NMHC (g/mi)
1	First catalyst (BASE LINE) No adsorber	1.357	-	0.023	-	0.065	0.098	0.096	0.085
2	First catalyst + Second catalyst No adsorber	1.400	-	0.017	-	0.066	0.092	0.098	0.081
3	No Hole	0.526	61%	1.02		0.000	0.07	0.094	0.079
4	1.91 cm Hole	0.581	57%	0.575	28.9	0.041	0.055	0.077	0.058
5	2.54 cm Hole	0.651	52%	0.48	35.3	0.043	0.06	0.076	0.056
6	2.54 cm Hole, with second catalyst moved 5 cm closer to adsorber	0.570	58%	0.702	13.7	0.036	0.05	0.082	0.068
7	5.08 cm Hole in 14.4 cm diameter adsorber	0.910	33%	0.145	72.7	0.024	0.059	0.068	0.058

TABLE 3: FTP RESULTS (g/mi) for REGULATED EMISSIONS

TEST NUMBER	SYSTEM	NMHC	CO	NOx
1	Base line (First catalyst)	0.085	0.962	0.110
2	Base-line with second catalyst	0.081	0.937	0.083
3	No Hole	0.079	0.751	0.124
4	1.9 cm hole	0.058	0.462	0.100
5	2.5 cm hole	0.056	0.442	0.092
6	2.5 cm hole, ads moved 5.1cm closer to adsorber	0.068	0.634	0.082
7	5.1 cm hole (14.4 cm diameter)	0.058	0.640	0.111

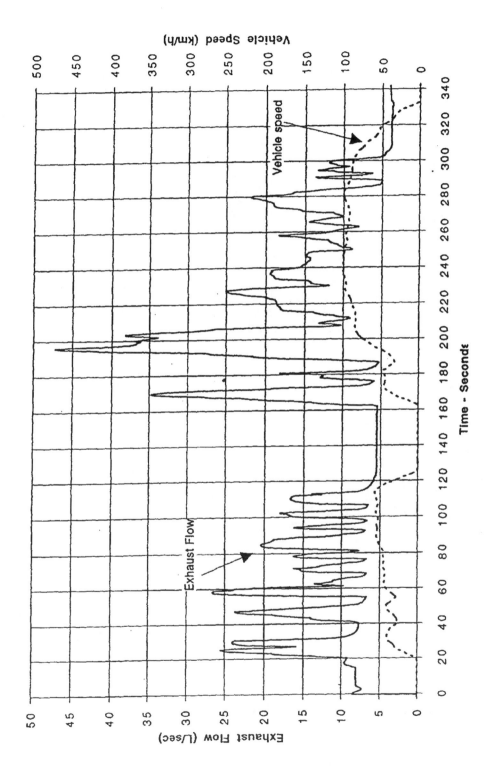

Figure 1: Exhaust flow velocities and vehicle speed duringearly part of FTP cycle

441

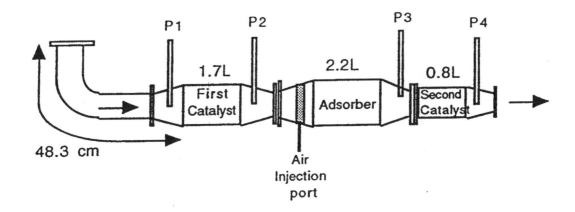

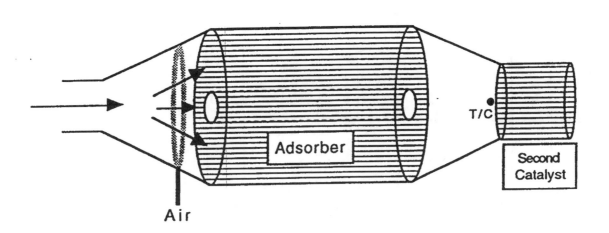

Figure 2: Schematic diagram of (top) hydrocarbon adsorber system and (bottom) on enlarged view of the adsorber and second catalyst. P1, P2, P3 and P4 are gas sampling ports for obtaining continuous HC, CO and O2 concentrations. T/C is second inlet thermocouple

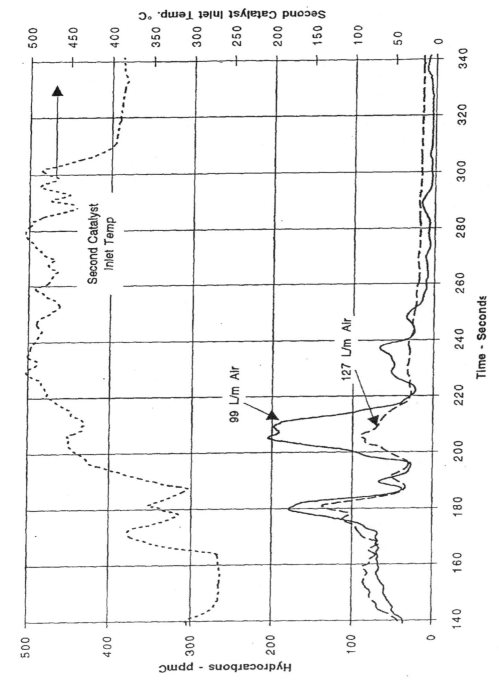

Figure 3: HC concentrations exiting the second catalystwith 99 L/m (3.5 scfm) and 127 L/m (4.5 scfm) injected air. The gas was sampled at probe P4. The second catalyst inlet temperature (upper plot) is also plotted.

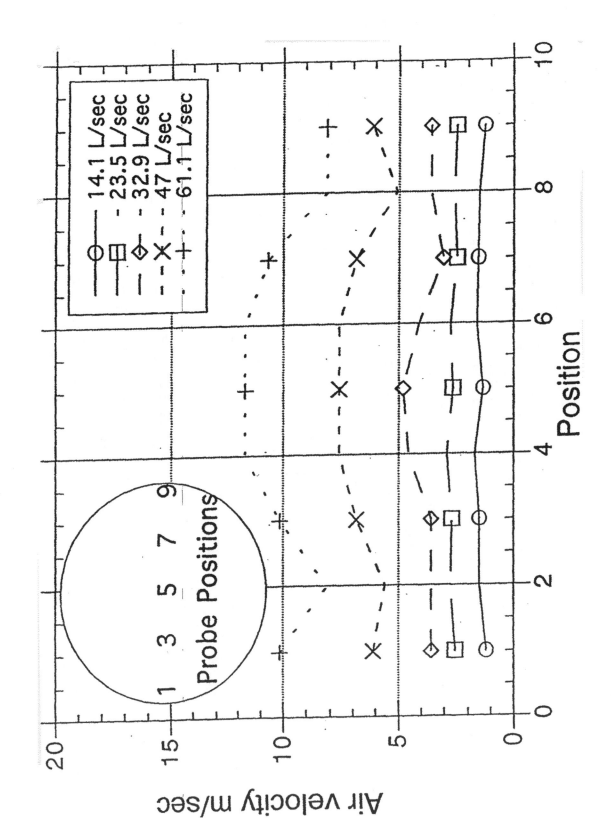

Figure 4: Flow Profile for 11.8 cm Substrate Without Hole for Five Different Inlet Air Flow Rates

Figure 5: Flow Profile for 11.8 cm Substrate With
1.27cm Hole for Five Differnet Inlet Air Flow Rates

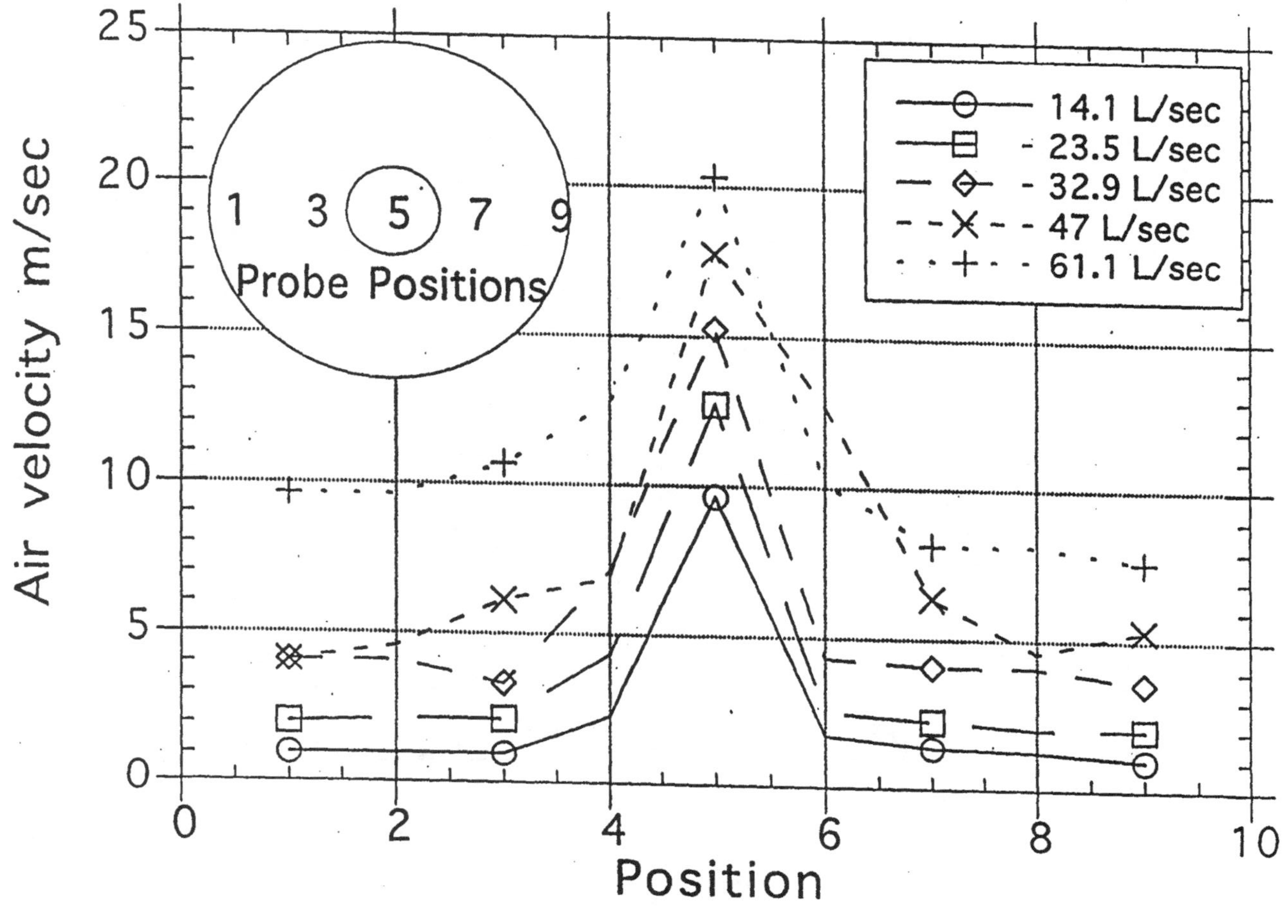

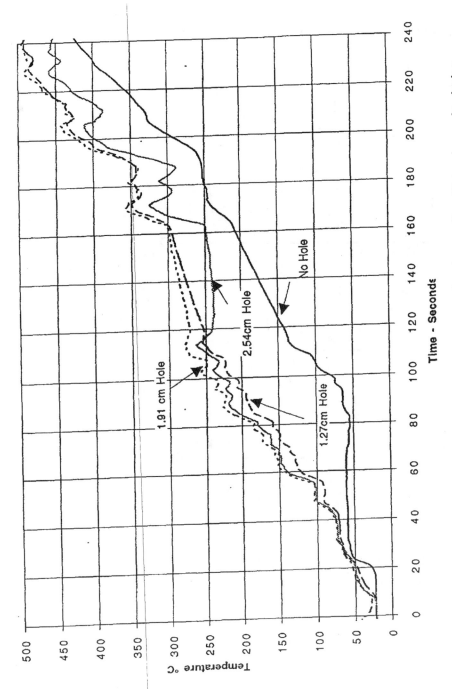

Figure 6: Second catalyst inlet temperatures for adsorber units with various size holes.

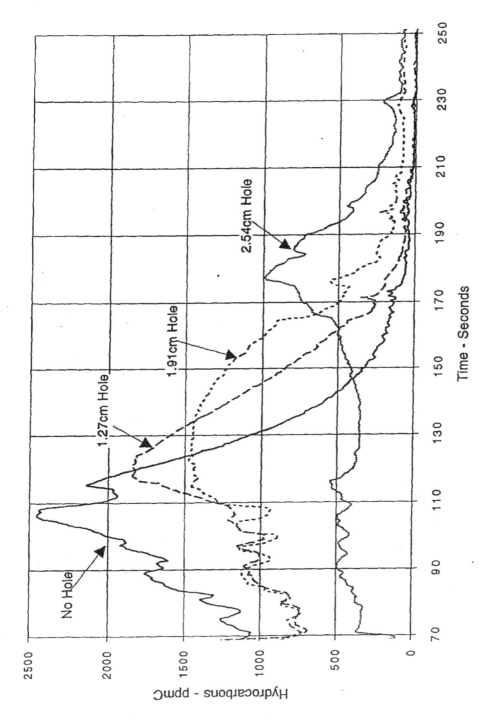

Figure 7: Continuous HC emissions sampled betweenadsorber with various hole sizes and second catalyst (probe P3).

Development of the Automotive Exhaust Hydrocarbon Adsorbent

Takaaki Kanazawa and Kazuhiro Sakurai
Toyota Motor Corp.

Copyright © 2001 Society of Automotive Engineers, Inc.

ABSTRACT

The hydrocarbon adsorption volume character of zeolite was studied. Specifically, the relationship between aluminum content and zeolite hydrocarbon adsorption was investigated, as a potential hydrocarbon adsorbent for exhaust gas. The study also analyzed the relationship between hole diameter and zeolite hydrocarbon adsorption. It was found that hydrocarbon adsorption increased with decreasing aluminum content. Zeolite with a pore size approximately 0.1nm greater than the diameter of hydrocarbon molecules showed the best performance. Zeolites with two different pore sizes were mixed, and succeeded in adsorbing hydrocarbons of carbon number 3 and above. Silver (Ag) ion exchanged zeolite was also used to increase the adsorption of exhaust gas hydrocarbons, including those of carbon number 2.

INTRODUCTION

Emission regulations are becoming increasingly stricter every year, and so, it is necessary to enhance the performance of 3-way catalysts. It is important, in particular, to develop a high-performance catalyst that can purify even at low temperatures. Conventional measures to cope with the regulations have consisted of enhancing the performance of catalysts and improving the engine (1). However, these approaches are insufficient to meet the recent SULEV and PZEV regulations. To satisfy the requirements imposed by these regulations, it is necessary to remove the hydrocarbons (HCs) exhausted immediately after engine startup, when the temperature has not yet reached the catalyst's activation level. It is, therefore, necessary to be able to adsorb HCs at a temperature below the catalyst's activation level, and then purify them with 3-way catalysts as they desorb with increasing temperature.

EXHAUST GAS HYDROCARBON COMPONENTS AND ADSORBENTS

Figure 1 shows a typical Hydrocarbon (HC) adsorber and catalyst system. The system will temporarily adsorb HCs that form just after engine startup. The gas flows into the catalyst only via the adsorbent immediately after engine

startup, when the direct access valve is closed. The valve is opened to allow purification of the detached HCs once the catalyst, located upstream of the HC adsorber, reaches activation temperature.

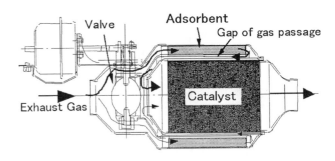

Figure 1 Structure of HC Adsorber

HCs in exhaust gases are comprised of paraffin, olefin, and aromatics. Each of these components contains HCs of various sizes, ranging from C1 to C11 (Figure 2). Effective HC adsorbents must adsorb all these HC sizes. Currently, there are several adsorbents, such as active carbon and zeolite, available. For this study, zeolite was selected due to its excellent heat resistance.

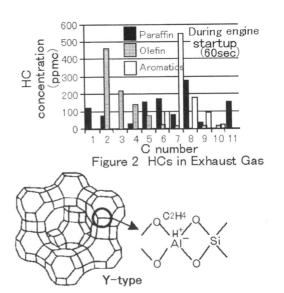

Figure 2 HCs in Exhaust Gas

Figure 3 Chemical Class to Zeolite and Al Section

Zeolite is a crystalline SiO_2 porous substance (2). It occurs in various forms, including ferrierite, ZSM5, mordenite and Y type. Zeolite exhibits physical adsorption capabilities, which trap HCs within its pores (3). It also exhibits chemical adsorption capabilities(4)(5), at points of unbalanced electric charges resulting from the addition of Al (Figure 3), which can bond the HCs. Factors affecting HC adsorption, include pore size, water content and Al quantity, were studied.

EXPERIMENTAL

Basic testing was preformed with the model gas device shown in Figure 4. A variety of HC adsorbents were prepared, each containing pellets 0.5-1.7mm in diameter. To determine the HC adsorption amount, 3000 ppmc HC gases were flowed through the apparatus at room temperature, and the HC concentrations of the inlet and outlet gases were measured. Gas flow rate is $10dm^3/min$ and Absorbent weight is 2.0g. Water vapor is added to the reactor by water vaporizer. The mole number of HC adsorbed per gram of adsorbent for each of the samples was compared (Figure 5).

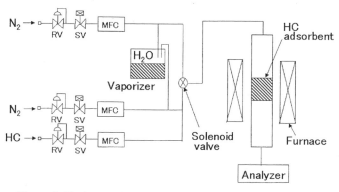

Figure 4 Equipment for model gas adsorbent evaluation

Pellet adsorbent
Adsorbent weight : 2.0g
Gas flow rate : $10dm^3/min$

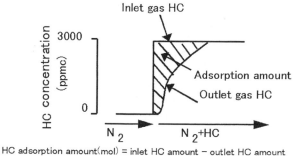

HC adsorption amount(mol) = inlet HC amount − outlet HC amount

Figure 5 HC Adsorption Measurement Method

In engine bench evaluations, a $0.7dm^3$ monolithic adsorbent was mounted to a $2.3dm^3$ engine. The monoliths ($0.7dm^3$ with 400cells/in^2) were prepared with washcoats of ZSM5 and Y-type(and ferrierite).The coat volume to monolithic is $200g/dm^3$.The performance of the adsorbent was evaluated 40-60 sec after engine startup(First Idle). The component of HC gas was analyzed by the gas chromatography. HC adsorption rate was determined as the ratio of inlet HC amount and outlet HC amount.

IMPROVEMENT OF PHYSICAL HC ADSORPTION CHARACTERISTICS

EFFECT OF WATER CONTENT

To study the effects of water content on performance, the C3 adsorption rates of ZSM5, both in a dry state and a 3 vol% water vapor atmosphere, were compared (Figure 6). Zeolite HC adsorption is known to be mostly of olefin type. Adsorption was found to be considerably restricted when water vapor existed, and adsorption of olefin and paraffin were at the same levels. Though olefin is known to adsorb onto the Al section, Fig 6 shows that, adsorption on the Al did not occurred in the presence of water vapor.

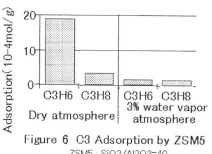

Figure 6 C3 Adsorption by ZSM5
ZSM5 : SiO2/Al2O3=40

Figures 7 and 8 show the change in adsorption of ZSM5 with varied Al contents. HC adsorption increases with increasing Al content in the dry atmosphere, and increases with decreasing Al content in the water vapor atmosphere. This tendency was also observed with mordenite and Y-type zeolites of a different pore sizes(Figure 9,10). These results suggest that, in the absence of H_2O, HCs adsorb on the Al section, while if H_2O is present it displaces the adsorption of HC on the Al section. Based on these findings, because automobile exhaust gas contains water, Al-free zeolite was selected for investigation as a potential exhaust gas purifier.

All discussions reported hereafter refer to analyses carried out in a water vapor atmosphere.

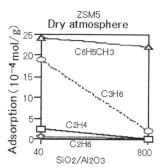

Figure 7 Al Content and Adsorption of Various HCs in Dry Atmosphere

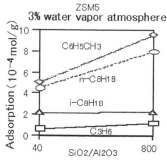

Figure 8 Al Content and Adsorption of Various HCs in Water Vapor Atmosphere

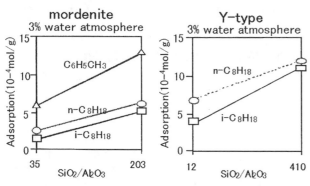

Figure 9 Al Content and Adsorption of Various HCs in Water Vapor Atmosphere mordenite

Figure 10 Al Content and Adsorption of Various HCs in Water Vapor Atmosphere Y-type

EFFECT OF PORE DIAMETER

The effect of HC size and zeolite pore size on adsorption was studied, using C_3H_6 and $C_6H_5CH_3$ gases, which occur in large quantities within one minute after engine startup. The results are shown in Figure 11. The figure shows that zeolites of any pore size adsorb HCs, and the pore size with the maximum adsorption is approximately 0.1nm greater than HC molecular diameter.

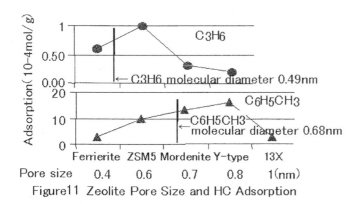

Figure11 Zeolite Pore Size and HC Adsorption

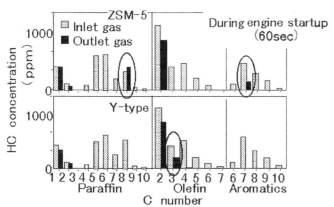

Figure12 HC Adsorption by ZSM5 and Y-type Zeolite in Actual Exhaust Gas

ZSM5 : SiO2/Al2O3=1900
Y-type : SiO2/Al2O3=400

The adsorption rate of various engine exhaust gas HCs was studied using ZSM5 and Y-type zeolites. The results are shown in Figure 12. ZSM5 was found to be inferior to the Y-type in adsorbing C8 paraffin, C7 aromatics, and other higher HCs (circled in the figure). The Y-type, on the other hand, could not adsorb C3 olefin. These results show that ZSM5 is effective for lower, C3 to C5, HCs, and Y-type is effective for higher, C6 and above, HCs. Neither adsorbent is effective for C2 and lower.

HEAT RESISTANCE

To examine the heat resistance of zeolite, HC adsorption rates were measured after varied heat treatments in 10% water vapor. The results are shown in Figure 13. Adsorption did not decrease for either ZSM5 or Y-type samples treated up to 1100 degC. This confirms that both ZSM5 and Y-type will resist the heat of automobile exhaust gas.

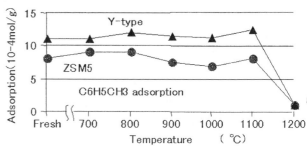

Figure13 HC Adsorption after durability Test

IMPROVEMENT OF CHEMICAL HC ADSORPTION CHARACTERISTICS

C_2H_4, which is a C2 olefin occurring in relatively large quantities in exhaust gases, is barely adsorbed by ZSM5, as shown in Figure 12. This is because the molecules are too small. A method to chemically adsorb C_2H_4 on Ag and Pd was investigated. Ag and Pd were added to ZSM5 and Al2O3, and C_2H_4 adsorption levels were compared. The results are shown in Figure 14. As the figure indicates, Ag and Pd impregnated Al2O3 did not adsorb HCs, and zeolite is the better Ag and Pd carrier. The behavior of Ag was examined via X-ray photoelectron spectrometry (XPS), and is shown in Figure 15. Ag impregnated in ZSM5 is close to being in an oxide state, while that in Al_2O_3 is metallic. This suggests that oxygen plays role in the chemical adsorption of olefin in Ag.

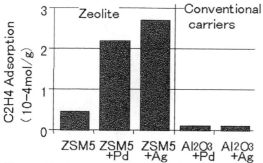

Figure 14 Improved C2H4 Adsorption through Adding Pd and Ag

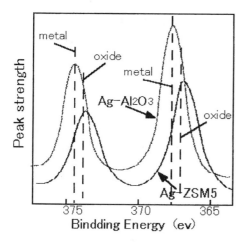

Figure15 Ag State (XPS)

The olefin adsorption rates, before and after Ag impregnation, were studied via adsorbing IR spectrum on C₃H₆. The results, in Figure 16, show that before Ag impregnation, CH₂ deformation vibration was minimal. For ZSM5 with Ag, C-O stretching vibration was observed in addition to CH₂ deformation vibration. This suggests that O and C in the HC are bonded together. The theorized C₃H₆ adsorbed structure on Ag is shown in Figure 17. This is based on the condition of the Ag revealed in the XPS analysis, and from the IR analysis results.

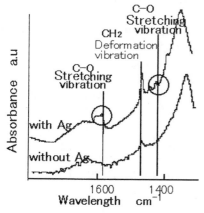

Figure 16 C3H6 Adsorption IR before and after Ag Addition (100 degC)

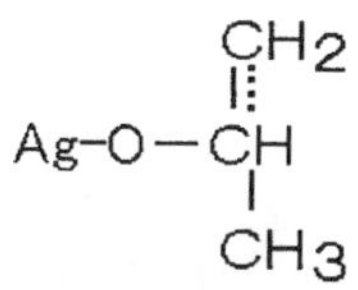

$$Ag-O-\underset{\underset{CH_3}{|}}{\overset{\overset{CH_2}{\|}}{CH}}$$

Figure 17 Chemical Adsorption of C3H6 by Ag

Figure 18 shows the relationship between Ag content and C₂H₄ adsorption for Ag impregnated ZSM5 (without Al). Initially, C₂H₄ adsorption increased with increasing Ag content. However, during durability testing, Ag's presence showed no effect on adsorption. This may due to the lack of Ag-zeolite bonding, allowing elemental cohesion of the Ag during heating, thus neutralizing its influence. In order to address this situation, an attempt was made to stabilize the Ag by attaching it to the Al-containing zeolite. This was done via ion-exchange, (see Figure 19) and fixing by electric charge.

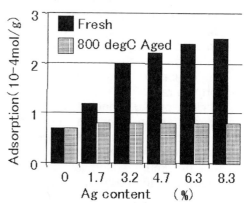

Figure18 C2H4 Adsorption by Ag Carrying ZSM5
ZSM5 : SiO2/Al2O3=1900

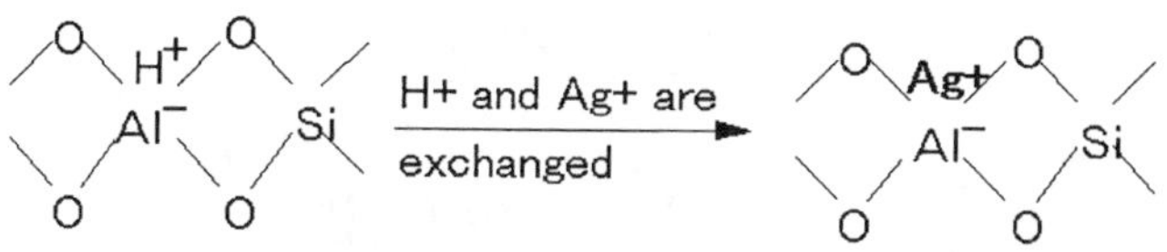

Figure 19 Ion-exchanged Ag

Ferrierite as a C₂H₄ adsorbent was also evaluated. This is a zeolite structure with ion exchange sites and 0.4nm pore size (Figure 20). Ion-exchanged Ag, on the Al section of zeolite, was found to be particularly thermally stable if the base is ferrierite. For this investigation, the Al addition was determined to be 3×10^{-4} mol/g for the ferrierite, and 7×10^{-4} mol/g for the ZSM5. These quantities coincide with the molecular weight of the adsorbed C₂H₄, verifying the feasibility of equal-mole adsorption by the ion exchanged Ag on the Al section.

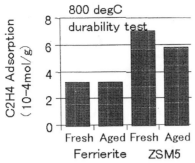

Figure 20   C2H4 Adsorption by Ag
Ion Exchanged Zeolite

Ferrierite : SiO2/Al2O3=62
ZSM5 : SiO2/Al2O3=40

CONFIRMATION WITH ENGINE EXHAUST GAS

HC adsorptions for ZSM5 and Y-type mixed zeolites were measured in order to optimize the mixing ratio. The highest adsorption was recorded for a mixing ratio of 3 : 1 (Figure 21). The adsorption did not decrease when the engine exhaust temperature reached 800 degC or after 100hr. durability testing. To adsorb C2 olefin that cannot be adsorbed by a ZSM5 and Y mixture, a 3-type mixed adsorbent was prepared. The 3-type adsorbent, containing Ag-ion exchanged ferrierite, was evaluated for HC adsorption rate. The results are shown in Figure 22. The addition of Ag enabled C2 adsorption. This improved adsorption rate was confirmed using actual vehicle evaluations. The adsorption rate did not fall when the adsorption cylinder temperature reached 500 degC (the maximum adsorption cylinder durability test temperature) or after 100hr durability testing (Figure 23).

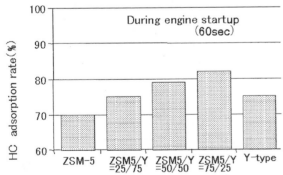

Figure 21   ZSM-5 and Y-type Combined Study

ZSM5 : SiO2/Al2O3=1900
Y-type : SiO2/Al2O3=400

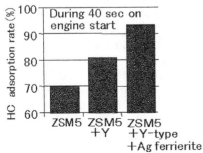

Figure 22 Effect of Adding Ag Zeolite

ZSM5 : SiO2/Al2O3=1900
Y-type : SiO2/Al2O3=400
Ferrierite : SiO2/Al2O3=62

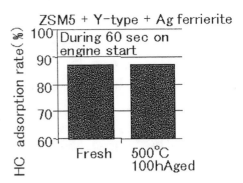

Figure 23 Ag Heat Resistance
in Exhaust Gas

In May 2000, the ZSM5 + Y-type zeolite adsorbent was applied in HC adsorbing cylinders in the US production Prius to meet the SULEV regulation.

CONCLUSION

(1) The factors for improving the physical adsorption characteristics of zeolite were identified. The effect of zeolite pore size and Al content were clarified.

- It is easiest to adsorb HCs of a molecular diameter that is equal to the zeolite pore size plus 0.1nm.

- The Al content in zeolite prevents HC adsorption in a water vapor atmosphere.

(2) HC adsorbents (ZSM5 and ferrierite) containing ion exchanged Ag, to improve chemical adsorption characteristics, are particularly thermally stable, and adsorb C2 olefin.

REFERENCES

1. M.Shinichi " Recent advances in automobile exhaust catalyst" Catalysis Surveys from japan 1(1997)111-117
2. Adsorption Technology Handbook, P643, N.T.S. (1999)
3. S.Namba and T.Yashima "Separation of p-isomers from disubstituted benzenes by means of shape-selective adsorption on mordenite and ZSM5 zeolites"Zeolites,4,77(1984)
4. H.Herden, W.D.Einicke,M.Jusec,U.Messow and R.Scrollner "Adsorption studies of n-olefin/n-paraffin mixtures on X- and Y-zeolotes" J.Colloid Int.Sci.,97,559(1984)
5. N.Y.Chen "Hydrophobic properties of zeolites" J.Phys.Chem.,80(1),64(1997)

CHAPTER 10

CATALYTIC CONVERTER SYSTEM MODELING

Tariq Shamim
The University of Michigan-Dearborn

10.1 OVERVIEW

Due to progressively stricter emission regulations, catalytic converter design and performance need to undergo continuous modification. In the past, much of the design and engineering process to optimize various components of engine and emission systems has involved prototype testing. The complexity of modern systems and the resulting flow dynamics and thermal and chemical mechanisms have increased the difficulty in assessing and optimizing system operation. Due to overall complexity and increased costs associated with these factors, modeling continues to be pursued as a method of obtaining valuable information supporting the design and development process associated with optimizing the exhaust emission system.

Mathematical models of catalytic converters have been employed for nearly 30 years. Modeling studies have generated new insight on catalyst performance during light-off, steady-state and transient driving conditions. Due to its various advantages, the use of modeling in designing and optimizing the catalytic converter system is now becoming a common practice. There are different approaches to simulate the phenomena of exhaust systems. These modeling approaches may be classified into two main categories: single-channel-based one-dimensional modeling, and multi-dimensional modeling of the entire catalytic monolith.

The single-channel-based 1-D modeling approach has been the most common and popular. It offers simplified, less computationally intensive 1-D handling and practically equivalent accuracy levels. In this approach, the non-uniform flow distribution at the monolith's face is generally neglected in order to simplify the mathematical model and for computational expediency. The number and type of species considered in different studies vary. However, most studies consider CO, NO and HC. Various hydrocarbon species in engine exhaust are generally categorized into two types: fast burning and slow burning. Some studies also consider an additional category of inert hydrocarbon.

Due to strong coupling of the individual channels of the monolith through heat transfer, and the inherent non-uniformity in flow distribution, the adequacy of single-channel-based 1-D models in designing and optimizing full-scale catalytic converters has been questioned. The flow maldistribution is due to the fact that the core area of the frontal face of a brick receives more than 50% of the emerging flow with higher velocity, and the edge of the brick only gets 5% to 10% of emerging flow. The effect is influenced by various parameters including inlet pipe bend, flow Reynolds number, brick resistance, and inlet velocity profile. This effect is even more significant over a driving cycle with aged catalysts.

While the importance of three-dimensional effects in catalyst modeling is well recognized, there are still a relatively small number of sophisticated models for simulating the entire catalytic monolith using simultaneously detailed models of transport and chemistry in each single channel. A major difficulty in three-dimensional modeling of a full-scale catalytic converter is the large amount of computation, which may take several weeks on a high-end processor. To circumvent this difficulty, many studies employ a mixed dimensional approach.

This approach exploits the fact that monolith channels are coupled to each other through heat transfer only, and thus it is possible to link one dimensional transport-reaction models within the monolith channels with three dimensional heat conduction models for the entire converter using prescribed heat and mass transfer coefficients to attain the desired coupling. This approach has been successful in modeling the global three-dimensional effects and predicting the effect of temperature non-uniformity on reaction kinetics and conversion efficiency. In another recent approach, the entire catalytic monolith is treated as an anisotropic porous medium, and sub-grid scale models are employed to represent the heterogeneous chemical reactions occurring at the solid-fluid interfaces within the monolith. This low computational cost approach allows full coupling between fluid flow, heat transfer, species transport, and heterogeneous chemical reactions through flux balance of species and energy at the solid-fluid interfaces.

An important component of all modeling studies is the simulation of catalyst chemical reactions. Various chemical reactions that occur in a catalyst may be classified into two categories: homogeneous reactions in a gas phase and heterogeneous reactions on the monolith surface. Due to low temperatures, the effect of homogeneous reactions in a gas phase is very small and is generally neglected. The heterogeneous chemical reactions, which mainly determine the catalyst performance, are generally modeled based on the classical work of Voltz et al. This work involved the measurement of pellet-type Pt catalyst performances and derived kinetic reaction rate expressions for the oxidation of CO and C_3H_6 under lean conditions. The expressions are of the Langmuir-Hinshelwood type and take into account the inhibition due to NO. These rate expressions with some modifications have been used for non-Pt catalysts due to lack of data for other types of catalysts.

Several reaction mechanisms appear in the literature that vary in details, accuracy and the convenience of use. Most mechanisms are based on global reaction rates in which adsorption, desorption and kinetic effects are lumped together. An alternate approach is the description of the chemical reactions by a set of elementary reaction steps. The reaction equations of the elementary steps describe the reactions on a molecular level, so the approach is much more accurate than a globally fitted kinetic. The main advantage of these detailed reaction mechanisms is their potential to predict the behavior of the chemical system at different external conditions. The disadvantage of using elementary chemical reactions is the large number of reaction equations, which demand a large computational capacity. Furthermore, the rate coefficients of all the single steps have to be known.

In addition to the pathways, which specify the kinetics over the noble metal sites, an additional kinetic mechanism is required to represent the oxygen storage capacity (OSC). The OSC is developed in the modern catalyst to improve catalyst performance under non-stoichiometric operating conditions. It allows storing the extra oxygen under fuel lean conditions and releasing it under rich conditions. The released oxygen may participate in the reactions with the reducing agents, thereby increasing the conversion of CO and HC in a rich exhaust-gas environment. The OSC is recognized as an important mechanism affecting catalyst behavior during vehicle acceleration and deceleration. There are various OSC mechanisms presented in literature. In a simple model, the oxygen storage mechanism is described by the oxidation and reduction of the cerium oxides present in the washcoat.

The transport mechanisms of heat and mass transfer from the exhaust gases to the catalyst surface also strongly influence the catalyst performance. The convective heat transfer between the exhaust gas and the catalyst substrate, and the heat generated during exothermic reactions are the main contributors to the heat transfer mechanisms in the catalytic converter. Due to low catalyst operating temperatures, the radiation heat exchange between the substrate and the surrounding walls is generally neglected. Most studies also include the effect of convective heat loss from the converter surface to the ambient. The mass transfer in the catalyst channels is due to the concentration gradients between the exhaust gas and the reactive washcoat. Most studies model the convective heat and mass transfer processes by using the simplified one-dimensional film model. In the film model, the dimensionless Nusselt (Nu) and Sherwood (Sh) numbers describe the heat and mass transfer rates. Without being very sophisticated, the film model is a good compromise to obtain a reasonable prediction of the converter efficiency. However, many uncertainties remain about the estimate of Nu and Sh numbers, which affect the accuracy of the modeling studies.

The effectiveness and predictive capabilities of catalyst models can be greatly improved if models can predict the catalyst behavior under dynamic operating conditions. The dynamic conditions, which occur during typical driving conditions as a result of the response lag of the engine air/fuel ratio control system, and acceleration and deceleration, make the catalyst behavior differ significantly from that under steady-state conditions. The modeling of catalyst dynamic behavior requires the consideration of all highly transient phenomena occurring in a catalytic converter, in that it differs from converter models that are developed for predicting cumulative emissions during legislated driving cycles and are generally quasi-steady. Some important transient phenomena that should be included in dynamic modeling include:

- The oxygen storage and release phenomena in the washcoat.
- Accumulation effects in the conservation of mass, species and energy.
- Water gas shift and steam reforming effects.
- Transient effects on reaction kinetics, adsorption and desorption.

Among these, the first three can easily be incorporated in the model. However, the inclusion of the last item is more challenging since it requires the development of transient kinetic, adsorption and desorption rates. The existing kinetic rates, which typically include the lumping effects of chemical kinetics, adsorption and desorption, are generally derived from steady or quasi-steady measurements. This limits their range of application in truly transient conditions.

However, the existing lumped kinetic rates may be used with caution to gain qualitative understanding of catalyst dynamic behavior.

10.2 SELECTED SAE TECHNICAL PAPERS

A Three-Dimensional Model for the Analysis of Transient Thermal and Conversion Characteristics of Monolithic Catalytic Converters

David K. S. Chen
ACRO Service Corp.
Livonia, MI

Edward J. Bissett
Mathematics Dept.
General Motors Research Laboratories
Warren, MI

Se H. Oh
Physical Chemistry Dept.
General Motors Research Laboratories
Warren, MI

David L. Van Ostrom
AC Spark Plug Div.
Flint, MI

ABSTRACT

A transient three-dimensional model has been developed to simulate the thermal and conversion characteristics of nonadiabatic monolithic converters operating under flow maldistribution conditions. The model accounts for convective heat and mass transport, gas-solid heat and mass transfer, axial and radial heat conduction, chemical reactions and the attendant heat release, and heat loss to the surroundings. The model was used to analyze the transient response of an axisymmetric ceramic monolith system (catalyzed monolith, mat, and steel shell) during converter warm-up, sustained heavy load, and engine misfiring. The simulation indicates that high solid temperatures are encountered during sustained heavy load or engine misfiring, while steep temperature gradients are developed during the converter warm-up period. Flow maldistribution and radial heat loss are major sources for the thermal gradients. The predicted temperature profiles provide a basis for the analysis of thermal stresses and fatigue in the monolith converter assembly.

THE CONVERSION EFFICIENCY and structural integrity of an automotive catalytic converter over its lifetime are complex functions of operating and design parameters. Traditional converter design relies heavily on empirical approaches, which can be very costly and time-consuming. Furthermore, such approaches can result in overdesign or underdesign of converters in terms of noble metal loading or structural strength. A comprehensive converter model enables one to explore many possible design options and promises to yield an optimum design within limited lead time if the model is carefully used and coordinated with tests.

A catalytic converter operates under highly transient conditions. At the time of a cold start, for example, a cold converter is suddenly exposed to exhaust gas at an elevated temperature. The temperature transients of a catalytic converter during its warm-up period have important implications in automobile emission control, because the government-prescribed emission test procedure requires that a test vehicle system be at room temperature for a minimum of 12 h prior to each test. Furthermore, monolithic catalysts are generally known to be prone to undesirable temperature excursions under certain transient driving conditions [1]. Therefore, a transient model is necessary for the design of monoliths with improved thermal and emission characteristics.

In addition to the conversion performance of a catalytic converter, its structural durability in the presence of the temperature transients is another important design factor. Among other causes, severe thermal gradients in monoliths during cold-start warm-up have been shown to initiate ring-off and longitudinal cracks [2]. Melting of monoliths is generally attributed to heat release accompanying the oxidation of high concentrations of unburned hydrocarbons [1,3]. An accurate mathematical model provides a convenient means of analyzing and predicting these potential thermally-induced structural failures.

The conversion performance and thermal response are coupled. The rates of chemical reactions in converters are highly nonlinear functions of temperature. Conversely, heat released during reactions contributes to the thermal response. This coupling implies that a realistic converter model must solve the chemical and thermal problems simultaneously. Complex physical/chemical phenomena occurring in a converter include the heat and mass transfer between the exhaust gas and catalyst surface, convective heat and mass transport, chemical reactions and the attendant heat release, heat conduction in the substrate, and heat loss to the surroundings. To further complicate the situation, the problem is three-dimensional in nature; the converter structure itself is three-dimensional and the exhaust gas flow is nonuniform at the front face. Previous efforts in the thermal and chemical modeling of monolithic converters are reviewed briefly in the following paragraph.

Most of the monolith modeling studies previously reported in the literature [4-9], with the exception of recent studies by Flytzani-Stephanopoulos et al. [10] and by Becker and Zygourakis [11], have focused on the behavior of adiabatic (i.e., completely insulated) monoliths exposed to a uniform flow distribution at the front face. In this case, temperature and concentration profiles in all channels of the monolith are the same, so that consideration of only one channel would be sufficient for the modeling purpose. However, actual automobile monolithic converters operate in a nonadiabatic mode under conditions where the gas flow is concentrated in the center of the monolith [12-14]. Consequently, different conditions prevail in each channel and interactions among neighboring channels through radial heat conduction require simultaneous consideration of the entire array of monolith channels. As will be shown in this report, a three-dimensional monolith model, such as the one developed here, is required to adequately predict the temperature gradients and the resulting thermal stresses.

This study was undertaken to provide guidance in the design of monoliths with improved performance and durability by mathematically analyzing their transient behavior under realistic flow conditions. To that end, a transient, three-dimensional monolith model is developed which accounts for all the important physical/chemical phenomena occurring in the catalyzed monolith, as well as in the surrounding materials, such as mat, steel shell and insulation (see Figure 1). The model is used to examine the thermal and conversion characteristics of an axisymmetric platinum-impregnated ceramic monolith for three different operating conditions: warm-up, sustained heavy load, and engine misfiring. Some insights into the temperature transients of monoliths gained from the simulation results will be discussed.

It should be noted that our monolith model is more comprehensive than the previously reported three-dimensional models of Flytzani-Stephanopoulos et al. [10], and of Becker and Zygourakis [11]; the former deals with only the heat transfer process in a non-reactive monolith, and the latter neglects the effects of heat conduction through the surrounding materials.

CHEMICAL REACTIONS AND KINETICS

Most catalytic converters are designed to operate at the stoichiometrically balanced air/fuel ratio for the simultaneous conversion of carbon monoxide, hydrocarbons, and nitrogen oxides in automobile exhaust. In this study, however, only the Pt-catalyzed oxidation reactions of CO, hydrocarbons and H_2 are considered. (At the present time, adequate prediction of the

CERAMIC MONOLITH CATALYTIC CONVERTER

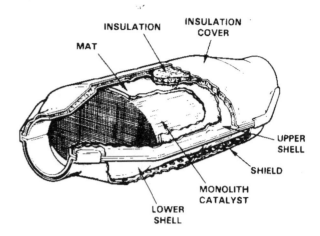

Fig. 1. Ceramic monolith catalytic converter.

conversion performance (including NO) of a multi-metallic commercial converter is hampered by the lack of reliable kinetic information over three-way catalysts.) Neglecting the exothermic heat associated with the reduction reactions of NO is expected to have a minor effect on temperature predictions in view of the low NO concentrations in exhaust (typically 500-1000 ppm).

In this model, the oxidation of hydrocarbons is represented by the reaction of propylene ("fast-oxidizing hydrocarbon") because "slow-oxidizing hydrocarbons," such as methane, have negligible effects on converter thermal response [8]. The oxidation reactions and their reaction heats for this model are [8]:

$$CO + 1/2\ O_2 \rightarrow CO_2$$
$$\Delta H = -2.832 \times 10^5 \quad J/mole \quad (1)$$

$$C_3H_6 + 9/2\ O_2 \rightarrow 3\ CO_2 + 3\ H_2O$$
$$\Delta H = -1.928 \times 10^6 \quad J/mole \quad (2)$$

$$H_2 + 1/2\ O_2 \rightarrow H_2O$$
$$\Delta H = -2.42 \times 10^5 \quad J/mole \quad (3)$$

The specific rates R_i (rates per unit Pt surface area) for the oxidation of CO, C_3H_6, and H_2 can be expressed as functions of concentrations c_i and temperature T:

$$R_{CO} = k_1 c_{CO} c_{O_2}/G \quad (4)$$

$$R_{C_3H_6} = k_2 c_{C_3H_6} c_{O_2}/G \quad (5)$$

$$R_{H_2} = k_1 c_{H_2} c_{O_2}/G \quad (6)$$

462

where

$$G = T (1 + K_1 c_{CO} + K_2 c_{C_3H_6})^2 \text{ x}$$

$$(1 + K_3 c_{CO}^2 c_{C_3H_6}^2)(1 + K_4 c_{NO}^{0.7}) \qquad (7)$$

$$k_1 = 6.699 \times 10^9 \exp(-12556/T)$$
$$k_2 = 1.392 \times 10^{11} \exp(-14556/T)$$
$$K_1 = 65.5 \exp(961/T)$$
$$K_2 = 2.08 \times 10^3 \exp(361/T)$$
$$K_3 = 3.98 \exp(11611/T)$$
$$K_4 = 4.79 \times 10^5 \exp(-3733/T)$$

From the stoichiometry, the reaction rate for oxygen must satisfy

$$R_{O_2} = 0.5 R_{CO} + 4.5 R_{C_3H_6} + 0.5 R_{H_2} \qquad (8)$$

BASIC EQUATIONS AND ASSUMPTIONS

As mentioned above, we are interested in the thermal and conversion characteristics of the entire converter structure illustrated in Figure 1. For the monolith catalyst, it is necessary to include energy and reaction species balances in the gas stream and the catalyzed wall during the exothermic chemical reactions. Outside the catalyzed monolith, however, the only phenomenon is heat conduction through the surrounding materials, such as mat, steel shell, and insulation.

The heat conduction equation for the surrounding materials can be expressed as

$$\rho \partial (CT)/\partial t = \lambda_x \partial^2 T/\partial x^2 + \qquad (9)$$
$$\lambda_y \partial^2 T/\partial y^2 + \lambda_z \partial^2 T/\partial z^2$$

subjected to the initial condition

$$T(x,y,z,0) = T_o(x,y,z) \qquad (10)$$

and the boundary condition for convection and radiation

$$\lambda_n \partial T/\partial n = h_c (T - T_a) + h_r (T^4 - T_a^4) \qquad (11)$$

where density ρ, specific heat C, and conductivity λ are functions of temperature and material with λ assumed to be orthotropic. Here, x is oriented along the monolith channel and y, z are the other two principal directions perpendicular to x. The boundary condition, eq. (11), accounts for the heat transferred from the converter external surface to ambient. In addition, temperature and radial heat flux are conserved across the radial interfaces between different materials.

For the reaction/transport phenomena in the catalyzed monolith, a transient three-dimen-

sional model has been developed by extending the one-dimensional model of Oh and Cavendish [8]. The resulting model includes the effects of the transverse heat flux and nonuniform gas flow distribution, which were neglected in the previous one-dimensional model. In this model, the mass and energy balances in the gas and solid phases are

$$\epsilon \rho_g \partial c_{g,i}/\partial t = - w \partial c_{g,i}/\partial x - \qquad (12)$$
$$\rho_g k_{m,i} S(c_{g,i} - c_{s,i})]_{y,z} \quad i = 1,\ldots,4$$

$$\epsilon \rho_g C_{pg} \partial T_g/\partial t = - w C_{pg} \partial T_g/\partial x \qquad (13)$$
$$+ h S(T_s - T_g)]_{y,z}$$

$$(M/\rho_g) a(x) R_i (\bar{c}_s, T_s) = \qquad (14)$$
$$k_{m,i} S(c_{g,i} - c_{s,i})]_{y,z} \quad i = 1,\ldots,4$$

$$(1-\epsilon)\rho_s \partial (C_s T_s)/\partial t = (1-\epsilon)(\lambda_x \partial^2 T_s/\partial x^2 \qquad (15)$$
$$+ \lambda_y \partial^2 T_s/\partial y^2 + \lambda_z \partial^2 T_s/\partial z^2) + h S(T_g - T_s)$$
$$+ a(x) \sum_{i=1}^{3} (-\Delta H)_i R_i (\bar{c}_s, T_s)$$

The unknowns $\bar{c}$ and T denote concentration and temperature, respectively. The subscripts g, s, and i refer to, respectively, gas phase, solid phase, and the species of interest: $i=1$, CO; $i=2$, C_3H_6; $i=3$, H_2; and $i=4$, O_2. The noble metal concentration $a(x)$, which depends on the noble metal loading and the extent of converter aging, is permitted to vary in the gas flow direction x only, although this assumption could easily be removed. Other parameters are defined in the NOMENCLATURE section.

In this model, both the axial diffusion of mass and heat, and the chemical reactions in the gas phase are neglected. Moreover, the time derivative terms in eqs. (12) and (13), $\partial c_{g,i}/\partial t$ and $\partial T_g/\partial t$, will be discarded because the time constants involved are typically much smaller than that of the solid thermal response [8]. Gas temperature, concentration, and mass flux ($w=\rho_g v$) can be functions of position, y and z, at the monolith front face. Heat exchange between the substrate and the surroundings at both the inlet and outlet faces of the monolith is neglected. With these assumptions, the appropriate initial condition (only for T_s) and boundary conditions are

Initial condition:

$$T_s(x,y,z,0) = T_{so}(x,y,z) \qquad (16)$$

At the inlet x_{in},

$$c_{g,i}(x_{in},y,z,t) = c_{g,i}^{in}(y,z,t) \qquad (17)$$
$$i = 1,\ldots,4$$

$$T_g(x_{in}, y, z, t) = T_g^{in}(y, z, t) \qquad (18)$$

$$\partial T_s/\partial x(x_{in}, y, z, t) = 0 \qquad (19)$$

$$w(x_{in}, y, z, t) = w^{in}(y, z, t) \qquad (20)$$

Since w is independent of x by conservation of mass, we immediately have

$$w(x, y, z, t) = w^{in}(y, z, t) \qquad (21)$$

for all x.

At the outlet x_{out},

$$\partial T_s/\partial x(x_{out}, y, z, t) = 0 \qquad (22)$$

On the radial boundary of the catalyzed monolith, both the temperature and the radial heat flux must be continuous at the interface of the monolith and the surrounding materials.

Heat and mass transfer coefficients, h and $k_{m,i}$, between the bulk gas stream and catalyst surface are estimated from the Nusselt and the Sherwood numbers for fully developed laminar flow with constant wall heat flux in the monolith channels [15]:

$$h = Nu_\infty \lambda_g/(2R_h) \qquad (23)$$

$$k_{m,i} = Sh_\infty D_i/(2R_h) \qquad (24)$$

The values of the molecular diffusivity of species i, D_i [16], and the thermal conductivity of the exhaust gas λ_g [8] are listed in Table 1. The heat and mass transfer coefficients calculated above neglect the variance within the developing laminar flow regime near the monolith inlet, because the length of this hydrodynamic entrance region is typically a small fraction of the total monolith length [17].

Table 1

MOLECULAR DIFFUSIVITIES D_i AND GAS THERMAL CONDUCTIVITY λ_g

D_{CO}	1.332	cm^2/s
$D_{C_3H_6}$	0.8095	cm^2/s
D_{H_2}	5.1863	cm^2/s
D_{O_2}	1.3541	cm^2/s
λ_g	$2.269 \times 10^{-6} T_g^{.832}$	$J/cm \cdot s \cdot K$

NUMERICAL SOLUTION OF THE EQUATIONS

Equations (9) and (12) through (15) represent a set of eleven coupled nonlinear, three-dimensional partial differential equations. The numerical solution of these equations presents a considerable challenge. A brief description of the methods employed to solve the equations is given below.

Equations (9) and (15) are actually simplified versions of the more general heat conduction equation with a heat source function F_1, namely,

$$\rho_s \partial(C_s T_s)/\partial t = \lambda_x \partial^2 T_s/\partial x^2 + \qquad (25)$$

$$\lambda_y \partial^2 T_s/\partial y^2 + \lambda_z \partial^2 T_s/\partial z^2 + F_1(T_s, T_g, \bar{c}_s)$$

For simplicity, we use the above equation to replace both the conduction equations, eqs. (9) and (15), with the parameters assumed to be temperature and material dependent. Notice that in the case of the catalyzed monolith, ρ_s and λ's in eq. (25), represent the volume-averaged (i.e., $(1-\epsilon)$ factored in) solid density and thermal conductivities, respectively. The thermal conductivity is assumed to be orthotropic (i.e., λ_x being the thermal conductivity of the solid in the flow direction and λ_y, λ_z in the other two principal directions). Other equations, eqs. (12) through (14), can be expressed in the following compact form

$$d\bar{c}_g/dx = \bar{F}_2(T_g, \bar{c}_g, \bar{c}_s)]_{y,z} \qquad (26)$$

$$dT_g/dx = F_3(T_s, T_g)]_{y,z} \qquad (27)$$

$$\bar{0} = \bar{F}_4(T_g, T_s, \bar{c}_g, \bar{c}_s)]_{y,z} \qquad (28)$$

The functions F_1, $\bar{F}_2$, F_3, and $\bar{F}_4$ in the above equations are defined in the NOMENCLATURE section.

We observe that the three-dimensional heat conduction problem, eq. (25), can be solved if T_g and $\bar{c}_s$ in F_1 are known. Conversely, eqs. (26) through (28) reduce to a multivariable initial value problem in x for each channel if the solid temperature field T_s is provided at a specified time t. This observation provides the basis for the numerical methods adopted here. Our approach to solving eqs. (25) through (28) was to break down the mathematical description of the monolith system into two problems. Problem 1 describes the three-dimensional heat conduction in the monolith substrate and the surrounding materials, such as mat, steel shell, and insulation with appropriate initial and boundary conditions. Problem 2 denotes the one-dimensional representation of the simultaneous processes of heat/mass transfer and chemical reaction in each of the monolith channels with the initial values specified at the monolith inlet. The two problems are linked together through T_s and F_1.

Figure 2 illustrates the basic scheme for the numerical solution. Problem 1 is solved by a nonlinear finite element solver for each iteration. The resultant temperature field T_s is then mapped onto Problem 2. With the provided T_s, a nonlinear finite difference solver for Problem 2 calculates the gas temperature T_g, the gas-phase concentrations $c_{g,i}$, the solid-phase concentrations $c_{s,i}$, and the heat source function F_1 for each monolith channel. The computed F_1 is then passed to Problem 1 for the next iteration. The convergence requirements, ϵ_1 and ϵ_2, for the two problems are set separately. For each time step, the solution is obtained when both of the two convergence requirements are satisfied.

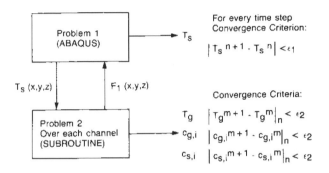

Fig. 2. Numerical solution scheme.

The finite element solver ABAQUS [18] was used in this model in view of its transient nonlinear capability and flexibility in interfacing with user-supplied subroutines (in this case the finite difference solver for Problem 2). The finite element algorithm for heat conduction is detailed in the ABAQUS manuals. ABAQUS adopts the two-point backward time integration scheme. The source term F_1 in Problem 1 is calculated at the midpoint between time t^n and t^{n+1} in Problem 2, which is described briefly in the following.

For each y, z position, eqs. (26) and (27) are solved as initial value problems in the x-direction with initial values supplied at the inlet from eqs. (17) and (18). Equation (28) is coupled to eq. (26), with O_2 concentration being dependent on the other three concentrations (see eq. (8)). Specifically, with T_s available and $\bar{c}_g(x=0)$ and $T_g(x=0)$, known from eqs. (17) and (18), the 4×4 nonlinear system, eq. (28), is solved by Newton's method to obtain $\bar{c}_s(x=0)$. Initial guesses for $\bar{c}_s(x=0)$ for Newton's method are obtained from $\bar{c}_s(x=0)$ at the previous time step, except in extreme cases requiring individual treatment. The solution proceeds from mesh point x_i (initially $x_i=x_{in}$) to mesh point x_{i+1} (finally $x_{i+1}=x_{out}$) as follows. First, eq. (27) is discretized by the midpoint

rule between x_i and x_{i+1} and the resulting nonlinear scalar equation is solved by Newton's method to obtain $T_g(x_{i+1})$. Equation (26) supplies $\bar{c}_g(x_{i+1})$ as a linear function of $\bar{c}_s(x_{i+1})$ which is then substituted into eq. (28) to obtain a 4×4 nonlinear system for $\bar{c}_s(x_{i+1})$ which is solved by Newton's method. The initial guess for Newton's method is simply $\bar{c}_s(x_i)$. This procedure is the same as that used by Oh and Cavendish [8], except that eq. (26) is substituted into eq. (28) to reduce the size of the nonlinear subproblem, and eq. (28) is here solved without truncation error at the nodes $\{x_i\}$, rather than at the midpoints.

ANALYSIS RESULTS

We applied the model to an axisymmetric converter system (which includes a cylindrical catalyzed monolith, mat, and steel shell) shown in Figure 3 together with the finite element mesh. The geometric dimensions and material properties are listed in Table 2. Three different operating conditions were simulated; namely, converter warm-up, sustained heavy load, and engine misfiring. For the warm-up simulations, we examined the converter behavior as a function of noble metal loadings for three different exhaust gas flow rates of 18, 25, and 40 g/s, which correspond to typical 4, 6, 8 cylinder engines, respectively. The two standard noble

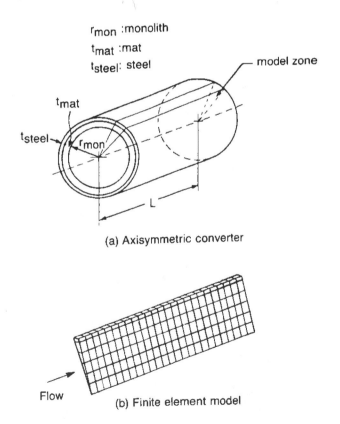

Fig. 3. Axisymmetric converter and finite element mesh; the steel shell is modeled by plate elements.

Table 2

CONVERTER DIMENSIONS AND MATERIAL PROPERTIES

r_{mon}	4.648	cm	
t_{mat}	0.33	cm	
t_{steel}	0.145	cm	
L	15.24	cm	
R_h	0.05085	cm	coated substrate
cell density	62	cells/cm^2	
ϵ	0.6408		coated substrate
ρ_{mon}	0.552	g/cm^3	volume averaged
ρ_{mat}	1.0	g/cm^3	
ρ_{steel}	7.8	g/cm^3	
C_{mon}	$1.071 + 1.56 \times 10^{-4} T_s$ $-3.435 \times 10^4 T_s^{-2}$	J/g•K	
C_{mat}	0.6	J/g•K	
C_{steel}	0.5023	J/g•K	
C_{pg}	1.089	J/g•K	
λ_{mon}	0.0053 0.00265	J/cm•s•K J/cm•s•K	λ_x volume averaged λ_y, λ_z volume averaged
λ_{mat}	0.0013 0.00265	J/cm•s•K J/cm•s•K	at 370 K at 617 K
λ_{steel}	0.26	J/cm•s•K	
h_c	4.84×10^{-3}	J/cm^2•s•K	
h_r	1.588×10^{-12}	J/cm^2•s•K^4	

metal concentrations adopted here are a = 537.9 and 268.95 cm^{2}Pt/cm^3 reactor, which are representative values for typical in-use converters. The upper limit for the noble metal concentration was taken to be a = 2500, which corresponds to the activity for fresh commercial converters with high noble metal loadings. Also, to simulate the behavior of a heavily deactivated monolith, we analyzed a case with a = 89.65, roughly 1/20 of the activity of low-loaded fresh commercial converters. To examine the thermal transients during sustained heavy load, a high flow rate (W=64 g/s) of exhaust gas containing relatively high reactant concentrations (1500 ppm C$_3$H$_6$ and 2.5% CO) was assumed to be fed to the inlet of a fully warmed-up converter with a = 537.9. The last examples considered were 12.5% and 25% misfiring conditions which, for example, represent 1 or 2 nonfiring cylinders in a V8 engine, respectively. The parameter values and inlet exhaust gas conditions for these test cases are summarized in Table 3. In the analysis, we have included the effects of nonuniform gas flow distribution and radial heat loss to ambient. A previously measured velocity profile at the monolith face, shown in Figure 4 [12], was selected for this study. The heat loss to ambient was governed by the convective and radiative boundary conditions on the steel shell surface with their coefficients listed in Table 2. The convective heat transfer (film) coefficient was evaluated based on an assumed

Table 3

PARAMETERS FOR SIMULATION

		Warm-up	Sustained Heavy Load	12.5% Misfiring	25% Misfiring
T_{so}	K	300	950	950	950
T_g^{in}	K	650	800	650	500
W	g/s	varies	64	40	40
a	cm^2Pt/cm^3	varies	537.9	537.9	537.9
$c_{g,i}^{in}$ % CO		2.0	2.5	2.0	1.5
% C_3H_6		0.045	0.15	0.60	1.2
% H_2		0.667	0.667	0.667	0.667
% O_2		5.0	1.5	3.5	5.0
% NO		0.05	0.05	0.01	0.01

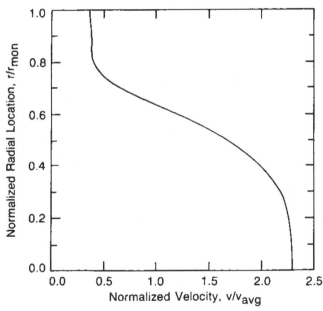

Fig. 4. Velocity profile at the inlet of the monolith.

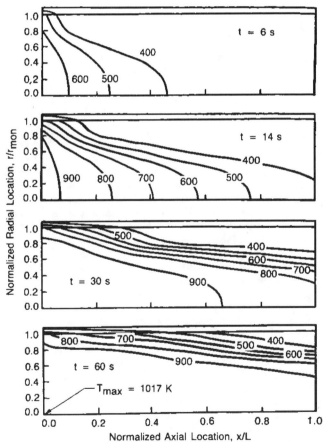

Fig. 5. Evolution of solid temperature contours (in K) during converter warm-up for a = 537.9 cm^2Pt/cm^3 and W = 18 g/s.

air velocity of 35 mi/h on the steel shell in the perforated converter housing.

Figures 5 through 7 illustrate the evolutions of the solid temperature contours during warm-up for the cases with a = 537.9 at various flow rates. The converter is hotter in the center region near the inlet, with the maximum temperature located at the center of the inlet. All the three sets of contours are similar. The maximum temperatures encountered during the

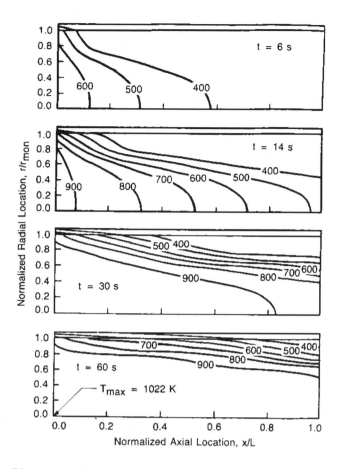

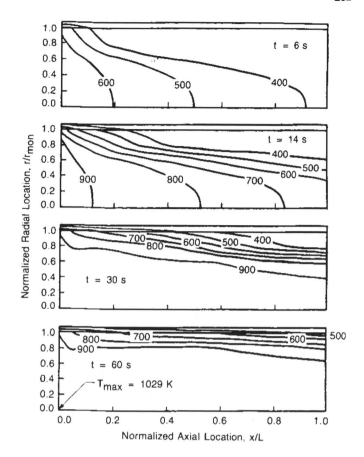

Fig. 6. Evolution of solid temperature contours (in K) during converter warm-up for a = 537.9 cm²Pt/cm³ and W = 25 g/s.

Fig. 7. Evolution of solid temperature contours (in K) during converter warm-up for a = 537.9 cm²Pt/cm³ and W = 40 g/s.

warm-up process are predicted to lie between 1017 and 1029 K (see t = 60 s). The most severe thermal gradients, as indicated by the most dense contour lines, occur at t = 30 s. From these contours, we observed that flow rate has minor effects on the basic characteristics of thermal response, except that higher flow rates result in somewhat steeper gradients and faster temperature rises. These effects can be attributed to a larger amount of convective heat flux coming into the converter and hence, quicker light-off.

Shown in Figures 8 is the thermal evolution during converter warm-up for the cases with a lower noble metal concentration of a = 268.95 at a flow rate of 40 g/s. Results of additional simultaions with lower exhaust flow rates, though not illustrated here, again show that flow rate variations have minor effects on the evolution of temperature contours. The maximum temperature (1028 K) and the severity of the thermal gradients in the late portions of the warm-up process in Figure 8 are generally similar to those of the case with a = 537.9. However, there is a temporary temperature peak (around 14 s), which develops somewhat downstream along the converter axis (peak position depends on flow rates) and then moves toward the

inlet. For this lower activity, the light-off does not occur at the monolith entrance because the reactant stream must be heated up to the reaction temperature in the inlet portion of the converter before being rapidly converted in the downstream section. The axial position and magnitude of the temperature peak are generally determined by the relative rates of heat generation by the chemical reaction and heat removal by the convective transport.

Figure 9 depicts the converter thermal evolution for the case with a = 2500 and W = 40 g/s. As can be anticipated from the high activity, the solid temperature at early times rises faster than that of the lower activity cases (compare with Figures 7 and 8). However, the maximum temperature (1031 K) and the magnitude of the thermal gradients in the late portions of the warm-up process are comparable to those of the previous cases. It is interesting to note that in this case of high activity there is no temperature peak developed during the warm-up period.

Our calculations show that the temperature profile at 60 s shown in each of the preceding figures closely approximates the steady-state profile attained at the end of the warm-up period. The highest temperatures encountered within the monolithic catalyst during converter warm-up are predicted to be 1015-1030 K.

In order to assess the importance of radial heat loss and nonuniform flow distribution during converter warm-up, we consider the following four cases: (a) no radial heat loss, uniform gas flow; (b) no radial heat loss, nonuniform gas flow; (c) radial heat loss, uniform gas flow; and (d) radial heat loss, nonuniform gas flow. The resulting model predictions are compared in Figure 10. The assumptions invoked in Case (a) lead to the previously developed one-dimensional model of Oh and Cavendish [8]. The radial temperature gradient at the monolith core for Case (b), involving nonuniform flow, is more severe than that for Case (c), accounting for the heat conduction through mat and steel and the heat loss to ambient. However, the reverse is true in the outer part of the monolith ($r/r_{mon} > 0.8$). These comparisons indicate that nonuniform flow distribution and radial heat loss are major sources for the thermal gradients. Accordingly, design changes aimed at more uniform gas flow and better insulation around the mat can reduce temperature gradients and thus thermal stresses.

Figures 11 through 13 depict the CO conversion predictions for an exhaust flow rate of 25 g/s at various noble metal concentrations. The four curves in each figure represent the four cases of monolith models discussed previously. It is apparent from the figures that all four models predict very similar light-off behavior. Notice that for the cases with a = 537.9 and 268.95, the four curves predict the same ultimate (i.e., steady-state) conversion performance of nearly 100%. However, for the very low activity case (a = 89.65, Figure 13), the insulated boundary assumption yields overestimates of the steady-state conversion efficiency. For the two cases with radial heat loss, the uniform gas flow yields a lower steady-state conversion efficiency compared with the nonuniform flow because in the latter case, most of the exhaust gas flows through the hot center zone, and hence, conversion of the reactants occurs more efficiently. We must bear in mind that in this study, the aging effect is assumed to be uniform within the entire monolith. In reality, nonuniform gas flow might poison the center region more heavily, thus reducing the overall conversion efficiency.

It can be seen from Figures 10 through 13 that the previously reported one-dimensional model of Oh and Cavendish [8], though adequate in describing the conversion performance of a monolith during its warm-up, is not capable of predicting the radial temperature profiles, and it significantly underestimates the severity of the temperature gradients.

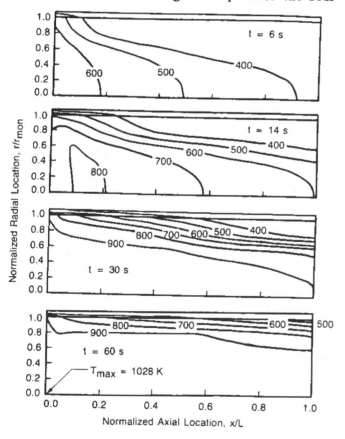

Fig. 8. Evolution of solid temperature contours (in K) during converter warm-up for a = 268.95 cm^{2}Pt/cm^3 and W=40 g/s.

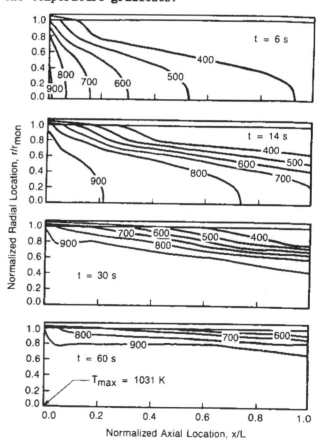

Fig. 9. Evolution of solid temperature contours (in K) during converter warm-up for a = 2500 cm^{2}Pt/cm^3 and W = 40 g/s.

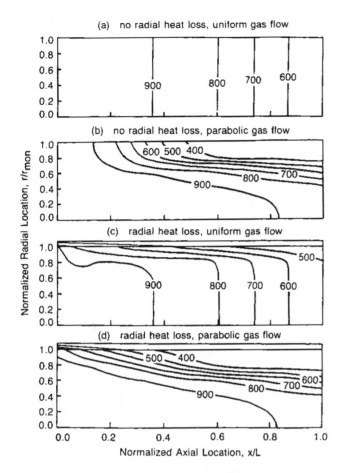

(a) no radial heat loss, uniform gas flow

(b) no radial heat loss, parabolic gas flow

(c) radial heat loss, uniform gas flow

(d) radial heat loss, parabolic gas flow

Normalized Radial Location, r/r_{mon}

Normalized Axial Location, x/L

Fig. 10. Comparison of the solid temperature contours (in K) predicted from various models; a = 537.9 $cm^2 Pt/cm^3$, W = 25 g/s, t = 30 s.

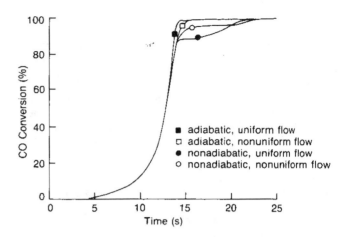

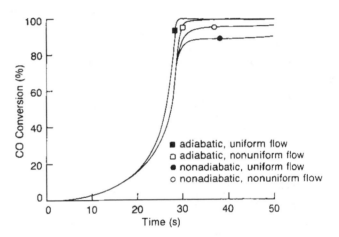

- ■ adiabatic, uniform flow
- □ adiabatic, nonuniform flow
- ● nonadiabatic, uniform flow
- ○ nonadiabatic, nonuniform flow

Fig. 12. Comparison of CO conversion efficiency during converter warm-up for a = 268.95 $cm^2 Pt/cm^3$ and W = 25 g/s.

- ■ adiabatic, uniform flow
- □ adiabatic, nonuniform flow
- ● nonadiabatic, uniform flow
- ○ nonadiabatic, nonuniform flow

Fig. 13. Comparison of CO conversion efficiency during converter warm-up for a = 89.65 $cm^2 Pt/cm^3$ and W = 25 g/s.

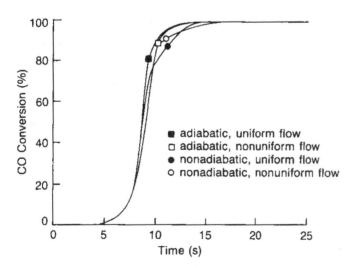

- ■ adiabatic, uniform flow
- □ adiabatic, nonuniform flow
- ● nonadiabatic, uniform flow
- ○ nonadiabatic, nonuniform flow

Fig. 11. Comparison of CO conversion efficiency during converter warm-up for a = 537.9 $cm^2 Pt/cm^3$ and W = 25 g/s.

Figure 14 compares the time variations of CO conversion during warm-up for different combinations of gas flow rates and noble metal concentrations for the nonadiabatic monolith with nonuniform flow distribution. The six curves in the figure fall into two separate categories depending on the noble metal concentrations, while flow rate has only minor effects on conversion performance in terms of percentage over the range of 18 to 40 g/s. Higher flow rates will, however, result in higher mass emissions of the exhaust species (i.e., in grams per mile).

It is worth mentioning that C_3H_6 and H_2 show very similar conversion characteristics (including similar light-off times) to CO as a result of kinetic coupling between the reactant species.

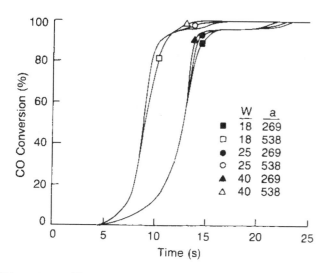

Fig. 14. CO conversion efficiency during the warm-up of a nonadiabatic monolith with nonuniform flow distribution.

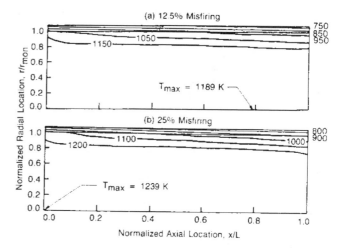

Fig. 16. Solid temperature contours (in K) at 60 s into engine misfiring conditions for a = 537.9 cm²Pt/cm³ and W = 40 g/s.

Figure 15 depicts the steady-state temperature contour within the converter during sustained heavy load. The analysis predicted a maximum temperature of 1069 K in the catalyzed monolith. In this case, the temperature gradient is relatively mild compared to those predicted for the converter warm-up conditions considered above.

Figure 16 illustrates the temperature contours for the cases of 12.5% and 25% misfiring engines at 60 s into the misfiring condition. The peak temperatures are predicted to be 1189 K and 1239 K for 12.5% and 25% misfiring, respectively. These temperatures are much lower than the melting point of the monolith substrate (about 1700 K). Neglecting the gas-phase reactions (which generally become significant above 1100 K) in our model is expected to have little effect on the interpretation of our simulation results here, because the concentration of unconverted reactants in the gas phase (and thus the extent of the homogeneous combustion) would become low by the time high gas temperatures are attained as a result of rapid catalytic reactions.

CONCLUDING REMARKS

In this study, we have developed a transient three-dimensional model for simulating the thermal response and conversion performance of nonadiabatic monolithic converters operating under flow maldistribution conditions. The model includes all the important physical/chemical phenomena, including convective heat and mass transport in the monolith channels, interphase heat and mass transfer between the exhaust gas and substrate, axial and radial heat conduction, chemical reactions and the attendant heat release, and heat loss to the surroundings. Numerical solutions of the equations were obtained based on a combination of finite element and finite difference methods.

We have applied the model to simulate the transient response of a monolith during converter warm-up, sustained heavy load, and engine misfiring. The model predicts that variations in exhaust flow rate have minor effects on the percent conversion efficiency and the nature of the temperature transients encountered during warm-up. However, the noble metal loading and the flow distribution of the inlet exhaust gas are shown to have a strong influence on the solid temperature profiles during converter warm-up. In addition to such parametric sensitivity analysis, the model has the capability of identifying operating conditions under which high temperatures or steep thermal gradients are encountered. Our simulation results indicate that steep radial and axial temperature gradients are developed during the converter warm-up period. Flow maldistribution and radial heat loss are major sources for thermal gradients; thus, design changes aimed at more uniform gas flow and better insulation would improve the structural durability of a monolith by reducing temperature gradients and the resulting thermal

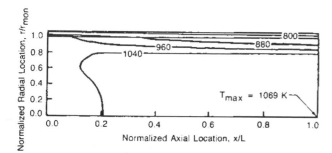

Fig. 15. Steady-state solid temperature contours (in K) during sustained heavy load for a = 537.9 cm²Pt/cm³ and W = 64 g/s.

stresses. The model also shows that high solid temperatures are encountered during sustained heavy load or engine misfiring.

One important model application area is optimum design of monoliths for improved structural and emission durability. The quantitative prediction of optimum monolith properties requires an evaluation of the validity of the model. Studies are underway for experimental verification of the monolith model developed here, and preliminary results indicate reasonable agreement for the converter warm-up conditions. The results of a thermal stress analysis aimed at assessing the structural impact of the temperature gradients predicted in this work are reported separately [19]. However, the ability of the model to predict FTP exhaust emissions (including NO) from modern catalytic converters is limited at this time by the lack of reliable kinetic expressions over commercial three-way catalysts.

NOMENCLATURE

$a(x)$: catalytic surface area per unit reactor volume, cm^2/cm^3

c_i : concentration of species i, mole fraction

$c_{g,i}$: concentration of species i in bulk gas stream, mole fraction

$c_{g,i}^{in}$: inlet concentration of species i, mole fraction

$\bar{c}_g$: vector with entries $c_{g,i}$, mole fraction

$\bar{c}_s$: vector with entries $c_{s,i}$, mole fraction

$c_{s,i}$: concentration of species i on the solid surface, mole fraction

C : specific heat, $J/g \cdot K$

C_{mat} : specific heat of mat, $J/g \cdot K$

C_{mon} : specific heat of monolith, $J/g \cdot K$

C_{pg} : specific heat of gas, $J/g \cdot K$

C_s : specific heat of solid, $J/g \cdot K$

C_{steel} : specific heat of steel, $J/g \cdot K$

D_i : diffusivity of species i in the reactive mixture, cm^2/s

F_1 : $hS(T_g - T_s) + a(x)\sum_{i=1}^{3}(-\Delta H)_i R_i (\bar{c}_s, T_s)$

$[\bar{F}_2]_i$: $-\rho_g k_{m,i} S(c_{g,i} - c_{s,i})/w$

F_3 : $hS(T_s - T_g)/wC_{pg}$

$[\bar{F}_4]_i$: $(M/\rho_g)a(x)R_i(\bar{c}_s, T_s) - k_{m,i}S(c_{g,i} - c_{s,i})$

G : quantity defined by eq. (7)

h : heat transfer coefficient, $J/cm^2 \cdot s \cdot K$

$(-\Delta H)_i$: heat of combustion of species i, $J/mole$

k_i : rate constant for reaction i, $mole \cdot K/cm^2 \cdot s$

K_i : adsorption equilibrium constant, (mole fraction)$^{-1}$

$k_{m,i}$: mass transfer coefficient for species i, cm/s

L : length of monolith, cm

M : molecular weight of reaction mixture, $g/mole$

Nu_∞ : limiting Nusselt number, $2hR_h/\lambda_g$

R_h : hydraulic radius of channel, 2(cross-sectional area)/(wetted perimeter), cm

R_i : specific reaction rate for species i, $mole/cm^2 Pt \cdot s$

r_{mon} : radius of monolith face, cm

S : geometric surface area per unit reactor volume, cm^2/cm^3

Sh_∞ : limiting Sherwood number, $2k_{m,i}R_h/D_i$

t : time, s

T : temperature, K

T_a : ambient temperature, K

T_g : gas temperature, K

T_g^{in} : inlet gas temperature, K

t_{mat} : thickness of mat, cm

T_s : solid temperature, K

T_{so} : initial solid temperature, K

t_{steel} : thickness of steel, cm

v : linear superficial velocity of gas, cm/s

w : mass flux of exhaust gas, $g/cm^2 \cdot s$

W : mass flow rate of gas, g/s

w^{in} : mass flux of inlet gas, $g/cm^2 \cdot s$

472

x : coordinate along flow direction, cm

y, z : principal axes perpendicular to flow direction, cm

GREEK LETTERS

ϵ : void fraction of the monolith

λ_g : thermal conductivity of gas, J/cm•s•K

λ_{mat} : thermal conductivity of mat, J/cm•s•K

λ_{mon} : thermal conductivity of monolith, J/cm•s•K

λ_s : thermal conductivity of solid, J/cm•s•K

λ_{steel} : thermal conductivity of steel, J/cm•s•K

ρ_g : gas density, g/cm^3

ρ_s : solid density, g/cm^3

ρ_{mat} : mat density, g/cm^3

ρ_{mon} : monolith density g/cm^3

ρ_{steel} : steel density, g/cm^3

REFERENCES

1. Morgan, C. R., Carlson, D. W., and Voltz, S. E., "Thermal Response and Emission Breakthrough of Platinum Monolithic Catalytic Converters," SAE Paper 730569, 1973.

2. Gulati, S. T. and Merry, R. P., "Design Considerations for Mounting Material for Ceramic Wall-Flow Diesel Filters," SAE paper 840074, 1984.

3. Mondt, J. R., "A Guard System to Limit Catalytic Converter Temperature," SAE paper 760320, 1976.

4. Heck, R. H., Wei, J., and Katzer, J. R., "Mathematical Modeling of Automotive Catalysts," AIChE J. 22, p. 477, 1976.

5. Young, L. C. and Finlayson, B. A., "Mathematical Models of the Monolith Catalytic Converter: Part I. Development of Model and Application of Orthogonal Collocation," AIChE J., 22, p. 331, 1976.

6. Lee, S. T. and Aris, R., "On the Effects of Radiative Heat Transfer in Monoliths," Chem. Eng. Sci., 32, p. 827, 1977.

7. Lee, S. T. and Aris, R., "Poisoning in Monolithic Catalysts," ACS Symp. Ser., No. 65, p. 110, 1978.

8. Oh, S. H. and Cavendish, J. C., "Transients of Monolithic Catalytic Converters: Response to a Step Change in Freestream Temperature as Related to Controlling Automobile Emissions," Ind. Eng. Chem. Prod. Res. Dev., 21, p. 29, 1982.

9. Oh, S. H. and Cavendish, J. C., "Design Aspects of Poison-Resistant Automobile Monolithic Catalysts," Ind. Eng. Chem. Prod. Dev., 22, p. 509, 1983.

10. Flytzani-Stephanopoulos, M., Voecks, G. E., and Charng, T., "Modeling of Heat Transfer in Nonadiabatic Monolith Reactors and Experimental Comparisons of Metal Monoliths With Packed Beds," Chem. Eng. Sci., 41, p. 1203, 1986.

11. Becker, E. R. And Zygourakis, K., "Monolith Catalyst Design Options for Rapid Light-Off," presented at the 1986 Annual Meeting of AIChE, Miami Beach, FL, Nov. 2-7, 1986.

12. Lemme, C. D. and Givens, W. R., "Flow Through Catalytic Converters - An Analytic and Experimental Treatment," SAE paper 740243, 1974.

13. Howitt, J. S. and Sekella, T. C., "Flow Effects in Monolithic Honeycomb Automotive Catalytic Converters," SAE paper 740244, 1974.

14. Wendland, D. W. and Matthes, W. R., "Visualization of Automotive Catalytic Converter Internal Flows," SAE paper 861554, 1986.

15. Shah, R. K. and London, A. L., "Laminar Flow Forced Convection Heat Transfer and Flow Friction in Straight and Curved Ducts - A Summary of Analytical Solutions," Technical Report No. 75, Stanford University, 1971.

16. Bird, R. B., Steward, W. E. and Lightfoot, E. N., "Transport Phenomena," John Wiley, New York, p. 505, 1960.

17. Sherony, D. F. and Solbrig, C. W., "Analytical Investigation of Heat and Mass Transfer Friction Factors in a Corrugated Duct Heat or Mass Exchanger," Int. J. Heat Mass Transfer, 13, p. 1455, 1970.

18. ABAQUS Theory Manual, Hibbitt, Karlsson and Sorensen, Inc. Providence, RI.

19. Verma, A. and Van Ostrom, D. L., "Thermal Stresses in Catalytic Converters," to be presented at the SAE International Congress & Exposition, Detroit, 1988.

920096

Modeling Current Generation Catalytic Converters: Laboratory Experiments and Kinetic Parameter Optimization - Steady State Kinetics

Clifford N. Montreuil, Scott C. Williams, and Andrew A. Adamczyk
Ford Motor Co.

ABSTRACT

An experimental data base of catalyst conversion efficiency was generated, using a tubular flow reactor which contained either a Pt/Rh (5:1; 40g/ft³) or a Pd/Rh (5:1; 40g/ft³) catalyst sample, for the purpose of updating the kinetic rate constants in the Ford TWC model. Steady-state conversion efficiency of CO, NO, C_3H_8, C_3H_6, H_2 and O_2 through these catalysts were determined for a variety of inlet species concentrations and inlet gas temperatures. These data were obtained for values of redox ratio between 0.5 (excess O_2) and 4.0, and inlet gas temperatures between 371°C and 593°C. All experimental details and modeling procedures utilized in obtaining an optimized set of kinetic parameters are included.

Results of these experiments show significant improvement in CO and NO conversion efficiency and an increase in NH_3 production for both catalyst formulations over previous generation catalyst formulations when redox ratio is greater than unity. These results required the reevaluation of several steady state kinetic parameters prior to any future adjustment of the parameters relating to the transient kinetic mechanism in the non-steady state version of our TWC model. The conversion efficiencies obtained with the Pt/Rh formulation were similar to those obtained with the Pd/Rh formulation over a wide range of conditions; the greatest differences occurred at low temperature or when a high concentration of "slow" burning hydrocarbons (C_3H_8) was present. Because of this similarity in performance, kinetic rate parameters were generated which describe the "averaged" performance of these two catalyst formulations in addition to the constants which describe the performance of either catalyst formulation.

INTRODUCTION

RECENT AMENDMENTS to the U.S. Clean Air Act and increasingly stringent clean-air legislation worldwide have led to a refocused effort to help meet the newly imposed standards. One aspect of this effort (largely research based) includes the building of computer models to predict emission system behavior for proposed emission-system geometries. These models may include descriptions of exhaust-system heat conservation and flow, catalyst behavior under steady state or transient operation, and/or emission/control system interactions.

In this report, we focus on the models which describe catalyst operation [see Otto,N.C. and LeGray,W.J.(1980); Oh, S.H.(1988a); Oh,S.H.,et al.(1988b); Kuo,J.W.C., et al.(1971); and Young,L.C. and Finlayson,B.R. (1976)] and the steps (both experimental and analytical) which are necessary to update their chemical kinetic submodels. Specifically, we will focus on the procedures used to update the parameters which describe the rates at which each kinetic step in the steady state kinetic mechanism proceeds. The procedures used to update the corresponding parameters for the transient kinetic mechanism are similar to these but additionally require a description of modulated input O_2 into the catalyst and dictate the use of the fully transient TWC model rather than its steady state counterpart; these procedures will be discussed in a later report. We also include, for completeness of the experimental description, many of the details which relate to reactor operation under transient conditions.

To update the model parameters which describe the chemical operation of the catalyst, we use the following approach. First, we determine experimentally the amount of chemical conversion (outlet concentration versus inlet concentration) of CO, NO, C_3H_6, C_3H_8, H_2, and O_2 through a representative catalyst sample. These experimental data are generated over a "wide" range of inlet temperatures, flow rates and species concentrations using a tubular flow reactor (see figure 1; discussed later) which contains an aged catalyst element. Second, we compare these experimentally determined results to model predictions of TWC performance at similar conditions. Third, we adjust the appropriate model parameters to yield the "best" overall agreement between the two results. Finally, we compare the predictions of the model, using the new parameter set, to experimental results obtained at conditions which lie outside the range of conditions used during parameter adjustment. One should note, however, that the kinetic mechanism generated in this manner is not necessarily unique but rather should be considered to be a mechanism which describes the behavior of the catalyst over a wide variety of relevant input conditions, while satisfying constraints imposed by the conservation of mass, momentum and energy in this reactive flow.

We start the procedure using the model parameters which were based on data obtained from the previous generation Pt/Rh catalyst, since the current Pt/Rh or Pd/Rh catalyst formulations are not thought to be radically different in terms of their overall kinetic mechanisms. However, we do note significant improvements in the steady state CO and NO conversion efficiency for this generation catalyst (redox > 1.0) and, therefore, begin our update with the steady state kinetic mechanism rather than moving directly to the full transient kinetic mechanism. For this report, we have determined the kinetic rate parameters for inlet temperatures above 371°C which will yield a good representation of catalyst performance after lightoff.

MODEL DESCRIPTION

The TWC mathematical model represents the one-dimensional conservation (both gas phase and surface) of mass, energy and momentum of a heterogeneous reacting flow along a surface which enhances catalytic reaction. Phenomena explicitly included in the model are:

1. The convective transport of energy and chemical species from the bulk flow to the surface using Nusselt and Sherwood number correlations to specify the lateral energy and mass transport,
2. The surface chemical reactions and corresponding heat liberation on the surface,
3. The adjustments to the mass transport rates for entrance effects (localized boundary layer development) as gases enter the honeycomb substrate, and
4. Thermal conduction along the substrate as well as energy loss to the atmosphere.

The model does not explicitly but does implicitly include many of the following:

1. The details of pore diffusion of species, as this effect is implicitly included in the kinetic rate expressions, (Note: this process was explicitly included in an earlier COC catalyst model but had been removed to reduce computation time of the TWC model),
2. Surface adsorption/desorption rates and gas-phase chemistry, as these factors are also "lumped" into the overall kinetic rate expressions,
3. The axial dispersion of gas-phase species, since the time scales associated with dispersive processes are considerably longer than the time scales for convection of material through the catalytic reactor
4. The lateral variations of gas-phase temperature, species concentration and velocity within the channels of the substrate material. Gas-phase values are specified as cross-sectional averages. Using these averages and Nusselt and Sherwood correlations [see Bird, et al. (1965)], the lateral mass and energy flux to the surface is specified as a potential difference between gas and surface conditions.

Therefore, using the definition of terms specified in nomenclature, the material and energy balances for the gas phase are:

$$e\frac{\partial X_g^i}{\partial t} + v\frac{\partial X_g^i}{\partial z} = -k_m^i G_A (X_g^i - X_s^i) \quad \textbf{(1)}$$

$$e\rho_g C_{P_g}\frac{\partial T_g}{\partial t} + v\rho_g C_{P_g}\frac{\partial T_g}{\partial z} = -hG_A(T_g - T_s) \quad \textbf{(2)}$$

The corresponding equations for the surface are:

$$(1-e)\frac{\partial X_s^i}{\partial t} = k_m^i G_A (X_g^i - X_s^i) - R_i(\overline{\mathbf{X}}_s, T_s) \quad \textbf{(3)}$$

$$(1-e)\rho_s C_{P_s}\frac{\partial T_s}{\partial t} = \lambda_s(1-e)\frac{\partial^2 T_s}{\partial z^2} + hG_A(T_g - T_s) -$$
$$-h_{amb}S_{ext}(T_s - T_{amb}) + \sum_{j=1}^{m}(-\Delta H)_j \overline{R}_j(\overline{\mathbf{X}}_s, T_s)$$
$$\textbf{(4)}$$

where the subscript i refers to the particular species of interest: i=1,CO; i=2,NO; i=3,NH$_3$; i=4,O$_2$; i=5,C$_3$H$_6$; i=6,H$_2$; i=7,C$_3$H$_8$, and X^i is the concentration of each species. (Note: In this TWC model, all hydrocarbon species are lumped into three general categories: fast burning hydrocarbons, slow burning hydrocarbons and inert hydrocarbons; and are characterized by propylene, propane and methane, respectively.) For this study of steady state TWC operation, the accumulation ($\partial/\partial t$) of mass and energy in the gas phase and upon the surface was neglected, and the conservation equations reduce to:

$$v\frac{dX_g^i}{dz} = -k_m^i G_A(X_g^i - X_s^i) \quad \textbf{(5)}$$

$$v\rho_g C_{P_s}\frac{dT_g}{dz} = -hG_A(T_g - T_s) \quad \textbf{(6)}$$

$$0 = k_m^i G_A(X_g^i - X_s^i) - R_i(\overline{\mathbf{X}}_s, T_s) \quad \textbf{(7)}$$

$$0 = \lambda_s(1-e)\frac{d^2 T_s}{dz^2} + \sum_{j=1}^{m}(-\Delta H)_j \overline{R}_j(\overline{\mathbf{X}}_s, T_s)$$
$$+hG_A(T_g - T_s) - h_{amb}S_{ext}(T_s - T_{amb}) \quad \textbf{(8)}$$

With this approximation, the inlet and boundary conditions for this two-point boundary value problem are:

$$X_g^i(0) = X_{in}^i \quad \textbf{(9)}$$

$$T_g(0) = T_{in} \quad \textbf{(10)}$$

$$\frac{dT_s(0)}{dz} = \frac{dT_s(L)}{dz} = 0 \quad \textbf{(11)}$$

where the boundary condition given in eq. 11 indicates that no energy from the substrate is exchanged with the flow at its inlet or outlet; however, provision is included so energy can be lost globally through the side of a catalyst element. Values of Nusselt, N$_u$, and Sherwood, S$_h$, numbers are obtained from conventional correlations with Reynolds, Prandtl and Schmidt numbers [see Bird, et al. (1965)]. The heat and mass transfer coefficients are calculated from

$$h = \frac{N_u \cdot \lambda_g}{2R_h} \quad \textbf{(12)}$$

$$k_m^i = \frac{S_h \cdot D_i}{2R_h} \quad \textbf{(13)}$$

where λ_g is the thermal conductivity in the gas, D_i represents the binary diffusion coefficients as calculated from Hirschfelder (1954) and R_h is the hydraulic radius (open area of the duct divided by its perimeter).

To determine the optimum values of the steady state kinetic rate parameters, we integrate eqs. 5-13 yielding values of species conversion efficiencies as inlet conditions, and kinetic parameters were varied in a prescribed manner to yield the "best" overall agreement between model and experiment. The procedure for parameter adjustment will be discussed later.

KINETIC MECHANISM

Since considerable effort has been expended in developing kinetic mechanisms for previous generation catalysts (Pt/Rh; 13:1; 40g/ft^3) and since the fundamental materials (precious metal type and substrate) are similar in the current generation catalyst formulations (Pt/Rh; 5:1; 40g/ft^3), we choose as a starting point the current kinetic mechanism (both kinetic pathways and rate expressions) developed for the 13:1 Pt/Rh catalyst. This kinetic description consists of 13 independent forward pathways for the oxidation of CO, H$_2$, C$_3$H$_6$, C$_3$H$_8$, and NH$_3$ with O$_2$ and NO as oxidizing agents, and is shown in appendix A. Their corresponding rich (R > 1; excess reductant) and lean (R < 1) kinetic rate expressions are shown in appendix B. See Gandhi(1976) for definition of R. Near stoichiometry (R=1), a linear blending function (discussed later) is used to obtain a smooth transition between lean-side rate expressions and their corresponding rich-side expressions.

476

This steady-state mechanism gives little importance to the water-gas-shift reaction, to the steam reforming reactions, or to the re-oxidation of NH_3 by NO, as they are not significant during steady-state operation [see Kummer (1980)], but will be included later in the kinetic mechanism which describes the operation of the catalyst under transient conditions. Also, as seen from the lead constant (C46 in appendix C) in the rate expression for ammonia oxidation by O_2, little importance is given to this pathway. The overall mechanism also indicates that saturated hydrocarbons are weak competitors with CO, H_2 and unsaturated hydrocarbons for the reduction of NO. In general, N_2O is considered equivalent to N_2, since it decomposes to N_2 at higher temperatures, and reaction pathways using N_2O are not present in this mechanism.

The corresponding rate expressions for this kinetic mechanism are shown in appendix B. They suggest the typical Langmuir-Hinshelwood form for heterogeneous catalysis with added production and inhibition terms. The sum total of parameters is approximately 95. However, upon testing the sensitivity of each kinetic parameter, the results of the sensitivity analysis suggest that there are approximately 20 parameters which strongly influence model predictions.

Historically, these rate expressions were developed through a "building block" approach in which rates were established for kinetic subsystems (i.e, binary systems, trinary, etc.) until a fully blended (CO, NO, C_3H_6, C_3H_8, H_2, O_2, NH_3, N_2, SO_2, and H_2O) inlet stream was tested. This approach initially enabled the adjustment of subsystem kinetic parameters without the complication of cross production or inhibition effects by other chemical species. Subsequently, additional species were added to the inlet stream and the appropriate cross kinetic terms added. Since this level of detail is already present in the current kinetic mechanism and since we hope it is similar in character to the mechanism which will describe this generation of catalyst, our current approach is to start these experiments with a fully blended inlet stream and adjust a minimum number of kinetic parameters to obtain adequate agreement between model and experiment. As additional information is needed of kinetic subsystems, it is acquired and the mechanism modified.

PARAMETER OPTIMIZATION

To obtain an accurate description of catalyst performance over a broad range of conditions, we generate an experimental data base to assess catalyst model performance, test its subprocess descriptions and adjust the important submodel parameters to yield "good" overall agreement between prediction and experimental result. In the TWC model, the most complex submodel is that which describes the chemical processes which occur in the catalyst; this submodel contains 95 parameters. To adjust all or even a fraction of these manually would be an impossible task and computer optimization techniques are used to adjust the majority of these parameters. We therefore initialize the baseline parameters (previous rate parameters), run the TWC model for initial conditions which correspond to the initial conditions of the experiments, compare the results from each and adjust the model parameters to yield a smaller difference between the two results.

To judge the error between model and experiment, we choose the following merit function

$$\sigma(P) = \frac{\sum_{i=1}^{m} W_i (\eta_{model_i} - \eta_{experiment_i})^2}{\sum_{i=1}^{m} W_i} \tag{14}$$

where η is the conversion efficiency of each species of interest, m the total number of comparisons and W_i, a weighting function, which can, in principle, be related to the local standard deviation of these data. In a more mundane way, we use the weight as a means of skewing the parameter optimization to yield "better" local agreement to areas of greater importance, e.g. near a redox ratio of 1.0.

For the optimization procedure, we choose the method of conjugate gradients [see Press, et al. (1988)] in multi-dimensions. In N-dimensional parameter space, the problem is to minimize $\sigma(P)$ where P represents the vector of parameter values. If we choose a vector direction, n from P, we can then proceed toward a minimum value for σ in this direction using a one-dimensional minimization algorithm. From this local minimum, we then choose another direction and again repeat the one-dimensional minimization. Most minimization techniques of this type only differ in how one chooses this "new" direction, remembering that they are not independent choices. In the conjugate gradient method, one chooses a new direction in such a manner that the change in the gradient of the function σ is perpendicular to n [see Press, et al. (1988)], and then continues this procedure until a new minimum is reached. In our case, we apply additional constraints to this minimization algorithm which prevent it from choosing parameters, such as reaction order, outside of reasonable limits. We also choose to optimize only a fraction of the total number of parameters during any single computer run with the parameters involved all relating to a specific kinetic subsystem, e.g. CO oxidation by NO.

Before proceeding with the automated process of model parameter optimization, a few select parameters are initially optimized manually. These parameters deal with the blending of lean/rich-side kinetic expressions near $R = 1.0$. Since different rate expressions were previously developed on either side of stoichiometry, a blending function is necessary to allow a smooth transition from one to the other. The rate expressions and blending functions, in general, take the form

$$Rate_{overall} = \beta_1 \cdot Rate_{lean} + (1 - \beta_1) \cdot Rate_{rich} \quad \textbf{(15)}$$

where β_1 is used to blend the rich/lean hydrocarbon and CO reactions (reactions 2,3,6,11,12 and 13) and β_2 is used to blend the rich/lean hydrogen reactions (reactions 8 and 9). For this purpose, β is bounded between 0 and 1, and is a linear function of the pseudo local redox ratio (local meaning at a position along the substrate) defined below.

$$\beta_1 = A_1 \cdot R_1 + A_2 \quad \textbf{(16)}$$

$$\beta_2 = A_1 \cdot R_2 + A_2 \quad \textbf{(17)}$$

and

$$R_1 = [CO] + 9.0[C_3H_6] + 1.5[NH_3] - [NO] - 2.0[O_2] - R_{flip} \quad \textbf{(18)}$$

$$R_2 = [CO] + 9.0[C_3H_6] + 1.5[NH_3] + [H_2] - [NO] - 2.0[O_2] - H_{flip} \quad \textbf{(19)}$$

where A_1, A_2, R_{flip}, and H_{flip} are constants and can be determined during the optimization of parameters.

EXPERIMENTAL DETAILS

Flow reactor system

The reactor system is shown schematically in Figure 1. The feed gas mixture originates as pure cylinder gases which are blended in a gas mixing manifold. Tylan mass flow controllers precisely control the various gas flows in the system. Before entering this manifold, however, CO passes through a heated trap to decompose any $Fe(CO)_5$ present. This forestalls its decomposition in the reactor and prevents the subsequent deposition of iron, with its possible adverse effects. CO_2, in a parallel stream, is routed through a steam generator and mixed with the water vapor produced in the correct proportions to simulate the level of water in vehicle exhaust. The CO_2/steam mixture is then mixed with the remaining manifold gases. Once blended, the gases may either be routed through a reactor by-pass line and then analyzed at the analytical console, or may be routed through the catalyst-reactor system to measure the extent of reaction.

The reactor tube assembly consists of three segments. The front tube is filled with quartz chips to ensure complete mixing and to act as a heat exchange medium, thereby ensuring uniform temperature throughout the flow. The central tube is a tee shaped arrangement which allows for pulse O_2 injection prior to the catalyst (Note: This feature is a necessity to determine the extent to which transient kinetic reactions contribute to the overall chemical conversion through the catalyst; however, it is not used in these experiments.). The final tube contains the catalyst element, which is a core sample measuring 2.54cm diameter x 3.81cm long. This catalyst sample is wrapped with Fiberfrax$^{(TM)}$ mineral wool to prevent gases from leaking past its edges.

Once through the catalyst sample, gases may either be routed directly to the analytical console, or are routed through a furnace containing a pelleted oxidation catalyst which quantitatively converts NH_3 to NO, and then analyzed. In general, prior to analysis, all gases are passed through two condensers connected in series which cool the gases and extract any water vapor present. This procedure is used to protect the various analyzers. The following is a more detailed description of various aspects of the reactor system.

Analytical console

The analytical system consists of the following units:

1. UTI mass spectrometer (primarily for H_2 analysis),
2. Beckman FID hydrocarbon analyzer (detects hydrocarbons as ppm C, cannot speciate),
3. Beckman Chemiluminescence NO/NO$_x$ analyzer,
4. Beckman NDIR CO analyzer,
5. Ametek O_2 analyzer (very precise, "our primary standard").
6. HP gas chromatograph for HC speciation.

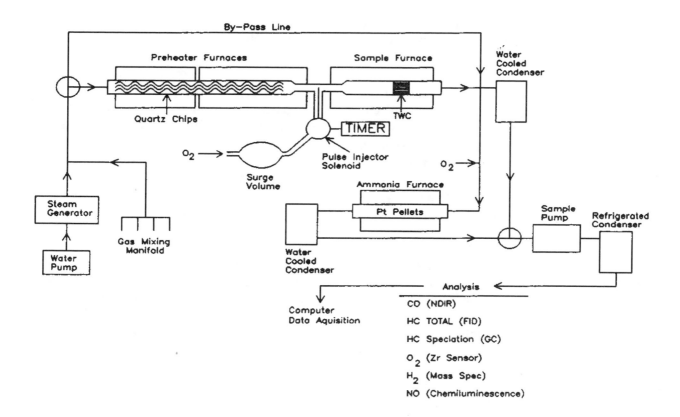

Fig. 1. Schematic diagram of the flow reactor system.

NH₃ analysis

Rather than utilizing the traditional method of analyzing NH_3 by wet chemical titration, we make use of the reaction $2 NH_3 + 2.5 O_2 \rightarrow 2 NO + 3 H_2O$ in the presence of an oxidation catalyst to quantitatively convert NH_3 to NO, with the subsequent increase in NO in the exhaust stream being measured by the chemiluminescence detector. We use the method as follows:

After analysis of NO in the exhaust stream, a valve is switched which routes a portion of the exhaust gas through a tube containing a pelleted Pt catalyst heated to 700°C. A small amount of O_2 is added upstream of this catalyst (usually sufficient to give an excess of 0.3-0.5% O_2 after reaction) and the resulting products are then passed to the analytical console. The increase of NO seen by the NO_x analyzer thus corresponds to the amount of NH_3 present in the exhaust stream from our test catalyst. This method has proven to be straightforward and quite reliable if precautions are taken and necessary conditions are met to assure overall conversion of NH_3 to NO. These conditions are the following:

1. It is important that the gases are sufficiently hot before they contact the catalyst; otherwise, the front pellets will cool to a temperature whereby a fraction of the NH_3 will be converted to N_2 rather than NO. To prevent this occurrence, the front of the NH_3 converter reactor tube contains quartz chips which act as a good heat exchange medium maintaining a high gas temperature. Using this arrangement, we find that there is a rather broad temperature plateau over which one obtains approximately 95% conversion of NH_3 to NO. This temperature range normally runs from 650-750°C, but is a function of space velocity into the oxidation catalyst and must be determined for each individual reactor system.

2. At extremely low space velocities, the reaction $2NH_3 + 3 NO \rightarrow 2.5 N_2 + 3 H_2O$ can become appreciable. Therefore, care must be taken to ensure that a sufficient space velocity exists to minimize this reaction pathway. As a rough

guide, we find that a space velocity of 10,000 hr^{-1} at 700°C is sufficient to eliminate this reaction.

Since ammonia present in the exhaust stream from a reactor is most commonly measured in our laboratory by the impinger/titration method rather than by the pelleted catalyst method mentioned above, we performed a few additional experiments to measure the amount of ammonia present, using both techniques to assure their agreement. In summary, the titration method involves routing a measured portion of the gas stream through an impinger containing a boric acid solution; the ammonia is collected in this solution; and the solution is then titrated using a known concentration of hydrochloric acid and indicator dye. For our comparison of these techniques, we first use a known unreacted blend of ammonia and nitrogen, with or without water present in the mixture. The results of this comparison are shown in figure 2 as the filled characters, and indicate exact correspondence between the two techniques. (Note: the dotted line in the figure corresponds to exact agreement between the titration method and the pelleted catalyst method). In addition, we also measured the amount of ammonia present in the outlet

gases from a heated Pd/Rh catalyst sample, when the full complement of reactive inlet gases is used. To vary the amount of ammonia in the exhaust stream, redox ratio to the catalyst sample was varied from 1.5 to 3. Figure 2 also shows the results of these measurements with the titration method measuring approximately 10% lower than the pelleted catalyst method, with the difference possibly being due to additional acidic species present in the product gases from the catalyst which would lower the results predicted by titration.

Steam generation (Inlet stream)

The steam generator system supplies 10% by volume of the total gas flow to the reactor. Water is supplied to the generator by an adjustable peristaltic pump which injects distilled water through a thin stainless steel tube at the entrance to the generator. The generator itself consists of a quartz tube which is heated in a "clamshell" type furnace. A strip of woven silica fabric (Siltemp$^{(TM)}$) with a very high capillary action is placed in the tube so that the leading end lies in a relatively cool zone, while the majority of the fabric lies in the hot zone. Water supplied to the cool section of the fabric immediately wicks to the hot zone, and is continuously vaporized. The resulting vaporized water is carried along by a stream of carbon dioxide which is routed through the tube and finally blends in with the remainder of the inlet gases. (Note: All lines downstream of the steam generator are heated to prevent condensation of the water vapor.)

Water removal (Exhaust stream)

Prior to analysis, it is necessary to remove water vapor from the gas stream. The formation of liquid droplets in certain analyzers would have effects ranging from dirtying optics to possibly cracking heated components. To dry the gas stream, two condensers are placed in series before the analytical console. The first condenser is water cooled and removes the bulk of the water vapor. A second condenser (refrigerant cooled to 1°C) is located just downstream of the sample pump which supplies the analytical console. These condensers remove sufficient water to ensure the reliability and durability of the analytical system.

Pulse injection (For future transient experiments)

In order to simulate the non-steady state nature of the inlet gas stream into a catalyst, we have added a gas-pulse injection system which enables us to run in

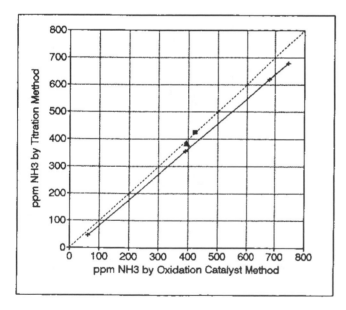

Fig. 2. - *Comparison of ammonia concentration as measured by the titration method and by the pelleted catalyst method. The dotted curve would represent exact correspondence between the two methods. The filled triangles represent results obtained using an unreacted blend of NH$_3$, N$_2$ and H$_2$O; the filled squares an unreacted blend of NH$_3$ and N$_2$. The crosses indicate results obtained while analyzing the product gases from a Pd/Rh catalyst sample with T$_{inlet}$ = 482°C.*

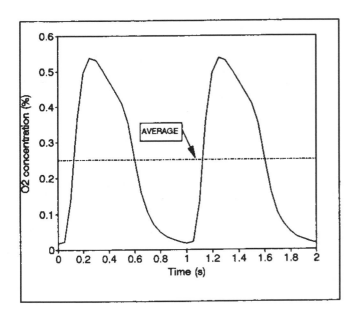

Fig. 3. - *Pulsed O_2 waveform shape. Total O_2 flow rate through injection system equals 40 cc/min. Base flow into reactor equals 16.09 l/min.*

either a modulated mode or a steady state mode. By running in a modulated mode, phenomena such as oxygen storage, water-gas shift and steam reforming may be observed and evaluated. The system injects a well characterized O_2 pulse into the steady state base flow directly upstream of the catalyst sample. The frequency and amplitude of the pulse are adjustable. Typically, we use a frequency of 1 Hz and an amplitude of 0.32% O_2.

The pulse injection system consists of the following components:

1. A mass flow controller which controls the total injected O_2 flow,
2. A small(150cc) surge volume to damp out pressure fluctuations and keep pulse shape constant,
3. A solenoid valve which injects the O_2 upstream of the catalyst by means of a small capillary,
4. An electronic counter circuit which controls the cycling of the solenoid valve.

A typical O_2 pulse is shown in Figure 3.

Data analysis (General)

The output from the analytical instruments and various detectors are recorded by an IBM PS-2/60 computer equipped with a Metrabyte data acquisition card capable of monitoring up to 16 channels of data

with a total throughput of 50KHz. The data can then be imported into a spreadsheet or other computer program for treatment as desired (i.e., calculation of mass balance, and conversion efficiencies). One use of the data acquisition system merits special mention. The O_2 pulse waveform must be thoroughly characterized; this is easily done by logging output from the O_2 analyzer at a sampling frequency of 100Hz.

Catalyst ageing

Prior to using the catalyst sample to obtain the experimental data base, it is put through an ageing process which attempts to simulate the initial period of rapid deterioration, with corresponding loss of activity, that a TWC experiences in actual operation. The treatment employed is patterned after the Thermal Deterioration portion of the Corporate Ageing Cycle for TWC's. Conditions are the following: Temperature = 788°C, space velocity = 25,000 hr^{-1}, duration = 24 hours. R=.7 (lean) with the following inlet gas concentrations: [CO]=1%; [H$_2$]=.33%; [C$_3$H$_6$]=500ppm; [NO]=1000ppm; [SO$_2$]=20ppm; [CO$_2$]=12.5%; [H$_2$O]=10%; and [O$_2$]=1.23%.

O_2 pulse characterization (For future transient experiments)

In order to know the magnitude of the imposed oscillations in redox ratio due to the injected O_2, it is necessary to accurately characterize the injected O_2 pulse. This measurement is performed on a monthly basis, since the pulse shape is quite constant with time. The waveform is analyzed using a fast response O_2 analyzer (Ametek S-3A) attached to a computerized data acquisition system which is set to acquire 100 data points/sec. The procedure is conducted using the following arrangement:

1. The normal total flow rate of gas (however, using N$_2$ only) is run through the catalyst sample. The O_2 pulsing system is then started and allowed to operate for a period of time in order to ensure that the surge volume is completely filled with O$_2$, thereby assuring a constant and uniform concentration.
2. A 1/8" stainless steel sampling tube is placed directly behind the center of the catalyst. The tube is placed at the exit of the reactor system and connects to a tee which routes the exiting gas sample directly to the O_2 analyzer cell. The excess

sample gas not used by the analyzer is dumped to atmosphere.

3. Sample gas is forced through the tube by means of the internal pressure of the reactor system, and is brought into the O_2 analyzer by a sample pump attached to the O_2 cell's outlet port.

The reason for this special plumbing arrangement for O_2 pulse characterization is to ensure that the waveform is not damped while flowing through a large amount of tubing or deformed due to pressure effects in the system. A number of cycles (100-1000) are then recorded, averaged and analyzed for their Fourier components. The average waveform is then incorporated into the model by an appropriate series representation.

Normal operating procedures

The procedures for steady-state and modulated operation are quite similar and are detailed below.

Steady-state operation

1. The gases (N_2, CO_2, H_2O) are blended and a constant flow of O_2 is added. These gases are then routed through the by-pass line and sent to the O_2 analyzer to precisely set the O_2 concentration. After determining the O_2 concentration without the presence of reductant, flow to the O_2 analyzer is stopped. This is necessary since O_2 in the presence of reductant would be combusted over the hot zirconium cell in the O_2 detector giving a reduced O_2 concentration, and possibly harming the cell.

2. The remaining gases are blended into the main stream and all other species are analyzed for their concentrations while by-passing the catalyst sample. Gases are then routed through the catalyst sample to determine catalyst conversion.

3. Sufficient time (it varies with inlet condition) is allowed to attain a steady-state condition (i.e. exiting species concentrations stabilize and remain constant). At this point, a portion of the sample is routed through the NH_3 conversion furnace and the NH_3 concentration is measured.

Modulated operation

1. The neutral gases are blended and routed through the catalyst. The pulse injection system is activated and gases are routed to the O_2 analyzer in the analytical console. By going through the full series of condensers, tubing etc., the injected O_2 pulse is damped to its average value by the time it reaches the console.

2. An additional steady stream of O_2 is then added to the gas flow at the main mixing manifold to attain the desired total average O_2 concentration into the catalyst sample.

3. At this point, the O_2 pulse injection system is stopped, gases are routed through the by-pass and the correct proportions of reductant are added to the blend.

4. Once the desired steady state blend is achieved, the full complement of gases is passed over the catalyst sample while simultaneously beginning pulse injection.

5. The reactor is allowed to stabilize and measurements of exhaust concentrations are made.

RESULTS AND DISCUSSION

The first step, in updating the kinetic parameters contained in the steady state TWC model, is to generate an experimental data base of CO, NO, NH_3, C_3H_6, C_3H_8, H_2 and O_2 conversion efficiencies (1.0 - [ ]$_{outlet}$/[ ]$_{inlet}$) as a function of inlet temperature, species concentrations and flow rate over a wide range of operating conditions. For these experiments, we used two catalyst formulations. The first formulation is a Pt/Rh (5:1) catalyst loaded to 40 g/ft^3. The second formulation is a Pd/Rh (5:1) catalyst also loaded to 40 g/ft^3. The cell density of the monolith was 400 cells/in^2. The inlet gas stream always contained 20ppm SO_2, 10% H_2O, 12% CO_2, and a balance of N_2 to yield a mixture of synthetic exhaust gas, in addition to the various concentrations of CO, NO, C_3H_6, C_3H_8, H_2 and O_2. 20ppm SO_2 is present since this is its average concentration level in automotive exhaust and since it can have a significant effect on the reducing power of H_2 and CO in Pt/Rh and Pd/Rh catalysts when R > 1 [see Kummer (1980)].

Conversion efficiencies for CO, NO, NH_3, C_3H_6, C_3H_8, H_2 and O_2 were determined for values of redox ratio (see appendix B for definition of redox ratio) between 0.5 and 4.5, at each of three inlet gas temperatures--371°C, 482°C, and 593°C. A synopsis of all experiments is listed in table I. For a complete redox/temperature sweep, 16 values of redox ratio were used, as were the three values of inlet temperature. This combination potentially yields 336 (7 species x 16 redox points x 3 temperatures) comparison points between model prediction and experimental result for one

482

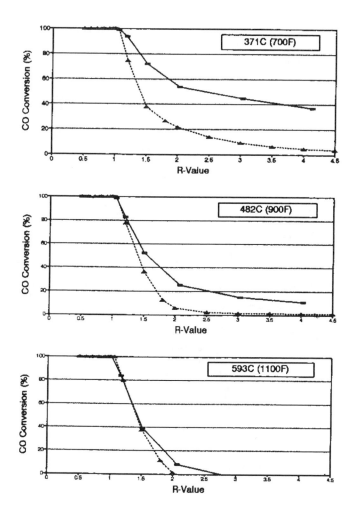

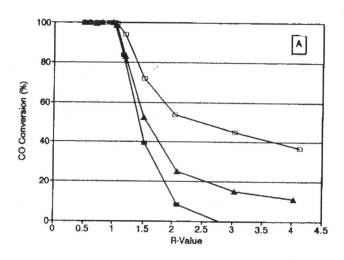

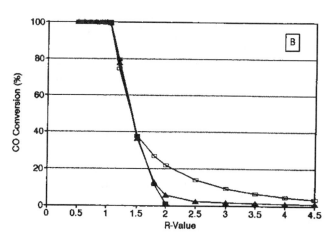

Fig. 4. - *Steady state CO conversion efficiency as a function of redox ratio. Triangles represent model predictions based on the previous kinetic mechanism for Pt/Rh catalysts. Squares represent experimental results using a current generation Pt/Rh catalyst formulation. Figure 4(a) shows results at 371°C; 4(b) at 482°C; and 4(c) at 593°C.*

Fig. 5. - *Steady state CO conversion efficiency: (A) from experiment; (B) from model predictions based on the kinetic mechanism for previous catalyst formulations. Unfilled squares represent data at 371°C; filled triangles at 482°C; and filled squares at 593°C.*

inlet stream composition. For the platinum catalyst sample, initial redox sweeps (tests 1-7 in table I) were necessary to determine performance differences when simpler kinetic subsystems were present. In addition, for this Pt catalyst, these first experiments were also necessary to continue the catalyst aging process, since the aging procedure, mentioned previously, was insufficient to pass the initial period of rapid catalyst deterioration. These experiments were not included in the final modeling procedure, but were used in a qualitative sense to determine differences in conversion efficiency as additional species were placed in the inlet stream. Results from experiments 8-13 (i.e. after the Pt catalyst sample stabilized) were used for modeling purposes as were the results from experiments 14 and 15 for the Pd/Rh catalyst sample.

(Note: For the Pd/Rh sample, the initial aging procedure, as described, was sufficient to stabilize the catalyst and no further aging was required.)

Comparison of Current Generation Catalyst Data and Model Predictions Base on Old Kinetic Mechanism

To define the differences between model predictions based on our previous kinetic mechanism and the results of experiments conducted using current Pt/Rh catalyst formulations, the TWC model was run using the previous kinetic mechanism for the conditions listed as test 13 in Table I. A comparison between the results from these experiments using a Pt/Rh catalyst sample and the corresponding model predictions is shown in Figures 4-9, with these comparisons being used to suggest potential changes to the kinetic mechanism.

As seen in figure 4, when R > 1.0, model predictions for CO conversion, based on the "old" kinetic mechanism, are lower than the experimental results generated from the current Pt/Rh catalyst sample at all three temperatures. For R < 1.0, the conversion efficiency is 100% for both model and experiment but this result yields little actual information, since breakthrough has not occurred at these conditions. These results are somewhat expected, since one design criteria for the new catalyst formulation was to significantly improve CO activity. At R = 2.0, and 371°C, the model predicts a conversion efficiency of 22% for CO, while experiment produced a 53% conversion efficiency. At 482°C, the difference between model and experiment was less; a conversion of 8% was obtained from the model and 25% from experiment; and, at 593°C, the model prediction was 1% CO conversion and the experiment yielded an 11% conversion. As shown in figures 5 (a) and (b), the model predictions of CO conversion do not exhibit the correct temperature dependence to describe the results obtained with this catalyst sample. In general, these results indicate that the parameters (see appendices B and C) associated with reaction steps 1, 3, 6, 7, and 11-13 (see appendix A) will require some modification when R > 1.0. The leading multiplicative constants may change, as well as the apparent activation energy or exponents on the inhibition terms.

Gross NO conversions are presented in Figure 6. As was expected, the new catalyst formulation has significantly improved rich-side NO activity. This is another designed-in feature of the new catalyst formulation over previous generation catalysts. In addition, to within experimental accuracy, lean-side NO conversions are comparable. As shown, when R > 1.0, the previous model predicts that the gross NO conversion will decrease as the redox ratio is increased and as temperature is raised. In contrast, the experimental results show little decrease in NO activity either as R increases or as inlet gas temperature is changed. For R > 1.0 the steady state NO conversion efficiency for the current Pt/Rh catalyst is always above 93%. At R = 4.0, the model predicts 80% NO conversion at 482°C and 57% conversion at 593°C. For this species, the old model kinetics are too sensitive to changes in temperature, and parameters associated with reactions 5-9 and 11-13 need some degree of modification. Figure 7 illustrates the need for adjustment of model parameters which govern the conversion of NO to ammonia. When R is greater than 1.5, the previous model predicts

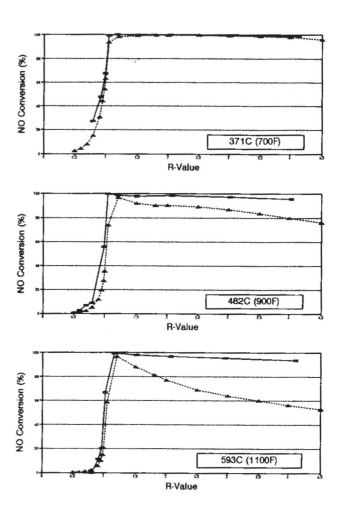

Fig. 6. - *Steady state NO conversion efficiency as a function of redox ratio. Triangles represent model predictions based on the previous kinetic mechanism for Pt/Rh catalysts. Squares represent experimental results using a current generation Pt/Rh catalyst formulation. The results shown in figure 6(a) were obtained at 371 °C; 6(b) at 482 °C; and 6(c) at 593 °C.*

less ammonia formation than was produced from this catalyst sample for 371°C and 482°C. For 593°C, the model predicts lower NH_3 levels over the entire range of redox ratio. Part of this difference can be accounted for, since the current Pt/Rh catalyst converts more NO than the previous generation catalyst formulation. However, in general, approximately 65% of the converted NO goes to NH_3 in the current Pt/Rh catalyst at a redox ratio of 4.0, while 45% was converted to NH_3 by the former Pt/Rh catalyst.

In our TWC model, all exhaust hydrocarbon species are placed into three representative groups -- "Fast" burning hydrocarbon species (olefins and aromatics), "Slow" burning species (small paraffinic hydrocarbon species) and "Inert" hydrocarbon species (methane). To assess HC conversion efficiency, we typically used a

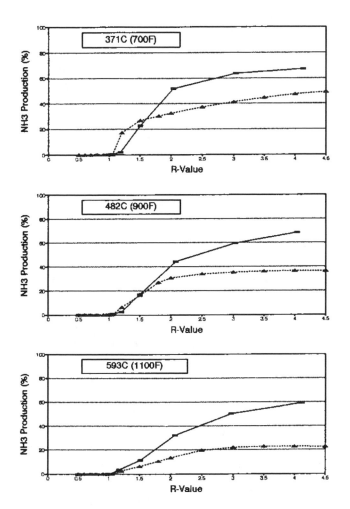

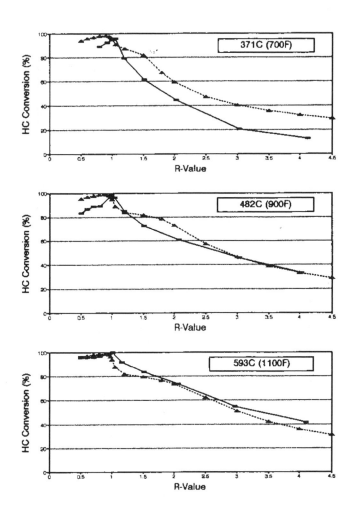

Fig. 7.- *Steady state production of NH₃ relative to inlet NO as a function of redox ratio. Triangles represent model predictions based on the previous kinetic mechanism for Pt/Rh catalysts. Squares represent experimental results using a current generation Pt/Rh catalyst formulation. The results shown in figure 7(a) were obtained at 371°C; 7(b) at 482°C; and 7(c) at 593°C.*

Fig. 8.- *Total steady state hydrocarbon conversion efficiency as a function of redox ratio. Triangles represent model predictions based on the previous kinetic mechanism for Pt/Rh catalysts. Squares represent experimental results using a current generation Pt/Rh catalyst formulation. The results shown in figure 8(a) were obtained at 371°C; 8(b) at 482°C; and 8(c) at 593°C.*

2:1 admixture of propylene (our characteristic "fast" burning hydrocarbon) and propane (our "slow" burning hydrocarbon). In addition, experiments with the Pt and Pd catalyst samples were performed using methane as the only reductant, and results from these experiments indicate that the conversion of methane was always less than 5% of its inlet level. Therefore, for these catalyst formulations, methane can still be treated as an inert species, and no additional kinetic steps are required to represent its oxidation characteristics.

Total hydrocarbon conversion efficiencies are shown in Figure 8 for an inlet stream containing 1000 ppm propylene and 500 ppm propane (see table I for full details of mixture). For R greater than 1.0, the results generated using the current Pt/Rh catalyst

show lower HC conversion at 371°C and approximately equal conversion at 482°C and 593°C. Differences between the curves (at 371°C) which represent model predictions and experimental results can be resolved by simple mass balance between CO, NO and hydrocarbon species, when R > 1.0, and essentially all oxidant is consumed, i.e., there is a tradeoff between CO and HC conversion. As an example, at 371°C with R equal to 2.0, the CO conversion efficiency of the "the current generation Pt catalyst formulation is 33% greater than the result predicted using the old model kinetics. Since all oxidant is consumed in either instance at this value of R, this increase must occur at the expense of another reductant species -- in this case the hydrocarbon species. For this blend of hydrocarbon species it takes approximately 3 times the amount of the oxidant,

485

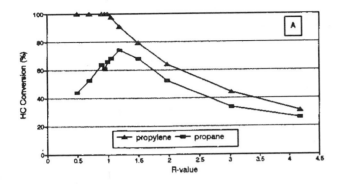

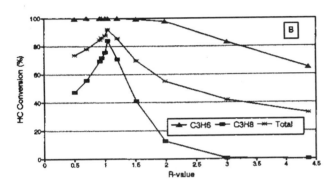

Fig. 9. - *Hydrocarbon conversion efficiency as a function of redox ratio. Figure 9(A) was generated using either 750 ppm propane or 750 ppm propylene in the inlet stream. Figure 9(B) was generated using a mixture of 750ppm propane and 750ppm propylene. T_{inlet} = 483°C; current generation Pt/Rh sample.*

NO, to convert a hydrocarbon molecule (represented as ppm'C_1') as it does to convert a CO molecule. Therefore, in principal, the hydrocarbon conversion should decrease by approximately 11% (33% CO increase/3). As seen in figure 8(a), the HC conversion efficiency decreases by approximately 15% at this value of redox ratio. In general, the differences in the other hydrocarbon conversion efficiency curves can be accounted for by the same procedure.

When R is less than 1.0, the results of our experiments show greater oxygen inhibition than the model predicts and, hence, at 371°C and 483°C, the total steady state hydrocarbon conversion efficiency is slightly lower for this new generation of catalyst. This inhibition is historically attributed to the presence of saturated hydrocarbon species [see Kummer(1980)], and this can be tested using a gas

chromatograph to differentiate between propane and propylene in the exhaust from our reactor or by placing only propane or propylene in the inlet stream. Figure 9(a) shows the resulting hydrocarbon conversion efficiency when either 750 ppm propylene or 750 ppm propane is used in the inlet stream as the total reductant. When only propylene is used, 100% propylene conversion occurs for R less than 1.0. When only propane is used, the maximum conversion is less than 100% (indicating lower reactivity) and, in addition, the propane conversion efficiency drops steadily as the O_2 concentration is increased.

When both 750 ppm propylene and 750 ppm propane are present in the inlet stream, the characteristic shape of the total hydrocarbon conversion efficiency versus redox curve is revealed, as shown in figure 9b (center curve). In addition, this figure also shows the results of speciating hydrocarbon molecules present in the exhaust stream. Both propane and propylene conversion efficiencies are shown and the results indicate, as expected, that the propylene is a significantly better competitor for oxidant than is the propane. This is clearly indicated by the fact that the conversion efficiency for propylene is nearly 100% to a redox ratio (based on both reductant) of 2.0. In essence, if one hydrocarbon species reacts much faster than its counterpart, it will use all available oxidant, its partner will have a much lower conversion efficiency, and its apparent stoichiometric proportion of oxidant will occur at a higher redox ratio when this redox ratio is based on the presence of both reductants in the inlet stream.

Comparison of Pt and Pd Catalyst Data

One specific intent of changing from previous catalyst formulations to current generation Pt/Rh or Pd/Rh formulations was to improve CO and NO conversion efficiency [Gandhi,H.S. (1990)], and Pd/Rh catalysts were designed to have equal or superior performance to their Pt/Rh counterparts. In general, these criteria have been met over a wide range of operating conditions although differences in catalyst performance do exist for specific inlet conditions. These differences may arise from fundamental differences in CO and HC oxidation rates for these two metals and may be accentuated in these tests, since 33% of the inlet hydrocarbon is propane which can be less active on Pd [see Kummer (1980)]. (Note: In general, the specific activity of Pd is greater than that of Pt for the oxidation of CO, olefinic hydrocarbons and methane; they are about equal for aromatics; and Pt is generally more active for the oxidation of paraffinic hydrocarbons.) This percentage of saturated hydrocarbon is greater

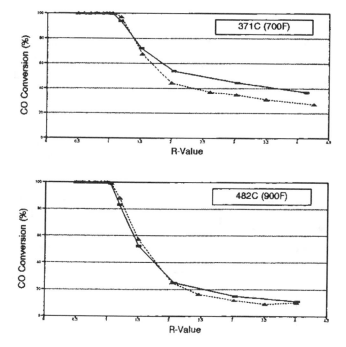

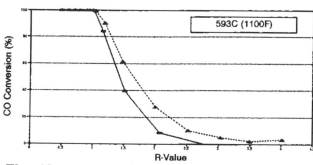

Fig. 10. - *Steady state CO conversion efficiency as a function of redox ratio. Triangles represent the results obtained using a current generation Pd/Rh catalyst sample; squares represent results obtained using a current Pt/Rh sample. Data in figure 10(A) are taken at 371°C; 10(B) at 482°C; and 10(C) at 593°C.*

than that which is normally used (14%) to model vehicle emission system performance. In these experiments, it was our intent to use this higher concentration of saturated hydrocarbons to amplify the differences in conversion between "fast" and "slow" burning hydrocarbon species and to clearly show the effect of oxygen inhibition on the conversion efficiency of saturated hydrocarbon species when R was less than 1.0. Also, when R is greater than 1.0, a difference in conversion efficiency of one reductant will alter the conversion efficiency of another, since O_2 and NO are depleted when using these catalyst formulations. Therefore, by simple mass balance, an increase in one reductant's conversion efficiency will arise at the expense of another's overall conversion. As an example, whenever the HC conversion efficiency (for R > 1.0) is greater in one

formulation, the CO conversion efficiency is usually lower in that formulation.

To show the differences in performance between these two catalyst formulations, the steady state conversion efficiencies of CO, NO, NH_3 and total HC have been plotted as a function of redox ratio in figures 10-13 for both catalyst formulations. These data were obtained using a full complement of inlet gases (NO, C_3H_6, C_3H_8, CO, H_2, O_2, SO_2, CO_2, H_2O, N_2; see table I for details) for temperatures between 371°C and 593°C. For many of these gases, the conversion efficiencies of the two catalysts were similar over a wide range of conditions with several of the major differences occurring when the temperature was lowest - - 371°C.

As mentioned previously, the steady state CO activity has increased for both current generation catalyst formulations over previous catalyst formulations. However, differences in CO activity do exist between

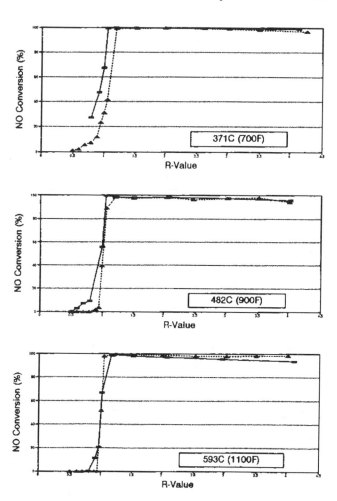

Fig. 11. - *Steady state NO conversion efficiency as a function of redox ratio. Triangles represent the results obtained using a current generation Pd/Rh catalyst sample; squares represent results obtained using a current Pt/Rh sample. Data in figure 11(A) are taken at 371°C; 11(B) at 482°C; and 11(C) at 593°C.*

the current generation Pt/Rh and Pd/Rh catalyst formulations and are shown in Figure 10. Their conversion efficiencies are identical at 482°C for all redox ratios tested, but differ at 371°C and 593°C on the rich side of stoichiometry. At 371°C, their CO conversion efficiencies differ by approximately 10% when R is greater than 2.0 (Pt having the higher conversion); at 593°C, the maximum difference in CO conversion efficiency (Pd having the higher conversion) is approximately 20% when $1.25 \leq R \leq 2.0$. Some of this difference in steady state CO conversion may relate to the large amount of "slow" burning hydrocarbon present in the inlet stream, since saturated hydrocarbon species have a lower activity on Pd than on Pt [see Kummer(1980)] or simply CO's greater activity on Pd. This lower activity for "slow" burning hydrocarbons may provide additional oxidant to extend CO activity. In either case, it represents the competition of each reductant for the available

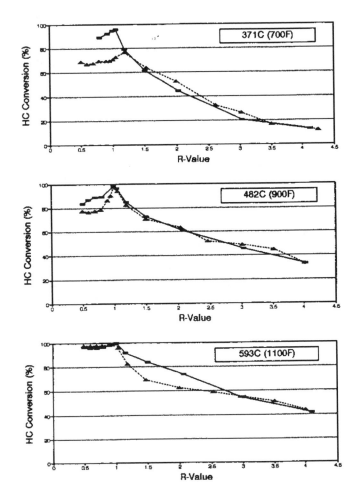

Fig. 13. - *Steady state hydrocarbon conversion efficiency as a function of redox ratio. Triangles represent the results obtained using a current generation Pd/Rh catalyst sample; squares represent results obtained using a current Pt/Rh sample. Data in figure 13(A) are taken at 371°C; 13(B) at 482°C; and 13(C) at 593°C.*

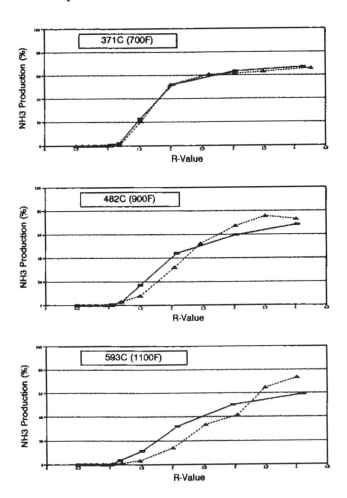

Fig. 12. - *Production of NH₃ relative to inlet NO as a function of redox ratio. Triangles represent the results obtained using a current generation Pd/Rh catalyst sample; squares represent results obtained using a current Pt/Rh sample. Data in figure 12(A) are taken at 371°C; 12(B) at 482°C; and 12(C) at 593°C.*

oxidant. Upon repeat experiments at R equal to 1.5, it was determined that the 1σ standard deviation in CO conversion efficiency about its mean value was approximately $\pm 3\%$ (conversion efficiency).

As with CO, the steady state NO conversion efficiency, as shown in Figure 11, has been substantially improved for both the current Pt/Rh and Pd/Rh catalysts over previous generation catalyst formulations. When R is greater than 1.0, the gross NO conversion efficiency for either current generation catalyst formulation is identical and is always above 95% conversion. In addition, at 482°C and 593°C, the steady state conversion of NO is essentially similar between these Pt/Rh and Pd/Rh formulations when R is less than or equal to 1.0. However, when the inlet gas temperature was lowered to 371°C, the Pt/Rh catalyst sample showed greater steady state NO conversion

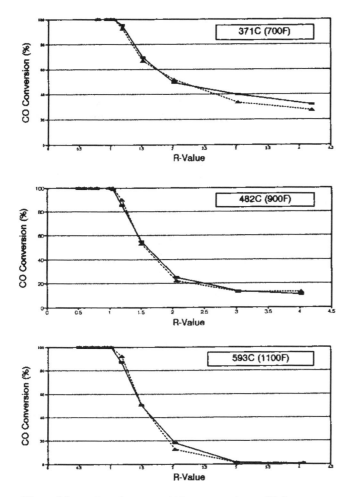

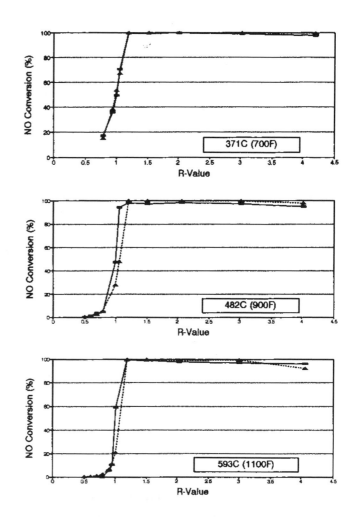

Fig. 14. - *Steady state CO conversion efficiency as a function of redox ratio. Triangles represent model predictions based on new kinetic parameters (see appendix D); squares represent averaged conversion efficiencies for Pt/Rh and Pd/Rh catalyst samples. Data in figure 14(A) were taken at 371°C; 14(B) at 482°C; and 14(C) at 593°C.*

Fig. 15. - *Steady state NO conversion efficiency as a function of redox ratio. Triangles represent model predictions based on new kinetic parameters (see appendix D); squares represent averaged conversion efficiencies for Pt/Rh and Pd/Rh catalyst samples. Data in figure 15(A) were taken at 371°C; 15(B) at 482°C; and 15(C) at 593°C.*

than its counterpart. Some of this difference in NO conversion efficiency at 371°C may be due to the exact placement of redox about a value of 1.0, since the drop in NO conversion is very steep, but, in general, for all temperatures tested, the steady state NO conversion was greater for the Pt catalyst as the level of excess oxygen was increased. (Note: some of this difference is reduced upon transient operation of the catalyst.)

The conversion of NO to ammonia is shown in Figure 12 for both catalyst formulations and all temperatures studied. At 371°C, their conversions are identical for all redox ratios tested. At 482°C and 593°C, the general trend of NO conversion to NH_3 is similar in either formulation, with the Pd catalyst producing less ammonia when R is less than 2.5 and a greater amount when R is near 4.0. At higher inlet

gas temperatures, the range in redox ratio over which Pd performance is better than the Pt catalyst is extended. However, as mentioned previously, both catalyst formulations produced a greater amount of ammonia than did previous catalysts under steady state conditions, and if ultimately used in combination with an additional downstream oxidation catalyst, this NH_3 may be reconverted to NO, thus reducing the overall improvement in gross NO conversion.

Total hydrocarbon conversions for the Pt/Rh and Pd/Rh catalyst samples are shown in figure 13. In general, for all values of redox ratio studied, their HC conversion efficiency increases as inlet gas temperature was increased. When R is greater than unity, they produce identical steady state conversions, except at 593°C. At this temperature, the total HC conversion is less in the Pd/Rh catalyst when $1 \leq R \leq 3$. This

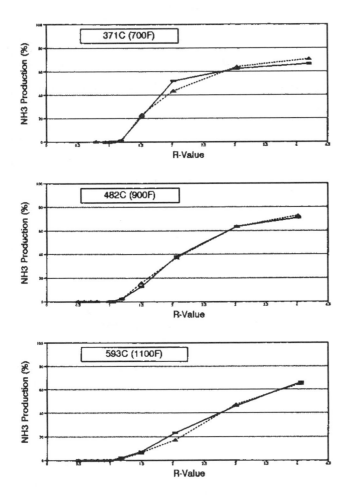

Fig. 16. - *Production of NH₃ relative to inlet NO as a function of redox ratio. Triangles represent model predictions based on new kinetic parameters (see appendix D); squares represent averaged conversion efficiencies for Pt/Rh and Pd/Rh catalyst samples. Data in figure 16(A) were taken at 371°C; 16(B) at 482°C; and 16(C) at 593°C.*

difference may be due to the level of saturated hydrocarbons in the inlet gas stream, since their activity on Pd is generally lower than on Pt; or, it may simply be due to the greater CO activity on Pd. As mentioned, since all oxidant is depleted when using these catalyst formulations with R greater than 1.0, an increase in conversion of one reductant arises at the expense of the other. When R is less than 1.0, the Pd/Rh catalyst is affected more strongly by oxygen inhibition at both 371°C and 482°C than the Pt catalyst. At 593°C, the oxidation rates of "slow" burning hydrocarbons are sufficiently rapid so little breakthrough occurs in either catalyst formulation. As shown earlier, in figure 9, the reduced hydrocarbon conversion efficiency is due to the presence of propane in the inlet gas stream, and is accentuated by our chosen level (33% rather than 14% used in modeling vehicle exhaust) of propane in

the inlet stream; this difference may be less when modeling a typical exhaust system.

Comparison of New Model and Experimental Data

To generate a new set of kinetic rate expressions, these experimental data are compared to model predictions at equivalent inlet conditions, and the appropriate parameters in the kinetic rate expressions are adjusted to yield "good" overall agreement between model and experiments, while ensuring that the value of any one individual parameter is within reasonable limits. Using these data, we have determined three sets of kinetic expressions: one set for the "averaged" performance of these two catalyst formulations (presented in appendix D); one set based on the results from the Pt/Rh sample (appendix E) and one set based on the results of the Pd/Rh sample (appendix F). In

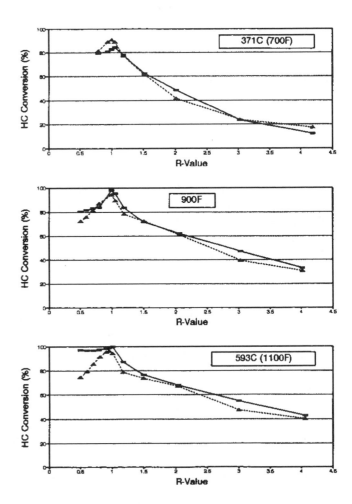

Fig. 17. - *Steady state hydrocarbon conversion efficiency as a function of redox ratio. Triangles represent model predictions based on new kinetic parameters (see appendix D); squares represent averaged conversion efficiencies for Pt/Rh and Pd/Rh catalyst samples. Data in figure 17(A) were taken at 371°C; 17(B) at 482°C; and 17(C) at 593°C.*

general, since the difference in their performance will be less than that indicated in figures 10-13 (i.e., the fraction of saturated hydrocarbons will be lower in vehicle exhaust) and in many cases may lie within experimental error, the rate constants based on their "averaged" performance may be sufficient for many of the modeling studies envisioned. Where specific differences between these formulations are important, the appropriate set of constants should be utilized.

As an example of the results produced by one set (based on the "averaged" catalyst performance) of these kinetic constants, CO, gross NO, NH_3, and HC conversion efficiencies are plotted in figures 14-17 for each temperature studied as a function of redox ratio. Also included are the corresponding experimental results. In general, the new constants reproduce the data to within experimental accuracy. The one feature which could still be improved is the oxygen inhibition at 593°C shown in figure 17. This may require the addition of another parameter to rate expression 13 or may ultimately require a separate rate expression for rich- and lean-side operation, as is done with many of the other rate expressions. In general, we believe that the error incurred by this difference is small when compared with other simplifications in the model or in experiments to analyze vehicle performance.

CONCLUDING REMARKS

An experimental data base consisting of CO, NO, NH_3, H_2, O_2 and HC conversion for two catalyst formulations was generated as a function of redox ratio between 0.5 and 4.0 and inlet gas temperatures between 371°C and 593°C. These data were used to update the parameters in our TWC model, which are used to describe the steady state chemical action of the catalyst. A set of kinetic parameters was generated for a Pt/Rh (5:1;40g/ft³) catalyst sample-- see appendix E--, a Pd/Rh(5:1; 40g/ft³) catalyst sample--see appendix F--, and for the results based on their "averaged" performance--see appendix D. These "new" kinetic parameters are being incorporated into the steady state TWC model and that portion of the kinetic mechanism in the transient version of the TWC model which describes its steady state base operation. In addition, work continues on determining the kinetic constants which describe the transient operation of the catalyst (O_2 storage effects, water gas shift and steam reforming reactions) for these catalyst formulations. Upon completion, they also will be incorporated into the TWC model which describes the full transient operation of the catalyst.

Nomenclature

A_1	=	kinetic rate blending function constant, 2500
A_2	=	kinetic rate blending function constant, 0.5
Cp_g	=	specific heat of gas, J/kg•K
Cp_s	=	specific heat of substrate, J/kg•K
D_i	=	diffusion coefficient, M^2/s
G_A	=	Geometric surface area, M^2/M^3
ΔH_j	=	heat of reaction j, J/mole
h	=	heat transfer coefficient between flow and substrate, J/M^2•s•K
h_{amb}	=	heat transfer coefficient between substrate and atmosphere, J/M^2•s•K
km^i	=	mass transfer coefficient for species i, M/s
L	=	length of substrate, M
m	=	total number of comparisons
N_u	=	Nusselt number
P	=	Vector of kinetic parameter values
R_j	=	reaction rate of jth reaction, mole/M^3•s
R_h	=	hydraulic radius, duct cross-sectional area/perimeter
S_{ext}	=	external surface to volume area ratio, M^2/M^3
S_h	=	Sherwood number
t	=	time, s
T_{amb}	=	ambient temperature, K
Tg	=	gas temperature, K
Ts	=	substrate temperature, K
v	=	average flow velocity through substrate channel, M/s
W_i	=	Weighting function
Xg^i	=	gas phase concentration, moles/M^3
Xs^i	=	surface concentration, moles/M^3
Xs	=	vector of surface concentrations, moles/M^3
z	=	axial length along substrate, M
β	=	linear kinetic blending function
ϵ	=	void volume fraction
η	=	conversion efficiency, {[ ]$_{inlet}$-[ ]$_{outlet}$}/[ ]$_{inlet}$
λ_g	=	thermal conductivity of gas, J/M•s•K
λ_s	=	thermal conductivity of substrate, J/M•s•K
ρ_g	=	gas density, kg/M^3
ρ_s	=	substrate density, kg/M^3
σ	=	weighted merit function to define error

Acknowledgement

We thank Dr. K. Otto for many discussions on catalyst action and Mr. D.R. Brigham for discussions and analysis of exhaust heat transfer. We also thank N.C. Otto for writing the original Ford steady-state and transient catalyst codes.

REFERENCES

Bird,R.B, Stewart,W.E. and Lightfoot, E.N., **Transport Phenomena**, John Wiley and Sons, Inc., New York, 1965.

Gandhi,H.S., Delosh,R.G., Piken,A.G. and Shelef,M., "Laboratory evaluation of three-way catalysts," SAE Trans. Sec. 2, Vol. 85, p.201, Society of Automotive Engineers, (1976).

Gandhi,H.S., Private communication (1990).

Hirschfelder, J.O., Curtiss, C.F. and Bird, R.B., **Molecular theory of gases and liquids**, John Wiley and Sons, Inc., New York, 1954.

Kummer,J.T., "Catalysts for automotive emission control," Prog. Energy Combustion Sci., **6**, p177-199, (1980).

Kummer,J.T., **Fuel Economy**, Hillard,J.C. and Springer,G.S. (Ed.), Plenum Publishing Corp., 1984.

Kuo, J.W.C., Morgan, C.R. and Lasson,H.G., "Mathematical modeling of CO and HC catalytic converter systems," S.A.E. paper no. 710289, The Society of Automotive Engineers, (1971).

Oh, S.H., "Thermal response of monolithic catalytic converters during sustained engine misfiring: A computational study," S.A.E. paper no 881591, The Society of Automotive Engineers, (1988a).

Oh, S.H., "A three-dimensional model for the analysis of transient thermal and conversion characteristics of monolithic catalytic converters," S.A.E. paper no. 880282, The Society of Automotive Engineers, (1988b).

Otto, N.C. and LeGray W.J., "Mathematical models for catalytic converter performance," S.A.E. paper number 800841, The Society for automotive engineers (1980).

Young, L.C. and Finlayson, B.A., "Part I. Development of model and application of orthogonal collocation," AIChE journal, Vol 22, No 2, p331, (1976).

Press, W.H., Flannery, B.P., Teuolsky, S.A., and Vetterling, W.T., **Numerical Recipes**, Cambridge University Press, Cambridge, England, 1988.

Table I
Steady State Flow Reactor Experiments

Inlet Gases[a, b, d]	Inlet Gas Temperature (°C)	Catalyst[h]	Test Number
1.33% CO, 1000ppm NO, O_2[g]	482	Pt/Rh	1
2% CO, 2000ppm NO, O_2[g]	482	Pt/Rh	2
2% CO, 500ppm NO[g]	482	Pt/Rh	3
1000ppm C_3H_6, 1000ppm NO, O_2[g]	482	Pt/Rh	4
1500ppm C_3H_6, 500ppm NO, O_2[g]	482	Pt/Rh	5
1500ppm C_3H_6, 1000ppm NO, O_2[g]	371, 482, 593	Pt/Rh	6
Simulated standard exhaust gas mixture[c, g]	371, 482, 593	Pt/Rh	7
900ppm CH_4, O_2[i]	482, 593	Pt/Rh	8
1500ppm C_3H_8, O_2[i]	482	Pt/Rh	9
1500ppm C_3H_6, O_2[i]	482	Pt/Rh	10
750ppm C_3H_6, 750ppm C_3H_8, O_2[i]	371, 482, 593	Pt/Rh	11
750ppm C_3H_6, 750ppm C_3H_8, 1000ppm NO, O_2[i]	482	Pt/Rh	12
Simulated standard exhaust gas mixture[c, e, i]	371, 482, 593	Pt/Rh	13
Simulated standard exhaust gas mixture[c, e, i]	371, 482, 593	Pd/Rh	14
Simulated standard exhaust gas mixture[c, f, i]	482, 593	Pd/Rh	15

[a]Space velocity = 50,000hr^{-1}. Redox set by varying the inlet O_2. All concentrations specified are before the addition of 10% H_2O by volume.

[b]Final gas mixtures contain: 10% H_2O vapor, 12% CO_2, 20ppm SO_2, plus balance nitrogen, by volume.

[c]Simulated standard exhaust mixture: 1% CO, 0.33% H_2, 1000ppm NO, 1000ppm C_3H_6, 500ppm C_3H_8, plus balance gas mixture specified in b.

[d]Full redox scans were performed unless otherwise noted. R = 0.5, 0.6, 0.7, 0.8, 0.9, 0.95, 1.0, 1.05, 1.2, 1.5, 2.0, 2.5, 3.0, 3.5, 4.0.

[e]Abbreviated redox scan: R = 0.8, 1.0, 1.05, 1.2, 1.5, 2.0, 3.0, 4.0.

[f]Mini redox scan: R equals 1.0, 1.5, 2.0.

[g]Experiments performed prior to full catalyst aging.

[h]Catalyst samples were 2.54cm dia.x3.81cm, 400 cell/in^2, 40g/ft^3 with a Pt/Rh or Pd/Rh ratio of (5:1). Aged by the procedure specified in the text.

[i]Results used to assess TWC model predictions and to adjust kinetic parameters.

APPENDIX A
Major Reactions

$$CO + \frac{1}{2}O_2 \rightarrow CO_2$$

$$CH_\alpha + M_1 \cdot O_2 \rightarrow CO_2 + \frac{\alpha}{2} \cdot H_2O \qquad (HC_F)$$

$$CH_\alpha + M_2 \cdot O_2 \rightarrow CO + \frac{\alpha}{2} \cdot H_2O \qquad (HC_F)$$

$$H_2 + \frac{1}{2}O_2 \rightarrow H_2O$$

$$NH_3 + \frac{3}{4}O_2 \rightarrow 1.5H_2O + \frac{1}{2}N_2$$

$$CO + NO \rightarrow CO_2 + \frac{1}{2}N_2$$

$$2.5CO + NO + 1.5H_2O \rightarrow NH_3 + 2.5CO_2 + \frac{1}{2}N_2$$

$$H_2 + NO \rightarrow H_2O + \frac{1}{2}N_2$$

$$2.5H_2 + NO \rightarrow NH_3 + H_2O$$

$$CH_\alpha + M_1 \cdot O_2 \rightarrow CO_2 + \frac{\alpha}{2} \cdot H_2O \qquad (HC_S)$$

$$\frac{1}{2M_2}CH_\alpha + NO \rightarrow \frac{1}{2M_2}CO + \frac{\alpha}{4M_2}H_2O + \frac{1}{2}N_2 \qquad (HC_F)$$

$$CH_\alpha + M_3 \cdot NO \rightarrow CO_2 + \frac{\alpha}{2} \cdot H_2O + \frac{M_3}{2} \cdot N_2 \qquad (HC_F)$$

$$\frac{2.5}{2M_2}CH_\alpha + NO + \frac{(3-\alpha)}{4M_2}H_2O \rightarrow NH_3 + \frac{2.5}{2M_2}CO \qquad (HC_F)$$

where α = Hydrogen-to-Carbon RATIO, and
$M_1 = [1 + (\alpha/4)]$; $M_2 = [\frac{1}{2} + (\alpha/4)]$; $M_3 = [2 + (\alpha/2)]$
HC_F = Fast burning HC; HC_S = Slow burning HC.

APPENDIX B
Steady-state rate expressions

$$RATE(1) = \frac{C_1 e^{-EA(1)/RT_s} X_{CO} X_{O_2}^{ex(1)}}{[(1 + C_4 X_{CO})^2 (1 + C_5 X_{C_3H_6})]}$$

$$RATE(2) = [C_2 e^{-EA(6)/RT_s} + (1-\beta_1) C_3 e^{-EA(3)/RT_s}] \cdot \frac{3 X_{C_3H_6} X_{O_2}^{ex(2)} (1 + C_{31} X_{C_3H_6} X_{CO})}{(1 + C_{35} X_{C_3H_6})^2 (1 + C_{32} e^{-EA(7)/RT_s} X_{CO})}$$

$$RATE(3) = \beta_1 \cdot [C_3 e^{-EA(3)/RT_s}] \cdot \frac{3 X_{C_3H_6} X_{O_2}^{ex(2)} (1 + C_{31} X_{C_3H_6} X_{CO})}{(1 + C_{35} X_{C_3H_6})^2 (1 + C_{32} e^{-EA(7)/RT_s} X_{CO})}$$

$$RATE(4) = \frac{C_6 e^{-EA(4)/RT_s} X_{H_2} X_{O_2}}{[(1 + C_7 X_{H_2})^2 (1 + C_4 X_{CO})^2 (1 + C_5 X_{C_3H_6})]}$$

$$RATE(5) = \frac{C_{46} X_{NH_3}^{ex(4)}}{(1 + C_{47} X_{H_2} + C_{48} X_{CO} + C_{49} X_{C_3H_6})^2}$$

$$RATE(6L) = \frac{(1-\beta_1) C_{15} X_{CO} e^{-EA(2)/RT_s} X_{NO}^{ex(10)}}{[(1 + C_{16} X_{O_2})^{ex(11)} (1 + C_{50} X_{CO})^2]}$$

$$RATE(6R) = \frac{\beta_1 \cdot C_{11} e^{-EA(5)/RT_s} X_{CO} X_{NO}^{ex(5)} (1 + C_{12} X_{O_2})^{ex(6)}}{[F1 + F2 + F3 - 2]^2}$$
$$F1 = (1 + CMILL \cdot X_{CO})^{ex(7)}$$
$$F2 = (1 + C_{14} X_{H_2})^{ex(8)}$$
$$F3 = (1 + C_{34} e^{EA(10)/RT_s} X_{C_3H_6})^{ex(23)}$$

494

$$RATE(7) =$$
$$\frac{\beta_1 \cdot C_{17} X_{CO} X_{NO}^{ex(13)} (1+CMILL \cdot X_{O_2})^{ex(14)}}{F4 \cdot F5 \cdot F6}$$
$$F4 = [1+C_{19}|T_S - C_{20}|^{ex(12)}]$$
$$F5 = [1+C_{13}X_{CO} + C_{33}X_{H_2}]$$
$$F6 = [1+C_{30}X_{C_3H_6}]$$

$$RATE(8L) = \frac{(1-\beta_2) C_{22} X_{H_2}^{ex(16)} X_{NO}^{ex(15)}}{(1+C_{44}X_{O_2})^2}$$

$$RATE(8R) = \frac{C_{21} X_{H_2} X_{NO}^{ex(9)} (1+C_{23}X_{O_2})^2 \cdot \beta_2}{[F7 + F8 + F9 - 2]^2}$$
$$F7 = (1+C_8 X_{CO})^{ex(7)}$$
$$F8 = (1+C_9 X_{H_2})^{ex(8)}$$
$$F9 = (1+C_{34}e^{EA(10)/RT_S}X_{C_3H_6})^{ex(23)}$$

$$RATE(9L) = (1.0 + \beta_2 \cdot CMILL \cdot X_{O_2})^{ex(18)} \cdot RATE(9R)$$

$$RATE(9R) = \frac{C_{25} X_{H_2} X_{NO}^{ex(17)}}{F10 \cdot F11 \cdot F12}$$
$$F10 = [1+C_{27}|T_S - C_{28}|^{ex(28)}]$$
$$F11 = [1+C_{24}X_{CO} + C_{18}X_{H_2}]$$
$$F12 = [1+C_{26}e^{EA(11)/RT_S}X_{C_3H_6}]$$

$$RATE(10) = \frac{3 C_{10}e^{-EA(8)/RT_S}X_{O_2} X_{C_3H_6}^{ex(26)}}{(1+C_{51}X_{O_2})^{ex(27)}}$$

$$RATE(11) =$$
$$\frac{\beta_1 6 C_{36} X_{C_3H_6} X_{NO}^{ex(24)} (1+C_{37}X_{O_2})^{ex(3)}}{(1+C_{39}X_{CO})[F1 + F2 + F3 - 2]^2}$$

$$RATE(12) = (1-\beta_1) 9 C_{42}e^{-EA(9)/RT_S}X_{NO}^{ex(22)} X_{C_3H_6}^{ex(25)}$$

$$RATE(13) =$$
$$\frac{\beta_1 \cdot 2.4 C_{38} X_{NO}^{ex(20)} X_{C_3H_6} (1+CMILL \cdot X_{O_2})^{ex(21)}}{[F13 \cdot F14 \cdot F15]}$$
$$F13 = [1+C_{40}|T_S - C_{41}|^{ex(19)}]$$
$$F14 = [1+C_{29}X_{C_3H_6}]^2$$
$$F15 = [1+C_{43}X_{CO} + C_{45}X_{H_2}]$$

where β_1 and β_2 are rich-to-lean blending functions;
$$0 \le \beta_1 \le 1 \text{ and } 0 \le \beta_2 \le 1$$

$$\beta_1 = 2500 R_1 + 0.5$$
$$\beta_2 = 2500 R_2 + 0.5$$

if $\beta < 0$, set $\beta = 0$
if $\beta > 1$, set $\beta = 1$

where

$$R_1 = X_{CO} + 9X_{C3H6} + 1.5X_{NH3} - X_{NO} - 2X_{O2} - R_{flip}$$

$$R_2 = X_{CO} + 9X_{C3H6} + 1.5X_{NH3} - X_{NO} - 2X_{O2} - H_{flip} + X_{H2}$$

R indicates rich-side (Redox > 1.0) rate
L indicates lean-side (Redox < 1.0) rate

$$Redox\ ratio = R = \frac{[CO] + [H_2] + 6 \cdot \left(1 + \frac{\alpha}{4}\right)[HC]}{[NO] + 2 \cdot [O_2]}$$

where α = the average hydrogen-to-carbon ratio of the fuel blend.

APPENDIX C
Previous generation Pt/Rh Kinetic constants

C1 = 2.04845E4	**EXPONENTS:**
C2 = 60.984	
C3 = 1.0601E6	EX1 = 1.2724
C4 = 303.53	EX2 = 1.0882
C5 = 14661	EX3 = 2.517
C6 = 2.5978E5	EX4 = 0.22484E1
C7 = 25.498	EX5 = 1.9079
C8 = 1.0577E06	EX6 = 2.075
C9 = 4.5825E06	EX7 = 1.3116
C10 = 184.34	EX8 = 1.2372
C11 = 3.6633E11	EX9 = 1.757886
C12 = 2.5655E04	EX10 = 0.0699
C13 = 1358.2	EX11 = 0.22347
C14 = 9.242E06	EX12 = 2.0
C15 = .051822	EX13 = 0.89458
C16 = 2.4175E07	EX14 = 0.77813
C17 = 1.2265E-3	EX15 = 0.99901
C18 = 6753.8	EX16 = 0.7082
C19 = 2.124E-6	EX17 = 0.73892
C20 = 1049.62	EX18 = 0.7116
C21 = 2.0755E10	EX19 = 2.0
C22 = 0.95396	EX20 = 1.0405
C23 = 61290.0	EX21 = 0.98519
C24 = 1579.3	EX22 = 1.0011
C25 = 1.4515	EX23 = 0.89309
C26 = 161.49	EX24 = 1.6036
C27 = 1.1572E-04	EX25 = 1.7273
C28 = 710.91	EX26 = 1.4428
C29 = 3170.7	EX27 = 2.2598
C30 = 11164.0	EX28 = 2.0
C31 = 0.13834E5	
C32 = 0.24075	**ACTIVATION ENERGIES:**
C33 = 7603.0	
C34 = 3.9355E09	EA1 = 10626.
C35 = 7474.8	EA2 = -8000.0
C36 = 5.8364E11	EA3 = 17185.8
C37 = 4132.5	EA4 = 12975.
C38 = 9.2382E-03	EA5 = 0.23097E4
C39 = 217.79	EA6 = 3699.8
C40 = 7.5254E-05	EA7 = -11278
C41 = 1503.6	EA8 = 2949.3E0
C42 = 1.265E07	EA9 = 14601.
C43 = 96.208	EA10 = 92.65
C44 = 558.70	EA11 = 5638.4
C45 = 5801.	
C46 = 0.0	CMILL = 1.0E06
C47 = .56488E5	RFLIP = 0.00057
C48 = 3.6744E4	HFLIP = 0.0
C49 = .22327E05	AH = 1.449E-06
C50 = 17.024	BH = 7.7E-10
C51 = 455.58	DIFCON = 6.4338E-07

APPENDIX D
Kinetic constants for Current Pt/Rh and Pd/Rh catalyst formulations (averaged performance)

C1 = 1.519E4	**EXPONENTS**
C2 = 539.1	
C3 = 8.794E5	EX1 = 1.225
C4 = 208.0	EX2 = 1.104
C5 = 7400	EX3 = 9.01E-4
C6 = 1.657E6	EX4 = 1.094
C7 = 1688	EX5 = 1.8960
C8 = 3.234E06	EX6 = 2.091
C9 = 3.034E06	EX7 = 1.295
C10 = 1067	EX8 = 1.180
C11 = 1.163E13	EX9 = 1.714
C12 = 3.103E04	EX10 = 1.140
C13 = 49.89	EX11 = 0.2203
C14 = 1.890E08	EX12 = 2.023
C15 = .01281	EX13 = 0.8885
C16 = 5.098E08	EX14 = 0.7949
C17 = 2.097E-4	EX15 = 0.8503
C18 = 1.599E5	EX16 = 0.8463
C19 = 9.639E-7	EX17 = 0.7040
C20 = 989.9	EX18 = 0.7187
C21 = 1.239E10	EX19 = 2.331
C22 = 64.73	EX20 = 1.068
C23 = 50450	EX21 = 1.023
C24 = 10210	EX22 = 0.8448
C25 = 20.58	EX23 = 1.937
C26 = 2288	EX24 = 1.533
C27 = 6.105E-5	EX25 = 1.733
C28 = 586.9	EX26 = 1.442
C29 = 390.2	EX27 = 3.118
C30 = 249.4	EX28 = 2.000
C31 = 5.611E4	
C32 = 0.02723	**ACTIVATION ENERGIES:**
C33 = 2.169E-5	
C34 = 4.420E05	EA1 = 10870
C35 = 7013	EA2 = -9747
C36 = 3.893E13	EA3 = 17400
C37 = 5.873E-6	EA4 = 13320
C38 = 1.898E-2	EA5 = 2622
C39 = 151.3	EA6 = 16270
C40 = 2.893E-5	EA7 = -12860
C41 = 1567	EA8 = 6989
C42 = 2.132E07	EA9 = 15820
C43 = 618.3	EA10 = 2.94E-3
C44 = 1033	EA11 = 5865
C45 = 2163	
C46 = 1.41E-16	CMILL = 3.0E06
C47 = 4.821E4	RFLIP = 5.7E-4
C48 = 7.057E4	HFLIP = 0.0
C49 = 2.343E05	AH = 1.45E-6
C50 = 19.17	BH = 7.7E-10
C51 = 592.0	DIFCON = 6.434E-7

<table>
<tr><td colspan="2">

APPENDIX E
Kinetic constants for Pt/Rh catalyst formulations

</td><td colspan="2">

APPENDIX F
Kinetic constants for Pd/Rh catalyst formulation

</td></tr>
</table>

APPENDIX E		APPENDIX F	
$C1 = 1.519E4$	**EXPONENTS**	$C1 = 1.734E4$	**EXPONENTS**
$C2 = 202.5$		$C2 = 1.084E3$	
$C3 = 8.966E5$	$EX1 = 1.225$	$C3 = 8.411E5$	$EX1 = 1.225$
$C4 = 208.0$	$EX2 = 1.104$	$C4 = 197.2$	$EX2 = 1.097$
$C5 = 7400$	$EX3 = 8.68E-4$	$C5 = 8171$	$EX3 = 1.03E-3$
$C6 = 1.556E6$	$EX4 = 1.094$	$C6 = 1.598E6$	$EX4 = 1.094$
$C7 = 1688$	$EX5 = 1.8960$	$C7 = 2563$	$EX5 = 1.887$
$C8 = 2.994E6$	$EX6 = 2.098$	$C8 = 1.849E06$	$EX6 = 2.075$
$C9 = 2.052E6$	$EX7 = 1.301$	$C9 = 4.737E05$	$EX7 = 1.259$
$C10 = 1048$	$EX8 = 1.180$	$C10 = 1483$	$EX8 = 1.147$
$C11 = 1.143E13$	$EX9 = 1.714$	$C11 = 2.101E13$	$EX9 = 1.474$
$C12 = 3.073E4$	$EX10 = 1.140$	$C12 = 3.169E04$	$EX10 = 0.07633$
$C13 = 45.60$	$EX11 = 0.2203$	$C13 = 13.75$	$EX11 = 0.8766$
$C14 = 1.937E8$	$EX12 = 2.031$	$C14 = 2.688E08$	$EX12 = 2.001$
$C15 = 7.015E-3$	$EX13 = 0.8891$	$C15 = 2.803E-3$	$EX13 = 0.8886$
$C16 = 5.098E8$	$EX14 = 0.7961$	$C16 = 5.098E08$	$EX14 = 0.7959$
$C17 = 2.253E-4$	$EX15 = 0.8503$	$C17 = 2.479E-4$	$EX15 = 0.8503$
$C18 = 6.308E4$	$EX16 = 0.8463$	$C18 = 1.222E3$	$EX16 = 0.8463$
$C19 = 1.105E-6$	$EX17 = 0.7034$	$C19 = 1.173E-6$	$EX17 = 0.7723$
$C20 = 955.2$	$EX18 = 0.7187$	$C20 = 870.1$	$EX18 = 0.7187$
$C21 = 2.061E10$	$EX19 = 2.465$	$C21 = 1.072E9$	$EX19 = 2.502$
$C22 = 69.67$	$EX20 = 1.091$	$C22 = 96.83$	$EX20 = 1.073$
$C23 = 49990$	$EX21 = 1.024$	$C23 = 75950$	$EX21 = 1.027$
$C24 = 10520$	$EX22 = 0.8448$	$C24 = 7217$	$EX22 = 0.8483$
$C25 = 36.01$	$EX23 = 1.987$	$C25 = 12.43$	$EX23 = 2.015$
$C26 = 2285$	$EX24 = 1.533$	$C26 = 2087$	$EX24 = 1.482$
$C27 = 6.107E-5$	$EX25 = 1.733$	$C27 = 6.37E-5$	$EX25 = 1.724$
$C28 = 597.9$	$EX26 = 1.442$	$C28 = 646.4$	$EX26 = 1.442$
$C29 = 250.8$	$EX27 = 3.047$	$C29 = 114.3$	$EX27 = 3.118$
$C30 = 230.5$	$EX28 = 2.001$	$C30 = 331.0$	$EX28 = 2.005$
$C31 = 5.094E4$		$C31 = 9.37E4$	
$C32 = 0.02559$	**ACTIVATION ENERGIES:**	$C32 = 0.03516$	**ACTIVATION ENERGIES:**
$C33 = 2.169E-5$		$C33 = 1.76E-8$	
$C34 = 4.896E05$	$EA1 = 10870$	$C34 = 4.65E5$	$EA1 = 10770$
$C35 = 6986$	$EA2 = -9718$	$C35 = 7549$	$EA2 = -7847$
$C36 = 2.135E14$	$EA3 = 17400$	$C36 = 4.17E13$	$EA3 = 17400$
$C37 = 1.450E-6$	$EA4 = 13320$	$C37 = 2.67E-6$	$EA4 = 13000$
$C38 = 2.544E-2$	$EA5 = 2.645E3$	$C38 = 1.98E-2$	$EA5 = 2.321E3$
$C39 = 412.7$	$EA6 = 16270$	$C39 = 70.39$	$EA6 = 7544$
$C40 = 1.344E-5$	$EA7 = -13190$	$C40 = 1.22E-5$	$EA7 = -12990$
$C41 = 1647$	$EA8 = 6492$	$C41 = 1544$	$EA8 = 7288$
$C42 = 2.097E7$	$EA9 = 14250$	$C42 = 2.48E7$	$EA9 = 15930$
$C43 = 439.2$	$EA10 = 2.01E-8$	$C43 = 633.5$	$EA10 = 3.34E-3$
$C44 = 1033$	$EA11 = 6073$	$C44 = 1033$	$EA11 = 3585$
$C45 = 2163$		$C45 = 902.8$	
$C46 = 2.42E-16$	$CMILL = 3.00E06$	$C46 = 2.69E-16$	$CMILL = 3.00E06$
$C47 = 4.821E4$	$RFLIP = 0.00057$	$C47 = 4.82E4$	$RFLIP = 0.00057$
$C48 = 7.057E4$	$HFLIP = 0.0$	$C48 = 7.06E4$	$HFLIP = 0.0$
$C49 = 2.343E5$	$AH = 1.449E-06$	$C49 = 2.34E5$	$AH = 1.449E-06$
$C50 = 19.17$	$BH = 7.7E-10$	$C50 = 19.17$	$BH = 7.7E-10$
$C51 = 452.5$	$DIFCON = 6.434E-7$	$C51 = 616.4$	$DIFCON = 6.4338E-07$

2000-01-1953

Comparison of Chemical Kinetic Mechanisms in Simulating the Emission Characteristics of Catalytic Converters

Tariq Shamim, Huixian Shen and Subrata Sengupta
The University of Michigan - Dearborn

Copyright © 2000 CEC and SAE International.

ABSTRACT

Engine exhaust systems need to undergo continuous modifications to meet increasingly stricter regulations. In the past, much of the design and engineering process to optimize various components of engine and emission systems has involved prototype testing. The complexity of modern systems and the resulting flow dynamics, and thermal and chemical mechanisms have increased the difficulty in assessing and optimizing system operation. Due to overall complexity and increased costs associated with these factors, modeling continues to be pursued as a method of obtaining valuable information supporting the design and development process associated with the exhaust emission system optimization.

Insufficient kinetic mechanisms and the lack of adequate kinetics data are major sources of inaccuracies in catalytic converters modeling. This paper presents a numerical study that investigates the performance of different chemical mechanisms in simulating the emission conversion characteristics of catalytic converters during both steady state and transient conditions. The model considers the coupling effect of heat and mass transfer with the catalyst reactions as exhaust gases flow through the catalyst. The heat transfer model includes the heat loss due to conduction and convection. The effect of radiation is assumed to be negligible and is not considered. The resulting governing equations based on the conservation of mass, momentum and energy are solved by a tridiagonal matrix algorithm (TDMA) with a successive line under relaxation method. The performance of different chemical kinetic schemes is reviewed by comparing the results of numerical model with the experimental measurements.

INTRODUCTION

Catalytic converters have been employed in vehicles for decades and have been proven to be very successful in reducing the exhaust emissions. However, with the worldwide trend of stringent emission regulations, their designs need to undergo continuous modifications. Mathematical modeling and numerical simulations play an important role in such design modification efforts. In addition to proper and accurate flow, heat and mass transfer models, the accuracy of numerical simulations depend on the accurate description of chemical kinetic mechanisms.

Several chemical reaction mechanisms are available in literature. Oh and Cavendish [1] used a three step reaction mechanism. Their model only considered oxidation processes of CO, HC, and H_2. They assumed that HC oxidation is represented by the reaction of propylene ("fast-oxidizing HC") and neglected other HC as insignificant influence on the converter thermal performance. Siemend et al. [2] used similar oxidation kinetic rates in their study of comparison between model simulations and experiments. In addition to oxidation, their reaction mechanism also includes the NO reduction. The rate expression for NO reduction by CO was taken from Subramanian and Verma [3,4]. In both models, all unburned hydrocarbons are represented by CH_y (y is the ratio of hydrogen to carbon in the fuel). The heat of formation of CH_y was assumed to be one-third of C_3H_6. Due to its simplicity and good accuracy, the 4-step mechanism is widely used in simulating the catalyst performance. However, this mechanism ignores the variation in the reaction rates of several HC species by lumping them into one category. It also ignores the effect of water gas shift and steam reforming effects, which become especially important under sever transient driving conditions.

An improvement in the reaction mechanisms may be made by lumping several HC species into more than one category. Koltsakis et al. [5] used a six-step chemical mechanism to describe the chemical reactions occurring on the surface of the automotive catalyst. Their model assumed the HC to be lumped into two categories: fast burning and slow burning. In addition to the oxidation and reduction processes, they also considered the steam reforming reaction. This reaction was attributed to high HC conversion efficiencies in fresh catalysts during operation in rich exhaust. Similar to most of the previous studies, their kinetic expressions were of the form proposed by Votz et al. [6]. They employed a tunable factor to match their NO reduction rate with the experimental measurements. A more detailed kinetic mechanism was proposed by Otto and LeGray (7), which was later modified to include an elaborate NOx mechanism and the formation of ammonia and its effect on NO oxidation under fuel-rich conditions. A methodology for updating steady-state kinetic data for this 13-step reaction mechanism was presented by Montreuil et al [8]. These kinetic rate expressions are to date the most detailed. In this mechanism, all the unburnt hydrocarbons are lumped into three categories: fast burning, slow burning, and inert. The mechanism has been found to simulate the performance of catalytic converters fairly well under both steady state and transient conditions [9]. However, the use of these detailed kinetic rate expressions require the knowledge of 97 constants, which requires detailed experimentation for different type of catalysts.

Even though all of the chemical reaction models mentioned above are applied to catalyst simulations, there exist differences in catalyst performance predictions because of variations in the chemical mechanisms. However, it is difficult to isolate the effect of chemical mechanism by simply comparing the above studies. As the differences in catalyst performance prediction may also be attributed to the differences in these studies related to test conditions, heat and mass transfer modeling, catalyst conditions, etc. The present study was motivated by recognizing the need for a systematic evaluation of the existing chemical reaction schemes under similar and realistic conditions. The study compares the performance of various mechanisms in simulating the catalyst operation during both steady and transient conditions. The model predictions were compared with experimental measurements. The transient conditions were simulated by considering the catalyst operation during the US Federal Test Procedure (FTP). In addition to the comparison, the study also identifies the salient feature of each mechanism. It is anticipated that this sensitivity study will lead to efforts in modifying or devising a better chemical catalyst mechanism.

MATHEMATICAL FORMULATION

GOVERNING EQUATIONS – The governing equations, as listed below, were developed by considering the conservation of mass, energy and chemical species.

The gas phase energy equation:

$$\rho_g C_{P_g} (\varepsilon \frac{\partial T_g}{\partial t} + v_g \frac{\partial T_g}{\partial z}) = -h_g G_a (T_g - T_s) \tag{1}$$

The surface energy equation:

$$(1-\varepsilon)\rho_s C_{Ps} \frac{\partial T_s}{\partial t} = (1-\varepsilon)\lambda_s \frac{\partial^2 T_s}{\partial z^2} + h_g G_a (T_g - T_s) - h_\infty S_{ext}(T_s - T_\infty)$$

$$+ G_a \sum_{j=1}^{n_{reaction}} R^j (T_s, C_s^1, \cdots, C_s^{n_{species}}) \cdot \Delta H^j \tag{2}$$

The gas phase species equation:

$$(\varepsilon \frac{\partial C_g^j}{\partial t} + v_g \frac{\partial C_g^j}{\partial z}) = -km^j G_a (C_g^j - C_s^j) \tag{3}$$

where the superscript j varies from 1 to 7 representing, respectively, following gas species: CO, NO, NH_3, O_2, C_3H_6, H_2 and C_3H_8.

The surface species equation:

$$(1-\varepsilon)\frac{\partial C_s^j}{\partial t} = km^j G_a (C_g^j - C_s^j) - G_a R^j (T_s, C_s^1, \cdots, C_c^{N_{species}}) \tag{4}$$

where superscript j varies from 1 to 7 representing the surface species in the same order as the gas phase species equation.

CHEMICAL REACTION MECHANISMS – As described in the Introduction, the present study considers several different chemical reaction mechanisms existing in literature. A brief description of these mechanisms is listed below:

3-Step Chemical Reaction Mechanism – This reaction mechanism, listed below, was proposed by Seh Oh and co-workers [1,10]

$$CO + \frac{1}{2}O_2 \rightarrow CO_2 \qquad Q_R = -2.832*10^5 \text{ (J/mol)}$$

$$C_3H_6 + \frac{9}{2}O_2 \rightarrow 3CO_2 + 3H_2O \qquad Q_R = -1.928*10^6 \text{ (J/mol)}$$

$$H_2 + \frac{1}{2}O_2 \rightarrow H_2O \qquad Q_R = -2.42*10^5 \text{ (J/mol)}$$

The expressions for reaction rates are as following:

$$R_1 = k_1 C_{CO} C_{O_2} / G$$

$$R_2 = k_2 C_{C_3H_6} C_{O_2} / G$$

$$R_3 = k_1 C_{H_2} C_{O_2} / G$$

where

$$k_1 = 6.699 \times 10^9 \exp(-12556/T_S) \qquad mol/cm^2 \cdot s$$

$$k_2 = 1.392 \times 10^{11} \exp(-14556/T_S) \qquad mol/cm^2 \cdot s$$

$$G = T_S (1 + K_1 C_{CO} + K_2 C_{C_3H_6})^2 (1 + K_3 C_{CO}^2 C_{C_3H_6}^2)$$
$$(1 + K_4 C_{NO}^{0.7})$$

$$K_1 = 65.5 EXP(961/T_S) \qquad \text{dimensionless}$$

$$K_2 = 2.08 \times 10^3 EXP(361/T_S) \qquad \text{dimensionless}$$

$$K_3 = 3.98 EXP(11611/T_S) \qquad \text{dimensionless}$$

$$K_4 = 4.79 \times 10^5 EXP(-3733/T_S) \qquad \text{dimensionless}$$

4-Step Chemical Reaction Mechanism

4-Step Chemical Reaction Mechanism – A widely used reaction scheme considers the above three reactions with the NO reduction by CO. The resulting 4-step mechanism has been shown to simulate the catalyst reactions with a reasonable accuracy [2]. The NO reduction used in this mechanism is as follows:

$$CO + NO = CO_2 + \frac{1}{2}N_2 \qquad Q_R = -3.73*10^5 \ (J/mol)$$

And the reaction rate expressions for the above reactions is:

$$R_4 = \frac{k_4 C_{CO}^{1.4} C_{O_2}^{0.3} C_{NO}^{0.13}}{T_S^{-0.17}(T + k_5 C_{CO})^2}$$

$$k_4 = 3.067 \times 10^8 EXP(-8771/T_S) \qquad mole/cm^2 \cdot s$$

$$k_5 = 1.2028 \times 10^5 EXP(653.5/T_S) \qquad K$$

Modified 4-Step Chemical Reaction Mechanism

Modified 4-Step Chemical Reaction Mechanism – This reaction mechanism is essentially similar to the 4-step mechanism described previously. The only difference is the modification in the rate expression of NO reduction. Based on the comparison with the experimental measurement, the authors modified the exponent of CO concentration in the rate expression R_4 from 1.4 to 1.9. The modified reaction rate of NO reduction is as following:

$$R_4 = \frac{k_4 C_{CO}^{1.9} C_{O_2}^{0.3} C_{NO}^{0.13}}{T_S^{-0.17}(T + k_5 C_{CO})^2}$$

5-Step Chemical Reaction Mechanism

5-Step Chemical Reaction Mechanism – This mechanism was obtained by adding the steam reforming reaction in the above modified 4-step reaction scheme. The steam reforming reaction and its rate expression used, as shown below, are similar to those used by Koltsakis et al. [5].

$$C_3H_6 + 3H_2O = 3CO + 6H_2 \qquad Q_R = 3.7346*10^5 \ (J/mol)$$

And the reaction rate expression:

$$R_5 = k_5 C_{C_3H_6} C_{H_2O}/G$$

$$k_5 = 1.7 \times 10^{12} EXP(12629/T_S)$$

13-Step Chemical Reaction Mechanism

13-Step Chemical Reaction Mechanism – This mechanism consists of 13 independent forward pathways for oxidation of CO, H_2, C_3H_6, C_3H_8, and NH_3 with O_2 and NO as oxidizing agents, and their corresponding rich and lean kinetic rate expressions. This chemical reaction scheme and kinetic data were originally presented by Otto and LeGray [7] and later presented with the modified kinetic data by Montreuil et al. [8]. The reaction scheme is shown as follows:

$$CO + \frac{1}{2}O_2 \rightarrow CO_2$$

$$CH_\alpha + M_1 \cdot O_2 \rightarrow CO_2 + \frac{\alpha}{2} \cdot H_2O \qquad (HC_F)$$

$$CH_\alpha + M_2 \cdot O_2 \rightarrow CO + \frac{\alpha}{2} \cdot H_2O \qquad (HC_F)$$

$$H_2 + \frac{1}{2}O_2 \rightarrow H_2O$$

$$NH_3 + \frac{3}{4}O_2 \rightarrow 1.5H_2O + \frac{1}{2}N_2$$

$$CO + NO \rightarrow CO_2 + \frac{1}{2}N_2$$

$$2.5CO + NO + 1.5H_2O \rightarrow NH_3 + 2.5CO_2 + \frac{1}{2}N_2$$

$$H_2 + NO \rightarrow H_2O + \frac{1}{2}N_2$$

$$2.5H_2 + NO \rightarrow NH_3 + H_2O$$

$$CH_\alpha + M_1 \cdot O_2 \rightarrow CO_2 + \frac{\alpha}{2} \cdot H_2O \qquad (HC_S)$$

$$\frac{1}{2M_2}CH_\alpha + NO \rightarrow \frac{1}{2M_2}CO + \frac{\alpha}{4M_2}H_2O + \frac{1}{2}N_2 \qquad (HC_F)$$

$$CH_\alpha + M_3 \cdot NO \rightarrow CO_2 + \frac{\alpha}{2} \cdot H_2O + \frac{M_3}{2} \cdot N_2 \qquad (HC_F)$$

$$\frac{2.5}{2M_2}CH_\alpha + NO + \frac{(3-\alpha)}{4M_2}H_2O \rightarrow NH_3 + \frac{2.5}{2M_2}CO \qquad (HC_F)$$

$$(5)$$

where α = Hydrogen-to-Carbon ratio, and

$$M_1 = [1 + \frac{\alpha}{4}]; \quad M_2 = [\frac{1}{2} + \frac{\alpha}{4}]; \quad M_3 = [2 + \frac{\alpha}{2}]$$

HC_F = Fast burning HC; HC_S = Slow burning HC.

The corresponding reaction rate expressions are listed elsewhere [9].

For different catalyst formulations, the coefficients in the reaction rate expressions are different. They also vary with the aging of catalytic converters. The present study uses a palladium-based catalyst. The coefficients in the reaction rate expressions were taken from Montreuil et al. [8] wherein they were appropriately adjusted using experimental flow reactor measurements.

OXYGEN STORAGE MECHANISM – The conversion efficiency of a three-way catalytic converter can be improved by storing the extra oxygen under fuel lean conditions and releasing it under rich conditions [11,12]. The released oxygen may participate in the reactions with the reducing agents, thereby increasing the conversion of CO and HC in a rich exhaust-gas environment [13-15]. Such an oxygen storage capacity (OSC) is developed in the modern catalyst by coating its substrate with a wash-coat material containing ceria.

The oxygen storage and release mechanism used in this study was modeled by a 9-step site reaction mechanism. This mechanism was developed by designating two kinds of sites that can be oxidized and reduced through a 9 step site reaction mechanism [14]. The reduced metal site on the surface is defined as $<S>$ and the oxidized site is defined as $<OS>$. This 9 step reaction mechanism is listed as following:

$$<S> + \frac{1}{2}O_2 \rightarrow <OS> \qquad \text{Site Oxidation}$$

$$<OS> + CO \rightarrow <S> + CO_2 \qquad \text{Site reduction by CO}$$

$$H_2O + CO \rightarrow H_2 + CO_2 \qquad \text{Water-Gas shift}$$

$$<OS> + H_2 \rightarrow <S> + H_2O \qquad \text{Site reduction by } H_2$$

$$CH_\alpha + H_2O \rightarrow CO + (1 + \frac{\alpha}{2}) \cdot H_2 \qquad \text{Steam reforming}$$

$$\frac{3}{2}CH_\alpha + <OS> \rightarrow <S> + \frac{1}{2}C + CO + (\frac{3}{4}\alpha)H_2$$

$$\text{Reduction by HC}$$

$$C + O_2 \rightarrow CO_2 \qquad \text{Coke Burn-off}$$

$$<S> + NO \rightarrow <OS> + \frac{1}{2}N_2 \qquad \text{NO Storage}$$

$$<OS> + \frac{2}{5}NH_3 \rightarrow <S> + \frac{2}{5}NO + \frac{3}{5}H_2O$$

$$NH_3 \text{ Site reduction}$$

where α is hydrogen-to-carbon ratio of the hydrocarbon.

Each reaction has two rate expressions, one being the fast site rate expression and another the slow site rate expression. The total sites are conserved for both fast site and slow site. Therefore,

$$S_{total,f} = <S>_f + <OS>_f$$

$$S_{total,s} = <S>_s + <OS>_s \qquad (7)$$

The rates of the transient reactions are of the form

$$R_{transient} = \frac{(OXSW) \cdot CTR^{-E/RT_s} \cdot \prod_{1}^{N_{specie}} (X_s^i)^{EX(i)}}{1 + \sum_{n=1}^{N_{specie}} k_n \cdot X_s^n} \qquad (8)$$

where

$\quad OXSW = 1$, if Redox ratio <1

$\quad OXSW = 0$, if Redox ratio >1

The corresponding coefficients for calculating the transient reaction rates are based on experimental data.

SOLUTION PROCEDURE – The governing equations were discretized by using a non-uniform grid and employing the control volume approach with the central implicit difference scheme in the spatial direction. Since more chemical reactions take place near the inlet, smaller grid spacing was used near the inlet and larger spacing near the exit. A standard *tridiagonal matrix algorithm* (TDMA) with an iterative successive line under relaxation method was used to solve the finite difference equations. Details of the solution procedure are described elsewhere [9].

RESULTS AND DISCUSSION

The performance of different chemical mechanisms was assessed by comparing the numerical predictions with the experimental measurements under both steady and transient conditions. The converters used for both steady state and transient performance assessment were palladium-based catalysts. For the steady state case, the catalyst has a length of 3.81 cm, cross-sectional area of 5.0671×10^{-4} m², cell density of 620,000 cell/m², and wall thickness of 1.88×10^{-4} m. And the feed gas composition was 1% CO, 0 ppm CH_4, 1000 ppm C_3H_6, 500 ppm C_3H_8, and 1000 ppm NO. The feed gas temperature was 371 °C, and space velocity of 50,000 hr^{-1}. For the transient case, the catalyst used has a length of 8.001 cm, cross-sectional area of 8.69254×10^{-3} m², cell density of 620,000 cell/m², and wall thickness of 1.905×10^{-4} m.

STEADY STATE PERFORMANCE – The steady state performance of the reaction mechanisms was assessed by comparing the model results with the experimental measurements of Montreuil et al. [8]. For this case, all transient terms in the governing equations were set to be zero. Figure 1 shows the comparison of the converter pollutant conversion efficiencies as predicted by different chemical mechanisms. Here, the conversion efficiencies are plotted as a function of redox ratio, which is defined as:

$$Redox\,Ratio = \frac{[CO] + [H_2] + 6 \cdot (1 + \frac{\alpha}{4})[HC]}{[NO] + 2 \cdot [O_2]} \qquad (9)$$

Figure 1a shows the results of the 3-step mechanism. The results depict a good agreement between the model predictions of CO and HC conversion and the experimental measurements. The agreement is particularly excellent in the range of low redox ratio. However, when the redox ratio increases, i.e., in rich mixture zone, the discrepancies between the model predictions and measurements increase. For a wide range of redox ratio greater than unity, the model underpredicts the CO conversion. The HC conversion is also initially slightly underpredicted. However, beyond redox ratio of 2.5, the model overpredicts HC conversion efficiency. The overall prediction of HC conversion is better than that of CO. The NO conversion, as mentioned earlier, is not included in this mechanism. The elimination of NO reaction may have some influence on the model prediction of CO and HC.

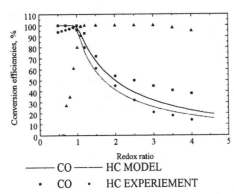

Figure 1a. Comparison of 3-step model results with experimental measurements under steady state condition

Figure 1b shows the results of the 4-step mechanism, which was obtained by adding the NO reduction reaction in the previous 3-step mechanism. For redox ratios greater than unity, the results show an excellent match between the model NO prediction and the experimental measurement. However, the NO prediction for redox ratio less than unity is very inaccurate. The addition of NO reaction also significantly influences the CO and HC conversion rates. It increases the conversion rates of both HC and CO. The results show large differences between the model predictions and the measurements. These discrepancies are due to inaccuracies in the model of NO reduction reaction.

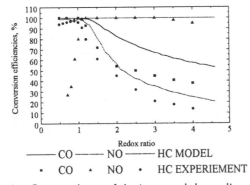

Figure 1b. Comparison of 4-step model results with experimental measurements under steady state condition

Figure 1c shows the results of the modified 4-step mechanism. Except for a slight modification in the reaction rate of NO, this mechanism including the reaction rates and other kinetic data is essentially similar to the previously discussed 4-step mechanism. The modification in the NO rate expression, as mentioned earlier, is made by changing the exponent of CO concentration from 1.4 to 1.9. The figure depicts a much better agreement between the model and the measurements. The prediction of NO conversion is particularly excellent over a wide range of redox ratio. The prediction of CO conversion also shows improvement. It matches very well with the measurements in the vicinity of redox ratio of unity. The CO conversion is slightly overpredicted between the

redox ratio of 1.1 and 2.5. Beyond this value, the conversion is underpredicted by the mechanism. The model prediction of HC conversion is similar to the 3-step mechanism, and has good agreement with measurements. The HC conversion efficiency is slightly overpredicted for redox ratio greater than unity.

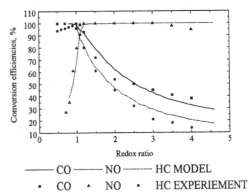

Figure 1c. Comparison of modified 4-step model results with experimental measurements under steady state condition

Figure 1d shows the results of the 5-step mechanism. This chemical scheme, as mentioned earlier, was obtained by adding a steam reforming reaction to the modified 4-step mechanism. The reaction rate expression for the steam reforming was obtained from literature [15]. The rate expressions for other reactions were similar to that of the modified 4-step. The results show that the model prediction of NO emission remains excellent and is not much influenced by steam reforming reaction. The major influence of steam reforming is found to be on HC emission. As expected, the results show an improvement in HC conversion efficiency due to steam reforming reaction. With the overprediction of HC conversion, the agreement between the model prediction and measurement is reduced. The prediction of CO conversion, however, is improved especially for the redox ratios between 1.1 and 2.5.

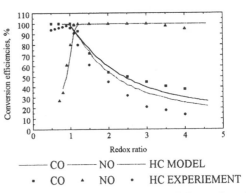

Figure 1d. Comparison of 5-step model results with experimental measurements under steady state condition

The results of the model with the 13-step chemical mechanism are shown in Figure 1e. The figure depicts that the model predictions, particularly at low redox ratios, are in good agreement with the measurements.

NO prediction is excellent over a wide range of redox ratios. The predictions of CO and HC, however, are not as good as those of 3 or 4 step mechanisms. For redox ratios greater than unity, the model significantly underpredicts the CO conversion and overpredicts the HC conversion. One reason for such a disagreement is the inaccuracy in chemical kinetic data, which are obtained by comparing with the actual measurements during engine operation. Since the engine operating conditions are in the close vicinity of redox ratio of unity, the constants and the rate expressions are tuned to better simulate the converter performance under these conditions. Hence the model can be used with a good accuracy to simulate the converter performance under normal engine operating conditions. However, a modification of its rate expressions will be required if the mechanism is employed to simulate the converter subjected to highly fuel rich conditions.

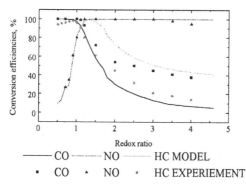

Figure 1e. Comparison of 13-step model results with experimental measurements under steady state condition

TRANSIENT PERFORMANCE – The steady state performance results clearly show that the modified 4-step mechanism is better than the 3-step mechanism. The modified mechanism also simulates the NO conversion in addition to embodying all the other features of the 3-step mechanism. Hence, the assessment of the transient performance of different mechanisms was limited to the modified 4-step and the 13-step mechanisms. The transient conditions were simulated by considering the converter performance during the US Federal Test Procedure (FTP).

Figure 2a shows the comparison of instantaneous HC emissions as determined by measurements and predictions by using the modified 4-step and the 13-step reaction schemes. The figure depicts a reasonable overall agreement between both model predictions and the measurements. Both of the mechanisms capture the major trends of HC emissions. The relatively large production of HC at the initial stage of cycle corresponds to the cold start conditions. This cold start behavior is also well simulated by both mechanisms. However, the model predictions have more spikes than those shown in measurements. Compared to the 13-step mechanism, the 4-step mechanism shows better agreement with the measurements. The more number of spikes in the

results of the 13-step mechanism may indicate a transient phenomenon that is not captured well by both the 4-step chemistry and the measurements that were limited by 1 Hz. The measurements with higher resolution are needed to clarify this point.

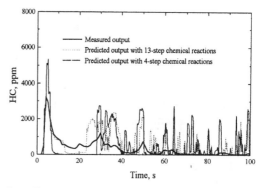

Figure 2a. Comparison of transient output HC emissions with experimental measurements

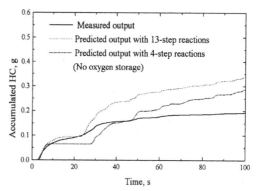

Figure 2b. Comparison of accumulated HC with experimental measurements

Figure 2b depicts the accumulated HC emission during the first 100 seconds of the FTP. The 4-step chemistry predicts less HC emission than the 13-step mechanism, and its predictions are closer to the experimental measurements. Both mechanisms, however, underestimate the catalyst HC conversion performance. A major reason for this discrepancy is that the simulations do not take into account the effect of oxygen storage capacity, which significantly improves the converter performance during the rich operating conditions. The effect of this mechanism is discussed in the later section. The converter HC conversion efficiency during the whole FTP cycle is measured to be 95.88%, compared to 80.37% and 85.14% predicted respectively by the 4-step and the 13-step mechanisms.

Figure 3 shows the instantaneous and accumulated CO emissions. The results show that the instantaneous CO emission is well predicted by both the mechanisms. These models capture the initial high CO production corresponding to cold start conditions, and simulate fairly well the various CO peaks during the FTP cycle. Similar to the HC emission prediction, the CO predictions of the 13-step mechanism have more spikes. The accumulated emission results depict a good agreement between the

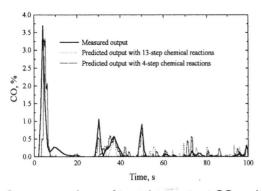

Figure 3a. comparison of transient output CO emissions with experimental measurements

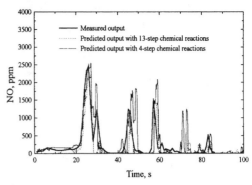

Figure 4a. Comparison of transient output NO emissions with experimental measurements

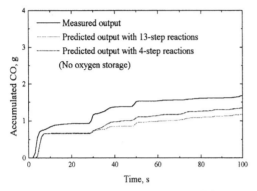

Figure 3b. Comparison of accumulated CO with experimental measurements

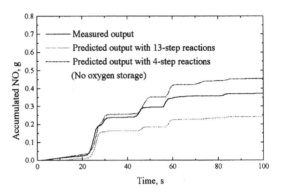

Figure 4b. Comparison of accumulated NO with experimental measurements

measurements and the predictions of the 4-step mechanism. Contrary to the case of HC conversion performance, the models overestimate the converter CO performance. The converter CO conversion efficiency during the whole FTP cycle is measured to be 91.53%, compared to 86.51% and 85.40% predicted respectively by the 4-step and the 13-step mechanisms.

The instantaneous and accumulated NO emissions are shown in Figure 4. The trend of instantaneous NO emission is fairly well predicted by both the 4-step and 13-step mechanisms. However, the 4-step mechanism shows larger spikes than those determined experimentally. Compared to the 4-step, the 13-step mechanism predictions are generally in closer agreement with the measurements. The accumulated NO emission results reveal that the 13-step overpredicts and the 4-step underpredicts the converter performance. The converter NO conversion efficiency during the whole FTP cycle is measured to be 92.03%, compared to 86.49% and 95.09% predicted respectively by the 4-step and the 13-step mechanisms.

EFFECT OF OXYGEN STORAGE CAPACITY – Most of the modern 3-way catalytic converters have oxygen storage capacity (OSC). This capacity, which is developed mainly by coating the catalyst substrate with a washcoat material containing ceria, allows oxygen storage under lean operating conditions and its release under rich conditions. The released oxygen improves the conversion of CO and HC during the rich cycle. During normal driving operation, the OSC plays an important role in improving the converter efficiency since there are continuous oscillations of rich and lean conditions due to rapid fluctuations in air-fuel ratio about stoichiometric conditions.

The 4-step and the 13-step mechanisms were compared by adding the OSC. The reaction scheme used for the OSC, as mentioned earlier, was a 9-step mechanism. Figures 5-7 show the accumulated HC, CO, and NO conversions. The results show that the model predictions of both mechanisms are improved by considering the OSC. The prediction of HC conversion especially shows significant improvement. The 13-step mechanism predictions are in excellent agreement with the measurements. The prediction of the 4-step mechanism is also improved, however, not as much as in the case of the 13-step mechanism. The HC conversion efficiency during the whole FTP cycle is 88.15% for the 4-step and 94.84% for the 13-step, compared to 95.88% as determined by measurements.

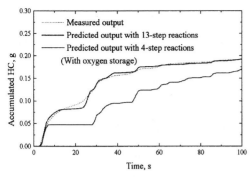

Figure 5. Comparison of accumulated HC with experimental measurements

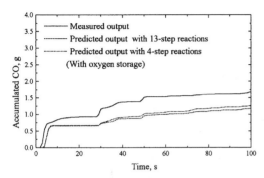

Figure 6. Comparison of accumulated CO with experimental measurements

Figure 6 shows that the addition of the OSC does not appreciably influence the CO conversion prediction of the 13-step reactions during the first 100 seconds of the FTP. The results of the 4-step mechanism with the OSC show an improvement of the CO conversion performance. However, since the mechanism with no OSC already overpredicts the CO conversion (see Figure 3), addition of the OSC makes the disparity between the model results and the experimental measurements larger. For the total FTP cycle, the OSC improves the CO conversion efficiency, which is underpredicted without OSC by both mechanisms. The CO conversion efficiency is 87.33% for the 4-step and 88.69% for the 13-step, as compared to 91.53% determined by measurements.

Similar to the case of CO emissions, there is not much influence of the OSC on the NO conversion prediction of the 13-step mechanism at both the beginning of the FTP (Fig.7) and during the whole FTP cycle (the total NO conversion efficiency changing to 95.93%, comparing with 95.09% of no OSC case). With the addition of the OSC, the 4-step mechanism predicts an increase in the NO and underpredicts the NO conversion performance (the total NO conversion efficiency decreasing to 75.38% from 86.49%). Compared to the 4-step, the 13-step mechanism is in closer agreement with the experimental measurements.

In summary, the results show that OSC improves the predictions of total conversion efficiencies of HC and CO during the whole FTP test, and the improvement of 13-

step mechanism is better than that of the 4-step mechanism. The prediction of NO conversion efficiency is not much influenced by the OSC for the 13-step mechanism, and it decreases significantly for the 4-step mechanism.

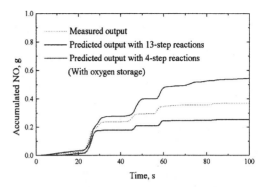

Figure 7. Comparison of accumulated NO with experimental measurements

CONCLUSIONS

A numerical study was carried out to investigate the performance of different chemical reaction mechanisms under both steady state and transient conditions. The results led to the following conclusions:

- The converter steady state performance can be simulated by several different chemical reaction mechanisms. The kinetic expressions of most of these mechanisms are generally tuned to yield optimum model performance near the stoichiometeric conditions. The 3-step mechanism, proposed by Seh Oh, which considers only the oxidation of CO, C3H6, and H2, gives satisfactory results of CO and HC conversion. The NO conversion performance can be obtained by adding the reaction of NO reduction by CO. The resulting 4-step mechanism is widely used in simulating the converter performance. A slight modification in the NO reduction rate expression is found to give the best results compared to measurements.

- The steam reforming reaction is found to mainly influence the HC emission by increasing its oxidation rate. It also affects CO oxidation but does not have much influence on NO reduction.

- The transient performance of reaction mechanisms studied is acceptable. The model predictions with 13-step mechanism have the best agreement with experimental measurements during the whole FTP test. The consideration of the oxygen storage mechanism improves the model predictions of HC and CO conversions for both 13-step and 4-step mechanisms. However, on the NO conversion predictions, it has almost no effect for 13-step mechanism and adverse effect for 4-step mechanism. In addition, the oxygen storage capacity has major influence on HC conversion.

ACKNOWLEDGMENTS

The financial support from the Ford Scientific Research Laboratory, Oak Ridge National Laboratory and the Center for Engineering Education and Practice (CEEP) of the University of Michigan-Dearborn is greatly appreciated..

REFERENCES

1. Oh, S.H. and Cavendish, J.C., "Transients of Monolithic Catalytic Converters: Response to a Step Change in Freestream Temperature as related to Controlling Automobile Emissions," Ind. Eng. Chem. Prod. Res.Dev., 21, p. 29, 1982.

2. S. Siemund, P. Leclerc, D. J. Schweich, M. Prigent and F. Castagna, "Three-way Monolithic Converters: Simulations versus Experiments," Chemical Engineering Science, Vol. 51, N0. 15, pp. 3709-3720, 1996.

3. Subramanian, B. and Varma, A., "Reactions of CO, NO, O2, and H2O on Three-way and Pt/AL2O3 Catalyst," Frontiers in Chemical Engineering Proceedings of the International Chemical Engineering Conference, Vol.1, pp. 231-240. 1984.

4. Subramanian, B. and Varma, A., "Reaction Kinetics on a Commercial Three-way Catalyst: the CO-NO-O2-H2O System," Ind. Engng Chem., Prod. Res. Dec. 24, pp. 512-516, 1985.

5. Koltsakis, G. C., Konstantinidis, P. A. and Stamatelos, A. M., "Development and Application Range of Mathematical Models for 3-way Catalytic Converters," Applied Catalysis B: Environmental, Vol, 12, No. 2-3, pp. 161-191, 1997.

6. Voltz, S. E., Morgan, C. R., Liederman, D. and Jacob, S. M., "Kinetic Study of Carbon Monoxide and Propylene Oxidation on Platinum Catalysts," Ind. Engng Chem. Prod. Res. Dev. 12, p. 294, 1973.

7. Otto, N. C. and LeGray W. J., "Mathematical Models for Catalytic Converter Performance," SAE paper No. 800841, 1980.

8. Montreuil, C. N., Williams, S. C., and Adamczyk, A. A., "Modeling Current Generation Catalytic Converters: Laboratory Experiments and Kinetic Parameter Optimization – Steady State Kinetics," SAE paper No. 920096, 1992.

9. Shen, H., Shamim, T., Sengupta, S., Son, S. and Adamczyk, A., "Performance Simulations of Catalytic Converters during the Federal Test Procedure," Proceedings of the 33rd National Heat Transfer Conference, August 15-17, 1999, Albuquerque, New Mexico.

10. Chen, D. K., Oh, S. H., Bisselt, E. J. and Van Ostrom, D. L., "A Three-dimensional Model for the Analysis of Transient Thermal and Conversion Characteristics of Monolithic Catalytic Converters," SAE paper 880282, 1988.

11. Gandhi, H. S., Delosh, R. G., Piken, A. G. and Shelef, M., "Laboratory Evaluation of Three-way Catalysts," SAE Transactions, Sec.2, Vol.85, 1976, p. 201.

12. Taylor, K. C., "Automobile Catalytic Converters", Springer-Verlag, Berlin, Heidelberg, 1984.

13. Herz, R. K., " Dynamic Behavior of Automotive Catalysts. 1. Catalyst Oxidation and Reduction." Ind. Engng Chem. Prod. Res. Dev.20, pp. 451-457, 1981.

14. Li, P., Adamczyk, A. A., and Pakko, J. D., "Thermal Management of Automotive Emission Systems: Reducing the Environmental Impact," The Japan-U. S. Seminar on Thermal Engineering for Global Environment Protection (A-3), 1994.

15. Koltsakis, G. C. and Stamatelos, A. M., "Catalytic Automotive Exhaust Aftertreatment," Progress in Energy and Combustion Science, v 23 n 1, 1997, pp. 1-39.

NOMENCLATURE

C_g^j = gas phase concentration, moles/M^3

C_s^j = surface concentration, moles/M^3

Cp_g = specific heat of gas, J/kg*K

Cp_s = specific heat of substrate, J/kg*K

EA, E = activation energy, Pa-m^3/g-mole

G_a = geometric surface to volume ratio, M^2/M^3

ΔH^j = heat of reaction j, J/mole

h_g = heat transfer coefficient between flow and substrate, J/M^2*s*K

h_∞ = heat transfer coefficient between substrate and atmosphere, J/M^2*s*K

km^j = mass transfer coefficient for species j, M/s

Q_R = reaction heat, J/mol

R^j = reaction rate of jth reaction, mole/M^3*s

S_{ext} = external surface to volume area ratio, M^2/M^3

t = time, s

T_∞ = ambient temperature, K

T_g = gas temperature, K

T_s = substrate temperature, K

V_g = gas flow velocity, m/s

X_s^i = mole fraction of species i in substrate

z = axial coordinate, m

α = hydrogen-to-carbon ratio in the fuel

ε = void volume fraction

λ_s = thermal conductivity of substrate, J/M*s*K

ρ_g = gas density, kg/M^3

ρ_s = substrate density, kg/M^3

2002-01-0065

Three-Dimensional Simulation of the Transient Behavior of a Three-Way Catalytic Converter

Joachim Braun, Thomas Hauber, Heike Többen, Julia Windmann and Peter Zacke
J. Eberspächer GmbH & Co.

Daniel Chatterjee[1], Chrys Correa[2], Olaf Deutschmann, Lubow Maier, Steffen Tischer and Jürgen Warnatz
University of Heidelberg

Copyright © 2002 Society of Automotive Engineers, Inc.

ABSTRACT

The ultimate goal in the numerical simulation of automotive catalytic converters is the prediction of exhaust gas emissions as function of time for varying inlet conditions, i.e. the simulation of a driving cycle. Such a simulation must include the calculation of the transient three-dimensional temperature-field of the monolithic solid structure of the converter, which results from a complex interaction between a variety of physical and chemical processes such as the gaseous flow field through the monolith channels, the catalytic reactions, gaseous and solid heat transport, and heat transfer to the ambience.

This paper will discuss the application of the newly developed CFD-code DETCHEM[MONOLITH] for the numerical simulation of the transient behavior of three-way catalytic converters that have a monolithic structure. The code combines the two-dimensional simulations of the reactive flows in a representative number of monolith channels with a transient simulation of the three-dimensional temperature field of the solid structure of the converter including insulation and canning. The chemical reactions are modeled by a multi-step heterogeneous reaction mechanism, which is based on the elementary pro??cesses on the platinum and rhodium catalysts used. The integration over the chemical conversion in the single channels leads to the total conversion in the converter as function of time.

This paper presents a numerical simulation of the start-up phase of an automotive catalytic converter for temporally varying inlet conditions. The variation of the temperature distribution in the solid structure and in the single channels as well as the species profiles are described. The numerically predicted time-dependent conversion of the combustion pollutants is compared with experimental data. The potentials and limitations of the models and computational tools are discussed.

INTRODUCTION

In the automotive industry, the development cycles are continuously reduced, whereas the guidelines of the legislation concerning the pollutant emission limits become more restrictive. The legislatively enforced emission limits can only be fulfilled by the optimization of the exhaust gas system. Not only the right choice of the single components but also their arrangement is important. Mathematical modeling and numerical simulation has been gaining significance in the design and optimization process, as extensive experimental setups are not only time consuming but also expensive. The numerical simulation of the exhaust system can already support catalyst design in an early stage of development. For the three-way catalyst, the major concern here is the correct prediction of the light-off behavior of the exhaust system, because the pollutant emitted in this period presents a large amount of the overall pollutant emission during the driving test cycle given by legislation.

In order to achieve reliable results, the numerical simulations have to be based on accurate models of all the significant chemical and physical processes in the catalytic converter. In the last years, several models were proposed for the numerical simulation of catalytic converters /1-4/ reaching from a one-dimensional up to a three-dimensional description of the temperature distribution in the monolith. Most of these studies have in common, that they use global reaction kinetics for the description of the chemical reactions.

---

[1] now DaimlerChrysler AG, Stuttgart
[2] now BASF AG, Ludwigshafen

Alternatively, the chemical reaction system can be described by a multi-step reaction mechanism based on the elementary steps occurring on a molecular level on the catalytic surface /5-7/. This approach requires more computational efforts because a large number of chemical species and reactions is considered. In particular, the description of the surface coverage of the catalyst with adsorbed species and the strongly non-linear reaction kinetics lead to a highly-nonlinear, stiff differential-algebraic equation system. Simultaneously, the modeling effort demands the knowledge of the reaction scheme and the rate coefficients for every single step. However, the effort is rewarded by the fact that the simulation can not only describe the system but also predict the system behavior under varying external conditions, which not have necessarily been investigated by experiments a priori. Therefore, such detailed models will lead to more accurate transient simulations of the behavior of an automotive catalytic converter.

This study continues our former work on modeling automotive catalytic converters /5/, in which steady-state simulations of the conversion process in a single monolith channel at constant channel wall temperature were carried out. Now we study the transient behavior of the entire monolith. In that former study, we used a sample gas containing the pollutants CO, NO, and C_3H_6 in the experiment and the simulation. Additionally, in the current simulation, CH_4 is considered as additional hydrocarbon species in order to map the behavior of a real exhaust gas. In the experiment, a real exhaust gas mixture coming from an SI engine is used.

Earlier simulations used FLUENT /8/ to calculate the reactive flow field in the channels. Those simulations were very time consuming and not well suited for a transient simulation of the catalytic converter. In order to speed up the single channel simulations, the recently developed computer program DETCHEM^{CHANNEL} /9-11/ was used. It solves the steady-state two-dimensional reactive flow in a straight channel based on a boundary-layer code and takes heterogeneous reactions into account.

The time-dependend temperature distribution in the two- or three-dimensional monolith structure is calculated by the code DETCHEM$^{MONOLITH\_2D}$ or DETCHEM$^{MONOLITH\_3D}$, respectively /10,11/. These codes are coupled with DETCHEM^{CHANNEL} to account for the behavior in the single channels. The DETCHEM^{MONOLITH} codes can also be linked to input/output data structures of the CFD-code FLUENT, which can be used to simulate the flow in front of and behind the converter.

In the present paper, we performed a numerical simulation of the start-up phase of an automotive catalytic converter for time varying inlet conditions. The variation of the temperature distribution in the solid structure and in the single channels as well as the species profiles are described. The numerically predicted time-conversion of the combustion pollutants is compar

with experimental data. The potentials and limitations of the models and computational tools are discussed.

MATHEMATICAL AND NUMERICAL MODEL

The numerical model for the simulation of the monolith consists of two parts. Since the time scales of the reactive channel flows and of the solid's thermal response are decoupled, time variations in the local monolith temperature can be neglected when calculating the fluid flow through a single channel. Thus, a time independent formulation is used to describe the gaseous flow in order to calculate heat source terms for a transient heat conduction equation for the solid /12/.

DETCHEM^{CHANNEL} models a single channel with cylindrical symmetry. Since the variation of the washcoat thickness over the channel walls has not been determined and a comparison with three-dimensional single channel simulations revealed only minor differences, a cylindrical channel model assuming a mean channel diameter and washcoat thickness was used. Given the inlet (velocity, temperature, density, species mass fractions) and wall conditions (axial temperature profile), the two-dimensional flow field of the fluid can be solved. The set of Navier-Stokes equations is the most accurate model for the description of the laminar flow of a chemically reacting fluid. However, due to their mathematical structure - in the time independent formulation they resemble a set of nonlinear elliptical partial differential equations - and their stiffness, a numerical solution is computationally expensive /13,14/. Therefore, simpler models such as plug-flow or boundary-layer models are frequently used /14,15/.

In the boundary layer of a fluid near a surface, the convection is mainly directed parallel to the surface. The diffusive transport in the same direction diminishes in comparison with the one perpendicular to the surface. This effect becomes more significant as the axial gas velocity is increased, i.e. for higher Reynolds numbers as long as the flow is laminar. The results achieved by the boundary-layer model can be as accurate as the results from the full Navier-Stokes model at high but laminar flow rates /12/. Mathematically, the character of the equations changes from elliptical to parabolic with a time-like coordinate along the channel axis. The set of equations consists of conservation equations for

Total mass

$$\frac{\partial(r\rho u)}{\partial z} + \frac{\partial(r\rho v)}{\partial r} = 0 \ , \tag{1}$$

Mass of species s

$$\frac{\partial(r\rho u Y_s)}{\partial z} + \frac{\partial(r\rho v Y_s)}{\partial r} = -\frac{\partial}{\partial r}(r j_s) \ , \tag{2}$$

Axial momentum

$$\frac{\partial(r\rho uu)}{\partial z} + \frac{\partial(r\rho vu)}{\partial r} = -r\frac{\partial p}{\partial z} + \frac{\partial}{\partial r}\left(\mu r\frac{\partial u}{\partial r}\right) , \quad (3)$$

And Enthalpy

$$\frac{\partial(r\rho uh)}{\partial z} + \frac{\partial(r\rho vh)}{\partial r} =$$

$$ru\frac{\partial p}{\partial z} + \frac{\partial}{\partial r}\left(\lambda r\frac{\partial T}{\partial r}\right) - \frac{\partial}{\partial r}\left(\sum_s rj_s h_s\right) \quad (4)$$

Given the inlet conditions, the boundary-layer equations are solved in a single sweep of integration along the axial direction by a method-of-lines procedure. The radial derivatives are discretized by a finite-volume method. The resulting differential-algebraic equation system is integrated using the semi-implicit extrapolation solver LIMEX /16/. The transport coefficients for radial diffusion (μ, λ) and the species diffusion flux j_s depend on temperature and species composition. Surface reaction source terms ($j_{s,surface}$) are modeled by elementary-step based reaction mechanisms as has been described in detail in /5,7,11/. The model of the catalytic reactions on the inner channel wall accounts for a varying surface coverage of adsorbed species along the channel. Furthermore, a model for pore diffusion inside washcoats /17/, which also depends on the local reaction rates, can be included when necessary.

In DETCHEM$^{MONOLITH\_2D}$ or DETCHEM$^{MONOLITH\_3D}$ /10,11/, the simulation of the thermal behavior of the entire monolithic structure, which is coupled with the single channel simulations, is modeled by a two- or three-dimensional temperature equation, respectively, i. e.

$$\frac{\partial T}{\partial t} = \nabla^2\left(\frac{\lambda T}{\rho C_p}\right) + \frac{q}{\rho C_p} . \quad (5)$$

The material properties (density ρ, heat capacity C_p and thermal conductivity λ) are functions of the local temperature and material (monolith, insulation and canning) and can also be specified as functions of the direction. Heat losses due to conduction, convection, and thermal radiation at the exterior walls of the monolith can be included. In order to obtain the source terms q in the temperature equation, the heat flux from the gas phase into the monolith bulk due to convection and chemical heat release is calculated for a representative number of channels. These single-channel simulations are carried out for each time step of the transient temperature simulation. They apply the actual local axial temperature profiles as boundary conditions and use the time-dependent initial flow conditions. Hence, time-varying inlet conditions can be specified as long as the conditions vary at a time scale that is larger than the residence time.

For the spatial discretization of the transient temperature equation, a finite volume approach is used. For the integration of the resulting ordinary differential equation system one of the solvers LIMEX /14/ or LSODE /18/ can be chosen from. Based on these models, the computational tool predicts the transient, two- or three-dimensional distributions of temperature and species concentrations.

CHEMICAL REACTION SYSTEM

A detailed multi-step reaction mechanism is used to model the catalytic reactions in a three-way catalytic converter that contains Pt and Rh as active catalysts. The surface coverage of the species on the catalytic material and the surface mass fluxes are also calculated as a function of the position in the channel. This approach has already been discussed in former publications /5,7,11/.

The mechanism includes only surface chemistry; gas phase chemistry can be neglected because of the low pressure and temperature, and the short residence time. In the simulation, the sample exhaust gas mixture is composed of C_3H_6, CH_4, CO_2, H_2O, CO, NO, O_2, N_2. The surface reaction scheme consists of 62 reaction steps among the 8 gas phase and further 29 adsorbed chemical species. It is assumed that all species are adsorbed competitively. The model also considers the different adsorption sites (platinum or rhodium) on the metallic catalyst surface. However, on rhodium, surface reactions are considered between NO, CO, and O_2 only. The kinetic data of the mechanism were taken either from literature or fits to experimental data. The mechanism is based on our former studies /5,7/. Here it was slightly extended by the introduction of CH_4 reactions on the platinum surface. The revised mechanism will be discussed in a forthcoming paper, when further experimental measurements and numerically predicted data are available for the validation of the mechanism.

The parameters of the catalyst used, e.g., metal composition and loading, dispersion, and so on, have taken as described in the preceding paper /5/; the experiment is carried out using the same catalyst. In the simulation, a simplified washcoat model is used with CO as species that determines the effectiveness factor /5,11/.

EXPERIMENTAL SETUP

For this study, experiments were carried out on an engine test bench. A 4-cylinder 1.6 l SI engine was used, which meets the EU-III pollutant emission standards. The design of the exhaust system was chosen in a way that guarantees a highly uniform flow distribution in front of the catalytic converter. Therefore, the exhaust system was provided with a long straight pipe and a long inlet cone before the monolith as shown in Figure 1.

A commercially available three-way catalyst was used, containing 50 g/ft^3 noble metal (Pt/Rh = 5:1) impregnated on a ceria stabilized γ-alumina washcoat. The washcoat was supported by a race-track shaped cordierit monolith with the dimensions 169.7 x 80.8 x 114.4 mm (6.68" x 3.18" x 4.5"), a cell density of 63 cells per cm^2 (400cpsi) and a wall thickness of 0.165 mm (6.5 mil). The length of the catalytically active monolith was varied keeping the total length constant. The results, presented in this study, were obtained with a catalytic converter of 57.2 mm (2.25") in length.

The surface temperature was measured at 8 points and gas-phase temperature was measured at 3 points along the exhaust system. The gas temperature was measured by 1.5 mm K type thermocouple wires. The mass flow rate was determined at the intake and converted to the corresponding data in front of the catalytic converter. Samples of the gas composition were taken at two positions as shown in Figure 1. The gas samples were then led through heated pipes to an exhaust gas analysis system from HORIBA /19/. The gas components CO, CO$_2$, NOx, O$_2$, and THC were detected. The sampling rate of all data was 1 Hz.

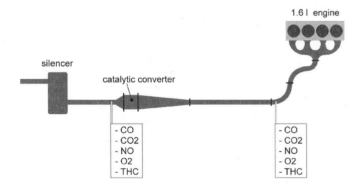

Figure 1 Sketch of the experimental set-up

RESULTS AND DISCUSSION

INLET AND BOUNDARY CONDITIONS

As an example, the experimental and numerical study of the cold start-up phase of a commercial three-way catalyst will be discussed. In Figure 2, the recorded profiles of the engine power and the rotational speed are shown. The engine is started at cold state and then phases of full and partial load alternate.

The corresponding mass flow rate was used to determine the axial inlet velocity at the front of the catalytic monolith, both shown in Figure 3. According to the experimental setup, the flow distribution was assumed to be spatially constant, even though the computational tool can handle spatially varying flow distributions at the monolith inlet. The inlet velocity was calculated on basis of the mass flow rate. Here, it was accounted for the density variation due to the higher gas temperatures and increase in volume by the combustion process.

The gas temperature, measured in front of the catalytic converter, was chosen as inlet temperature for the simulation. However, due to the thermal mass and thereby slow thermal response of the chosen thermocouples, a reverse transformation was applied to the temperature signal in order to approximate the actual gas phase temperature. Figure 4 shows not only the measured inlet and outlet gas temperature but also the corrected gas inlet temperature and the numerically predicted average gas-phase temperature at the catalyst exit.

The species concentrations at the monolith inlet were chosen according to the experimental data measured 5 cm in front of the straight pipe (Figure 1). The unburnt hydrocarbons were distributed among C$_3$H$_6$ (95% C -mol) and CH$_4$ (5% C -mol), neglecting the fact that the composition of the exhaust gas is much more complex /20/. This real gas effect is studied in on-going research.

Due to the fact that the model does not include oxygen storage effects (see also paragraph Outlook) and stoichiometric conditions were assumed, the oxygen concentration at the inlet was adjusted to unity redox ratio /5/.

Thermal convection and radiation at the exterior walls was taken into account, when calculating the temperature distribution in the entire monolith. The emissivity of the external surface of the catalytic converter was assumed as 0.8. Convection was taken into account by a heat transfer coefficient of 80 W/m^{2}K.

The material properties were determined by experiments (thermal conductivity of the monolithic structure) or taken from the literature and product information from the manufacturers.

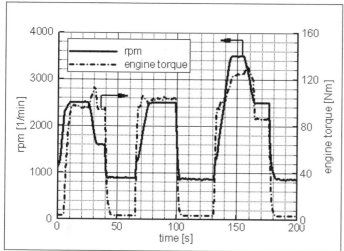

Figure 2   Engine torque and rotational speed profiles

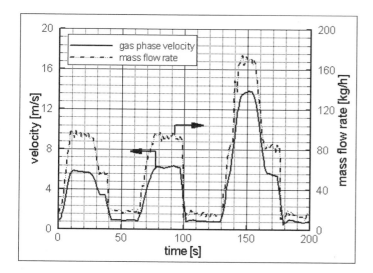

Figure 3   Mass flow rate at the inlet manifold and velocity magnitude
before the catalytic converter

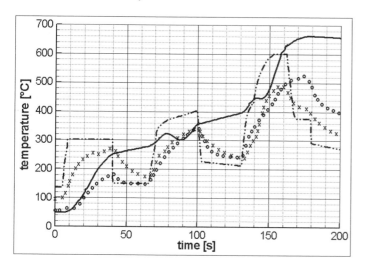

Figure 4   Measured and calculated temperature data:
measured gas temperature before catalytic converter (o)
measured gas temperature after catalytic converter (x)
approximated actual inlet gas temperature (- · · -)
calculated gas temperature after catalytic converter (——)

TEMPERATURE PROFILES

Using the computational tools and these inlet and boundary conditions, the transient behavior of the catalyst monolith was simulated for a cylindrical converter with elliptical cross-section as illustrated in Figure 5. DETCHEM$^{MONOLITH\_3D}$ is also capable of modeling converters with more complex cross-section.

Figures 6 – 9 reveal the solid phase temperature distribution of the monolith including insulation and canning at times of 10 and 41 seconds after start-up, respectively. The incoming hot exhaust gas heats up the monolithic structure. At the exterior wall, steep temperature gradients occur due to external heat loss. However, these gradients mainly occur inside the insulation and canning. Hence, the temperature of the catalytic channels vary only slightly over the cross-section.

At the early stage of operation, the heat is primarily provided by the heat capacity of the incoming exhaust gas. Heat release due to chemical reactions does not play a significant role, which is also revealed by the conversion as shown in Figures 20 – 23. After the converter has reached its operating temperature, the exit gas temperature exceeds the incoming gas temperature (Figure 4) caused by exothermic reactions.

In Figure 10 simulation results of the 2D and the 3D model are contrasted. As the presented cross-section of the 3D simulation result implies the maxim radius of the elliptic profile (maximum diameter: 169.7 mm), but the 2D model is based on the mean diameter of the monolith (mean diameter: 122.9mm), the radial ranges differ. The simulation results correspond very well.

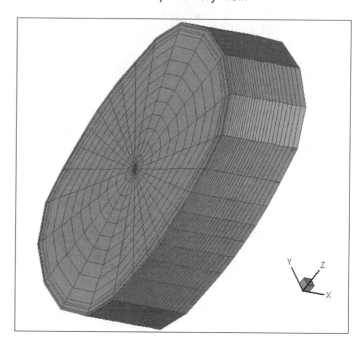

Figure 5   The 3-dimensional calculation grid. The flow runs in the positive x-direction.

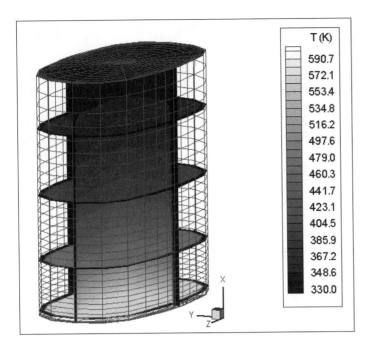

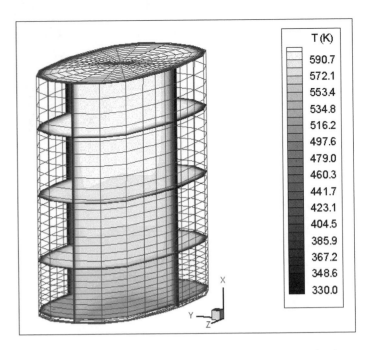

Figure 6   Temperature distribution in parallel slices, 10 seconds after start-up. The flow runs in the positive x-direction that is stretched by a factor of 3 for visual clarity; only the catalytic active part is shown

Figure 8   Temperature distribution in parallel slices, 41 seconds after start-up. The flow runs in the positive x-direction that is stretched by a factor of 3 for visual clarity; only the catalytic active part is shown

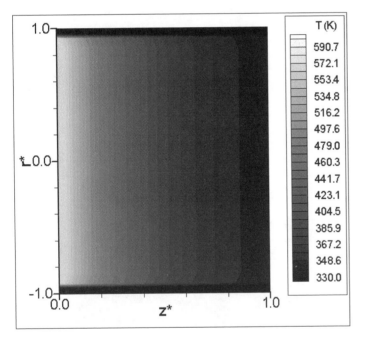

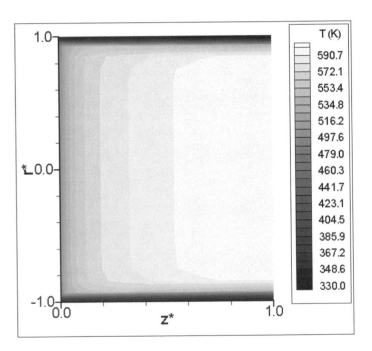

Figure 7   Temperature distribution in the monolith 10 seconds after start-up. Cross-section through the 3-dimensional temperature field in x-z-plane. x* is the normalized x-coordinate in the flow direction, r* is the normalized radial coordinate; only the catalytic active part is shown

Figure 9   Temperature distribution in the monolith 41 seconds after start-up. Cross-section through the 3-dimensional temperature field in x-z-plane. x* is the normalized x-coordinate in the flow direction, r* is the normalized radial coordinate; only the catalytic active part is shown

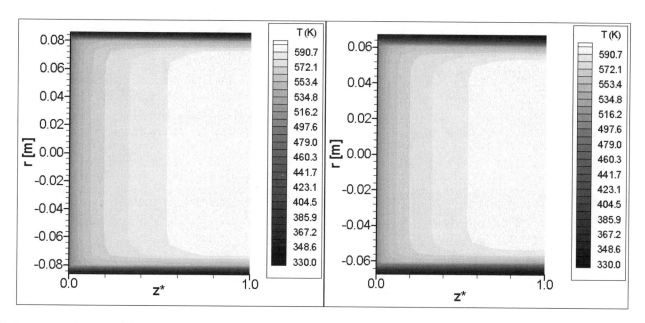

Figure 10: Comparison between the predicted temperature distributions in the monolith 41 seconds after start-up of the 3D (left hand-side) and the 2D simulation (right hand-side). The cross-section for the 3D simulation result is in the x-z-plane, z* is the normalized coordinate in flow direction; only the catalytic active part is shown

Figures 11 – 13 present the two-dimensional gas phase temperature profiles inside a channel located in the center of the monolith. 10 seconds after start-up, an axial temperature variation of approximately 230 K occurs. The hot exhaust gas is cooled down very quickly due to radial heat conduction upon entering the cold monolith. A few seconds later, the axial temperature profile flattens (Figures 8,9 and 12). Here, the monolith structure has almost reached the exhaust gas temperature. At that time, significant chemical conversion has started (Figures 20 – 23). The light–off temperature is already reached and due to the exothermical reactions the gas phase temperature in the channel increases in flow direction.

140 seconds after start-up, the catalyst temperature is sufficient for complete conversion. In the experiment as well as in the simulation (Figure 4), the exit gas temperature exceeds the incoming gas temperature. Here, chemical heat release leads to the temperature increase as the exhaust gas flows through the monolith channels.

We are aware of the fact that the predicted temperature difference between the inlet and outlet gas phase temperature does not agree well with the measured temperature difference (Figure 4). Further experimental investigations shall answer the question whether the deviations are due to configuration of measurement or due to discrepancies in simulation parameters.

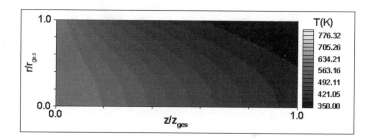

Figure 11 Temperature distribution in a single channel, 10 seconds after start-up

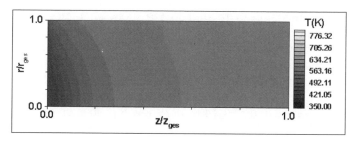

Figure 12 Temperature distribution in a single channel, 41 seconds after start-up

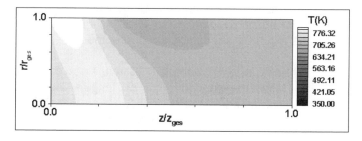

Figure 13 Temperature distribution in a single channel, 140 seconds after start-up.

513

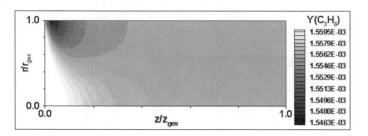

Figure 14   C$_3$H$_6$ mass fraction in a single channel, 10s after start-up.

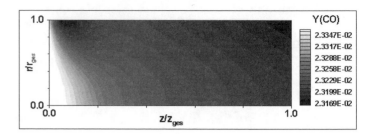

Figure 17   CO mass fraction in a single channel, 10s after start-up.

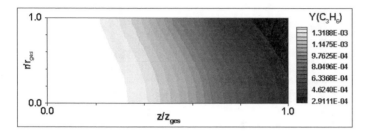

Figure 15 C$_3$H$_6$ mass fraction in a single channel, 41s after start-up.

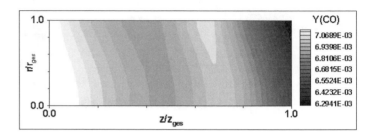

Figure 18   CO mass fraction in a single channel, 41s after start-up.

Figure 16 C$_3$H$_6$ mass fraction in a single channel, 140s after start-up.

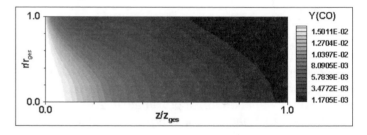

Figure 19   CO mass fraction in a single channel, 140s after start-up.

CHEMICAL SPECIES PROFILES

The conversion is related to the monolith temperature and the composition of the inlet feed. 10 seconds after start-up, when the catalyst is still cold, almost no chemical reactions take place. Only in the first section of the catalyst, where the temperature is already above 500K, less than 1% of the chemical species are converted on the catalyst. Figures 14 – 19 show the species profiles for C$_3$H$_6$ and CO.

Forty-one seconds after start-up, about 50% of the relevant emission species are converted, i.e. the reaction is lit off. A relatively good agreement between measured and calculated conversion is achieved. In Figures 20 – 23 results of the 2D simulation are presented.

After light-off, almost complete conversion can be achieved. Now, the conversion is no longer kinetically limited but controlled by mass transport, i.e., diffusion of reactants to the wall and products back into the fluid. Larger radial gradients are developed (Figure 16 and Figure 19).

Even though, the general trend of the experimentally observed conversion as function of time could be well predicted by the numerical simulation, some deviations occur. The conversion behavior of NO during the first 70 seconds is well predicted, but then the quality of the correlation decreases. This is likely caused by the assumption of unity redox ratio.

The experimentally determined decrease of the conversion rate of the hydrocarbon species at about 75 seconds is well matched, but the following increase is over-predicted. This might be a hint that the calculated temperature profile in the monolith is over-predicted, too. This might also be the reason for the over-predicted decrease of the CO conversion rate at 70 seconds and the over-predicted re-increase.

The reaction mechanism has been primarily developed using laboratory steady-state experiments with a well-defined simulation gas, including only propane and propylene as unburned hydrocarbons /5, 7, 17/. Methane oxidation on Pt and Rh has already been extensively studied /13,14,21/. However, the complex exhaust gas mixture contains a much wider variety of hydrocarbons.

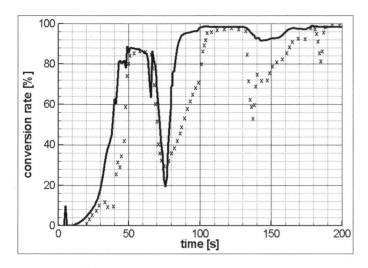

Figure 20 Measured (x) and numerically predicted (—) THC conversion.

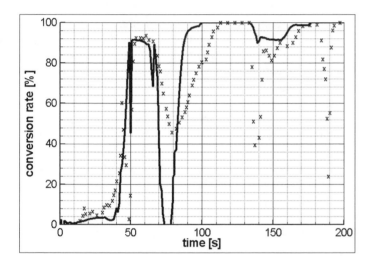

Figure 21 Measured (x) and numerically predicted (—) CO conversion.

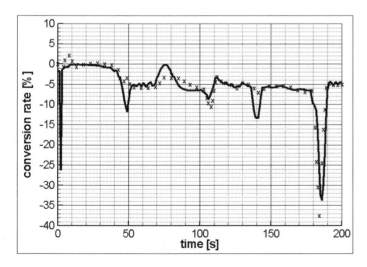

Figure 22 Measured (x) and numerically predicted (—) CO2 conversion.

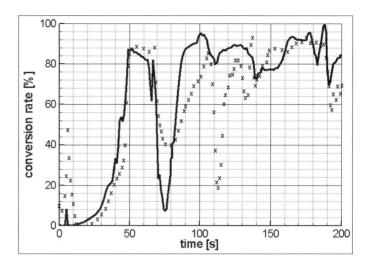

Figure 23 Measured (x) and numerically predicted (—) NO conversion.

Their representation by only two simple species may be one source of deviations. The surface reaction mechanism also needs further validation. For instance, no storage model has been implemented yet, as already mentioned.

On the experimental side, accurate measurements of the hydrocarbon concentration and composition have turned out to be a challenge.

2D VERSUS 3D MONOLITH SIMULATION

The DETCHEM^{MONOLITH} software can be used for the simulation of three- and two-dimensional converter geometries. The computation is time-consuming, especially if a large number of single channels needs to be analyzed to represent the monolith behavior. A typical CPU-time needed for the 3D simulation of the case studied takes about 24 - 48 hours depending on the hardware, if the parallel version of the code is used, in which each channel is simulated on a single processor. Therefore, the question arises whether or not a two-dimensional simulation is sufficiently accurate and how many channels need to be simulated.

A comparison of a 2D and a 3D simulation concerning the exit exhaust gas temperature is shown in Figure 24. In the 2D simulation the two-dimensional elliptic cross-section was replaced by a single averaged spatial coordinate. In the 2D simulation 3 channels were simulated. A comparison with simulations with more channels showed no qualitative difference to the simulation results presented in this paper. In the 3D simulation only 2 channels were simulated.

Over a wide range, the temporal temperature profiles do not show any difference and neither do the total species conversions (not shown here) (see also Figure 10). In case of a non-monotonic temperature profile in radial direction inside the monolith it seems to be necessary to consider a larger quantity of channels. Such intervals can be found at 45 – 55 seconds and 120 – 130 seconds. However for spatially varying inlet conditions, more channels have to be simulated.

A two-dimensional model considering more than 2 channels is appropriate in the case discussed in this study, since spatially constant inlet conditions were assumed.

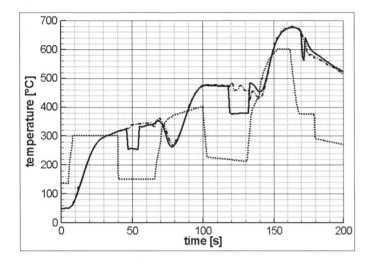

Figure 24 Comparison of the outlet fluid temperature for a 2D (- · -) and a 3D (—) simulation; for the 3D simulation the coated part only was taken into account. Additionally the considered calculation inlet gas temperature is plotted (····)

CONCLUSION

The recently developed computational tool DETCHEM^MONOLITH has been applied for the transient three-dimensional simulation of a cold start-up of an automotive three-way catalytic converter. The numerical code is based on a two-dimensional description (boundary-layer approximation) of the flow field in the single channels of the monolith coupled with a detailed multi-step surface reaction mechanism and a washcoat model. All the models were based on the physical and chemical processes occurring in the catalytic converter. To our knowledge, it is the first time that this detailed approach has been used for the transient numerical simulation of catalytic converters.

The simulation describes the heat-up of the catalytic monolith and the onset of the chemical reactions. The predicted time-dependent exit temperature and conversion are compared with experimental data. The light-off behavior is described well, as well as the conversion behavior in general, while deficiencies still exist in the temperature prediction. In the model, major uncertainties still remain concerning the exhaust gas composition and its representation by a limited number of chemical species.

Nevertheless, a computational tool is now available that allows predictive numerical simulations of the pollutant emissions during a driving cycle within a passable amount of time. This tool can be used for converter design and optimization.

The inlet conditions for DETCHEM^MONOLITH can either be taken from CFD-simulations or experimental data, and thus, the influence of the exhaust system design can be examined. Furthermore, the impact of the velocity distribution in front of the catalytic converter on conversion can be studied. The interactions between the thermal and geometrical properties and the light-off behavior of the monolith can also be examined.

OUTLOOK

We are currently working on the following aspects:

- Reaction schemes of more complex hydrocarbon species. Detailed gas analyses showed that a wide variety of HC species is present in the exhaust gas [20]. The influence of some of these species on the reaction system in a 3-way catalytic converter will be examined.

- Implementation of a model of oxygen storage effects into the numerical codes.

- Further tests of the surface reaction mechanism, also under transient conditions.

- Further improvements of the numerical algorithms (speed-up).

REFERENCES

/1/ S.-J. Jeong and W.-S. Kim, "Three-Dimensional Numerical Study on the Use of Warm-up Catalyst to Improve Light-Off Performance"; SAE 2000-01-0207, (2000)

/2/ G.C. Koltsakis, P.A. Konstantinidis and A.M. Stamatelos, "Development and application range of mathematical models for 3-way catalytic converters"; Appl. Catal. B: 12, 161 - 191, (1997)

/3/ T. Kirchner and G. Eigenberger, "On the dynamic behavior of automotive catalysts"; Catal. Today, 28, 3 (1997)

/4/ A. Onorati, G. Ferrari and G. D'Errico, "1D Unsteady Flows with Chemical Reactions in the Exhaust Duct-System of S.I. Engines: Predictions and Experiments"; SAE 2001-01-0939, (2001)

/5/ J. Braun, T. Hauber, H. Többen, P. Zacke, D. Chatterjee, O. Deutschmann and J. Warnatz, "Influence of Physical and Chemical Parameters on the Conversion Rate of a Catalytic Converter: A Numerical Study"; SAE 2000-01-0211, (2000)

/6/ S. Sriramulu, P. D. Moore, J.P. Mello and R. S. Weber, "Microkinetics Modeling of Catalytic Converters"; SAE 2001-01-0936, (2001)

/7/ D. Chatterjee, O. Deutschmann, J. Warnatz, "Detailed surface reaction mechanism in a three-way catalyst", Faraday Discussions119 (2001), in press

/8/ FLUENT 5.5, Fluent Inc., Lebanon, NH (2001)

/9/ S.Tischer and O.Deutschmann, "Modeling of catalytic partial oxidation reactors", Chem. Eng. Sci.(submitted)

/10/ S. Tischer, C. Correa, O. Deutschmann, "Transient three-dimensional simulations of a catalytic combustion monolith using detailed models for heterogeneous and homogeneous reactions"; Catalysis Today (in press)

/11/ DETCHEM, Version 1.4.2, O. Deutschmann, C. Correa, S. Tischer, D. Chatterjee, S. Kleditzsch, J. Warnatz, IWR, Universität Heidelberg

/12/ R. Jahn, D. Snita, M. Kubicek, M. Marek, "3-D modeling of monolith reactors", Catalysis Today 38, 39 (1997)

/13/ O. Deutschmann and L.D. Schmidt, "Modeling the Partial Oxidation of Methane in a Short-Contact-Time Reactor", AIChE J. 44, 2465 (1998)

/14/ L.L. Raja, R.J. Kee, O. Deutschmann, J. Warnatz and L.D. Schmidt, "A Critical Evaluation of Navier-Stokes, Boundary-Layer and Plug-Flow Models of the Flow and Chemistry in a Catalytic-Combustion Monolith", Catalysis Today 59, 47 (2000)

/15/ H. Schlichting and K. Gersten, "Boundary-Layer Theory", 8th ed., Springer-Verlag, Heidelberg (1999)

/16/ P. Deuflhardt, E. Hairer and J. Zugk, "One-Step and Extrapolation Methods for Differential-Algebraic Systems", Num. Math. 51, 501 (1987)

/17/ D. Chatterjee, Modellierung von Autoabgaskatalysatoren, Dissertation (PhD thesis), Naturwissenschaftlich - Mathematische Gesamtfakultät der Ruprecht-Karls-Universität Heidelberg, 2001

/18/ A.C. Hindmarsh, "ODE pack, a systematized collection of ODE solvers", in: Scientific Computing, R.S. Stepleman et al. (eds.), North-Holland, Amsterdam, 1983, pp. 55-64

/19/ MEXA - 7000 - Serie Abgas-Analysesystem Bedienungsanleitung 4. Ausgabe, Version 1.2, HORIBA Europe GmbH, Steinbach / Ts., Germany 1996

/20/ W.O. Siegl, E.W. Kaiser, A.A. Adamczyk, M.T. Guenther, D.M. DiCicco and D. Lewis, "A Comparison of Conversion Efficiencies of Individual Hydrocarbon Species Across Pd- and Pt-Based Catalysts as a Function of Fuel-Air Ratio"; SAE 982549, (1998)

/21/ O. Deutschmann, F. Behrendt, and J. Warnatz, „Modelling and Simulation of Heterogeneous Oxidation of Methane on a Platinum Foil"; Catalysis Today 21 (1994), 461

NOTATION

$c_{p,i}$	specific heat at constant pressure of species i
h_i	enthalpy of species i
h	enthalpy of the mixture
p	pressure
q	heat flux
j_{ir}	diffusive flux
r	radial spatial coordinate
T	temperature
t	time
u	axial velocity
v	radial velocity
Y_i	mass fraction of species i
z	axial spatial coordinate
λ	thermal conductivity
μ	viscosity
ρ	density

CHAPTER 11

EVAPORATIVE EMISSIONS REDUCTION

Christopher Hadre and Jenny Spravsow
DaimlerChrysler Corporation

11.1 OVERVIEW

Evaporative emissions are defined as emissions from a vehicle when parked. These emissions can be divided into two categories: fuel based and non-fuel based. The fuel based evaporative emissions typically are caused by permeation through fuel system components, such as fuel tanks and vapor harnesses, as well as emissions from fuel tank breathing from the canister vent. The non-fuel based evaporative emissions are largely derived from other vehicle systems. Air conditioning refrigerant R134a and washer solvent are two main sources of non-fuel based emissions.

The California Air Resources Board (CARB) has created the California Low Emissions Vehicle II (LEV II) standards for evaporative emissions. The passenger car standard is 0.50 g and the light duty truck standard is 0.90 g for the three day test procedure. This legislation also comes with stricter tailpipe standards.

CARB has also mandated that by 2005 10% of its fleet will have zero evaporative emissions. These vehicles, typically fuel cell or electrically powered, are considered to have inherently zero emissions. In lieu of this requirement, manufacturers can develop and sell Partial Zero Emission Vehicles (PZEV) to earn credits. The PZEV standard is 0.0 g per day which is accepted as 0.054 g fuel based evaporative emissions, and must be demonstrated on a vehicle carcass basis.

The key to successfully demonstrating LEV II and PZEV emission levels is to control the permeation and the containment of any vapors that are generated in the fuel system. To control permeation, the proper materials must be specified for the fuel tank, pump module flange and seal, and vapor tubes. Connections in the fuel system should have properly designed features to ensure adequate sealing. Use of multilayer HDPE (High Density Polyethylene) fuel tanks with a minimum number of welded components, or steel tanks with proper corrosion coatings, and fluorocarbon (FKM) o-rings and seals are some of the proven technologies available for LEV II and PZEV.

There are several ways to control the generation of vapor from the fuel in the fuel tank. In one concept, positive pressure is applied to the fuel system to prevent any vapors generated from loading the canister. A smaller canister can be used, saving space, weight and potentially improving the drivability of the vehicle. This concept lends itself to steel tanks with remote vapor domes very well. Another way to reduce the formation of vapor is to prevent the tank from having vapor space. This is done using a "bladder", or flexible barrier, within a fixed-wall structure for the fuel tank. As fuel is consumed by the engine, the bladder decreases in size. If only a very small amount of vapor is generated, a very small canister is needed. A third type of fuel and vapor control system is the one that allows free breathing of the canister and tank. The vapors are allowed to generate and are contained by the canister. This system is less complex than the previous systems. During refueling, vapor generation is reduced when some of the gasoline vapor-saturated air is recirculated to the top of the fill tube.

One area of research is in canister technology. Developing different, more effective carbons and canister designs are being done by many companies. As fuel vapor loads into the canister, it condenses on the carbon pellets in the canister bed. The carbon holds on to the fuel vapor, allowing fresh air to pass through. When the engine is on and the sensors detect a need to purge the canister, air is drawn through the canister, pulling off the evaporating hydrocarbon molecules. There must be enough volume of carbon in the canister to contain all the hydrocarbon fuel vapors that the tank generates; typically the largest load occurs during an ORVR (Onboard Refueling Vapor Recovery) refueling event. Canister "bleed" emissions must be tightly controlled in a PZEV environment. Bleed emissions are those hydrocarbon molecules that diffuse through the canister into the fresh air. In a strict standard such as PZEV, even milligrams of bleed emissions can cause a vehicle to fail the standard. In addition to specialized carbons, canister geometry can be modified to improve bleed emissions.

Developing carbons that adsorb emissions from the air induction system in the engine is a high priority in the LEV II and PZEV process. Fuel vapor concentration can be higher in the cylinders and intake manifold, due to the proximity to the fuel injectors. This vapor can migrate past the throttle body and through the fresh air intake out to the atmosphere in the SHED (Sealed Housing for Evaporative Determination) test. A "trap" may be required to control those emissions. Several concepts have been researched, from a honeycomb brick of carbon, to specially designed carbon-impregnated air filters. A significant challenge to these traps is the amount of pressure drop across them in terms of a decrease in engine horsepower.

Due to the increasingly challenging emissions standards, more resources have been focused on fuel system and engine evaporative emissions. There are many effective ways to improve evaporative emissions from fuel system components. Choosing the proper materials for components and seals, understanding different strategies for controlling vapor generation in the fuel tank, and major strides in canister bleed emissions control have made the seemingly impossible task of LEV II and PZEV standards achievable.

11.2 SELECTED SAE TECHNICAL PAPERS

2000-01-1099

Fuel Permeation Study on Various Seal Materials

Hélène Aguilar
Acadia Polymers

Ron G. Kander
Virginia Tech

Copyright © 2000 Society of Automotive Engineers, Inc.

ABSTRACT

The advent of low emission regulations on fuel systems has made conventional sealing materials such as acrylonitrile butadiene rubber (NBR) unfit for sealing most fuel systems. Therefore, it is imperative to look beyond conventional rubbers and towards more exotic materials to seal such applications. In this study, the permeation characteristics and the change in physical properties of several elastomeric materials (NBR, hydrogenated NBR (HNBR), and fluorocarbon elasto-mers (FKM)) as well as various poly(tetrafluoroethylene) (PTFE) composites were evaluated with four different fuel mixtures. The sealing materials were tested using vaporimeter cups. The results are discussed as a function of the materials' nature, composition, and filler content.

INTRODUCTION

For many years, automotive fuel systems have been sealed using materials such as acrylonitrile butadiene rubber (NBR). However, with the advent of low emission regulations, these conventional materials are not enough to keep these same fuels from emitting into the atmosphere. More exotic materials have been studied by many researchers to replace the conventional materials in such applications.[1-11] However, these studies mostly involve extruded materials for hose applications[1-7] and have not focused on the materials used to seal fuel tanks or fuel pumps. The materials in those studies range from hydrogenated acrylonitrile butadiene rubbers (HNBR) to highly fluorinated fluorocarbon elastomers (FKM). In this paper, long-term permeability results of these materials and glass fiber reinforced poly(tetrafluoroethylene) (PTFE) composites are reviewed and compared with a reference NBR material.

EXPERIMENTAL PROCEDURE

MATERIALS – Table 1 gives the matrix of elastomeric materials used in this study, while Table 2 shows the composition and thickness of the different PTFE composite sheets. FKM-1 and FKM-3 elastomers are production compounds mixed in an internal production mixer, while the other four elastomers were mixed on a laboratory mill. All elastomeric compounds were then compression molded into sheets according to ASTM D3182 and post-cured to optimize cross-link density according to rubber manufacturer's recommendations. The PTFE sheets were made of PTFE powder from Custom Compounding, a Division of Dyneon LLC. These PTFE sheets are skived tapes from billets.

Four different fuels were used in this study and are described in Table 3. All four fuels are based on reference fuel C, which is an ASTM standard and is made of 50% by volume isooctane and 50% by volume toluene. Fuel CM contains an oxygenate, methyl-*t*-butyl ether (MTBE), which is currently used in most gasolines to help the gas burn in a cleaner and more efficient manner. MTBE has been under investigation for contaminating groundwater, but thusfar it has not been subjected to replacement.

FUEL PERMEATION TESTING – Testing was performed according to ASTM D814. The rubber and PTFE sheets were marked along the grain of the material to identify the mechanical testing direction. Circular samples were cut from the provided slabs of rubber and PTFE. The original weight of each sample was then taken to the nearest 0.1 mg and recorded. 100 ml of the desired testing fuel was poured into a clean aluminum testing cup. The samples were placed over the mouth of the cup and fastened down tightly with thumbscrews. Due to the reduced thickness of the PTFE films (0.38 mm to 0.64 mm), rubber seals were placed around the edges of the cup to ensure total sealing with no leakage at the edge. No rubber seals were used for the elastomeric materials, since the elastomeric sheets were thick enough. After complete assembly, the total weight of the testing cups was measured to the nearest 0.01 g and recorded.

Table 1. Elastomeric materials used in this fuel permeation study

Compound #	Materials type	Fluorine content
NBR	NBR/PVC blend	N/A
HNBR	HNBR	N/A
FKM-1	Dipolymer FKM	66%
FKM-2	GFLT FKM	67%
FKM-3	Terpolymer FKM	70.2%
FKM-4	Terpolymer FKM	73%

Table 2. PTFE composite systems used in this fuel permeation study

Sample #	Glass fiber content	Sheet thickness
1	5%	0.508 mm (0.020 in)
2	15%	0.635 mm (0.025 in)
3	25%	0.381 mm (0.015 in)
4	25%	0.508 mm (0.020 in)
5	25%	0.635 mm (0.025 in)

Table 3. Fuel mixtures used in this fuel permeation study

Fuel name	Ingredients	Percentages
Fuel C	Reference Fuel C	100%
Fuel CE15	Fuel C	85%
	Ethanol	15%
Fuel CM15	Fuel C	85%
	Methanol	15%
Fuel CM	Fuel C	62%
	Methanol	30%
	Methyl-*t*-butyl ether (MBTE)	8%

The assembled test cups were turned upside down to create contact between the fuel and the sealing film and were placed in a fume hood at room temperature for the duration of the experiment. Periodic weight measurements were performed to monitor the permeation of the fuel through the sealing material. At the end of the testing time (1008 hours), the test cups were disassembled and the samples removed. The samples were immediately weighed to the nearest 0.1 mg and placed in airtight individual plastic bags for a maximum period of 30 minutes until mechanical testing was performed. The amount of fuel remaining in the test cup was measured with a graduated cylinder and recorded.

MECHANICAL TESTING – Promptly after removal from the testing cups, small dog-bone shaped specimens were cut from the samples using a Dewes-Gumbs Die Company manual die press. The cuts were made along the length of the material (grain) for the elastomers and along the length and width of the material for the PTFE composites, using the marks made on the sample before exposure to the fuel as a guide. A Polymer Laboratories MINIMAT Miniature Materials Tester was used to pull the samples in tension at a speed of 20 mm/min. Three dog-bones were tested per film. Load and extension data were recorded, from which stress vs. strain data were calculated. Finally, graphics software was used to present the information. Only the median stress vs. strain curve from the three dogbones was used for analysis.

THERMAL ANALYSIS – Thermogravimetric analysis (TGA) was performed on the NBR compound (original and exposed to fuel CM) using a Perkin Elmer TGA 7. A sample weighing 10 ± 2 mg was cut from the middle of the material slab and was tested at temperatures between 30 and 900 °C at a heating rate of 20 °C/min. Nitrogen purge was used below 600 °C, while oxygen purge was utilized above 600 °C to burn off all the remaining organic matter in the material.

RESULTS

ELASTOMERS

Fuel permeation – Figures 1-4 give the change in weight of the total system (cup, fuel, and sealing sample) over 1008 hours at room temperature for all the elastomers studied in fuels C, CM15, CE15, and CM, respectively. From these figures it is possible to extract the following information:

1. The order of aggressiveness of the four testing fuels is (by increasing aggressiveness):

   Fuel C << Fuel CM15 << Fuel CE15 << Fuel CM

2. The system sealed with NBR displays a large weight loss when the cup is filled with fuel CM (this elastomer was not tested with the other three fuels for 1008 hours). Thus, this nitrile compound is very permeable to fuel CM.

3. The HNBR compound also shows a very large decrease in the total weight of the system over 1008 hours of testing in all four fuel systems studied. Therefore, this HNBR compound is not suited for fuel applications.

4. The higher the fluorine content, the heavier the compound and thus the entire system.

5. In general, the higher the fluorine content in the fluorocarbon elastomers, the less permeable the materials are to fuel systems.

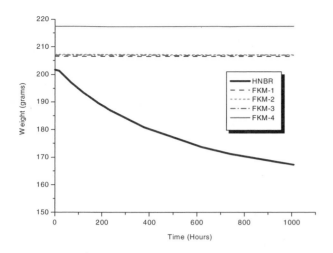

Figure 1. Weight change of vaporimeter cups filled with fuel C and sealed with various elastomers for 1008 hours

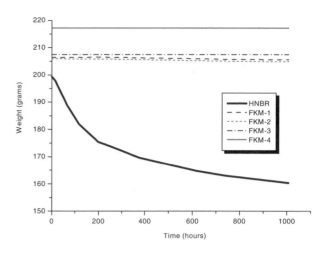

Figure 3. Weight change of vaporimeter cups filled with fuel CM15 and sealed with various elastomers for 1008 hours

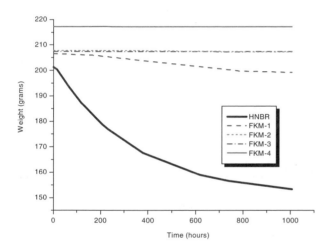

Figure 2. Weight change of vaporimeter cups filled with fuel CE15 and sealed with various elastomers for 1008 hours

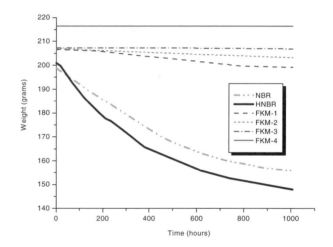

Figure 4. Weight change of vaporimeter cups filled with fuel CM and sealed with various elastomers for 1008 hours

<u>Tensile properties</u> – Figure 5 gives the stress-strain curves for the elastomers before exposure to any fuels. The tensile properties shown on this figure can be summarized by calculating three important parameters: Young's modulus, maximum strain, and maximum strength. These three parameters are given in Table 4 for all six elastomers before exposure to fuels.

Figures 6-8 show the change in all three tensile parameters (Young's modulus, maximum strain, and maximum strength, respectively) for the elastomers exposed to all four fuels for 1008 hours at room temperature. From these three figures, it is possible to observe the following trends:

1. The order of aggressivity of the fuels towards the tensile properties of the elastomers is consistent with that observed during the permeation study.

2. The NBR and HNBR compounds do present a relatively low degradation of their tensile properties, compared to their fuel permeation rates. That is, the NBR and HNBR compounds display a smaller change in tensile properties than the lower fluorine content FKM compounds (*i.e.*, compounds FKM-1 and FKM-2).

3. In general, the lower the fluorine content in the FKM compounds, the more degradation in their tensile properties.

Table 4. Original tensile properties for all six elastomers: Young's modulus, maximum strain, and maximum strength.

Compound #	Modulus (MPa)	Max. strain (%)	Max. strength (MPa)
NBR	7.29	382.16	16.01
HNBR	6.84	395.40	23.16
FKM-1	6.46	313.39	14.49
FKM-2	5.33	276.00	14.00
FKM-3	4.58	355.30	13.53
FKM-4	11.71	372.78	23.57

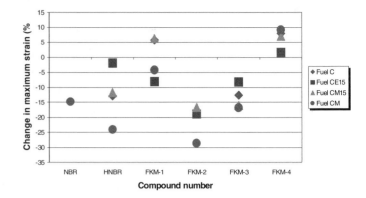

Figure 7. Change in maximum strain of elastomers exposed to various fuels for 1008 hours at room temperature

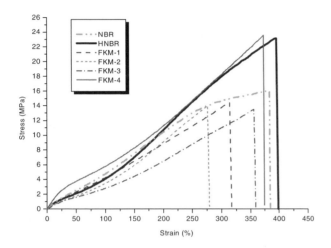

Figure 5. Stress vs. strain curves of elastomers before exposure to fuels

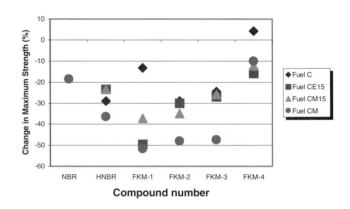

Figure 8. Change in maximum strength of elastomers exposed to various fuels for 1008 hours at room temperature

When working with elastomers, it is common to work with moduli at various strains (*i.e.*, modulus at 50% strain, modulus at 100% strain, etc). These moduli correspond to the strength values at the specified strains. When plotting the change of these moduli after exposure to the four fuel mixtures, the above-mentioned trends were also observed.

Sample weight uptakes – Figure 9 presents the weight uptake of the elastomeric samples after exposure to the different fuels for 1008 hours. The major surprising result from this figure is the very low weight uptake of the NBR compound, with only 0.538% weight uptake in Fuel CM. All the other weight uptakes follow the same trends as the tensile properties described above.

This fuel permeation experiment was repeated for most elastomeric materials for lower periods of time (*i.e.*, 70 and 504 hours) and the weight uptake was monitored to follow the weight gain of the sealing samples over time.

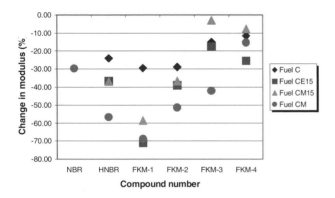

Figure 6. Change in Young's modulus of elastomers exposed to various fuels for 1008 hours at room temperature

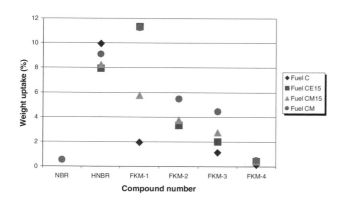

Figure 9. Weight uptake of elastomers after exposure to various fuels for 1008 hours

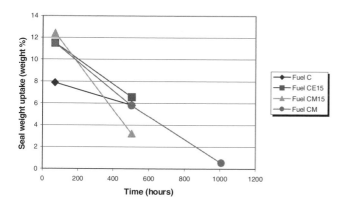

Figure 10. Periodic weight uptake of NBR compound exposed to four fuel mixtures for periods of time up to 1008 hours

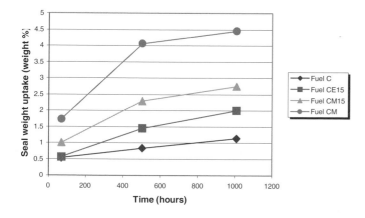

Figure 11. Periodic weight uptake of FKM-3 compound exposed to four fuel mixtures for periods of time up to 1008 hours

Figures 10 and 11 show the weight uptake of NBR compound and FKM-3 compound, respectively. The major

difference between those two compounds is that the weight uptake trends are reversed. That is, the NBR compound displays a weight uptake that decreases over time, while the FKM compound has a weight uptake that increases over time. The latter result (*i.e.*, weight uptake increasing with fluid exposure time) is the expected response of a material when exposed to a fluid. All the other FKM compounds behaved like compound FKM-3. No periodic weight uptake results were obtained for the HNBR compound.

The NBR compound's behavior will be discussed in further detail in the analysis section of this paper.

PTFE COMPOSITES

Fuel permeation – The order of aggressivity of the different fuels on the PTFE composites is the same as with the elastomeric materials; however, the difference in aggressivity among the various fuels is more subtle with PTFE than with the elastomers. The two major factors affecting the permeability of the fuels through the PTFE films were (1) the percent glass fiber reinforcement and (2) the thickness of the films.

Figure 12 presents the change in weight of the total system over 1008 hours at room temperature for all PTFE composites exposed to fuel CM15. As shown on this figure, the higher the glass fiber reinforcement and/or the thinner the PTFE film, the more permeable the sealing material is to fuel systems. These results will be discussed further in the analysis section.

Figure 13 shows the final fuel volume that remained in the cups after the 1008 hours of testing. As expected from Figure 12, the lower the glass fiber content and/or the thicker the films, the more fuel remained in the cups. Also, some of the samples (25% glass filled at various thicknesses) let most or all the fuels evaporate, which means that these materials are very poor fuel sealant materials.

Tensile properties – Figure 14 displays the stress-strain curves of the unexposed PTFE composites. It can be noted that as the glass fiber content increases, the material becomes less tough (*i.e.*, less energy is required to break the material). Table 5 gives the three tensile parameters for the original PTFE sheets, while Figures 15-17 give the change in those parameters after exposure to the four fuels mixtures.

After fuel exposure, these PTFE films present (1) a decrease in modulus (78.6% of the samples), (2) an increase in maximum strain (92.9% of the samples), and (3) an increase in maximum strength (78.6% of the samples). This general behavior indicates a *plasticization* of the composites by the fuels.

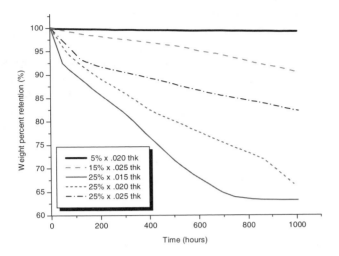

Figure 12. Change in weight of total system for PTFE composites exposed to fuel CM15 for 1008 hours at room temperature

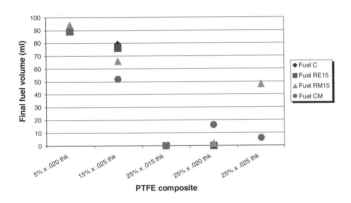

Figure 13. Final fuel volume remaining in the cups after the 1008 hours of testing

There does not seem to be any trend on the effect of the filler content and the thickness of the material on this plasticization. This is probably due to the disparity in the tensile results one would expect with this kind of testing. Also, as discussed above, some of the PTFE films let the whole fuel amount evaporate so that the film was drying for several days or weeks before mechanical testing. This would give different tensile properties than if the film had stayed in contact with the fuel until just before mechanical testing.

Sample weight uptakes – Figure 18 shows the weight uptake of the PTFE composites after exposure to the four fuel mixtures for 1008 hours. As expected from the per-meability results, the more reinforcement and the thinner the film, the higher the weight uptake of the sample. Also, the weight uptake does not always follow the order of aggressivity of the fuels (i.e., the more aggressive the fuel, the more weight uptake of the sample). This is probably due to the fact that, as mentioned above, the dif-ference in aggressivity among the various fuels is not as defined with PTFE as it is with elastomers.

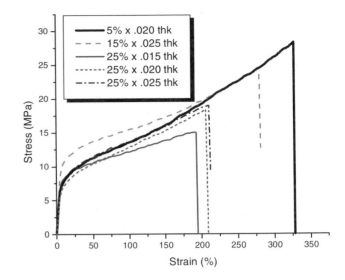

(a)

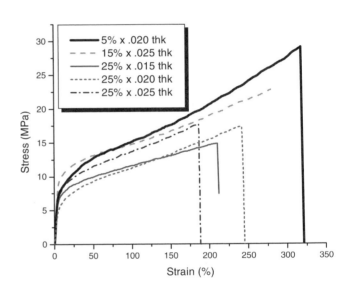

(b)

Figure 14. Stress vs. strain curves of unexposed PTFE composites. (a) along length of material film, and (b) along width of material film.

Table 5. Tensile properties of the original PTFE films

PTFE film	Modulus (MPa)		Max. strain (%)		Max. strength (MPa)	
%/thk	length	width	length	width	length	Width
5/.020	164.23	168.72	327.02	318.08	28.31	29.06
15/.025	187.23	199.47	279.34	279.21	24.53	22.79
25/.015	112.77	135.52	235.94	214.98	13.29	14.33
25/.020	134.49	118.50	206.06	241.91	18.42	17.34
25/.025	159.89	144.14	208.95	186.57	18.95	17.69

530

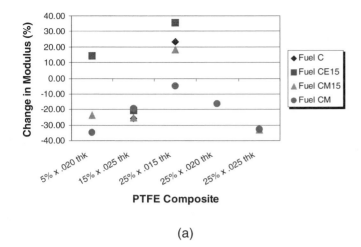

(a)

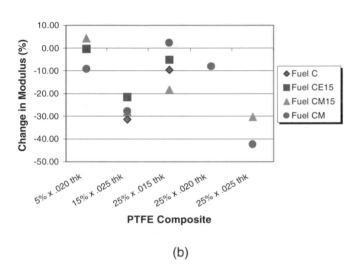

(b)

Figure 15. Percent change in modulus of the PTFE composites exposed to various fuels for 1008 hours – (a) along length of material and (b) along width of material

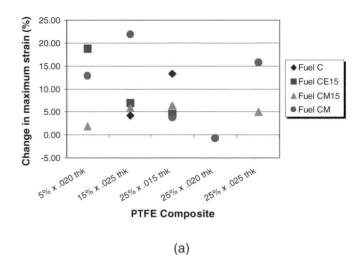

(a)

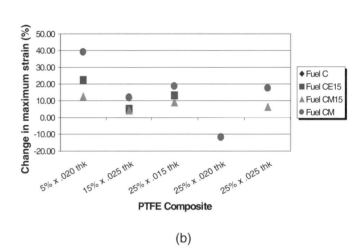

(b)

Figure 16. Percent change in maximum strain of the PTFE composites exposed to various fuels for 1008 hours – (a) along length of material and (b) along width of material

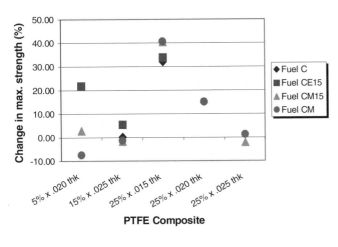

(a)

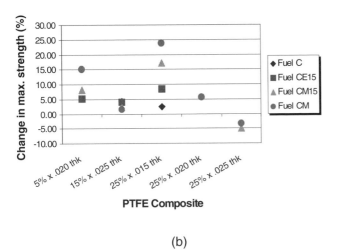

(b)

Figure 17. Percent change in maximum strength of the PTFE composites exposed to various fuels for 1008 hours – (a) along length of material and (b) along width of material

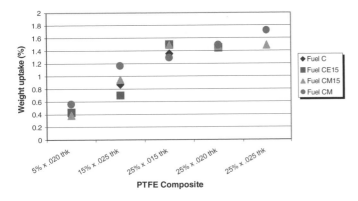

Figure 18. Weight uptake of the PTFE composites after exposure to the four fuel mixtures for 1008 hours

Figure 19. Schematic of nitrile butadiene rubber (NBR) and hydrogenated nitrile butadiene rubber (HNBR) chemical structures

ANALYSIS

ELASTOMERS – The NBR and HNBR compounds tested in this study did not perform well with any of the four fuel mixtures studied. Although these materials have been used for years to seal most fuel applications, there are better materials to control emission as this study suggests. Surprisingly, the HNBR material permeated more fuel than the NBR material. Since HNBR has a more saturated backbone than NBR as shown in Figure 19, HNBR should have more chemical resistance than NBR. The poor fuel resistance of NBR has been observed by Sargent et al.[10], while Stahl and Stevens[12] have shown that HNBR does permeate fuel at a higher rate than FKM materials. Dunn and Vara[13] have shown that permeation and swelling in NBR compounds were reduced by increasing acrylonitrile amount in the polymer or blending with PVC. The acrylonitrile content of the NBR compound in this study is 40%, while that of the HNBR compound is 36%. Also, our NBR compound is an NBR/PVC blend. Therefore, the reasons why the NBR compound is less permeable to fuel than the HNBR compound are probably that (1) the acrylonitrile amount is higher in the NBR compound than in the HNBR compound and (2) the NBR compound is a PVC/NBR blend.

One of the worst effects a fluid can have on an elastomeric material is to change the material's composition and thus its inherent properties. This has occured with the NBR compound as it displays a weight loss during extended exposure to fuels. This weight loss may be due to some plasticizer materials being extracted out of the elastomers because of the presence of the fuel mixtures.

Thermogravimetric analysis (TGA) was performed on the original NBR compound and on the same compound exposed to fuel CM for 1008 hours. Figure 20 shows the TGA curves for the two samples. As shown on these

TGA curves, exposure to the fuel does decrease the amount of plasticizer in the material, *i.e.* the weight loss associated with plasticizer materials (between 200 and 400 °C) disappears after the material has been exposed to fuel CM for 1008 hours. According to the TGA curves, the original compound contains 8.3% by weight of plasticizer materials, while the exposed compound has 2.5% by weight of plasticizer materials. Thus the exposure to fuel CM has reduced the plasticizer content by 5.8%.

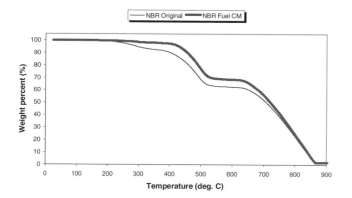

Figure 20. Thermogravimetric analysis of NBR compound before and after exposure to fuel CM for 1008 hours

Another consideration one has to take is that the experiment described here does not provide for large loss of fuel, where the composition of the fuel will change over time since toluene, ethanol, and methanol evaporate at a higher rate than isooctane. This is the case for both NBR and HNBR compounds. This could also explain the fact that the NBR compound loses weight overtime. However, this change in fuel composition would not happen in a real fuel system, which has a much larger supply of fuel and can be considered an infinite reservoir of fuel. Therefore, the NBR behavior may or may not occurr in a real fuel sealing application.

All the FKM compounds tested did show a substantially lower permeability to the four fuel mixtures, with improved chemical resistance as the fluorine content increased. This trend was also observed by Balzer and Edmonson in their FKM hose compound study.[6] However, there are major drawbacks to using very high fluorine content FKM compounds: (1) added cost, (2) rigidity at subambient temperatures, and (3) poor processability. This study shows that a 70.2% terpolymer FKM compound, such as compound FKM-3, would be sufficient to seal most fuel systems as its permeability rate is very low with all four fuel mixtures studied and its physical properties are not altered greatly after fuel exposure.

PTFE COMPOSITES – As mentioned above, the higher the glass fiber reinforcement and/or the thinner the PTFE film, the more permeable the sealing material is to fuel systems. The effect of the thickness of the PTFE film is well known. According to Crank and Park[14], if the material is the same, the flux should be inversely proportional to the thickness.

On the other hand, the effect of the filler content on the fuel permeation (*i.e.*, the more the glass fibers, the more permeable the material) requires a more in-depth explanation. The glass fibers do act as disruptive elements on the PTFE morphology.

First, since PTFE has very low surface energy, it is difficult to wet and bond to reinforcing glass fibers.[15] Most of the sizing materials (*i.e.*, coating materials used on reinforcing fibers to promote bonding between the fibers and the matrix material) available commercially are not compatible with PTFE so there is no real bond between the PTFE matrix and the glass fibers.[15-16] Poor adhesion leaves voids between the fibers and the matrix material providing paths where fuel can permeate.

Second, PTFE is a semi-crystalline material which is very resistant to most chemicals. The crystals in a thermoplastic contribute the most to chemical and thermal resistance, while the amorphous fraction is more vulnerable to chemicals and heat. Introducing foreign objects into a material, such as glass fibers, disrupts the crystallinity of the material. In general, the higher the glass fiber content, the lower the degree of crystallinity. Therefore, the higher fuel permeation rate for the most filled PTFE composites could also be due to an increased fraction of amorphous phase in the material.

Finally, if there were any bond between the fibers and the matrix, the fiber reinforcement usually introduces "interphase" material between the fibers and the matrix material. This interphase material constitutes the bond between the two materials and has a different crystalline structure than the matrix material. The more glass fibers in the matrix, the more interphase material there is in the overall material, which could contribute to increased fuel permeation.

CONCLUSION

Permeation of four different fuel mixtures of various aggressivity levels was studied using six elastomers and five PTFE composites. Permeation rate, ultimate tensile properties, and final structure were analyzed. The following conclusions were drawn:

1. Both nitrile-based elastomers (NBR and HNBR) are very permeable to the fuel mixtures studied. Their composition is changed during exposure to aggressive fuels (*i.e.*, the plasticizer material is extracted out of the compound)

2. Fluorocarbon elastomers are relatively impermeable to all four fuels studied. The higher the fluorine content, the less permeable they become to fuels.

3. In PTFE composites, the higher the glass fiber reinforcement and/or the thinner the PTFE film, the more permeable the sealing material is to fuel systems.

FUTURE WORK

In order to answer some of the questions raised above, we are currently performing the following tests:

1. Fuel permeation tests with hollow and full glass fiber reinforced PTFE.

2. Use of other exotic materials such as graphite seals and THV material (a polymer of tetrafluoroethylene, hexafluoropropylene and vinylidene fluoride) to seal off the vaporimeter cups.

3. Investigating osmosis *vs.* fuel contact tests, where the cups are left with the mouth up and not upside down. This osmosis test has already been performed by other authors with other kinds of materials.[17,18] In our case, osmosis may or may not accelerate fuel permeation.

4. Investigating the use of a controlled environmental chamber. Placing the vaporimeter cups in a fumehood tends to accelerate permeation of the fuels through the seals as there is more air flow around the cups. Even if the fumehood experiment gives us trends to compare different materials and simulates better a real fuel system environment, it is more difficult to control than a chamber environment, where temperature and humidity are easily monitored.

ACKNOWLEDGMENTS

The authors would like to thank Ms. Melinda Guinn, Ms. Melanie Lashus, Ms. Michelle Jensen, and Ms. Julie Martin for their help in performing the tests presented in this paper. Particular thanks to Mr. Gabe Willams from Acadia Polymers for providing the PTFE samples and to Ms. Nena Serios from Dyneon, LLC for her excellent advice in conducting this study.

REFERENCES

1. Vara R. G., Dunn J.R.; ACS, Rubber Division Meeting; Orlando, Fl.; 26th-29th Oct. 1993; Paper 13; p 25 (1993)

2. Brit. Plast. Rubb.; Jan. 1998; p 2 (1998)

3. Ries H.; Plast' 21; no 56; pp 61-63 (1996)

4. Knights M.; Plast. Technol.; vol. 43; no. 9; p 32-34 (1997)

5. O'Neill M.; Mod. Plast. Int. ; vol. 27; no. 2; pp 56-58 (1997)

6. Balzer J. R., Edmonson A. L.; SAE International Congress and Exposition; Detroit, MI; Feb. 25 – Mar. 1, 1991; paper no. 910106 (1991)

7. Plast. Rubb. Weekly; no. 1663; 22nd Nov. 1996; p 10 (1996)

8. De Gaspari J.; Plast. Technol.; vol. 42; no. 5; pp 15-19 (1996)

9. Brullo R. A., Sohlo A. M., SAE International Congress and Exposition; Detroit, MI; Feb. 29 – Mar. 4, 1988; paper no. 880021 (1988)

10. Sargent D. H., Bielawski C.; Summary report Jun. 30-Oct. 31, 1969; Defense Technical Information Center; 30 pages (1969)

11. Landwehr R. S.; IISRP 31st Annual Meeting Proceedings; Capetown; 23rd-27th April 1990; p 179-185 (1990)

12. Stahl W.M., Stevens R.D.; SAE International Congress and Exposition; Detroit, MI; paper no. 920163 (1992)

13. Dunn J. R., Vara R.; *Elastomerics*; vol. 29 (1986)

14. Crank J, Park G.S.; *Diffusion in Polymers*; Academic Press; New York (1968)

15. Wang X.Q., Han J.C., Du S.Y., Wang D.F.; Journal of Reinforced Plastics and Composites; vol. 17; no. 17; pp1496-1506 (1998)

16. Voss H., Friedrich K; *Journal of Materials Science Letters*; vol. 5; pp 569-572 (1986)

17. Nulman N., et al.; SAE International Congress and Exposition; Detroit, MI; paper no. 981360 (1998)

18. Brahmi A., Wolf R.; SAE International Congress and Exposition; Detroit, MI; paper no. 1999-01-0380 (1999)

CONTACT

Dr. Hélène Aguilar obtained her Ph.D. in Materials Engineering and Science in May 1995 from Virginia Tech. Her dissertation was on The Effect of Intermediate Solvents on Poly(Ether Ether Ketone). Dr. Aguilar joined the R&D team at Acadia Polymers in July 1996, where her work has encompassed projects involving adhesives development, surface analysis work, elastomers compatibility with automatic transmission fluids, and rubber compounding. Dr. Aguilar can be reached at helene.aguilar@acadiapolymers.com or visit Acadia at www.acadiapolymers.com.

DEFINITIONS, ACRONYMS, ABBREVIATIONS

HNBR: Hydrogenated acrylonitrile-butadiene rubber (ASTM D1418 nomenclature)
FKM: Fluorocarbon elastomer (ASTM D1418 nomenclature)
MTBE: Methyl-*t*-butyl ether. An oxygenate used in fuel systems to make gasoline burn cleaner and more efficiently.
NBR: Acrylonitrile-butadiene rubber (ASTM D1418 nomenclature)
PTFE: Poly(tetrafluoroethylene)
TGA: Thermogravimetric analysis. An instrument that records the loss of mass in a material while being heated through a defined range of temperature.

2000-01-0895

Studies on Carbon Canisters to Satisfy LEVII EVAP Regulations

Hideaki Itakura, Naoya Kato and Tokio Kohama
Nippon Soken.Inc.

Yoshihiko Hyoudou, Toshimi Murai
Toyota Motor Corp.

Copyright © 2000 Society of Automotive Engineers, Inc.

ABSTRACT

Recently, the California Air Resources Board (CARB) has proposed a new set of evaporative emissions and "Useful Life" standards, called LEVII EVAP regulations, which are more stringent than those of the enhanced EVAP emissions regulations. If the new regulations are enforced, it will become increasingly important for the carbon canister to reduce Diurnal Breathing Loss (DBL) and to prevent deterioration of the canister. Therefore, careful studies have been made on the techniques to meet these regulations by clarifying the working capacity deterioration mechanism and the phenomenon of DBL in a carbon canister.

It has been found that the deterioration of working capacity would occur if high boiling hydrocarbons, which are difficult to purge, fill up the micropores of the activated carbon, and Useful Life could be estimated more accurately according to the saturated adsorption mass of the activated carbon and the canister purge volume. As a result, it is presumed that a more adaptable, longer Useful Life can be realized by providing a sufficient purge.

It has been also found that the butane diffusion in a carbon canister during vehicle parking which is loaded to the canister during the DBL test, is the main cause of evaporative emissions from the canister. To prevent such diffusion, it is effective to divide the carbon bed into separated segments and insert some "labyrinth" between such carbon beds. Compared with the conventional canister, the improved canister was able to reduce DBL by half._Furthermore it became clear that DBL is reduced to approximately 1/3 when the gasoline fuel vapor is loaded to the canister instead of butane, which is the main cause of DBL. It was also concluded that the evaluation method should be reconsidered to account for real world conditions.

INTRODUCTION

CARB has proposed for 2004, which is when both the enhanced EVAP emissions regulations and On-Board Refueling Vapor Recovery (ORVR) regulations will once be settled, a new set of LEVII regulations including a more stringent set of evaporative emissions and Useful Life standards as compared to those of the enhanced EVAP emissions regulations. In these regulations CARB also has required the evaporative emissions to be restricted to 1/4th of the parameter stated in the 1995 regulations and the Useful Life to be extended from 10 to 15 years.

The sources of evaporative emissions generated in DBL test have already been investigated and it is clarified that a canister is one of the typical sources of the evaporative emissions. To meet the LEV II regulations, it has become more important for the canister to reduce DBL and to extend its useful life span.

The useful life span of the canister depends on that of the activated carbon which adsorbs the hydrocarbons(HC). Therefore, several studies have been made on the deterioration of adsorption performance of the carbon canister withdrawn from the vehicles after long usage and the change in adsorption performance during the adsorption/desorption cycle tests in the laboratory. However, little research analyzing the relation between characteristics of activated carbon and its deterioration phenomenon has been done, and a general method for estimating canister life span has not been developed.

Furthermore the effects on DBL by external factors such as temperature and the length of parking have been clarified, and this information has led to the formulation of the present Federal Test Procedure. Recent studies show the effects of the design factors of canister configuration and purge amount on DBL. However, detailed analysis on the adsorption states of activated carbon in the canister and on the mechanism of DBL have been few.

Therefore, the authors tried to clarify the mechanisms for canister life span and DBL. In addition, the authors demonstrated by a simple model, and tried to make clear a canister design that would comply with the LEVII regulations.

EXPERIMENTAL METHODS

The test equipment used in the experiment is shown in Figure 1. The equipment continuously measures the canister conditions, as recording the adsorption mass in the canister, the breakthrough concentration from the canister, fuel temperature, and fuel tank pressure, and then logs these data onto a computer file.

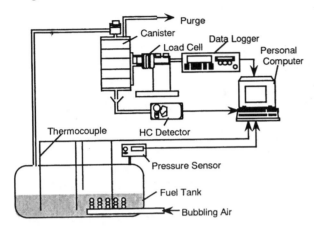

Figure 1.   Test Equipment

DETERIORATION ANALYSIS – The adsorption/ desorption cycles test procedure is shown in Figure 2. By using a 0.12 liter, small-sized canister enclosed with activated carbon, gasoline temperature was adjusted to 35°C and gasoline fuel vapor was generated by bubbling dry air through gasoline in the fuel tank. 1.0-1.2g/min HC were loaded by breakthrough at a room temperature of 25°C. Then the canister was purged with dry air at 10 liter per minute for 3.6 minutes(300 bed volume). This adsorption/desorption was carried out for 240 cycles. Reid Vapor Pressure_(RVP) of the gasoline used for this test was 62KPa, and it was exchanged with new gasoline every thirty cycles. The gasoline working capacity of the canister was prescribed by the amount of adsorption of each cycle. Residual HC components adsorbed in the activated carbon were extracted by using dichloromethane (CH_2Cl_2) as a solvent, and were then analyzed by gas chromatography.

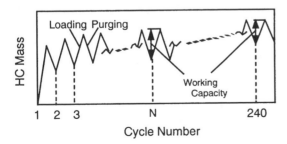

Figure 2.   Test Pattern (Adsorption/Desorption Cycles)

DBL ANALYSIS – The diurnal test procedure is shown in Figure 3, and the conditions of each process are as follows.

1. Canister stabilization: The gasoline fuel vapor generated by bubbling dry air through gasoline was loaded by breakthrough to the 2.0 liter canister which enclosed activated carbon. Then the canister was purged with dry air at 20 liters per minute for 30 minutes(300 bed volume). This adsorption/ desorption_was carried out for 6 cycles at 25°C.

2. Canister loading and purging: Loading was conducted by 2 grams breakthrough using a 50/50 percent by volume mixture of n-butane and nitrogen at a flow rate equal to 40g butane/hour. Then the purging was conducted with dry air at 20 liter per minute for 15 minutes(150 bed volume).

3. Environmental conditions at parking: After soaking for 12 hours at 25°C, the canister went through two 24-hour EPA diurnal temperature cycles.

The canister was divided into 5 layers, as shown in Figure 4, and the activated carbon in each layer was removed.   After measuring the adsorption mass of the activated carbon removed from each layer, the components adsorbed in the activated carbon were analyzed by gas chromatography.

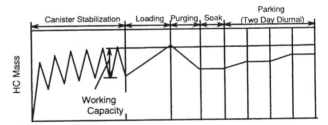

Figure 3.   Diurnal Test Procedure

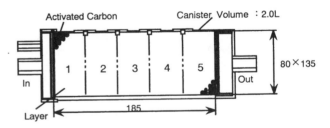

Figure 4.   Canister Configuration Used in Diurnal Test

CANISTER PERFORMANCE DETERIORATION MECHANISM

After long usage, canister performance may occasionally show reduced working capacity which could sometimes result in aggravated DBL as well. This is called "deterioration". In this experiment, the canister was repeatedly loaded and purged. The results of compositional analysis of the fuel and its vapor applied in this adsorption test are shown in Figure 5.

Figure 6 shows the transitions of the adsorption mass. In the first cycle of adsorption/desorption, approximately 1/3 of the adsorption mass that was unable to desorb remained after purging. The residual HC continued to climb after every adsorption/desorption cycle, but after the 10th cycle, this trend leveled out considerably. The adsorption mass at such time was also found to be stable. After such stable conditions continued for a while, the adsorption mass began to reduce. This is called a "deterioration phase". This entire transition span of working capacity from the beginning of a stable state to a definite deteriorating state is counted as one span of the "Useful Life".

Figure 7 shows the transitions of mass and components of the residual HC after purging in the canister. The residual HC in the initial aging phase was mostly dominated by C4,C5,C6. Likewise, in the stabilized phase, all such components were replaced with high boiling hydrocarbons of C7 and over for a long time. In the deterioration phase, C4, C5 were not apparent in the residuals, but hydrocarbons of C7 and over continued to increase.

Figure 8 shows the transitions of pore conditions of the activated carbon. In the initial aging phase, micropores over 13 angstroms decreased whereas those of approximately 9 angstroms to 11 angstroms increased. In the stabilized phase, the pores of 9 angstroms to 11 angstroms continued to decrease. The decrease in micropores of larger size was small. In the deterioration phase, micropores of the size equivalent to hydrocarbon molecules in the gasoline fuel vapor began to decrease.

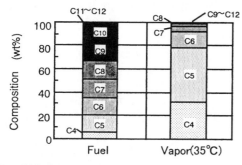

Figure 5. HC Components of Fuel and Vapor Applied in Adsorption Cycle

Thus, based on the above-mentioned results, it can be concluded that deterioration is caused when high boiling hydrocarbons of C7 and over begin to accumulate in, and fill up those pores effective for adsorption, subsequently causing the working capacity to drop. (See Fig.9)

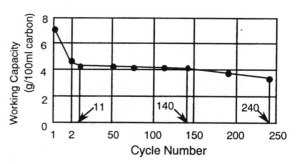

Figure 6. Transition of Working Capacity during Adsorption/Desorption Cycles

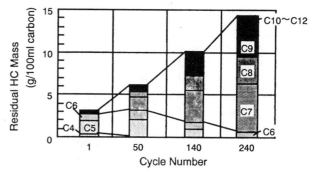

Figure 7. Transition of Mass and Components of Residual HC in Canister

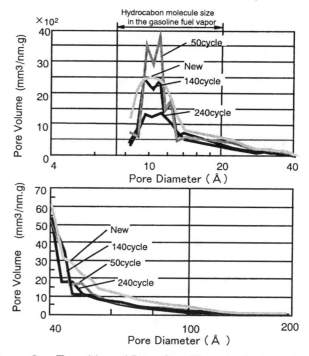

Figure 8. Transition of Pore Conditions in Activated Carbon

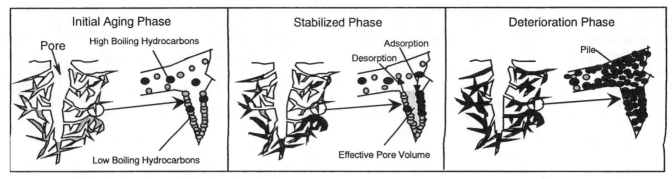

Figure 9. Activated Carbon Deteriorating Mechanism

USEFUL LIFE ESTIMATION – Based on the above-mentioned results, the canister adsorption performance is assumed to deteriorate when the residual mass of the_high_boiling_hydrocarbons reaches a specific amount. Such mass is proportional to the saturated adsorption mass_which is the adsorption mass in the case that all pores are filled by gasoline. According to Figure 7, on completion of Useful Life (140 cycles), the residual mass of high boiling_hydrocarbons was found to be 8.2g/100ml carbon. Depending on the pore volume per 100ml and the test gasoline density, the saturated adsorption mass was calculated to be 26.5g/100ml carbon. Therefore, the residual rate of 0.30(8.2/26.5) was assigned as the standard value of Useful Life. The rate of accumulated pore volume of micropores under 20 angstroms in activated carbons used for automotive canisters is almost 30%.(Fig.10) Therefore, it is assumed that the deterioration occurs when the high boiling hydrocarbons fill up the micropores equal to those molecule size.

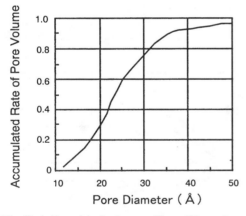

Figure 10. Relationship between Pore Diameter and Accumulated rate of Pore Volume

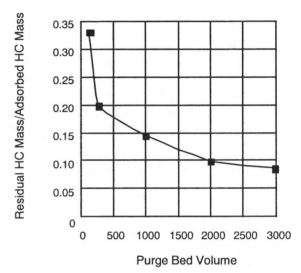

Figure 11. Purge Influence on the Residual HC

Saturated adsorption mass varied according to types of activated carbon, and purge amount influenced the residual mass of high boiling hydrocarbons. The temperature change of the canister installed in close proximity to the fuel tank was not large enough to influence the desorption ability of high boiling hydrocarbons of C7 and over, but the purge amount was noteworthy. In Figure 11, the purge amount influence_to_the_residual_high boiling_hydrocarbons in the stabilized_phase is shown. It became clear_that 90% of the high boiling hydrocarbons of C7 and over could be desorbed by increasing number of purge bed volumes.

A_formula for estimating the canister Useful Life was devised. The flow mass of C7 and over for each refueling is Vr(g), and the flow mass of C7 and over flowing into the canister in the state of driving and parking, until the next refueling is Vd(g). R(P) is the rate of residual mass of C7 and over according to the purge amount P as indicated in Figure 11, and is set at 0.1. If Q(L) represents carbon volume, then the residual mass Rm (g/100ml carbon) of C7 and over upon one refueling can be expressed as the following:

$$Rm = \frac{(Vr + Vd) \times 0.1}{Q \times 10} \quad (1)$$

Provided that a given vehicle can travel a distance of 300 miles after refueling, by the point the vehicle accomplishes the designated DD (driving distance) miles, it will have been refueled DD/300 times. Even if the residual mass of components of C7 and over at such time-point has not reached the standard value of Useful Life, the canister Useful Life is guaranteed all the way up to such driving distance.

$$\frac{DD}{300} \times Rm \leqq Mmax \times 0.30 \qquad (2)$$

Mmax :Saturated Adsorption Mass

Following this estimation formula, an adsorption mass of the canister in every refueling cycle that would guarantee a Useful Life after 150,000 miles of driving was deduced. The factors for this calculation are presented in Table 1. Compared with the previously mentioned activated carbons, the pore volume on these activated carbons was increased by 30%.

Table 1. Parameters and Conditions

Gasoline RVP	: 62KPa
Fuel Temperature	: 35℃
Saturated Adsorption Mass	: 34. 2g/100ml carbon
Carbon Volume	: 2. 5 Liter

Calculations show that (Vr+Vd) should be under 5.1 grams. As shown in Figure 5, since the proportion of components of C7 and over in the gasoline fuel vapor is 6.6%, the adsorption mass of the canister at the refueling interval should be under 78(5.1/0.066) grams. The calculated value(5.1g) requires a purge amount in excess of 2000 bed volume. This is equivalent to a purge amount of 6.7 bed volume per 1 mile. Based on such results, it is clear that by guaranteeing a reasonable purge amount, canister life can be extended, or in other words, deterioration can be more restrained.

Next, in order to verify the DBL reduction effects of the anti-deterioration efforts, DBL of a canister in the Useful Life phase and another canister in the deteriorating phase were compared. The amount of DBL from the canisters are shown in Figure 12. Compared with the canister in the Useful Life phase, the canister in the deterioration phase proves to have two days worth of DBL. Furthermore, DBL of the second day exceeded the amount of the first day. As a result, it was confirmed that restrained deterioration leads to greater assurance of DBL during Useful Life.

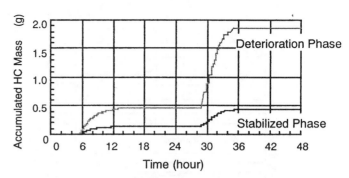

Figure 12. Comparison od DBL Performance of Canister in Stabilized Phase and in Deterioration Phase

CANISTER DBL MECHANISM

Figure 13 shows an example of DBL measurement and results that are apparent under a condition lower than the working capacity of the canister. The weight of the canister increased when the fuel temperature rose during the diurnal cycle. On the contrary, when the fuel temperature dropped, the weight of the canister reduced. This is because the fuel tank breathes according to the change in the fuel temperature. The total vapor flow mass upon two day diurnal was 52g. In spite of the fact that it was lower in capacity than the 80g working capacity of the canister, DBL already occurred in the first cycle. An interesting point is that evaporative emissions occurred only upon rising of the fuel temperature. Figure 14 shows the composition of the evaporative emissions. They were of low boiling points under gasoline fuel vapor such as C4and C5.

In order to clarify the causes of such phenomenon, the changing trends of HC mass and composites inside the canister were examined. The results are shown in Figures 15 and 16. Low boiling hydrocarbons of C4 and C5 have been found to diffuse slightly on the outlet end of the canister under the state of soak and in the diurnal cycle performed inside the canister.

Thus, based on the above results, the DBL mechanism can be explained as in Figure 17. Low boiling hydrocarbons, which adsorbed on the canister inlet side, diffuse to the outlet side of the canister during soak and during the diurnal cycle. The hydrocarbons in the gasoline fuel vapor, which flow into the canister upon rising of the fuel temperature in the diurnal cycle, adsorb on the canister inlet side. Therefore only the air flows into the outlet side of the canister. By this air, low boiling hydrocarbons are desorbed, and thus leakage occurs.

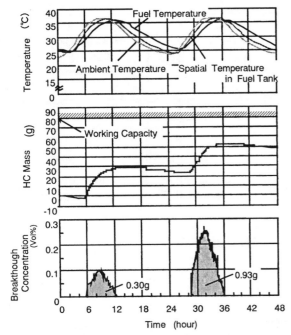

Figure 13. Temperature, HC Mass, and Breakthrough HC Concentration Profiles during Two Day Diurnal

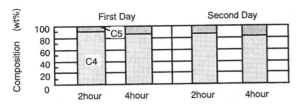

Figure 14. Breakthrough HC Components

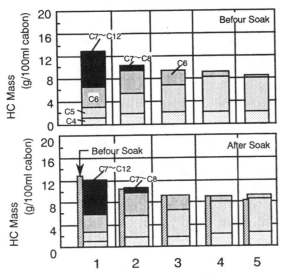

Figure 15. Transition of HC Mass and Components of Each Layer during Soak

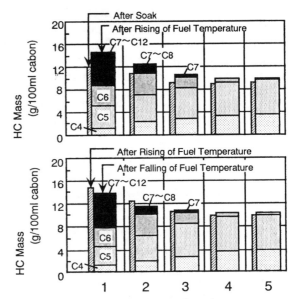

Figure 16. Transition of HC Mass and Components of Each Layer during First Diurnal

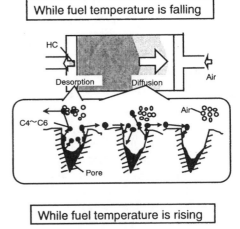

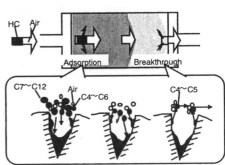

Figure 17. Mechanism of Diurnal Emissions from Canister

CANISTER DBL REDUCTION TECHNIQUES

REDUCTION OF DBL BY IMPROVING CANISTER CONFIGURATION – It has been verified that the main factor causing DBL in the canister is caused by the in-canister diffusion of low boiling hydrocarbons such as C4 and C5 that are present slightly in the residual HC after purge and in the HC adsorbed during the DBL test. For this reason, reducing the residual HC after purge and HC diffusion inside the canister have been found to be effective techniques for the DBL reduction. Because the amount of DBL increased along with the number of days, restraining HC diffusion inside the canister was particularly important. An equation explaining this diffusion is as follows:

$$\text{Amount of Diffusion} = -D \times A \times \partial C / \partial X \quad (3)$$

$$D = Ds \times (T/273)^n \times 760/P \quad (4)$$

D: Coefficient of Diffusion
A: Cross-Sectional Area
$\partial C / \partial X$: Concentration Gradient
Ds: Coefficient of Diffusion at Standard Conditions
  〈0℃, 1atm〉
T: Temperature
P: Pressure

According to EQ 3, for preventing diffusion it is effective to minimize D, A, and $\partial C / \partial x$. However, A should not be minimized because the pressure loss of the canister could be increased. Therefore, a low pressure loss and an anti-diffusion technique was developed. Figure 18 shows the pattern of HC diffusion between activated carbons.

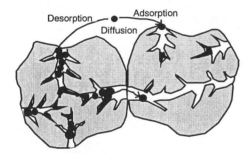

Figure 18. HC Diffusion Pattern between Activated Carbons

Therefore, even if a massive amount of HC is adsorbed in the micropores of activated carbon, provided that the HC concentration in the ambience of such activated carbon is low, it then becomes easy for the HC to desorb from the micropores. Furthermore, if the adsorption HC mass in other activated carbon micropores is small, then adsorption of HC in the ambient space by such pores becomes possible. In this way, the HC seemingly diffuses as if it migrates among the activated carbon micropores.

In order to restrict such diffusion, activated carbon with large adsorption and that with small adsorption should be separated as much as possible. For confirmation, a canister with two activated carbon beds were separated by an air bed along the vapor flow direction was prepared. The length of the air bed was made to be variable. Only one carbon bed was adsorbed with HC, then aged afterwards inside a thermostatic chamber for a specific amount of time. Finally, the mass of the two activated carbon beds were measured. The results are shown in Figure 19. The HC migration to another carbon beds was found to be smaller when the air bed was longer and when the temperature was lower. At the same temperature, diffusion mass between carbons separated by an air bed of 20 cm was found to be only 1/4 the diffusion mass of carbons without an air bed.

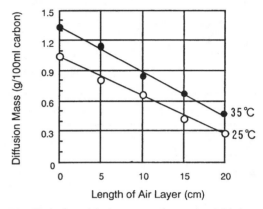

Figure 19. Relationship between Length of Air Layer and Diffusion Mass

Based on these results, the improved canister which has a labyrinth air bed was designed as taking into account the adsorption mass upon actual DBL test, and its DBL became 1/2 of a conventional canister. Such results are shown in Figure 20.

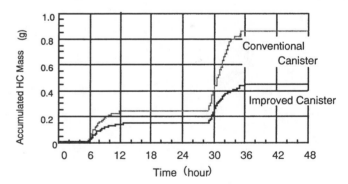

Figure 20. Comparison of DBL between Improved Canister and Conventional Canister

REVISION OF DBL TESTING CONDITIONS – As seen in the diurnal test procedure (Fig. 3), there are several cycles of adsorption test prior to the DBL test. More officially speaking, this is done to adsorb the gasoline fuel vapor, but the butane loading is widely conducted as an optional method.

It has been shown by this investigation that DBL occurs only because of the diffusion of low boiling hydrocarbons such as C4 and C5. For reference, in Table 2 the diffusion coefficients indicate how easily the diffusion of saturated hydrocarbons can take place. Based on the fact that the smaller the amount of carbon in components, the faster the speed of their diffusion will be, it can be understood that the difference of HC adsorbed before the diurnal cycle can no longer be ignored. Only in the canister loading stage, the gasoline fuel vapor loading and the butane loading could be studied comparatively. All other adsorption in the canister stabilization stage had to be tried with the gasoline fuel vapor. As seen in Figure 21, the gasoline fuel vapor loaded canister showed 1/3 the DBL amount for 2 cycles compared to that of butane loaded canister. Also, with the gasoline fuel vapor loaded canister, the DBL fluctuation appeared to be more reduced in comparison with the butane loaded one.

Table 2. Coefficient of Diffusion at Standard conditions (0°C, 1atm)

Saturated Hydrocarbon	Coefficient of Diffusion
n-Butane	0.0806
n-Hexane	0.0677
n-heptane	0.0600
n-Octane	0.0589
n-Decane	0.0474

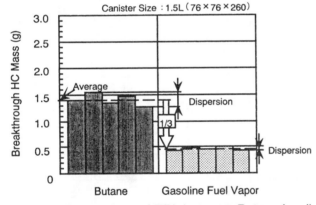

Figure 21. Comparison of DBL between Butane Lording and Gasoline Fuel Vapor Lording

Outside of the laboratory, DBL actually occurs after the gasoline fuel vapor loading, and such results are usually more emphasized. If a canister must be designed according to that result gained by employing a butane loading method, an extra performance margin ends up being created in order to compensate for that aggravated DBL and its fluctuations. To resolve this situation, one idea is to clarify the amount of such margin caused by the butane loading method, which consequently leads to the belief that gasoline fuel vapor loading will definitely guarantee higher precision than the butane method.

CONCLUSION

By a detailed investigation and analysis of carbon adsorption/desorption states, the deterioration mechanism and DBL mechanism were clarified.

- Deterioration of adsorption abilities is caused when the micropores effective for adsorption are closed due to accumulation of high boiling hydrocarbons of C7 and over which are present in low concentrations in the gasoline vapor.

- The main cause of evaporative emissions from the canister is butane-diffusion occurring inside the canister while the car is parked.

Based on this mechanism a technology responding to extended Useful Life and enhanced evaporative emissions was clarified and the following points were acknowledged

- Deterioration of a current carbon canister can be restrained by increasing the purge amount and thus Useful Life can be extended

- To design a canister reasonably , canister loading in real world conditions must also be taken into account for evaluation.

REFERENCES

1. Harold M. Haskew and William R. Cadman, "Evaporative Emissions Under Real Time Conditions", SAE Paper No. 891121,1989.

2. Harold M. Haskew , William R. Cadman, and Thomas F. Liberty, "The Development of a Real-Time Evaporative Emission Test", SAE Paper No.901110,1990.

3. Harold M. Haskew , William R. Cadman, and Thomas F. Liberty, "Real-Time Non-Fuel Background Emissions", SAE Paper No.912373,1991.

4. J. E. Urbanic, E. S. Oswald, N. J. Wagner, and H. E. Moore, "Factors Affecting the Design and Breakthrough Performance of Evaporative Loss Control Systems for Current and Future Emissions Standards", SAE Paper No.890621,1989.

5. H. R. Johnson and R. S. Williams, "Performance of Activated Carbon in Evaporative Loss Control Systems", SAE Paper No.902119,1990.

6. Ames A. Grisanti, Ted R. Aulich, and Curtis L. Knudson, "Gasoline Evaporative Emissions – Ethanol Effects on Vapor Control Canister Solvent Performance", SAE Paper No.952748,1995.

7. Michael E. Payne, Jack S. Segal, Matthew S. Newkirk, and Lawrence R. Smith, "Use of Butane as an Alternative Fuel Emissions from a Conversion Vehicle Using Various Blends", SAE Paper No.952496,1995.

8. Philip J. Johnson, James R. Jamrog, and George A. Lavoie, "Activated Carbon Canister Performance During Diurnal Cycles: An Experimental and Modeling Evaluation", SAE Paper No.971651,1997.

9. Philip J. Johnson, Roger J. Khami, Jeffrey E. Bauman, Thomas D. Goebel, Vernon L. Clark, David L. Hirt, and Paul J. Luft, "Carbon Canister Development for Enhanced Evaporative Emissions and On-Board Refueling", SAE Paper No.970312,1997.

10. Marek C. Lockhart, "Predicting Tank Vapor Mass for On-Board Refueling Vapor Recovery", SAE Paper No.970308,1997.

11. George A. Lavoie, Philip J. Johnson, and Jeffrey F. Hood, "Carbon Canister Modeling for Evaporative Emissions: Adsorption and Thermal Effects", SAE Paper No.961210,1996.

Development of Vapor Reducing Fuel Tank System

**Tomoyasu Arase, Takashi Ishikawa, Masahide Kobayashi,
Yoshihiko Hyodo, Masahiro Kasai and Nobuhiro Nakano**
Toyota Motor Corp.

Copyright © 2001 Society of Automotive Engineers, Inc.

ABSTRACT

In succession to the world-first introduction of a mass production gasoline hybrid passenger car into the Japanese market in 1997, Toyota also has introduced an enhanced version of the above to the US and European markets in 2000.

Upon introduction of Toyota Hybrid System (THS) into the US market, a drastic reduction of gasoline vapor evaporation from the fuel tank was necessary, in order to meet the most stringent exhaust emission (SULEV) and evaporative emission standards in the world.

In order to meet this requirement, a fuel tank system named "Vapor Reducing Fuel Tank System" was developed. This is the first commercial application in the world to use a variable tank volume to drastically reduce gasoline vapor generation.

INTRODUCTION

Proposals have been made in recent years to intensify automobile exhaust/evaporative emission standards in the US, Japan and Europe in relation to global environmental issues.

Taking the opportunity of introduction into the US market, the mass production of gasoline hybrid passenger car was materialized, with advanced features to meet the world most stringent exhaust emission standard and the gasoline evaporative emission standard of California in the US ahead of their enforcement. In line with this mass production of vehicle, a fuel tank system with a new structure has been also developed.

Table 1 shows the current standard values and the target values of the most stringent emission standard proposed for development by California in the US.

FEATURES OF TOYOTA HYBRID SYSTEM

Various measures are implemented in THS for the reduction of Carbon dioxide (CO2) by reducing the emission levels and increasing the fuel economy. A particularly significant feature utilizing the hybrid characteristics is the capability to select a proper pattern to stop the engine and to run the vehicle by means of electric motor in the low power range, aiming at the enhancement of overall efficiency.

REQUIREMENTS FOR FUEL TANK

The following problems occurred in an attempt to ensure compatibility between the requirements of exhaust emission (SULEV) and evaporative emission standards at a high level for THS having the features stated above. Namely, precise control of air/fuel ratio (hereafter called A/F) is necessary to reduce exhaust emissions, whereas the gasoline vapor evaporated from the fuel tank through the canister is purged and burned in the engine as the evaporated gas. Hence the precise control of A/F is impeded by the influence of evaporated gas that cannot be measured at that time, which makes it difficult to ensure conformity with the SULEV standard.

For the reduction of evaporative emission, on the other hand, it is necessary to purge the gasoline vapor adsorbed in the canister. However, the amount of purging air becomes much smaller than the conventional system due to the necessity to reduce the intake manifold vacuum pressure for the enhancement of engine efficiency, as well as the frequent on-off engine operations stated earlier. In order to meet such conflicting requirements at the same time, it became necessary to develop a fuel tank capable of drastic reduction of gasoline vapor generation.

Table.1 Present standard & target values new

US California		NMOG g/mile	CO g/mile	NOx g/mile	HCHO g/mile
Present Standards	ULEV/50kmiles	0.040	1.7	0.2	0.008
	/100kmiles	0.055	2.1	0.3	0.011
Target	SULEV/120kmiles (2004MY~	0.010	1.0	0.02	0.004

CONCEPT OF DEVELOPMENT

It is difficult to reduce the amount of gasoline vapor generation markedly for a conventional fuel tank having a fixed capacity, as the amount of gasoline vapor a function of the total vapor volume of the tank. The vapor volume increases as gasoline is consumed. It was, therefore, decided to develop a fuel tank (Vapor Reducing Fuel Tank System) with a structure entirely different from those of conventional systems, with a feature that varies its capacity in line with the amount of the fuel remaining in the tank (Figure 1).

Toyota has succeeded in its mass production of the fuel tank system for THS, and the details of structure will be described in the following

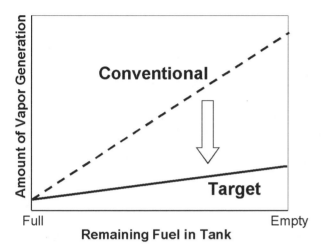

Fig.1 Relationship between Amount of Vapor Generation & Remaining Fuel

VAPOR REDUCING FUEL TANK

The main components of the newly developed fuel tank an be roughly divided into the following;
(1) Fuel storage unit; (2) fuel sending and level gauge unit; and (3) outer tank to support and contain those units. The outline of each component is described below.

FUEL STORAGE (BLADDER TANK)

A structure that can expand or shrink in proportion to the amount of remaining fuel is necessary to reduce the vapor volume in the fuel tank. Rubber which is generally elastic may be considered as the material to allow the expansion and shrinkage, but it is extremely difficult to use it in practice, considering the long period of service in automobile under severe operating conditions, and the permeation of fuel through the rubber material itself.

Therefore, the multi-layer resin membrane shown in Figure 2 (hereafter called "bladder membrane") is employed, because of high durability and excellent performance against fuel permeation as the membrane of fuel storage unit. The thickness and the shape are optimized as shown in Figure 3, with which it has become possible to reduce the vapor volume and to ensure the necessary reliability. Reliability is confirmed by various durability tests including vibration, heat and c operation.

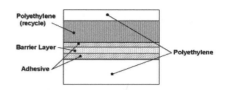

Fig.2 Structure of Multi-layer Membrane
(With Six Layers and Four Kinds of Materials)

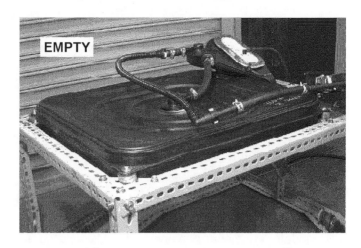

Fig.3 Shape of Bladder Membrane

FUEL SENDING AND LEVEL GAUGE UNIT

A conventional fuel tank consists of a fuel pump to feed the fuel into the engine and a fuel gauge to detect the amount of fuel remaining in the tank. If such a fuel pump and a fuel gauge are installed in the capacity variable bladder membrane described in the foregoing, the expansion and shrinkage of the bladder membrane would be impeded and the vapor volume would be increased. Hence, a structure to accommodate such parts in a container (sub-tank) separate from the fuel tank is employed in the new system. Moreover, a cylinder type fuel gauge is also employed for the downsizing, and all components that used to be installed in a conventional fuel tank, such as the fuel filter, control valve, temperature sensor, etc. are installed in the sub-tank.

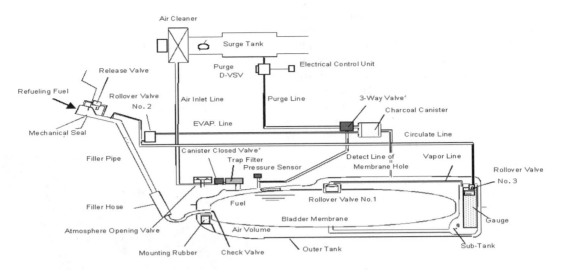

Fig.4 Show the Outline of the System

OUTER TANK

A double enclosed structure is employed to accommodate the capacity variable bladder membrane and the sub-tank in the outer tank to ensure reliability and safety. The outer tank plays not only the role of supporting the components described in the foregoing, but also the role of providing a space layer to contain fuel permeation from the membrane and the discharge of break through vapor from the canister so as to prevent their escape outside of vehicle, in order to an increase in evaporative emission even in extremely severe operating environments. Owing to this space layer, it is possible to maintain clean conditions of the system by causing the flow of air purge through the space layer while the vehicle is running.

OUTLINE OF VAPOR REDUCING FUEL TANK SYSTEM

Figure 4 shows the outline of the system.

PURGE CONTROL

A purge control system similar to that of a conventional fuel tank is employed, but more precise A/F control has become possible as the amount of gasoline vapor is extremely small.

OBD FOR EVAPORATIVE EMISSION CONTROL SYSTEM

On-Board Diagnosis (OBD) for the system is performed to detect two kinds of failures. One is the evaporated fuel treatment system failure detected by negative pressurizing, and the other is the bladder membrane failure detected by the hydrocarbon (HC) concentration measuring. The HC concentration method is developed for the bladder membrane leak check, that is, when HC concentration in the space layer increases due to

some hole(s) in the bladder membrane, the presence of such hole is detected by measuring higher HC concentration.

The HC concentration is measured by using the exhaust oxygen (O2) sensor. The sampling is through a tube connecting the air volume to the engine.

MECHANISM TO DETECT REMAINING FUEL

The remaining fuel is measured by a fuel level gauge in the sub-tank. The newly developed remaining fuel detection system is shown in Figure 5.

The remaining amount of fuel is measured with the gauge signal, but the temperature in the outer tank, slant of vehicle, and the fuel injection rate are also used for correction.

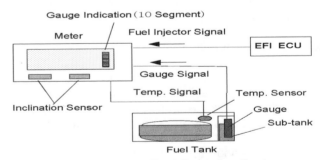

Fig.5 Remaining Fuel Detection System

RESULTS

EFFECT OF VAPOR REDUCTION

The amount of gasoline vapor with the conventional system and that with the new system are shown respectively in Figures 6, 7 and 8. Those data are measured on the certification test methods in the US.

Drastic reductions in amount of vapor during refueling (Onboard Refueling Vapor Recovery Test), parking(Diurnal Breathing Loss Test and Hot Soak Loss Test) and vehicle running (Running Loss Test) are achieved in comparison to those with equivalent conventional fuel tank systems.

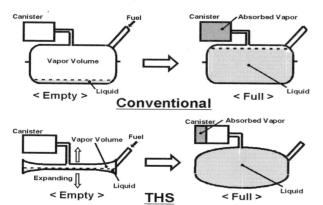

Conventional

THS

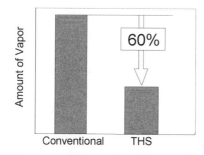

Fig.6 Amount of Gasoline Vapor Generated (Refueling)

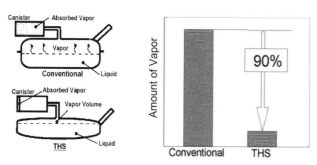

Fig.7 Amount of Gasoline Vapor Generated (Parking)

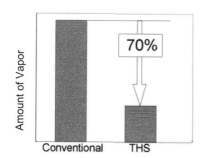

Fig.8 Amount of Gasoline Vapor Generated (Running)

The necessary capacity of canister is also reduced to one-third of the conventional one, and a major reduction of the amount of purge air is also achieved (Figure 9).

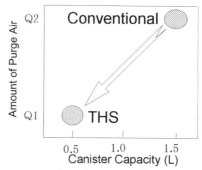

Fig.9 Relationship between Canister Capacity & Amount of Purge Air

EFFECT ON EXHAUST EMISSION REDUCTION

Precise control of A/F is required in order for THS to conform to the exhaust SULEV standard, but the A/F ratio tends to fluctuate by purging the canister because an unmeasurable amount of fuel goes into the engine, which gives an adverse effect on the exhaust emissions.

With this new system, however, the emission levels do not increase owing to the small amount of purging air Q1.

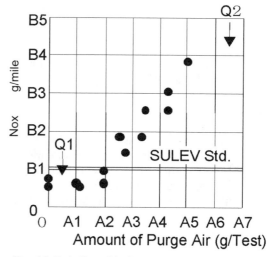

Fig.10 Relationship between Amount of Purge Air & Exhaust Emission

Figure 11 shows the results of exhaust emissions. Compatibility between the purging performance and the conformity to the exhaust SULEV emission standard is thus established

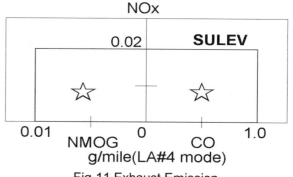

Fig.11 Exhaust Emission

547

SUMMARY

The cleanest mass production vehicle in the world, meeting the most stringent California emission standard in the US, has been materialized by using this new system as described below.

1. Even with the small amount of purging air, the vehicle has cleared the hurdles of the exhaust SULEV emission standard and the LEVII evaporative emission standard in California, USA.

2. For the first time in the world, the fuel tank system, that expands and shrinks according to the amount of fuel remaining in the fuel tank, has been developed.

3. By using the multi-layer resin membrane, the bladder membrane has been developed and the capability to expand or shrink according to the amount of fuel remaining in the tank.

4. A reduction in the amount of gasoline vapor from the fuel tank has been achieved and required canister capacity is 1/3 what is required in conventional system.

5. The failure diagnosis system has been developed, with the capability to check on the fuel leak from the bladder membrane by measuring the HC concentration in the space layer.

6. The system to detect the amount of remaining fuel in the capacity variable fuel tank has been developed

ACKNOWLEDGMENTS

Sincere appreciation is expressed to people from Aisan Industry Co., Tokai Rubber Industries, Horie Metal Co., Yazaki Corporation and the divisions/departments concerned at Toyota, who have extended cooperation to the development of the fuel tank system with a new structure

REFERENCES

1. Toshio Inoue, Hirose, Sakurai, Ishikawa: "Challenge for the Cleanest Car Part 2," JSAE Spring Convention Proceedings 2000-5 20005188 (in Japanese)

2. Toshio Inoue, Kusada, Kanai, Hino, Hyodo: "Improvement of a Highly Efficient Hybrid Vehicle and Integrating Super Low Emission," SAE Paper 2000 Fall Fuels & Lubricants Meeting & Exposition 2000-01-2930

2000-01-2926

Development of a Gasoline-Fueled Vehicle with Zero Evaporative Emissions

Hideyuki Matsushima, Akio Iwamoto, Masahiro Ogawa, Tomoyuki Satoh and Katsunori Ozaki
Nissan Motor Co., Ltd.

Copyright © 2000 Society of Automotive Engineers, Inc.

ABSTRACT

...Technologies for reducing evaporative emissions generated from gasoline vapors have been developed. To reduce evaporative emissions, both permeation from fuel and vapor lines and breakthrough from the evaporative canister need to be diminished. Fewer fuel line connections are used and hose and valve materials have been modified to reduce permeation. Component test results confirm that permeation is substantially reduced from the level of previous parts. A new type of activated charcoal, which has a high specific heat characteristic and improves adsorption and desorption performance, has been applied to reduce canister breakthrough. Additionally, the amount of purge air has been increased by applying purge control using an air-fuel ratio sensor. The problem of canister breakthrough has thus been resolved by the new evaporative canister combined with increased purge flow to the engine.

Endurance mode tests equivalent to 15 years/150,000 miles of driving were conducted on the fuel evaporative system parts and the results confirmed that this level of emission durability could be assured. The new system has been developed for the Sentra CA sold in the U.S. market. This vehicle satisfies the zero evaporative emission regulations as a result of adopting this combination of new technologies.

INTRODUCTION

An evaporative emission regulation was established for the first time in California in 1970. Evaporative emission standards have since been tightened and compliance with an evaporative emission regulation that simulates the states when a vehicle is being driven, is stationary and is parked has been mandatory in the United States since 1995.

With the aim of reducing evaporative emissions even further, the California Air Resources Board (CARB) proposed the Low Emission Vehicle (LEV)-II zero and near zero evaporative emission standards, which would take effect in 2004. These proposed standards were adopted by the state of California as official regulations in autumn of 1999.

This study of technologies for reducing evaporative emissions was carried out against such a background. This paper describes the technologies incorporated in the zero evaporative emission system that has been developed as a result of this study.

DEVELOPMENT AIM

OVERVIEW OF EVAPORATIVE EMISSION REGULATIONS –

The zero and near zero evaporative emission regulations proposed by CARB for enforcement in 2004 are summarized in Table 1. Running loss (RL) test is aimed to simulate evaporative emission generated during driving. Hot soak loss (HSL) test and Diurnal breathing loss (DBL) test is for vehicle being stationary. The new regulations mainly concern evaporative emissions that occur while a vehicle is stationary or is parked.

Evaporative emission of HSL test is measured for one hour just after RL test is finished, and 72 hours of DBL test is conducted after the HSL test. Ambient temperature is varied from 18°C to 41°C during DBL test, which simulates vehicle being parked on hot summer days.

The regulation for the useful life of a vehicle is 15 years or 150,000 miles, which has been increased from the current requirement of 10 years or 100,000 miles. There are two different hydrocarbon (HC) standards for zero evaporative emission regulation of HSL+DBL test. One is a requirement for vehicle and the other is for fuel and evaporative systems, and 0.0 gram is specified for the latter as it is called.

California Evaporative Emission Standard		Hydrocarbon Standards		Useful Life
		Running Loss Test (g/mile)	Hot Soak Loss + Diurnal Breathing Loss (g/test)	
LEV-I	Current	Vehicle: RL=0.05	Vehicle: HSL+DBL=2.0	10 years / 100,000 miles
LEV-II	Near Zero	Vehicle: RL=0.05	Vehicle: HSL+DBL=0.50	15 years / 150,000 miles
	Zero	Vehicle: RL=0.05	Vehicle: HSL+DBL=0.35 / Fuel+Evap. Systems = 0.0	

Table 1 : Evaporative Emission Standards

ZERO EVAPORATIVE EMISSION TARGETS –

Using a U.S.-specification Sentra fitted with a four-cylinder inline engine as a base vehicle, a study was made of a system for reducing evaporative emissions. First, the respective contribution of the fuel system and fuel evaporative system components to evaporative emissions was investigated. Technologies for reducing HC emissions were examined in order from the places where large HC releases occurred.

To verify the results of that investigation, evaporative emission tests were conducted using a Variable Temperature-Sealed Housing for Evaporative Determination (VT-SHED) to identify the locations and quantities of HC emissions. An outline of the VT-SHED equipment is shown in Figure 1. This VT-SHED equipment is of the closed type so that daytime air temperature conditions can be simulated. Evaporative emissions from the test vehicle were measured with an HC analyzer and the quantity of HCs emitted was calculated.

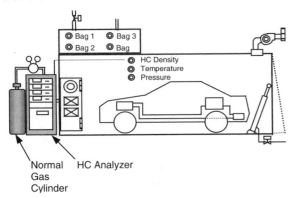

Figure 1 : Outline of VT-SHED Equipment

The target set for the development of this evaporative emission reduction system was to reduce the emission level to 1/10 of the conventional evaporative emission test standard required for vehicle, i.e., to a value of 0.20 g/test for the HSL + DBL test, compared with a value of 2.0 g/test at present. The aim was to reduce the level to less than the California zero evaporative emission regulation value of 0.35 g/test for the HSL + DBL test. A further goal was to comply with the more stringent emission durability requirement resulting from an extension of the useful life to 15 years/150,000 miles from 10 years/100,000 miles at present.

TECHNOLOGIES FOR ACHIEVING ZERO EVAPORATIVE EMISSIONS

The technologies for reducing evaporative emissions from vehicles can be classified under three categories.

1. Technologies for reducing the quantity of HC permeation from the fuel system and fuel evaporative system

2. Improvement of evaporative canister for reducing HC breakthrough

3. Purge control technology for increasing purge air volume

It had been thought that it would be difficult to develop such evaporative emission reduction technologies for gasoline-fueled vehicles. However, improvements to the system components and advances in control technology have made such technologies feasible. HC permeation from the fuel and fuel evaporative systems was dramatically reduced, and improved control technology for increasing the amount of engine purge air was added to the new system so as to make full use of the improvement made to the internal structure of the evaporative canister and the enhanced adsorption capacity of the activated charcoal.

TECHNOLOGIES FOR REDUCING HC PERMEATION –

HC permeation from the fuel system mainly occurs from parts made of plastic and rubber. The first step taken to reduce such HC permeation was to identify the components that are large contributors to evaporative emissions. The places where HC permeation occurs in the fuel system and the fuel evaporative system are shown in Figure 2. Large permeation occurs from rubber parts such as fuel hoses, filler hose, and diaphragm of pressure control valve.

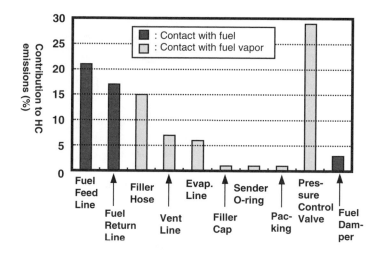

Figure 2 : Contribution to HC emissions by parts

The factors influencing permeation of fuel from plastic and rubber parts include a difference in pressure before and after the permeation film, a difference in concentration, solvent (gasoline in this case) and a difference in the solubility coefficient of the permeation film. Among these factors, concentration differences cannot be controlled, but differences in pressure and in the solubility coefficient are controllable. Measures for reducing the quantity of HC permeation were taken from the three perspectives described below. The configuration of the fuel system and the fuel evaporative system is shown in Figure 3. The measures adopted were aimed at reducing HC permeation from parts coming in contact with the fuel and fuel vapors.

<u>Reduction of the Number of Parts and Connections –</u>

A return-less fuel delivery system is adopted and the location of fuel filter is moved inside of fuel tank to eliminate the chance of HC permeation. Also the number of connections was reduced by re-examining the system and simplifying the parts so as to reduce the quantity of permeation from the O-rings used for sealing pipe connections and from hose terminals. The number of pipe connections in the fuel line was reduced from ten places in the previous system to three places in the new system.

<u>Change to Low Permeation Plastic and Rubber –</u>

The plastic and rubber materials used to make various components of the fuel system and the fuel evaporative system were changed to ones displaying a low level of fuel permeation. The amount of permeation can be reduced by using materials having a solubility coefficient that is greatly different from that of gasoline. Specifically, fluorine rubber (FKM) was adopted for O-rings, packing and valve diaphragms, and a plastic rubber compound was adopted for the filler hoses along with using three-layer plastic for fuel hoses. Cross-sectional views of the previous rubber fuel hose and the improved multi-layer plastic fuel hose are shown as examples in Figure 4. The use of fluorine plastic for the innermost layer has reduced the quantity of permeation to less than 1/100 of the level for the previous fuel hose

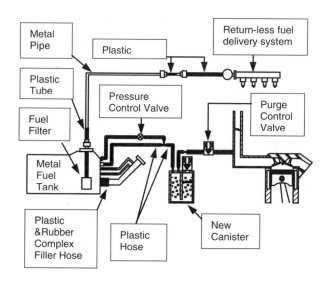

Figure 3 : Fuel & Evaporative Systems for Sentra CA

Item	Previous rubber fuel hose	Improved plastic fuel hose
Structure	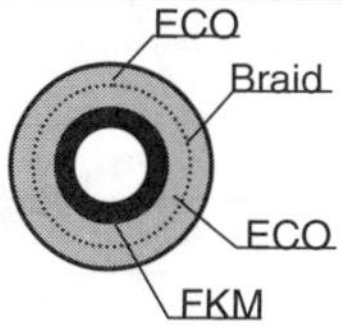 ECO Braid ECO FKM	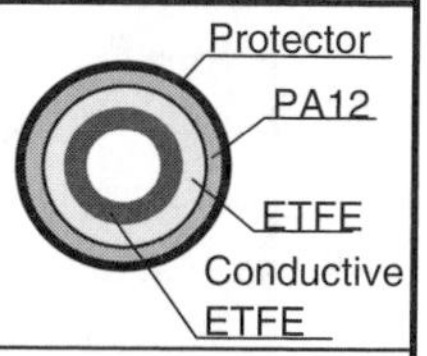 Protector PA12 ETFE Conductive ETFE
Comparison of permeation under same test conditions	100%	Less than 1%

ECO : Epichlorohydrin rubber

FKM : Fluorcarbon rubber

ETFE : Polyethylene-Tetrafluoroethylene copolymer

PA12 : Polyamide 12

Figure 4 : Low Permeation Fuel Hose

Reduction of Fuel Tank Pressure –

It was found that the quantity of permeation from connections and plastic and rubber parts of fuel tank was strongly influenced by differences in internal and external pressures. Consequently, the vapor control pressure inside the fuel tank was lowered to 0.4 kPa from about 1.5 kPa for the purpose of reducing the quantity of permeation.

IMPROVEMENT OF EVAPORATIVE CANISTER FOR REDUCING HC BREAKTHROUGH

Activated charcoal inside the evaporative canister adsorbs fuel vapors. In general, fuel vapors are adsorbed at low temperature and desorbed when the temperature is high. Adsorbed vapors are easily affected by temperature and tend to diffuse from places of high HC concentration to low concentration areas at high temperature. The characteristics of the activated charcoal were studied with the aim of preventing diffusion of fuel vapors adsorbed by the charcoal. While activated charcoal can adsorb a large quantity of fuel vapors, canister breakthrough tends to occur when the residual HC adsorption concentration increases. This makes it necessary to control suitably the HC residual adsorption concentration of fuel vapors by the activated charcoal. To increase the adsorption capacity of the activated charcoal, the residual HC adsorption concentration was made 5% before HSL+DBL test, as indicated in Figure 5.

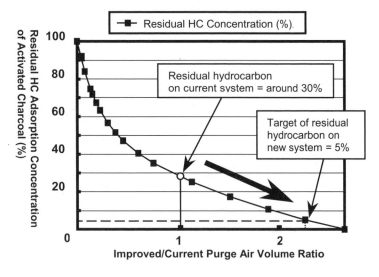

Figure 5 : Residual HC adsorption concentration vs. Purge air volume ratio

The internal structure of the canister was examined with the aim of allowing fuel vapors adsorbed by the activated charcoal to be purged as much as possible during driving. The internal structure of the canister adopted in the new system is shown in Figure 6.

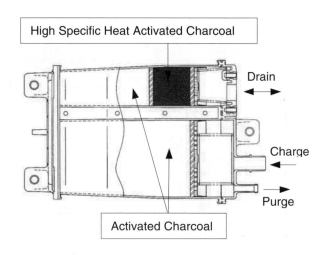

Figure 6 : Internal Structure of New Canister

The activated charcoal used previously in the evaporative canister provided high adsorption performance in order to meet the regulation for on-board refueling vapor recovery (ORVR). As part of the improvement made to the canister this time, activated charcoal with a high specific heat characteristic was adopted on the drain side, as indicated in Table 2. HC adsorption by the activated charcoal involves an exothermic reaction, whereas HC desorption is an endothermic reaction. The use of activated charcoal with a high specific heat characteristic promotes HC desorption in the area where the two types of activated charcoal are mixed, owing to the specific heat effect. Figure 7 shows the temperatures of the high specific heat activated charcoal during HC desorption as a function of the purge air volume.

Item	Previous Canister	New Canister
Charge side	Activated charcoal	Activated charcoal
Drain side	Activated charcoal	Activated charcoal + High specific heat activated charcoal
Shape	Pelletized	Pelletized
Density (g/cm³)	0.35	0.35 for Activated charcoal 0.66 for High specific heat charcoal
Particle diameter (mm)	2.2	2.2 for Activated charcoal 2.5 for High specific heat charcoal

Table 2 : Comparison of canister specifications

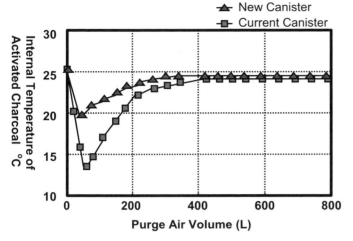

Figure 7 : Activated Charcoal Temperature vs. Purge Air Volume

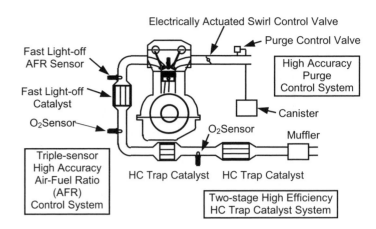

Figure 8 : Sentra CA Exhaust System

The internal temperature of the high specific heat activated charcoal showed a smaller decline in this experiment, thereby confirming the improved promotion of HC desorption. Because the residual HC level of the high specific heat activated charcoal can be kept low, it works to prevent diffusion of fuel vapors inside the activated charcoal and reduces HC breakthrough.

PURGE CONTROL TECHNOLOGY FOR INCREASING PURGE AIR VOLUME

From the LEV-II regulations an exhaust emission standard has been more stringent as well as evaporative emissions, which means that more accurate purge control system is required. A technology for controlling a large volume of purge air was achieved while giving full consideration to the impact on tailpipe emissions. Purge control technology was improved to accommodate the increased quantity of purge air. The configuration of the exhaust system is shown in Figure 8. A new purge control valve was developed that allows highly accurate control of a large volume of purge air, making it possible to achieve improved control of the increased purge air flow rate.

A fast-response feedback control system was examined for more accurate air-fuel ratio control through the adoption of a fast light-off air-fuel ratio sensor. The use of this A/F ratio sensor has remarkably improved feedback control of exhaust emissions. The amount of purge air supplied to the canister was increased to utilize the adsorption capacity of the activated charcoal fully while taking the influence on tailpipe emissions into consideration. The effect of large purge air volume control on reducing HC breakthrough is shown in Figure 9 in a comparison between the standard and new canisters. The data confirm that the increased purge air volume and canister improvement reduce the HC breakthrough.

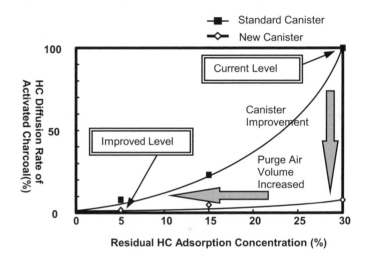

Figure 9 : Residual HC Adsorption Concentration vs. HC Diffusion Rate

ATTAINMENT OF ZERO EVAPORATIVE EMISSIONS

An evaporative emission durability test procedure was examined for the extended emission durability distance of 150,000 miles, increased from 100,000 miles at present. A system test procedure corresponding to 150,000 miles of driving was proposed to CARB. The newly developed system was subjected to this evaporative emission durability test equivalent to 150,000 miles of operation and attained a level of 0.20 g/test for the HSL + DBL test. The results of this 150,000-mile evaporative durability test are shown in Figure 10. The data indicate that the emission level following the durability test remained unchanged from the initial state.

553

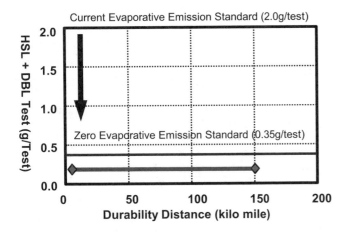

Figure 10 : Evaporative Emission Durability Test

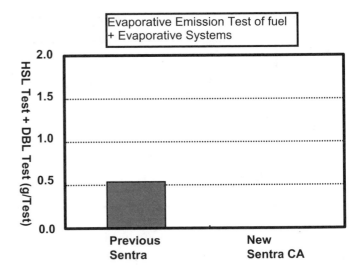

Figure 12 : Comparison of Evaporative Emission Test Results for Current and New System

The comparison of the evaporative emission reduction on reducing the HC level for the vehicle as a whole is shown in Figure 11. HC emissions were reduced to the level that satisfies the zero evaporative emission standard, which was the aim of this work. It should be noted that this evaporative emission test level is equivalent to an electric vehicle. Moreover, an evaporative emission test was also conducted on the fuel system and the fuel evaporative system. Under a test condition equivalent to 150,000 miles of driving, a value of 0.0 g/test was achieved for the HSL test + DBL test. The effect of the evaporative emission reduction from the fuel system and the fuel evaporative system on reducing the HC emission level is shown in Figure 12.

CONCLUSION

A zero evaporative emission system capable of meeting the zero evaporative emission regulation has been developed as a result of improving existing technologies. The results of this study are summarized below.

1. The amount of HC permeation was reduced substantially by improving the materials of the fuel system and the fuel evaporative system and by decreasing the number of parts and connections.

2. The purge air flow rate was significantly increased by improving purge air control through the adoption of a new A/F ratio sensor.

3. HC breakthrough from the evaporative canister was greatly reduced by improving the internal structure of the canister (i.e. through the use of high specific heat activated charcoal on the drain side) and increasing the purge air flow rate.

4. As a result of these improvements, a gasoline-fueled vehicle has been developed that complies with the zero evaporative emission regulation.

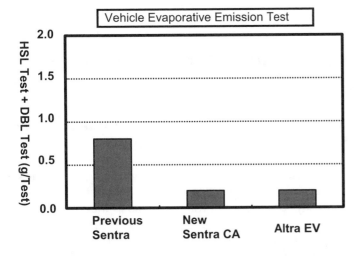

Figure 11 : Comparison of Evaporative Emissions on Test Vehicles

REFERENCES

(1) David Parker, Susan O'Connor, Stephan Lemieux and Kathleen Nolan, "CARB Evaporative Emission Test Program," SAE Paper 1999-01-3528.

(2) Kimiyoshi Nishizawa, Sukenori Momoshima, Masaki Koga and Hirofumi Tsuchida, "Development of New Technologies Targeting Zero Emissions for Gasoline Engines," SAE Paper 2000-01-0890

1999-01-0771

Reduction Technologies for Evaporative Emissions in Zero Level Emission Vehicle

Yoshio Nuiya, Hajime Uto and Takeshi Suzuki
Honda R&D Co., Ltd.

Copyright © 1999 Society of Automotive Engineers, Inc.

ABSTRACT

We have conducted technical research to achieve levels of evaporative emissions very close to zero. In the conventional fuel tank system, the internal pressure in the fuel tank and vapor lines varies between positive and negative values due to the outside temperature, rendering emission control and performance in the actual vehicle very difficult. We have developed a control system utilizing the negative pressure in the engine intake manifold to maintain a negative pressure (vacuum) in the fuel tank at all times. By always maintaining such a vacuum, the variation in internal pressure that depends on diurnal variation in ambient temperature and the rise in internal pressure that depends on engine load, have been eliminated, enabling emissions from the fuel system to be reduced significantly. This paper gives an outline of the technology mentioned above.

INTRODUCTION

Although improvements are being made to resolve the problem of automobile air pollution in recent years, several issues remain to be addressed. Emissions from automobiles include exhaust gas, evaporative gas and evaporative gas emitted during refueling of the vehicle. Although more stringent EVAP regulations since 1995 M/Y and evaporative emission regulations for refueling since 1998 M/Y have been introduced, further reinforcement of the evaporative emission regulations are being considered for inclusion in LEV-II and Tier-II standards.

We have carried out technical research on a completely new vacuum type sealed tank system for reduction of evaporative emissions as part of the technology for full control of evaporative fuel emissions, and have obtained good results. The outline of our work and the results are reported here.

1. STUDIES ON NON-EVAPORATIVE EMISSION SYSTEMS

The main source of evaporative emissions today are:

1. Breathing action due to pressure changes in the fuel tank
2. Permeation of fuel from tubes, joints, etc.
3. Emissions from paints on the vehicle body, rubber components, oils and greases

From the above, (1) and (2) are evaporative emissions related to fuel. To drastically reduce these emissions, we decided to combine technology for a sealed type fuel system with measures against permeation. As shown in Table 1, the next step in the investigation was to select the type of sealed tank. We selected the vacuum type (negative pressure) sealed tank considering the ease of fitting such a tank in the vehicle, and the permeation reduction effect by vacuum, explained later in this report.

Table 1.

Available options / Cause of HC emission	Sealed type with no pressure control	Sealed type with vacuum control	Bladder type tank system	Bag type tank system
Permeation	△	○	△	△
Ease of fitting in vehicle	○	○	△ Size is large compared to others	○
Range of pressure variation in tank (gauge)	-500mmHg ~800mmHg	-300mmHg ~ 0mmHg	0mmHg	0mmHg

○: Effect of reducing HC emission is large
△: Possibly some reduction of HC emission

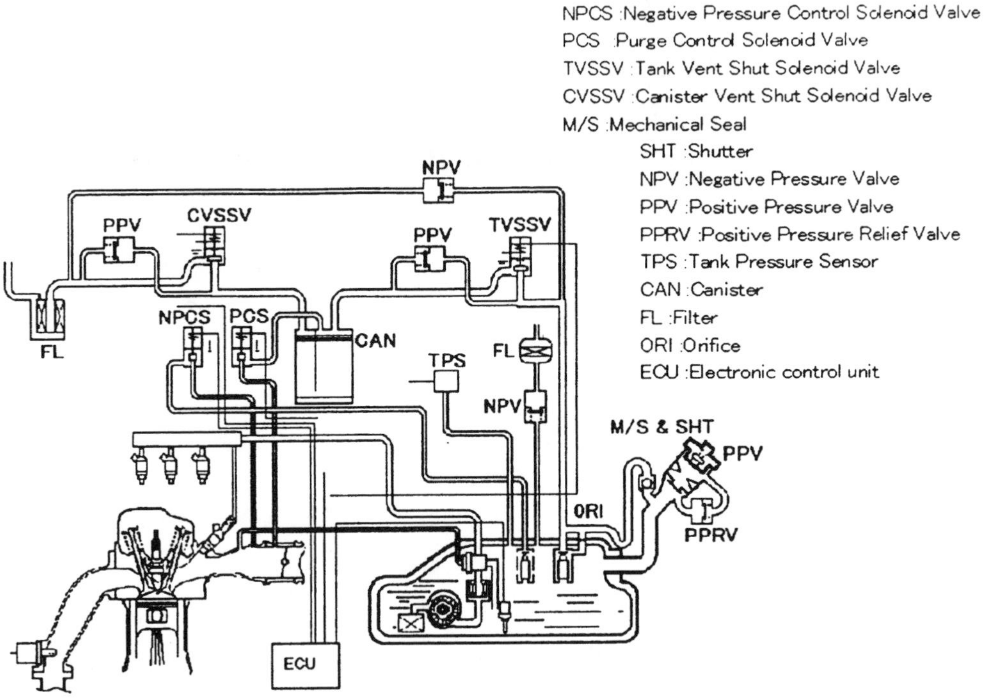

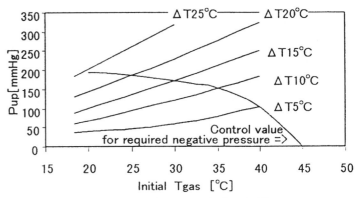

Figure 1. The vacuum type sealed tank system

NPCS :Negative Pressure Control Solenoid Valve
PCS :Purge Control Solenoid Valve
TVSSV :Tank Vent Shut Solenoid Valve
CVSSV :Canister Vent Shut Solenoid Valve
M/S :Mechanical Seal
SHT :Shutter
NPV :Negative Pressure Valve
PPV :Positive Pressure Valve
PPRV :Positive Pressure Relief Valve
TPS :Tank Pressure Sensor
CAN :Canister
FL :Filter
ORI :Orifice
ECU :Electronic control unit

2. CONCEPT OF VACUUM TYPE SEALED TANK SYSTEM

Figure 1 shows a sketch of the vacuum type sealed tank system. The objective of this system is to maintain a vacuum in the sealed tank during both running and at rest conditions of the vehicle. The concept of this system is given below.

1. When the vehicle is in motion, the vacuum in the engine breather pipe is used to control the vacuum in the sealed pressure tank by duty control of vacuum solenoid (NPCS).

2. When the vapor is adsorbed during refueling and at all times other than purging of the canister, the canister drain is closed completely by the canister vent shut-off solenoid valve (CVSSV).

3. The canister is purged by duty control of the purge control solenoid (PCS) when the engine is running.

4. At times other than refueling, the tank is closed completely by the tank vent shut-off solenoid valve (TVSSV). The signal for closing the TVSSV is activated by the sensor in the lid when the fuel lid is closed after for refueling.

5. Evaporative emission during refueling is reduced based on the ORVR system fitted in the 1998 Accord. The refueling gun is sealed mechanically, restricting the emission of vapor from the refueling port.

6. Tank pressure abnormalities are avoided when the vehicle is at rest or when any part has failed, by means of positive pressure relief valves and negative pressure relief valves (1) to (6), ensuring reliability of the tank.

3. TARGET PRESSURE IN THE TANK

To maintain vacuum in the tank with the vehicle at rest, the control values of the target pressure of tank when the vehicle is in motion are set.

3-1. RISE IN TEMPERATURE OF GASOLINE WHEN THE VEHICLE IS AT REST AND RANGE OF RISE IN TANK PRESSURES – Fig.2 shows the pressure increase characteristics after vehicle stop. The pressure in the tank rises when the temperature of gasoline (Tgas) increases by DT(amount of temperature increase) soon after the vehicle has come to a stop. The maximum outside temperature when the vehicle has come to a stop, is expected to be 45°C. If the temperature of gasoline is also assumed to increase to 45°C, the control value for obtaining the required negative pressure can be obtained from the figure.

Figure 2. Rise in Ptank due to variation in Tgas (max. 45°C assumed)

556

3-2. RISE IN PRESSURE AFTER STOPPING NEGATIVE PRESSURIZATION – In practice, if negative pressure is established in the tank, the pressure in the tank (Ptank) starts increasing gradually after the negative pressurization is stopped, as shown in Fig. 3. This is because vapor is supplied so that the saturated vapor pressure of gasoline corresponding to the temperature of the gas is reached in the sealed tank in which negative pressure has been achieved. This point also has to be taken into consideration before negative pressurization of the tank. This rise in pressure of the tank is taken as ΔPup, and the required negative pressure is shown in Fig. 4-(2).

3-3. SETTING THE TARGET PRESSURE OF THE TANK – As mentioned above in 3-1 and 3-2, if the required drop in pressure when the vehicle is in motion is established during engine operation, the negative pressure in the sealed tank can be maintained even when the vehicle is at rest. Consequently, the sum of these pressures becomes -300 mmHg approximately, as shown in Fig. 4, and the temperature of the gas, Tgas, controls the target pressure of the tank, 3 of Figure 4.

4. TEST RESULTS

4-1. EVAPORATIVE EMISSION – The SHED test was carried out using the hardware system shown in Fig. 1. From the current US test modes, the 3DBL test and ORVR refueling test modes were followed. To avoid background effects of the car, the test was conducted with the fuel system of Fig. 1 installed in the SHED house. Consequently, hot soak loss and running loss modes were not evaluated. For the sake of comparison, the 1998 Accord (US specification) currently in mass production, was used. The line from the fuel pump to the injector has a pressure of approximately 3 kg/cm^2 even when the vehicle is at rest. This pressure does not affect the negative pressure in the tank. Various measures are necessary to minimize the losses of this positive pressure system.

4-1-1. Permeation reduction effect due to negative pressure – Fig. 5 shows the results of permeation measurements in the FKM hose. The tests were carried out after filling a long length hose with gasoline. From this figure, it is observed that the HC permeation varies depending on the pressure in the hose. HC is detected even in a hose not filled with gasoline. However, because of the negative pressurization, similar data was obtained even when no gasoline was present in the hose. Accordingly, if negative pressure (vacuum state) is maintained, the HC permeation of fuel can be eliminated.

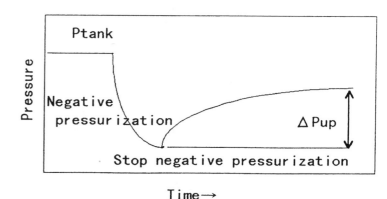

Figure 3. Pressure in the tank is increasing gradually after the negative pressurization stopped

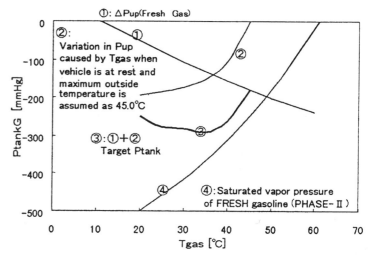

Figure 4. Required negative pressurization in sealed tank

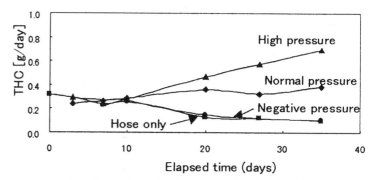

Figure 5. Confirmation of HC penetration in FKM hose (hose length 10m)

4-1-2. SHED values around the tank – From the effects indicated in Fig. 5, the SHED values around the tank including the canister are as shown in Fig. 6. No HC permeation due to fuel can be observed for the vacuum type sealed tank system.

4-1-3. HC permeation of high pressure line – The line from the fuel pump to the injector had a positive pressure of approximately 3 kg/cm^2 even when the vehicle was at rest. This pressure does not affect the negative pressure in the tank. Various measures are necessary to minimize the effects of high pressure. Measures include using plastic hose and metal sealing of various joints. Prospects of reduction of HC permeation after adopting the measures are as shown in Fig. 7.

4-2. CONTROLLABILITY OF PRESSURE IN TANK – In sealed type vacuum tank systems, the most demanding driving modes are very hot running and high altitude running for the reasons given below.

1. If the temperature rise of gas is high, the target pressure of the tank may not be reached.

2. When hill climbing, the engine load increases, therefore, the negative pressure in the intake manifold, which is the basis of the vacuum, decreases.

3. When hill climbing, the atmospheric pressure (Pa) decreases, gauge pressure in the sealed tank increases, and the load to maintain the vacuum increases.

4. At high altitudes, even at the same running mode, the negative pressure in the intake manifold decreases, and the frequency of building up the negative pressure decreases.

In view of the above reasons, hot running up hill and high altitude operating conditions were investigated.

4-2-1. Hot running up a slope – Fig. 8 shows data for continuous running from a full tank condition to empty tank condition of fuel in tests carried out at Death Valley, where the outside air temperature rises to a level of 50°C. Hill climbing conditions can be added even in Death Valley. The vehicle was driven up hill to an altitude of 1600 m and the vacuum state was always maintained during the run. Next, the running of the vehicle in Death Valley was indicated in terms of absolute pressure of the tank. Fig. 9 shows the transition of temperature of gas and pressure in the tank during operation. According to these figure, PtankG(gauge pressure of tank)would vary considerably due to the effect of reduction of Pa(atmospheric pressure) when climbing up the slope. However, when the same data is observed in terms of absolute pressure, the transition of pressure moves on the line of the vapor pressure curve. This is a vapor pressure line for a gasoline. There isn't air inside the tank since the tank space is filled with gasoline vapor. The transition of pressure is to the right of the vapor pressure curve. This transition is due to the effect of evaporated gas. And the vapor pressure moved in the direction of the dotted line of Fig. 9 representing a used gasoline vapor pressure curve.

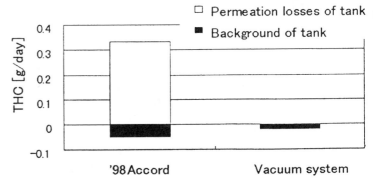

Figure 6. Unit total DBL SHED test value around tank

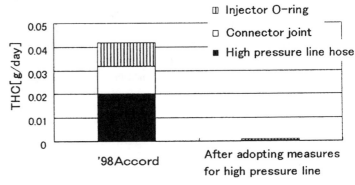

Figure 7. HC permeation of fuel line (DBL mode)

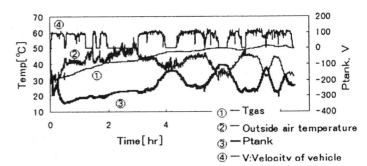

Figure 8. One – tank continuous hit running (Death Valley)

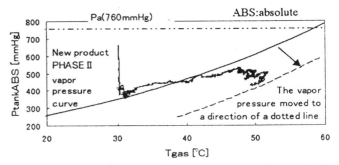

Figure 9. Transition of Ptankabs during running tests in Death Valley

- The judgement on whether the vacuum state can be maintained or not when the vehicle is at rest is made as given below.
- If the temperature of the gasoline immediately after vehicle comes to a stop is greater than the maximum temperature of the gasoline when the vehicle is at rest.
- Due to a drop in the temperature of the gas when the vehicle is at rest, vacuum buildup is automatic. Therefore the vacuum state is maintained.
- If the temperature of the gasoline immediately after vehicle comes to a stop is less than the maximum temperature of the gasoline when the vehicle is at rest.
- Up to the point where the vapor pressure curve and the Pa value at that location intersect in Fig. 9, the vacuum state is maintained even if the temperature of the gas rises.

Accordingly, the vacuum could be maintained at all times even in the running conditions in Death Valley.

4-2-2. Running tests of the vehicle at high altitude locations – Running tests were carried out after selecting Lapland Pass (3500 m, Pa 500 mmHg) as the high altitude test location. As shown in Fig. 10, the vehicle was refueled at the foot of the slope before the vehicle was driven up the slope, and the vacuum state was retained all the time during the run. If the phases are expressed in terms of Ptankabs(absolute pressure of tank), similar to Fig. 9, then Fig. 11 is obtained. The target pressure of the tank (-300 mmHg) at low altitude, in terms of absolute pressure, falls below the vapor pressure curve of gasoline due to the reduction in Pa at high altitude. Consequently, Ptankabs does not move below the vapor pressure curve, and it cannot reach the target pressure of the tank, Ptank (-300 mmHg gauge). However, since the vacuum state can be maintained above the vapor pressure curve, it means that this state can be maintained at an atmospheric pressure of 500 mmHg up to 42°C, taking the maximum temperature of the gas when the vehicle is at rest. Accordingly, the vacuum state can be maintained in a high altitude environment when the vehicle is at rest.

4-3. MAINTAINABILITY OF VACUUM STATE – Fig. 12 shows the results of long term exposure tests carried out after fitting the fuel system in Fig. 1 on an actual vehicle. According to this figure, the maintainability of vacuum state in the tank is satisfactory. No rise in pressure due to leakage was observed. Consequently, prevention of HC emission during long term exposure of the vehicle may be anticipated since the sealed, less than atmospheric pressure condition can be maintained over a long period.

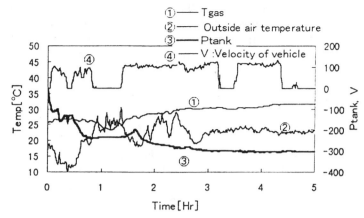

Figure 10. High altitude running
(summer, Lapland Pass, 3500m class)

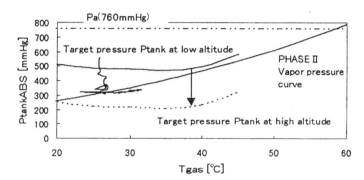

Figure 11. Lapland Pass running (3500 m class)

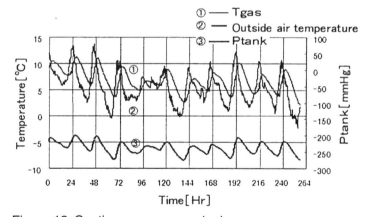

Figure 12. Continuous exposure test

5. CONCLUSION

We conclude that the prospects are bright for practically achieving zero level emissions by using a sealed negative pressure type tank system. Since the vacuum state can be maintained even in very adverse environments, we believe that this system can be used effectively in the market. Henceforth, we will carry out further research on vacuum controls and confirm the applicability of this system to the market.

Using existing SHED emissions measurement equipment for these low emission systems are likely to bring up limitations in measurement accuracy. Evaluation by unit MINI SHED tests and evaluation by design criteria are anticipated.

CONTACT

YOSHIO NUIYA
e-mail:yoshio_nuiya@n.t.rd.honda.co.jp
ENGINEERING DEVELOPMENT DEPARTMENT 3
HONDA R&D CO., LTD. TOCHIGI R&D CENTER
4630 Shimotakanezawa Haga-machi Haga-gun Tochigi
JAPAN
Telephone (028) 677-3311 Telefax (028)677-7114

CHAPTER 12

ON BOARD DIAGNOSTICS

Glenn Zimlich, Kathleen Grant and Timothy Gernant
Ford Motor Company

12.1 OVERVIEW

On Board Diagnostics (OBD) is a set of diagnostic algorithms in the powertrain control strategy that detects when an emission-critical component has failed or reduced efficiency. When an emission-related component is detected to have failed, the malfunction indicator light (MIL) must be illuminated. In addition, the OBD system must record the status and test results of the various diagnostic tests so they can be reviewed at I/M (inspection maintenance) stations. The diagnostics work to improve air quality by detecting when a vehicle produces more emissions than allowed. The OBD II system also aids the service community in identifying and isolating vehicle faults.

The diagnostics monitor any component that affects the emission system performance or is an input or output to the electronic powertrain control module. The monitoring systems are affected by new low emission standards (LEV II) because diagnostic failures are in most cases related to an increase in vehicle emissions measured on the Federal Test Procedure (FTP) emission drive cycle. The emission constituents of interest are carbon monoxide (CO), non-methane organic gases (NMOG) and oxides of nitrogen (NOx). There are cases when the malfunction of an emission control device does not lead to an OBD emission failure. The manufacturer is required to monitor the component for function only (functional monitor). If the deterioration of a component leads to an OBD emission failure, the component monitor must be calibrated to illuminate the MIL at the specified OBD II emission threshold (threshold monitor). Beyond the functional/threshold monitor types, the OBD II monitors can be further separated into two categories, continuous and non-continuous. Continuous monitors, like circuit checks (open, short, out-of-range), execute constantly during engine operation. The non-continuous monitors execute a specific test within a given set of entry conditions once per trip. The latest regulations stipulate performance metrics on the frequency of execution of some of the non-continuous monitors. The OBD II system will now calculate and track an in-use performance ratio. The numerator of this ratio is defined as the number of times the correct window of opportunity has been encountered such that each monitor would have detected a malfunction. The denominator is defined by the number of times the vehicle has been operated in a CARB (California Air Resource Board) defined manner that monitors should have completed.

All of the monitors, except the CCM (Comprehensive Component Monitoring), catalyst and evaporative system monitors, must illuminate the MIL before the vehicle exceeds 1.5 times the applicable tailpipe emission standard (for CO, NMOG, or NOx). There is relief for the Super Ultra Low Emission vehicles (SULEV II / PZEV). The emission failure thresholds increase to 2.5 times the SULEV II emission standards. The Evaporative System Monitor is not calibrated to a tailpipe emission standard. The monitor detects a leak of a prescribed orifice diameter in the evaporative system (ex. fuel tank, fuel vapor lines, fittings/connectors). The Catalyst Monitor is a special case in that the monitor must detect a 1.75 times (2.5 times SULEV II) emission degradation of NMOG or NOx constituents only.

The diagnostic system is affected by low emission technology in primarily two ways. First OBD emission standards are linked (multiples of 1.5 times for most monitors) by the regulatory tailpipe standard to which the vehicle is certified. A second direct impact of low emission vehicles is the increased hardware that is required to achieve lower emissions standards. The complexity of the control system and the ability to monitor new components must be managed to provide a complete diagnostic system.

The fuel system monitor on low-emission vehicles is of particular interest. The fuel system must be capable of adapting to fuel bias quickly and accurately. The reduction of catalyst efficiency due to non-adaptation can be crucial to meeting low emission levels. A fuel-biased vehicle system must nearly replicate the air-fuel of a non-biased vehicle. The OBD fuel monitor must be capable of detecting these shifts.

The catalytic converter is one of the primary means of reducing tailpipe emissions. In the past, the catalyst monitor had only been required to detect a failure of the catalyst before the emissions exceeded 1.75 times the NMOG standard. For LEV II applications, the monitor must also detect an increase in NOx emissions. The incremental requirement is being phased in with the LEV II requirements. OBD catalyst monitoring involves correlating the depletion of catalyst oxygen storage capacity (OSC) to the degradation of tailpipe emissions. The catalyst OSC provides a mechanism to store excess oxygen during lean operation and release oxygen during rich conditions. As the catalyst OSC is depleted, there is an increase in the air-fuel fluctuations that leads to increased voltage activity on the oxygen sensor mounted after the monitored catalyst. The increased voltage activity is then correlated to the increase in tailpipe emissions. A catalyst with greater OSC tends to have much less voltage swings than the catalyst with lower OSC. The correlation of tailpipe HC performance to catalyst OSC during warm engine steady-state conditions is challenging because the degradation of vehicle emissions during warm operation is small.

Vehicular emissions are traditionally thought of as those relating to the incomplete products of combustion within the engine. However, unburned hydrocarbons can also enter the atmosphere through evaporation from the fuel tank. Instead of venting the fuel tank to the atmosphere, the vapors pass through an activated charcoal canister. The charcoal adsorbs the unburned hydrocarbons while the vehicle is not in operation. Later, the vapors are drawn through the canister and into the intake manifold to undergo combustion in the engine. Most of the emissions-based OBDII regulations dictate that the MIL must illuminate when the tailpipe emissions exceed a 1.5 times tailpipe standard in terms of grams per mile. Those dealing with the evaporative system monitor handle things differently by specifying a given leak size within the evaporative emissions system that must illuminate the MIL. The most stringent requirement dictated by CARB, requires the OBD II system detect a 0.020" diameter leak.

One common method for leak detection uses the same system hardware as the vehicle uses to control evaporative emissions, with the addition of a valve to seal the vent at the carbon canister. By sealing the system, while actively pulling vapors into the intake manifold, a vacuum is pulled on the tank. Once the system reaches a target vacuum, the system is also sealed from the intake manifold, and the pressure is monitored over time. In a vehicle with good system integrity, the pressure will remain nearly constant over time. However, if the system has a leak, the influx of air from the atmosphere into the system will cause the pressure to rise back to atmospheric over time. The rise in system pressure is correlated to leak size to determine when to illuminate the MIL. This method does have several pitfalls, such as dynamic vehicle operation, vehicle grades, fuel volatility, and large tank volumes. One of the methods employed to overcome several of these obstacles is to incorporate the use of additional system hardware, in the form of an electronically controlled pump. The pump is used to apply a positive pressure to the sealed system. Initial systems would apply a positive pressure until the system pressure came into equilibrium with the internal spring load of the pump. A leak-free system would maintain the target pressure. The same noise factor associated with fuel volatility that plagues the vacuum decay method is present on the pressure decay systems, and as expected, has the opposite effect. Previously, where volatile fuel would create a false detection, the pressure rise associated with fuel evaporation could now cause the diagnostic to yield a false pass. Additionally, the application of a positive pressure to the fuel system will cause unburned hydrocarbons to be forced out of the system through any leaks that are present in the system. Another method used to achieve the 0.020" standard, has its basis in the ideal gas law. Instead of using either intake manifold vacuum or a pump, this method uses the natural thermodynamics of the fuel system following vehicle operation. Assuming that the volume of the system is held constant, the system pressure will change with the associated temperature changes brought about during the soak period after vehicle operation. Initially, after shut down, heat will be transferred to the tank in the form of waste heat from other vehicle components. Assuming a constant volume, this heat addition will cause a pressure rise in the tank. Eventually the vehicle will begin to cool, and bring with it a corresponding drop in fuel system pressure. Although the magnitude of pressure changes generated in the tank are not as large as methods that use a pump or the intake manifold as a vacuum source, the change in pressure can be viewed over a much longer time scale by allowing the test to run while the engine is off.

There are several new or revised monitoring requirements for LEV II type vehicles. One new requirement is the Cold Start Emission Reduction Control Strategy Monitor. Engine start-up control strategies that reduce emissions are to be monitored. In 2006, the secondary air system monitoring requirements are also modified, the system must be monitored while it is being used (i.e. cold start) and it must be emission-based. Due to these considerable changes a more direct measurement of the secondary air system may be required. Beginning in 2006 MY on LEVII applications, the regulations expand the degree to which the VVT systems must be monitored. The impact on emissions of both target error and response rate of the system will need to be monitored

On-board vehicle diagnostics are an ever-expanding regulatory effort and the regulations are continuing to evolve. The expansion is not limited to the U.S.: Europe and Japan have implemented diagnostic standards (EOBD and JOBD respectively). The trend will most likely continue into other countries. It is clear the regulations will continue to evolve, both on the quest for lower vehicle emissions as well as the refinement of the diagnostic requirements.

12.2 SELECTED SAE TECHNICAL PAPERS

OBD-II Performance of Three-Way Catalysts

J. S. Rieck
Johnson Matthey, Wayne, PA

N. R. Collins and J. S. Moore,
Johnson Matthey, Royston, UK

Copyright © 1998 Society of Automotive Engineers, Inc.

ABSTRACT

The current method for on-board monitoring of catalyst performance involves detecting the degradation in the oxygen storage capacity as the catalyst ages. Inherent in this method is the need to correlate the deactivation in HC perform-ance with oxygen storage capacity. However, as HC standards become more stringent, light off becomes the key factor impacting HC emission levels, and it is increasingly difficult to detect failures in HC performance based on OSC deactivation. A possible approach to address this challenge is to include catalyst formulation as a variable in performing OBD-II calibrations. This study explores the potential for tailoring the OBD-II performance of a catalyst by customizing the PGM/OSC component to give the desired degree of thermal stability. The effects of sulfur and aging conditions are also investigated. The potential for independent adjustment of OSC and HC performance is discussed.

INTRODUCTION

The dual oxygen sensor method has become the industry standard for on-board monitoring of catalyst deactivation (1). With this technique, an oxygen sensor is placed down stream of the three-way catalyst. Perturbations in the air/fuel ratio of the feed gas to the catalyst are recorded with the upstream (controlling) sensor. Due to the oxygen storage capacity (OSC) of the catalyst, the perturbations in the feed gas AFR will be dampened in the exhaust downstream of the catalyst. Therefore, comparison of the upstream and downstream signals provides a measurement of the OSC of the catalyst.

As HC emission standards have been tightened, it has become increasingly difficult to correlate HC emission failures with the deact-ivation in OSC (2). HC light off tends to be the controlling parameter in reaching LEV and ULEV HC emissions. However, most OBD-II systems measure OSC during warmed-up steady-state operating conditions, and the catalyst perform-ance in this state does not necessarily correlate with light off performance.

For ULEV systems, a close-coupled Pd-only catalyst is commonly used to achieve rapid HC light off, and an oxygen sensor can be placed immediately downstream of the close-coupled catalyst (2,3). In Pd-only catalysts, the OSC component greatly influences CO and NOx performance but has little effect on HC light off (4). Since the key function of the close-coupled catalyst is to provide HC conversion, the potential exists to modify the OSC of the catalyst to provide characteristics which may be desirable for OBD-II calibrations.

Hepburn, et al. (5) presented a laboratory technique for investigating the relationship between OSC and HC performance. In addition, they demonstrated that the catalyst washcoat can be engineered to alter this relationship. The present study investigates the effect of the OSC component on the relative deactivations in HC conversion and OSC. The effect of sulfur and aging conditions are also investigated.

EXPERIMENTAL

Catalyst formulations were coated on 55 in^3 ceramic substrates with 400 cells/in^2 and 6 mil wall thicknesses. For aging and evaluation, 1 inch diameter by 3 inch long cylindrical cores were removed from the catalyst blocks. Agings were performed in a Lindberg furnace equipped with a quartz tube to allow control of the aging atmosphere. Agings were performed either in flowing dry air or in a cycled lean/rich atmos-phere. The cycled aging gas composition consisted of 0.5% O_2 (lean) or 1.0% CO (rich) with 10% H_2O and balance N_2. The aging times and temperatures were varied to achieve various degrees of catalyst deactivation.

After aging, the catalysts were evaluated in a synthetic exhaust mixture which cycled be-tween lean and rich conditions at a frequency of 1 Hz. The reactor was equipped with a UEGO sensor downstream of the catalyst. The gas compositions are shown in Table 1. The total flow rate was 25.33 standard l/min which corresponds to a space velocity of 39,360 hr-1.

The catalysts were loaded in the reactor at room temperature, and the temperature was ramped to 450°C at 30°C/min in the absence of sulfur. The temperature was allowed to stabilize at 450°C for 5 minutes prior to recording the conversions and UEGO sensor response. SO_2 was then introduced to the catalysts, and the conversions were recorded after 30 minutes of sulfur exposure. Prior to each series of tests, a blank cordierite substrate was tested to measure the background conversions and UEGO sensor response.

Table 1. Synthetic Exhaust Composition

Component	Rich Stream Concentration	Lean Stream Concentration
C_3H_8	200 ppm	200 ppm
C_3H_6	200 ppm	200 ppm
NOx	500 ppm	500 ppm
CO_2	14.0%	14.0%
H_2	0.17%	0.17%
H_2O	10.0%	10.0%
CO	2.64%	0.50%
O_2	0.50%	1.57%
N_2	Balance	Balance

For these experiments, the OSC of the catalysts is reported as the ratio of the amplitude of the perturbation in the HEGO sensor voltage measured downstream of the catalyst to that measured downstream of a blank cordierite substrate. A thorough discussion of this technique has been presented by Hepburn, et al. (5). The UEGO response for a typical catalyst with high OSC is shown in Figure 1.

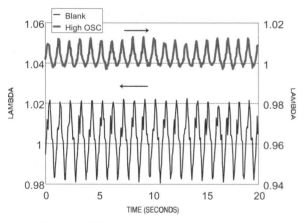

Figure 1. Typical UEGO response for blank and high OSC catalysts.

An engine based evaluation method was also developed to directly measure the oxygen storage capacity of three-way catalysts. For this test, signals were recorded from UEGO sensors mounted directly in front of and directly behind the test catalyst. The engine was operated to give a catalyst inlet temperature of approximately 350°C. The fuel system, operating open-loop, was adjusted by

the control algorithm to give a standard lean air/fuel ratio (Lambda = 1.068), and this operating point recorded. A standard rich (Lambda = 0.932) condition was then determined in the same way. Forced oscillation of lambda between these values was then started, and after a ten-minute preconditioning period, data from ten one-minute cycles were recorded at a rate of one reading per second. (Figure 2) The procedure was then repeated at an engine condition giving a catalyst inlet temperature of approximately 450°C.

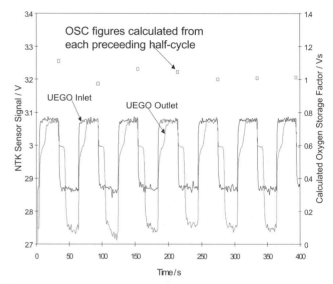

Figure 2. Procedure for engine measurement of OSC.

To give a final figure for OSC, the area between the two UEGO signal traces was determined for each lean half-cycle. These areas were then averaged for all ten cycles, giving an overall figure for each of the two engine load conditions. These figures are a measure of the oxygen storage capacity of the sample at different temperatures. The result is in Volt-seconds, but given an approximate calibration of the UEGO sensor from voltage into % O_2, and an air-flow figure for each condition, it is possible to convert this into mass of oxygen stored on the sample (see Appendix). The result shown in Figure 2, which was recorded at an air mass flow of 25 kg·h⁻¹, with an OSC of 1 Vs, shows that the test catalyst was storing approximately 701mg O_2.

This method can be combined with catalyst aging on the same engine to give an oxygen storage life history or OBD fingerprint of a catalyst formulation in a twenty hour continuous test cycle. The aging involves running the engine rich to give 7% CO engine out and injecting air to give lambda = 1 at the catalyst inlet. Catalyst bed temperatures of up to 1250°C result from such conditions, thus giving a rapid thermal deacti-vation of the catalyst. The

[1] Jordan K., Moore J.S., Collins N.R., Brogan M.S., Chandler G.R., Webster D.E., Wilkins A.J.J., Brisley R.J., Modern Catalytic Coatings - Systems and Test Methods, Haus der Technik Congress, Essen, Germany, 1996

catalyst aging is halted every hour and a oxygen storage capacity measurement performed.

A range of OSC components were evaluated in the course of this study as listed in Table 2. Type A is an unstabilized ceria; while, in types B-E, the ceria is stabilized with zirconia.

Table 2.   OSC Components

Component	Composition
A	ceria
B	ceria-zirconia (low Zr/Ce)
C	ceria-zirconia
D	ceria-zirconia (high Zr/Ce)
E	ceria-zirconia

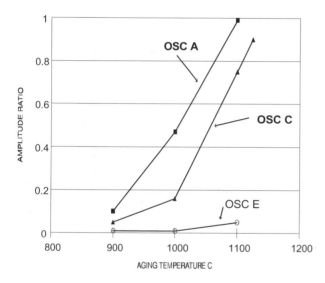

Figure 3.   Effect of aging temperature amplitude ratio of OSC A (0.61 g/l Pt, 0.15 g/l Rh), C (0.61 g/l Pt, 0.15 g/l Rh), and E (1.91 g/l Pd, 0.21 g/l Rh)

RESULTS AND DISCUSSION

A. EFFECT OF CERIA LOADING – The effect of ceria loading on OSC was evaluated for Pt/Rh catalysts (0.61 g/l Pt, 0.15 g/l Rh) containing OSC components A and C, and a Pd/Rh catalyst (1.91 g/l Pd, 0.21 g/l Rh) containing OSC component E. All agings were performed in air for this study. The effect of aging temperature on the amplitude ratio is shown in Figure 3 for the three catalysts at a fixed OSC component loading. The stability of the OSC is seen to correlate with the degree of ceria stabilization E > C > A. Because of the varying stabilities of these materials, the effect of ceria loading on the amplitude ratio was evaluated for aging temperatures of 900°C for component A, 1000°C for component C, and 1100°C for component E. These results are shown in Figure 4.

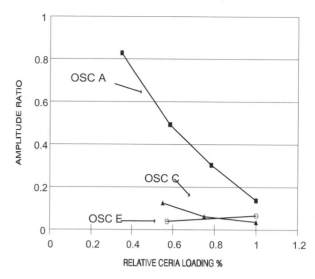

Figure 4.   Effect of OSC loading on amplitude ratio of OSC A (0.61 g/l Pt, 0.15 g/l Rh), C (0.61 g/l Pt, 0.15 g/l Rh), and E (1.91 g/l Pd, 0.21 g/l Rh)

Figure 4 shows that, with increasing stability of the OSC material, the amplitude ratio is less affected by Ce loading. Dropping the Ce loading in half for the Pd/Rh-E catalyst has little effect on the amplitude ratio. In contrast, halving the Ce content of the Pt/Rh-A catalyst dramatically increases the amplitude ratio in spite of the 200°C lower aging temperature. This is not surprising, since the mixed oxides formed in the stabilized cerias have been observed to have more capability for oxygen storage than pure ceria (6).

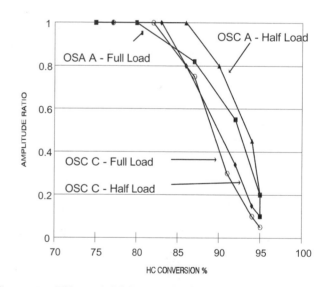

Figure 5.   Effect of OSC loading on amplitude ratio vs HC conversion for OSC A and C (0.61 g/l Pt, 0.15 g/l Rh).

The effect of OSC component loading on the relation between amplitude ratio and HC conversion are shown in Figure 5 for the Pt/Rh catalysts containing OSC components A and C. In both catalysts, the OSC component loading was halved. The results show that the loading has little effect on the amplitude ratio vs. HC conversion curve for either catalyst. This occurs for differing reasons for the two catalysts. For the Pt/Rh-A catalyst, the OSC deactivates very rapidly relative to the HC conversion. As a result, changing the ceria loading can have little effect on the shape of the curve. For the Pt/Rh-C catalyst, the OSC is more durable relative to the HC conversion. However, as shown in Figure 4, the amplitude ratio is not a strong function of Ce loading for this catalyst. These results show that altering Ce loading alone is not an effective way to affect the OBD-II performance of catalysts made using these components.

B. INFLUENCE OF OSC COMPONENT

Pd-only catalysts are commonly used in a close-coupled position in LEV/ULEV systems to achieve rapid HC light off (3). Provided that the key function of the close-coupled catalyst is HC conversion, the potential exists to modify the OSC component in this catalyst to provide desirable characteristics for an OBD-II calibration.

Figure 6 shows the amplitude ratios and HC conversions for two Pd-only catalysts containing 10.59 g/l of Pd following agings at 1000°C and 1100°C in air. The first catalyst contains no OSC component, and the second catalyst contains OSC component C. As expected, the catalyst with no OSC has an amplitude ratio of 1; in contrast, the second catalyst has a very low amplitude ratio. However, the HC conversions of the two catalysts are very similar. This demonstrates that, in this situation, the OSC package may be altered without affecting HC performance. In light of this, components D and E were investigated as potential additives for adjusting the OBD-II characteristics of Pd-based catalysts.

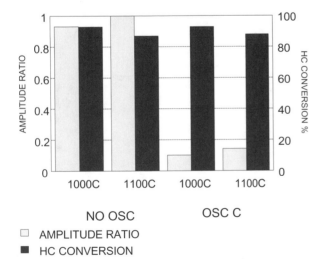

Figure 6. Effect of presence of OSC component on HC conversion and amplitude ratio (10.59 g/l Pd)

Figure 7 shows the amplitude ratio for various aging temperatures for catalysts made with OSC components D and E containing a PGM at a loading of 0.21 g/l. The data show that a varying quantity of OSC can be attained for a given aging temperature depending on the choice of PGM/OSC component. Component E is more durable than component D, and OSC E/ PGM1 is the most durable PGM/OSC com-ponent combination. Based on these results, OSC E/PGM1 and OSC D/PGM2 were selected for further study since Figure 7 shows these are the most durable and least durable OSC components, respectively. Catalysts were formualted containing 7.06 g/L Pd and incorporatiang one of these PGM/OSC components

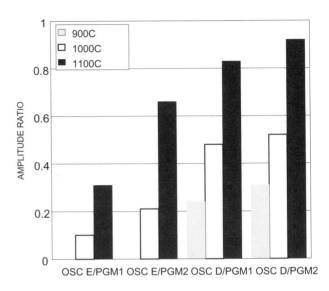

Figure 7. Effect of OSC composition on amplitude ratio (0.21 g/l PGM).

The relation between amplitude ratio and HC conversion for these catalysts is shown in Figure 8. The data show that the curve for the catalyst containing OSC D/PGM2 has a steeper slope due to a relative faster rate of deactivation in OSC relative to HC conversion, and the OBD-II characteristics of the catalyst have been effectively altered. This indicates that proper engineering of the OSC material should prove effective at tailoring the OBD-II characteristics.

It should be noted that, as more attention in the U.S is focussed on reducing NOx emissions with LEV-II standards, the feasibility of radically altering catalyst formulation strictly to achieve desirable OBD-II characteristics may decrease, since altering the OSC package of a Pd-only catalyst can affect NOx performance (4). However, since most systems for ULEV are two brick sysems, the rear brick can be formulated such that no detriment in NOx performance for the overall system is observed.

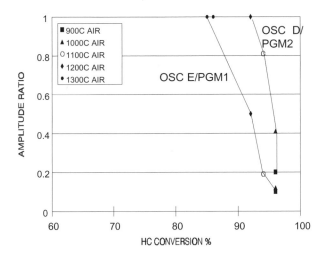

Figure 8. Relation between amplitude ratio and HC conversion for 7.06 g/l Pd catalyst containing OSC D/PGM2 or OSC E/PGM1

The engine test procedure was utilized to evaluate the oxygen storage decay characteristic of two palladium/ rhodium fully formulated three-way catalysts, containing components B and E. The catalysts were loaded at 100 gft[-3], 14:1 (Pd:Rh) on 4" x 6" 400 cpsi, 0.006 in wall thickness substrate. Figure 9 shows the decay curves of the two catalysts on aging; the oxygen storage capacity of catalyst containing oxygen storage component B deteriorates at a much faster rate than that of the catalyst containing component E. Indeed, the measurable oxygen storage of the B formulation has become undet- ectable with this test after 9 hours aging in contrast to that of the component E formulation at this aging point which is still significant. This increased robustness in oxygen storage has been achieved by altering the chemistry of the oxygen storage component and its interaction with the platinum group metals present. Catalysts containing oxygen storage component E will have a greater lifetime of monitorablility by an OBD system as once the oxygen storage of the catalyst is lost the dual sensor system cannot detect further deactivation of three-way activity. This may well be important for proposed European Stage III OBD legislation where levels of catalyst deactivation (0.4 g/km, over European Stage III drive cycle) required to be measured are relatively high compared to California LEV, ULEV threshold limits.

C. EFFECT OF AGING CONDITIONS – Thermal agings in air are commonly used to produce deactivated catalysts which can be used for calibrating OBD-II systems (7). These thermal agings sinter the PGM as well as the base metal oxides present in the washcoat leading to a loss in both HC performance and OSC. However, exhaust stoichiometry is likely to have an effect on the relative deactivation of the PGM and the OSC material. Redispersion of PGMs can occur depending on the reducing/ oxidizing characteristics of the gas stream, and the phase stability of mixed oxides can also be greatly affected by the stoichiometry. In light of this, experiments

were conducted to determine the relative effects of thermal aging in air and aging in a cycled lean/rich exhaust stoichiometry.

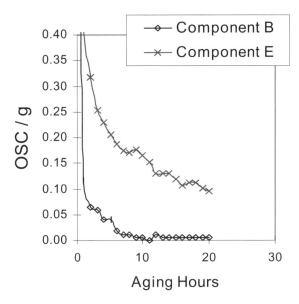

Figure 9. Decay in OSC with engine aging (3l30 g/l (Pd), 0.24 g/l Rh)

Figure 10 shows the relation between amp-litude ratio and HC conversion for a Pd/Rh catalyst (1.91 g/l Pd, 0.21 g/l Rh) containing OSC component E. A general trend between OSC and HC conversion is followed regardless of the aging environment, although the cycled aging is seen to be less severe for both OSC and HC conversion than the aging in air at a given temperature. Figure 11 shows the depen-dence of amplitude ratio on HC conversion for a catalyst of the same PGM composition containing OSC component B. In this case, the cycled aging results in a much greater deact-ivation in the OSC of the catalyst and less deactivation in HC performance relative to aging in air. As a result, the aging environment has a great effect on the relationship between OSC and HC performance over this catalyst.

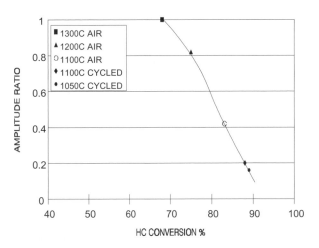

Figure 10. Effect of aging conditions on relation between amplitude ratio and HC conversion for OSC E (1.91 g/l Pd, 0.21 g/l Rh)

For both catalysts, HC conversion was better under a cycled aging condition relative to air aging. Under severe oxidizing conditions, Rh is known to be deactivated due to reaction with alumina, and a severe reducing condition can reverse this deactivation (8). As a result, the higher HC conversions noted following the cycled agings may result from greater contribution of the Rh to HC performance. It should be noted that typical procedures for OBD-II agings involve a preconditioning following the air aging, and this preconditioning can regen-erate the activity of the Rh.

The radically differing effect of the cycled aging on the OSC of the catalysts may be explained by the stability of the OSC materials. Component E is a stabilized ceria which has been shown to be very stable in cycled stoichi-ometries (4), while component B has been shown to be less stable. These results demonstrate that the OSC component can have a large effect on the durability of the catalysts under varying aging environments, and could potentially affect the validity of using air agings to pro-duce catalysts for use in OBD-II calibrations. However, the results also show that the more advanced OSC com-ponents are unaffected by aging environment.

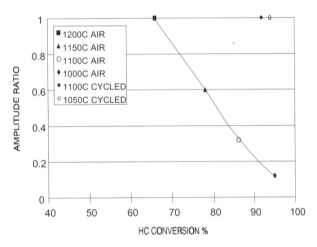

Figure 11. Effect of aging conditions on relation between amplitude ratio and HC conversion for OSC B (1.91 g/l Pd, 0.21 g/l Rh)

D. SULFUR EFFECTS – A major issue regarding the accuracy of OBD-II systems in accurately detecting fail-ures in catalyst performance revolves around vari-ability in fuel sulfur levels. For severely deacti-vated catalysts near the failure threshold, it has been demonstrated that high sulfur levels can severely inhibit the OSC and the HC perfor-mance of the catalyst (5,9). As a result, an OBD-II system calibrated in clear fuel could potentially trigger a false failure in the presence of high sulfur.

The effect of sulfur was studied by measuring the HC conversions and amplitude ratios in the presence of 70 ppm SO_2 after allowing for 30 minutes of equilibration. This corresponds to a fuel sulfur level of approx-imately 1000 ppm. Blank experiments showed that the HEGO response itself was not affected by sulfur. The recovery from the effects of the sulfur exposure was also evaluated

by removing sulfur from the gas stream, and measuring the performance after 30 minutes. Figure 12 shows the effect of sulfur on the relationship between HC conver-sion and amplitude ratio for a Pt-Rh catalyst (0.61 g/l Pt, 0.15 g/l Rh) containing OSC component A. The data shows that, for a given aging condition, the OSC is much more severely degraded than the HC conversion. In addition, it was observed that removing sulfur from the gas stream did not result in a recovery of OSC. This sug-gests that, consistent with previous observations (5), the presence of high fuel sulfur levels could cause the OBD-II system to trigger a false failure.

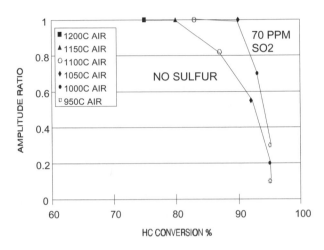

Figure 12. Effect of sulfur on relation between amplitude ratio and HC conversion for OSC A (0.61 g/l Pt, 0.15 g/l Rh)

Previous studies have shown that sulfur can potentially have a much greater effect on the performance of Pd-only catalysts relative to Pt-Rh catalysts (10). In addition, the nature of the OSC component and the inclusion of promoters in the washcoat can have an influence on the relative effect of sulfur (4). To investigate the potential ramifications for OBD-II, the effects of sulfur were deter-mined for two Pd-only catalysts. The first contained OSC component C along with an La promoter, which is com-monly used to promote the activity of Pd-only catalysts (11). The second catalyst contained OSC component E with an alternate promoter package. In a previous study (4), this catalyst was shown to be less affected by sulfur. Both catalysts contained 3.88 g/l of Pd.

Figure 13 shows the effect of sulfur on the La-promoted catalyst containing OSC compo-nent C. In contrast to the results for the Pt-Rh catalyst, both the OSC and the HC conversion are severely degraded by the presence of sulfur for a given aging condition. Similar to the results for the Pt-Rh catalyst, the sulfur poisoning was also not reversible. The data also shows that sulfur deactivates the OSC more than the HC conversion - the relation between OSC and HC conversion shifts so that, for a given HC conversion, a higher amplitude ratio will be measured in high sulfur. This indicates the potential for a false OBD-II failure in the presence of high sulfur levels.

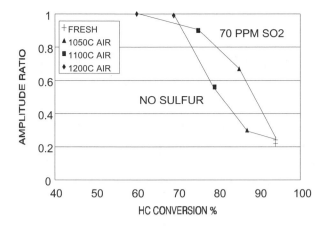

Figure 13. Effect of sulfur on relation between amplitude ratio and HC conversion for OSC C (1.91 g/l Pd, 0.21 g/l Rh)

Figure 14 shows the effect of sulfur on the Pd-only catalyst containing OSC component E with an alternate promoter package. Sulfur also has a negative impact on both the HC con-version and OSC of this catalyst. However, comparison with Figure 13 shows that this catalyst is less affected by sulfur, and the effect of sulfur on the catalyst was largely reversible in agreement with previous studies (4). In contrast to the other catalysts studied, the sulfur results in a greater deactivation in HC conversion relative to OSC - for a given HC conversion, a lower amplitude ratio is measured in high sulfur. This would be expected to minimize the potential for a false OBD-II failure due to high sulfur fuel. To the contrary, this could potentially result in a false passage as seen in other studies (12).

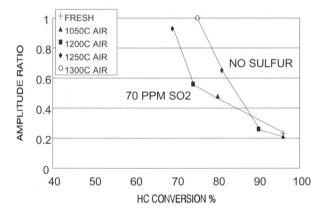

Figure 14. Effect of sulfur on relation between amplitude ratio and HC conversion for OSC E (1.91 g/l Pd, 0.21 g/l Rh)

SUMMARY

The relationship between OSC and HC performance can be affected by altering the catalyst formulation variables. For Pd-only catalysts, OSC can be altered without a large impact on HC conversion. The inclusion of an OSC component with the desired thermal durability can potentially have a greater impact on this relationship than altering the quantity of OSC for a given component. The nature of the OSC component can also impact the effect of sulfur on the OBD-II response of the system. Depending on the properties of the OSC component, the aging environment may alter the relationship between the deactivation in OSC and HC conversion.

REFERENCES

1. W.B. Clemmens, M.A. Sabourin, T. Rao, SAE Paper 900062, 1990.
2. J. Hepburn, T. Ghanko, G. Zimlich, SAE Paper 972847, 1997.
3. R.J. Brisley, G.R. Chandler, H.R. Jones, P.J. Andersen, P.J. Shady, SAE Paper 950259, 1995.
4. P.J. Andersen, J.S. Rieck, SAE Paper 970739, 1997.
5. J.S. Hepburn, D.A. Dobson, C.P. Hubbard, S.O. Guidberg, E. Thanasiu, W.L. Watkins, B.D. Burns, H.S. Gandhi, SAE Paper 942057, 1994.
6. G.J.J. Bartley, P.J. Shady, M.J. D'Aniello,Jr., G.R. Chandler, R.J. Brisley, D.E. Webster, SAE Paper 930076,
7. J. Hepburn, T. Chanko, J. McKenzie, R. Jerger, and D. Dobson, SAE Paper 972853, 1997.
8. K. Otto, W.B. Williamson, H.S. Gandhi, Ler. Eng. Sci. Proc., 2, (1981).
9. D.D. Beck, T.W. Silvis, and S.T. Mahan, SAE Paper 941057, 1994.
10. J.E. Thoss, J.S. Rieck, and C.J. Bennett, SAE Paper 970737, 1997.
11. H. Muraki, H. Shinjoh, H. Sobukawa, K. Yokota, Y. Fujitani, Ind. Eng. Chem. Prod. Res. Dev., 25, p. 202 (1986).
12. J. Hepburn, M. Sweppy, Z. Zaghati, SAE Paper 972855, 1997.

APPENDIX

Derivation of oxygen storage figure from test data

Mass of oxygen stored = $\int DO_2 \, dt$ over one lean half-cycle, (where DO_2 is the difference in mass flow rate of O_2 from catalyst inlet to outlet).

But DO_2 =

mass flow$_{exh}$ x [(vol% O_2 in feed gas - vol% O_2 in tailpipe gas)/100] x mol wt O_2/mol wt exh.

Empirically, variation in vol% O_2 is approximately linear with sensor voltage from lambda = 1 to 1.05, and is approximately 0.11 V per vol%O_2 in this range, taking into account also that the analyser measurement is of dry gases.

So DO_2 = mass flow$_{exh}$ x (Volts$_{feed\ gas\ sensor}$ - Volts$_{tailpipe\ sensor}$)/11 x 32/28.8, (taking molecular weight of exhaust gas to be 28.8 and of O_2 to be 32) and hence

Mass of oxygen stored (kg) =

mass flow$_{exh}$ (kg.s^{-1}) x $\int$(Volts$_{feed\ gas\ sensor}$ - Volts$_{tailpipe\ sensor}$)dt x 0.101

Or mass of oxygen stored (mg) =

mass flow$_{exh}$ (kg.h^{-1}) x $\int$(Volts$_{feed\ gas\ sensor}$ - Volts$_{tailpipe\ sensor}$)dt x 0.101 x 10^6/3600 =

mass flow$_{exh}$ (kg.h^{-1}) x $\int$(Volts$_{feed\ gas\ sensor}$ - Volts$_{tailpipe\ sensor}$)dt x 28.05.

Development and Benchmarking of Leak Detection Methods for Automobile Evaporation Control Systems to Meet OBDII Emission Requirements

Paul D. Perry and J. P. Gilles Delaire
Siemens Automotive Ltd.

Copyright © 1998 Society of Automotive Engineers, Inc.

ABSTRACT

This paper describes the development and benchmarking of two 'automobile fuel tank evaporation control system' leak detection methods, which include 1) Positive pressure decay and 2) Negative pressure decay. In the past, negative pressure decay was the least expensive method that met the current OBDII requirement for a 1.0mm leak but these systems exhibit deficiencies when attempting a 0.5mm leak test. Positive pressure systems overcome most of these deficiencies but respond too slowly for evaporative control strategies of the future. Testing was done to compare the ability of each system to detect a 0.5mm fuel tank leak under various environmental conditions. It was found that both systems exhibit similar leak detection capability if a specific degree of stability is attained with respect to tank pressure.

INTRODUCTION

Development of leak detection systems at Siemens started in 1992. The first system developed was a vacuum system. Subsequently, several positive pressure systems were developed. The chronology of products discussed here include:

1. VLDS = Vacuum Leak Detection System
2. LDP1.0 = Leak Detection Pump (vacuum activated)
3. LDPxHF = High Flow Pressure Leak Detection
4. VLDSII = Second generation Vacuum System
5. LDPxST = Stabilized Pressure Leak Detection

The Vacuum Leak Detection System was the earliest attempt at leak detection and is widely used today in the auto industry. This system involves evacuating the fuel tank and measuring the time required for pressure to rise as a result of a leak, hence the terms 'negative pressure decay' or 'vacuum decay'. This system is more economical to implement than current 'positive pressure' systems but smaller leaks are more difficult to detect with this system due to limitations in transducer sensitivity. False leak detection can also occur because evaporating fuel will increase vacuum decay. It is, however, a relatively quick test. Typical test times are under 30 seconds.

Another characteristic of a vacuum leak detection system is that during the test period, fresh air is being drawn in through any leak present in the system. The system is therefore considered to be more environmentally friendly.

Positive pressure decay systems, on the other hand, are less prone to the false leak detection caused by fuel vapor pressure. Excessive fuel vapor pressure in the fuel tank will cause a positive pressure decay system to underestimate the size of a leak.

The first commonly employed positive pressure system was LDP1.0, which contains a diaphragm pressure pump, a mechanical vent valve and a reed switch feedback mechanism. The test time for this system was relatively long at about 3 minutes.

When OBDII legislation introduced the necessity for a 0.020" leak test, it was felt that some effort was needed to develop a second generation system, which would address the three main aspects of interest in this science, namely "cost", "speed" and "accuracy".

A new system arising from this research effort was designated 'LDPx'. The LDPx concept utilized an electric turbine pump to pressurize the fuel tank, a differential pressure switch to detect pressure decay and a canister vent solenoid (CVS). Test times for this system were typically less than 30 seconds.

In order to compare the performance of vacuum and pressure systems for 0.5mm leak detection, a Benchmark Study was initiated. The Benchmark Study included hot and cold environmental chamber testing, and a road trip to Arizona (Hot Trip) in June 1997 to do field-testing.

The VLDSII and LDPxST systems were developed as a result of the Benchmark Study.

LEGISLATION

In 1994, the California Air Resources Board (CARB) and the U.S. Environmental Protection Agency (EPA) enacted OBDII, a set of on-board diagnostics to detect emission-related malfunctions on selected vehicle models. Within OBDII, the 'evaporative system leak detection' section called for detection of a 1.0mm diameter leak by 1996. The EPA is currently evaluating its OBDII requirements with respect to a further reduction of the leak threshold to 0.5mm. Current dialog suggests that the 'Green States' and California will be required to comply with a 0.5mm evaporative leak test by the year 2000.

DESCRIPTION OF 3 SYSTEMS

VACUUM LEAK DETECTION SYSTEM (VLDS) – The VLDS system uses a transducer to measure fuel tank pressure. The transducer consists of a piezo-resistive element with a range of +/- 380mm H_2O. Figure 1 is a schematic diagram of this system. Figure 2 shows the operating cycle, which consists of the following sequence.

At time zero, the canister vent solenoid (CVS) is closed and an initial pressure reading P_A is made. A period of time t_1 passes before a second pressure reading P_B is made. The difference

$$dP_G = P_A - P_B \qquad (Eq. 1)$$

is an indication of fuel vapor activity. At this point, duty cycle purge is initiated and held constant until a target vacuum is reached. The purge is then ramped down. A short period of time is needed for the system to stabilize before recording the pressure P_C. Time t_2, which is equal to t_1, passes before measuring P_D. The difference

$$dP_{INC} = P_D - P_C \qquad (Eq. 2)$$

is the primary leak information. The diagnostic output can now be calculated as

$$dP_{DIAG} = dP_{INC} - dP_G \qquad (Eq. 3)$$

which corrects for fuel vapor pressure. This diagnostic output must also be corrected for fuel tank volume using fuel gage information.

It is important to note that the vapor pressure check also measures ambient pressure change (i.e.: during up or down hill grades), since the tank transducer is referenced to atmospheric pressure.

LEAK DETECTION PUMP (LDP 1.0) – Figure 3 shows a simplified cross section drawing of the LDP 1.0. Figure 4 is a schematic diagram of the LDP 1.0 system. The LDP 1.0 is a spring and vacuum powered diaphragm pump, which is energized by engine vacuum under electrical control to pressurize the evaporative emission system for leak detection.

The pump cycle is started by a short pulse of engine vacuum, which raises the diaphragm and compresses the spring, drawing air into the volume under the diaphragm through a check valve. Simultaneously, the integral canister vent valve is closed. At the end of the pulse, the vacuum is released and the force of the spring pushes the diaphragm down, pumping the air into the evaporative emission system through a second check valve. As the diaphragm reaches the end of the down stroke, a reed switch assembly provides an electrical timing signal to the control system, which re-initiates the cycle.

If there is no leakage in the system, the pressure rises after a number of cycles to the point at which the spring load and the system pressure are in balance, and the pumping cycle stops. If there is a leak, the cycling time or "pulse duration" stabilizes at a rate, which replaces the leakage loss. The pulse duration is measured by the control system, which determines whether or not the leak exceeds a certain threshold.

It should be noted that the LDP 1.0 could detect leaks as small as 0.25 mm. In order to have reasonable capability it is necessary to do the leak test with the vehicle stationary. This will avoid inaccuracies due to fuel slosh, elevation changes, etc.

In addition, small leaks cause the LDP 1.0 to cycle very slowly, which in turn results in relatively long test duration. To compensate for the extra time, attempts were made to optimize the charge phase. Two approaches included:

1. Modification of the pump cycle duration.
2. Development of an "early detection" algorithm.

1.Pump Cycle – Two changes were explored in the pressurization cycle of the LDP. In the first of these, knowledge of the manifold vacuum level was used to modify the time allowed for the diaphragm to reset. By minimizing the reset time, especially when manifold vacuum was high, the pump cycle was optimized and pressure was established faster.

In the second development, the pump return stroke delay was progressively decreased as the fuel tank pressure increased.

Both of these changes provided a faster charge time and therefore, shorter test duration.

2.Early Detection Algorithm – Here, an attempt was made to predict the final pump pulse duration using a 3-point fit of a modified logistics curve based on the response of the LDP. The first data point was the minimum pulse duration experienced during initial pumping. The following two points were obtained by measuring the pulse duration at two fixed subsequent time intervals.

The resulting pulse duration could be predicted with an error of less than 15% within 100 seconds. This was a significant decrease over the previous test time of 3 minutes.

ADVANCED LEAK DETECTION (LDPX) – In this system the fuel tank was pressurized to a specified level (200mmH$_2$O), then the pump flow was interrupted and the pressure decay was measured using a differential pressure switch.

Various methods were employed to accomplish the task of gagging the system pressure decay. One method, known as LDPxPRESS used a combination of high and low flow vent valves to 'modulate' the pressure in the tank (see figure 5a). By alternately charging and discharging the pressure across the control points of the differential switch, a duty cycle, proportional to system leak, could be calculated. Since this method is truly ratiometric in nature, the resulting duty cycle was independent of system volume. While this method was largely successful, the method proved to be too slow for practical application.

A much faster method was designated as 'LDPxHF' (see figures 5b and 6). This technique resulted in a single pressure decay measurement after suitable conditioning of the fuel tank.

Figure 5b shows the schematic diagram of this system. The cycle is illustrated in figure 6. The objective was to condition the air in the fuel tank headspace by holding the pressure at a given level for a period of time. The pressure decay rate was then measured with the CVS closed and the blower off.

First, the tank was pressurized to 200mm H$_2$O as indicated by the differential pressure switch. It was then over-pressurized for a constant time t_A. The pump was then de-energized and the pressure allowed to drop down to the lower (190mm H$_2$O) limit. At this point the pump was again started and the pressure rose to the upper limit again. A time t_1 was then recorded. t_1 was considered to be a reasonable indicator of system volume. The pressure was then allowed to over-shoot again for a time period t_B, which was proportional to t_1.

This proportionality ensures that the tank volume is taken into account when estimating the over-pressurization required for system stability before starting the test. The pump was stopped again and pressure decay time was monitored. The time taken for the pressure to decay from the upper limit to the lower limit t_2 establishes the leakage rate. A pass/fail time threshold was calculated using the t_1 volume indicator.

BENCHMARK STUDY

The purpose of the Benchmark Study was to compare vacuum methods with pressure methods in order to try and recommend a system for 0.5mm leak detection. The main areas of interest were:

- Static vs. dynamic capability of each system.
- Effect of system volume.

- Effect of altitude and road grade.
- Effect of temperature.
- Effect of leak location.

To address worst-case real-world conditions, the majority of the field tests were conducted during mid-summer conditions in Phoenix Arizona (Hot Trip).

ENVIRONMENTAL CHAMBER TESTING – Chamber testing was conducted at -10° and 0° Celsius to determine the effect of temperature on the components of the leak detection systems. A vehicle test chamber was used to ensure that the entire system was subjected to the same conditions. The vehicle was preconditioned for 10 hours at temperature before test initiation.

The effects of temperature on calibration of the vacuum sensor and pressure switches were of particular interest. The electric air pumps used in the leak detection systems can also experience performance shifts at extreme temperatures. These effects were important for sensitivity and accuracy of the leak detection system.

HOT TRIP TESTING

While bench results are relatively easy to obtain repeatedly, real-world tests are prone to error because environmental conditions cannot be controlled to the same extent. Considerable effort was required to attain statistical validity

To limit the effect of fuel vapor, the test vehicles used on the Hot Trip were fitted with both functioning (production) fuel tanks and non-functioning or auxiliary tanks. The auxiliary tanks were filled with an ethylene glycol mixture. A transfer pump within the auxiliary tank allowed the liquid level to quickly be changed without disturbing the thermal or mechanical equilibrium of the system.

During the study, testing was also performed on the production evaporative system. It was found however, that only the first few tests of the day were valid due to progressive heating of the evaporative system. These results were not included in the statistical evaluation of each system, but were used as supportive data.

In general, 10 repetitions (minimum) were made under each condition to ensure statistical validity. A time of approximately five minutes was allowed between diagnostic tests to allow the evaporative system to settle.

RESULTS OF THE HOT TRIP

The most significant results of the hot trip are shown in figures 7 through 10, which contain the results for the VLDS and LDPxHF systems. These figures compare the ability of the two systems to distinguish between leak sizes with different fuel tank volumes.

Figure 7 shows the average vacuum decay of the VLDS system for 3 tank volumes and 5 leak orifice diameters. The vacuum decay for a given leak diameter increased with fuel tank volume. Correction was required for tank volume when estimating leak size. The most striking characteristic was the shallow slope of the curve. The shallow slope, combined with the wide data scatter in figure 8, made it difficult to distinguish the 0.5mm leak from the others. Figure 8 shows the cumulative spread of all test data for the same 5 leak orifice diameters.

Figure 9 is the LDPxHF equivalent of figure 7. The abscissa units in figures 7 and 9 reflect the nature of each system. It was found that the LPDxHF system had greater sensitivity to small leak sizes as a result of the increased slope of the curves. The scatter for the LDPxHF cumulative data of figure 10 shows more separation between the 0.25mm data group and the 0.5mm data group than found in figure 8. The curves for larger volumes would have higher decay times if full stabilization had been reached.

SUMMARY OF THE BENCHMARK STUDY

VLDS

Major Findings

- VLDS system meets time criteria
- VLDS not sensitive enough for 0.5mm leak detection.
- Some false leak detection occurred in a sealed system due to evaporation of fuel.

Static vs. Dynamic

- Leak detection of 0.5mm is not practical even in static conditions
- Leak detection of 1.0mm is feasible under dynamic conditions.

Temperature

- No effect.

Front vs. Rear Leak

- No effect.

Altitude

- Increased altitude reduces signal resolution.
- During a dynamic test the t1 measurement can compensate for a steady grade (up or down).

LDPXHF

Major Findings

- LDPxHF system meets time criteria
- LDPxHF is sensitive enough for 0.5mm leak detection.
- Some false leak detection occurs in a sealed system due to lack of system stability.

Static vs. Dynamic

- Leak detection of 1.0mm is feasible under both static and dynamic conditions.

- Leak detection of 0.5mm is feasible under static but not dynamic conditions.

Temperature

- No effect.

Front vs. Rear Leak

- No effect.

Altitude

- No effect in static mode.
- Grades will cause inaccuracy in dynamic mode.

POST HOT TRIP DEVELOPMENT

The hot trip data revealed that some decay characteristic exists in a vehicle fuel system, even when the system is essentially 'sealed'. This characteristic was experienced with both the pressure and vacuum systems. It was observed that the longer the fuel tank was held at pressure, the more stable the leak-down values became.

The physics behind this stabilization phenomenon are not understood at this point but investigations are underway to mathematically isolate thermal effects, which are suspected of being the cause.

In order to evaluate system stability, a series of bench tests were conducted to determine precisely how much time must be allowed for a vehicle fuel tank to fully stabilize.

Figure 11 summarizes these findings for different leak sizes. For example with the 0.25mm orifice, approximately 120 seconds of 'stabilization hold time' was required in order to obtain repeatable or stable decay times, while with the 0.375mm and 0.5mm leaks, it took only about 40 or 50 seconds. This suggests that the longer the pressure is held, the more separation is evident in the decay values and hence, the more sensitive the system.

Using this information the vacuum and pressure systems were modified to become VLDSII and LDPxHF, respectively.

VLDSII – Figure 12 shows the VLDSII system schematic. A vacuum regulator was added to simplify the task of achieving a stable test pressure. Figure 13 shows the VLDSII test cycle, which is described as follows.

After starting the engine and achieving closed loop fuel control, the Canister Vent Solenoid (CVS) was closed to seal the fuel tank. The vacuum regulator was now active. As normal duty cycle purge was started, the fuel tank was held at a constant vacuum (perhaps 250mm H_2O) by the regulator. The system was allowed to stay in this state for t_{S1} seconds while the system stabilized. This phase may be either static or dynamic, but must be of a minimum duration to ensure stability. It was found that typically a stabilization time of 120 seconds was required. When the vehicle ECU determined that the vehicle had come to a stop an additional time t_{S2} was allowed to give any fuel slosh a chance settle. At this point the CVS was opened for a short time ($t_{SEAT} \sim 50$ ms) to drop the vacuum slightly and force the regulator diaphragm to seal. At the same time the purge was discontinued. The vacuum decay was then recorded over duration t_L. The slope (d_1 / t_L), which indicates the leak size is dependent upon the volume of the tank and the rate of vapor being produced by evaporation of the fuel. This necessitated two additional checks to finalize the determination.

System volume was measured by recording the time required for the system to vent with the CVS open (d_2). The vacuum deviation d_2 was used to modify the decay duration t_L to account for system volume.

Once the vacuum was released and the system returned to atmospheric pressure, the fuel vapor activity was measured by closing the CVS for the period t_v and recording the corresponding increase in system pressure. The pressure change d_3 was then used to modify the decay duration t_L to account for vapor pressure or changes in atmospheric pressure during the test period.

Figure 14 shows the scatter of the final data for the VLD-SII. This method shows greater separation between 0.25mm and 0.5mm leaks than the earlier VLDS method (figure 8).

LDPXST – The LDPxST algorithm uses essentially the same hardware as the earlier LDPxHF concept (figure 5b). However, the accuracy of the system was improved by using a 'dual contact' switch, which provided more control of pressure level. The system also used a longer stabilization period.

Figure 15 illustrates the LDPxST test cycle. After starting the engine and achieving closed loop fuel control, the pump was activated to bring the system up to pressure. The motor was then modulated using the upper level (250mm H2O) of the differential pressure switch as feedback to keep the pressure constant. The tank was held in this state for the required 120 seconds (minimum) to ensure stability of the system.

When the vehicle came to a complete stop, and the engine was confirmed to be at idle, the leak detection algorithm was initiated. If at any time the vehicle started to move, the test would drop back into the stabilization mode.

The CVS was then energized to seal the canister vent and the pump was stopped. By eliminating the pressure source, any leak in the CVS seal would also be detected. When the upper switch level was reached the leak-down timer was activated. The resulting leak down time t_L was a measure of system leakage but it is dependent on system volume.

To decrease the test duration, an extrapolation technique was employed to predict extended t_L times. When time t_L exceeded 10 seconds, purge pulses were introduced to force the tank pressure down to the lower switch point.

The pulse count PC1 was a measure both of remaining pressure and tank volume.

To cancel out the effect of volume, the tank was first re-pressurized for several seconds (ts2) then discharged again across the entire switch range. The resulting pulse count PC2 could then be used in conjunction with PC1 to extrapolate the predicted value of t_{LP} using the formula:

$$t_{LP} = 10 \text{ seconds} \times (PC2/PC2\text{-}PC1) \qquad \text{(Eq. 4)}$$

where t_{LP} was the predicted leak down time.

A final calculation was made to compensate the real or extrapolated leak-down times for tank volume. This was accomplished with PC2, a parameter that is essentially a measure of system volume. The equation is:

$$t_{LC} = t_{LP} - k_1(PC2\text{-}k_2) \qquad \text{(Eq. 5)}$$

where k1 and k2 are parameters specific to the vehicle system.

Figure 16 shows the scatter of the final data for LDPxST. The separation of the groups of data make it possible to distinguish between the 0.25mm and 0.5mm leaks more clearly than with the LDPxHF (see figure 10).

CONCLUSIONS

1. Prior to establishing the need for system stability, pressure methods performed better than vacuum methods in their ability to reliably detect 0.5mm leakage.

2. After implementing algorithms to exploit the effect of system stability, the new vacuum method (VLDSII) was as sensitive as the new pressure method (LDPxST).

3. Practical methods have been developed to determine and correct for system volume.

4. Vacuum methods are prone to indicate a false leak detection due to fuel evaporation.

5. Both the VLDSII and the LDPxST systems will satisfy OBDII requirements into the next millennium.

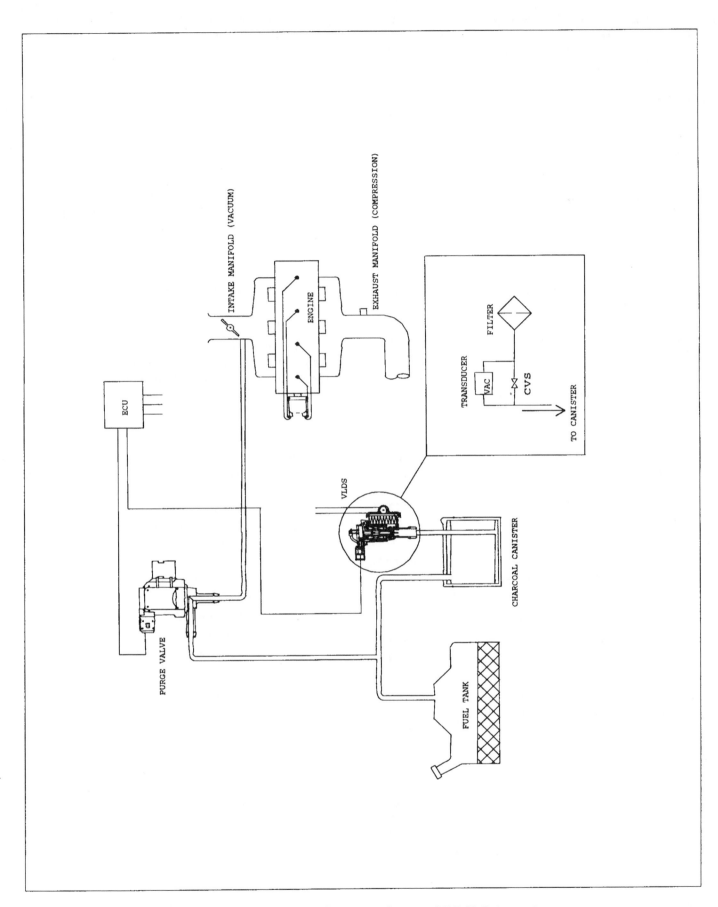

Figure 1. Vacuum Leak Detection System (VLDS) Schematic

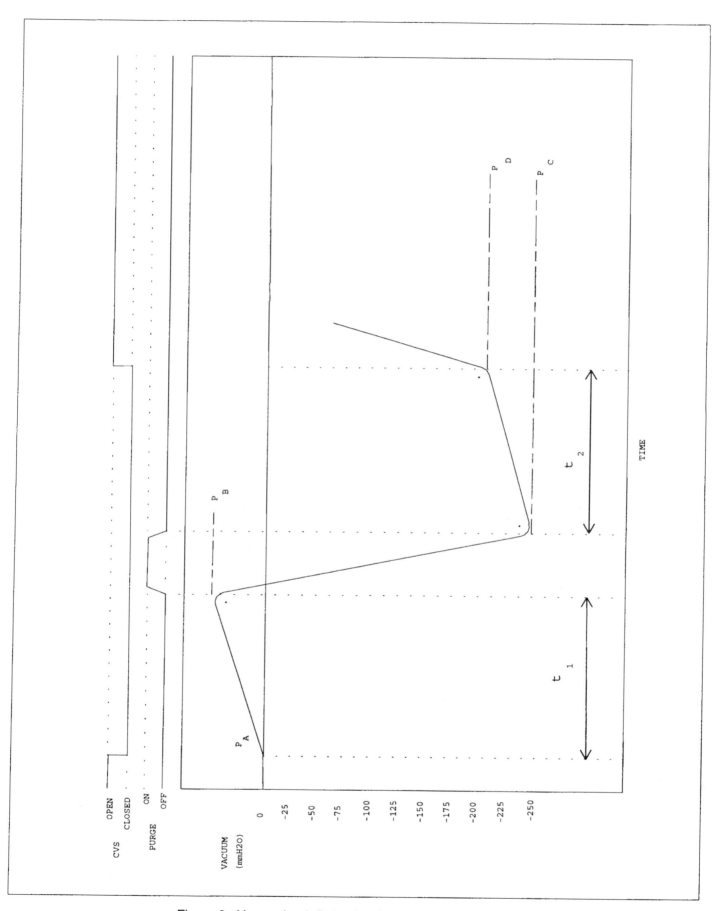

Figure 2. Vapour Leak Detection System (VLDS) Test Cycle

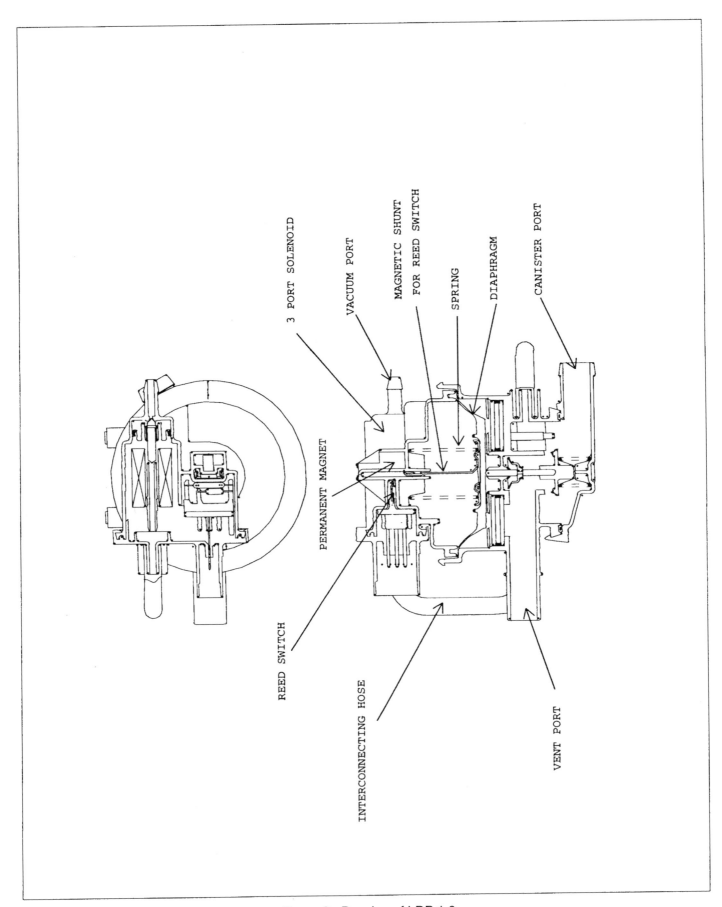

Figure 3. Drawing of LDP 1.0

584

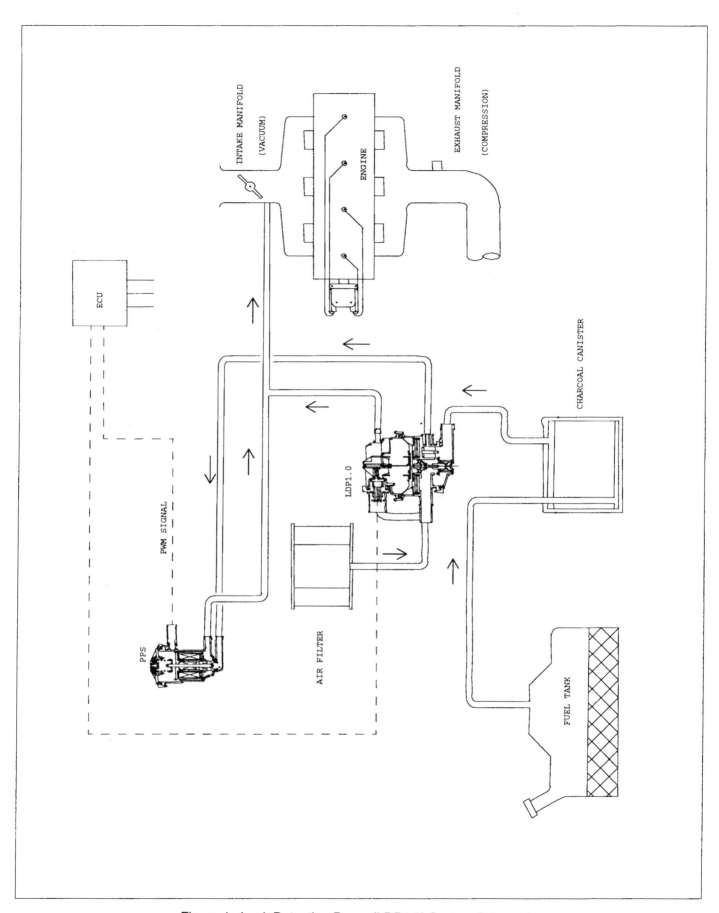

Figure 4. Leak Detection Pump (LDP1.0) System Schematic

585

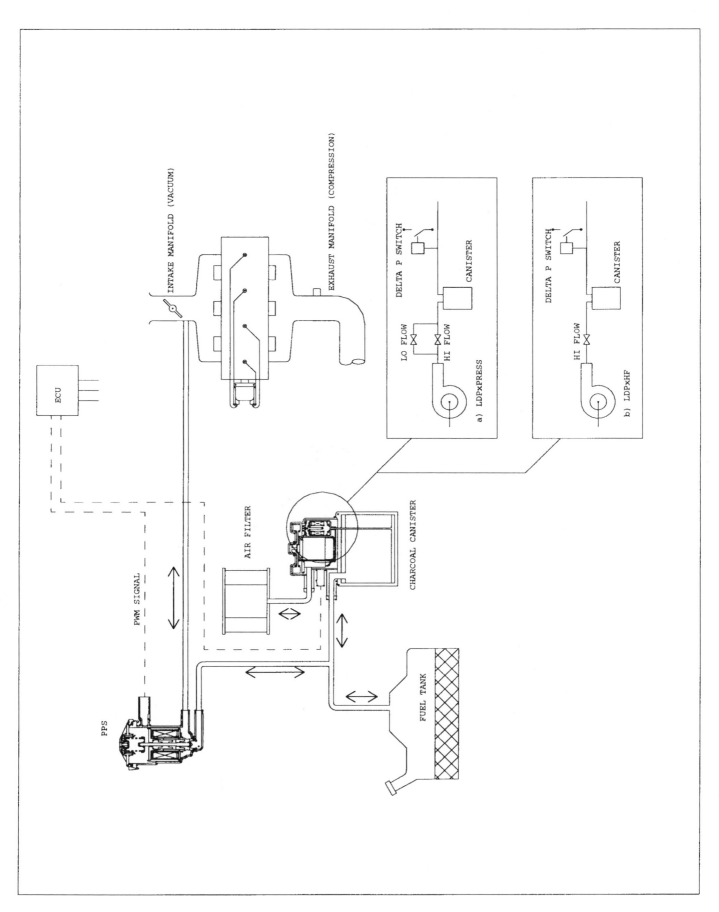

Figure 5. (A&B) - Advanced Leak Detection Pump (LDPx) System Schematic

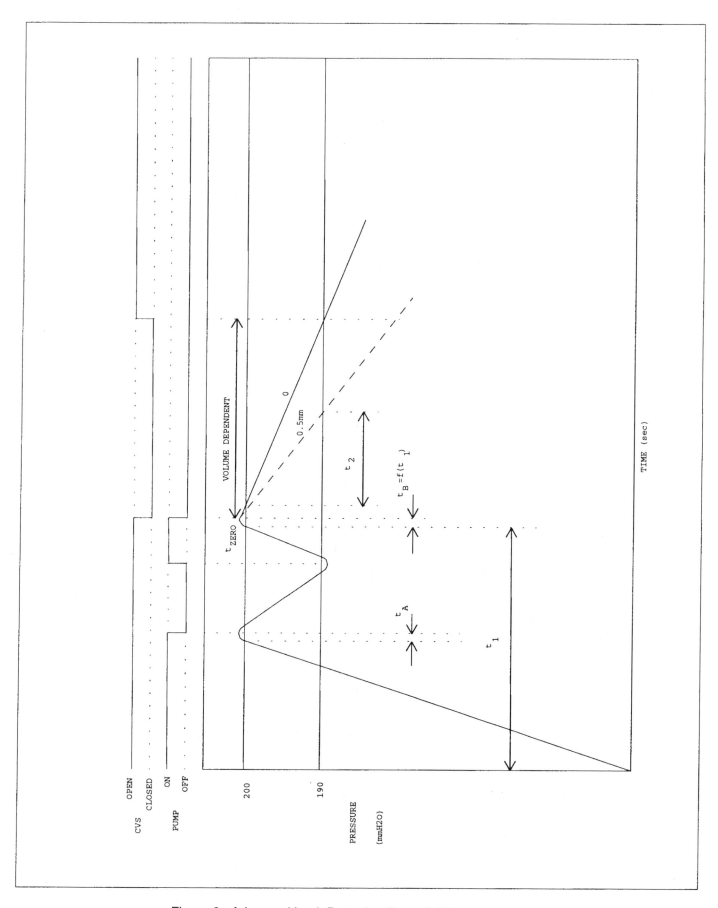

Figure 6. Advanced Leak Detection Pump (LDPxHF) Test Cycle

587

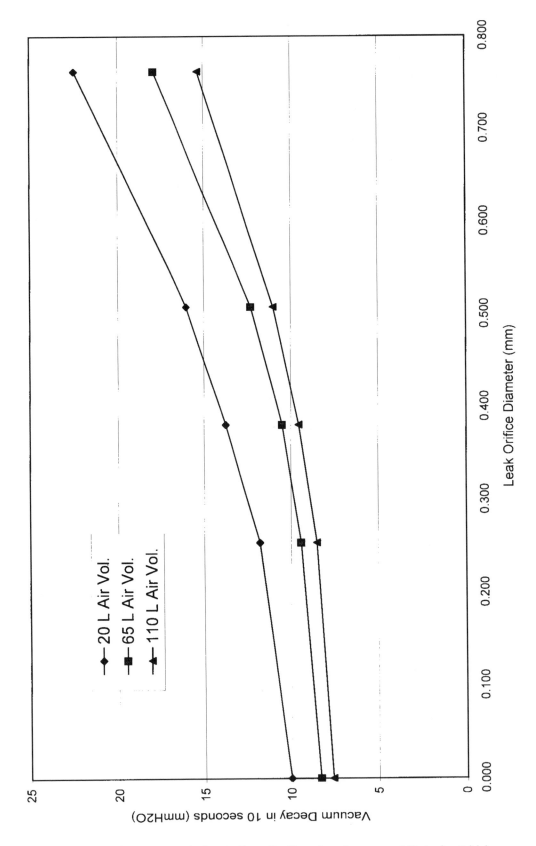

Figure 7. VLDS Benchmark Study Results Showing Average of Data for 3 Volumes

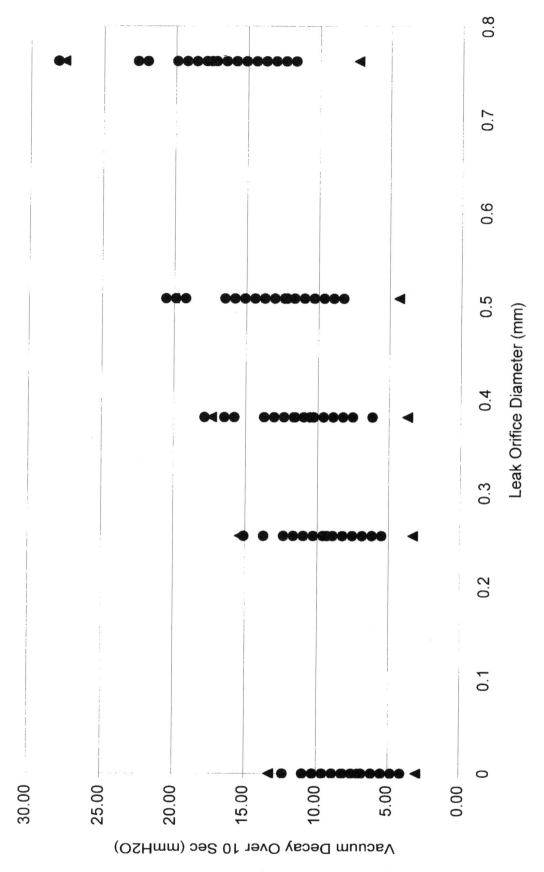

Figure 8. VLDS Benchmark Study Results Showing Spread of Cumulative Data from All Combinations
(upper triangles=+3sigma, lower triangles=-3sigma)

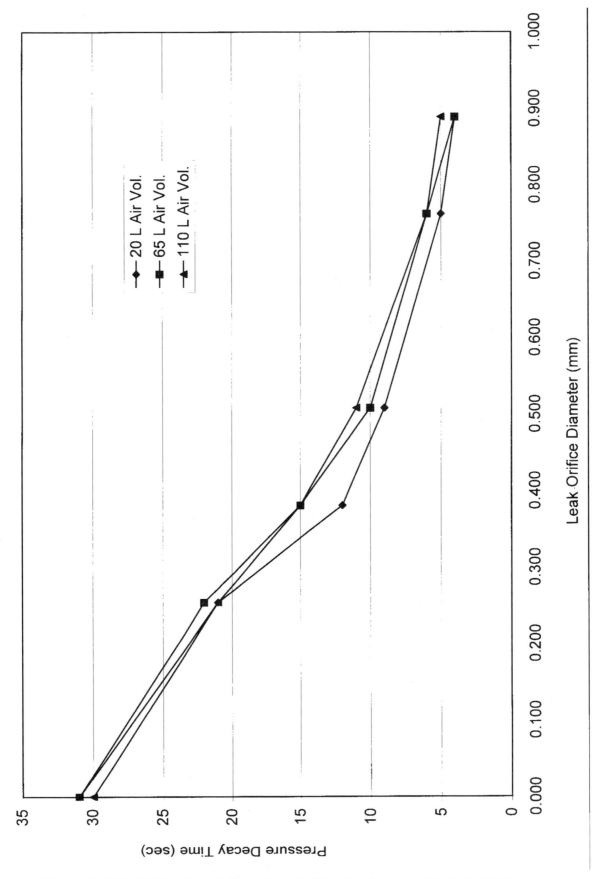

Figure 9. LDPxHF Benchmark Study Results Showing Average of Data for 3 Volumes

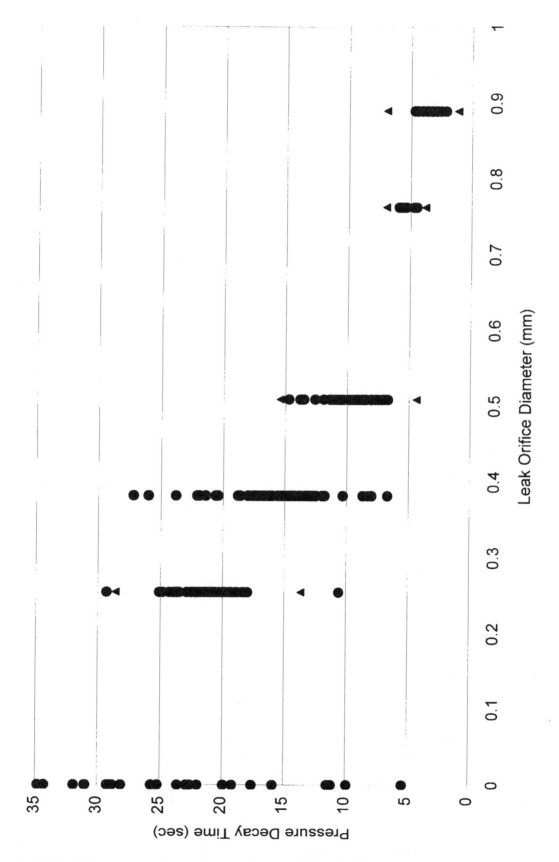

Figure 10. LDPXHF Benchmark Study Results Showing Spread of Cumulative Data from All Combinations
(upper triangles=+3sigma, lower triangles+-3sigma)

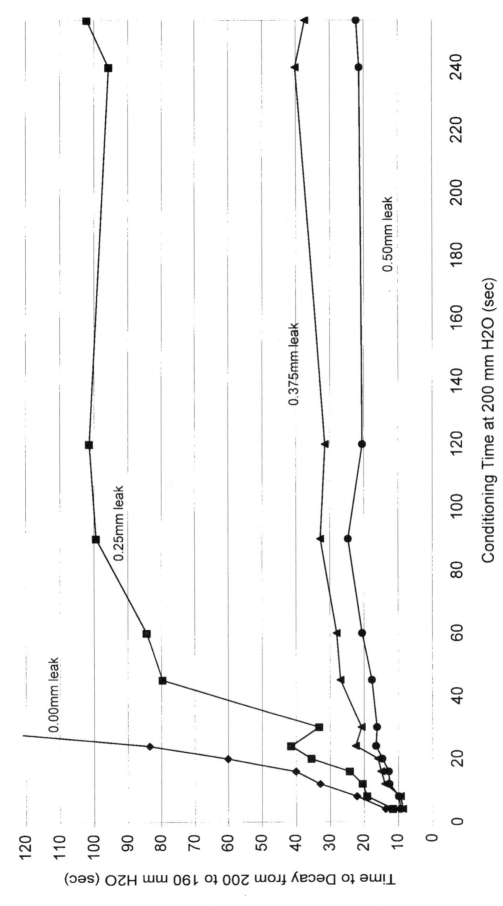

Figure 11. System Stability Study Results 85 L Volume, Static Engine Off

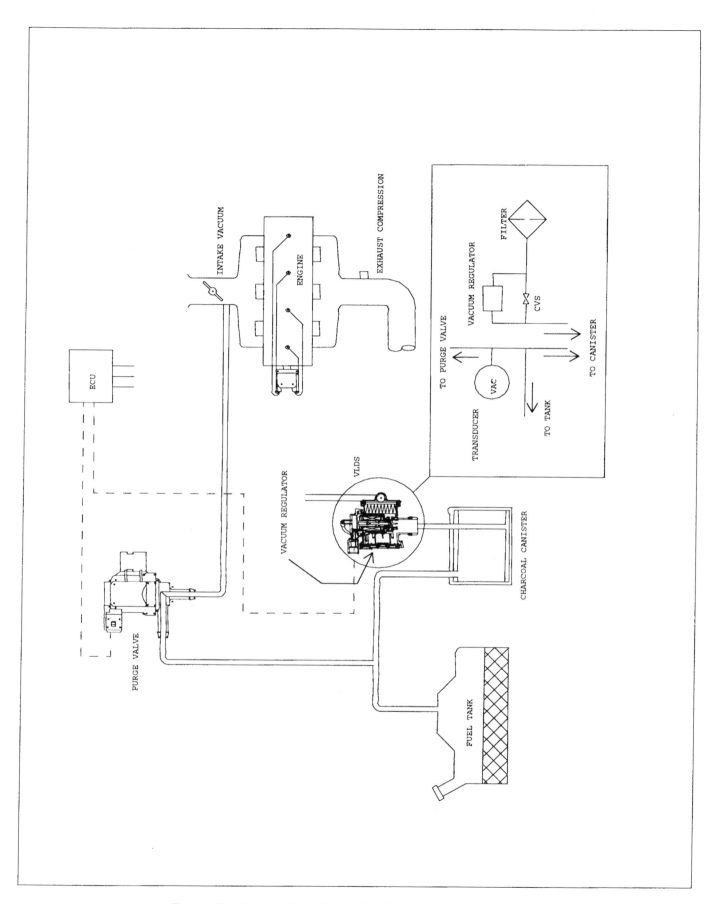

Figure 12. Vacuum Leak Detection System (VLDS) Schematic

593

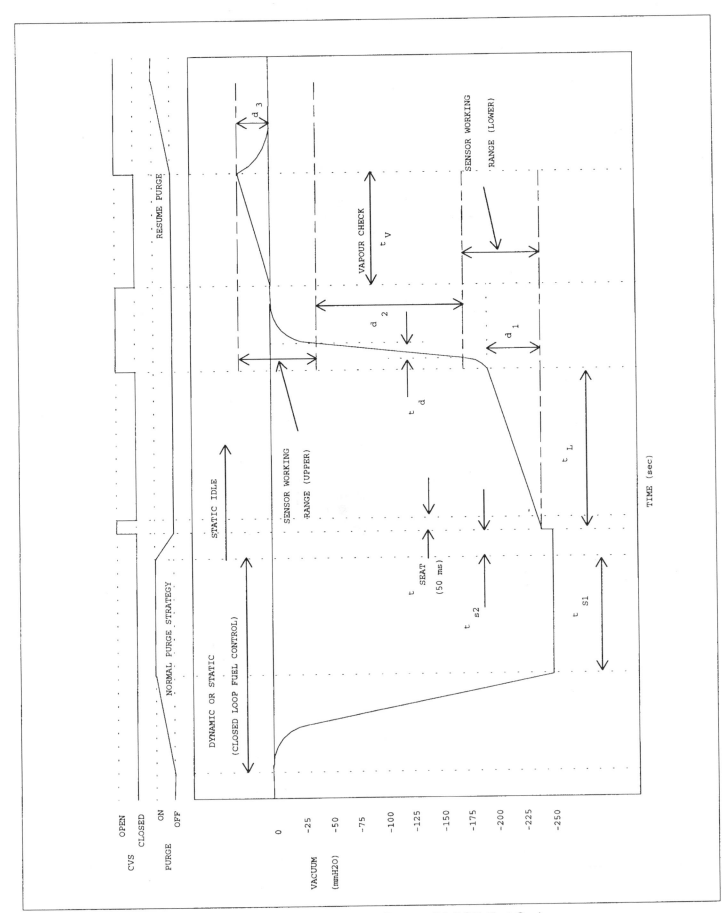

Figure 13. Vapor Leak Detection System (VLDSII) Test Cycle

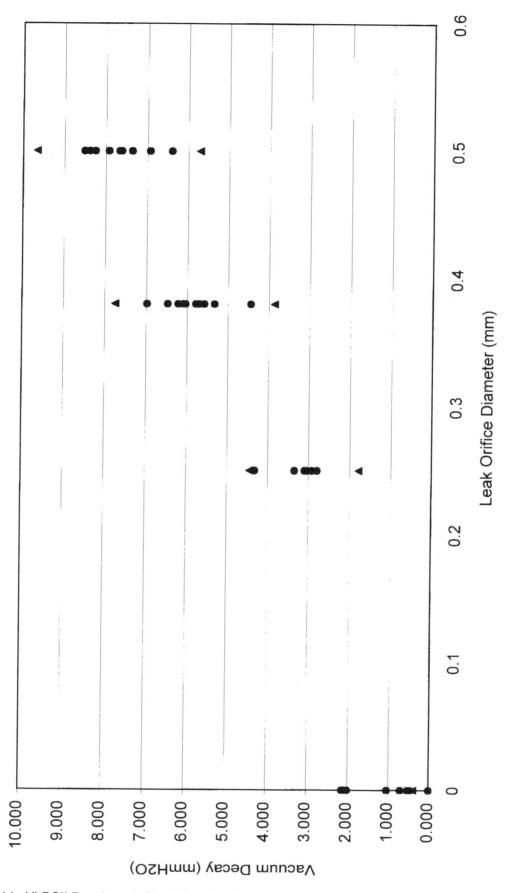

Figure 14. VLDSII Benchmark Study Results Showing Spread of Cumulative Data from all Data
(upper triangles=+3sigma, lower triangles=-3sigma)

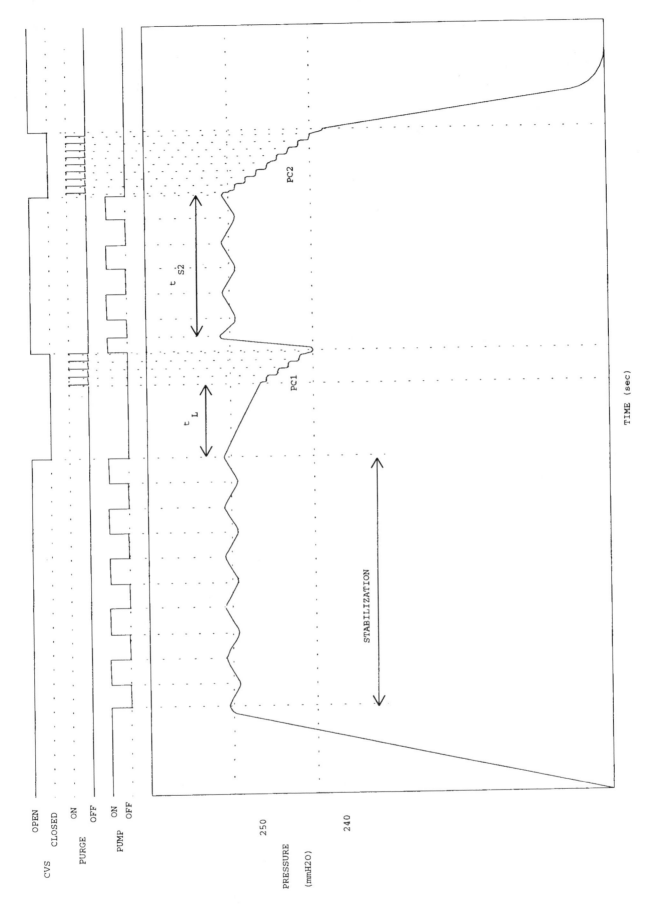

Figure 15. Advanced Leak   n Pump (LDPxST) Test Cycle

596

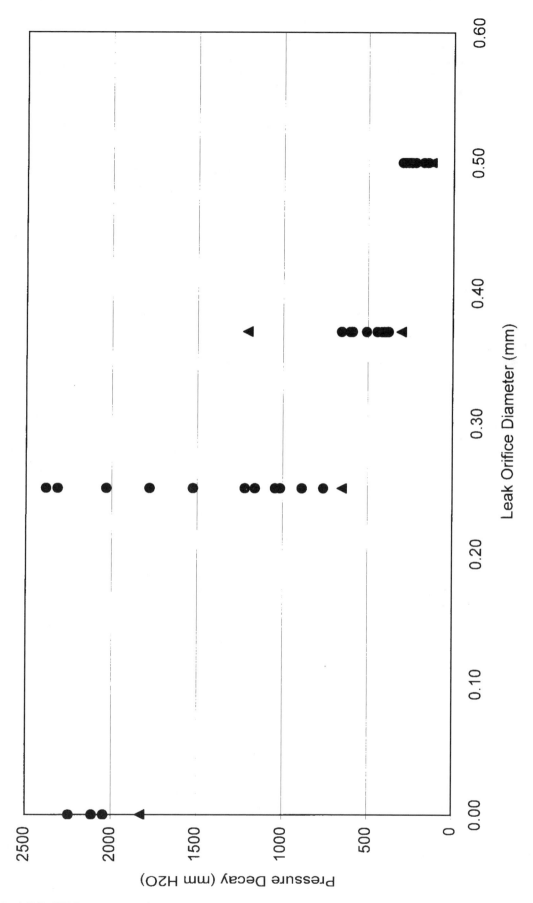

Figure 16. LDPxST Benchmark Study Results Showing Spread of Cumulative Data from all Combinations
(upper triangles=+3sigma, lower triangles=-3sigma)

The Development and Implementation of an Engine Off Natural Vacuum Test for Diagnosing Small Leaks in Evaporative Emissions Systems

Michael DeRonne, Greg Labus, Chad Lehner,
Marc Gonsiorowski, Bill Western and Kevin Wong
General Motors

Copyright © 2003 SAE International

ABSTRACT

This paper discusses an approach to detecting small leaks in an automobile's evaporative emissions systems that is a technique based upon ideal gas laws. It does this by monitoring pressure in the system while the vehicle's engine is off. This low cost solution can be easily implemented on General Motors vehicles using existing components. The topics covered in this paper include details on the background of the problem and the technique, the underlying thermodynamics of the technique, a description of the algorithm, testing and data collection considerations.

INTRODUCTION

Evaporative emission system leak detection has been a growing challenge for the automotive industry since it was mandated by the California Air Resources Board (CARB) in 1996. The introduction of the 0.51mm (0.020") diameter leak detection requirement in 2000 necessitated the creation of new leak detection methods that might be needed for many vehicle applications (large tanks, plastic tanks, etc.). The challenge for a easily integrated, low cost, yet robust leak diagnostic has led General Motors to invent a new approach which is based on the ideal gas law and referred to as the natural vacuum technique.

The fundamentals of the natural vacuum technique are based on the natural gas laws. The natural gas laws dictates that a temperature change for a gas in a sealed system will have a corresponding pressure change. A considerable amount of heat energy is transferred into a vehicles fuel tank while it is operating. When the vehicle is shut down, a change in vapor temperature is observed. Corresponding pressure changes in the vapor space are directly monitored by the diagnostic to make a judgment on the leak integrity of the system. With a 0.51mm (0.020") leak in the system, the amount of pressure change observed is significantly less than that of a sealed system.

The natural vacuum technique is an extremely robust approach to small leak detection. It demonstrates capability across a wide range of tank and platform configurations and lends itself to more common calibrations. It is robust to many of the things that affected traditional leak detection methods such as tank size, tank geometry and flexibility, and fuel slosh. Data shows that the natural vacuum 0.51mm (0.020") leak detection performance shows great potential in terms of meeting the small leak detection regulatory requirements.

BACKGROUND

The system described in this paper is a method to detect small leaks in an automobile's evaporative emissions control system. Many gasoline powered automobiles are equipped with evaporative emissions control systems (EVAP systems). The primary objective of these systems is to prevent hydrocarbon (HC) vapors from escaping out of the vehicle's fuel handling system to the environment. HC vapors are generated as liquid fuel evaporates. Release of these vapors to the environment has a negative impact on ambient air quality. Since the 1996 model year. the CARB regulations required that the on-board diagnostic monitor the EVAP system for leaks with effective areas equivalent to a 0.040" diameter orifice, or larger. Starting with the 2000 model year, the requirement was reduced to a 0.020" diameter orifice.

Automakers have implemented a variety of on-board diagnostic schemes. Perry and Delaire surveyed a number of these schemes in their SAE paper 980043. One popular algorithm is the "vacuum decay method" that GM has used since the 1996 model year. It executes while the engine is running. It uses the engine as a vacuum source and manipulates the EVAP system valves and monitors the response of the pressure sensors in the system. Based on the behavior of the system pressure, the on-board diagnostic can determine whether or not there are any leaks or malfunctions in the system.

An difficulty with this approach is that it is affected by fuel movement motion. Harsh movement of the liquid fuel can cause it to evaporate. The evaporation affects the pressure in the system, which disturbs the results of the leak test. One can limit the test to steady cruise or idle operation; however, this limits how often the test can execute. See SAE paper 1999-01-8061 for a description of how this can be done. Regulations published in 2002 now require certain minimum test frequencies. Thus, this may not be a good solution.

Another issue is fuel tank stiffness. GM's vacuum decay small leak test generates a significant amount of vacuum in the EVAP and fuel handling system (> ~1.25 kPa vacuum) in order to achieve a good signal-to-noise ratio for the test. Vehicles produced in the 1996 model year mostly had fuel tanks made of steel, which was stiff enough to resist deflection under these vacuum levels. However, fuel tank design has been switching to the use of plastic. Plastic is less rigid than steel and thus deflects more under these vacuum levels. The deflection in the tank material translates into a reduced sensitivity for the small leak test.

Lastly, the signal-to-noise ratio of the engine-run small leak test reduces as the fuel tank capacity gets larger. It is more difficult to detect small leaks in a 55 gallon tank than a 15 gallon tank (all else equal).

Given the above mentioned issues with the existing engine-run small leak test along with the more stringent leak detection requirement for the 2000 model year, General Motors needed to develop a new approach that could perform 0.51mm (0.020") leak detection across all of its applications. This investigation lead to an approach known as the engine-off natural vacuum (EONV) method. As indicated by its name, it executes while the engine is off. Thus, the issue of fuel movement is all but eliminated. Also, as the name implies, it generates pressure changes in the EVAP system via the thermodynamics naturally available while the engine is off (as opposed to using the engine an its vacuum source). In addition, inherent to the way that EONV works, the system generates lower vacuum and pressure levels than the engine-run scheme. Thus, it is less susceptible to fuel tank flexibility. Lastly, EONV does not have problems with large fuel tank capacities.

THE HISTORY OF EONV DIAGNOSTICS

Dr. Sam Reddy of General Motors Research Labs first documented the concept of monitoring pressure in the EVAP system during engine-off periods for the purpose of detecting leaks [US Patent #5,263,462, Nov. 1993]. All other schemes subsequent to this invention are based on the fundamental philosophy described in this patent. The embodiments described in the invention were mostly mechanical in nature. It used pressure and temperature switches to capture the occurrences of certain pressure levels and temperature changes. It also used a tank pressure control valve to control the flow of HC vapor from the fuel tank through the canister and eventually out of the canister fresh air vent. Even

inventions, there were issues with its specific implementation that rendered it less useful to current model year vehicles (2003 model year).

Current model year vehicles are different in three significant ways. First, most fuel tanks are now fabricated from plastic as opposed to steel. Steel is a much better thermal conductor than plastic; and, thus allowed for much better heat transfer into and out of the fuel system. Second, many fuel systems have "returnless" fuel delivery systems. These systems do not have the warm fuel re-circulated from the fuel rail at the engine back to the fuel tank. Rather they have either no fuel return line ("demand fuel delivery") or a short re-circulation loop that is contained within the fuel tank itself (e.g. not coming from the fuel rail and thus not being warmed by the engine). The combination of both of the differences described above (and some other changes such as better heat shielding) means that the thermal changes in the fuel system are less pronounced than they were in older model year vehicles. The third significant issue was the tank pressure control valve used in the original patent. This valve would have kept the EVAP and fuel handling system sealed during most of the vehicle soak period. The canister's vent to atmosphere would have only opened when the pressure exceeded the tank pressure control valves blow-off threshold. Thus, if there was a leak in the fuel tank then while pressure was building the HC vapors would be forced out of the leak instead of migrating to the canister where they are adsorbed by treated carbon. This characteristic would have made it difficult for the system to meet future EVAP emissions standards such as those described (LEV2, PZEV and ORVR regulations). In addition, the tank pressure control valve would not have worked well during fueling situations with on-board vapor recovery (ORVR) systems.

Another issue would have been the calibration of the mechanical switches. As described above, the thermal changes are less pronounced in current model year vehicles. Thus, the pressure changes are less pronounced. This requires sensing devices with more precision over smaller pressure ranges which is difficult to do. In addition, the pressure switch threshold would have to be calibrated individually for each vehicle application (or at least for each significantly different fuel handling system architecture). This would have driven a proliferation of part numbers and cost.

The next development in engine-off EVAP system leak diagnostics came in 1994 by Greg Rich of General Motors North American Operations Chassis Center. This scheme was referred to as the "Hot Soak Vacuum" approach. This invention was an enhancement of Dr. Sam Reddy's original work. It included a thermistor in the fuel tank (as opposed to a temperature change switch) for more precise temperature measurement. In also had a pressure/vacuum relief control valve combination which included two mechanical control valves (one for vacuum, one for pressure) and a 3-port solenoid to separately control the backpressure control valve. This system was obviously more complicated and involved more devices (and cost). However, it was more compatible with ORVR systems and better able to meet the stricter evaporative emissions standards.

Still there were two remaining issues with both of the previously described inventions. First, neither had a way to command the canister fresh air vent to the closed position. Thus, vehicles using these schemes would not be compliant with mode $08 portion of the SAE J1979 standard. This portion of the standard was put in place so that EVAP systems could be sealed for testing during a state Inspection and Maintenance test. The second issue was that neither scheme took advantage of an available fuel level input. The fuel level input can be useful for a number of reasons. It can be used to determine the vapor space in the fuel tank, which can help determine what the expected pressure or vacuum change would be. It can also be to disable the system under very low and very high fuel levels when diagnostic results may not be reliable. In addition, it can be used to determine if a refueling event has occurred thus allowing one to abort the test results since they would be unreliable after a refueling event.

The next development in the EONV diagnostics was described in a patent by Reddy et al. [US Patent #6,321,727, Nov. 2001]. The first significant enhancement was the use of a solenoid-controlled valve to control the canister fresh air vent. This allowed more precise control of when the system would be sealed or allowed to vent to atmosphere as the newer algorithm has various stages requiring transitions between sealed and vented. In addition, this allowed the system to be compliant with the mode $08 portion of the SAE J1979 standard.

The next significant enhancement was the use of a precision vacuum/pressure sensor instead of a vacuum switch. This allowed for a more precise measurement and thus a more reliable test. It also allowed the test to monitor for either pressure or vacuum changes. In addition, the sensor could be used to identify abrupt pressure changes associated with refueling events. The test could then be aborted during these unreliable conditions. Lastly, the use of a pressure/vacuum sensor allowed for one pressure sensing part that could be used on all applications (as opposed to the mechanical switches that would require custom adjustments for each significantly different fuel handling system architecture). Any differences could be accommodated via software based calibration changes within a vehicle's controller module.

THE EONV DIAGNOSTIC

The latest development in EONV diagnostics the subject of this paper. Recent developments have yielded two basic benefits: improved robustness and quicker

execution which reduces the diagnostics impact on battery life.

PHYSICAL DESCRIPTION OF THE SYSTEM

EONV System Hardware

General Motors implements the same basic evaporative emissions hardware architecture on all vehicle applications. A major advantage of evolving from a vacuum decay diagnostic was that EONV could be implemented with the existing hardware and only minor wiring changes. Figure 1 shows the major components used to perform EONV.

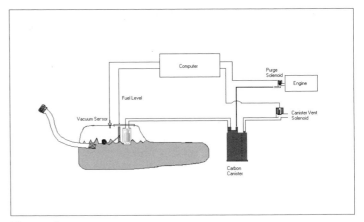

Figure 1

Normally when a vehicle is parked, the fuel tank vapor space is vented to the atmosphere through the carbon canister. The engine burns hydrocarbons captured by the carbon canister by using manifold vacuum to draw the hydrocarbons out of the carbon canister through the purge solenoid. During the EONV diagnostic, the canister vent and purge solenoids seal the system from the atmosphere. The dynamic temperature change and mass flux between the liquid fuel and the vapor space drive pressure change in the system. Very small or no pressure changes indicate that the system is leaking. Larger pressure changes indicate that it is sealed. The vehicle computer measures these pressure changes and processes the results. It also monitors the fuel level sensor and tank pressure sensor for any change that would indicate a refueling event. Once the diagnostic is finished, the controller opens the canister vent and then powers down.

System Thermodynamics

The driving force behind the EONV diagnostic is heat transfer into and work performed on the fuel tank vapor space. Heat transfer can come from several different sources including convection, conduction and radiation. Convection currents account for the most heat transfer into the tank. Sources of convective heat include waste engine and transmission heat, heat from the road, ambient heat, and exhaust heat. Conduction heat sources include heat from the fuel pump, return fuel that has been warmed by the engine, ambient air, and exhaust heat. Radiation energy from the sun can transfer a significant amount of heat to the tank. Work energy comes from the action of the fuel pump on the liquid fuel.

When a vehicle is shut down and parked, the temperature in the fuel vapor space is usually greater than the ambient temperature. This difference is dependent on a number of factors, and is typically in the range of 5 degrees C. In almost all cases the temperature continues to build for some time after the vehicle is shut down due to the loss of convective currents which carry large amounts of heat away from the vehicle. Typically the temperature will peak and begin to decay within the first 30 minutes of the diagnostic; however, the onset of temperature decay has been observed to be as long as three hours. When the system is sealed, the dynamic temperature change inside of the system generates pressure changes that are measured to determine if any leaks are present.

The EONV phenomena can be described using the ideal gas law (Equation 1).

$$P \cdot V = m \cdot r \cdot T$$

Equation (1)

Where P is the pressure, V is vapor space volume, m is the mass in the vapor space, r is the specific gas constant for the vapor mixture, and T is the vapor space temperature.

In differential form, Equation 2 shows the relationship between pressure, mass flux and temperature for a dynamic system at constant volume.

$$\frac{\partial P}{\partial t} = \frac{R}{V}\left[\left(\frac{\partial m}{\partial t}_{\text{fuel phase transformation}} + \frac{\partial m}{\partial t}_{\text{leak}}\right)T + \frac{\partial T}{\partial t}m\right]$$

Equation (2)

The mass flux is broken up into two parts. The first part is a result of phase transformation of the fuel. The second term measures the rate of mass transfer into and out the system when the system is leaking.

Ideally, the diagnostic would have the capability to measure the temperature, pressure, and mass flux due to phase transformation in order to solve for the mass transfer rate due to leaking. Using one dimensional flow calculations, a leak size could then be calculated. With an accurate leak size, the diagnostic could sort out any leak under the 0.51mm (0.020") requirement, and only set a service light for leaks greater than or equal to the 0.51mm (0.020") requirement. Mass flux proved to be difficult to measure or predict on a production vehicle as compared to temperature and pressure. In some cases

the mass flux has little to no effect on the system and there is a direct correlation between temperature and pressure as shown in Figure 2.

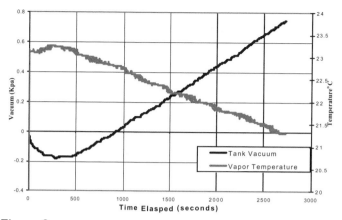

Figure 2

In other cases the mass flux has a noticeable effect on pressure generation as shown in Figure 3.

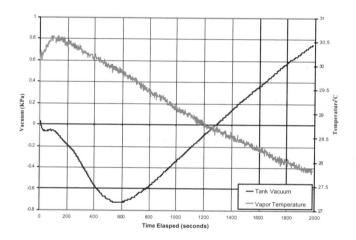

Figure 3

Figure 3 shows a more typical case where the pressure and temperature do not correlate. It can be seen that the pressure lags the temperature trend by almost 400 seconds. In this case the mass flux was still positive (fuel vaporization) after the temperature peak, this overrode the temperature decay, causing continued pressure build. The pressure peak occurs at approximately 600 seconds and indicates the point where mass flux has either stabilized (gone to zero) or has become so small that it is overridden by temperature. From this point on, a direct correlation between temperature and pressure is observed.

The most difficult part about quantifying the effects of fuel volatility is the fact that volatility varies greatly. Winter fuels contain higher quantities of light hydrocarbons (more volatile) opposed to summer fuels that contain lower quantities. Depending when and where a vehicle is fueled, the volatility can change drastically making a prediction almost impossible. This made it virtually impossible to determine if mass flux in the system was the result of volatility or leaks. Since this prevented the diagnostic from being able to calculate a leak size a more empirical approach was taken.

All of the previously mentioned inventions reflect the development work done by General Motors on EONV approaches for small leak detection in EVAP systems. Other developments outside of General Motors include work by Daimler-Chrysler and Siemens. Detailed descriptions of their inventions are outside the scope of this paper. See US patents #6,314,797 B1, #6,073,487 A1 and #6,089,081 for more details on these systems.

DEVELOPMENT OF THE ALGORITHM

Running the diagnostic while the engine is off has it's own unique challenges. Keeping the vehicle computer and the appropriate sensors alive, and energizing the canister vent solenoid can consume up to 1.5 amps. Since the vehicle charging system is not operating, the battery is providing all of the necessary power. To preserve battery life and maintain the necessary state of charge, constraints were placed on the frequency and duration of the EONV diagnostic. The operation was limited to 40 minutes, allowing one complete test per day with no more than two attempts.

Testing in Arizona demonstrated that the original algorithm using fuel vapor temperature was highly influenced by sun/heat load variations a different approach was needed. Test data showed that waiting for the fuel vapor temperature peak detection could extend the diagnostic test time beyond the 40 minute time limit. Also, the vapor pressure often lagged temperature to a degree where temperature was not useful in predicting the onset of a vacuum build.. Thus, the algorithm was changed to allow for a pressure measurement without waiting for system cooling; and total pressure change over both phases of the test is used to determine system integrity.

Both Figures 2 and 3 show an initial pressure build prior to pressure decay. Using the pressure build in addition to vacuum allowed the diagnostic to complete in a timely manner and often times shortened the diagnostic to just a few minutes. This method provided better results with difficult applications such as dual tanks, tanks with packaging issues, and very large tanks; and it allowed the diagnostic to run at higher ambient temperatures. If the test cannot make a decision during the pressure phase, then it continues on to execute the vacuum phase.

Early test data showed that the pressure generated during the test was influenced by different volatility fuels with winter grade fuel being the worst case. This data also showed that winter grade fuel at higher ambient temperatures could build considerable amounts of pressure in the system with the canister vent open. An algorithm was developed to qual

volatility at the beginning of the EONV test. The volatility is represented by an approximation of the area under the pressure curve as shown in Figure 4. Since the Pressure Phase of the diagnostic was influenced by the volatility of the fuel, it was also influencing the test results. Fuel volatility was classified into 3 states, Low, Medium and High. High Volatility aborts the diagnostic since pressure generated with the canister vent open would make a 0.51mm (0.020") leak undetectable with the vent closed. Medium is factored into the test results for better leak predictions. Low volatility allows the diagnostic to run without any corrections.

ALGORITHM DESCRIPTION

The empirical method is split into 4 major phases; the volatility phase, the pressure phase, the vacuum phase and the analysis phase. The basic concept of the empirical method is that after the vehicle has been driven for a period of time, the liquid and vapor fuel in the tank, as well as the components nearby the tank, would have acquired a different temperature than the ambient environment. This would then be followed by a natural cooling or heating effect on the fuel liquid and vapor in the tank. As a result of the heating or cooling, a corresponding pressure or vacuum can be monitored in the system to determine if the system is sealed or leaking.

Volatility Phase

After a drive cycle has been completed, the canister vent valve is left open, and the test enters a timed volatility phase. One of two conditions are typically encountered: either zero delta pressure results or a positive pressure is measured by the tank pressure sensor. In the case of zero delta pressure, the fuel is estimated to be low volatility, and the test continues to the next phase (pressure phase). In the case of moderate pressure builds, the fuel is estimated to be medium volatility and a correction factor is determined to be used later in the results. The test would then continue to the pressure phase. In the case of high pressure builds during the volatility phase, the test is aborted because the vapor generation is too high and will overcome the leak size and yield false results. (Figure 4)

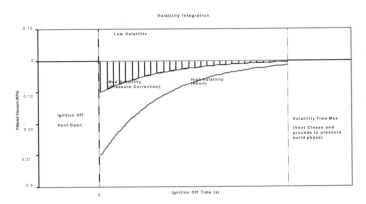

Figure 4

Pressure Phase

If the test passes the volatility phase, it will proceed to the pressure phase. Five scenarios are possible in the pressure build phase:

A. The pressure build reaches a pressure threshold before a test timer expires. In this case the test proceeds to the reporting results phase. (Figure 5)
B. Pressure builds to a peak and begins to drop off by a calibrated value. In this case the test proceeds to the vacuum build phase. (Figure 5)
C. Vacuum begins to build and exceeds a pressure phase vacuum threshold. In this case the test proceeds to the vacuum build phase. (Figure 6)
D. Very little pressure or vacuum is built and the sensor reading stays within two pressure levels, a zero vacuum max and a zero pressure min. While the pressure reading is between these two thresholds, a fast expire timer is incremented. If a fast expire timer threshold is exceeded, the test proceeds to the vacuum build phase. This would likely be the case if a leak were present in the EVAP system. (Figure 6)
E. Pressure continues to build (without peaking), but does not reach target pressure before the test timer expires. An upper limit on time is required due to test time constraints (mainly on battery life) (Figure 5)

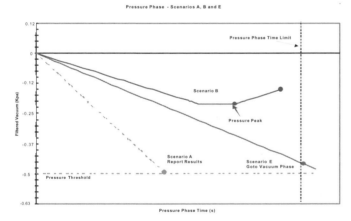

Figure 5

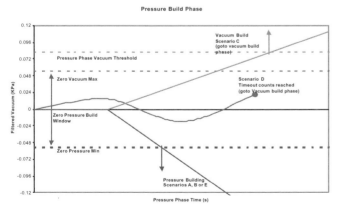

Figure 6

Vacuum Phase

In the scenarios where the pressure phase has resulted in proceeding to the vacuum phase, the timed vacuum build phase now executes. The vacuum phase works similarly to the pressure build phase and has six possible scenarios.

F. Pressure begins to build and reaches a pressure threshold. In this case, the canister vent is re-opened to vent the pressure through the canister, a time is waited (in scenario K), and then the vent is resealed. This cycle can be repeated several times as time allows. (Figure 7 and 9)
G. Very little pressure or vacuum is built and the sensor reading stays within two vacuum levels, a zero vacuum max and a zero pressure min. While the pressure reading is between these two thresholds, a fast expire timer is incremented. If a fast expire timer threshold is exceeded, the test proceeds to the analysis phase. This would likely be the case if a leak were present in the system. (Figure 7)
H. The vacuum build continues but the vacuum phase max time is exceeded. In this case the test proceeds to the analysis phase. (Figure 8)
I. Vacuum builds to a peak and begins to fall off a calibrated delta. In this case the test proceeds to the analysis phase. (Figure 8)
J. Vacuum builds and reaches a threshold which is a calculated value that depends on the corrected results achieved in the pressure build phase. The vacuum build threshold is the pressure threshold minus the corrected pressure built during the pressure build phase. In the case where the vacuum exceeds the vacuum threshold, the test proceeds to the analysis phase. (Figure 8)
K. If pressure has built to a pressure threshold and the vent has be re-opened in scenario F, a time is waited until the vent is resealed so that the vacuum phase can continue. If this happens multiple times and a maximum number of attempts has been exceeded, this scenario proceeds to the analysis phase. (Figure 7 and 9)

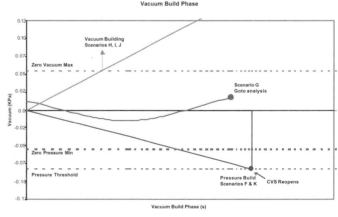

Figure 7

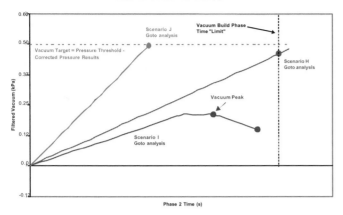

Figure 8

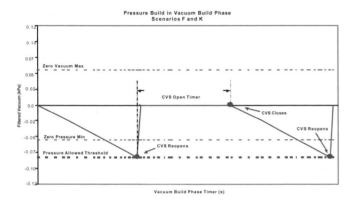

Figure 9

Analysis Phase

The first step that the analysis phase performs is to combine the results from the pressure and vacuum phases to determine the total pressure change. Thus, a statistical filter was used to process the test results before pass/fail determination. Before filtering the data, the algorithm preprocesses the data by normalizing it to the expected total pressure change for the current test conditions. The normalized result is a rational number between zero (0) and one (1). Zero (0) representing a "perfectly passing" system and one (1) representing a "perfectly failing" system. (Equation 3)

$$Normalized\ \ Result = 1 + \frac{Vacuum\ Results - Pressure\ Results}{Pressure\ Threshold}$$

Equation 3

Next, the normalized result is processed by an exponentially weighted moving average (EWMA) filter. The primary purpose of the filter is to reduce the variation caused by factors such as fuel volatility, fuel level, driving patterns, and ambient conditions. (Equation 4)

$$EWMA_{New} = Filter * (Results_{Normalized} - EWMA_{Prev}) + EWMA_{Prev}$$

Equation 4

The output of the EWMA filter is what is ultimately used for pass/fail determination. Fail thresholds are typically between 0.4 and 0.7. When the EWMA filter output goes above the fail threshold, a failure will be logged in the engine controller and the "service engine soon" light on the vehicle dash will be illuminated. The following diagram illustrates typical results for a system with a 0.51mm (0.020") leak installed.

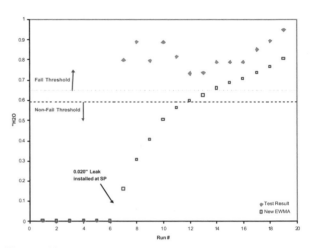

Figure 10

ALGORITHM DIAGRAM

Figure 10 depicts the major function blocks of the algorithm. It is not intended to show the complete functional algorithm nor the control of the canister vent solenoid which is inherent to each function. The algorithm is patent pending.

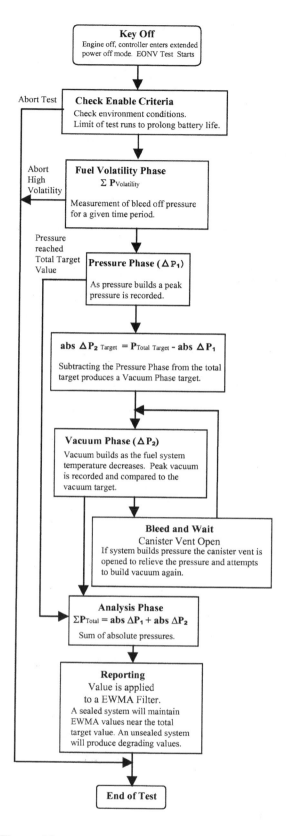

Key Off
Engine off, controller enters extended power off mode. EONV Test Starts

Abort Test

Check Enable Criteria
Check environment conditions.
Limit of test runs to prolong battery life.

Abort High Volatility

Fuel Volatility Phase
$\Sigma\,P_{Volatility}$
Measurement of bleed off pressure for a given time period.

Pressure reached Total Target Value

Pressure Phase (ΔP_1)
As pressure builds a peak pressure is recorded.

abs $\Delta P_{2\,Target}$ = $P_{Total\,Target}$ - abs ΔP_1
Subtracting the Pressure Phase from the total target produces a Vacuum Phase target.

Vacuum Phase (ΔP_2)
Vacuum builds as the fuel system temperature decreases. Peak vacuum is recorded and compared to the vacuum target.

Bleed and Wait
Canister Vent Open
If system builds pressure the canister vent is opened to relieve the pressure and attempts to build vacuum again.

Analysis Phase
ΣP_{Total} = abs ΔP_1 + abs ΔP_2
Sum of absolute pressures.

Reporting
Value is applied to a EWMA Filter.
A sealed system will maintain EWMA values near the total target value. An unsealed system will produce degrading values.

End of Test

Figure 10

TESTING AND DATA COLLECTION CONSIDERATIONS

TESTING WITHOUT CONTROL SOFTWARE

Initial EONV fuel tank testing can be done without the algorithm coded into engine controller software. Since the basic algorithm is actually simple, testing can be done without complex equipment. If the vehicle has an evaporative emissions system with a fuel tank pressure sensor and a vent valve that can be manually controlled to seal the fuel tank, then an EONV test can be run. The vent solenoid control can be achieved with a service tool that communicates with the Powertrain Control Module (PCM), or by wiring a switch directly to the vent valve to control it manually.

Initial GM testing and algorithm development was performed this way on a variety of vehicles. Proof of concept was done on a Grand Prix, Seville, TransSport, Corvette and a Suburban. These vehicles were selected for their diversity in fuel system design.

VEHICLE COMBINATIONS FOR DEVELOPMENT

For the 2003 vehicle model year, GM chose to implement EONV on 19 different fuel system designs. These fuel systems are represented on a wide variety of mid-sized and full-sized pick-ups and utilities and light duty service (delivery) vehicles. These vehicles have fuel tanks that range in useable fuel capacity from 66.2 L (17.5 gal) to 227 (60 gal). Fuel system designs included both metal and plastic fuel tanks, and included single and dual tank designs. In addition to developing EONV on unleaded gasoline vehicles, it was also implemented on ethanol flexible fueled vehicles (E0 to E85). This diversity of fuel system configurations illustrates the flexibility of the EONV algorithm. Many of the final calibrations are common across the population of vehicles while many more are common within a vehicle body style.

EONV DEPENDENCIES

There are many environmental factors that may effect the pressure in a fuel system during an EONV test. The factors considered were lumped into two types of variables. One type of factor is a variable that can be measured or estimated in the PCM. The other type is a variable that is essentially unknown to the PCM. It is this second type of variable that may create minor variation in the results of the EONV tests.

Fuel volatility

Fuel with higher volatility will generate pressure easier than low volatility fuel. With the elevated temperature in the tank, the higher the volatility, the more fuel will boil and generate vapor. In the sealed system, this vapor will generate pressure. High volatility fuels will create situations that make detecting leaks harder, while lower volatility fuels will create situations that make detecting a sealed system more difficult. Since the volatility of fuel delivered to fueling stations is regulated based upon geographical location and season, the probability of using a given volatility can be considered in developing the test matrix and vehicle calibration. The EONV algorithm does have features that will allow for detection of high volatility fuel situations. Other than this feature, fuel volatility is essentially a variable that is unknown to the PCM. As a result of this issue, fuel volatility was significantly considered in the test matrix.

Fuel Level

The amount of fuel in the tank affects the test in two ways. A lower fuel level has a larger vapor volume and a smaller mass of liquid fuel. Large vapor volumes provide the ability to create higher pressure changes during the test. This, however, is time dependent. As the vapor is generated, the pressure rises very slowly until the large volume becomes saturated with vapor. Once the space is filled, the momentum of the pressure rise will allow the system to achieve the higher pressure levels rather easily. Some vehicles show this trend more than others. On the other hand, a greater mass of liquid fuel provides for a significant heat sink. More time is required to heat the fuel during the drive before the test. If the system has a small temperature difference between the ambient and the bulk fuel, then it will be harder to achieve the targeted pressure threshold within the time limit for the sealed system. To protect against false results due to this phenomena, the test is only required to run with fuel levels between 15% and 85% of usable fuel capacity. Figure 12 shows a vehicle configuration that has a dependency on fuel level.

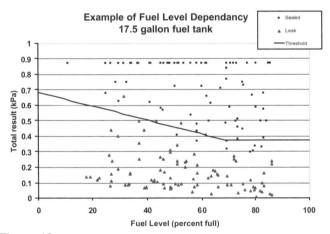

Figure 12.

Ambient Temperature

The difference in temperature between the ambient environment and the localized fuel system drives the pressure and vacuum changes. The ambient temperature is also monitored for limit conditions that can create unpredictable results for the EONV test. At temperatures near freezing, there is the possibility that ice will form on the tank vent. This would stop the tank from venting and create unknown results for the test. At high temperatures (>30C), the possibility that the volatility of the fuel will generate high pressure and mask a leak is great. For these reasons, the test is only

enabled within an ambient temperature window. At higher temperatures, the fuel will tend to boil and generate pressure for long periods of time. Because of this, the time for the volatility phase and the target for the pressure thresholds are based upon the ambient temperature. Because ambient temperature is a significant factor, an algorithm to estimate ambient air temperature was created (patent pending). This was done using data from the intake air temperature sensor, the vehicle speed sensors and the mass airflow sensor. This algorithm allows for ambient temperature to be estimated during the drive, no matter how long the drive is. Because the estimation is based on induction air temperature, the calibration is dependent on vehicle induction and front fascia design, not on tank configuration.

Preconditioning Drive

The length and type of drive before the EONV test runs, affects the test. The purpose of the prep drive is to create the temperature differentials between the fuel tank and the environment that will then drive the pressure and vacuum builds. If the drive is too short, there will not be enough heat imparted to the fuel tank to achieve proper results. The type of the drive is also a variable. Different speeds and loads create different temperature profiles. It was determined that a minimum engine run time and drive mileage are required to effectively prepare the fuel system. Short drives (long enough to create results) will tend to use phase one of the test more. This is because there is a lot of temperature around the fuel tank that has not been soaked into the tank yet. This will cause temperature rise during the test. Drives that are long enough to warm up the entire system, will use phase two of the test more because the system will cool down sooner. In general, large capacity fuel systems will require longer drives because of the amount of heat that will be required to create a driving temperature difference in large masses of liquid fuel. Some of the trucks have high GVWs. This can create significantly different speed and engine load requirements. Testing was done with delivery trucks in unloaded and loaded conditions. This created variation that was similar to other drive types. Overall, different drives create different temperature profiles and the test runs differently. However, the overall results are predictable as long as the minimum drive requirement is met.

Vehicle Configuration

There are almost an infinite number of items that can influence the EONV test from a vehicle standpoint. The major factors need to be understood to manage change control during development. The thing that drives the EONV test is temperature. This means that anything in the vehicle that can influence the temperature in the tank, may influence the EONV test. Therefore, changes to any of these parts may change the results of the tests. There are many factors on this list, but only significant differences look to affect the results of the tests. This also means that similar vehicles can have similar calibrations and will have similar results. A sample of these factors are: tank size, location, and construction, material and items around the tank (floor above, pickup bed, exhaust, axle), ground clearance, fuel system vent restrictions, etc. Of the 19 configurations that were developed, many of them fell into a few different groups that have very similar calibrations. Figure 13 is data from the old vacuum pull down EVAP test. It shows the difference between a normal plastic tank and a tank that has been stiffened. Figure 14 is the same two tanks run with EONV. There is no result differences between the tanks.

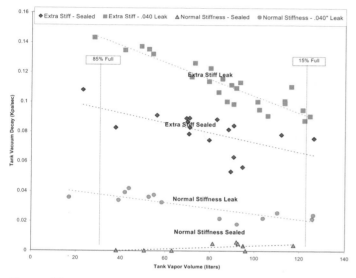

Figure 13

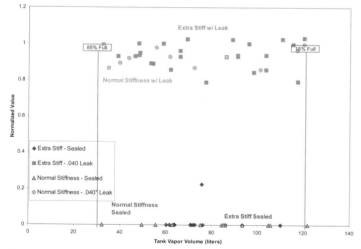

Figure 14

Ethanol and Gasoline

The EONV test will function without modification on a vehicle that will run on a mix gasoline and ethanol (up to 85% ethanol by volume). For the purposes of the EONV test, the major difference between gasoline and E85 is the fuel volatility. The RVP of E85 is relatively low.

Typically, E85 will act similar to 7 RVP gasoline. Because the volatility of fuel can vary and is unknown to the EONV test, calibration of the EONV test covers a wide range of fuel volatility that includes E85. Therefore, E85 has not been a significant factor within the test matrix. Performance on E85 is not different than low volatility gasoline.

Altitude

The underlying principle of the EONV test is based upon the temperature and pressure differentials. When the test starts, the fuel system is vented to atmosphere. This could be at sea level or high in some mountain pass. For a sealed system, it has not matter what the barometric pressure is during the test as long as it is greater than 74 kPa and it is relatively steady. There is a possibility that the weather could change during the test, but the test is fast relative to the weather. Extreme cases of weather change are considered rare, and thus are filtered out by using statistical processing with an EWMA. Testing was done at altitude. Although no major concerns were found, higher altitude has the ability of making the fuel look more like a slightly more volatile on a leaking system. This phenomena is accounted for in the legislation of fuel volatility within geographic region and season as referred to earlier in the section on fuel volatility.

Other Uncontrolled Variables

In the real world environment, there are many factors that can drive differences in the EONV pressure and test results. Several of these factors remain unknown to the PCM. Several of these factors were considered during development of EONV. Some of the major factors that were considered included sun load on different vehicle colors, radiated heat from different types of surfaces under the vehicle during the test, wind and rain. A study of vehicle color was done. Data was compared for a dark blue pick up truck and a white pick up truck of the same model. For identical test conditions, the data proved to be within test to test variation. In addition, during data collection throughout the development phases of EONV, weather and parking location was logged. This data was analyzed to look for trends. There were a few tests that showed differences associated with wind and rain and parking surface, but there was not enough evidence in the data to conclude that there was reason to be concerned about those factors. There are many factors that drive variation in test results, but that is why a significant amount of testing was done: to include the variation in the data analysis. With further development, more factors can be understood and possibly controlled or compensated for in the algorithm.

CONCLUSION

The natural vacuum technique is an extremely robust approach to small leak detection. It demonstrates capability across a wide range of tank and platform configurations and lends itself to more common calibrations. It is robust to many of the things that affected traditional leak detection methods such as tank size, tank geometry and flexibility, and fuel slosh. Data shows that the natural vacuum 0.51mm (0.020") leak detection performance shows great potential in terms of meeting the small leak detection regulatory requirements.

ACKNOWLEDGMENTS

Sam Reddy, Greg Rich, Craig Kemler

REFERENCES

[1] Perry, P.D. and Delaire, J.P., "Development and Benchmarking of Leak Detection Methods for Automobile Evaporation Control Systems to Meet OBDII Emission Requirements," SAE paper No. 980043, Feb. 1998.

[2] Majkowski S.F., et al., "Development and Validation of a 0.020" Evaporative Leak Diagnostic System Utilizing Vacuum Decay Methods," SAE paper No. 1999-01-0861, March 1999

[3] Reddy, S.R., "System and Method for Detecting Leaks in a Vapor Handling System," U.S. Patent No. 5,263,462, Nov. 1993.

[4] Reddy, S.R., et al., "Leak Detection for a Vapor Handling System," U.S. Patent No. 6,321,727, Nov. 2001.

CONTACT

Michael DeRonne – Diagnostic Strategist: Premium V8 programs – michael.deronne@gm.com

Chad Lehner – Diagnostic Development Engineer: EVAP – chad.lehner@gm.com

Gregory E. Labus – General Motors, EVAP Diagnostic Technical Expert – gregory.labus@gm.com

ADDITIONAL SOURCES

Here are any additional sources. This is an optional section.

DEFINITIONS, ACRONYMS, ABBREVIATIONS

Calibration: collection of values that can be changed and control the execution of the software algorithm

E85: 85% denatured ethanol, 15% gasoline mix

EONV: Engine Off Natural Vacuum

EVAP: Evaporative emissions control system

EWMA: Exponentially weighted moving average

PCM: Powertrain Control Module

RVP: Reid Vapor Pressure – used to measure fuel volatility

CHAPTER 13

CERTIFICATION OF
ULTRA-CLEAN VEHICLES

Michael Akard
Horiba Instruments, Inc.

13.1 OVERVIEW

The subject of emission testing for ultra-clean gasoline-powered vehicles is an extensive topic. The testing procedures can be broken down into two basic topics: exhaust emissions and evaporative hydrocarbon emissions.

Exhaust emissions for ultra-clean gasoline-powered vehicles have been driven lower by regulations. The first major market to require a significant portion of the new vehicles sold to be ultra-clean gasoline-powered vehicles is California. Currently, the lowest exhaust emissions in the California market are termed SULEV (super ultra-low emission vehicles). When evaporative emissions are also reduced to defined zero level, the cars are classified as PZEV (partial zero emission vehicles). Emissions for other major markets like the rest of the United States, the European Union and Japan are also being regulated to very low emissions in the near future.

For SULEV-level testing, vehicle emissions must be quantified that are very close to ambient air concentrations. This testing is complicated by the fact that ambient air concentrations vary with location, time of day, season and nearby emission sources. In order to successfully measure SULEV emissions, testing variability must be reduced dramatically from the original testing methodology first introduced in the United States. There are many different sources of variation during testing that must be addressed.

The fundamental challenge of SULEV testing stems from the fact that the automotive emissions must be reduced to concentrations very close to ambient air concentrations for three of the four regulated emissions. The ambient air not only provides intake air for the engine but traditionally supplies the dilution gas for the constant volume sampler (CVS) system used to collect the tailpipe emissions. The most variable ambient air concentration of the regulated emissions is the total hydrocarbons (THC). Carbon monoxide and the nitrogen oxides (nitric oxide and nitric dioxide) are generally at low concentrations and stable unless a nearby emission source contaminates the ambient air. Carbon dioxide is the only emission that remains well above ambient concentrations.

The CVS collects the entire tailpipe emissions for a test cycle and dilutes the exhaust with dilution air in order to fill a sample bag. While the transient dilution ratio changes as the exhaust flow of the vehicle changes, an overall dilution factor can be determined. This dilution factor is used to calculate the exhaust emissions from the difference between the sample and ambient bag concentrations in order to give mass emissions (g/km) for a vehicle test. As the tailpipe emissions for the hot-start emission cycles are lowered closer to the concentrations in the ambient air, the difference between the sample and ambient bag becomes very small relative to the measured concentration. The measurement of small differences between relatively large values is analytically difficult and prone to error due to variability in the testing. This problem requires variations on the basic CVS concept for successful testing. There have been three different fundamental approaches to SULEV testing in order to improve the measurement accuracy and reproducibility.

The first successful approach was developed in Japan. In order to use the CVS with the same dilution rates, the ambient air is scrubbed of background hydrocarbons, carbon monoxide and nitrogen oxides. By diluting with cleaned ambient air, the variability in the dilution gas concentrations is significantly reduced. In addition, the effect of the dilution factor error is minimized in the emission calculation by multiplying the term with the dilution factor by a lower ambient concentration. The net effect of this approach is to measure small differences between small numbers. While this is analytically easier, this approach requires lower-range analyzers.

The second approach to successfully enhance the CVS includes various improvements designed to reduce sampling variability and dilution ratios. The primary factor determining the dilution factor is the need to avoid condensation in the sampling handling system or the sample bag. In this approach, the sample and ambient bags can be heated to allow the use of lower dilution rates and to keep the bags clean. The dilution air is also dried, heated and run through a charcoal filter to help stabilize the dilution gas properties. Other CVS enhancements include separate lines for clean and dirty lines and pre-treating bags. The net effect of this approach is to increase the difference between the ambient and sample bags, measuring a larger difference between two large numbers. The analyzers used in this method do not need ranges as low as the first approach.

The third approach to successfully test SULEV emissions was developed by American Industry/Government Emissions Research (AIGER). This approach does not use a CVS at all but a bag mini-diluter (BMD). The BMD does not collect all the tailpipe exhaust. Instead, the BMD measures the tailpipe and collects a portion of the exhaust. This exhaust is diluted with refined dilution gas with a constant dilution ratio on a transient basis. This technique is also known as partial flow dilution as opposed to the full flow dilution technique of the CVS. The advantages of this approach include the clean background of the first approach and the lower dilution rates of the second approach. The measurement is reduced to a large difference between small numbers. This approach also requires analyzers with lower ranges. While this approach has not been written into any regulations, the United States Environmental Protection Agency (USEPA) has issued an acceptance letter to the automotive manufacturers allowing the BMD to be used for certification.

Using the BMD requires a fast and accurate exhaust flow measurement. This measurement is necessary to control proportionality and to calculate the total emissions in mass per kilometer. The proportionality for the test is dependent upon the speed of the measurement and the control of the bag flow. In addition to the exhaust flow measurement, the BMD must maintain a constant dilution ratio. An advantage of this technique is to eliminate the dilution factor error.

The next step in certifying an ultra-clean gasoline-powered vehicle is to test evaporative emissions. Evaporative emissions fall into two categories, diurnal and running loss. There is not a definitive zero evaporative emissions paper focused solely on the measurement technology and therefore this chapter does not include an evaporative emissions paper. The SAE paper 2000-01-2926 by

Matsushima et. al. included in Chapter 11 covers some of the concepts for this testing. Papers have addressed micro-SHED (SAE 980402) or mini-SHED (SAE 2003-01-3156) used for the testing of automotive components. A general paper regarding evaporative emissions from the California Air Resources Board (CARB) describes diurnal emissions measurement (1999-01-3528).

13.2 SELECTED SAE TECHNICAL PAPERS

A Study of a Gasoline-Fueled Near-Zero-Emission Vehicle Using an Improved Emission Measurement System

Akira Tayama, Kazuhiko Kanetoshi, Hirofumi Tsuchida and Hiroshi Morita
Nissan Motor Co., Ltd.

Copyright © 1998 Society of Automotive Engineers, Inc.

ABSTRACT

This paper concerns research on an emission control system aimed at reducing emission levels to well below the ULEV standards. As emission levels are further reduced in the coming years, it is projected that measurement error will increase substantially. Therefore, an analysis was made of the conventional measurement system, which revealed the following major problems.

1. The conventional analyzer, having a minimum full-scale THC range of 10 ppmC, cannot measure lower concentration emissions with high accuracy.

2. Hydrocarbons are produced in various components of the measurement system, increasing measurement error.

3. Even if an analyzer with a minimum full-scale THC range of 1 ppmC is used in an effort to measure low concentrations, the 1 ppmC measurement range cannot be applied when the dilution air contains a high THC concentration. This makes it impossible to obtain highly accurate measurements.

4. Since the conventional CVS has a constant flow rate, there are some test phases with an excess dilution rate, which lowers the exhaust gas concentrations, resulting in larger measurement error.

Improvements were made to the conventional measurement system with the aim of resolving these problems. As a result, a system has been established that can measure emission levels of around one-tenth of the ULEV standards with much greater accuracy than the previous system.

This improved measurement system is now being used to develop an exhaust gas aftertreatment system that integrates an electrically heated catalyst (EHC), a three-way catalyst (TWC) and an HC adsorber. In addition, the individual performance of the HC adsorber and the TWC has also been improved. Although this aftertreatment system is still at the research stage, preliminary test results indicate that it has the potential to reduce emission levels to one-tenth of the ULEV limits.

INTRODUCTION

More stringent exhaust emission regulations have been enforced in recent years amid the increased attention paid to environmental issues. In the U.S., the state of California is even discussing possible regulations that would require exhaust emission levels lower than the ultra-low emission vehicle (ULEV) standards. Since it is possible that tougher emission regulations will be adopted in the future, research is under way on emission control systems aimed at reducing the constituents of gasoline engine exhaust to atmospheric levels, i.e., the attainment of near-zero emissions. However, this further reduction of emission concentrations is expected to lead to substantially larger measurement error with existing measurement equipment.

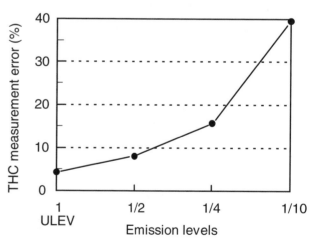

Figure 1. Calculated Measurement Accuracy
Calculated based on the guaranteed accuracy of a conventional analyzer

To investigate this possibility, estimates were made of the measurement error relative to various emission levels, based on the guaranteed accuracy of a conventional exhaust gas analyzer. The results shown in Fig. 1 indicate that measurement error increased as the emission levels decreased. At a level equal to one-tenth of the ULEV regulations, measurement error rose to around 40%.

Since it was projected that measurement accuracy would be reduced even further by the addition of other measurement error apart from that of the analyzer, a study was undertaken to find ways of improving accuracy. This paper presents the results of that study and discusses the potential of a system now being researched that combines an electrically heated catalyst (EHC), a three-way catalyst and an HC adsorber. This system is presented as one example of a future exhaust gas aftertreatment system.

IMPROVEMENT OF MEASUREMENT EQUIPMENT

Various techniques have been proposed for measuring vehicle emissions, including a modal mass method [1] and a mini diluter method [2]. In this research, however, an effort was made to improve the accuracy of the constant volume sampler (CVS) that has traditionally been used in measuring exhaust emissions.

It has been pointed out that the measurement of total hydrocarbons (THC) entails more problems than the measurement of carbon monoxide (CO) or nitrogen oxides (NOx) [3]. Since the greatest error tends to occur in the measurement of THC, the discussion here will deal with THC measurement. The following four points were the focus of a study aimed at improving measurement accuracy.

1. An analyzer capable of measuring low concentration emissions with high accuracy.
2. Preservation of gas concentrations in the process from sampling to measurement.
3. Concentrations of target components in the dilution air.
4. CVS dilution ratio.

The problems that occurred in the previous system are explained here along with the improvements that have been made based on this study.

(1) ANALYZER – The exhaust gas analyzer is subject to random measurement variation as a result of being used in repeated measurements, the occurrence of drift and other factors. Consequently, measurement error tends to increase as the concentrations of emission components are reduced. This means that an analyzer capable of measuring lower concentrations with little variation is required in order to prevent any further increase in measurement error. Since the error of an ordinary analyzer is determined by the measurement range, it is necessary to obtain accurate measurements in a range of lower emission concentrations.

The minimum measurement range of the bag concentration analyzer used in our laboratory at the Nissan Research Center is 10 ppmC for THC, 10 ppm for NOx and 100 ppm for CO. Accordingly, if an attempt is made to measure THC at levels below 10 ppmC, all measurements are made at the full-scale range of 10 ppmC.

Since the full-scale analyzer error is ±1%, i.e., ±0.1 ppmC, measurement of THC at, say, 1 ppmC would lead to large measurement error of ±10%. Measurement error would be even larger at lower emission concentrations. For example, in the case of emissions at one-tenth of the ULEV levels, it is estimated that the bag concentrations (when THC is calculated assuming zero background) in phases 2 and 3 would be less than 1 ppmC, and even that of phase 1 would be less than 2 ppmC. At such low levels, large error would occur with the conventional analyzer, making it impossible to obtain highly accurate measurements.

In order to be able to measure low emission concentrations accurately, it was decided to implement a bag concentration analyzer with a minimum range of 1 ppmC for THC, 1 ppm for NOx and 10 ppm for CO. That would make it possible to use the 1 ppmC range in case we wanted to measure THC at the 1 ppmC level. Accordingly, since the full-scale error of the analyzer would be 1%, or 0.01 ppmC, measurement error would be reduced to one-tenth of the level of the conventional analyzer. The use of an analyzer with a low concentration range would therefore improve measurement accuracy for low concentration emissions.

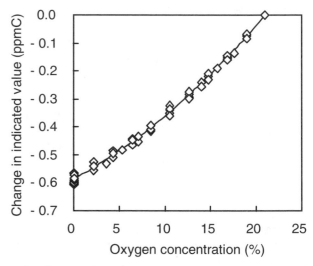

Figure 2. Oxygen Interference in Flame Ionization Detector
Measured concentrating rang: 0.0-0.9 ppmC

It was also predicted that oxygen interference in the flame ionization detector (FID) generally used to measure THC would increase as the emission concentrations to be measured decreased. Therefore, a study was made of oxygen interference at emission concentrations of less than 1 ppmC. The results are shown in Fig. 2. It should be noted that the indicated value at an oxygen concentration of 21% was taken as the baseline value. As seen in the figure, the indicated value decreased with a decreasing oxygen concentration, regardless of the THC concentration. It was found that oxygen interference would have an exceptionally large effect on the measured values at extremely low concentrations of less than 1 ppmC.

Oxygen concentrations differ between the sample bags and the background bag, and the oxygen concentration in the sample bags also varies depending on the dilution rate. For these reasons, the change in the indicated value due to oxygen interference is not constant. Therefore, we adopted a procedure for reducing measurement error due to oxygen interference by correcting the measured values on the basis of the oxygen concentration in the bags. Another possible measure for improving measurement accuracy would be to reduce the amount of oxygen interference by adjusting the quantity of fuel supplied to the FID. It is planned to examine this measure in future work.

As the foregoing discussion has indicated, measurement accuracy for low concentration emissions has been improved by implementing an analyzer capable of measuring low concentrations more accurately and by correcting THC values measured with the FID so as to compensate for oxygen interference.

(2) PRESERVATION OF GAS CONCENTRATIONS – The preservation of exhaust gas concentrations in the process from sampling to measurement was examined, because changes in component concentrations during this interval can lead to measurement error. It was found that HCs are produced in different parts of the measurement system through which the exhaust gas flows, causing measured concentrations to vary [3]. Various sources can be considered for such HCs, including oils or organic solvents that collected on the components at the time of manufacture, adhesives used in the components and also the materials of the components themselves. Accordingly, all components coming in contact with the exhaust gas from sampling to measurement were examined, and the materials of components that were confirmed to be sources of HCs were changed, as outlined in Table 1.

Figure 3 compares the HC concentrations measured in the previous system and in the improved system following the material changes. These measured data represent the HC concentrations that occurred when the systems were supplied with air having a high degree of purity. Changing the component materials greatly reduced the generation of HCs in the system, thereby improving the preservation of the gas sample concentrations.

The improved system was then used to measure the change in the system-derived HC concentration relative to the concentration of the gas passed through the system. Measurements were made for a total of six sample bags and background bags combined, including their respective lines. The results shown in Fig. 4 are expressed as the result of a linear regression showing the measured concentration as a function of the gas concentration that was passed through the system. It is seen that the slope has a value of nearly 1 and that the mea-

sured concentration is larger than the supplied value by the intercept regardless of the gas concentration. This means that the system-derived HC concentration was virtually constant to a concentration of 1 ppmC, irrespective of the concentration of the gas passed through the system. On the basis of this result, it was decided to calculate the emission concentration by subtracting the system-derived HC concentration from the measured value. The same correction values were used for both the gas sample lines and the background line, as no significant difference was observed between the HC concentrations produced in each line.

Table 1. Material Changes for Measurement System Components

Component	Previous material	Improved material
Gas piping	Teflon	Stainless steel
Sample bags	Tedlar	Fluororesin
Filters	Silica filled	Sintered stainless steel
Solenoid valve seat	Viton	Teflon coated
Pressure regulator	With Viton rubber	Without Viton rubber

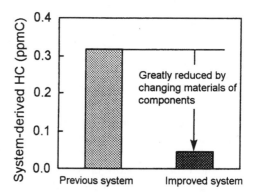

Figure 3. Reduction of System-derived HC Concentration by Changing Materials

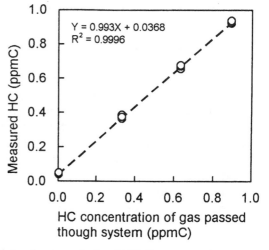

$$Y = 0.993X + 0.0368$$
$$R^2 = 0.9996$$

Figure 4. System-derived HC Concentration

619

The foregoing results showed that measurement error can be corrected and reduced because the system-derived HC concentration is virtually constant, regardless of the concentration of the gas flowing through the system. However, the system-derived HC concentration can vary due to differences in the lines or in the sample bags as well as to random variation that occurs with repeated measurements. Such variation gives rise to measurement error.

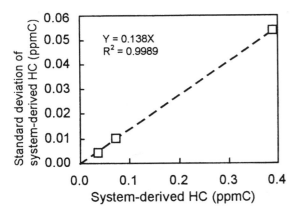

Figure 5. Variation in System-derived HC Concentration

Figure 5 shows the standard deviation of the HC concentration relative to the average system-derived HC concentration. Variation in the HC concentration was measured using lines and bags containing identical foreign substances. It should be noted that the measurement variation of the analyzer was removed by calculation. The results indicate that there is a correlation between the average HC concentration and the degree of variation. This variation in the system-derived HC concentration must be controlled in order to reduce measurement error. It became clear that one effective way to accomplish that was to reduce the HC concentration produced in the system itself. Accordingly, changing the materials of the system components so as to reduce the occurrence of HCs also had the effect of suppressing the degree of variation, which in turn reduced measurement error.

As the next measure, a capability was added for purging the sample bags and the transport lines from the sampling point to the analyzer. That was done to prevent changes in measured concentration (i.e., so-called hang-up) due to contamination that collects on the surfaces of the system during use. While ambient air is generally used for purging purposes, it has a THC content of 2-3 ppmC, which would reduce the purging effect. Accordingly, purging was accomplished with purified air the THC content of which was lowered to less than 0.1 ppmC by using an air purifier, as will be explained later.

Another measure taken to avoid hang-up was to construct the gas flow lines, from the sample lines and bags to the analyzer, as two independent systems so that the parts coming in contact with the gas stream are separate. One line is used for measuring high-concentration exhaust gas, such as engine-out emissions, and the other line is used for measuring low concentration exhaust gas. Additionally, a low-concentration line has been provided specifically for phase 1, which involves a relatively high-concentration gas stream, thereby preventing the occurrence of hang-up in phases 2 and 3.

There was still a possibility that the HC concentration produced in the system would change during the course of use, even though material substitutions had been made, a purging capability had been added and the sampling lines had been separated.

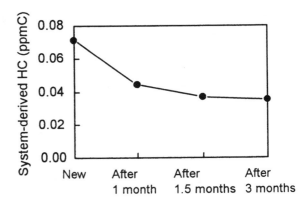

Figure 6. Change in System-derived HC Concentration Over Time

Figure 6 shows the change in the HC concentration that was measured over a three-month period. The system-derived HC concentration decreased with elapsed time, which means that measurement error also decreased. This result is thought to indicate a reduction in the quantity of HCs produced by oils and organic solvents that collected on the surfaces of the components at the time of manufacture.

Since no significant difference has been observed at this point between the HC concentrations that occur on the sample bag side and on the background bag side, it is concluded that hang-up has been effectively prevented. In order to maintain measurement accuracy in the future, it will be necessary to determine periodically if there is any change in the system-derived HC concentration and also to perform regular system maintenance.

The components of the standard gas supply line were also examined from the standpoint of preserving the gas concentration. The standard gas supply line has a pressure regulator for the storage tank gas.

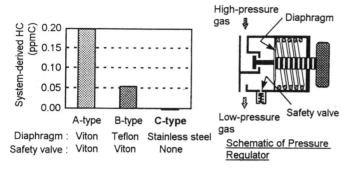

Figure 7. Pressure Regulator

As shown in Fig. 7, the regulator incorporates a diaphragm for adjusting the pressure and a safety valve for releasing the pressure in the event of a component failure somewhere in the system. It was found that a regulator that used Viton rubber for the diaphragm and the safety valve produced approximately 0.2 ppmC of HCs, and one that used Viton rubber only for the safety valve produced 0.05 ppmC, or only one-fourth as much. This variation in the HC concentration according to the number of places where such rubber was used suggested that the rubber was one source of the HCs produced in the system. This level of HC concentration had little effect on the results in the measurement range of the conventional analyzer, but it would significantly affect the measured values in the low-concentration measurement range. For that reason, a clean pressure regulator that does not incorporate any rubber was adopted.

The piping was changed from the previous Teflon tubes to bright annealed (BA) stainless steel tubes. Before use, the piping was purged with hot N$_2$ gas to prevent the generation of HCs from oils or solvents on the inner surfaces of the tubes. The use of these supply line components has made it possible to control the change in THC concentration in the system during standard gas transport from the storage tank to the analyzer.

(3) CONCENTRATIONS OF TARGET COMPONENTS IN DILUTION AIR – The ambient air that is generally used as the dilution air contains 2-3 ppmC of THC. Following dilution, the THC concentration in the exhaust gas tends to exceed 1 ppmC, making it impossible to use an analyzer with a full-scale range of 1 ppmC. This gives rise to the problem that low concentration emissions cannot be measured with high accuracy.

Furthermore, when calculating emission levels, the component concentrations in the dilution air must be subtracted from the gas sample concentrations in order to take into account the dilution factor. Since the dilution factor also contains measurement error and calculation error [4], there is the problem that larger error can occur in the subtraction process when the dilution air contains high concentrations of the emission components to be measured.

These problems were resolved by using an air purifier to reduce the THC content of the dilution air to less than 0.1 ppmC. This purified air is also used to purge the lines from the constant volume sampler (CVS), including the bags, to the analyzer.

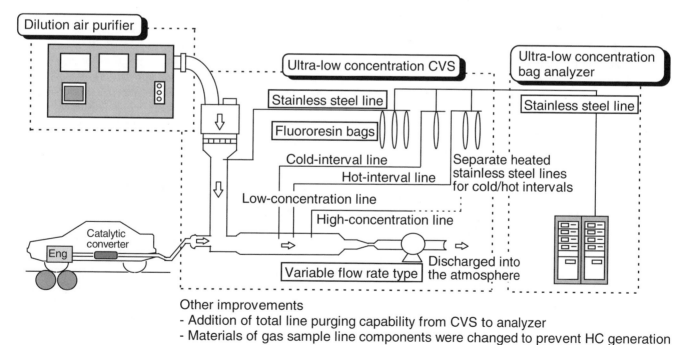

Other improvements
- Addition of total line purging capability from CVS to analyzer
- Materials of gas sample line components were changed to prevent HC generation

Figure 8. Exhaust Emission Measurement System with Low Concentration Capability

(4) CVS DILUTION RATE – An excess dilution rate in the CVS leads to lower exhaust gas concentrations following dilution, giving rise to the problem of larger measurement error. The optimum CVS flow rate is the minimum flow rate that does not cause water condensation in the system from the time the samples are taken until the analysis is performed. However, because the conventional CVS has a constant flow rate, it is not possible to set the optimum CVS flow rate for each phase [5].

To overcome that problem, a variable flow CVS was adopted [6-7]. This CVS is equipped with a venturi flowmeter (VFM) that measures the CVS flow rate and uses a blower to control the flow rate to the set value. This makes it possible to obtain the optimum CVS flow rate for each phase. During measurement, the CVS flow rate was set to the minimum level that did not produce water condensation in the bags at room temperature at the conclusion of sampling. Water condensation in the sample lines was prevented by heating the lines, and water condensation in the sample bags was prevented by operating the air-conditioning system to keep the room temperature above 25°C[8].

INVESTIGATION OF MEASUREMENT ERROR

Measurement error was investigated using a measurement system that incorporated the improvements noted above (Fig. 8). This investigation was conducted by creating an emission model based on experimental data and calculating what effect measurement error at a 95% confidence level would have on the tailpipe emission mass. The measurement error taken into account in the investigation included that of the analyzer and measurement error related to preservation of the gas samples. It should be noted that the measurement error of the standard gas concentrations and the measurement error of the CVS flow rate were not taken into account.

Figure 9 shows the calculated results for THC measurement error at emission levels of one-tenth of the ULEV regulations. The effects of the improvements explained in subsections (1)-(4) above are also broken down. Improving the preservation of the gas concentrations had the largest effect on reducing measurement error, resulting in a 66% improvement in measurement accuracy. That was followed by the increased accuracy of the analyzer and the use of purified dilution air, which together improved measurement accuracy by 29%. Finally, optimizing the CVS flow rate improved measurement accuracy by 1%. Since the effect of optimizing the CVS flow rate varies according to the ratio of the optimized flow to the previous constant flow, the degree of improvement obtained may be greater than 1% in some cases [9].

The combined result of all the improvements explained in subsections (1)-(4) above is shown in Fig. 10.

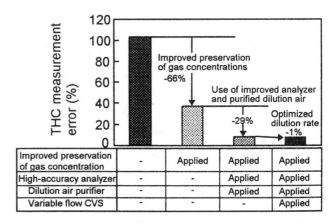

Figure 9. Increased Measurement Accuracy Resulting from Improvements
For measurement of emissions at 1/10 of the ULEV standards

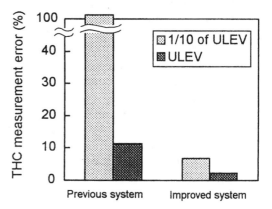

Figure 10. Measurement Accuracy

Measurement error of emissions at levels equal to one-tenth of the ULEV regulations has been greatly reduced from approximately 103% with the previous system to about 7% with the improved system. That is even lower than the ULEV measurement error of the previous system, which was approximately 11%. However, maintaining this level of accuracy requires extremely complex measurement and maintenance procedures, which will have to be simplified in future work.

INVESTIGATION OF THE ULTIMATE GASOLINE ENGINE EMISSION CONTROL SYSTEM

The improved measurement system explained in the preceding section is now being used in research into technologies for reducing gasoline engine emissions to extremely low levels. The test vehicle being used in this research is based on the Nissan Altima and is fitted with a modified version of the 2.4-liter inline 4-cylinder engine used on this production model. The specifications of the test vehicle are given in Table 2.

Table 2. Specifications of Test Vehicle

Vehicle	Model	Nissan Altima
	Transmission	4-Speed Automatic
	Inertia Weight	1,474 kg (3,250lbs)
Engine	Model	KA24DE-Modified
	Type	Inline 4-Cylinder, 4Valves/Cyl.
	Bore × Stroke	89.0 × 96.0 mm
	Displacement	2,389 cm^3
	Compression Ratio	9.2:1
	Fuel Supply	Multi-Point Injection
		2-Hole Injector

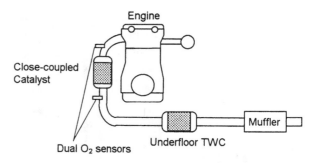

Figure 12. Reference Emission Control System

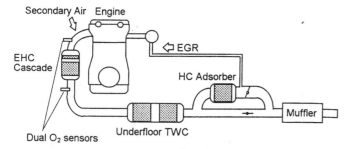

Figure 11. Exhaust Gas Aftertreatment System

An example of an exhaust gas aftertreatment system that was investigated in the present work in shown in Fig. 11. The system uses an EHC and secondary air and is designed to reduce cold-start HC emissions by shortening the light-off time of the catalyst. The EHC is located directly after the manifold in order to make effective use of exhaust gas heat. An HC adsorber is also used to reduce HC emissions that are not converted in the interval before the catalyst reaches its operating temperature. Since the HC adsorber begins to desorb HCs as the temperature rises, it is positioned in a relatively low temperature location downstream of the underfloor catalyst. The HC adsorber is installed in a secondary flow channel that allows the exhaust flow to be switched between the adsorber and the muffler. This arrangement keeps the HC adsorber temperature from rising, thereby preventing desorption of HCs that are adsorbed in the initial interval after a cold start. A large-capacity three-way catalyst (TWC) is installed in the underfloor location to reduce emissions following engine warm-up.

Measurements were made of FTP-test emissions from the test vehicle equipped with this exhaust gas aftertreatment system. For the sake of comparison, emission measurements were also made using a test vehicle fitted with the reference system shown in Fig. 12, incorporating a close-coupled catalyst and an underfloor TWC.

When fresh catalysts were used, the NMOG emission level measured with the aftertreatment system in Fig. 11 was 5 mg/mile, which was approximately one-fourth of the level obtained with the reference system and about one-eighth of the ULEV standard after 50,000 miles (Fig. 13). Further reductions will be needed to attain the goal of reducing the concentrations of exhaust components to atmospheric levels. The figure also compares the NMOG emission quantities for the cold interval (until the first peak) and the hot interval (from the second to the twenty-third peak) of the FTP-test. It is seen that there was no large difference in emission levels, indicating that further reduction of emissions in both the cold and hot intervals is desired. To accomplish that, the HC adsorber was improved to reduce cold-interval emissions and the underfloor TWC was improved to reduce hot-interval emissions.

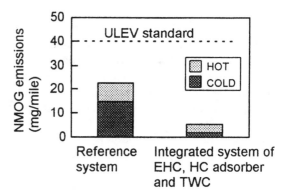

Figure 13. Results of Emission Evaluation
With fresh catalysts

(1) IMPROVEMENT OF HC ADSORBER – Figure 14 shows the adsorption rates found for different carbon numbers (CN) of HC species before and after the HC adsorber was improved.

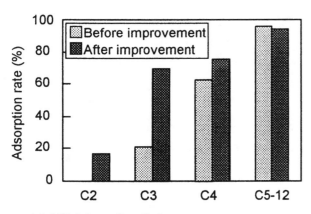

Figure 14. HC Adsorption Rates
With fresh catalysts

Before the HC adsorber was improved, it achieved high adsorption rates for HC species having a CN of 5 or higher, but the adsorption rate declined with a decreasing CN. This indicated that improving the adsorption rate for low CN species would be effective in increasing overall HC adsorption. Improvements were then made to the HC adsorber to increase its ability to adsorb low CN species. As a result, compared with the previous adsorber, the improved HC adsorber achieves higher adsorption rates for low CN species without any decline in the adsorption rate for high CN species.

(2) IMPROVEMENT OF UNDERFLOOR TWC – The HC species flowing into the underfloor TWC were analyzed for the purpose of improving this catalyst system. As shown in Fig. 15, the results indicated that large quantities of alkanes were present in the exhaust gas at the inlet to the underfloor TWC. The exhaust gas entering the underfloor TWC contains NMOG components that were not converted by the manifold catalyst system. Since the underfloor TWC adopted a catalyst having the same specifications as the manifold catalyst, it was assumed that high conversion rates could not be expected for those NMOG components.

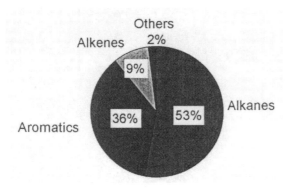

Figure 15. NMOG Components at Inlet of Underfloor
TWC
Phase 2&3 data (with fresh catalysts)

Therefore, the underfloor TWC was improved to increase its conversion efficiency for alkanes that were not converted by the manifold catalyst system.

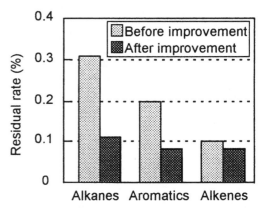

Figure 16. Comparison of Conversion Efficiencies for HC
Species
Phase 2&3 data (with fresh catalysts)

A comparison was made of the residual rates of HC species in phases 2 and 3 (6th to 23rd peak) before and after improving the underfloor TWC. The results shown in Fig. 16 indicate that the improvement reduced alkanes to approximately one-third of the previous level. It was also confirmed that the improvement did not lower the conversion performance for other NMOG species.

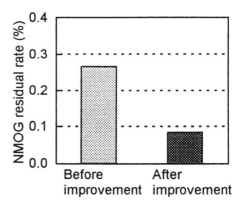

Figure 17. Comparison of Residual Rates in Hot Interval
(2nd to 23rd peak)

Measurements were then made with a fresh improved TWC. As shown in Fig. 17, it was found that the residual rate of NMOG in the hot interval (2nd to 23rd peak) was reduced to less than 0.1% and that the hot-interval emission level was reduced to approximately one-third of that prior to improvement of the underfloor TWC.

Figure 18 shows the results of emission measurements that were made with the improved HC adsorber and the improved underfloor TWC. When new catalysts were used, the NMOG emission level was reduced to approximately 2 mg/mile, i.e., to about one-third of the level measured with the previous devices. The NMOG emission level measured with aged catalysts was approximately 4 mg/mile, indicating roughly a twofold increase in emissions compared with the results for fresh catalysts. Nonetheless, the value is approximately one-tenth of the ULEV limit.

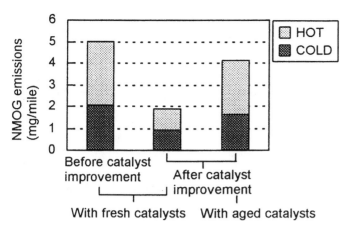

Figure 18. Emission Levels Before and After Catalysts Improvement

It should be noted that this exhaust gas aftertreatment system is still at the research stage. There are various issues that must be addressed with respect to ease of installation, cost, durability, component reliability and other aspects before it can be implemented on production vehicles.

SUMMARY

This paper has described the improvements made to the conventional CVS measurement system in the following four areas:

1. adoption of an analyzer capable of measuring low concentration emissions,

2. improvement of the preservation of gas concentrations,

3. reduction of concentrations of target components in the dilution air, and

4. optimization of the CVS dilution rate.

As a result of these improvements, a measurement system has been established that can measure emissions concentrations of around one-tenth of the ULEV standards with higher accuracy than that obtained with the previous measurement system at ULEV levels.

An exhaust gas aftertreatment system has also been developed that integrates an electrically heated catalyst (EHC), a three-way catalyst (TWC) and an HC adsorber. This system uses an HC adsorber that achieves higher adsorption rates for low carbon number species and a TWC that provides higher conversion efficiency for alkanes. Although this aftertreatment system is still at the research stage, recent test results indicate that it has the potential to reduce emission levels to around one-tenth of the ULEV regulations.

ACKNOWLEDGMENTS

The authors would like to thank various individuals at Horiba, Ltd. for their invaluable cooperation in connection with this research.

REFERENCES

1. J. Baronick, et al., "Modal Measurement of Raw Exhaust Volume and Mass Emissions by SESAM," SAE 980047, 1998.
2. J. McLeod, et al., "A Sampling System for the Measurement of PreCatalyst Emissions from Vehicles Operating Under Transient Conditions," SAE 930141, 1993.
3. M. Sawano, et al., "A Study of Low-Emission Measurement Techniques," Journal of the Society of Automotive Engineers of Japan, 9631010, 1996.
4. T. C. Austin and L. S. Caretto, "Improving the Calculation of Exhaust Gas Dilution During Constant Volume Sampling," SAE 980678, 1998.
5. F. Black and R. Snow, "Constant Volume Sampling System Water Condensation," SAE 940970, 1994.
6. J. Saito, "LEV Exhaust Emission Measurement," Journal of the Society of Automotive Engineers of Japan, 9631029, 1996.
7. W. M. Silvis, "Constant Volume Sampler CVS-7000 Series-Variable Flow CVS-7600 Type Based on Sub-Sonic Venturi Flowmeter," Readout Horiba Technical Reports, No. 11, pp. 35-42, 1995.
8. J. F. Hood and W. M. Silvis, "Predicting and Preventing Water Condensation in Sampled Vehicle Exhaust for Optimal CVS Dilution," SAE 980404, 1998.
9. J. Velosa, "Error Analysis of the Vehicle Exhaust Emission Measurement System," SAE 930393, 1993.

2002-01-0048

Studies on Enhanced CVS Technology to Achieve SULEV Certification

H. Behrendt and O. Mörsch
DaimlerChrysler AG

C. T. Seiferth
BMW AG

G. E. Seifert
Porsche AG

J. W. Wiebrecht
Audi AG

Copyright © 2002 Society of Automotive Engineers, Inc.

ABSTRACT

For the measurement of exhaust emissions, Constant Volume Sampling (CVS) technology is recommended by legislation and has proven its practical capability in the past. However, the introduction of new low emission standards has raised questions regarding the accuracy and variability of the CVS system when measuring very low emission levels.

This paper will show that CVS has the potential to achieve sufficient precision for certification of SULEV concepts. Thus, there is no need for the introduction of new test methods involving high cost.

An analysis of the CVS basic equations indicates the importance of the Dilution Factor (DF) for calculating true mass emissions. A test series will demonstrate that, by adjusting the dilution and using state of the art analyzers, the consistency of exhaust results is comparable with those of LEV concepts, measured with conventional CVS systems and former standard analyzers.

Blank tests and vehicle emission tests demonstrate that emissions at SULEV level can be measured with high accuracy.

Finally, this paper discusses the influence of individual components of the CVS system on exhaust results. It suggests an enhanced CVS system that detects SULEV limits with high precision.

INTRODUCTION

Since CVS technology was recommended in 1972 by legislation of the USA /1/ including California for the determination of exhaust emissions, this technology has established itself as a standard worldwide. All emission limits defined since then are checked by using the CVS technique. All measures introduced to tighten the regulations were taken on this basis, and consequently SULEV, as the most stringent exhaust emission limit currently in the USA, is also based on this definition.

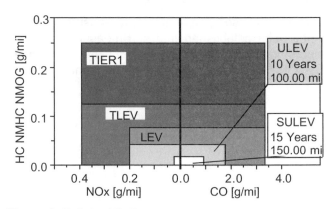

Figure 1: Pollutant limits

The limit for hydrocarbons HC/NMHC/NMOG, i.e. 0.01 g/mi, corresponds to 0.28% of the original value from 1972 (3.6 g/mi). While the exhaust emission limits have been reduced to a fraction as can be seen in Fig. 1, nothing has been changed in the basic formulas or in the measuring technology principle since introduction of CVS technology in the seventies. Requesting the measurement of the extremely low emission limits, CVS technology is now the subject of discussion. On the one hand, there is doubt about the attainable measuring accuracy since vehicles in some cases produce pollutant concentrations in the exhaust gas below those in the ambient air. On the other hand, the calculation of the emitted pollutant mass may lead to negative results. A question that arises here is whether these are measurement errors or results based on the principle of CVS technology. For this reason alternative methods are currently discussed as replacements for CVS technology /2,3,4/. However, new methods not only have to display the necessary measuring accuracy, but also verification that their results are equivalent to those of CVS technology. This applies both to determination of the mass of the pollutant emissions and to fuel economy.

To clarify the potential of CVS technology and the calculation of negative mass emissions a theoretical analysis will be conducted here with a derivation of the basic equations. An error analysis and resulting enhancement options will be presented. Improvements of CVS technology have already proven effective in practice for ULEV concepts /5/. Further studies for enhancement of CVS technology for SULEV were conducted and are reported on here. The enhancement capabilities determined in this process must be examined with an eye to technical feasibility, suitability in practice as well as cost aspects in order to maintain worldwide acceptance.

CVS-PRINCIPLES

The US Federal Register defines the pollutant masses indicated in the limits as the "true mass" (absolute emitted mass) at the tailpipe of the vehicle /6,7/. CVS technology is specifically described in the US /7/, European /8/ and Japanese /9/ test regulations for determination of pollutant emissions. The calculation equations for determination of the pollutant mass are described explicitly in the Federal Register /10/. The equations indicated there contain simplifying assumptions that may be noticeable in mass calculations at the SULEV level. This will be shown in the following analysis of the CVS basic equation.

ANALYSIS OF BASIC CVS EQUATION

The CVS equations are derived from the basic principle of conservation of mass within a control zone. The CVS system can be analyzed as a control zone with incoming and outgoing masses, differentiated for the components i, Fig. 2.

For the CVS system control zone the following balance can be drawn up for the exhaust gas component i at the *entry*. The inflowing pollutant mass $m_{i\ entry}$ of the component i is determined on the basis of the dilution air volume V_{air}, the pollutant concentration $c_{i\ air}$ of the dilution air, as well as its density ρ_i and the incoming undiluted

exhaust gas volume V_e with the respective concentration $c_{i\ e}$ and density ρ_i:

$$m_{i\ entry} = V_{air} \cdot c_{i\ air} \cdot \rho_i \ + V_e \cdot c_{i\ e} \cdot \rho_i \qquad \textbf{Eq. 1}$$

The significance of the designations will be pointed out in the following balance analyses.

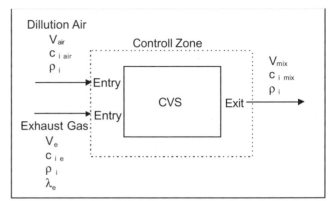

Figure 2: Control zone: CVS system

At the *exit* of the CVS system the pollutant mass $m_{i\ exit}$ is determined through the CVS volume V_{mix}, the respective pollutant concentration $c_{i\ mix}$ and the density ρ_i, based on:

$$m_{i\ exit} = V_{mix} \cdot c_{i\ mix} \cdot \rho_i \qquad \textbf{Eq. 2}$$

Under the condition that a change in mass for the component i in the CVS system due to further chemical reactions is ruled out, the following applies:

$$m_{i\ entry} = m_{i\ exit} \qquad \textbf{Eq. 3}$$

i.e. the incoming component mass $\mathbf{m_{i\ entry}}$ and the outflowing component mass $\mathbf{m_{i\ exit}}$ are equal. Furthermore, it must be ensured in the CVS system that the condition $\mathbf{V_e \le V_{mix}}$ is met at all times.

The following variables are measured by means of current CVS technology:

- CVS volume $\mathbf{V_{mix}}$,
- Diluted pollutant concentration $\mathbf{c_{i\ mix}}$,
- Dilution air concentration $\mathbf{c_{i\ air}}$.

The volume of the dilution air $\mathbf{V_{air}}$ or of the undiluted exhaust gas $\mathbf{V_e}$ is not measured. CVS technology circumvents this problem as follows:

By using **Eq. 1 – 3**, the following results:

$$V_e \cdot c_{i\ e} \cdot \rho_i = V_{mix} \cdot c_{i\ mix} \cdot \rho_i - V_{air} \cdot c_{i\ air} \cdot \rho_i \qquad \textbf{Eq. 4}$$

According to the CVS principle, the following applies:

$$V_{air} = V_{mix} - V_e \qquad \textbf{Eq. 5}$$

For the absolute pollutant mass follows:

$$m_{ie} = V_{mix} \cdot c_{imix} \cdot \rho_i - (V_{mix} - V_e) \cdot c_{i\,air} \cdot \rho_i \qquad \textbf{Eq. 6}$$

The unknown exhaust gas volume V_e is shown with the dilution ratio DR as follows:

$$DR = \frac{V_{mix}}{V_e} = \frac{V_{mix}}{V_{mix} - V_{air}} \qquad \textbf{Eq. 7}$$

Accordingly, the absolute pollutant mass $m_{i\,e}$ is determined as calculated "true mass" by using Eq. 6 and Eq. 7 as follows,

$$m_{ie} = V_{mix} \cdot \rho_i \cdot (c_{i\,mix} - c_{i\,air} \cdot (1-1/DR)) \qquad \textbf{Eq. 8}$$

as it also continues to be used for ULEV-SULEV limits in /10/ for determination of the mass.

For the dilution ratio DR, Eq. 7, the Federal Register for gasoline fueled vehicles (H/C ratio of the fuel = 1.85) gives the following definition DF as an approximation of DR /10/:

$$DF = \frac{134000}{CO_{2\,mix} + CO_{mix} + HC_{mix}} \qquad \textbf{Eq. 9}$$

with HC_{mix} in ppm C1, CO_2 and CO in ppm

A derivation for this definition is not indicated in the CFR, it has not changed since the introduction of DF in the early 70th.

ANALYSIS REGARDING DILUTION FACTOR DF

DF is calculated according to Eq. 9 from the relationship between the theoretically formed CO_2 concentration in the humid, undiluted exhaust gas of an ideal combustion and the measured diluted concentrations of the hydrocarbon products (CO_2, CO, HC).

The basic assumption behind this DF calculation is that for actual combustion, only the stochiometric amount of air is used.

$$C_8H_{14,8} + 11{,}7 \cdot (O_2 + 3{,}77 \cdot N_2) \rightarrow 8\,CO_2 + 7{,}4\,H_2O + 44{,}14N_2$$
$$(H/C = 1{,}85;\ m = 111g) \qquad 13{,}4Vol\% + 12{,}4Vol\% + 74{,}2Vol\%$$
$$\textbf{Eq. 10}$$

For other fuels, such as methanol, LPG and CNG, other DF definitions are given in /10/. They are also based on stochiometric equations.

In the case of secondary air feed, fuel cutoff in the overrun, etc. the exhaust gas volume actually emitted by the engine, V_e, is not exactly determined by Eq. 9. It is reduced to stochiometric conditions.

Another simplification in Eq. 9 is that the measured diluted concentrations used without correcting the influence of dilution air background level of 0.04 Vol%.

To determine the dilution ratio, Eq. 7, V_e must be measured directly or indirectly via $V_e = V_{mix} - V_{air}$ through measurement of V_{air}.

The calculation of masses according to Eq. 8 with DF, Eq. 9, and DR, Eq. 7, have to lead to different results even with identical CVS measured values. Using Eq. 8 and Fig. 3 the mass difference

$$\Delta m_{ieDF-DR} = m_{ieDF} - m_{ieDR}$$

is expressed by:

$$\Delta m_{ieDF-DR} = \rho_i \cdot V_{mix} \cdot c_{i\,air} \cdot (1/DF - 1/DR) \qquad \textbf{Eq. 11}$$

with

$$DF = V_{mix}/V_{eDF}, \quad DR = V_{mix}/V_e$$

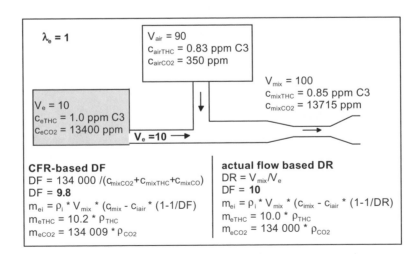

Figure 3: Comparison of the mass calculation according to DF and DR, example 1

and the approximation, appendix 1

$$V_{eDF} \approx V_e/\lambda_e$$

follows:

$$\Delta m_{ieDF-DR} \approx \rho_i \cdot V_e \cdot c_{i\,air} \cdot (1-\lambda_e) \qquad \textbf{Eq. 12}$$

Eq.12 shows that only for $\lambda_e \neq 1$ a mass difference proportional to V_e, $c_{i\,air}$ and $1 - \lambda_e$ appears. There is no influence of $c_{i\,mix}$ in mass difference $\Delta m_{ieDF-DR}$. That means mass difference is in good approximation independent on emission level.

To illustrate this, examples 1 – 3 are calculated with DF and DR, Fig. 3 and Appendix 2.

1. Exhaust gas values of a spark ignition engine operated on an average $\lambda_e = 1$
2. Exhaust gas values of a spark ignition engine with an average $\lambda_e = 1.2$
3. Exhaust gas values of a spark ignition engine with an average $\lambda_e = 1.2$ and complete exhaust gas purification

Mass calculations by DF and DR, for examples 1 - 3 are carried out for component HC (low emission limit, dilution air and diluted exhaust gas concentrations within same range) and will be examined against true mass in undiluted exhaust.

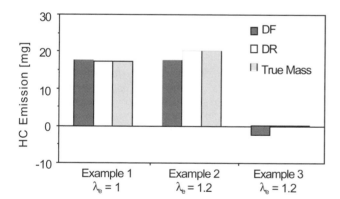

Figure 4: Comparison of HC mass emission with calculation according to DF and DR (example 1-3)

The results of this comparison are shown in the Fig. 4. The calculation of the mass with the real dilution factor DR corresponds to the "true mass". Whereas in case 1 the difference between calculation with DF and DR is not significant and is only due to neglecting CO_2 background, in case 2 and 3 differences can be seen.

To summarize the result:

- CVS-Mass calculation formula, Eq.8, includes an approximation for Dilution Ratio DR by Dilution Factor DF. The calculation of DF by diluted exhaust

concentration, Eq. 9, is a basic assumption and has not changed since introduction of CVS- Technology for emission testing. /1/.

- Eq. 11 and 12 explain a difference in mass-calculation by Eq. 8 if DF is not equal DR. In the case of extremely low-emission vehicles like SULEV (example 3) the simplifying assumptions in mass-calculation, Eq.8 , will lead to negative emissions.

ERROR ANALYSIS OF MASS CALCULATION

With CVS Technology true mass is calculated from CVS-measurement values. It must be expected in general for all measured values that they are not free of measurement-errors. To examine the error in the calculation of mass, an error propagation calculation based on Eq. 8 is carried out.

Although SULEV represents an NMOG limit value, a more far-reaching analysis is conducted only for HC and not for NMOG. The reason for this is that NMHC as the basis for NMOG can be determined in different ways and one single standardized mathematical error analysis is thus not possible. Instead, typical HC values for SULEV vehicles are examined here. This also maintains comparability to previously conducted error analyses regarding Tier1, LEV I and LEV II /5,11 /.

Based on Eq. 8, the error is made up of $dm_{i\,e} = f(dV_{mix}$ ($V_{mix} = f(Q,t)$), $dc_{i\,air}$, $dc_{i\,mix}$, dDF, dd). For the distance d, time t and flow rate Q the legally permissible deviations were taken as the basis and for the concentration measurement the manufacturer's specifications for the analyzers (Appendix 3).

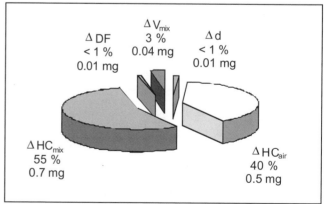

Figure 5: Composition of the calculated probable error (ΔmHC e $= \pm1.3$ mg/mi $\equiv 100\%$) at weighted mass emission of 6 mg/mi in FTP75

For a SULEV a typical weighted HC emission in an FTP75 test is 6 mg/mi. From the theoretical error calculation, Eq. A7, the probable error for HC in this case is ±1.3 mg/mi (= 22%). The decisive sources of errors in HC-mass calculation are given by concentration measurement of air and exhaust sample bags, Fig. 5.

Further error variables are negligible. The error for DF shown here is merely the random error that results through uncertainty in the measurement of the concentration of CO_2 (and CO, HC). It has nothing to do with the systematic error due to the simplifications of DF that were discussed in the previous section, Eq. 11 and Eq. 12.

ENHANCING CVS- TECHNOLOGY

To reduce the statistical error in the CVS based mass calculation the following possibilities result from the above considerations:

- Reduce diluted exhaust gas volume V_{mix} for a higher pollutant concentration in diluted exhaust gas and a reduced mass calculation error by an optimized dilution (Eq. A3 – A5),
- Improve accuracy of measurement of q_{mix} and q_{air} (Fig. 5, Eq. A3 – A5).
- Reduce background concentration $c_{i\,air}$ (Eq. A3 – A5)

OPTIMIZED DILUTION - As shown before, a reduction of V_{mix} means an increase of the share of exhaust gas volume V_e in the diluted exhaust gas V_{mix} and thus in the exhaust sample bag. As the error analysis shows before, Eq. A3 - A5, a reduction of V_{mix} proportionally reduces the mass calculation error.

In the following section, the theoretical minimum dilution for an FTP75 will be derived. The CVS-principle requires:

- $V_{mix} \geq V_e$ (DF $\geq$ 1) at all times,
- no condensation of water in the sampling system at any time,
- no condensation of water in the sample bags at any time.

As DF-calculations based on CO_2-measurements show, Fig. 6, the lowest possible average dilution factor that ensures DF $\geq$ 1 at all times during phase 1 of an FTP 75 is typically DF $\approx$ 4. The same applies for phase 2 and 3. Under these extreme conditions the minimum DF at t $\approx$ 200 seconds of cold start is around 1.2. This is almost the theoretical limit of the constant volume sampling with respect to the optimized dilution. No measurements were done under these extreme conditions.

As described in Eq. 10, at stochiometric combustion of gasoline with air the raw exhaust gas includes approx. 12.4 volume percent water vapor. The conditions in the CVS system are to be designed such that this water vapor cannot condense at any time during emission measurement. This means that for a given average DF the temperature of the sampling system has to be kept above the dew point at the point of maximum water vapor content in V_{mix}. Time-resolved measurements show that, apart from the cold start phase, the necessary information on the water content in the diluted exhaust gas is

obtained through the CO_2 curve over time. In Fig. 7 the maximum humidity is reached at t $\approx$ 200 s with an absolute humidity of 17 $g_{H2O}/kg_{dry\,air}$.

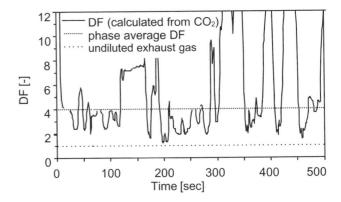

Figure 6: Time resolved dilution factor for an average DF of 4 (calculated from CO_2), Phase 1 FTP75

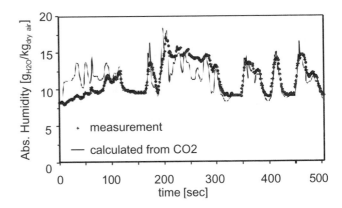

Figure 7: Abs. humidity in mixed exhaust gas at sample venturi in phase 1 of the FTP-75, comparison of a measurement (capacitive humidity sensor) and values calculated from CO_2 (DF = 13)

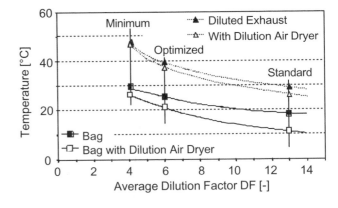

Figure 8: Temperature to avoid water vapor condensation in diluted exhaust gas lines and bags as function of average dilution factor DF

The corresponding minimum temperature of the sample lines and the exhaust gas bags for an FTP75 with a DF = 13 can be taken from Fig. 8. It also shows the required minimum temperature in the diluted exhaust gas lines and in the exhaust gas bags with and without application of dilution air dryer has been calculated as function of the average dilution factor. The values are calculated from the CO_2 data. Background humidity was assumed to be 2 $g_{H2O}/kg_{dry\ air}$ with dilution air dryer and 8 $g_{H2O}/kg_{dry\ air}$ without.

In the sample bag the exhaust gas, which is very moist for a short time, is mixed with less moist exhaust gas. As a consequence, the moisture in the exhaust gas bags and the required minimum temperature is considerably lower than for the sampling lines.

Based on these considerations the minimum DF of 4 requires heating of the sampling system to T ≈ 50 °C and the sample bags to T ≈ 30 °C. The theoretical DF of 4 is not feasible in practice for several reasons, such as:

- unknown vehicle emissions before test,
- actual water concentration in cold start differs slightly from theoretical value, Fig 7,
- safety margin required for a sound process.

For an FTP75 an average DF = 6 in each phase is, therefore, recommended here. This can be achieved by using a multiple critical flow venturi or a PDP-type CVS with multiple speed operation. Sample bags should then be heated up to 30 °C and diluted exhaust gas lines to 50 °C. The sample lines can be heated directly. The bulk flow can be heated by using a dilution air heater. By using a dilution air dryer, the bag temperature can be reduced (Fig. 8). With respect to standard CVS dilution of DF = 13 a DF of 6 reduces V_{mix} to about 50 per cent in phase 1 and 75 per cent in phase 2, respectively, table 1. As stated above this decreases the statistical error of mass calculation accordingly.

FTP-75 Test cycle	Standard CVS sampling	DF-optimized CVS sampling	Increase in exhaust gas volume in exhaust gas bag
Phase 1	DF = 13	DF = 6	2–fold
Phase 2	DF = 22	DF = 6	4–fold
Phase 3	DF = 16	DF = 6	3–fold

Table 1: DF of a standard and optimized CVS setting

Equivalent considerations for special tests, such as SC03 or US06, are presented in /5/.

ACCURACY OF ANALYZERS - For the exhaust gas measurement of extremely low emission vehicles besides standard analyzers, nowadays more expensive analyzers that have been specially optimized for very low measurement ranges (low-emission analyzers) are also available.

In Fig. 9 the zero and span stability of modern standard analyzers based on zero scatter and full-scale scatter is compared to low-emission analyzers. Zero and full-scale scatter are defined via mean and standard deviation of the difference between zero and span calibration before and zero and span check after the bag analysis. The data were obtained through evaluation of exhaust gas tests over a period of approx. 4 months. The measurement ranges 10 ppm C3 for HC and 10 ppm for NOx were used. HC and NO_x LE analyzers are slightly improved by lower zero-scatter against standard analyzers in practice though a systematic upward deviation exists with the HC full-scale value. The full-scale scatter is significantly less than the manufacturer's specification (± 0.1 ppm C3) with all analyzers. However, the standard analyzers lie slightly outside the specification (± 0.004 ppm) for the zero scatter, though the above described approach displays the real scatter in the measurement and does not correspond to the determination of the verifiable limit according to specification.

Since particularly the full-scale deviation of state of the art standard analyzers provides considerably better values in practice than the specification, there is no decisive advantage in using the significantly more expensive LE analyzers examined here for extremely low measurement ranges. In addition, with the HC analyzer concentrations of less than 0.7 ppm C3 do not occur anyhow because of the background.

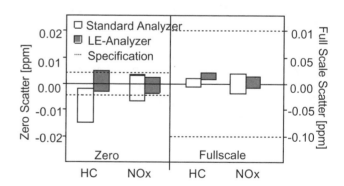

Figure 9: Re-zero and re-span scatter of state of the art standard and LE analyzers

SAMPLE BAG OUT-GASSING - Emissions of hydrocarbons (mostly DMA = NN-dimethylacetamid and phenole /12/) from the bag material (e.g. Tedlar) are a known phenomenon /13/ and certainly of special interest for measurement of HC/NMHC/NMOG - emissions at SULEV level. To quantify the out-gassing, the bags were filled with synthetic air and in some tests additionally with CO2 and water vapor. The bag temperature was 35°C. The HC concentration in the bag was determined with the standard FID of the analyzer bench after 10, 25 and 40 minutes. With this procedure contamination from the sampling system is included.

Fig. 10 shows the results. In the case of the filling with synthetic air, a HC concentration in the bag of approx. 0.06 ppm C3 results after 10 minutes. The differences in the HC concentrations between synthetic air and CO2 in synthetic air are not significant here. For CO2 and H2O in synthetic air, however, considerably higher values result, i.e. approx. 0.1 ppm C3, which means the out-gassing of HC is reinforced by moisture (DMA is soluble in water).

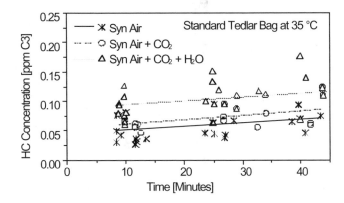

Figure 10: Out-gassing from Tedlar bag at 35 °C for different gas compositions

Through pretreatment of the tedlar bags /14/ out-gassing can be reduced from 0.06 to 0.02 ppm C3, Fig. 11. As discussed above, 30 °C is sufficient for an FTP 75 test. Out-gassing is than further reduced to approx. 0.005 ppm C3 which is almost identical to the out-gassing from an unheated bag (24 °C).

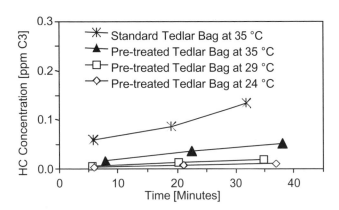

Figure 11: Out-gassing from Tedlar bags at different temperatures (filled with synthetic air)

CVS SYSTEM TEST (BLANK TESTS) - Repeatability, zero and mass-noise on zero-emission level of the entire CVS system (CVS unit, analyzer system, measuring procedures) can be estimated through so-called blank tests. A blank test is defined such that an emission test (FTP75) is carried out without vehicle. The exhaust inlet is closed and exhaust and ambient sample bags are filled with dilution air only. By Eq. 9 and 8 mass emission is calculated using CVS - measurement values and a DF = 8. Because no emitting vehicle is operated in blank test, for all pollutants mass-emission zero is expected.

Blank tests were carried out on a SULEV test cell with heated bags (35°C). NMHC as in all following figures was determined measuring CH_4 with a gas chromatograph. For HC and NMHC almost identical values were found with NMHC scatter only slightly larger. Both components showed a scatter of ± 0.1 mg/mi with an offset of −0.2 mg/mi. For NO_x the scatter is ± 0.05 mg/mi, an offset was not observed, Fig. 12.

For more realistic blank tests, sample flow to diluted exhaust bag was moistened in an additional series of measurements to a absolute humidity of 18 $g_{H2O}/kg_{dry\ air}$. Compared to the blank test without moistening, a detectable increase in scatter to ± 0.35 mg/mi results for HC and NMHC, which corresponds to 3.5% of the SULEV emission limit. Together with the offset of 0.2 mg/mi the maximum deviation, i.e. 0.6 mg/mi, is still only 6% of the SULEV emission limit. The observed shift in the mean value is due to increased out-gassing in the exhaust gas bag due to moisture. Out-gassing in sample and ambient bag in this case is not completely compensated. Scatter of NO_x in blank tests is not influenced by increased humidity in diluted exhaust, Fig. 12.

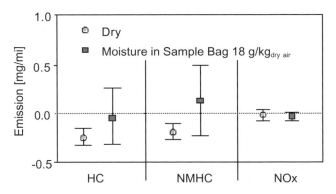

Figure 12: Mean value and standard deviation of 10 FTP75 blank tests with and without moisture in the exhaust gas bag (bag heated to 35°C)

SCATTER OF SULEV EMISSION TESTS - Most important for suitability of CVS-technology is scattering in vehicle tests at SULEV level. Fig. 13 shows emission results of FTP75 exhaust gas tests with 4 different SULEV development cars in an emission test cell configured as shown in Fig. 17. Each vehicle was tested 4-times on a FTP75-cold test. For NMHC an average scatter of approx. ± 1 mg/mi or approx. 10% of the SULEV emission limit was detected. The scatter of the NO_x emission results is around ± 2 mg/mi, which also corresponds to 10% of the SULEV emission limit.

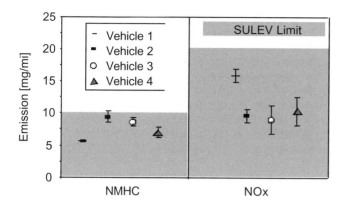

Figure 13: Mean value and standard deviation NMHC and NOx of 4 successive FTP75 cold tests on 4 SULEV development cars in each case (NMHC shown instead of NMOG)

If one compares the scatter occurring in these vehicle tests to the scatter from the above described blank test, Fig. 14, it can be seen that the total scatter of the FTP75 tests in phase 1 is one order of magnitude above the scatter in the blank test. In phase 2, in which the vehicle produces practically no HC or NOₓ emissions, the scatter of the blank test is close to that of the FTP75 exhaust gas test. This means that the scatter in phase 1 is not significantly caused by the measuring technique. A significant influence on the emission result therefore stems from the vehicle itself or from not known influences in the test procedure.

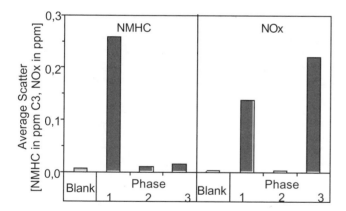

Figure 14: Comparison of the scatter in the blank test from Fig. 12 to the average scatter of the FTP75 from Fig. 13

HANG-UPS - An example of how the test handling influences the emission result concerns so-called hang-ups. These represent contamination of the measuring system partially released in the following exhaust gas test and thus falsifying the emission result. To study the extent to which these hang-ups in the exhaust gas test have an impact on the following test, three blank tests were conducted subsequent to the exhaust gas tests, Fig. 15.

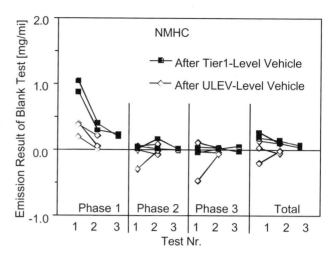

Figure 15: Effect of previous test on the result of blank tests (hang-ups)

Depending on the emission level of the previously tested vehicle, a significantly increased value, which may amount to over 1 mg/mi results in phase 1 during the following test. Since this effect is only observed in phase 1, it involves contamination in the sampling line and in the exhaust sample bag. Although the influence on the total emission result, i.e. a maximum of about 0.3 mg/mi, is still acceptable in the cases shown, a SULEV emission test after a high-emission vehicle may be strongly influenced by hang-ups. To minimize the effects of the previous test, it is useful to equip the test cell with an additional sampling line for higher emissions (dirty line and dirty bag). Additional improvements can be achieved by purge procedures (e.g. blank test).

DISCUSSION

CVS technology has clearly proven to be effective for the certification of vehicles with exhaust gas limits according to ULEV in the past. This is substantiated by the EPA, among others /12/.

For HC emissions which are typical for SULEV (6 mg/mi) a calculated probable error of over 20 % was determined on the basis of a theoretical error analysis. Consequently the measuring accuracy for SULEV certification by means of CVS technology initially appears questionable. Measurements show that the actual mass error is considerably smaller than the theoretical. This is shown by Fig. 16 which compares the calculated error with HC scatter in the blank test (± 0.1 mg/mi HC), and in FTP75 phase 2 vehicle tests (± 0,2 - 0,5 mg/mi). Both are significantly below the scatter (probable error) from the error calculation, which is of ± 2.4 mg/mi for phase 2, appendix 2. Reasons for this:

- The error in the measurement of c_{mix} and c_{air} are not independent as assumed by the error calculation. By using the same analyzer (identical calibration,

zero/span adjustment, etc.) the influence of errors on the mass calculation will be compensated such as:
- Zero and full-scale drift
- Non linearity (especially at very low emission levels since the concentrations of sample and ambient bag are almost identical)
- State of the art analyzers measure significantly more reproducible than the manufacturer's specifications state.

In addition to the theoretical errors discussed above, the CVS principle mostly compensates out-gassing from bag material (same bag material, aging and temperature).

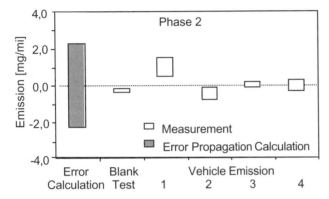

Figure 16: FTP 75, Phase 2, comparison of calculated error (appendix 2), blank test scatter and vehicle measurement scatter, Fig. 14.

Since out-gassing is not completely compensated due to different moisture in exhaust and air bag during emission tests, excessive heating above 30 °C has to be avoided.

In test series with SULEV development cars a HC scatter of approx. 10% of the SULEV emission limit was verified. The same applied to NO$_x$, Fig. 13. This shows that the SULEV emission limit can be verified with a high degree of certainty with CVS technology.

For vehicles with internal combustion engines with $\lambda_e > 1$, the existing definition of the DF in the legislation may lead to the calculation of negative total emissions, especially in phase 2 and 3. This can be avoided by determining DR by measurement of the dilution air volume V_{air} or by dilution air refinement. However, the typical error is negligible even at SULEV level. It is further reduced when negative emission results are set to zero as stipulated by the EPA /15/.

Fig. 17 shows a SULEV test cell with components that are meaningful also from a cost/benefit point of view.

In particular, heating and drying of the dilution air, heating of gas lines up to 50°C and bags up to 30°C is recommended to avoid condensation with optimized dilution.

Application of catalytic conditioning of the dilution air can be a possibility to further increase accuracy and to eliminate the systematic difference due to Eq. 12. This studies show, that SULEV certification without DAR can be done with high accuracy.

The commonly applied charcoal filter predominantly serves to smooth the background /16/.

A separate sampling line for high emissions (e.g. the cold start phase) and the use of pretreated bags reduce the problem of hang-ups and out-gassing.

The heat exchanger in the CVS provides for a nearly constant volume flow, which makes its determination considerably more accurate. Through the adjustable flow rate (multi-venturi or PDP) the dilution can be optimized for each test phase.

State of the art analyzers with a high degree of

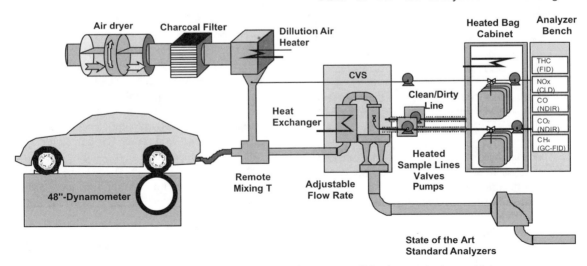

Figure 17: Recommended equipment for a test cell for low-emission vehicles

reproducibility are required, but more expensive analyzers specially optimized for very low measuring ranges (HC < 3 ppm C3; NOx < 10 ppm) are not necessary. On the one hand, because appropriate calibration gases are not available and, on the other hand, because for HC, for example, no concentrations less than 0.7 ppm C3 occur.

Furthermore, a precisely defined test procedure is decisive for less scatter in vehicle measurement, i.e.:

- Use SULEV test cell only for extremely low-emission vehicles
- Purge procedure
- Defined vehicle preconditioning
- Defined vehicle state (e.g. fuel composition, etc.)
- Precisely defined ambient conditions

CONCLUSIONS

The CVS technology is explicitly proposed by the legislature. A departure from the basic calculation formulas in the CFR would lead to a change in the mass emissions in comparison to the LEV I and LEV II definition. A discussion of the fundamental assumptions of CVS technology for the calculation of pollutant mass emission showed:

- The basic calculation formulas in the CFR can lead in some cases to negative mass emission results.
- For vehicle concepts with $\lambda_e \sim 1$ these are negligible even at SULEV level.

The results of a theoretical error analysis were compared to blank tests and SULEV emission tests. This showed that the real measuring uncertainty is considerably less than the uncertainty estimated via the error analysis.

Practical studies show:

- Due to the differential measurement principle of CVS technology, some errors are almost compensated.
- Through reduction of the dilution to DF=6, the concentration difference between ambient air and exhaust sample bag and, hence, the mass calculation accuracy can be nearly doubled.
- Vehicle measurements at SULEV level resulted in a scatter of ± 10% of the emission limit which shows that the SULEV emission limit can be verified with a high degree of certainty.
- Meticulous test preparation and execution is absolutely imperative for reproducible measurement results.
- SULEV test cells need enhanced CVS:
  - Charcoal filter
  - Heating dilution air
  - Remote mixing T
  - Separation of sampling lines (clean/dirty line)
  - Heat exchanger in the CVS
  - Heated sampling lines

- pre-treated bags
- State of the art analyzers

A reduction of DF additionally requires:
- Drying of dilution air
- Optimized dilution
- Pre-treated bags heated up to 30 °C
- Sample Lines heated up to 50 °C

CVS technology has been tested for many years and studies show that an enhanced CVS test cell has the required accuracy for certification of SULEV concepts. As a consequence, the worldwide comparability of the exhaust gas results is maintained.

REFERENCES

1 Mondt, J. Robert: Cleaner cars: the history and technology of emission control since the 1960; Copyright SAE, Warrendale; PA 15096-001 USA; ISBN 0-7680-0222-2

2 AIGER Bag Mini Diluter Technical Exchange Meeting , August 22-24,2000, Detroit.

3 W. M. Silvis, R. N.I Harvey, and A. F. Dageforde, A CFV Type Mini-dilution Sampling System for Vehicle Exhaust Emissions Measurement, SAE 1999-01-0151

4 K. Guenther, T. Henney, W.M. Silvis, S. Nakatani, and Dien-Yeh Wu, Improved Bag Mini-diluter Sampling System for Ultra-low Level Vehicle Exhaust Emissions, SAE 2000-01-0792

5 A. Gifhorn, H. Tieber, J. W. Wiebrecht, R. Ballik, Basic Investigations into the use of the CVS System for Ultra- low-emission Vehicles with Gasoline Engines, MTZ 61(2000)2, Pg. 106, MTZ Worldwide 61 (2000)2, Pg. 17

6 40 CFR 86.082-2 Definitions

7 40 CFR 86.109-94 Exhaust gas sampling system; Otto-cycle vehicles not requiring particulate emission measurement

8 70/220/EEC;1999/102 EC; Appendix 8

9 Automobile Type Approval Handbook For Japanese Certification; Edited as of December 31,1993; Japan Automobile Standards Internationalization Center; 5 Technical Standard for 10-Mode and 11-Mode Exhaust Emission Measurement for Gasoline-Fueled Motor Vehicles.

10 40 CFR 86.144-90 Calculations; exhaust emissions.

11   M. T. Sherman, R. E. Chase, and K. M. Lennon, Error Analysis of Various Sampling Systems, SAE 2001-01-0209

12   E.I. Sun, and W. N. McMahon, Evaluation of Fluorocarbon Polymer Bag Material for Near Zero Exhaust Emission Measurement, SAE 2001-01-3535

13   Sawano, Yoda, and Uchida, A Study of Low-Emission Measurement Techniques, JSAE 9631010, 1996

14   HORIBA Europe procedure, Sulzbach, Germany

15   Correcting Negative Calculated Emission Levels to Zero, United States Environmental Protection Agency; February 8, 2001

16   J. Heckelmann, Analyse und Verification verfahrenstechnischer Apparate zur Bereitstellung von stabiler und getrockneter Verdünnungsluft bei der CVS Abgasmessung, Diplom Thesis Fachhochschule Heilbronn, Institut für Verfahrenstechnik, 2001

CONTACT

Harald Behrendt and Dr. Oliver Mörsch
DaimlerChrysler AG
HPC D606
D-70546 Stuttgart
Germany
email: Harald.Behrendt@daimlerchrysler.com
email: Oliver.Moersch@daimlerchrysler.com

Christa T. Seiferth
BMW AG
Petuelring 130
D-80788 München
Germany
email: Christa.Seiferth@bmw.de

Dr. Gerd E. Seifert
Dr. Ing. h.c. F. Porsche AG
Porschestraße
D-71287 Weissach
Germany
email: Seiferge@porsche.de

Dr. Jörg W. Wiebrecht
AUDI AG
N/EA-521
D-74172 Neckarsulm
Germany
email: Joerg.Wiebrecht@audi.de

SYMBOLS

c	concentration
c_{iair}	dilution air pollutant concentration.
c_{ie}	exhaust gas pollutant concentration
c_{imix}	diluted pollutant concentration
CO	carbon monoxide concentration
CO_2	carbon dioxide concentration
HC	total hydrocarbon concentration
d	distance
DR	dilution ratio
DF	dilution factor
k	k-factor
m	mass
$m_{i\,eDF}$	Mass calculated using DF
$m_{i\,e}$	Mass calculated using DR
$m_{i\,entry}$	incoming component mass
$m_{i\,exit}$	outflowing component mass
M	weighted mass
p	pressure
Q	CVS flow rate
T	temperature
t	time
V	volume
V_{air}	dilution air volume
V_e	undiluted exhaust gas volume
$V_{e\,DF}$	Ve calculated using DF
V_{mix}	CVS volume
δ	differential
Δ	difference
λ	air/fuel ratio
λ_e	average air/fuel ratio over test
ρ_i	density

ABBREVIATIONS

CFR	Code of Federal Register
CVS	Constant Volume Sampler
CFV	Critical Flow Venturi
DAR	Dilution Air Refinement
D. L.	Detection Limit
FTP	Federal Test Procedure
LEV	Low Emission Vehicle
MR	measuring range
MV	measured value
NMHC	non-methane hydrocarbon
NMOG	non-methane organic gas
PDP	Positive Displacement Pump
SULEV	Super Ultra Low Emission Vehicle
ULEV	Ultra Low Emission Vehicle

Appendix 1:

The stochiometric combustion of one mole of the theoretical molecule $C_8H_{14,8}$ representing gasoline fuel with H/C = 1,85 according the formula:

$$C_8H_{14,8} + 11,7 \cdot (O_2 + 3,77 \cdot N_2) \rightarrow 8\ CO_2 + 7,4\ H_2O + 44,14N_2$$

yields 59.5 moles of exhaust gas. Therefore, $V_{eDF} = 59.5$. The table shows the exhaust volume in moles for different λ_e values. Assuming ideal gas behavior, the total is equal to the real exhaust volume V_e. It can be seen that the assumption $V_e = \lambda_e\ V_{eDF}$ is a good approximation.

λ [–]	1	1,2	1,5	2	3
CO_2 [mole]	8	8	8	8	8
H_2O [mole]	7.4	7.4	7.4	7.4	7.4
N_2 [mole]	44.1	52.9	66.15	88.2	132.3
O_2 [mole]	0	2.3	5.85	11.7	23.4
total =V_e [mole]	59.5	70.7	87.4	115	171
$\lambda\ V_eDF$ [mole]	59.5	71.4	89.3	119	179
Error [%]	0	1.0	2.1	3.1	4.2

Appendix 2:   Comparison of calculation DF/DR

The examples 1- 3 are theoretical analysis on the CVS system as a control zone. Based on the specifications of the engine emissions V_e and $c_{e\ i}$ as well as the dilution air V_{air}, $c_{air\ i}$, the mass calculations can be carried out according to the CFR with the dilution ratio DF (§86.544-94) and according to the real dilution ratio DR. It is shown that the mass calculation with the real dilution ratio is identical to the "true mass".

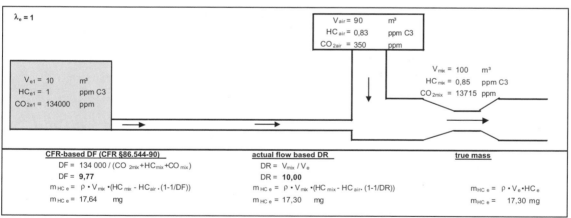

Example 1: $\lambda_e = 1$, SULEV emission level

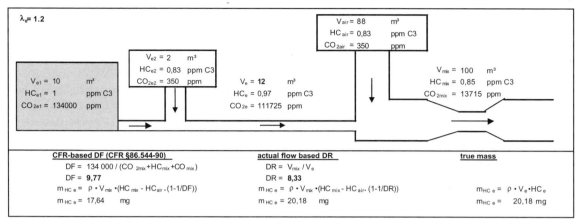

Example 2: λ_e = 1.2, SULEV emission level

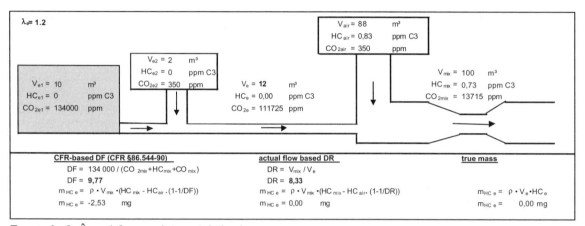

Example 3: λ_e = 1.2, complete catalytic clean up.

For λ_e = 1 there is very good consistency with the "true mass" in the HC determination of mass with DF and DR. The difference of 0.36 mg is due to the fact that the DF calculation is based only on the sample bag concentrations without correction for CO_2 background. The difference between calculation with DF and DR for λ_e = 1.2 is 2.53 mg. In the case of catalytic clean up, this leads to negative values. Since negative emission is set to zero, the final results for DF and DR in example 3 are zero in both cases.

Appendix 3: Error Calculation

1 CVS Volume V_{mix}

$$V_{mix} = Q \cdot t \qquad\qquad \textbf{Eq. A1}$$

$$\Delta V_{mix} = \sqrt{\left(\frac{\delta V_{mix}}{\delta Q} \cdot \Delta Q\right)^2 + \left(\frac{\delta V_{mix}}{\delta t} \cdot \Delta t\right)^2}$$

$$\Delta V_{mix} = \sqrt{(t \cdot \Delta Q)^2 + (Q \cdot \Delta t)^2}$$

$\Delta Q/Q = \pm 0.5\ \%$ $\qquad \Rightarrow$ legislator § 86.119-90

$\Delta t = \pm 0{,}05s$ $\qquad \Rightarrow$ legislator: §86.119-90

example:

for $\qquad$ Q = 9 m^3/min; $\qquad$ t = 505 s

$\qquad\qquad \Delta Q = 0.09$ m^3/min; $\qquad \Delta t = 0.1s$

$V_{mix} = 75.75$ m^3

$\Delta V_{mix} = 0{,}757$ m^3 => $\Delta V_{mix}/V_{mix} = 1\%$

2 Dilution Factor DF

$$DF = \frac{134000}{CO_{2mix} + CO_{mix} + HC_{mix}} \qquad\qquad \textbf{Eq. A2}$$

$$\Delta DF = \sqrt{\left(\frac{\delta DF}{\delta CO_{2mix}} \cdot \Delta CO_{2mix}\right)^2 + \left(\frac{\delta DF}{\delta CO_{mix}} \cdot \Delta CO_{mic}\right)^2 + \left(\frac{\partial DF}{\partial HC_{mic}} \cdot \Delta HC_{mix}\right)^2}$$

$$\frac{\delta DF}{\delta CO_{2mix}} \cdot \Delta CO_{2mix} = -\frac{134000}{\left(CO_{2mix} + CO_{mix} + HC_{mix}\right)^2} \cdot \Delta CO_{2mix}$$

$$\frac{\delta DF}{\delta CO_{mix}} \cdot \Delta CO_{mix}\ ;\ \frac{\delta DF}{\delta HC_{mix}} \cdot \Delta HC_{mix}\ : \text{analogous}$$

$\Delta\ CO_{2mix}/\ CO_{2mix} = 1.49\%$ (see table under 7.)

errors of CO and HC measurement are negigible, because the absolut values are small in comparision to CO_2

$\Rightarrow \Delta DF/DF \approx 1{,}5\%$

3 Mass Emission per Phase

$$m_i = \rho_i \cdot V_{mix} \cdot \left[c_{i\,mix} - c_{i\,air} \left(1 - \frac{1}{DF} \right) \right]$$

Eq. A3

ρ: constant

$$\Delta m_i = \sqrt{\left(\frac{\delta m_i}{\delta V_{mix}} \cdot \Delta V_{mix} \right)^2 + \left(\frac{\delta m_i}{\delta DF} \cdot \Delta DF \right)^2 + \left(\frac{\delta m_i}{\delta c_{i\,mix}} \cdot \Delta c_{i\,mix} \right)^2 + \left(\frac{\delta m_i}{\delta c_{i\,air}} \cdot \Delta c_{i\,air} \right)^2}$$

$$\frac{\delta m_i}{\delta V_{mix}} \cdot \Delta V_{mix} = \rho_i \cdot \left[c_{i\,mix} - \left(1 - \frac{1}{DF} \right) \cdot c_{i\,air} \right] \cdot \Delta V_{mix}$$

$$\frac{\delta m_i}{\delta DF} \cdot \Delta DF = -\rho_i \cdot V_{mix} \cdot c_{i\,air} \frac{1}{DF^2} \cdot \Delta DF$$

$$\frac{\delta m_i}{\delta c_{i\,mix}} \cdot \Delta c_{i\,mix} = \rho_i \cdot V_{mix} \cdot \Delta c_{i\,mix}$$

Eq. A4

$$\frac{\delta m_i}{\delta c_{i\,air}} \cdot \Delta c_{i\,air} = -\rho_i \cdot V_{mix} \cdot \left(1 - \frac{1}{DF} \right) \cdot \Delta c_{i\,air}$$

Eq. A5

$\Delta V_{mix} / V_{mix} = 1\%$; $\Delta DF / DF = 1{,}5\%$
$\Delta HC_e / HC_e$, $\Delta HC_{air} / HC_{air} \approx 3{,}5\%$ (see table under 6)

4 Weighted Mass M (g/mi)

$$M = 0{,}43 \frac{m_1 + m_2}{d_1 + d_2} + 0{,}57 \frac{m_2 + m_3}{d_2 + d_3}$$

Eq. A6

d_1, d_2, d_3: distance in each phase
$d_1 + d_2 = d_2 + d_3 = 7{,}48$ mi
$\Delta d = 0{,}0003$ mi
(EPA:Specification for electric chassis dynamometers, attachment A RFD c 100081T1, 1991)

$\Rightarrow$ measurement uncertainty ΔM:

$$\Delta M = \sqrt{\left(\frac{\delta M}{\delta m_1} \cdot \Delta m_1 \right)^2 + \left(\frac{\delta M}{\delta m_2} \cdot \Delta m_2 \right)^2 + \left(\frac{\delta M}{\delta m_3} \cdot \Delta m_3 \right)^2 + \left(\frac{\delta M}{\delta d_1} \cdot \Delta d_1 \right)^2 + \left(\frac{\delta M}{\delta d_2} \cdot \Delta d_2 \right)^2 + \left(\frac{\delta M}{\delta d_3} \cdot \Delta d_3 \right)^2}$$

Eq. A7

$$\frac{\delta M}{\delta m_1} \cdot \Delta m_1 = \frac{0{,}43}{d_1 + d_2} \cdot \Delta m_1$$

$$\frac{\delta M}{\delta m_2} \cdot \Delta m_2 = \frac{1}{d_1 + d_2} \cdot \Delta m_2$$

$$\frac{\delta M}{\delta m_3} \cdot \Delta m_3 = \frac{0{,}57}{d_2 + d_3} \cdot \Delta m_3$$

$$\frac{\delta M}{\delta d_1} \cdot \Delta d_1 = -\frac{0{,}43(m_1 + m_2)}{(d_1 + d_2)^2} \cdot \Delta d_1$$

$$\frac{\delta M}{\delta d_2} \cdot \Delta d_2 = -\frac{0{,}43(m_1 + m_2) + 0{,}57(m_2 + m_3)}{(d_1 + d_2)^2} \cdot \Delta d_2$$

$$\frac{\delta M}{\delta d_3} \cdot \Delta d_3 = -\frac{0{,}57(m_2 + m_3)}{(d_2 + d_3)^2} \cdot \Delta d_3$$

5 Measuring error of modern Equipment

$$\text{total error} = \sqrt{(\text{linearity})^2 + (\text{reproducib ility})^2 + \left(\frac{\text{drift}}{4}\right)^2}$$

	FID	CO_{low}	CO_2
detection limit	6 ppb für MR:10 ppm	50 ppb für MR: 50 ppm	15 ppm für MR: 20%
drift	< 0,5% MV + 2xD .L./h	<1,0% MV + 2xD .L./h	<1,0% MV + 2xD .L./h
linearity	< 1,0% MV + 2xD .L.	<1,0% MV + 2xD .L.	< 1,0% MV + 2xD .L.
reproducibility	< 0,5% MV + 2xD .L.	< 0,5% MV + 2xD .L.	< 0,5% MV + 2xD .L.

D. L. Detetion Limit
MV: measured value
MR: measuring range

6 Error Calculation in Bag Concentration Measurement

test cycle : FTP 75	HC ppm C3	CO ppm	NO$_x$ ppm	CO$_2$ ppm
Sample bag phase 1	1.677	5.125	0.474	13600
phase 2	0.887	0.185	0.042	13287
phase 3	0.877	0.165	0.078	17040
Ambient bag phase 1,2,3	0.980	0.130	0.033	420

measuring uncertainty for concentrations

		single error			total. error	total. error.
Sample bag		lin.	repro.	drift	Δc ppm	$\Delta c/c$ %
phase 1:HC$_{mix}$	ppm	0.029	0.020	0.014	0.038	2.26
phase 2:HC$_{mix}$	ppm	0.021	0.016	0.013	0.030	3.34
phase 3:HC$_{mix}$	ppm	0.021	0.016	0.013	0.030	3.37
phase 1:CO$_{2\,mix}$	ppm	166	98	64	203.12	1.49
Ambient bag						
HC$_{air}$ (all phases)	ppm	0.022	0.017	0.013	0.031	3.12

7 Total Error in Calculation of Mass

ρ_{HC} = 576.8 g/m³
ρ_{CO2} = 1835 g/m³

		phase1	phase2	phase3
flow rate Q	m³/s	0.15	0.15	0.15
time t	s	505	869	505
Volume V$_{mix}$	m³	75.75	130.4	75.75
distance d	mi	3.59	3.89	3.59
Dilution Factor DF	-	9.85	10.08	7.86
emission				
m$_{(HC)}$	mg	98	0.9	2.8
$\Delta m_{(HC)}$	mg	6	9.1	5.2
$\Delta m_{(HC)}$ /m$_{(HC)}$	%	6.0	965	184
$\Delta m_{(HC)}$	mg/mi	1.7	2.4	1.5

Weighted Total HC Emission		
M	mg/mi	6.0
ΔM	mg/mi	1.3
$\Delta M/M$	%	22

2002-01-0046

Advanced Emissions Test Site for Confident PZEV Measurements

Mark Guenther, Michael T. Sherman and Mike Vaillancourt
Correlation Engineering, Ford Motor Company

Dan Carpenter
Product Analysis and Verification, Ford Motor Company

Rick Rooney and Scott Porter
Horiba Instruments, Inc.

Copyright © 2002 Society of Automotive Engineers, Inc.

ABSTRACT

As automakers begin to develop and certify vehicles that meet the California Air Resources Board LEV II and Environmental Protection Agency Tier II Regulations, emissions test cells must be designed and implemented that are capable of accurate low-level measurements. A new test cell has been installed at Ford Motor Company for use in testing vehicles that meet the stringent Partial Zero Emission Vehicle tailpipe requirements (NMOG = 10 mg/mile, NOx = 20 mg/mile). This test cell includes a redesigned Bag Mini-Diluter (BMD), improved analytical benches, an ultrasonic exhaust flow meter with an integrated tailpipe pressure control system, a conventional constant volume sampler (CVS), and a moveable electric dynamometer. The Bag Mini-Diluter will be used as the primary sampling system for the tailpipe measurements. The moveable electric dynamometer enables the test cell to be configured so that the vehicle is moved to the test equipment rather than moving the test equipment to the vehicle.

INTRODUCTION

A new test site has been installed at Ford Motor Company to test vehicles designed to meet stringent California Partial Zero Emission Vehicle (PZEV) tailpipe emissions requirements [1]. The success of this project and return on investment will be measured by the increased confidence in exhaust emissions measurements below PZEV levels, i.e., below 10 mg/mile NMOG (Non Methane Organic Gases) and 20 mg/mile NOx (Oxides of Nitrogen). Several upgrades were necessary to each of our traditional systems that make up the test installation in order to specifically address the difficulties in making measurements at these low levels.

To help visualize the task, exhaust emissions measurements are normally made with diluted samples in the parts per million regime. These concentrations are then resolved in accordance with the regulations for final gram per mile determinations. For the new PZEV requirements, however, it is now more convenient to reference part per billion (ppb) concentrations and milligram per mile determinations to better communicate the low-level performance and accuracy necessary. For example, a systemic error of 50 ppbC applied to all three bags of the FTP (Federal Test Procedure) could amount to an error of 1 mg/mile to the final determination. To ensure accurate measurements, the requirement for a "clean" sampling system was determined to be a system contribution of no more than 30 ppbC to any measurement.

When using the conventional CVS (Constant Volume Sampling) system and testing PZEV units, it is not uncommon for the tailpipe exhaust to be cleaner than the ambient air used in the dilution process. This creates a problem for the laboratory and for those of us concerned with measurement accuracy and variability. Significant upgrades were needed to the traditional systems and methodologies normally used for emissions testing to ensure confident measurements for NMOG emissions below 10 mg/mile and NOx emissions below 20 mg/mile. For example, a Bag Mini-Diluter (BMD) was selected as the optimal sampling system for the PZEV test cell.

Several other considerations in the exhaust measurement process were also examined and refined. Apart from the test vehicle, the variability or uncertainty of an exhaust emissions test can be grouped for consideration in five different categories. These were evaluated for cost effective improvement with our suppliers and PZEV performance specifications were jointly established. Our considerations for the dynamometer, bag bench (gas analyzers used to measure diluted exhaust gas collected in sample bags), sampling system, facilities, and testing processes are presented here.

PZEV MEASUREMENT CONSIDERATIONS

DYNAMOMETER - Standard electric single roll dynamometer installations are required to meet the industry acceptance procedures established by expert representatives of AAM (Alliance of Automobile Manufacturers), AIAM (Association of International Automobile Manufacturers), EPA and ARB [2]. Additionally, for the PZEV Test Site, a movable dynamometer was specified to allow consistent vehicle positioning relative to the cooling fan, climate control system, and emissions measurement equipment. For additional accuracy, we specified a ½pound (2.2 N) tolerance for simulated load throughout the speed range. This was achieved by incorporating a high precision load sensor with a new mounting system, a redesigned motor bearing lubrication system, and improved dynamometer control algorithms.

GAS ANALYZERS (BAG BENCH) - The basic requirements for low-level emissions measurements are met with the low emission (LE) type instrumentation [3]. For this PZEV test cell application, additional improvements were made to the FID (Flame Ionization Detector) for total hydrocarbon measurement, GC (Gas Chromatography) for methane measurement, Chemi-luminescent detector for NOx measurement and NDIR (Non-Dispersive Infrared) for CO measurement. Signal to noise ratio was improved in each case by a factor of two, except for the CO instrument, where an improvement factor of nearly 100 was realized.

Previously, the noise associated with the analyzer output signal near the zero point made ultra-low concentration measurements difficult as the signal to noise ratio influenced the repeatability of these measurements. Various methods were utilized to stabilize this signal. Stabilization of the high voltage power in the analyzer detector and heat isolation of electrical components resulted in the greatly improved zero signal noise levels.

The new series of instruments developed for this project is now available as super low emission (SLE) type instrumentation. A schematic configuration of this system is shown in Figure 1.

In addition to the technical refinements of the redesigned instruments, the bench was critically examined and engineered to minimize sample surface area and the potential for hydrocarbon adsorption. Where appropriate, Teflon was replaced with electropolished, chemically passivated stainless steel,

and new materials were selected to reduce hydrocarbon desorption.

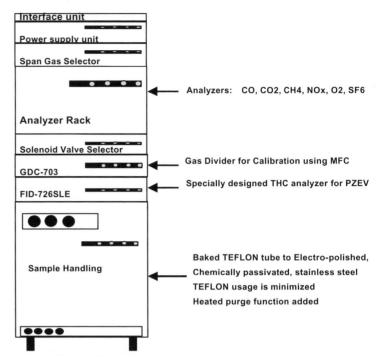

Figure 1. Bag bench for low-level emissions measurement

A heated purge function was added to address all components and lines in contact with the hydrocarbon sample stream. The sequence for the heated purge of the bag read line includes the gradual ramping of temperature to 100°C over a time interval of 11 minutes, while flowing high-purity nitrogen or zero air. This encourages the release of hydrocarbons from the internal surfaces of the tubing, valves and other components. As illustrated in Figure 2, the temperature is then gradually reduced to ambient temperature over a time period of 9 minutes. Rather than maintain a constant temperature of 100°C, the purge begins and ends at ambient temperature to allow the gas analyzer to monitor purge gas concentrations under conditions similar to actual testing.

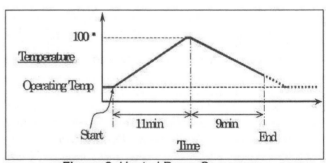

Figure 2. Heated Purge Sequence

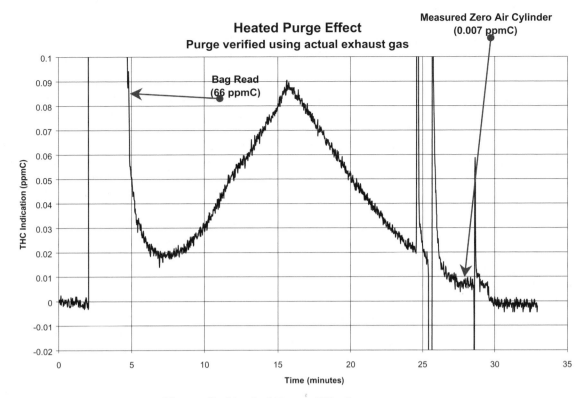

Figure 3. Heated Purge Effectiveness

Experimental tests have shown that above 100°C, there is minimal improvement in hydrocarbon reduction in the sample line. Purge gas temperatures up to 150°C showed negligible concentration differences when compared to concentration of purge gas temperatures at 100°C. The time to achieve sample line concentrations below 0.01 ppmC was approximately 240 seconds at purge gas temperatures of both 100°C and 150°C.

An experiment was performed to optimize the purge time and determine the effectiveness of the heated purge function. Two sample bags were filled; one with exhaust gas and the other with high-purity zero air. The THC concentration in the bag read line was monitored during the following sequence:

1. Zero the gas analyzer
2. Read the bag filled with diluted exhaust gas
3. Perform heated purge function
4. Initialize a system reset
5. Read the bag filled with zero-air
6. Zero the gas analyzer

Figure 3 summarizes the results of the heated purge experiment. A direct correlation exists between the temperature of the purge gas and the amount of hydrocarbons removed from the system tubing. As the temperature increases, the concentration of the gas sample increases proportionally until the hydrocarbons are purged from the system.

For this experiment, the exhaust gas in the sample bag consisted of the concentrations listed in Table 1.

Exhaust Gas Constituent	Sample Bag Concentration
THC	66.00 ppmC
CH_4	12.53 ppm
NOx	0.55 ppm
CO	48.90 ppm
CO_2	2.14%
O_2	17.96%

Table 1. Sample Bag Composition

All instruments in the SLE bag bench are operated on their high gain setting, which limits the full dynamic range but allows better resolution for low-level measurements. To better understand the sampling system performance and to support test site diagnostics, oxygen, SF_6 and humidity instruments compliment the standard bag bench analyzers. We are currently evaluating each of these instruments to determine whether they should be added to future bag benches on PZEV-capable test sites.

The oxygen instrument provides a Quality Assurance measure to monitor the dilution process and the integrity of the emissions test. The test site software allows the operator to select the oxygen instrument for bag measurements or to operate the system without the oxygen instrument.

Due to a lower dilution ratio than the CVS system, the Bag Mini-Diluter sample bags have a lower oxygen concentration in the sample bags. As the total hydrocarbon (THC) concentration in the bag approaches zero, the oxygen concentration in the sample bag has a greater effect on the analyzer reading. The oxygen concentration in the sample gas adds to the flame air supply to determine the air/fuel ratio in the flame. Variations in this air/fuel ratio will vary the temperature and size of the flame and affect the ionization efficiency producing a quenching effect. While the analyzer air supply can be regulated, the amount of oxygen introduced in the sample stream can vary (e.g. CVS sample bags versus Bag Mini-Diluter sample bags).

In order to eliminate oxygen quenching, zero and span gases with the same oxygen concentration as the sample gas could be used. This is not practical for many applications because of the unique requirements for these zero/span gases. For BMD applications, another approach would be to use nitrogen as the diluent instead of zero air. This approach was considered but not adopted due to concerns with potential effects on CO_2 measurements [6].

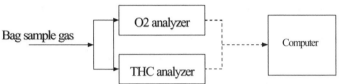

Figure 4. Total Hydrocarbon (THC) Compensation for O_2 Concentration

Figure 4 illustrates the method that was adopted to compensate the THC analyzer reading based on the oxygen content in the sample. Corrections for the oxygen quench can be classified into two categories, span and zero corrections. Recently, it was demonstrated that the O_2 quenching of the THC signal is not only significant for span gas quenching, but also for the zero reading.

The following formula is used to correct for the oxygen quenching effect:

THC indication = THC detected concentration - coefficient * O_2 detected concentration

Where the coefficient is represented as a "ZERO shift " and calculated from O_2 interference at zero.

Though it is recognized that changes in oxygen concentration may have a minor influence on low-level measurements [4], the current strategy is to monitor and control the range of oxygen manufactured by the zero air systems. Investigations are underway for an improved, dedicated, zero air generation system to compliment the unique demands of the BMD test site.

A relative humidity sensor has been adopted on the bench to determine the water content in each sample

bag. The water and oxygen content is recorded for each sample bag to increase understanding of the NDIR interferences by both of these components [5]. Finally, since SF_6 can be injected into the vehicle tailpipe during an FTP test and not be altered by the gaseous emissions, it is planned to trace SF_6 through the system and into the bag for a "dynamic" mass recovery test [6].

The modal benches (engine out, mid-bed and tailpipe) are equipped with the standard analyzers except for two revisions: each line includes a methane analyzer and the tailpipe bench includes an additional SF_6 analyzer.

SAMPLING SYSTEM - We believe that the sampling system is the major contributor to hydrocarbon measurement uncertainty and recognize that a number of alternatives are available to improve the standard CVS approach. The CVS process gathers all of the vehicle's exhaust during a prescribed driving cycle on the dynamometer (see Figure 5). The exhaust is diluted with ambient air and a Critical Flow Venturi then meters the dilute stream. The diluted samples are collected in bags for subsequent analyses. This dilution process is typically on the order of 10:1 ambient air to exhaust, and corrections are made to account for the ambient air concentrations when determining the tailpipe exhaust emissions.

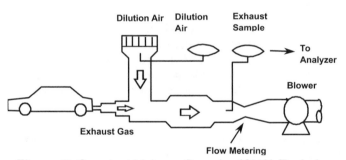

Figure 5. Constant Volume Sampler (CVS) Technique

The CVS dilution process is variable. Too much dilution takes place during idle conditions and too little occurs during accelerations. Furthermore, ambient air is no longer suitable as a diluent.

For this project, Dilution Air Refinement and heated CVS systems were considered. Both offer significant improvement and are attractive because they fit within the confines of the existing Code of Regulation. However, we elected the Bag Mini-Diluter approach as being the most cost effective, accurate, and capable technology with great potential for technical improvement in the future [7]. The Bag Mini-Diluter features a constant dilution process whereby dilution is the same for all portions of the test and is fixed by design (see Figure 6). Dry zero grade air is used for the diluent and ambient measurements are not required in the calculation of mass emissions.

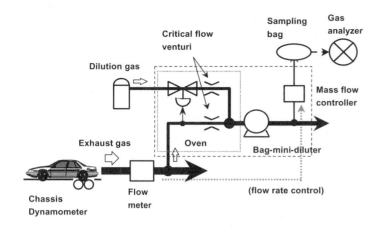

Figure 6. Schematic of Bag Mini-Diluter Technique

Bag Mini-Diluter Modifications - The new "production version" of the Bag Mini-Diluter (see Figure 7) includes many new features to enhance the quality of the PZEV measurements. The BMD improvements are summarized in Table 2.

For manufacturing of the new BMD, experience was gleaned from the semi-conductor industry to help define new manufacturing methods and to select better materials. Fittings were avoided wherever possible.

Electro-polished, chemically passivated, stainless steel tubing with large radius bends was used to minimize microscopic fissures (and nucleation sites) internal to the tubing. Minimum use of Teflon was permitted only at pump connections, and Latex gloves were worn during the assembly process.

As a result this fastidious "attention to detail", the total hydrocarbon "outgas" specification of 30 parts per billion carbon (ppbC) was satisfied with no individual hydrocarbon compound in excess of 10 ppbC. This performance specification was verified by Gas Chromatography with a dry zero air bag fill. More stringent diagnostics are being evaluated.

Traditional Tedlar® sample bags, especially when new, are known to outgas N,N-dimethylacetamide and may not be suitable for low-level hydrocarbon applications. While investigations continue to identify better materials, for the interim, a TFM (Dyneon) material has been selected for the BMD sample bags. This material is a blend of two different Teflon materials that minimizes hydrocarbon contamination/outgassing and reduces permeation of CO_2. In our permeation experiments with conditioned bags at room temperature, bags were filled with zero air and with 2% CO_2. We then made measurements before and after a 1-hour soak. The Dyneon bag outgassing and permeation data is shown in Table 3.

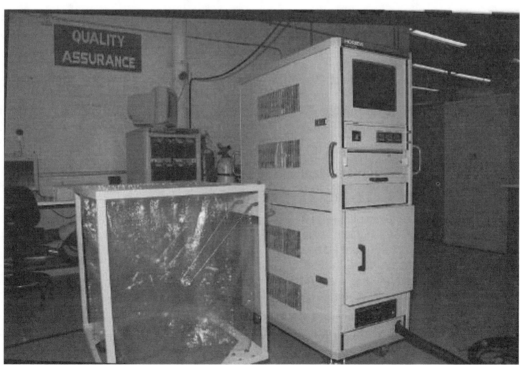

Figure 7. PZEV Test Cell Bag Mini-Diluter

BMD Improvement	Expected Benefit
1. Use semiconductor grade, electro-polished, 316 stainless steel (Ra<10)	Reduce HC hang-up and outgassing
2. Upgrade sample bags: 70 liter TFM crinkle w/o internal plumbing	Eliminate outgassing, and minimize CO2 permeation
3. Eliminate Viton	Reduce HC outgassing
4. Minimize internal volumes for sample stream	Reduce likelihood of sample contamination during gas transportation
5. Adopt miniature valve and manifolds	Minimize internal volume and reduce gas transportation delays
6. Use custom KNF Neuberger Pump w/ Teflon coated heads	Improve reliability and reduce outgassing
7. Use VCR fittings in critical areas	Improve sealing and maintenance access
8. Improve oven design	Reduce size and Improve access
9. Improve H_2O detector	Better response time (<10 seconds)
10. Increase heating coil	Improve temperature condition at CFV
11. Adopt PC control system	Improve compatibility and data storage
12. Adopt NI LabView Software	Incorporate user-friendly software
13. Adopt NI data boards	Increase to 16-bit resolution
14. Incorporate Windows 2000 System	Improve compatibility and ease-of-use
15. Eliminate process controllers	Improve compatibility and ease-of-use
16. Add diluent (zero air) bag	Ensure diluent purity and monitor contamination
17. Adopt 100% heated purge	Sweep residual HC's from BMD
18. Improve system and bag leak check	Improve system integrity
19. Add on-board read pump	Localize critical BMD components
20. Add venturi temperature thermocouples	Monitor CFV performance and improve ability to control dilution ratio at fixed rate

Table 2. PZEV Bag Mini-Diluter Upgrades

For the BMD application, the bags are not heated, and a consistent 5.5:1 dilution ratio is maintained. Further reduction in dilution is not practical due to gas composition effects and the desire to observe CO_2 correlation to the CVS technique. The PZEV Test Cell also includes a traditional CVS system for paired emissions measurements, ensuring that correlation is maintained between the BMD and CVS CO_2 measurements.

Dilution Ratio Accuracy - To maintain a consistent dilution ratio in the Mini-Diluter, the gas temperatures and pressures at the inlet to the critical flow venturis must be held constant and near the same value [7]. To improve the control of the gas temperature at the inlet to the critical flow venturis, the diluent and exhaust gases pass through coils of tubing which are contained in an oven in the Mini-Diluter prior to reaching the venturis (see Figure 8). The oven is maintained at 70°C to prevent condensation of the exhaust gas sample. One tube coil is used for the diluent gas and one is used for the exhaust gas.

As the gases pass through the coiled tubing, heat is transferred between the gases and the air in the oven, so that the temperature of the gas leaving the coil is near the oven temperature. Figure 9 shows the performance of the venturi inlet temperature control for the diluent and exhaust gases using as PZEV vehicle during phase 1 of an FTP test cycle. The exhaust (sample) gas temperature remains relatively constant throughout the test cycle, but the diluent gas temperature declines slightly throughout the phase. The slight decline in temperature is attributed to rapid expansion of the diluent gas through the inlet pressure regulator at high flow, thus slightly decreasing the diluent inlet temperature.

	Tedlar	Dyneon	Teflon PTFE
Outgassing ppmC	2.90	0.04	0.03
CO_2 Permeation (% change per hour)	-0.12	- 0.20	- 0.47

Table 3. Bag Material Comparison

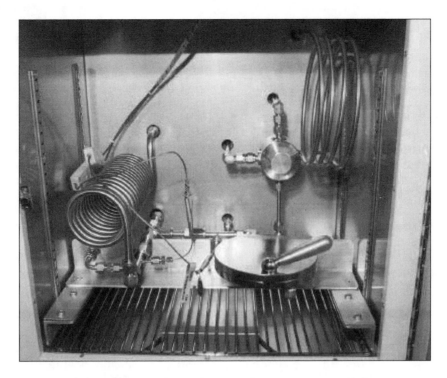

Figure 8. Heat transfer coils in BMD oven for stable venturi inlet temperatures

A faster relative humidity sensor (improved from the prototype BMD) with a T90 response time of less than 10 seconds was also implemented on the BMD. This relative humidity sensor is used to calculate the water content of the BMD sample. The BMD water content is then used to account for the slight influence of water on the sample venturi during the Cold Start portion of the FTP test cycle. As a result, a water adjustment factor (typically less than 2%) is commanded to the bag fill mass flow controller to compensate for the small change in the BMD dilution ratio during the Cold Start. A fast humidity sensor response time is important in order to implement compensation as close to "real-time" as possible.

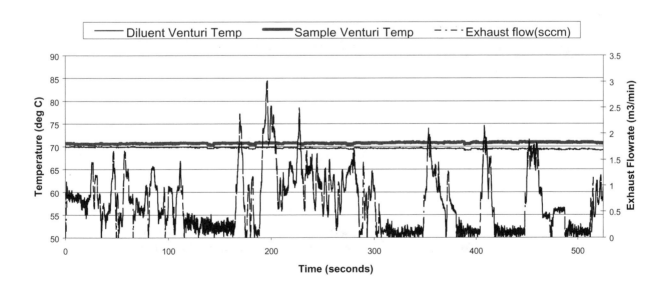

Figure 9. Venturi Inlet Temperatures during FTP, Phase 1 on Bag Mini-Diluter (ULEV Test Vehicle)

Diluent Bag - For the latest version of the Bag Mini-Diluter, a separate bag was added to the system to monitor the purity of the diluent gas that is mixed with the vehicle exhaust gas. Typically, subtraction of the low diluent gas concentrations are not required when using Bag Mini-Diluter technology, as is the case when using a CVS. The Diluent Bag may be used as a quality control tool to ensure on a test-by-test basis that the diluent gas is truly "zero" grade. This diluent bag is filled with the diluent gas at the same time as a sample bag during an emission test, and then analyzed separately to determine the purity level of the diluent gas. Improperly labeled gas bottles, excessive load on the zero air supply system, or contaminated sample read lines might be detected using this quality control method.

Bag Fill Proportionality - The BMD time delay [6] must be determined to account for the gas transport time and the response time of the bag-fill mass flow controller (MFC). A Fast Flame Ionization Detector (FID) was used to determine the time between hydrocarbon concentration changes during propane injections while a dynamic blower "stepped" between simulated exhaust flow rates. Using this technique, a time delay of 1.4 seconds was determined for the BMD system. This time delay, along with the MFC response time, is used in the

BMD control software to ensure proportionality between the bag fill MFC and the exhaust flow signal. On the new BMD design, the mass flow controller demonstrated a T90 response time of less than 0.4 seconds.

Several revisions in the BMD control software have been adopted to improve MFC proportionality. The software has been modified to provide the actual flow curve from the MFC, as determined by a flow standard, and an auto-zero function has been incorporated to provide better accuracy. In addition, the internal MFC "auto-linearize" function has been turned off so that the actual flow throughout the full-scale range will be output by the MFC (rather that a flow output that was forced through an internal MFC linearization scheme).

100% Heated Purge - A six-stage heated purge sequence was introduced to the new BMD design to remove residual hydrocarbons from the bag fill and bag analysis lines. The purge sequence is performed during both pre-test and post-test operations. The six stages encompass 100% of the system sample and read bag flow circuits. These stages consist of the following sequences, whereby a purge gas flows through the system while being heated by the sample lines:

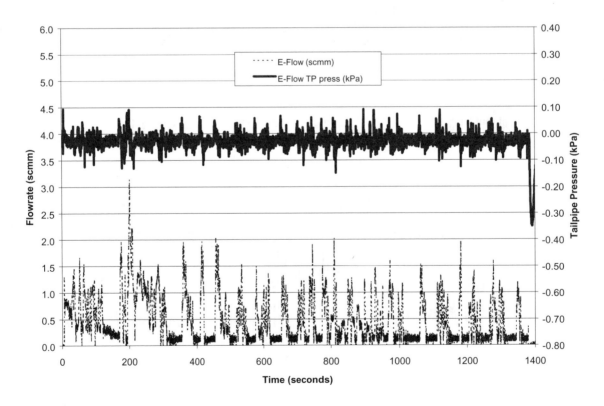

Figure 10. Tailpipe Pressure Control, 350 scfm (9.91 m^3/min), Phase 1 and 2 of FTP, ULEV Test Vehicle

650

1. Purge sample line (backflush).
2. Purge sample-fill line to the bag fill manifold.
3. Purge sample-fill valves to the samples bags.
4. Purge the read-bag manifold
5. Purge the read-bag valves
6. Purge bags

The system incorporates two user-configurable purge times for each stage. A shorter purge-time is currently used pre-test to perform a quick flush of the system to guarantee the system is clean prior to sampling. Another longer purge time is used post-test to remove residual hydrocarbons from the system. Effectiveness of the purge strategy is demonstrated by routine contamination checks (see data in "Process" section).

Ultrasonic Exhaust Flow Measurement - To complete the system, a modified E-Flow Direct Exhaust Volume Measurement System is provided with the ultrasonic exhaust measurement section aligned vertically to improve stratification under low flow conditions. The E-Flow also acts as a Mixing Tee for ambient air dilution to a standard multiple venturi CVS system connected in series. The CVS provides a selection of 14 flow rates for the diluted exhaust flow.

This flow meter contains an integral tailpipe pressure control system to maintain a range of ±1.0 inches W.C. (±0.25 kPa) during a Federal Test Procedure (FTP) test cycle. The pressure control system uses closed-loop control whereby the tailpipe pressure is measured by a transducer at the vehicle tailpipe. A loop controller processes the pressure transducer signal and provides a set point for a variable frequency drive, which in turn controls the motors for the dilution air blowers. Figure 10 demonstrates the capability of the pressure control system to maintain the pressure within the defined limits throughout flow rate excursions experienced during the first two phases of a FTP test using a PZEV test vehicle. For this test, the pressure set point was biased at -0.25 inches W.C (-0.06 kPa).

The response time of the E-Flow to a step change in flow rate was measured using a dynamic blower controller containing a hot-wire anemometer. The blower was used to inject air into the E-Flow while simultaneously varying the flow rate in step changes. Figure 11 demonstrates the E-Flow response compared to the hotwire anemometer under ascending flow conditions. The data was collected at 200 Hz, which provides ample resolution to determine the response of the E-Flow.

The initial response time of 0.390 seconds was recorded for ascending flow. Ascending flow was measured when the flow rate injected into the E-Flow was ramped from 3 scfm to 113 scfm (0.09 scmm to 3.20 scmm). Additionally, when the flow rate was ramped from 113 scfm to 4 scfm (3.20 scmm to 0.11 scmm), the initial descending flow response is slightly slower at 0.460 seconds (Figure 12).

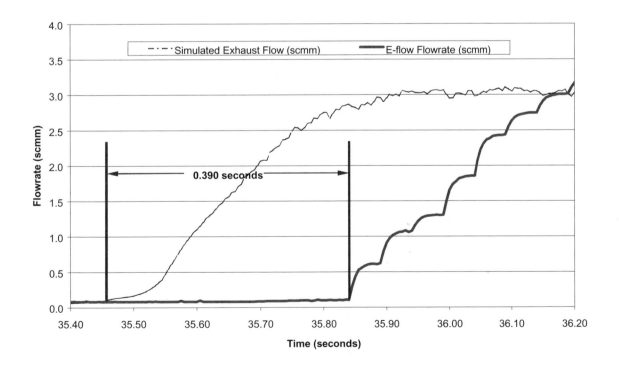

Figure 11. E-Flow Response Time – Ascending

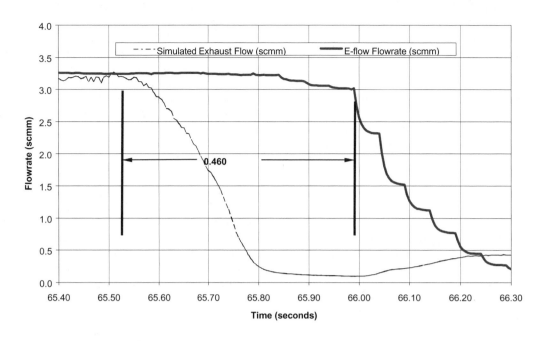

Figure 12. E-Flow Response Time – Descending

The T90 time was measured using the same technique. A step change in flow resulted in a 90% response time of 0.400 seconds (see Figure 13). This is less than the manufacturer's stated response of 0.500 seconds.

The critical factors for PZEV emissions measurement have been limiting contamination and improving instrument accuracy. However, for fuel economy, the key element is exhaust flow measurement. Prior to shipment, the E-Flow was calibrated by the manufacturer using a bell prover. Verification of the calibration was performed using Laminar Flow Elements (LFEs) and a blower to push ambient air into the inlet of the E-Flow.

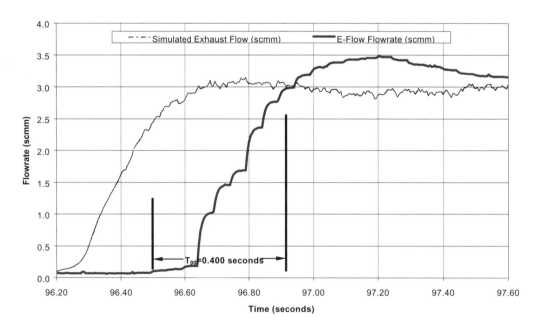

Figure 13. E-Flow T90 Response Time

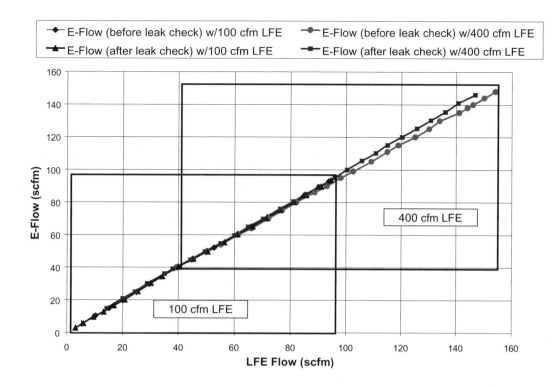

Figure 14. E-Flow Calibration Using "Overlapping" LFEs

Two different LFEs were used during this calibration in order to have the capability to span the full-scale range of the E-Flow (325 cfm or 9.20 cmm) while maintaining accuracy at the lower 20% of the range. A laminar flow element with a full-scale range of 100 cfm (2.83 cmm) was used for calibration of the E-Flow from 3 to 94 scfm (0.08 scmm to 2.66 scmm). A second LFE with a full-scale range of 400 cfm (11.33 cmm) was used to calibrate the E-Flow from 40 to 145 scfm (1.13 scmm to 4.11 scmm). This dual LFE method allowed for overlap of several data points to gain further confidence in the flow data (see Figure 14). Due to limitations on the blower supplying air to the E-Flow and LFE, 145 scfm was the maximum flow rate achieved for this calibration verification. The maximum flow rate of 145 scfm satisfies the flow range of PZEV vehicles currently under test.

Initial calibration data was collected with the E-Flow 'as-is'. Discrepancies between the factory calibration and the LFE calibration were discovered as the flow rate increased, indicating cause for investigation. Investigation of the interconnect piping from the calibration equipment to the E-Flow and internal components in the flow meter uncovered leaks at junction points. As the flow rate of the calibration blower increased, the pressure in the system piping increased producing more apparent leaks. The calibration procedure was repeated after these leaks were sealed, with improved results.

After repairing the leaks, data collected using the 100 cfm LFE demonstrated that the E-Flow calibration is accurate to within ±1.5% of point for flows between 3

and 30 scfm (0.08 and 0.85 scmm). Above 30 scfm, both the 100 cfm LFE and the 400 cfm LFE agreed with the E-Flow to within ±1.0% of point. Verification of the E-Flow calibration offers confidence in the measurement of the exhaust flow rate used to calculate the mass emissions of PZEV vehicles.

ENVIRONMENTAL OR FACILITY ISSUES - The PZEV Test Cell is temperature and humidity controlled to tight tolerances along with the aisle leading into the testing area. The cell is closed to the aisle during testing to help stabilize the ambient and to ensure front-to-rear airflow. A Road Speed Modulated (RSM) fan (45,000 cfm or 1,275 cmm) is provided along with the industry standard Hartzell. For US06 testing, the RSM fan is maintained at 15,000 cfm (425 cmm), which corresponds to an air speed of 20 mph (32 kph). Ample space in the test cell is needed to reduce variability due to air handling and airflow differences from car line to car line. The cell is approximately 1900 square feet by 19 feet in height (176.5 square meters by 5.8 meters in height) with 60 air exchanges per hour.

The layout of the equipment within the test site is for performance and not convenience. Plumbing lengths are kept to a minimum. The gas cylinders used for the zero, span and linearization functions are closely coupled to the bag bench by housing the cylinders in special cabinets within the test cell. The BMD sampling system and bag rack are within 8 feet of the bag bench, with a heated line between the BMD and bag bench. The moveable dynamometer allows the sampling equipment to remain stationary, while ensuring that the

BMD is no more than 12 feet from the vehicle's tailpipe at any time.

Also included in the "facility" category is the manufactured zero air generation system which is routinely tested for oxygen stability and hydrocarbon background level. The current system, though monitored, supplies 45 other test sites in the laboratory in addition to the PZEV facility. We believe there may be an opportunity for further improvement in this area and are experimenting with an independent system with part per billion (ppb) purification capabilities to determine the quantitative improvement to the testing data.

SAES Pure Gas manufactured the alternative on-site air purification system. Their MegaTorr Clean Air Supply (CAS) purifier is capable of providing zero air for BMD dilution with impurities at the ppb level. The purification is accomplished through three steps. First, gross moisture is removed using a water trap. This is followed by heated catalytic conversion of hydrocarbons, CO and H_2. Finally, NO_X, CO_2 and H_2O are removed through ambient adsorption. The purification process will also reduce the variability in oxygen content experienced with the current zero air supply system. Further testing, including continuous measurements of hydrocarbons,

CO_2, NO_X, and oxygen, will be conducted over the next several months.

PROCESS - By "process" we refer to the operations involved within the test cell, the quality control of the equipment, and the change in culture needed to operate consistently at PZEV levels. While every precaution has been made to minimize the risk of hydrocarbon contamination, our intent is to further restrict the usage of the test cell to "clean" vehicles only. The suppliers, technicians, and skilled trades involved with the installations and operation of the test cell have an understanding of the "part per billion" influences on PZEV emissions. Of course, smoking is strictly prohibited.

Quality Assurance tools have been developed that are unique for the BMD sampling system, but perhaps the most significant development is the Vehicle Exhaust Emissions Simulator (VEES, See Figure 15 for photograph and subsequent data in "Test Site Comparison"). The VEES allows consistent monitoring of test site performance at the PZEV levels by simulating the vehicle's second by second mass emissions with gases injected directly into the BMD or CVS system [8].

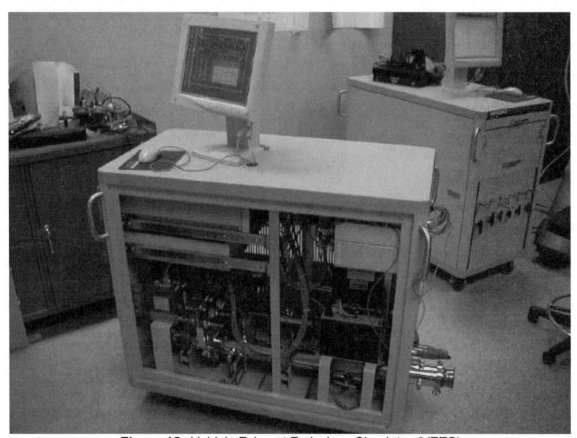

Figure 15. Vehicle Exhaust Emissions Simulator (VEES)

Other quality assurance techniques are being evaluated for routine usage in the PZEV Test Cell, including dynamic propane injections, SF$_6$ injections, contamination (or "zero") checks, and dilution ratio checks.

Dynamic Propane Injections - During the Dynamic Propane Injection, propane is injected into the sampling system using a critical flow orifice (CFO) for metering. The propane is mixed with ambient air that follows a dynamic profile generated by a blower and a computer-controlled flow valve. A tolerance of ±2% is maintained between the mass of propane recovered by the BMD and the calculated mass from the CFO. This technique has proven to be a useful quality assurance tool for the BMD measurement system because it ensures accuracy of the exhaust flow meter and the system delay times.

Sulfur Hexaflouride Injections - To ensure the accuracy and measurement integrity of the test cell analysis equipment, a known mass of Sulfur Hexaflouride (SF$_6$) gas may be injected into the sample stream and collected in the BMD and CVS sample bags. A special gas analyzer calibrated to read SF$_6$ is used to read the sample bags and calculate the injected mass of the gas. SF$_6$ has several advantages over the traditional propane mass recovery method.

The first advantage is that SF$_6$ is not typically present in the test cell background air. Correction of background propane concentrations when calculating injected mass is a large source of variability and error in the propane recovery method. Because SF$_6$ is not present in the background air, subtraction of background concentrations is not necessary, thus eliminating this source of variability.

Another major advantage of the SF$_6$ mass recovery method is that the SF$_6$ gas may be injected during an actual vehicle test, without interfering with the emissions test results. This technique may then be used to verify the analysis system under actual operating pressures, temperatures, and flow rates. The accuracy of the Bag Mini-Diluter, exhaust flow measurement system, CVS, and gas analysis system may all be verified using this recovery technique. Due to time constraints, we have not collected sufficient data to report on the usefulness of SF$_6$ injections for quality assurance.

Contamination Checks - As reported previously [6 and 7], contamination checks are a valuable tool to determine the cleanliness of the measurement system. During the contamination check, no exhaust gas is introduced into either the BMD or CVS system. The BMD sample probe is flushed with high-purity zero air and the exhaust inlet to the E-Flow is capped while the timing of sample collection/analysis for a typical emissions test is duplicated. Three separate phases are performed for each contamination check to mimic the Federal Test Procedure.

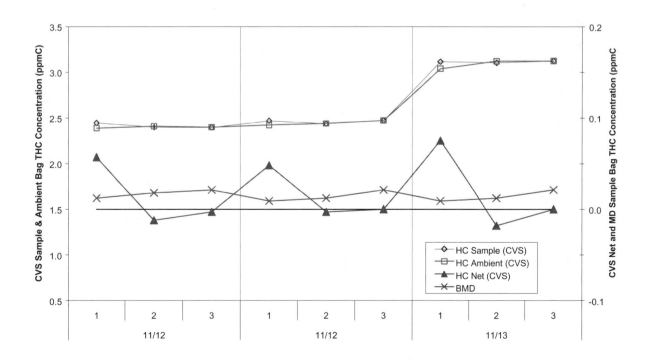

Figure 16. BMD and CVS Contamination Check Comparison on PZEV Test Cell

655

	Average THC Bag Concentration (ppmC)	Standard Deviation of THC Bag Concentration (ppmC)	Estimated THC Impact on FTP (mg/mi)
BMD	0.015	0.0017	0.087 ± 0.012
CVS Net	0.016	0.0077	0.086 ± 0.087

Table 4. Results from Three Contamination Checks on PZEV Test Cell

The results of three contamination checks on the PZEV Test Cell, which were interspersed with vehicle tests, are presented in Figure 16. Both the BMD and CVS demonstrated significant improvements when compared to results from the prototype equipment in Reference [7]. Table 4 includes a summary of the average concentration, standard deviation, and estimated impact of the contamination checks on the emissions measurements from the BMD and CVS. We currently perform contamination checks on a weekly basis.

Dilution Ratio Checks - The nominal dilution ratio for the critical flow venturis used in the BMD was determined to be 5.51:1 (i.e. one part of exhaust gas is diluted by 5.51 parts of zero air) using laminar flow elements as a flow standard and dry synthetic air as the calibration gas. Weekly dilution ratio checks, using the technique outlined in Reference [6], are performed using 4% CO_2 in nitrogen calibration gas to ensure that the BMD dilution control system is working properly. The current requirement for this check is that the dilution ratio measured using the 4% CO_2 calibration gas must agree with the nominal dilution ratio, as determined by LFEs, to within $\pm 1.0\%$. Venturi adjustment factors are used to account for the small gas composition effects on the flow through the "exhaust" venturi when exposed to the different gases.

TEST SITE COMPARISON

A test program was performed using the VEES to compare emissions measurements from different test cells in our laboratory, including the PZEV Test Cell. The VEES was programmed to simulate the phase-by-phase FTP emissions profile from a representative PZEV test vehicle (3 phases were completed for each "composite FTP" simulation). Table 5 below compares the performance of various test cells with different vintages of equipment. At least four simulated FTP tests were conducted on each test cell.

The results from the PZEV Test Cell are unique in that the BMD and CVS are sampling in parallel with bag measurements taken by the same analyzer bench. The use of "E-CVS" in Table 5 below indicates that the CVS in the PZEV Test Cell has been enhanced by numerous improvements as compared to the older vintage CVS systems, including purge strategies, multiple venturi flow rates, and electropolished stainless steel plumbing. The data indicates that for low-level emissions measurement, modern analyzers are extremely important and further accuracy is achieved with a BMD sampling system.

	PZEV Cell	PZEV Cell	Site A	Site B	Site C
Sampling System	BMD	E-CVS	CVS	CVS	CVS
Analyzers	7000-LE	7000-LE	Analog	Analog	Analog
THC target (mg/mi)	7.10	7.10	7.10	7.10	7.10
THC average (mg/mi)	7.07	6.01	6.45	8.02	5.88
THC COV (%)	0.24%	1.07%	10.29%	15.19%	25.50%
CH_4 target (mg/mi)	1.20	1.20	1.20	1.20	1.20
CH_4 average (mg/mi)	1.21	0.90	-0.39	2.36	1.25
CH_4 COV (%)	0.42%	6.10%	>100%	27.59%	57.61%
NOx target (mg/mi)	2.40	2.40	2.40	2.40	2.40
NOx average (mg/mi)	2.47	2.44	3.04	2.99	3.03
NOx COV (%)	0.55%	2.85%	67.57%	8.09%	8.75%

Table 5. Composite FTP Results from PZEV Experiments with VEES (COV = Coefficient of Variation)

	Sampling System	THC (mgs)	CH$_4$ (mgs)	NO$_x$ (mgs)	CO (mgs)	CO$_2$ (gms)
Phase 1	BMD	70.9	16.6	29.8	1227.8	1346.3
	CVS	69.1	16.6	28.4	1228.5	1338.9
Phase 2	BMD	2.1	0.0	0.3	268.1	1314.8
	CVS	-1.7	0.1	1.6	310.5	1330.4
Phase 3	BMD	1.7	0.9	0.5	165.8	1140.2
	CVS	-1.3	1.0	0.8	191.5	1138.2
Highway	BMD	1.3	0.6	1.1	358.0	2310.4
	CVS	-3.8	1.2	-1.7	403.0	2314.2

Table 6. Paired FTP and Highway Data Obtained From PZEV Test Vehicle

The data from the VEES experiment can be analyzed further to evaluate the measurement limits of the sampling and analytical equipment. During the 2nd Phase of the simulated FTP test, the VEES was programmed to inject just 2.0 mgs of THC (1 mg of propane plus 1 mg of methane), 1.0 mg of CH$_4$, and 1.0 mg of NOx into the sampling systems in the different test cells. On the PZEV Test Cell, the BMD recovered an average of 1.7 mgs of THC, 1.0 mg of CH$_4$, and 1.1 mgs of NOx. The Coefficient of Variation (COV) from these measurements was 0.08% for THC, 2.76% for CH4, and 2.97% for NOx. The 2nd Phase CVS measurements of THC and CH$_4$, regardless of the test cell, were typically negative. The COV during the 2nd Phase for all three emissions constituents (THC, CH$_4$, and NOx) on the CVS systems varied extensively, but was consistently much larger than 10%.

VEHICLE TESTING

Although the PZEV Test Cell is still in the commissioning stage, a variety of vehicles have been tested in the facility to ensure proper operation and to demonstrate correlation with other certification test cells. Table 6 includes an example of paired (i.e. both BMD and CVS measurements are taken in parallel) emissions results on a PZEV-compliant vehicle during the EPA City (FTP) and Highway test cycles. The composite FTP emissions from this test vehicle, as measured by the BMD, were 3.6 mgs/mile of NMHC and 1.8 mgs/mile of NOx. The PZEV test vehicle had a 2.0-liter, four-cylinder engine and a test weight of 3000 pounds (1361 kgs). It is important to note that only BMD results will be reported on the PZEV Test Cell in the future. The CVS results are only necessary to demonstrate correlation, and will be analyzed on a periodic basis by laboratory-support engineers.

CONCLUSION

The technical evolution of the Bag Mini-Diluter sampling system continues at a rapid pace. As discussed in various papers in the past, the BMD has several inherent advantages over the traditional Constant Volume Sampling (CVS) System.

- Dilution is at a fixed ratio (the CVS is a variable dilution process)
- Dilution with dry zero air obviates the ambient bag and corrects dilution factor assumptions
- Gaseous composition of the bag is consistent for water, Oxygen and CO2 (at stoichiometry)
- Considerations for heating, purging, reducing outgassing levels, etc. are simpler with the inherently smaller BMD sampling system.

Concerns with past prototype systems have been with measuring the gas transport delay time (from the probe to the bag), providing a proportional bag fill with respect to the exhaust flow, and achieving sufficient accuracy and/or response for the exhaust flow measurement system. We believe these concerns have been addressed and that appropriate tools exist to ensure traceable operation of the BMD system. We look forward to working with the new equipment, driving further improvement, and sharing further experiences as data is collected and analyzed.

ACKNOWLEDGMENTS

The authors wish to thank the following people for their contributions to this paper:

- Noelle Baker, Richard E. Chase, Darius Harrison, Travis Henney, Steve Hunter, Kim Isbrecht, Ed Kulik, Terry Laskowski, Mark Polster, Andre Welch from Ford Motor Company.
- Dave Balaka, Al Dageforde, Bob Gierada, Neal Harvey, Karl Oestergaard, Dan Whelan, Denny Wu from Horiba Instruments.
- Mike Wusterbarth from Flow Technologies, Inc.

REFERENCES

1. California Air Resources Board, "California Exhaust Emission Standards and Test Procedures for 2003 and Subsequent Model Zero-Emission Vehicles, and 2001 and Subsequent Model Hybrid Electric

Vehicles, in the Passenger Car, Light-Duty Truck and Medium-Duty Vehicle Classes", May 30, 2001

2. California Air Resources Board, "Dynamometer Performance Evaluation and Quality Assurance Procedures", March 17, 2000

3. M. Sherman, R. Chase, A. Mauti, Z. Rauker, W. Silvis, "Evaluation of Horiba MEXA-7000 Bag Bench Analyzers for Single Range Operation", SAE paper 1999-01-0147

4. A. Tayama, K. Kanetoshi, H. Tsuchida, H. Morita, "A Study of a Gasoline-Fueled Near-Zero-Emission Vehicle Using an Improved Emission Measurement System", SAE Paper 982555

5. K. Inoue, M. Ishihara, K. Akashi, M. Adachi, K. Ishida, "Numerical Analysis of Mass Emission Measurement Systems for Low Emission Vehicles", SAE Paper 1999-01-0150

6. M. Guenther, T. Henney, W. Silvis, S. Nakatani, D. Wu, "Improved Bag Mini-diluter Sampling System for Ultra-low-level Vehicle Exhaust Emissions", SAE Paper 2000-01-0792

7. M. Guenther, K. Brown, M. Landry, M. Sherman, D. Wu, "Refinement of a Bag Mini-Diluter System", SAE 2001-01-0212

8. M. Landry, M. Guenther, K. Isbrecht, G. Stevens, "Simulation of Low Level Exhaust Emissions for Evaluation of Sampling and Analytical Systems", SAE Paper 2001-01-0211

CONTACT

Author Information:
1. Mark Guenther, Senior Technical Specialist, Correlation Engineering, Vehicle Environmental Engineering, Ford Motor Company, Allen Park Test Laboratory, Suite 3W-100, 1500 Enterprise Drive, Allen Park, MI, 48101-2053, phone: (313) 594-2054, fax: (313) 594-2044, e-mail address: mguenthe@ford.com

2. Michael T. Sherman, Senior Technical Specialist, Correlation Engineering, Vehicle Environmental Engineering, Ford Motor Company, Allen Park Test Laboratory, Suite 3W-100, 1500 Enterprise Drive, Allen Park, MI, 48101-2053, phone: (313) 594-2056, fax: (313) 594-2044, e-mail address: msherma1@ford.com

3. Mike Vaillancourt, Principal Engineer, Correlation Engineering, Vehicle Environmental Engineering, Ford Motor Company, Allen Park Test Laboratory, Suite 3W-100, 1500 Enterprise Drive, Allen Park, MI, 48101-2053, phone: (313) 322-5228, fax: (313) 594-2044, e-mail address: mvaillan@ford.com

4. Daniel Carpenter, Product Development Engineer, Product Analysis and Verification, Ford Motor Company, Allen Park Test Laboratory, 1500 Enterprise Drive, Allen Park, MI, 48101-2053, phone: (313) 621-2396, fax: (313) 390-3089, e-mail address: dcarpent@ford.com

5. Rick Rooney, System Engineer, Engine Measurement Division, Horiba Instruments, Inc., 5900 Hines Drive, Ann Arbor, MI, 48108, phone: (734) 213-6555 x596, fax (734) 213-6525, e-mail address: rick.rooney@horiba.com

6. Scott Porter, System Engineer, Engine Measurement Division, Horiba Instruments, Inc., 5900 Hines Drive, Ann Arbor, MI, 48108, phone: (734) 213-6555 x869, fax (734) 213-6525, e-mail address: scott.porter@horiba.com

Evaluation of the Bag Mini-Diluter and Direct Vehicle Exhaust Volume System for Low Level Emissions Measurement

Donald B. Nagy, Jeffrey Loo, Jim Tulpa and Pat Schroeder
Powertrain Div., General Motors Corporation

Rick Middleton and Craig Morgan
DaimlerChrysler Corporation

Copyright © 2000 Society of Automotive Engineers, Inc.

ABSTRACT

With the adoption of the California Low-Emission Vehicle Regulations and the associated lower emission standards such as LEV (Low-Emission Vehicle in 1990), ULEV (Ultra-Low-Emission Vehicle), and LEV II (1998 with SULEV-Super Ultra Low Emission Vehicle), concerns were raised by emissions researchers over the accuracy and reliability of collecting and analyzing emissions measurements at such low levels. The primary concerns were water condensation, optimizing dilution ratios, and elimination of background contamination. These concerns prompted a multi-year research program looking at several new sampling techniques. This paper will describe the cooperative research conducted into one of these new technologies, namely the Bag Mini-Diluter (BMD) and Direct Vehicle Exhaust (DVE) Volume system.

INTRODUCTION

The objective of the study was to design and develop a bag mini-diluter system, then compare its performance to that of a conventional test site CFV using vehicle-generated emissions. Early in the development of the device, the following performance requirements were established in "Specifications For Advanced Emissions Test Instrumentation" book, dated February 1994, under Specification 1.0 "Mini-Diluter Technology", and Specification 2.0 "Vehicle Exhaust Volume Measurement". An excerpt of some important performance requirements from these specifications were:

1. Carbon dioxide (CO_2) measurement agreement to conventional test site technology (CFV) to within 1%.

2. Total hydrocarbon (HC), carbon monoxide (CO), methane (CH_4) and oxides of nitrogen (NO_x) measurement agreement to conventional test site technology (CFV) to within 3%.

3. Time response of BMD "system" – 0.1 second.

Clearly, good correlation with today's test sites using contemporary emission level vehicles was important. However it was also recognized that significant unknowns existed at very low emission levels and the two technologies (BMD system versus the CFV) would probably differ at very low levels. If this anticipated difference occurs, then the question would be which system is reading emissions levels correctly.

DISCUSSION-TODAY'S MEASUREMENT SYSTEM PROBLEMS

To understand the low level measurement concerns, requires to first look at the conventional technology used today. Most vehicle manufacturers and regulatory agencies utilize a single venturi, "cold" CFV exhaust sampling system. This system was originally developed for automotive testing back in the 1970's. Although incremental improvements were made over the past 30 years, the technology has remained basically the same. A simplified drawing of the CFV system is shown in Figure 1 below:

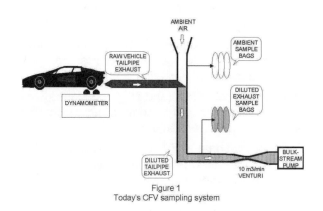

Figure 1
Today's CFV sampling system

The theory behind this sampling system is to dilute the vehicle's entire raw exhaust gases with enough ambient air such that water condensation does not occur in the diluted exhaust mixture (a typical dilution ratio is 15:1 for an average vehicle). The way this mixing occurs is to pull air through the diluted exhaust pipe using a large sample pump ("Bulkstream Pump"), and measure how much air is being pumped using a CFV nozzle ("10 m³/min Venturi"). A small sample of this diluted exhaust is then diverted into a set of three bags ("Diluted Exhaust Sample Bags"), and filled continuously over the entire test schedule. The three bags represent the three main phases of the Federal Test Procedure (FTP), that is "cold start" (or Phase 1), "hot stabilized" (or Phase 2), and "hot start" (or Phase 3). To account for ambient air contamination of the diluted exhaust, a similar set of three bags ("Ambient Air Sample Bags") are filled with ambient air, one bag per test phase. After the test is completed, both sets of bags (ambient and diluted exhaust bags) are analyzed for concentrations of total hydrocarbons (HC's), methane (CH_4), carbon monoxide (CO), carbon dioxide (CO_2) and oxides of nitrogen (NO_x.). Using these measured concentrations, plus knowing the diluted exhaust volume from the CFV, the masses of each emission constituent can be determined and then compared to the appropriate standard to verify conformance to the regulations. Simplified versions of the equations used are:

Mass Emissions = Corr Conc * Dil Exh Vol * Density (1)

Where:

Corr Conc = Dil Bag Conc–Amb Bag Conc*(1–1/DF) (2)

Where Dilution Factor (DF) is:

$$DF = \frac{\text{Carbon Mole Fraction of the Fuel}}{\text{Carbon Mole Fraction of the Exh Gases}} \quad (3)$$

As previously mentioned, this system has worked very well for many years, but it does have some intrinsic problems. In summary these are:

1. Current dilution ratios will be too low to prevent water condensation for future testing needs, such as US06 schedules or for alternatively fueled vehicles like ethanol, methanol or compressed natural gas (CNG).

2. Current dilution ratios are too high for future low emissions measurements; thereby reducing the concentration of diluted exhaust emissions, possibly below the systems' detection limits.

3. Ambient hydrocarbon (and NO_x) concentrations are not controlled and will become a significant interference at the new lower standards.

4. The conventional CFV system relies on certain assumptions about the chemistry of the exhaust gases to accurately measure the mass emissions. The assumptions may not always be true for future vehicle technology and testing requirements. Each of these concerns will be discussed in further detail below.

DILUTION RATIOS TOO LOW TO PREVENT WATER CONDENSATION – As previously mentioned, today's CFV system pulls a nominal fixed flowrate of 10 m³/min through a CFV. Since this diluted exhaust flowrate remains essentially constant throughout the FTP cycle, but the vehicle's raw exhaust flowrate fluctuates for the various modes of the test (i.e., idle, acceleration, deceleration, cruise), basic physics dictates that the dilution ratio must vary during the FTP test. To avoid water condensation, the fixed flowrate of the CFV must be set high enough so that a minimum of 6:1 dilution is maintained during peak exhaust volume modes (accelerations/cruises).

For the FTP schedule, and for most vehicles, a CFV flowrate of nominally 10 m³/min has traditionally been used. However with the advent of the Supplemental FTP test cycle (US06), much higher vehicle accelerations are produced, which leads to much higher vehicle exhaust volume flowrates. This, in turn, requires the CFV flowrates to be increased to 20 m³/min (or maybe as high as 35 m³/min) in order to maintain that minimum dilution ratio.

Another factor affecting the minimum dilution ratio is alternative fuels. For oxygenated fuels such as methanol or ethanol, or CNG fuels, there is much more water being produced in the combustion process. This extra water appears in the vehicles' raw exhaust, thereby forcing higher dilution ratios in the CFV system, again, to avoid water condensation. Some studies have indicated that, with various combinations of vehicle size, coupled with high speed driving cycles, like US06, and with alternative fuels, the industry may have to go to CFV flowrates of 60 m³/min or greater.

Of concern in this water condensation issue are the effect of liquid water on emissions measurements. Of the traditional emission constituents, NO_x (mostly the NO_2 component) and hydrocarbons are somewhat soluble in water. Obviously, if they go into solution with water, they probably will not be present in the diluted sample bags for analysis. This has been a small but recognized problem in the past. However with the advent of the new California LEV (and LEV II) regulations, more complex compounds like methanol, ethanol and formaldehyde, are required to be measured. These new compounds, which are all highly soluble in water, will again require raising the dilution ratio in order to avoid water condensation.

DILUTION RATIOS TOO HIGH FOR LOW LEVEL MEASUREMENTS – To avoid water condensation in the future with the CFV system requires increasing the CFV flowrate and, thereby, will raise the dilution ratio. However, the higher the dilution ratio, the lower the emission concentrations in the diluted exhaust stream, which makes it harder to accurately measure the emission levels present. To improve the measurement of these new lower concentrations requires lowering the dilution ratios to their theoretical minimum value (e.g., 6:1 for conventional gasoline). These two requirements are totally contradictory, and unfortunately with the conventional CFV, it is difficult to satisfy both simultaneously. This dilution ratio issue will be a significant measurement problem for measuring emissions at ULEV and LEV II SULEV standards.

AMBIENT AIR CONTAMINATION INTERFERING WITH LOW LEVEL EMISSIONS MEASUREMENT – As lower vehicle hydrocarbon emissions are achieved, they will approach and may even be lower than ambient levels. As the raw exhaust concentrations approach ambient levels, it will become increasingly difficult to differentiate the vehicle's emissions from ambient contamination. This is particularly true at higher dilution ratios, because proportionally there is much more ambient air than exhaust gases in the diluted exhaust bag. One of the measurement problems this causes is computing vehicle emissions includes the taking small difference between two similar numbers. Thus taking the difference is very small, which causes the variability, uncertainty and accuracy of the results to increase. Statistically, it can be shown to cause an increase in measurement variability by $\sqrt{2}$, or approximately 41%.

Using electronics terminology, the signal-to-noise ratio decreases dramatically as the vehicle exhaust concentration approaches ambient. Similarly, if the ambient pollutant level increases significantly, as a result of local factors like where the laboratory is physically located, nearby construction or fueling area vapors, then the measurement effect will be similar to an increase in the dilution ratio.

Yet another issue is the practice of filling the ambient bag set at a constant rate. Theoretically, the rate at which the ambient bags are filled should vary based on the amount of dilution air being put into the sample bags at any instant in time, and is proportional to the amount of ambient air being used to dilute the raw exhaust. Although this is generally thought to be a small correction, at SULEV levels it may be of greater concern.

Related to this is non-representative ambient bag sampling. This is fairly difficult to quantify, but the concept is simple. The premise is that the ambient bag may not completely represent the dilution air that is mixed with the raw exhaust gases. Again, today's equations assume that the ambient air going into the ambient bags is of the same composition as the air diluting the raw exhaust and going into the diluted exhaust sample bags. Examples of how these two air streams could be different are: (1) spatial non-uniformity of the ambient air probe causing non-representative sampling; (2) contamination in the bulkstream after the ambient air probe; (3) contamination in the ambient air bag or sampling system; and (4) sampling system leaks.

CFV RELIANCE ON EXHAUST GAS CHEMISTRY – When the CFV system was adopted, by necessity, it made certain assumptions about the physical measurement being performed. As previously mentioned, one of these assumptions was that the ambient sample bag concentration accurately reflects the make up dilution air used in the diluted exhaust sample bag.

As an example, the dilution factor equation being used assumes that there is no ambient air contamination, which there always is, unless an artificial ambient air source is used. Second, the numerator of the dilution factor equation assumes that the vehicle is always operating at stoichiometry throughout the entire test. Vehicles don't necessarily operate at stoichiometry during all modes of the test and, in fact, some vehicle designs may utilize non-stoichiometric technologies such as lean-burn engines or deceleration fuel shutoff. Any of these conditions will affect the actual dilution factor, as compared with the idealized dilution factor equation (reference SAE Paper # 980678; "Improving the Calculation of Exhaust Gas Dilution During Constant Volume Sampling, Thomas C. Austin and L.S. Caretto, Sierra Research, Inc).

A third concern, although not directly related to the CFV system, is the use of CO_2 tracer techniques to determine instantaneous exhaust volume. This technique was used for many years for modal (real-time, non-regulatory) emissions analysis. The fundamental principle in this technology is that CO_2 is present in the tailpipe exhaust and could be "traced" throughout the system. Again, the measurement (exhaust volume) fundamentally relies on exhaust gas composition assumptions that may not be true in the future.

ALTERNATIVE SOLUTIONS TO THE TRADITIONAL CFV ISSUES

Many different approaches have been proposed to resolve the previously mentioned issues. One thought is to heat the analyzers and sampling systems to above the diluted (or raw) exhaust's dew point. This sounds plausible, but heating the sampling systems tends to create significant maintenance problems. Additionally, there is the question of which analyzers must be heated. Not all current analyzers work properly at elevated temperatures. Also, heating lots of sample lines, analyzers, pumps, benches, bags, etc is very difficult to do without creating local cold and hot spots. Localized cold spots could cause water condensation inside the sampling system, which will effect emissions measurements.

Another possibility is to vary the entire diluted exhaust volume flowrate by using a multi-venturi (switchable) CFV system or sub-sonic nozzle. The principle here is to adjust the diluted exhaust flowrate to match vehicle, fuel, and test schedule requirements, and, thereby optimize the dilution ratio for the specific test conditions. This is a step in the right direction. However, it still does not optimize the dilution ratio to the theoretical minimum because of the varying vehicle exhaust volume flowrates. The diluted flowrate will still need to be sized for the maximum vehicle exhaust volume flowrate in order to eliminate the water condensation issue. Past this condensation concern, ambient air contamination still remains an issue. To address this, one could attempt to clean the entire bulkstream flowrate (up to 10, 20, 35 or 60 m³/min), but the additional equipment costs and size requirements are prohibitive to some laboratories.

A third possibility is to perform modal emissions testing. Although it sounds good, today's modal systems use sample conditioning units (scu's) containing refrigeration units to remove water. Unfortunately, along with the water, some portions of the exhaust emissions are also removed. To solve this, either a modal mini-diluter system could be used or heat all of the analyzers (previously mentioned problems). Another issue is that, with some of the more exotic constituents, there are either limited or no modal methodologies. Examples are ethanol, methanol, aldehydes, speciated hydrocarbons, and even methane. Therefore, these constituents require batch (or bag) sample analyses. Another issue with modal analysis is accurate vehicle exhaust volume determination. The current CO_2 tracer technique has some known accuracy problems. To solve this issue, some type of direct vehicle exhaust volume measurement technology would be needed. As one explores the concept of modal measurements for certification, further analysis finds it falls short of some required measurements like NMHC, NMOG, alcohols, particulate, HC speciation, etc. This leads to the realization that sample bags or "batch" analysis will still be required for the foreseeable future. Since modal analysis would still need many of the advanced technologies being used in BMD systems (DVE and modal mini-diluters) the perceived advantage of simpler modal analyses for certification (and development) testing appears nebulous.

BAG MINI-DILUTER – THEORY OF OPERATION

THEORY – The approach investigated during this study was a technology called Bag Mini-Diluters (BMD). This technology, when coupled with Direct Vehicle Exhaust (DVE) volume measurement, shows the best promise for future low level emissions measurements of technologies tested thus far. A schematic of this technology is illustrated in Figure 2 below:

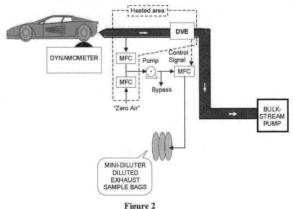

Figure 2
Bag Mini-Diluter plus Direct Vehicle Exhaust Volume Measurement

This technology has several parts. First, the vehicle's entire raw exhaust flow is precisely measured using a "Direct Vehicle Exhaust" volume instrument. Several different methodologies for exhaust volume measurement are being explored by a variety of different manufacturers. For this report, a commercially available ultrasonic unit was used to do this measurement.

Note in Figure 2 that water condensation is avoided by heating the vehicle exhaust gases up to and including the DVE volume system. Also the bag mini-diluter sample lines are heated up to and including the dilution point. Past this dilution point, the diluted exhaust gas dew point temperature is low enough to avoid water condensation in the sample plumbing.

For the purposes of this study, "zero air" is purified ambient air in which hydrocarbons, carbon monoxide and water have been removed by a conventional zero air generator. In the future, zero nitrogen could also be used for this purpose, which might further improve the emissions measurement; particularly for NO_x (in N_2, less NO will be converted to NO_2).

Knowing the exhaust volume, the second function the mini-diluter system performs is to precisely dilute a small portion of the exhaust gas with "zero air". This is accomplished using two small mass flow controllers (MFC). One MFC meters the raw exhaust sample flow and a second MFC meters the zero air flow. Again, the raw exhaust lines and all of the plumbing up to and including the mix point are heated to avoid water condensation. The ratio between the zero air flowrate and the raw exhaust flowrate is set to the minimum dilution to avoid water condensation, ideally 6:1 for conventional gasoline (Indolene). Typical flowrates through the dilution mass flow controllers are 6 liters per minute (L/min) for the raw exhaust MFC, and 35 L/min for the zero air MFC. The diluted flow of 41 L/min is enough to properly fill a conventional CFV bag through all the various accelerations and cruises. Lower flowrates could be used if smaller bags were to be used. Once set, this dilution ratio remains constant throughout the entire vehicle test. Note that this ratio could be set to any value, but the minimum dilution ratio (6:1 for Indolene) is desired.

The last function of the mini-diluter is to fill the sample bags with this diluted exhaust mixture. This is done with the third MFC, whose flowrate is controlled proportionally with the vehicle's exhaust volume flowrate. The direct vehicle exhaust volume instrument provides this control signal. Because the flowrate through the third mass flow controller varies, but the diluted exhaust flow remains fixed, the excess flow is bypassed, as shown in Figure 2

With this architecture, all four of the previously mentioned CFV problems are addressed. First the dilution ratio is now programmable and set to eliminate water condensation. Second, this dilution ratio is optimized to the lowest value, thereby maximizing the diluted exhaust concentrations. Third, the dilution air is free of contaminants because ambient air is not being used; instead zero grade air or nitrogen is utilized which has minimal contamination. And fourth, the system is less sensitive to the vehicle's exhaust chemical composition; instead, laws of physics and pneumatic principles are used to properly make the measurement.

Note that Figure 2 is a simplified schematic for illustrative purposes only. In reality, the mini-diluter has many more components to properly route the various gases and provide appropriate diagnostics for calibration and quality control. A prototype system was manufactured for performance testing as described in later in this study. With the bag mini-diluter system, a simplified version of the mass equations become:

$$\text{Mass Emissions} = \text{Conc} * \text{Exh Vol} * \text{DR} * \text{Density} \quad (4)$$

Where: Conc=Concentration, Exh Vol=Exhaust Volume
DR=Dilution Ratio, Density=Constituent Density

ADVANTAGE OF MINIMAL CALIBRATION – The calibration of the BMD itself was fairly simple and is done in two steps. First, the dilution ratio was nominally set by electronically adjusting the raw exhaust and diluent mass flow controllers to the desired values. Typically this ratio was set to 10:1, but could be set to any value. To determine the exact dilution ratio, a known calibration gas mixture of CO_2 and N_2 was injected into the raw exhaust MFC, thereby simulating a vehicle exhaust gases. The bag bench CO_2 analyzer then read this diluted mixture. Knowing the concentration of the calibration gas injected into the BMD on and the bag bench analysis of the diluted mixture, one can accurately calculate the actual dilution ratio. A nominal 13.5% CO_2 calibration gas was chosen to minimize CO_2 interference's during the actual vehicle test.

The second part of the BMD calibration is to establish proportional flow through the third MFC. Here the importance is proportionality with the vehicle exhaust volume. The accuracy of this third MFC really doesn't matter; the consequences of mis-adjustment are either over or under filling the sample bags. Proportionality is the key. Both the direct vehicle exhaust volume instrument and the MFC's are very linear, therefore proportionality is basically achieved by definition. However, one parameter which can effect this, is zero offsets either in either of these two devices. To prevent this error in the bag mini-diluter system, a check and adjustment of the zero offset to zero was part of the calibration.

The DVE's calibration was checked periodically via a smooth approach orifice, which was traceable to NIST. In addition several of these units have been sent to NIST for direct calibration with good correlation results

WATER AND CO_2 EFFECTS AND THEIR COMPENSATION

CO_2 EFFECTS ON MASS FLOW CONTROLLERS PERFORMANCE – The effect of CO_2 on mass flow controllers is well documented and can be verified by normal scientific experiments. The effect is due to the diabatic-flow process that the MFC's are based on. This effect can be mathematically calculated. Experiments were performed to analytically confirm the theoretical effects of CO_2.

Basically, as the CO_2 level increases, the mass flow controller's actual flowrate will decrease. This error in flow is a small, but not insignificant in a raw exhaust stream. In rough terms, for every 4% of CO_2 concentration increase, the MFC's flowrate will decrease by -1%. With today's vehicles operating basically at stoichiometry, the tailpipe CO_2 concentration will be approximately 13.5%, which will yield about a -4% error in the MFC's flowrate. Although small, if left uncorrected, this would bias the bag mini-diluters results by the same amount, about -4%. To compensate for this effect, the bag mini-diluter has been calibrated on a nominal 13.5% CO_2 calibration gas in order to set the actual dilution ratio. Doing this CO_2 injection into the raw exhaust MFC accurately compensates for the CO_2 effect. The degree to which this compensation works is demonstrated with actual vehicle testing. In the attached vehicle test data, the reported CO_2 mass of the bag mini-diluter system agrees very well with that of the CFV (less than +0.35% difference).

In some vehicle operating modes the tailpipe CO_2 levels are lower than stoichiometric conditions. Examples of this are modes during the initial start of the test (e.g., crank, idle) and during modes where deceleration fuel shutoff is used. During these times the tailpipe CO_2 levels can be significantly less than those at stoichiometic conditions. Even though this will affect the MFC's performance, the effect appears to be minimal because the time increment that the vehicle is away from stoichemetry is small compared to the total test time. In an attempt to correct for this effect at crank, the BMD used in this paper's testing had a compensation circuit built in which adjusted the MFC's flowrate for the first few seconds of the test. Although a technically correct thing to do, in reality, this compensation probably did little good, again due to the small time interval involved.

In addition, for deceleration fuel shutoff modes, experience has shown that the tailpipe CO_2 levels rarely hit 0%; rather they ramp down to some value, then ramp back up. Because there is some CO_2 in the exhaust; this tends to minimize the error.

WATER VAPOR EFFECTS ON MASS FLOW CONTROLLERS – The effect of water vapor on MFC's is not well documented nor apparently well understood. Manufacturers of MFC's disagree on this topic, with some saying there is an effect and others saying there isn't. Using the same diabatic-flow process theory of MFC's, there should be a small but (again) not insignificant effect due to water vapor. This water vapor effect should be in the same direction as the CO_2 case (i.e., creates lower flows through the MFC's) and have a magnitude of approximately half of the CO_2 effect. Note that the water vapor concentration in raw exhaust is basically the same as CO_2 at stoichiometry.

The contradiction of water vapor effects on MFC's mentioned above may be due to the fact that the experimental apparatus and methodology needed to measure water vapor interference is difficult to operate correctly. The reason this measurement is difficult is due to the potential of water condensation introducing measurement errors. In doing a water vapor effect experiment on MFC's, there is a requirement to keep all parts of the apparatus at an elevated temperature and above the dew point of the gases involved.

Several experiments were run to try to quantify the effect of water vapor on MFC's, but some of these experiments had suspect results (data showed opposite effect predicted by theory) and therefore questionable data. This may have been due to water vapor condensing in cold spots, therefore effecting the data. However in a few experiments it appeared that water vapor (47 degC dew point) did have an adverse effect on the MFC's flowrate by about -1 to -1½ % (lower flowrate). It is very difficult to quantify this error further, since the test apparatus used probably had accuracy and repeatability errors of the same magnitude or higher. However this –1 to –1½ % error was not apparent in the vehicle data, so either it is (1) not real, or (2) something else is compensating for it, or (3) is so small that it is less than the test sites ability to measure it. Options (1) & (3) may be very likely scenarios, and further research could resolve the issue. However option (2) was investigated further because it seemed to hold the most promise in explaining the issue.

WATER VAPOR AND CO_2 EFFECTS THE DIRECT VEHICLE EXHAUST VOLUME INSTRUMENT – Based on the above analysis, it was desirable to determine if anything else in the BMD system could be compensating for the predicted –1 to –1½ % effect due to water on MFC's. One possibility was the DVE volume instrument because exhaust volume is a direct multiplier to determine mass emissions. Several DVE ultrasonic instruments were sent to the National Institute of Standards and Technology (NIST). In these evaluations, the accuracy of the DVE were measured under a variety of test conditions including changing the composition of the gases flowed. In one test, simulated exhaust gas was used which contained concentrations of heated H_2O

vapor, CO_2, N_2 and O_2 that closely resembled those of raw exhaust gas. In reviewing the NIST evaluation reports of several DVE ultrasonic meters, it was discovered that heated simulated exhaust gas appears to have an effect on the DVE. When heated simulated gas was injected into the DVE meter, its measured flow values appeared to be artificially higher than that of ambient air, by approximately +1%.

Further study indicates that theory may support the introduction of a small error in ultra-sonic flow measurements due to exhaust gas composition. The error is due to air vector velocity profiles changing as a function of Reynolds numbers. This effect needs to be further studied, but if true, it may explain how a –1% water effect on the raw exhaust MFC could be compensated by a raw exhaust gas effect on the ultrasonic meter. Both the MFC effect (mini-diluter dilution ratio) and the ultrasonic meter effect (exhaust volume) are direct multipliers in the mass equations and therefore may tend to cancel each other out. Again, vehicle data would tend to support this premise.

COST EFFECTIVENESS – RETROFIT ABILITY OF BAG MINI-DILUTERS TO TODAY'S TEST SITES

The BMD system was designed such that one could retrofit an existing exhaust volume sampler with this technology and avoid the cost of buying all new equipment. A simplistic view of the BMD is a replacement of today's small bag fill venturis. The BMD itself is a small device, measuring about 0.5 meters on a side. Retrofitting an existing exhaust volume sampler allows the reuse of existing bag fill, read, and purge/evac hardware. This approach also allows for the normal routing of vehicle exhaust gases out of the test site and provides a mechanism to continue closed loop tailpipe pressure control if used.

Of course the DVE device would be a new instrument to the test cell and would have to be located elsewhere in the test site.

In the process of retrofitting an existing exhaust volume sampler with bag mini-diluter technology, one could also make a switchable device and utilize both technologies, i.e., has the capability to switch between the bag-mini-diluter technology and the regular bag fill venturis. This has successfully been done using a concept of a heated, hybrid CFV.

The bag mini-diluter also has a low procurement cost and uses minimal utilities to operate. Because the bag mini-diluter system is miniaturized, it uses much less diluent gas than a normal exhaust volume sampler. For example, the bag mini-diluter uses about 30 liters/minute (0.03 m^3/min) of diluent flow to properly dilute the raw exhaust sample. As previously mentioned, a conventional exhaust volume sampler has a bulkstream diluent air flowrate of 10, 20 even 35 m^3/min. Taking a simple ratio of these numbers, the bag mini-diluter uses between 1/100th to 1/10,000th less diluent gas than the conventional CFV does. If this diluent needs to be conditioned such that it is dehumidified and purified of contaminants, then from a utilities perspective, the bag mini-diluter would obviously cost much less to operate.

OTHER MINI-DILUTER APPLICATIONS

Although this report will focus on one prototype bag mini-diluter system used for chassis dynamometer testing, it is important to note that many mini-diluters with similar technology have been successfully implemented in multiple applications for years. As an example, during the Auto-Oil Air Quality Improvement Research Program (early 1990's), bag mini-diluter systems were employed to do engine out speciation testing. This work is documented in SAE technical paper series 930141 "A Sampling for the Measurement of Pre-Catalyst Emission from Vehicles Operated Under Transient Conditions", published March 1st, 1993, by Jon McLeod et al.

On a larger scale, modal mini-diluters have been successfully implemented for vehicle development emissions measurements. They replaced the older technology sample conditioning units (SCU's) which condense water out of the raw exhaust gases by cooling the sample (4 degrees C typical). Modal mini-diluters eliminate the need to chill the raw exhaust sample (removing water) by diluting the raw exhaust sample with enough pure nitrogen to lower its dew point below sample line temperatures. This dilution ratio is fixed throughout the test. Architecturally this technology is identical to the dilution portion of the bag mini-diluter system, but the bag mini-diluter has a further functionality of taking this diluted sample and proportionally filling sample bags.

This modal application of the mini-diluter technology has been successfully used in vehicle development testing since the early to mid 1990's, and currently dozens of these systems are in routine production testing.

Another application is for diesel particulate measurements. Although for particulate measurements the mini-diluter configuration is somewhat different to avoid interferences with the particulate collection, the principle is the same. Other applications could be stack gas and engine dynamometer emissions testing.

DIRECT VEHICLE EXHAUST VOLUME INSTRUMENT (DVE)

The direct measurement of the vehicle exhaust volume has never been required to certify automobiles for the Environmental Protection Agency (EPA) or the California Air Resources Board (CARB). However, in order to optimize the vehicle emission control packages in real time, second-by-second results are needed. The vehicle exhaust volume combined with the emission concentrations provides mass results throughout the entire test

sequence. Since the early 1970's the exhaust volume has been measured using the CO_2 dilute ("tracer") method. This method requires measuring the CO_2 concentrations at the raw tailpipe and the diluted sample in the CFV simultaneously. As noted above, the drawbacks with the CO_2 tracer method are issues like time alignment of analyzers and problem modes like crank and initial idle.

As mentioned earlier, exhaust volume measurement was identified in the 1994 Specifications document. This has been a very challenging task since vehicle exhaust has a wide temperature range, high percentage of water, changing concentrations of components, and is very corrosive.

A commercially available DVE flowmeter was used in this study. The meter is a differential transit time, ultra sonic meter and uses two ultra sonic sensing units to measure velocity of the exhaust gas in a pipe. Each sensing unit consists of an ultrasonic transmitter and receiver. The sensors are placed diametrically opposed to each other and at an angle to the flow so that one sensor is up stream of the other.

During operation the transmitters emit alternating synchronized ultrasonic waves. The instrument measures the time required for the sound to travel across the fluid. In the absence of flow sound waves would travel between the sensors at the speed of sound in vehicle exhaust. In the presence of flow the velocity of the sound wave travelling downstream would be accelerated, while the velocity of the sound wave travelling upstream would be decreased. The difference between the two measurements is directly proportional to the velocity of the fluid. Since it is the difference between the two measurements that is used in the calculation of flow, fluid properties are not required to measure fluid velocity. Pressure and temperature transducers are located in the flow stream. Temperature and pressure, along with pipe diameter are used to calculate volumetric flow.

The flowmeter has provisions for both heating and cooling the exhaust gas. Heating blankets are installed over piping in contact with the exhaust gas to prevent condensation inside the meter. A shell and tube heat exchanger coupled with a radiator provides cooling.

The flowmeter has an internal computer that collects the pressure, temperature, and ultrasonic sensor output and calculates flow, both actual and standard. The meter has analog outputs for standard flow, actual flow, temperatures at various points within the meter, and pressure. Digital outputs provide several error messages, plus a pulse count of standard flow.

The specifications on the meter state a response time of 0.5 seconds to 90% of full scale flow, an accuracy of 1.0% of point down to a Reynolds Number of 3000 and 1.0% of full scale below that. Repeatability is 0.5% of point. Several years of experience with the meter indicate that its performance is in line with manufacturer specifications.

Other DVE technologies were investigated for this study, but the ultra sonic appeared to have the best performance characteristics for this studies purpose. As the need for direct vehicle exhaust volume increases, it is anticipated that other measurement technologies will be further developed.

PATENT AWARD

The uniqueness of the bag mini-diluter technology was recognized by the United States Patent Office by granting a patent to the Environmental Research Consortium in July of 1997 (patent # 5,650,565).

DATA - CFV TO BAG MINI-DILUTER COMPARISON

APPARATUS USED TO TEST VEHICLES – To verify the mini-diluter's operation required the comparison of the bag mini-diluter results to that of a conventional CFV. At higher emissions levels, such as Tier I, the two systems should agree very well. However, at lower levels, such as ULEV and SULEV, the previously mentioned problems inherent in the CFV system should become apparent.

There are several ways to compare the two sampling systems. One way is to run vehicles first on the CFV system, and then repeat the tests on the bag mini-diluter system. This testing sequence makes it difficult to perform a comparative analysis of just the sampling systems, since vehicle test-to-test repeatability may play a major role in the magnitude of the perceived differences. Therefore, it was decided to measure the vehicle's emissions both ways on all tests (i.e., run the bag mini-diluter and the CFV in parallel). This cuts down on the number of tests needed and removes many of the vehicle test repeatability concerns.

A valid way to compare the results from the two different sampling systems is to analyze the differences in the results stemming from vehicle tests run with the parallel sampling methods. Replicate tests are run for a number of different vehicles. The difference data are then collected from each vehicle and statistical information is generated. This treatment helps to put a "real world" face on the results. For example, it could be said that the cold start (Phase 1) HC's from a LEV run on a 48" single roll dynamometer on a 10 m^3 CFV site are X% higher from the BMD than from the CFV.

Another way of looking at the data is to combine the results from all of the vehicle tests and generate statistics on the entire data set. In this way, the vehicle is taken out of the picture in the data interpretation. For example, if a certain data trend is true for a particular bag sample from the LEV, the same data trend may hold for all other samples with the same constituent mass level, regardless of which vehicle created it. Note, however, that a constituent data trend at a particular mass level might or might not hold true for higher or lower mass levels. For example, let's assume that the BMD, on average, reads higher for HC's than does the CFV. It may not be safe to assume that the BMD reads higher than the CFV at very low levels, as well as at very high levels. Likewise, it may not be safe to assume that the reason that the BMD reads higher than the CFV for a particular mass range is the same for all other mass ranges, higher and lower than the original. There may be different physical mechanisms at play, which may affect the relative magnitude of the measured differences. However, it may be reasonable to say that, if there is bias in the measured data, the physical reasons for the bias in a particular sample are likely to be the same in a different sample of the same concentration range, assuming that the testing conditions were nominally the same.

Grouping the data by constituent mass level allows the use of a randomized paired t-test comparison, as described by Box, Hunter, & Hunter in Statistics for Experimenters, Chapter 4 (Wiley & Sons, 1978). Within given mass range intervals for each constituent, the difference between the data from the BMD and the CFV for all results can be averaged and a sample standard deviation determined. For easy comparison, the average of the CFV masses within a given mass range interval is calculated and the "relative average difference," with respect to the average of the CFV mass values is calculated. Then, for each range, the null hypothesis is tested, to see if the average difference is significantly different from zero. In this case, an insignificant null hypothesis result means that the perceived difference between the BMD and the CFV for the given mass range is not statistically different from zero. It should be noted that this statistical test cannot be used for the entire range of data results. The use of a randomized paired t-test comparison assumes that data is gathered randomly. The simple act of combining vehicle data into separate mass range "bins" creates some order to the data, but is unavoidable in order to easily interpret the results. Creating wider mass range "bins" or combining the results from adjacent "bins" makes the sampling less and less random, which invalidates the statistical test. By keeping the mass range relatively narrow, the statistical test is more able to determine true significant differences between the BMD and the CFV sampling systems.

Figure 3 shows the test setup (again) in simplified format. Note that the mini-diluter sample probe to the raw exhaust pipe was actually taken after the direct vehicle exhaust volume instrument. This was done to eliminate any possible effects due to exhaust emission compounds reacting with the DVE device, unlikely as it is.

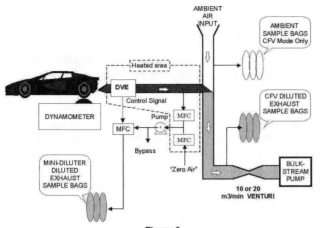

Figure 3
Bag Mini-Diluter / CFV Evaluation System

Picture 1 shows a picture of the prototype bag mini-diluter system that was used for these vehicle tests.

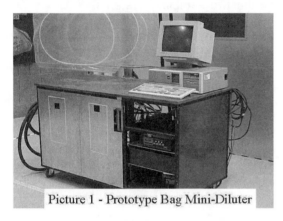

Picture 1 - Prototype Bag Mini-Diluter

Picture 2 shows a close-up picture of the production bag mini-diluter device.

Picture 2 - Production Bag mini-Diluter

CFV-TO-BMD CORRELATION TEST RESULTS

1996 VEHICLE TESTING – Four different vehicles were run on the parallel sampling systems in 1996. These vehicles consisted of a 4-cylinder passenger car, a 6-cylinder car, an 8-cylinder car, and a correlation vehicle which was modified to produce higher emissions levels

for test equipment diagnostic work (designated Repca VIII). The data and statistics are shown in Table 1. Each vehicle test was an EPA II test (the first two phases of the FTP). All testing was done on a twin-roll dynamometer site operating at nominally 10 m^3/min. Although a 6:1 dilution setting is the theoretical minimum dilution ratio to eliminate the possibility of water condensation, the BMD was set to a 10:1 dilution. This was done to allow a generous safety margin for the vehicle tests. Data from each phase, plus a two-phase composite, are shown. As can be seen from the coefficient of variance (COV) values, the variability in the test results is dominated by the run-to-run variability of the vehicles themselves. It is worth noting that the COV for the CFV and for the BMD were virtually identical for all components and for all vehicles. This indicates that, although the vehicles were themselves quite variable from run to run, the two different sampling systems tracked each other very well and were operating properly. Tests on both the 4-cylinder vehicle and Repca showed excellent agreement between the CFV and the BMD for all components. There was a slight offset (about 5% higher for the BMD with respect to the CFV) between the BMD and the CFV for HC's and non-methane HC's (NMHC's) for the 6-cylinder and the 8-cylinder vehicle. However, the CO_2 values agreed well for tests on these two vehicles, indicating that the source of the HC offset is probably not the vehicles or the proper functioning of the sampling systems. It is more likely to be from a fundamental sample handling difference between the two sampling systems.

1997 VEHICLE TESTING – Vehicle testing in 1997 was done on the same four vehicles as in 1996, with the addition of a lower-level laboratory "repeater" car, designated Repca IX. As before, the BMD was run at a 10:1 dilution ratio. The data and statistics are shown in Table 2. As can be seen by comparing the Repca VIII results with those from Repca IX, the HC AND CO variabilities for Repca IX are much higher. However, this is to be expected, since the HC mass values are lower for Repca IX by about a factor of six and the CO mass values are almost halved. Even with the higher variability, both vehicles continued to give results indicating good agreement between the BMD and the CFV, again meaning that the sampling systems are operating properly. There was the same offset between the BMD and the CFV for HC's and NMHC's for the 6-cylinder and the 8-cylinder vehicle, although the magnitude of the offset was slightly higher (7-9% higher for the BMD). Again, since the CO_2 values agreed so well for all vehicles, the reason for the HC offset is likely to be a difference in sampling system design.

1998 VEHICLE TESTING – Comparative vehicle testing on parallel CFV/BMD sampling systems was done exclusively on ULEV-class vehicles in 1998. Two similarly equipped commercially available ULEV's were used (identified as ULEV # 1 and ULEV # 2). Full three-phase

FTP testing was done both on a 10 m³/min twin-roll dyna-mometer site and on a 20 m³/min single-roll site. Additionally, US06 testing was done on the 20 m³/min single-roll site. The higher CFV flowrate was needed to avoid water condensation for a US06 test schedule. To better ensure accurate measurements for the lower emission level vehicles, the BMD was run at an 8:1 dilution ratio. The vehicle data is summarized in Table 3. In general, the COV values for the two sampling systems are very similar, indicating that the sampling systems tracked each other well and are probably operating properly. The data where the COV values diverged tended to be cases where the emission levels were at the extreme low end of the analyzer ranges. Noteworthy was that the offset between the BMD and the CFV for HCs and NMHCs for the ULEV # 1 vehicle doubled for the data from the 20 m³/min site, as compared to the data from the 10 m³/min site. Combining this with the fact that the CO_2 values agreed very well for both sets of data leads to the conclusion that there is a sample concentration effect to the system design difference of the two sampling systems, as previously mentioned. The performance of the two sampling systems for the US06 testing was not markedly different from their performances for the FTP testing.

CFV-TO-BMD CORRELATION TEST RESULTS – CUMULATIVE CONSTITUENT RESULTS

TOTAL HYDROCARBONS – The HC results are shown in Figure 4. Using the statistics of the randomized paired t-test, there is a high degree of confidence (generally greater than 95%) that the BMD system read higher than the CFV system for bags of about 0.6 g/mile or less. For higher mass bags, the BMD generally read higher, although the differences were not as statistically significant. The overall average percent difference was +2.7%. More importantly, the data shows an asymptotic trend at high mass levels towards about +2%. These results are consistent with the design theories of the BMD system. As mentioned previously, the fact that, at high mass levels, the differences between the two sampling systems become smaller (and reach a nearly constant level) is as expected. Also predicted is the fact that the percent differences become very large towards very low emission levels. This trend tends to support the idea that the completely heated sample system and lower dilution ratios attainable by the BMD make it more capable of truly measuring low-level hydrocarbons.

CH_4 – The CH_4 results are shown in Figure 5. To a high degree of confidence (better than 99%), the BMD system read lower for bags of 0.045 g/mile or less. For the samples analyzed, this mass level covered virtually all of the available data from the vehicle tests. The overall average percent difference was –2.9% and the high mass asymptote was at about +0.03%. This data trend is not altogether unexpected. For gasoline-fueled vehicles, CH_4 is not a major exhaust constituent. For very low CH_4 emission levels, it is expected that the ambient bag concentration would be nominally the same as the sample bag

concentration. For extreme cases (such as in much of the ULEV testing), when there is virtually no CH_4 emission, the error and variability in the ambient bag subtraction would tend to result in non-zero calculated emissions for the CFV. Because there is no ambient to subtract, the BMD would tend to correctly report the zero emissions.

NMHC – The NMHC results are shown in Figure 6. To a high degree of confidence (better than 99%), the BMD system read higher for bags of 0.55 g/mile or less. For higher mass bags, the BMD also read higher, although the differences were not as statistically significant. The overall average percent difference was +3.2% and the high mass asymptote was at about +2%. Since NMHC is the difference between the HC and the CH_4 values, these results for NMHC are not surprising. Of particular interest is that, at LEV levels and higher, the regression line indicates a mass percent difference of less than 5% between the two sampling systems. It is at ULEV and SULEV levels where the differences become dramatic and almost exponential. It should be noted that the various emission level ranges shown in the top of the chart are "best case scenarios." The calculation of the range boundaries assumed that there are NO EMISSIONS in any sample bag other than the Phase 1 bag. Since, in most cases, there are measurable emissions in the Phase 2 and 3 bags, those emissions will contribute to the total values and force the range boundaries towards lower emission levels (i.e., towards the left of the chart). This will make the differences between the BMD and the CFV more pronounced at a lower emission level.

CO – The CO results are shown in Figure 7. The CO differences were more random than any of the HC results. There were a few bag mass ranges that gave highly significant statistical differences. However, those mass ranges were interspersed within ranges that did not show significant differences. On the whole, the overall average percent difference was +1.6% and the high mass asymptote was about +1.3%. Just as for CH_4, at extremely low levels, the differences became dramatically large. However, since the BMD generally read higher than the CFV, there is likely a different reason for the divergence.

NO_x – The NO_x results are shown in Figure 8. The NO_x differences were only slightly more biased towards the BMD than the CO values. However, the confidence limits for the ranges were generally less than 95%. The overall average percent difference was +1.6% and the high mass asymptote was about +0.1%. The mechanism for the extremely low-level regression trend is probably the same as for CO.

CO_2 – The CO_2 results are shown in Figure 9. These comparative differences were the most unbiased of all of the species measured. Although there were some mass intervals that showed significant differences, on the whole, the overall average percent difference was only +0.3%. There was no significant trend to the differences at high or at low emission levels.

CFV-TO-BMD CORRELATION TEST RESULTS – TRIANGLE PLOTS

The emissions composite average and COV data from Table 1 (1996 Testing), Table 2 (1997 Testing) and Table 3 (1998 Testing), averaged by vehicle, was graphed using triangle plotting techniques. Figures 10 through 15 show the 1996 composite mass data, plotted by vehicle type. The high and low points on each triangle represent the averaged +COV and –COV (respectively), and the third corner of the triangle (on the right) reflects the average. For each vehicle, the CFV and the bag mini-diluter data are plotted using separate triangles. Likewise Figures 16 through 21 show the data from 1997 vehicle testing, and figures 22 through 27 show the data from 1998.

This data represents a culmination of 170 vehicle tests, run on 7 vehicles, over the 1996 through the 1998 time frame. Again, the data shows very good quantitative agreement between the CFV and the bag mini-diluter. In addition, the COV's are very similar. The slightly higher COV of the bag mini-diluter on CO_2 emissions could be predicted, and probably due to the variability of the direct vehicle exhaust volume instrument. Even then, the 0.5% increase is less than normal variability between two identical test sites.

CONCLUSION

The design and development of the bag mini-diluter system was successfully completed. Many revisions to the prototypes were made over the past seven years to improve performance and reliability. The system was found to be fairly rugged in construction, even though some of the high technology elements did require a learning curve to properly operate.

The accuracy requirements (bag mini-diluter versus conventional CFV) were basically met for all constituents for Tier 1 level vehicles. The time response of the DVE instrument was approximately 0.5 seconds and the 3rd MFC (or proportional MFC) time response was 1.0 seconds. This was slower than the cited performance specification of 0.1 second, however this longer time response did not appear to have an adverse effect on the data. Although measurement speed is always a desired feature in test equipment, there doesn't appear to be a demonstrated need (today) for faster time response.

One interesting trend that the data set showed was that, in general, the bag mini-diluter system reported statistically higher hydrocarbon (and non-methane hydrocarbon) emissions at very low levels (ULEV/SULEV). Again, the fact that the two systems would differ at lower levels was anticipated when the study was started. Also, some of the hydrocarbon data taken suggests that the bag mini-diluter system may also exhibit lower variability in hydrocarbon measurements at these low emission levels, as compared to the CFV. This was a predicted benefit in

the theory of the device. Note however that lower variability with the bag mini-diluter has not been definitively established in this study because of the significantly higher variability of the vehicles themselves at these very low emission levels. It is believed that vehicle variability may be masking the true variability of the two measurement systems.

Another interesting conclusion was that the bag mini-diluter also typically reported lower methane emissions especially at the lower levels. This phenomenon was not anticipated but one speculation as to the cause is that it appears to demonstrate the CFV's inability to correctly differentiate vehicle-produced methane gas versus ambient contamination. Although speculative, there is some data supporting this premise. The above methane measurements were made using the on-site, dedicated gas chromatograph methane analyzer. As an independent measurement, some of these vehicle tests were further analyzed using hydrocarbon speciation techniques, as specified in the California Non-Methane Organic Gas Test Procedures. The results of this hydrocarbon speciation confirmed that at very low hydrocarbon levels (phase II bags from the ULEV vehicles), there was little or no methane emissions found in the mini-diluter bag, yet the CFV bag did report methane as part of the vehicles' exhaust emissions.

THE REAL INVENTION OF THE BAG MINI-DILUTER SYSTEM is its high technology approach to solving tomorrow's vehicle emission measurement issues. "High technology" does not mean exotic or high risk parts either since the BMD system utilizes components; that have been commercially used for years. The BMD advantages are:

1. **Theoretically correct approach** to solving the issues of water condensation, dilution ratio and background contamination. Clean slate view.

2. **One technology to solve multiple problems** which can be applied to bag, modal, and NMOG speciation measurement systems. In addition, some instrument manufacturers have successfully used similar mini-diluter technology to do raw exhaust diesel particulate sampling.

3. **Minimize calibration gases** by sharing span gases. When bag mini-diluters are coupled with modal mini-diluters, this allows the sharing of span gases since the analyzer ranges can be the same.

4. **Minimized capital cost and small physical size**. Does not require huge test sites to be constructed to house new (and large) test equipment. Also much lower operating utility costs as compared to alternative technologies. This reduction is both in operating electricity of the BMD itself but also in the minimization of facility heating and air conditioning costs by minimizing the exhausting of expensive conditioned air to the outside.

5. **Minimized calibration and maintenance**

FUTURE STUDIES – Further studies are planned to look into very low level (SULEV) emissions measurements with the bag mini-diluter / CFV combination and comparing results. This will be done using both correlation vehicles and simulators. In addition low level NO_x will be looked at to determine if any new issues arise. It is hoped that studies like these will advance the technology further and allow for better understanding of low level measurement issues.

DEFINITIONS, ACRONYMS, ABBREVIATIONS

BMD: Bag Mini-Diluter
CARB: California Air Resources Board
CH_4: Methane
CFV: Critical Flow Venturi
CNG: Compressed Natural Gas
COV: Coefficient of Variability
CO: Carbon Monoxide
CO_2: Carbon Dioxide
H_20: Water
DVE: Direct Vehicle Exhaust Volume System
EPA: Environmental Protection Agency
ERC: Environmental Research Consortium
FTP: Federal Test Procedure
HC: Hydrocarbons
LEV: Low Emission Vehicle
MFC: Mass Flow Controller
MMD: Modal Mini-Diluter
NIST: National Institute of Standards & Technology
NMHC: Non-Methane Hydrocarbon
NMOG: Non-Methane Organic Gas
NO_x: Oxides of Nitrogen
O_2: Oxygen
SAE: Society of Automotive Engineers
SULEV: Super Ultra Low Emission Vehicle
ULEV: Ultra Low Emission Vehicle
w.r.t: with respect to

ADDITIONAL SOURCES

SAE technical paper series 930141 "A Sampling for the Measurement of Pre-Catalyst Emission from Vehicles Operated Under Transient Conditions", published March 1st, 1993, by Jon McLeod et al.

SAE technical paper series 1999-01-0151 "A CFV Type Mini-Dilution Sampling System for Vehicle Exhaust Emissions Measurement", published March, 1999, by William M. Silvis et al.

SAE technical paper series 1999-01-1460 "A New Proportional Collection System for the Extremely Low Emission Measurement in Vehicle Exhaust", published March, 1999, by Kimikazu Yoda et al.

AUTHORS

D.B. Nagy, J.F.Loo, J.G.Tulpa and F.P. Schroeder (ret.) all are employed at the Vehicle Emission Laboratories, Powertrain Division, at the Milford Proving Grounds. Their mailing address is:

General Motors Proving Grounds - Powertrain
3300 General Motors Road
M/S483-331-000
Milford Michigan, 48380

Rick Middleton and Craig Morgan are employed at the Emissions Laboratory, at the DaimlerChrysler Proving Grounds in Chelsea, Michigan. Their mailing address is:

DaimlerChrysler Corporation
Chelsea Proving Grounds
CIMS 422-01-11
3700 S. M-52
Chelsea, Michigan 48118

APPENDIX (ATTACHED, CONTENTS BELOW)

Table 1 – 1996 Vehicle Correlation Tests

Table 2 – 1997 Vehicle Correlation Tests

Table 3 – 1998 Vehicle Correlation Tests

Figures 4 through 9, Scatter Plots of 1996-1998 Vehicle Correlation Tests

Figures 10 through 27, Triangle Plots of 1996-1998 Vehicle Correlation Tests

Table 1 CFV-to-BMD Correlation
1996 Testing

4-Cylinder Vehicle - 1996 Testing (g/mi) [n=35]
CFV Setting = 350 scfm
BMD Setting = 10:1 dilution ratio

Constituent	Phase 1					Phase 2					Composite				
	CFV_{avg}	BMD_{avg}	%Δ wrt CFV	COV_{CFV}	COV_{BMD}	CFV_{avg}	BMD_{avg}	%Δ wrt CFV	COV_{CFV}	COV_{BMD}	CFV_{avg}	BMD_{avg}	%Δ wrt CFV	COV_{CFV}	COV_{BMD}
THC	0.449	0.448	0.0%	19%	19%	0.014	0.014	-1.9%	17%	20%	0.224	0.224	-0.1%	18%	19%
CO	5.54	5.39	-2.7%	10%	11%	0.47	0.47	-0.2%	44%	44%	2.92	2.85	-2.5%	11%	12%
NOx	0.516	0.519	0.6%	9%	10%	0.033	0.035	4.7%	33%	33%	0.267	0.269	0.9%	9%	10%
CO2	319.2	313.5	-1.8%	5%	6%	329.0	331.4	0.7%	8%	8%	324.3	322.7	-0.5%	6%	7%
CH4	0.035	0.033	-4.7%	11%	13%	0.007	0.006	-11.3%	19%	22%	0.020	0.019	-5.9%	10%	12%
NMHC	0.414	0.416	0.4%	20%	20%	0.007	0.008	7.4%	22%	25%	0.204	0.205	0.5%	19%	19%

6-Cylinder Vehicle - 1996 Testing (g/mi) [n=32]
CFV Setting = 350 scfm
BMD Setting = 10:1 dilution ratio

Constituent	Phase 1					Phase 2					Composite				
	CFV_{avg}	BMD_{avg}	%Δ wrt CFV	COV_{CFV}	COV_{BMD}	CFV_{avg}	BMD_{avg}	%Δ wrt CFV	COV_{CFV}	COV_{BMD}	CFV_{avg}	BMD_{avg}	%Δ wrt CFV	COV_{CFV}	COV_{BMD}
THC	0.303	0.318	5.1%	26%	26%	0.053	0.055	4.5%	56%	55%	0.174	0.182	5.0%	24%	23%
CO	3.01	3.19	5.8%	35%	36%	0.20	0.21	3.9%	54%	51%	1.56	1.65	5.7%	34%	35%
NOx	0.350	0.366	4.6%	21%	21%	0.012	0.013	12.7%	64%	69%	0.175	0.183	4.9%	21%	22%
CO2	434.0	432.4	-0.4%	2%	3%	455.6	468.3	2.8%	3%	4%	445.2	450.9	1.3%	3%	3%
CH4	0.024	0.024	-0.7%	21%	22%	0.011	0.010	-6.4%	36%	38%	0.017	0.017	-2.6%	19%	20%
NMHC	0.279	0.295	5.6%	27%	26%	0.042	0.045	7.3%	65%	62%	0.156	0.166	5.8%	25%	24%

8-Cylinder Vehicle - 1996 Testing (g/mi) [n=36]
CFV Setting = 350 scfm
BMD Setting = 10:1 dilution ratio

Constituent	Phase 1					Phase 2					Composite				
	CFV_{avg}	BMD_{avg}	%Δ wrt CFV	COV_{CFV}	COV_{BMD}	CFV_{avg}	BMD_{avg}	%Δ wrt CFV	COV_{CFV}	COV_{BMD}	CFV_{avg}	BMD_{avg}	%Δ wrt CFV	COV_{CFV}	COV_{BMD}
THC	0.505	0.533	5.5%	21%	21%	0.078	0.081	3.3%	18%	20%	0.285	0.299	5.2%	17%	18%
CO	5.92	6.12	3.3%	11%	10%	0.65	0.65	0.8%	30%	29%	3.20	3.29	3.0%	11%	10%
NOx	0.248	0.252	1.5%	29%	30%	0.036	0.037	0.4%	110%	110%	0.139	0.141	1.3%	37%	38%
CO2	503.0	498.5	-0.9%	3%	2%	541.0	547.4	1.2%	3%	4%	522.6	523.8	0.2%	3%	3%
CH4	0.042	0.042	-0.8%	11%	12%	0.021	0.019	-5.6%	12%	13%	0.031	0.030	-2.5%	9%	11%
NMHC	0.463	0.491	6.1%	23%	23%	0.058	0.061	6.5%	21%	23%	0.254	0.269	6.1%	19%	19%

Repca VIII - 1996 Testing (g/mi) [n=6]
CFV Setting = 350 scfm
BMD Setting = 10:1 dilution ratio

Constituent	Phase 1					Phase 2					Composite				
	CFV_{avg}	BMD_{avg}	%Δ wrt CFV	COV_{CFV}	COV_{BMD}	CFV_{avg}	BMD_{avg}	%Δ wrt CFV	COV_{CFV}	COV_{BMD}	CFV_{avg}	BMD_{avg}	%Δ wrt CFV	COV_{CFV}	COV_{BMD}
THC	0.705	0.700	-0.6%	4%	5%	0.840	0.843	0.4%	2%	5%	0.774	0.774	0.0%	3%	5%
CO	3.74	3.71	-0.8%	2%	4%	3.92	3.97	1.3%	3%	3%	3.83	3.84	0.3%	2%	3%
NOx	1.772	1.730	-2.4%	4%	5%	0.950	0.946	-0.4%	7%	8%	1.347	1.325	-1.7%	5%	6%
CO2	389.9	382.2	-2.0%	1%	3%	469.3	471.0	0.4%	2%	4%	431.0	428.1	-0.7%	1%	3%
CH4	0.030	0.029	-2.2%	4%	4%	0.041	0.041	-1.2%	5%	3%	0.036	0.035	-1.6%	4%	3%
NMHC	0.675	0.671	-0.6%	4%	6%	0.798	0.802	0.5%	3%	5%	0.739	0.739	0.0%	3%	5%

Table 2 CFV-to-BMD Correlation
1997 Testing

4-Cylinder Vehicle - 1997 Testing (g/mi) [n=5]
CFV Setting = 350 scfm
BMD Setting = 10:1 dilution ratio

Constituent	Phase 1					Phase 2					Composite				
	CFV_{avg}	BMD_{avg}	%Δ wrt CFV	COV_{CFV}	COV_{BMD}	CFV_{avg}	BMD_{avg}	%Δ wrt CFV	COV_{CFV}	COV_{BMD}	CFV_{avg}	BMD_{avg}	%Δ wrt CFV	COV_{CFV}	COV_{BMD}
THC	0.457	0.464	1.5%	16%	17%	0.031	0.029	-7.7%	13%	14%	0.237	0.239	0.9%	14%	16%
CO	5.46	5.26	-3.7%	24%	28%	1.06	1.05	-1.0%	26%	27%	3.19	3.09	-3.2%	18%	22%
NOx	0.493	0.512	4.0%	6%	6%	0.032	0.031	-3.9%	18%	11%	0.255	0.263	3.5%	6%	6%
CO2	315.1	318.9	1.2%	2%	2%	328.2	340.8	3.9%	4%	3%	321.9	330.2	2.6%	3%	2%
CH4	0.037	0.036	-4.0%	14%	16%	0.013	0.013	-4.7%	16%	16%	0.025	0.024	-4.2%	11%	14%
NMHC	0.419	0.428	2.0%	16%	17%	0.018	0.016	-10.0%	14%	12%	0.212	0.215	1.5%	15%	17%

6-Cylinder Vehicle - 1997 Testing (g/mi) [n=4]
CFV Setting = 350 scfm
BMD Setting = 10:1 dilution ratio

Constituent	Phase 1					Phase 2					Composite				
	CFV_{avg}	BMD_{avg}	%Δ wrt CFV	COV_{CFV}	COV_{BMD}	CFV_{avg}	BMD_{avg}	%Δ wrt CFV	COV_{CFV}	COV_{BMD}	CFV_{avg}	BMD_{avg}	%Δ wrt CFV	COV_{CFV}	COV_{BMD}
THC	0.253	0.272	7.5%	6%	6%	0.021	0.026	26.3%	22%	38%	0.133	0.145	9.0%	4%	5%
CO	2.57	2.70	5.4%	23%	24%	0.15	0.15	-1.2%	46%	49%	1.32	1.38	5.0%	20%	21%
NOx	0.331	0.362	9.4%	13%	12%	0.019	0.016	-17.0%	27%	31%	0.170	0.183	7.9%	11%	11%
CO2	443.5	457.4	3.1%	1%	2%	458.6	466.9	1.8%	0%	2%	451.3	462.3	2.4%	1%	2%
CH4	0.023	0.023	-0.6%	5%	7%	0.006	0.005	-14.3%	5%	8%	0.014	0.014	-3.7%	3%	4%
NMHC	0.230	0.249	8.3%	6%	6%	0.014	0.020	44.5%	30%	49%	0.118	0.131	10.6%	4%	5%

8-Cylinder Vehicle - 1997 Testing (g/mi) [n=5]
CFV Setting = 350 scfm
BMD Setting = 10:1 dilution ratio

Constituent	Phase 1					Phase 2					Composite				
	CFV_{avg}	BMD_{avg}	%Δ wrt CFV	COV_{CFV}	COV_{BMD}	CFV_{avg}	BMD_{avg}	%Δ wrt CFV	COV_{CFV}	COV_{BMD}	CFV_{avg}	BMD_{avg}	%Δ wrt CFV	COV_{CFV}	COV_{BMD}
THC	0.549	0.600	9.3%	15%	17%	0.060	0.053	-11.9%	16%	20%	0.296	0.317	7.1%	15%	15%
CO	7.12	7.67	7.7%	9%	7%	0.73	0.69	-5.7%	15%	14%	3.82	4.06	6.4%	9%	7%
NOx	0.294	0.308	4.7%	11%	11%	0.021	0.022	2.7%	55%	22%	0.153	0.160	4.6%	13%	10%
CO2	524.4	533.2	1.7%	1%	1%	564.0	556.6	-1.3%	1%	2%	544.9	545.3	0.1%	1%	1%
CH4	0.054	0.057	4.5%	8%	6%	0.019	0.016	-12.7%	10%	11%	0.036	0.036	-0.1%	9%	6%
NMHC	0.495	0.544	9.9%	16%	18%	0.041	0.036	-11.6%	19%	29%	0.260	0.282	8.1%	16%	16%

Repca VIII - 1997 Testing (g/mi) [n=10]
CFV Setting = 350 scfm
BMD Setting = 10:1 dilution ratio

Constituent	Phase 1					Phase 2					Composite				
	CFV_{avg}	BMD_{avg}	%Δ wrt CFV	COV_{CFV}	COV_{BMD}	CFV_{avg}	BMD_{avg}	%Δ wrt CFV	COV_{CFV}	COV_{BMD}	CFV_{avg}	BMD_{avg}	%Δ wrt CFV	COV_{CFV}	COV_{BMD}
THC	0.757	0.766	1.2%	3%	4%	0.923	0.935	1.2%	3%	4%	0.843	0.853	1.2%	3%	4%
CO	3.71	3.73	0.5%	7%	9%	3.94	4.05	2.9%	7%	4%	3.83	3.90	1.8%	5%	5%
NOx	1.809	1.822	0.7%	5%	5%	1.070	1.100	2.8%	7%	7%	1.427	1.449	1.5%	5%	5%
CO2	363.7	363.5	-0.1%	2%	3%	441.8	450.0	1.8%	2%	2%	404.1	408.2	1.0%	2%	3%
CH4	0.034	0.033	-2.7%	3%	5%	0.049	0.048	-1.6%	2%	3%	0.042	0.041	-2.1%	2%	3%
NMHC	0.723	0.733	1.3%	3%	4%	0.874	0.887	1.4%	3%	4%	0.801	0.812	1.4%	3%	4%

Repca IX - 1997 Testing (g/mi) [n=11]
CFV Setting = 350 scfm
BMD Setting = 10:1 dilution ratio

Constituent	Phase 1					Phase 2					Composite				
	CFV_{avg}	BMD_{avg}	%Δ wrt CFV	COV_{CFV}	COV_{BMD}	CFV_{avg}	BMD_{avg}	%Δ wrt CFV	COV_{CFV}	COV_{BMD}	CFV_{avg}	BMD_{avg}	%Δ wrt CFV	COV_{CFV}	COV_{BMD}
THC	0.154	0.157	1.7%	16%	16%	0.133	0.134	0.7%	14%	13%	0.144	0.145	1.2%	11%	11%
CO	2.63	2.68	1.9%	25%	25%	1.65	1.67	1.2%	18%	19%	2.13	2.16	1.6%	18%	19%
NOx	0.188	0.188	0.0%	10%	9%	0.156	0.160	2.6%	11%	12%	0.171	0.173	1.3%	8%	7%
CO2	378.1	378.0	0.0%	2%	2%	452.8	458.2	1.2%	5%	6%	416.7	419.4	0.7%	3%	4%
CH4	0.028	0.028	-1.7%	12%	13%	0.023	0.024	1.3%	20%	21%	0.026	0.026	-0.3%	12%	11%
NMHC	0.126	0.129	2.5%	17%	17%	0.110	0.111	0.5%	15%	15%	0.118	0.120	1.5%	12%	12%

Table 3 CFV-to-BMD Correlation
1998 Testing

1998 MY ULEV#1 - 1998 Testing (g/mi) [n=3]
CFV Setting = 350 scfm
BMD Setting = 8:1 dilution ratio

Constituent	Phase 1					Phase 2					Phase 3					Composite				
	CFV_{avg}	BMD_{avg}	%Δ wrt CFV	COV_{CFV}	COV_{BMD}	CFV_{avg}	BMD_{avg}	%Δ wrt CFV	COV_{CFV}	COV_{BMD}	CFV_{avg}	BMD_{avg}	%Δ wrt CFV	COV_{CFV}	COV_{BMD}	CFV_{avg}	BMD_{avg}	%Δ wrt CFV	COV_{CFV}	COV_{BMD}
THC	0.129	0.135	5.2%	5%	8%	0.002	0.004	86.0%	32%	11%	0.003	0.005	56.1%	60%	56%	0.029	0.032	10.1%	5%	8%
CO	1.20	1.27	5.9%	10%	9%	0.05	0.05	-9.2%	23%	23%	0.05	0.06	17.9%	159%	169%	0.29	0.31	5.0%	17%	18%
NOx	0.223	0.227	1.8%	4%	1%	0.001	0.001	43.0%	45%	49%	0.048	0.047	-2.4%	53%	54%	0.060	0.060	1.1%	15%	12%
CO2	371.0	382.4	3.0%	1%	1%	369.6	371.0	0.4%	1%	4%	315.3	321.0	1.8%	2%	2%	355.0	359.7	1.3%	1%	2%
CH4	0.009	0.011	25.9%	5%	10%	0.000	0.000	-100.0%	0%	#DIV/0!	0.001	0.001	17.7%	21%	109%	0.002	0.002	8.7%	6%	13%
NMHC	0.120	0.127	5.5%	5%	7%	0.002	0.004	128.4%	39%	11%	0.002	0.004	83.2%	64%	43%	0.026	0.030	11.7%	4%	7%

1998 MY ULEV#1 - 1998 Testing (g/mi) [n=4]
CFV Setting = 700 scfm
BMD Setting = 8:1 dilution ratio

Constituent	Phase 1					Phase 2					Phase 3					Composite				
	CFV_{avg}	BMD_{avg}	%Δ wrt CFV	COV_{CFV}	COV_{BMD}	CFV_{avg}	BMD_{avg}	%Δ wrt CFV	COV_{CFV}	COV_{BMD}	CFV_{avg}	BMD_{avg}	%Δ wrt CFV	COV_{CFV}	COV_{BMD}	CFV_{avg}	BMD_{avg}	%Δ wrt CFV	COV_{CFV}	COV_{BMD}
THC	0.119	0.129	7.7%	5%	8%	0.003	0.007	104.3%	44%	51%	0.001	0.006	883.5%	71%	45%	0.027	0.032	19.7%	7%	14%
CO	1.10	1.24	12.5%	12%	15%	0.03	0.02	-35.5%	36%	45%	0.02	0.01	-54.3%	43%	6%	0.25	0.27	7.9%	12%	15%
NOx	0.282	0.301	6.7%	10%	7%	0.004	0.006	48.1%	91%	52%	0.094	0.096	2.2%	13%	13%	0.087	0.092	6.4%	2%	3%
CO2	353.2	355.4	0.6%	0%	2%	347.8	333.9	-4.0%	1%	5%	301.0	307.2	2.1%	1%	8%	336.1	331.0	-1.5%	1%	3%
CH4	0.008	0.007	-3.1%	4%	4%	0.000	0.000	-100.0%	69%	#DIV/0!	0.001	0.000	-55.6%	28%	37%	0.002	0.002	-16.7%	10%	5%
NMHC	0.110	0.119	8.4%	4%	8%	0.003	0.007	129.8%	54%	51%	0.000	0.006	#DIV/0!	#DIV/0!	48%	0.024	0.030	23.2%	6%	13%

1998 MY ULEV#2 - 1998 Testing (g/mi) [n=6]
CFV Setting = 700 scfm
BMD Setting = 8:1 dilution ratio

Constituent	Phase 1					Phase 2					Phase 3					Composite				
	CFV_{avg}	BMD_{avg}	%Δ wrt CFV	COV_{CFV}	COV_{BMD}	CFV_{avg}	BMD_{avg}	%Δ wrt CFV	COV_{CFV}	COV_{BMD}	CFV_{avg}	BMD_{avg}	%Δ wrt CFV	COV_{CFV}	COV_{BMD}	CFV_{avg}	BMD_{avg}	%Δ wrt CFV	COV_{CFV}	COV_{BMD}
THC	0.135	0.138	2.9%	11%	12%	0.008	0.009	16.1%	81%	62%	0.007	0.007	-8.2%	66%	47%	0.034	0.035	3.8%	19%	18%
CO	1.33	1.40	5.0%	15%	16%	0.02	0.01	-24.6%	68%	87%	0.02	0.02	-8.3%	113%	120%	0.29	0.30	3.7%	16%	16%
NOx	0.250	0.270	7.9%	25%	25%	0.006	0.007	27.1%	49%	45%	0.028	0.028	0.6%	34%	35%	0.062	0.067	7.9%	23%	21%
CO2	368.6	371.3	0.7%	1%	3%	362.8	343.3	-5.4%	2%	4%	315.8	310.4	-1.7%	1%	2%	351.1	340.1	-3.1%	1%	3%
CH4	0.009	0.009	-4.2%	16%	18%	0.000	0.000	-100.0%	113%	#DIV/0!	0.001	0.000	-84.4%	36%	143%	0.002	0.002	-15.1%	19%	18%
NMHC	0.125	0.129	3.4%	11%	12%	0.008	0.009	20.0%	81%	62%	0.007	0.007	-2.4%	70%	47%	0.032	0.033	5.1%	20%	18%

1998 MY ULEV#1 - 1998 Testing (g/mi) [n=4]
CFV Setting = 700 scfm
BMD Setting = 8:1 dilution ratio

Constituent	US06				
	CFV_{avg}	BMD_{avg}	%Δ wrt CFV	COV_{CFV}	COV_{BMD}
THC	0.021	0.025	16.6%	32%	44%
CO	14.90	15.10	1.3%	13%	12%
NOx	0.031	0.031	1.1%	14%	14%
CO2	332.8	336.8	1.2%	0%	1%
CH4	0.009	0.009	-2.7%	32%	34%
NMHC	0.012	0.016	30.7%	32%	50%

1998 MY ULEV#2 - 1998 Testing (g/mi) [n=9]
CFV Setting = 700 scfm
BMD Setting = 8:1 dilution ratio

Constituent	US06				
	CFV_{avg}	BMD_{avg}	%Δ wrt CFV	COV_{CFV}	COV_{BMD}
THC	0.027	0.026	-6.2%	47%	38%
CO	16.27	15.40	-5.4%	22%	19%
NOx	0.017	0.020	16.3%	23%	27%
CO2	344.5	345.5	0.3%	1%	1%
CH4	0.011	0.010	-10.7%	43%	36%
NMHC	0.017	0.016	-3.4%	50%	39%

Figure 4 - Comparison of Sampling Systems
BMD-to-CFV THC Mass Differences wrt CFV

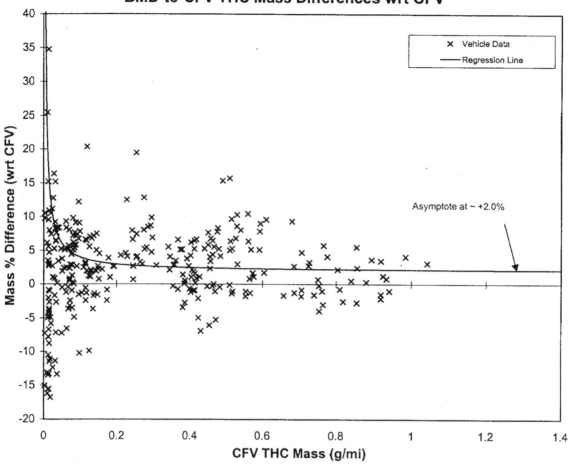

Mass Range (g/mi)		n	Average Difference	Difference COV	±1 sd Difference (g/mi)			Rel Ave Diff wrt Ave of Mass Range	% Probability* BMD ≠ CFV
0.00	to	0.05	89	0.0010	307.7%	-0.0021	to	0.0040	5.9%	99.9
0.05	to	0.10	50	0.0016	390.8%	-0.0048	to	0.0081	2.2%	95
0.10	to	0.15	29	0.0031	205.8%	-0.0033	to	0.0096	2.5%	99
0.15	to	0.20	10	0.0028	127.8%	-0.0008	to	0.0064	1.6%	97.5
0.20	to	0.25	9	0.0119	67.8%	0.0038	to	0.0200	5.1%	99.9
0.25	to	0.30	15	0.0204	55.2%	0.0091	to	0.0316	7.4%	> 99.95
0.30	to	0.35	4	0.0121	43.7%	0.0068	to	0.0174	3.6%	99
0.35	to	0.40	17	0.0059	179.5%	-0.0047	to	0.0164	1.6%	97.5
0.40	to	0.45	20	0.0051	302.6%	-0.0104	to	0.0206	1.2%	90
0.45	to	0.50	20	0.0174	134.7%	-0.0060	to	0.0408	3.7%	99.75
0.50	to	0.55	13	0.0305	83.4%	0.0051	to	0.0559	5.9%	> 99.95
0.55	to	0.60	13	0.0199	118.7%	-0.0037	to	0.0435	3.5%	99
0.60	to	0.70	6	0.0217	149.3%	-0.0107	to	0.0540	3.3%	90
0.70	to	0.80	13	0.0049	471.2%	-0.0183	to	0.0282	0.7%	75
0.80	to	0.90	9	0.0081	290.5%	-0.0154	to	0.0316	1.0%	75
0.90	to	1.05	9	0.0122	186.3%	-0.0105	to	0.0350	1.3%	90

| 0.00 | to | 1.05 | ## | 0.0072 | 218.5% | -0.0085 | to | 0.0228 | 2.7% | |

* The value (% Probability, also the Confidence Limit) is the probability that an average difference of the value stated for the given mass interval could <u>not</u> be expected from pure chance. Or restated, it is the probability that the results from the BMD for the given mass interval <u>are statistically different</u> from those for the CF\ Tabulated using the randomized paired t-test comparison, as described by Box, Hunter, & Hunter, 'Statistics for Experimenters, Ch. 4, Wiley & Sons, 1978.

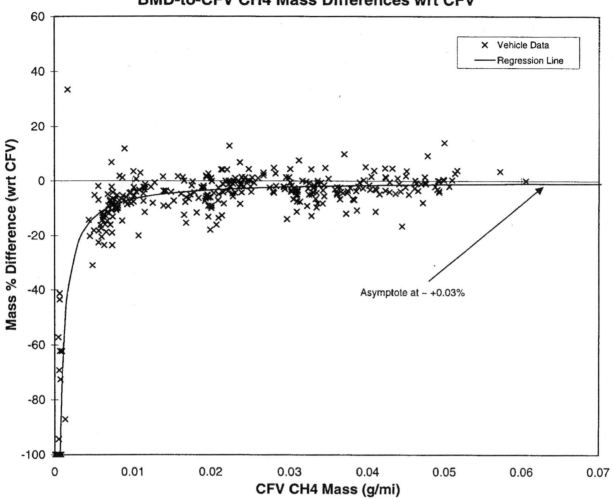

Figure 5 - Comparison of Sampling Systems
BMD-to-CFV CH4 Mass Differences wrt CFV

Mass Range (g/mi)		n	Average Difference	Difference COV	±1 sd Difference (g/mi)			Rel Ave Diff wrt Ave of Mass Range	% Probability* BMD ≠ CFV
0.000 to 0.005		29	-0.0004	-87.5%	-0.0008	to	-0.0001	-42.0%	>99.95
0.005 to 0.010		57	-0.0007	-67.2%	-0.0012	to	-0.0002	-9.3%	>99.95
0.010 to 0.015		25	-0.0005	-111.2%	-0.0011	to	0.0001	-4.4%	>99.95
0.015 to 0.020		32	-0.0010	-95.6%	-0.0020	to	0.0000	-5.5%	>99.95
0.020 to 0.025		47	-0.0005	-228.6%	-0.0016	to	0.0006	-2.2%	99.75
0.025 to 0.030		21	-0.0003	-338.4%	-0.0015	to	0.0008	-1.2%	90
0.030 to 0.035		43	-0.0011	-119.6%	-0.0025	to	0.0002	-3.5%	>99.95
0.035 to 0.040		23	-0.0012	-145.2%	-0.0029	to	0.0005	-3.2%	99.75
0.040 to 0.045		22	-0.0009	-245.2%	-0.0030	to	0.0013	-2.0%	95
0.045 to 0.050		21	-0.0005	-337.0%	-0.0021	to	0.0011	-1.0%	90
0.050 to 0.065		7	0.0016	160.4%	-0.0010	to	0.0042	3.0%	90

Mass Range (g/mi)		n	Average Difference	Difference COV	±1 sd Difference (g/mi)			Rel Ave Diff
0.000 to 0.065		##	-0.0007	-188.9%	-0.0019	to	0.0006	-2.9%

* The value (% Probability, also the Confidence Limit) is the probability that an average difference of the value stated for the given mass interval could <u>not</u> be expected from pure chance. Or restated, it is the probability that the results from the BMD for the given mass interval <u>are statistically different</u> from those for the CF   Tabulated using the randomized paired t-test comparison, as described by Box, Hunter, & Hunter, 'Statistics for Experimenters, Ch. 4, Wiley & Sons, 1978.

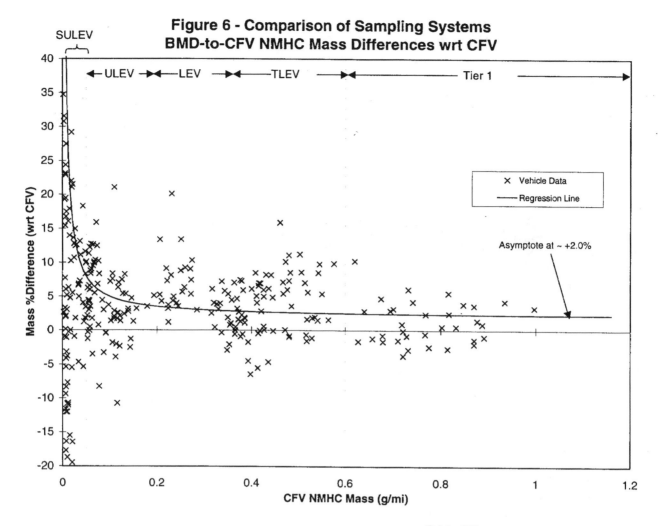

Figure 6 - Comparison of Sampling Systems
BMD-to-CFV NMHC Mass Differences wrt CFV

Mass Range (g/mi)			n	Average Difference	Difference COV	±1 sd Difference (g/mi)			Rel Ave Diff wrt Ave of Mass Range	% Probability* BMD ≠ CFV
0.00	to	0.05	104	0.0017	171.5%	-0.0012	to	0.0047	11.3%	> 99.95
0.05	to	0.10	42	0.0027	236.9%	-0.0036	to	0.0089	4.0%	99.5
0.10	to	0.15	31	0.0035	175.0%	-0.0026	to	0.0097	2.9%	99.75
0.15	to	0.20	3	0.0071	36.0%	0.0045	to	0.0096	4.0%	97.5
0.20	to	0.25	15	0.0152	71.2%	0.0044	to	0.0261	6.7%	> 99.95
0.25	to	0.30	9	0.0204	38.4%	0.0126	to	0.0283	7.7%	> 99.95
0.30	to	0.35	13	0.0104	92.3%	0.0008	to	0.0200	3.1%	99.75
0.35	to	0.40	26	0.0063	212.2%	-0.0070	to	0.0196	1.7%	97.5
0.40	to	0.45	20	0.0146	121.5%	-0.0031	to	0.0323	3.4%	99.9
0.45	to	0.50	13	0.0322	80.5%	0.0063	to	0.0581	6.8%	> 99.95
0.50	to	0.55	14	0.0223	100.3%	-0.0001	to	0.0447	4.2%	99.75
0.55	to	0.60	2	0.0327	103.8%	-0.0012	to	0.0667	5.7%	75
0.60	to	0.70	8	0.0123	213.6%	-0.0140	to	0.0387	1.9%	75
0.70	to	0.80	12	0.0028	786.3%	-0.0194	to	0.0250	0.4%	60
0.80	to	0.90	13	0.0099	216.6%	-0.0115	to	-0.0312	1.2%	90
0.90	to	1.00	2	0.0364	12.9%	0.0317	to	0.0411	3.8%	> 99.95

| 0.00 | to | 1.00 | 327 | 0.0078 | 190.0% | -0.0070 | to | 0.0226 | 3.2% | |

* The value (% Probability, also the Confidence Limit) is the probability that an average difference of
the value stated for the given mass interval could not be expected from pure chance. Or restated, it
is the probability that the results from the BMD for the given mass interval are statistically different
from those for the CFV.   Tabulated using the randomized paired t-test comparison, as described by
Box, Hunter, & Hunter, 'Statistics for Experimenters, Ch. 4, Wiley & Sons, 1978.

676

Figure 7 - Comparison of Sampling Systems
BMD-to-CFV CO Mass Differences wrt CFV

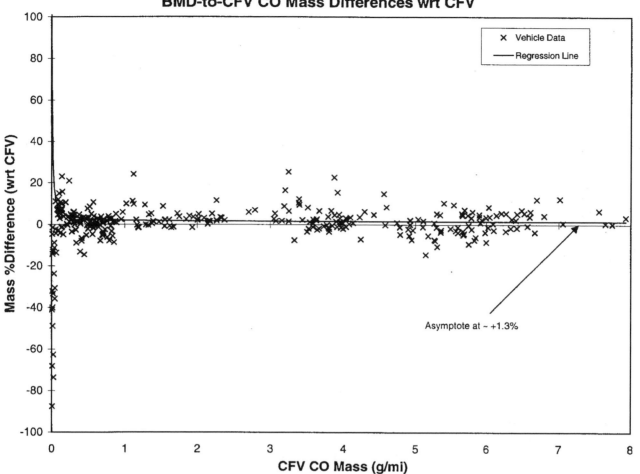

Mass Range (g/mi)			n	Average Difference	Difference COV	±1 sd Difference (g/mi)			Rel Ave Diff wrt Ave of Mass Range	% Probability* BMD ≠ CFV
0.0	to	0.5	90	0.0021	817.2%	-0.0148	to	0.0189	1.1%	75
0.5	to	1.0	51	0.0003	8898.5%	-0.0276	to	0.0282	0.0%	< 60
1.0	to	1.5	18	0.0474	158.9%	-0.0279	to	0.1227	3.8%	99
1.5	to	2.0	14	0.0343	144.6%	-0.0153	to	0.0839	2.0%	99
2.0	to	2.5	14	0.0711	86.7%	0.0095	to	0.1328	3.3%	> 99.95
2.5	to	3.0	2	0.1760	13.9%	0.1516	to	0.2005	6.4%	95
3.0	to	3.5	14	0.2255	115.2%	-0.0343	to	0.4853	6.9%	99.5
3.5	to	4.0	32	0.0768	272.3%	-0.1324	to	0.2860	2.0%	97.5
4.0	to	4.5	10	0.0689	231.4%	-0.0905	to	0.2282	1.7%	75
4.5	to	5.0	13	0.0104	2650.2%	-0.2649	to	0.2857	0.2%	< 60
5.0	to	5.5	16	-0.1010	-329.9%	-0.4342	to	0.2322	-1.9%	75
5.5	to	6.0	25	-0.0094	-3193.7%	-0.3112	to	0.2923	-0.2%	< 60
6.0	to	6.5	15	0.1415	164.6%	-0.0915	to	0.3744	2.3%	97.5
6.5	to	7.0	7	0.2557	132.2%	-0.0824	to	0.5937	3.9%	95
7.0	to	8.0	6	0.2810	120.2%	-0.0568	to	0.6188	3.8%	95

Mass Range (g/mi)			n	Average Difference	Difference COV	±1 sd Difference (g/mi)			Rel Ave Diff	
0.0	to	8.0	327	0.0400	469.9%	-0.1478	to	0.2278	1.6%	

* The value (% Probability, also the Confidence Limit) is the probability that an average difference of the value stated for the given mass interval could not be expected from pure chance. Or restated, it is the probability that the results from the BMD for the given mass interval are statistically different from those for the CFV.   Tabulated using the randomized paired t-test comparison, as described by Box, Hunter, & Hunter, 'Statistics for Experimenters, Ch. 4, Wiley & Sons, 1978.

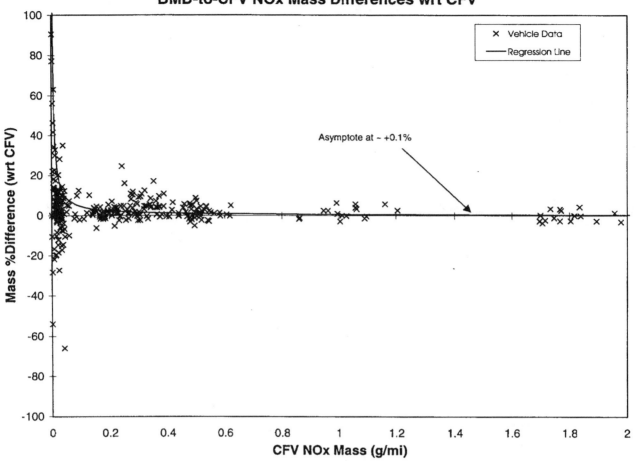

Figure 8 - Comparison of Sampling Systems
BMD-to-CFV NOx Mass Differences wrt CFV

Mass Range (g/mi)			n	Average Difference	Difference COV	±1 sd Difference (g/mi)			Rel Ave Diff wrt Ave of Mass Range	% Probability* BMD ≠ CFV
0.00	to	0.05	129	0.0005	740.8%	-0.0035	to	0.0045	2.6%	90
0.05	to	0.10	10	0.0022	226.9%	-0.0028	to	0.0073	3.1%	90
0.10	to	0.15	9	0.0034	128.9%	-0.0010	to	0.0077	2.5%	97.5
0.15	to	0.20	26	0.0014	279.9%	-0.0025	to	0.0054	0.8%	95
0.20	to	0.25	21	0.0068	199.4%	-0.0067	to	0.0202	3.0%	97.5
0.25	to	0.30	23	0.0114	117.5%	-0.0020	to	0.0249	4.1%	> 99.95
0.30	to	0.35	16	0.0144	97.0%	0.0004	to	0.0283	4.4%	> 99.95
0.35	to	0.40	13	0.0187	99.1%	0.0002	to	0.0372	5.0%	99.75
0.40	to	0.45	6	0.0066	169.6%	-0.0046	to	0.0179	1.6%	75
0.45	to	0.50	19	0.0071	242.8%	-0.0102	to	0.0244	1.5%	95
0.50	to	0.55	16	0.0080	188.4%	-0.0071	to	0.0231	1.5%	95
0.55	to	0.60	4	0.0052	86.9%	0.0007	to	0.0097	0.9%	90
0.60	to	0.65	3	0.0105	167.8%	-0.0071	to	0.0282	1.7%	75
0.65	to	1.30	16	0.0169	178.2%	-0.0132	to	0.0471	1.7%	97.5
1.30	to	2.00	16	-0.0074	-646.6%	-0.0550	to	0.0402	-0.4%	60

Mass Range (g/mi)			n	Average Difference	Difference COV	±1 sd Difference (g/mi)			Rel Ave Diff wrt Ave of Mass Range	
0.00	to	2.00	327	0.0047	347.8%	-0.0117	to	0.0212	1.6%	

* The value (% Probability, also the Confidence Limit) is the probability that an average difference of the value stated for the given mass interval could <u>not</u> be expected from pure chance. Or restated, it is the probability that the results from the BMD for the given mass interval <u>are statistically different</u> from those for the CFV.   Tabulated using the randomized paired t-test comparison, as described by Box, Hunter, & Hunter, 'Statistics for Experimenters, Ch. 4, Wiley & Sons, 1978.

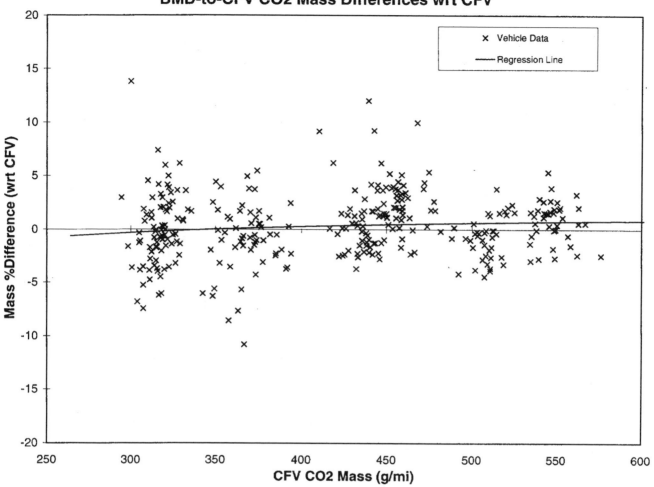

Figure 9 - Comparison of Sampling Systems
BMD-to-CFV CO2 Mass Differences wrt CFV

Mass Range (g/mi)			n	Average Difference	Difference COV	±1 sd Difference (g/mi)			Rel Ave Diff wrt Ave of Mass Range	% Probability* BMD ≠ CFV
250	to	300	2	1.94	493.6%	-7.64	to	11.52	0.7%	< 60
300	to	325	70	-0.81	-1356.5%	-11.85	to	10.23	-0.3%	60
325	to	350	19	-1.12	-1029.5%	-12.65	to	10.41	-0.3%	60
350	to	375	36	-2.12	-625.5%	-15.35	to	11.12	-0.6%	75
375	to	400	17	-3.39	-213.0%	-10.61	to	3.83	-0.9%	95
400	to	425	8	5.90	290.9%	-11.26	to	23.06	1.4%	75
425	to	450	54	3.62	360.4%	-9.42	to	16.66	0.8%	97.5
450	to	475	38	10.65	103.7%	-0.40	to	21.70	2.3%	> 99.95
475	to	500	9	-0.56	-1769.9%	-10.44	to	9.33	-0.1%	< 60
500	to	525	34	-4.15	-252.8%	-14.66	to	6.35	-0.8%	97.5
525	to	550	27	4.59	181.3%	-3.73	to	12.91	0.8%	99.5
550	to	600	13	3.03	344.7%	-7.41	to	13.46	0.5%	75

| 250 | to | 600 | 327 | 1.39 | 875.1% | -10.81 | to | 13.60 | 0.3% | |

* The value (% Probability, also the Confidence Limit) is the probability that an average difference of the value stated for the given mass interval could <u>not</u> be expected from pure chance. Or restated, it is the probability that the results from the BMD for the given mass interval <u>are statistically different</u> from those for the CFV.   Tabulated using the randomized paired t-test comparison, as described by Box, Hunter, & Hunter, 'Statistics for Experimenters, Ch. 4, Wiley & Sons, 1978.

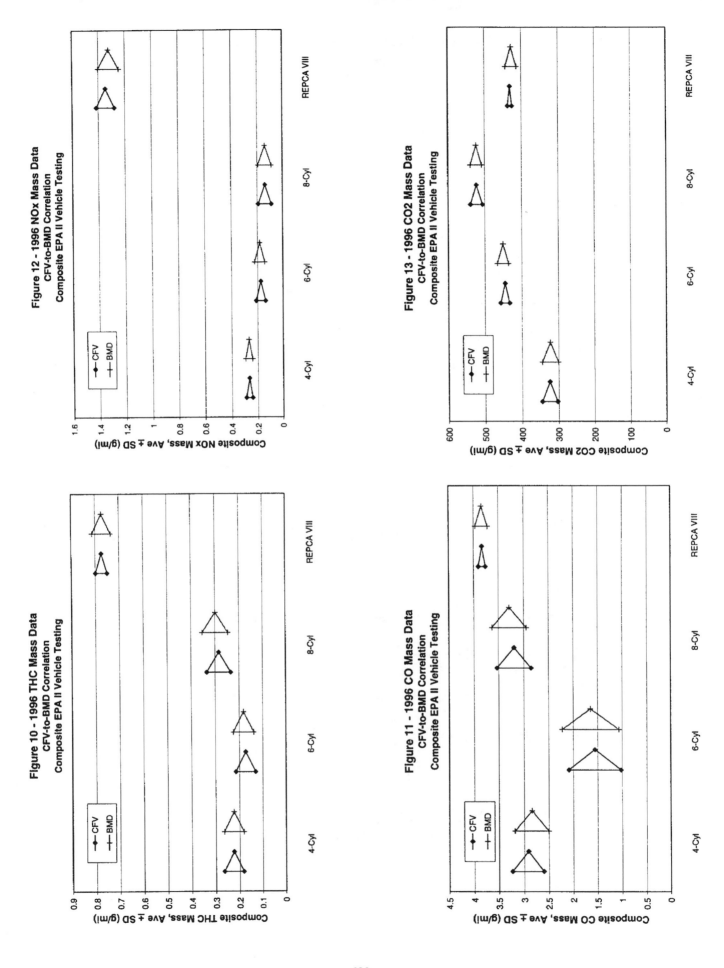

Figure 10 - 1996 THC Mass Data
CFV-to-BMD Correlation
Composite EPA II Vehicle Testing

Figure 11 - 1996 CO Mass Data
CFV-to-BMD Correlation
Composite EPA II Vehicle Testing

Figure 12 - 1996 NOx Mass Data
CFV-to-BMD Correlation
Composite EPA II Vehicle Testing

Figure 13 - 1996 CO2 Mass Data
CFV-to-BMD Correlation
Composite EPA II Vehicle Testing

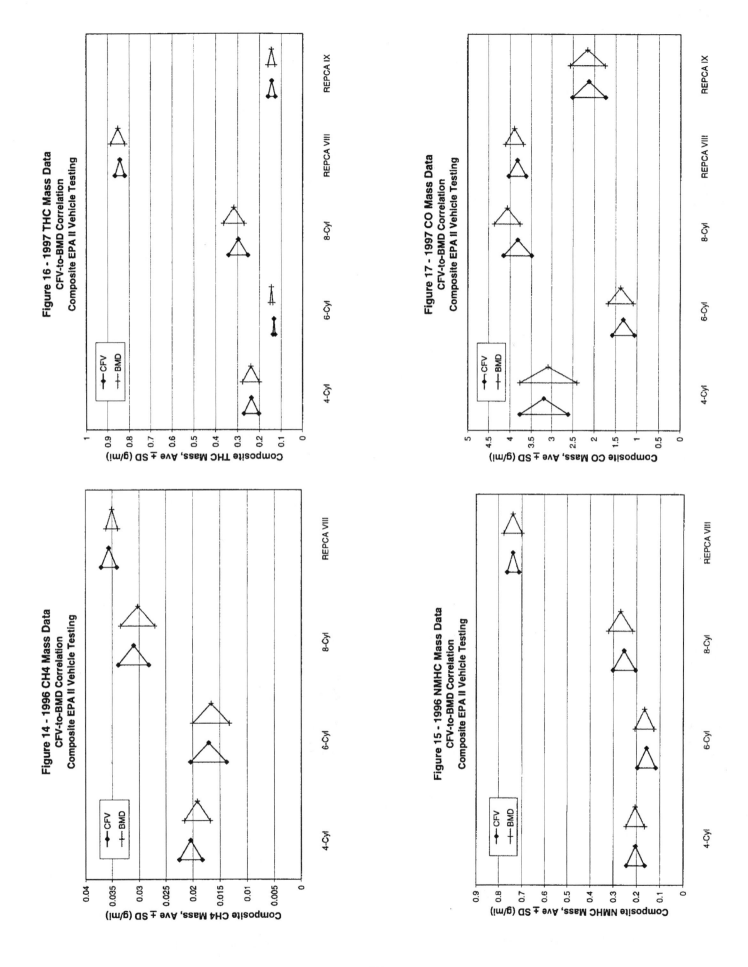

Figure 16 - 1997 THC Mass Data
CFV-to-BMD Correlation
Composite EPA II Vehicle Testing

Figure 17 - 1997 CO Mass Data
CFV-to-BMD Correlation
Composite EPA II Vehicle Testing

Figure 14 - 1996 CH4 Mass Data
CFV-to-BMD Correlation
Composite EPA II Vehicle Testing

Figure 15 - 1996 NMHC Mass Data
CFV-to-BMD Correlation
Composite EPA II Vehicle Testing

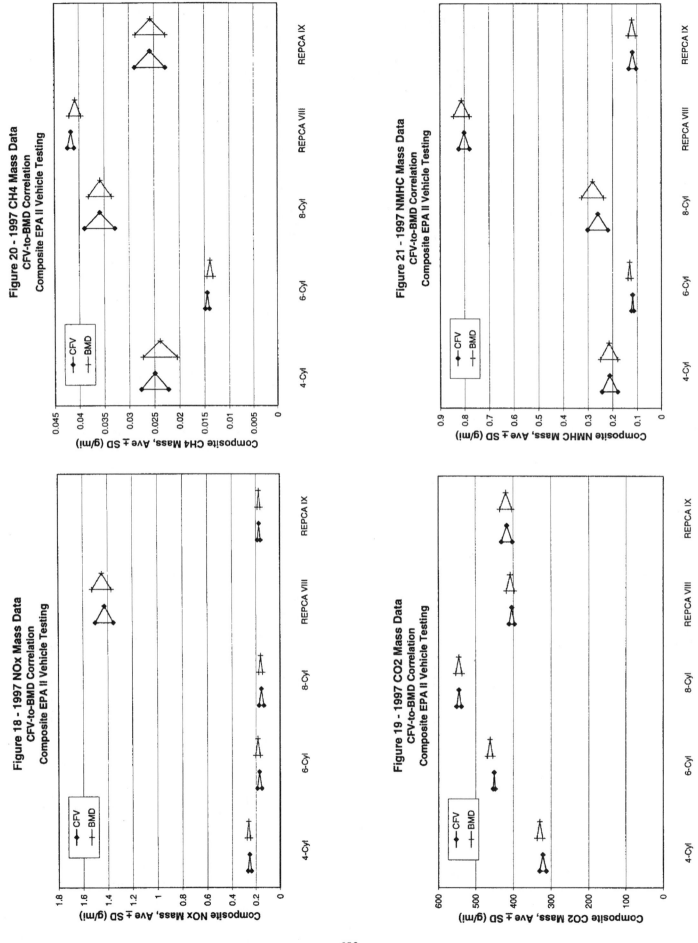

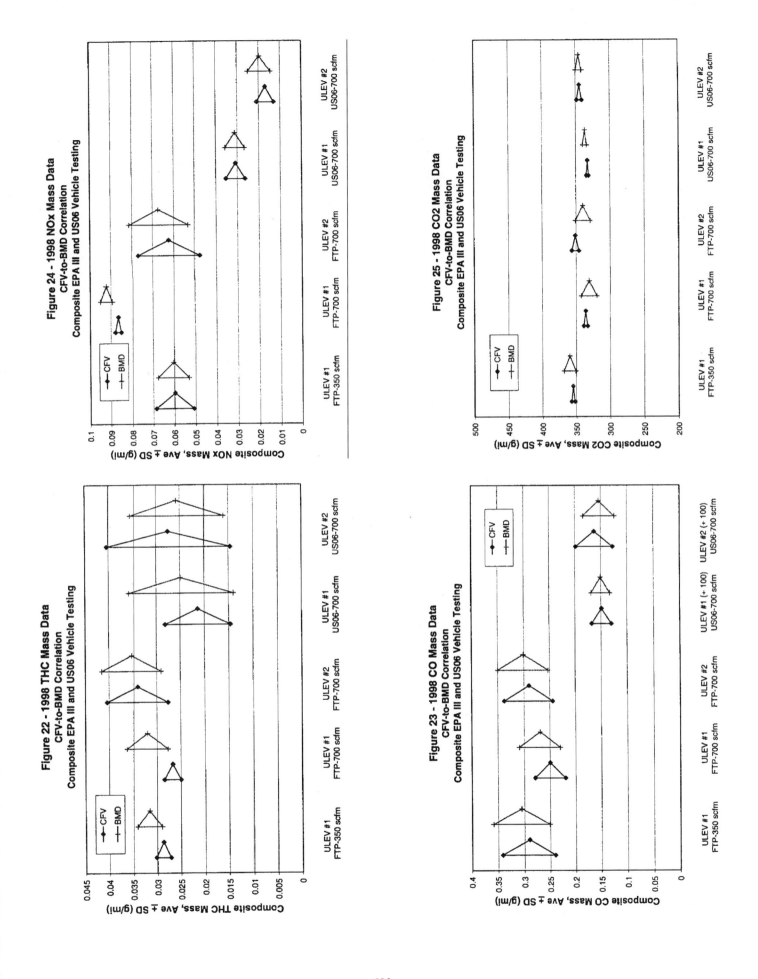

Figure 22 - 1998 THC Mass Data
CFV-to-BMD Correlation
Composite EPA III and US06 Vehicle Testing

Figure 24 - 1998 NOx Mass Data
CFV-to-BMD Correlation
Composite EPA III and US06 Vehicle Testing

Figure 23 - 1998 CO Mass Data
CFV-to-BMD Correlation
Composite EPA III and US06 Vehicle Testing

Figure 25 - 1998 CO2 Mass Data
CFV-to-BMD Correlation
Composite EPA III and US06 Vehicle Testing

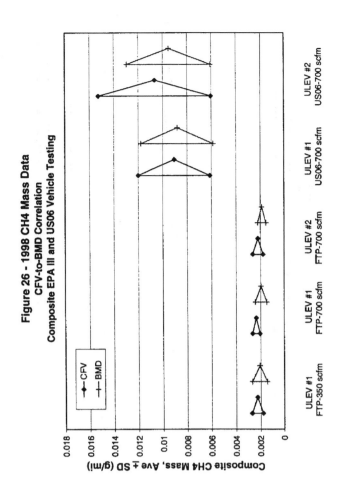

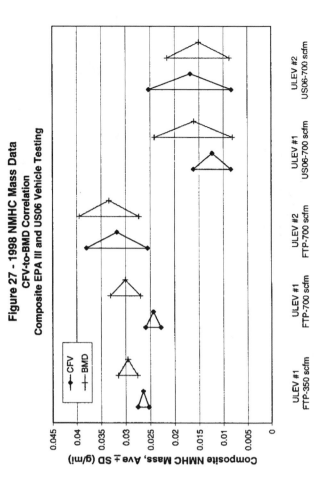

684

CHAPTER 14

ULTRA-CLEAN VEHICLE SYSTEMS

Fuquan (Frank) Zhao
DaimlerChrysler Corporation

14.1 OVERVIEW

The ultra-clean emission standards such as PZEV (partial zero emission vehicle) requires a reduction of more than 90% in tailpipe emissions of HC, NOx and CO from the current vehicle level. Lower tailpipe emissions are only part of the emissions reduction requirements for those ultra-clean gasoline-powered vehicles. Due to the fact that the tailpipe emission is getting lower and lower day by day, evaporative emissions are becoming an increased fraction of the total vehicle system emissions. The PZEV vehicles must also meet zero evaporative emissions requirements. As a result, zero fuel-based evaporative emissions are allowed for a PZEV vehicle where 0.054g per test of fuel-based evaporative emissions or less rounds to zero. The remaining evaporative emissions are non-fuel-based, which are usually derived from paint, tires, seats, washer fluid, sealants and adhesives. The ultra-clean vehicles must also meet the OBD requirements for detecting deterioration in emission control systems. The challenge for the PZEV OBD system is to develop the capability of detecting the difference between an adequately functioning emission control system meeting the emission standard and one with a deterioration such that the emissions are about to exceed the required threshold, defined as a multiple of the emission standard. This multiple is 2.5 for the PZEV vehicles. In addition, the PZEV vehicles are also required to meet the durability requirement of either 15 years or 150K miles for all emissions-related components. Above all of this, the PZEV vehicles must retain all of the customer attributes such as rapid starting in all ambient conditions, stable idle, good driveability, high fuel efficiency and good performance with an affordable price. Clearly all these requirements put a tremendous challenge on developing such a vehicle for mass production.

Since many vehicle design and operating parameters affect emissions, a system approach must be taken to meet the increasingly stringent emission standards: optimization of the engine system, control system and catalyst system to reduce engine-out emissions and improve catalyst conversion efficiency. For a cold engine, the emphasis should be on reducing cold-start emissions and improving catalyst light off.

All engine design and operating parameters that determine mixture preparation, combustion, emission formation and catalyst conversion efficiency must be carefully examined from the system perspective to minimize engine-out emissions while enhancing catalyst light-off performance. Those parameters include, but are not limited to, fuel preparation, spray characteristics (droplet size distribution, cone angle, targeting), fuel injection timing, valve timings, air flow, cranking speed, spark timing, burn rate, idle speed, engine warm-up rate, coolant distribution around the intake valves, crevice volume and exhaust manifold design. The benefits of various types of injectors such as single-fluid, multi-stream, air-assisted and heated injectors must be assessed to take advantage of each technology. The piston ring top land crevice must be minimized to reduce HC emissions. The excess coolant on the intake side must be minimized to enhance mixture preparation process during the cold start and warm-up period.

The exhaust manifold design for enhancing the mixing process between the cylinder events and reducing the heat loss need careful consideration, particularly when implementing the secondary air injection strategy. The engine start-up and shut-down strategies must also be carefully investigated to maximize the gain of each strategy.

Many factors affect catalyst performance, including heat flux to the catalytic converter, heat loss from the engine to the exhaust system, air-fuel ratio and its control, catalyst formulation, catalyst loading, catalyst volume, catalyst aging (thermal degradation and chemical poisoning), substrate configuration for reduced thermal mass, catalyst configuration, catalyst utilization, operating temperature and air-fuel ratio control with an aged catalyst. The catalyst type and the amount of catalyst loading have a significant impact on catalyst light-off temperature. The thermal mass of the catalyst substrate configuration also significantly affects the catalyst light-off time. An ultra-thin wall substrate with a high cell density can significantly reduce the time constant to light off the catalyst. As the thermal resistance of the catalysts is improved, chemical poisoning primarily from the phosphorus in the engine oil is increasingly important. Reducing engine oil consumption is getting particularly important in developing ultra-clean gasoline-powered vehicles. The reduced phosphorus loading on the catalyst is effective in reducing catalyst light-off time.

The most important factor that determines the catalyst light-off time is the engine-out heat flux to the converter during cold start. Strategies that are used to achieve a rapid catalyst heating can be classified into two approaches. One is the lean starting approach that typically requires a slight enleanment of the air/fuel ratio in combination with a significantly retarded spark timing. How to maintain good combustion stability with a slightly lean mixture and an extremely retarded spark timing is a daunting task, particularly with a high DI (Driveability Index) fuel. The other is the rich starting approach, which is achieved by operating the engine with an extremely rich mixture and injecting secondary air into the exhaust port via an air pump during cold start. Rich combustion produces a large amount of engine-out emittants of H_2, CO and HC, which will react with the secondary air inside the exhaust manifold to reduce the converter-in emissions and also significantly boost catalyst light-off performance. Since there is no need to over-retard the spark timing, this rich starting approach is more robust and tends to be less sensitive to fuel properties.

Fuel-based evaporative emissions must also be reduced on a system approach that should be focused on both the powertrain, and fuel storage and delivery sources. The powertrain source includes those from both the base engine and the air induction system. Evaporative emissions of fuel storage and delivery systems are from fuel tank, fuel lines, vapor lines, charcoal canister, fuel filler cap and control valves. To get rid of nearly all fuel-based evaporative emissions, a high rate of purging of vapor emissions is necessary during nearly all operating conditions. In addition to providing high purging rates, new materials are also a mandate for reduced permeability and leakage under the required lifetime of durability.

Several OEMs have succeeded in launching mass-produced vehicles that are certified to meet the California PZEV standard. Two representative systems are outlined here to explain the basic elements in the PZEV system.

Nissan PZEV System is based on a lean starting approach to achieve the catalyst light-off performance in combination with an HC trap to adsorb HC emissions before the catalyst is lit off (Sources: SAE 2000-01-0890; SAE 2001-01-1310; SAE 2003-01-0816).

- Engine System: 4-valve, four-cylinder engine; fine-spray injector; electrically-actuated swirl control valve; stainless steel dual-wall exhaust manifold for the first and second generations of PZEV (inner pipe has a thinner wall thickness of 0.8 mm); cast iron exhaust manifold for the third generation of PZEV; high-speed starter; electrically-actuated EGR valve for the first and second generations of PZEV only; continuously variable valve timing for the third generation of PZEV to replace the EGR system.
- Control System: fast light-off air-fuel ratio sensor; improved air-fuel ratio control; increased purging air for the canister.
- Catalyst System: fast light off catalyst substrate (900 cell density substrate); ultra-thin-wall (2 mm) catalyst substrate for the first and second generations of PZEV; the wall thickness of the third generation PZEV is 1.8 mm; two-stage HC trap; three three-way catalysts (one close-coupled catalyst, two two-stage HC traps) for the first generation PZEV; two three-way catalysts (first HC trap integrated inside the close-coupled catalyst; second trap integrated inside the underfloor catalyst) for the second and third generations of PZEV.
- Catalyst-coated radiator for reducing ozone.

Ford PZEV System is based on a rich starting approach via secondary-air injection to achieve the catalyst light-off performance (Sources: O. Kunde et al. at the GPC 2003 Conference, 2.3L PZEV Ford Focus, Proceedings of Advanced Engine Design & Performance, pp. 179-186; W. Wade at the ASME Fall ICE Conference, Near Zero Emission Internal Combustion Engines, Paper No. ICEF2003-775, 2003).

- Engine System: 2.3L, four-valve, four-cylinder engine; 12-hole, low-leakage injector with fine droplets; 90% of the fuel targeted on intake valve (goal); 2-position charge motion control valve (CMCV); 4 mm piston ring top land height; EGR system; dual-wall exhaust manifold; electric thermactor air pump; air induction system HC trap; coil-on-plug ignition system; iridium spark plugs; new horning specification for improving bore cylindricity and wall surface finish; reduced piston ring tension; zero-permeability gaskets/joints.
- Control System: heated exhaust gas oxygen (HEGO) sensor for pre- and after-catalyst oxygen control; mass air flow sensor; x-Tau transient fuel compensation is used for air/fuel ratio control; +/-0.5% A/F ratios for

cylinder-to-cylinder air flow distribution (goal); +/-0.5% A/F ratio (warmed up) for air/fuel ratio control; open loop air/fuel ratio control adaptively updated based on most representative speed/load cell in closed loop; oxygen sensors prior to/after catalyst are used for high mileage air/fuel ratio control; EGR valve is controlled by stepper motor; purge control via vapor management valve and purge compensation for air/fuel ratio control during purging.

- Catalyst System: close-coupled catalyst with two bricks (first brick: 900/2.5 and 0.69L; second brick: 400/4.0 and 0.69L); underfloor catalyst with two bricks (first brick: 400/4.0 and 0.84L; second brick: 400/4.0 and 0.84L).

Even though great success has been achieved in developing the ultra-clean gasoline-powered vehicles to meet the world's most stringent regulation of PZEV, the application is limited to small-displacement, 4-cylinder-engine-powered vehicles. Applying these technologies to larger-displacement-engine-powered vehicles presents significant challenges. Continued efforts must be directed towards improving the existing PZEV technologies and exploring new enablers for simplifying the system and enhancing its capability for a broader range of applications.

14.2 SELECTED SAE TECHNICAL PAPERS

2000-01-0890

Development of New Technologies Targeting Zero Emissions for Gasoline Engines

Kimiyoshi Nishizawa, Sukenori Momoshima, Masaki Koga and Hirofumi Tsuchida

Nissan Motor Co., Ltd.

Copyright © 2000 Society of Automotive Engineers, Inc.

ABSTRACT

This paper describes new technologies for achieving exhaust emission levels much below the SULEV standards in California, which are the most stringent among the currently proposed regulations in the world. Catalyst light-off time, for example, has been significantly reduced through the adoption of a catalyst substrate with an ultra-thin wall thickness of 2 mil and a catalyst coating specifically designed for quicker light-off. A highly-efficient HC trap system has been realized by combining a two-stage HC trap design with an improved HC trap catalyst. The cold-start HC emission level has been greatly reduced by an electronically actuated swirl control valve with a high-speed starter. Further, an improved Air Fuel Ratio (AFR) control method has achieved much higher catalyst HC and NOx conversion efficiency.

INTRODUCTION

Environmental issues have been the subject of discussion the world over in recent years and concern about the environment has been rising steadily. In California the Low Emission Vehicle (LEV) regulations will be tightened to LEV 2 regulations in 2004, which are the world's most stringent standards (Table.1). The Super Ultra Low Emission Vehicle (SULEV) standards in LEV 2 is a new regulation category that would mandate a 75% reduction from the Ultra Low Emission Vehicle (ULEV) levels. In addition to that, California requires the introduction of Zero Emission Vehicles (ZEVs). Only vehicles that satisfy all the following requirements will be recognized as clean vehicles and can be partially counted as ZEVs.

1. Meeting SULEV standard up to 150,000mile.
2. Fulfilling the OBD requirement.
3. Meeting the zero evaporative emissions standard.

Moves to tighten exhaust emission standards further are not limited to the U.S., but rather are accelerating throughout the world.

We have been developing new technologies for reducing exhaust emissions to levels even lower than these standards. The aim of this effort is to achieve exhaust emissions as clean as the surrounding air, a level that would not pose any HC related problem to the environment. Through those development efforts, the 2000 Nissan Sentra CA has become the world's first gasoline-powered vehicle to qualify for partial ZEV credits. This paper describes new technologies, mainly in the area of exhaust emission control, which are applied to the 2000 Nissan Sentra CA.

Table 1   California LEV2 Regulations

	NMOG Exhaust Emissions (g/mile)			Evaporative Emissions
	~ 50,000 miles	~120,000 miles	~150,000 miles	
LEV	0.075	0.090	—	—
ULEV	0.040	0.055	—	—
SULEV	—	0.010	—	—
P-ZEV	—	—	0.010	Zero Evaporative Emission is Required

OVERVIEW OF EMISSION REDUCTION SYSTEM

The concept of the emission reduction technologies described here is outlined in figure 1 in relation to the change in the HC emission concentration following engine start. Attaining exhaust emissions as clean as the surrounding air requires a reduction of cold-start HCs emitted in the interval from engine start to catalyst light-off and also hot-phase HCs emitted after the catalyst has reached its working temperature. Reducing cold-start HC emissions requires such approaches as a reduction of the engine-out HC level, quicker catalyst light-off and the adoption of an HC trap system. Meanwhile, reducing hot-phase HCs requires a reduction of the residual rate of hot-phase emissions through improvements to catalyst performance and to AFR control. In order to achieve zero evaporative emissions, we increased the purge air amount for the canister, because the canister should be cleaned thoroughly during driving. This requires much more precise AFR control.

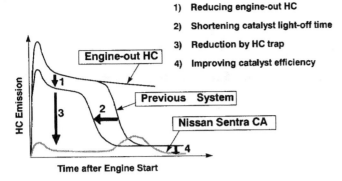

1) Reducing engine-out HC

2) Shortening catalyst light-off time

3) Reduction by HC trap

4) Improving catalyst efficiency

Fig. 1   Concept of Emission Reduction Technologies

The emission reduction system developed in this work is shown schematically in figure 2. This new system incorporates four major improvements.

The first improvement is a substantial reduction of catalyst light-off time as a result of adopting measures to warm up the catalyst faster. One such measure is the use of a double-walled exhaust manifold with a thin inner pipe wall. Another is the adoption of an ultra-thin catalyst substrate having a wall thickness of only 2 mil. In addition, the method of coating the catalyst was improved in order to develop a catalyst designed specifically for quicker light-off.

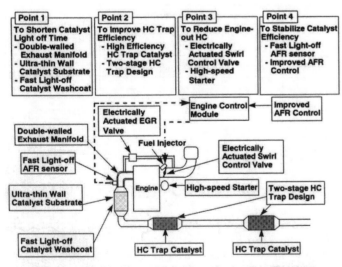

Fig. 2   Overview of Emission Reduction Strategy

The second improvement concerns the adoption of HC trap catalysts, which increase the conversion efficiency of trapped HCs and a two-stage trap system. The trap has been designed in two stages with a heat mass placed in between. Since the downstream HC trap again adsorbs HCs that have yet to be converted, the conversion performance of the trap system is significantly increased.

The third improvement concerns an increase in the engine-out exhaust temperature and a reduction of HC emissions as a result of improving combustion stability

especially right after engine start. There is a particularly strong need to improve catalyst light-off performance when the intake air quantity is small. To accomplish that, an electronically actuated swirl control valve has been adopted, which allows more accurate control for improved combustion stability when the airflow rate into the cylinders is small. As a result, it is possible to raise the exhaust gas temperature by substantially retarding the ignition timing. Additionally, the engine cranking speed has been increased and the fuel supply at engine start has been optimized. These measures are designed to reduce and stabilize cold-start emission levels.

The fourth improvement concerns the adoption of a control procedure for increasing the conversion efficiency of the catalyst. To stabilize the onset of catalyst activity, a fast light-off AFR sensor is employed. The accuracy of AFR control has been improved for the primary purpose of reducing and stabilizing nitrogen oxide (NOx) emissions.

The combined use of these emission reduction technologies enables an in-line 4-cylinder gasoline engine to achieve emission levels much below the proposed SULEV standards.

INDIVIDUAL EMISSION REDUCTION TECHNOLOGIES

EARLIER CATALYST LIGHT-OFF – Increasing the temperature rise characteristic and improving low-temperature activity are effective ways of achieving earlier catalyst light-off. In the newly developed system, the heat mass of the substrate has been reduced to improve the temperature rise characteristic of the catalyst. That has been accomplished by using a substrate with an ultra-thin wall thickness of 2 mil. A double-walled exhaust manifold has also been adopted. The inner pipe has a thinner wall thickness of 0.8 mm, which reduces its heat mass. The combined use of this substrate and exhaust manifold has made it possible to quicken the temperature rise of the catalyst compared with the previous system. The effect of this substrate with its 2 mil wall thickness is shown in figure 3.

As seen in the figure, reducing the substrate wall thickness to 2 mil from the previous 4 mil works to quicken the catalyst light-off time by 5 sec. The effect of the dual-walled exhaust manifold is shown in figure 4. The use of a dual-walled exhaust manifold shortens by 2 sec. the time needed for the exhaust gas upstream of the catalyst to reach a temperature of 300°C.

Furthermore, the low-temperature activity of the catalyst was enhanced by improving the coating method. As shown in figure 5, the improvement combining the above mentioned three items has the effect of reducing the catalyst light-off time to less than one-half that of the previous system and the light-off time is shortened to as little as 12 seconds.

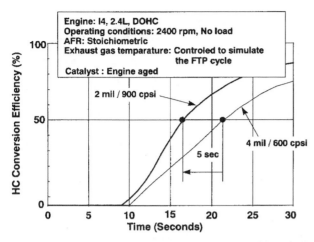

Fig. 3 Effects of Ultra-Thin Wall Catalyst Substrate

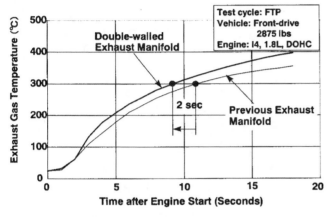

Fig. 4 Effects of Double-walled Exhaust Manifold

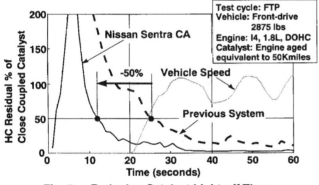

Fig. 5 Reducing Catalyst Light-off Time

IMPROVED HC TRAP SYSTEM EFFICIENCY – An HC trap temporarily stores HCs on an adsorber, such as zeolite, in the low temperature region below the catalyst's working temperature. Then, when the exhaust gas temperature rises, it releases the adsorbed HCs for conver-

sion. Heretofore, HC trap catalysts have provided sufficiently high trapping levels of 80-90%, but the rate of converting desorbed HCs has not been sufficient, generally being no higher than around 30%.[1]

A typical emission pattern of an HC trap system is shown in figure 6. In this example, nearly 100% of the HCs are trapped during the initial interval of approximately 20 sec. Subsequently, the HC level at the trap outlet exceeds that at the trap inlet, indicating that HCs are desorbed without being completely converted.

The first step taken in this work was to adopt newly developed HC trap catalysts to improve the conversion level of desorbed HCs. The resulting improvement is shown in figure 7. The conversion level of desorbed HCs was around 20-30%. Improving the wash coat layer raised the conversion level to 30-40%. As a result, the total quantity of HCs in the first bag of the FTP cycle was reduced by 30%.

The next step taken to improve the conversion level of the HC trap catalyst further was to adopt a two-stage trap system. The resulting improvement is also shown in figure 7. While the conversion level of desorbed HCs was around 30-40% with a single-stage trap, the adoption of a two-stage system improved conversion performance to 60%. As a result, the total quantity of HCs in the first bag was reduced by 50%. In this two-stage HC trap system, heat capacity exists between the two HC traps, resulting in a temperature difference between them. Hydrocarbons are initially trapped and released from the first HC trap, and at that time the second HC trap is still at a low temperature and can trap HCs released from the first HC trap. This mechanism improves system performance significantly.

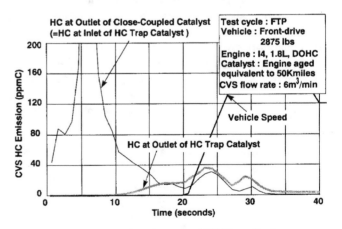

Fig. 6 Emission Pattern of HC Trap System

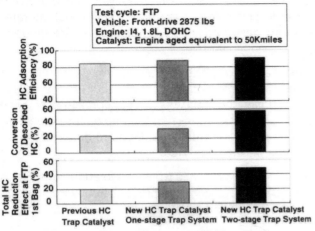

Fig. 7   Effect of Improved HC Trap System

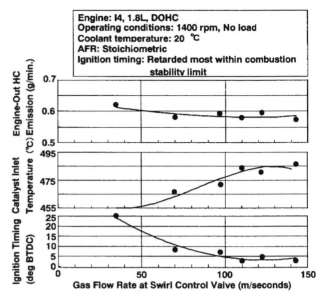

Fig. 8   Effect of Gas Flow Control
with Swirl Control Valve

REDUCTION OF ENGINE-OUT HC LEVEL – Figure   8 shows the effect of the gas flow rate on increasing the exhaust gas temperature and reducing the HC emission level, when the intake air flow is throttled by the swirl control valve. This throttling increases the gas velocity at one of the two ports of each cylinder. It induces swirl gas motion in the combustion chamber and has the effect of improving combustion stability, enabling the ignition timing to be retarded and thus resulting in a higher exhaust temperature and lower HC emissions. The results indicate that the air flow velocity must be raised above a certain level in order to obtain a sufficient exhaust temperature increase and HC reduction. Additionally, the catalyst temperature must be raised especially under a condition of a small intake air volume. Therefore, in order to increase the intake air flow velocity, it is necessary to throttle the intake air further. The application of an electrically actuated swirl control valve provides a variable capability that allows the air flow to be throttled more. The resulting effect combined with an optimally shaped and positioned swirl control valve is shown in figure 9. This figure indicates that throttling the air flow with the swirl control valve achieves amply stable combustion even if the ignition timing is retarded further, thereby making it possible to raise the catalyst inlet exhaust temperature.

Figure 10 shows the cold-start HC emission peak as a function of the engine starting speed. As seen in the figure, increasing the starting speed has the effect of reducing the HC peak at engine start. By increasing the speed from 200 to 250 rpm, the HC emission peak was reduced by 10%.

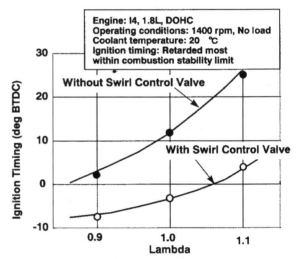

Fig. 9   Expansion of Ignition Retard
by Swirl Control Valve

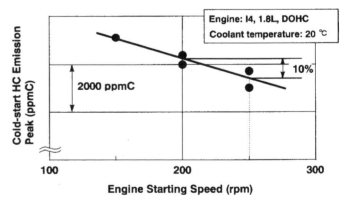

Fig. 10   Effect of Reducing HC Peak at Engine Start

IMPROVEMENT OF A/F RATIO CONTROL FOR IMPROVED CATALYST CONVERSION PERFORMANCE – In this newly developed system, the catalyst lights off at around 12 seconds after engine start. Because emission levels do not stabilize unless AFR feedback control is applied prior to catalyst light-off, a fast light-off AFR sensor is employed in the new system. An improved AFR control strategy is also used in the new system. In this strategy, short-term feedback is provided by the front AFR sensor and the long-term feedback is provided by the rear oxygen sensor. The accuracy and speed of the short-term feedback are significantly improved by the use of the AFR sensor. Figure 11 compares the NOx and HC conversion levels obtained with the previous control procedure and with the improved control technique adopted in the new system. Using the improved AFR control technique makes it possible to reduce NOx to a sufficiently low level without any accompanying degradation of CO and HC performance.

Improved AFR control also made it possible to increase the amount of purge air for the canister, which is required to realize zero evaporative emissions, without sacrificing exhaust emission performance.

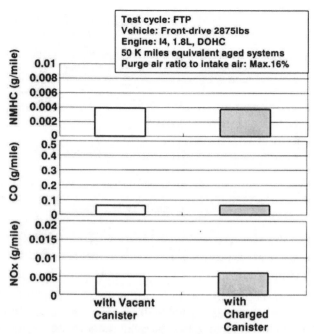

Fig. 12   Zero Evaporative Emissions

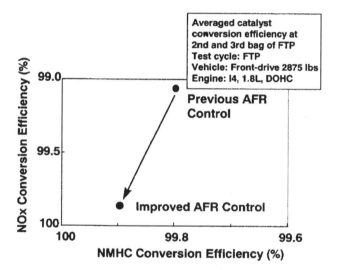

Fig. 11   Effect of Improved AFR Control

Figure 12 shows a comparison between emissions with a vacant canister and with a fully charged canister. The results show that the fuel-rich air from the canister has little effect on exhaust emissions, even with a large amount of purge air.

PERFORMANCE OF NEWLY DEVELOPED SYSTEM

The emission degradation characteristic of the new system incorporating the four features explained above is shown in figure 13. It is clear from this figure that minimal emission degradation occurs up to 150,000 miles and that the emission levels are still fully below the SULEV standards even after 150,000 miles.

In order to investigate variation in catalyst durability, the change in emission levels was examined when the catalyst durability temperature was raised by 50°C above that of ordinary use. As the results shown in figure 14 indicate, the HC emission level changed by less than 10%, which verifies the high stability of the new system.

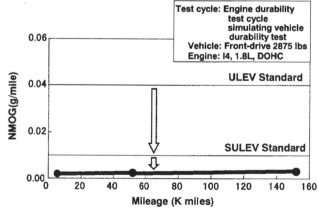

Fig. 13   Durability Performance of Nissan Sentra CA

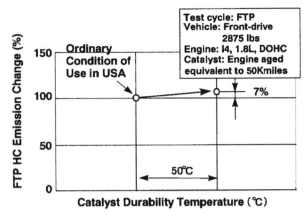

Fig. 14 Effect of Increased Durability Temperature on HC Emission

The THC emission pattern obtained with this system is shown in figure 15. It is seen that the HC concentration in the exhaust gas is higher than the atmospheric level immediately after engine start, but it decreases substantially below the atmospheric level after the engine is warmed up. As a result, the average concentration is nearly the same as that of the surrounding air. However, the level of NMHC emissions is still higher than the atmospheric concentration. It is planned to combine another ozone reduction method, such as the PremAir®, with the technologies described here for obtaining near zero emissions, with the aim of achieving virtually no increase in the ozone level. PremAir® is the catalyst coating on radiators and it acts to convert ozone contained in the air going through a radiator to harmless oxygen. Some calculations show that the ozone reduction effect per radiator with PremAir® is equal to a 0.01g/mile reduction of NMOG from vehicle exhaust.[2] As the vehicle described in this paper has much less exhaust NMOG emission than 0.01g/mile, the total effect on ozone formation falls below zero, if the effect of PremAir® is added.

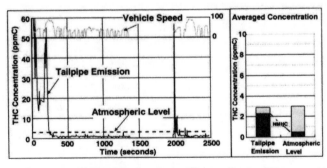

Fig. 15 THC Emission from Nissan Sentra CA

CONCLUSION

1. Steps were taken to reduce catalyst light-off time to less than one-half that of previous systems. This has been accomplished in part by reducing the heat mass through the use of a double-walled exhaust manifold with a thinner inner pipe wall and an ultra-thin catalyst substrate. Another contributing factor is the development of a catalyst coating specifically designed for quicker light-off.

2. A newly developed HC trap catalyst and a two-stage HC trap were adopted. These measures have more than tripled the trap HC conversion efficiency compared with the performance of previous traps.

3. The cold-start HC emission level has been reduced and stabilized through the combined use of an electronically actuated swirl control valve and a high-speed starter.

4. An improved AFR control strategy was applied for greater control accuracy, thereby improving the conversion performance of the catalyst. This precise AFR control enables a large amount of purge air for the canister, which is required to realize zero evaporative emissions.

5. A new system incorporating the measures noted in (1)-(4) achieves clean emissions at levels much below the SULEV standards and satisfies all the requirements for the partial ZEV credit. Moreover, the system shows high stability with respect to catalyst durability and variation in emission control performance.

6. Since the technologies described here attain near zero emissions, there is the potential to achieve virtually zero or minus effect on the ozone level by combining these technologies with other ozone reduction techniques, such as the PremAir®.

REFERENCES

1. S. Mitsuishi, K. Mori, K. Nishizawa and S. Yamamoto, "Emission Reduction Technologies for Turbocharged Engines" SAE Paper 1999-01-3629.

2. Jeffrey B. Hoke, Ronald M. Heck and Terry C. Poles, "Premair Catalyst System-A New Approach to Cleaning the Air" SAE Paper 1999-01-3677.

2003-01-0817

Development of PZEV Exhaust Emission Control System

Toru Kidokoro, Koichi Hoshi, Keizo Hiraku, Koichi Satoya, Takashi Watanabe and Takahiko Fujiwara
TOYOTA Motor Corporation

Hideki Suzuki
DENSO Corporation

Copyright © 2003 SAE International

ABSTRACT

A new exhaust emission control system has been developed which complies with the world's most severe emission standard: CARB PZEV. Leaner combustion in cold condition was enabled and rapid warm-up of a close-coupled catalyst was realized by utilizing a newly developed Intake Air Control Valve (IACV) system and hyper-atomization fuel injector. In addition, the newly developed HC adsorbing type 3-way catalyst realized cold HC reduction at lower cost.

For further reduction of the exhaust emission, the Variable Valve Timing-Intelligent (VVT-i) system was positively operated immediately after the cold start. By the suitable operation of Variable Valve Timing (VVT), the blow-back from the cylinder enhanced the fuel atomization and re-burning of remaining unburned hydrocarbons (HCs), and increased in-cylinder residual gas reduces NOx. In addition, to pursue a better catalyst conversion rate after warm-up, the oxygen (O_2) sensor downstream of the close-coupled catalyst was improved to a new type. We have thus developed an exhaust emission control system with less emission variability to comply with the 15 years or 150k miles warranty.

INTRODUCTION

With the increasing concern for the earth's environment protection in recent years, social demand for cleaner emissions and increased fuel economy has been rising all over the world. Particularly, in California, Zero Emission Vehicle (ZEV) regulation has been mandated for 2003 and subsequent model years. The regulation requires an obligatory introduction of ZEV for 10% of the sales volume in California. However, ZEVs consist of Electric Vehicles (EV), Fuel Cell Electric Vehicles (FCEV), and a new category called Partial-ZEV (PZEV) which, after being multiplied by a prescribed allowance, can account for 6% of ZEV sales. Fig.1 and Fig.2 show the qualifications of the PZEV standard. It calls for the tailpipe emission to be less than 1/5 for

NMOG and less than 1/15 for NOx of the ULEV (LEV-I/100k miles) standards. Moreover, the PZEV standard requires zero fuel oriented evaporative emissions as shown in Fig. 2. The PZEV standard also requires the vehicle to be equipped with an onboard diagnostic system (OBD-II). In addition, the emission reduction system is required to provide a long and quite severe warranty for 15 years or 150k miles.

Since 2000, auto-manufacturers[1][2] have introduced vehicles complying with SULEV and PZEV standards. We have introduced the SULEV "Prius"[3], the world's first production type hybrid vehicle with the Toyota HC Adsorber and Catalyst System (Toyota-HCAC-System).

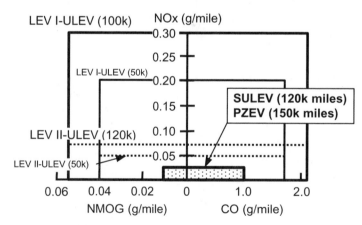

Fig.1 Tailpipe Emission Standards

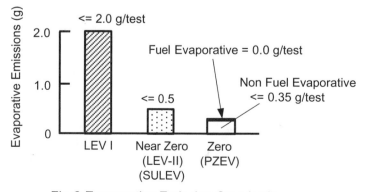

Fig.2 Evaporative Emission Standards

EMISSION REDUCTION TECHNOLOGY AND MAJOR ISSUES

As the emission standards lower, reducing HC and NOx emission in cold condition, including cold starting which account for more than 95% of the total emission, becomes quite important. Conventionally, leaner air-fuel ratio (A/F) by reduced fuel enrichment and improved catalyst warm-up performance by retarding the ignition timing have been applied to reduce emissions. Recently, however, compatibility is demanded between driveability with low-vaporizing fuel and the provision of margin for detecting misfire as required by the onboard diagnostic system (OBD-II), from the view point of "a vehicle as a whole" product quality. These circumstances make compliance for the controls all the more difficult.

We kept a conventional fundamental emission reduction concept, and we have developed a device to improve combustion in cold condition with minimal engine modifications. Another focus was to reduce the wall wetting on the intake port and the cylinder wall surface. The IACV-system we adopted not only improved the combustion in cold condition, but also reduced wall wetting on the intake port and cylinder wall surface. We also aimed at improved conversion rate in the catalyst during warm-up condition. Through the application of the aforementioned combustion improvement and wall wetting reducing technology, we were able to achieve a leaner A/F during the engine warm-up process. HC reduction was accomplished by promoting an oxidation reaction. In addition to activating the catalyst within a limited period, cold engine out emission reduction before catalyst activation is very important.

Fig. 3 illustrates a PZEV emission reduction system. First, we have adopted a high-efficiency close-coupled catalyst integrated with a double-walled compact exhaust manifold for improving catalyst warm-up performance. Next, we have adopted the HC adsorbing type 3way-catalyst. Located under the floor, the catalyst functions to temporarily adsorb HCs that flow through the close-coupled catalyst in cold condition and convert these HCs after the catalyst is activated. We also have adopted a fast light-off planar A/F sensor for earlier stable A/F control in lean condition during fast idling, for effective HC conversion. An O_2 sensor, newly developed for the PZEV system, is located down stream of the close-coupled catalyst. It detects the oxygen concentration in the exhaust gas accurately and contributed greatly to optimizing the catalyst conversion performance.

The top section in Fig. 3 shows a simple explanation of the emission reduction concept during the cold start and the engine warm-up condition. At respective timing, each emission control is sequentially executed to control

the emission below the PZEV level. Table 1 indicates the engine and the emission control system specifications in comparison with the previous year model (LEV I-ULEV).

Table.1 Engine and Exhaust Emission Control System for PZEV (ULEV)

	'03MY (PZEV)	'02MY (ULEV)
Displacement	2362 cc	→
Bore × Stroke	88.5 × 96.0 mm	→
Number of Cylinder	4	→
Number of Valves	Intake (IN) 2, Exhaust (EX) 2	→
Valve Timing (IN Open/Close, EX Close/Open)	VVT-i $\frac{3}{60}\frac{3}{45} \leftrightarrow \frac{46}{17}\frac{3}{45}$ (deg.)	VVT-i $\frac{-4}{60}\frac{3}{45} \leftrightarrow \frac{46}{10}\frac{3}{45}$
Compression Ratio	9.6	→
Fuel Injector	295cc/min (Improved 12 holes)	265cc/min (12 holes)
Intake Manifold	With IACV	Without IACV
Exhaust Manifold	Stainless Steel Compact Double Walled	Stainless Steel Single Walled
Close-Coupled Catalyst	1.1L, Ceramic 2mil-900cpsi	1.1L, Ceramic 3mil-600cpsi
Underfloor Catalyst (UFC)	1.3L, Ceramic 3mil-600cpsi (HC Adsorbing 3-Way Catalyst)	0.9L, Ceramic 4mil-400cpsi
Air-Fuel Ratio (A/F) Sensor	Planar Type (Fast light-off)	Cup Type
Sub-Oxygen (O_2) Sensor	Super Stability Type	Normal Type

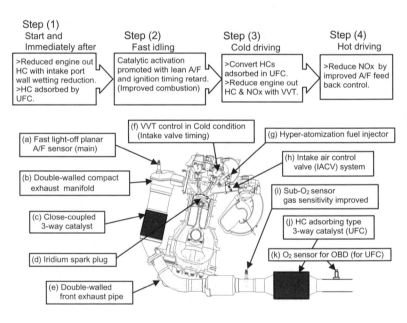

Step (1) Start and Immediately after	Step (2) Fast idling	Step (3) Cold driving	Step (4) Hot driving
>Reduced engine out HC with intake port wall wetting reduction. >HC adsorbed by UFC.	Catalytic activation promoted with lean A/F and ignition timing retard. (Improved combustion)	>Convert HCs adsorbed in UFC. >Reduce engine out HC & NOx with VVT.	>Reduce NOx by improved A/F feed back control.

(a) Fast light-off planar A/F sensor (main)
(b) Double-walled compact exhaust manifold
(c) Close-coupled 3-way catalyst
(d) Iridium spark plug
(e) Double-walled front exhaust pipe
(f) VVT control in Cold condition (Intake valve timing)
(g) Hyper-atomization fuel injector
(h) Intake air control valve (IACV) system
(i) Sub-O_2 sensor gas sensitivity improved
(j) HC adsorbing type 3-way catalyst (UFC)
(k) O_2 sensor for OBD (for UFC)

Fig.3 Exhaust Emission Control System for PZEV

EMISSION REDUCTION TECHNOLOGY DEVELOPMENT THROUGH IMPROVED COMBUSTION AND REDUCED WALL WETTING IN THE COLD CONDITION

Fig. 4 shows our basic concept for emission reduction in cold condition. To improve combustion, the IACV-system is adopted since it only requires modification on the intake system, which improves fuel atomization and also reduces wall wetting on the intake port and on the cylinder in cold condition. These enable slight lean A/F combustion and significant ignition timing retard immediately after the cold start. For further reduction of engine out emissions, a minor modification has been made to the VVT. At the most retarded VVT timing, the valve timing is overlapped, and the increased blow-back gas promotes fuel atomization. In addition, VVT operation in cold condition also reduces the engine out HCs and NOx. We also have improved the fuel injector atomization. We will introduce the states of development of respective components.

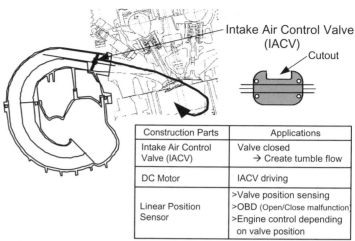

Construction Parts	Applications
Intake Air Control Valve (IACV)	Valve closed → Create tumble flow
DC Motor	IACV driving
Linear Position Sensor	>Valve position sensing >OBD (Open/Close malfunction) >Engine control depending on valve position

Fig.5 Intake Air Control Valve (IACV) System

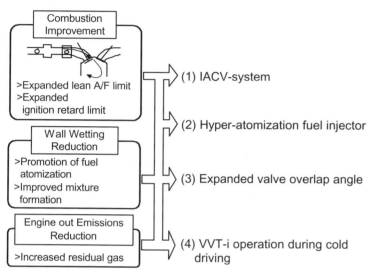

Fig.4 Cold Emissions Reduction Concept

(1) Development of IACV-system

Fig. 5 shows the outline of the IACV-system developed for the PZEV. The intake air control valve is located in the intake manifold passage, just upstream of the intake port. A cutout (approximately 15% of the whole valve area) is provided at the top of the valve. During cold fast idling, the valve is closed to only allow airflow from the cutout. This airflow forms a fierce tumble flow in the cylinder, improving the combustion with the turbulence. In addition, fuel wet on the intake port wall and the cylinder wall is drastically reduced and improves the mixture formation.

Fig. 6 shows "the required fuel amount" at cold start with and without the IACV-system. The required fuel amount refers to a 1.3 times of fuel amount of the misfiring lean limit injection amount, obtained injection by injection sequentially for each cylinder. As Fig. 6 is self-explanatory, the effect of the IACV-system is observed even under the engine starting condition. When compared with a cumulative amount for 4 cylinders, approximately 24% of fuel amount was reduced. This is quite effective for reducing HC emissions with improved driveability at immediately after the cold start, when the maximum HCs are emitted before the catalyst is activated.

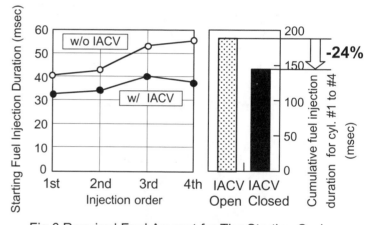

Fig.6 Required Fuel Amount for The Starting Cycle

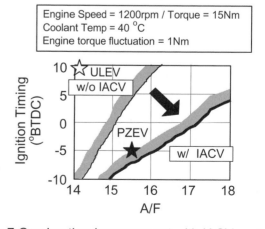

Fig.7 Combustion Improvement with IACV-system

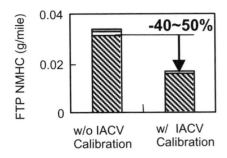

Fig.8 Comparison of Tailpipe HC emission
with/without IACV-System

To verify the atomizing effect of the improved fuel injector, we investigated the relationship between A/F and engine torque fluctuation under the fast idling condition. These results are shown in Fig. 10. With closed IACV-system, the lean limit A/F at the same torque fluctuation was greatly increased compared with opened IACV-system. Further expansion of the lean A/F limit was obtained with the hyper-atomization fuel injector; the improvement was approx. one A/F unit. The degradation in driveability with leaner A/F was greatly reduced by the improved fuel injector and IACV-system.

Fig. 7 shows the engine torque fluctuation in cold condition with and without the IACV-system. The IACV-system enabled not only much leaner A/F, but also retarded ignition timing. A darkened star (★) shows fast idling A/F and ignition timing applied to PZEV system.

Fig. 8 shows the effect of the IACV-system for the HC emission under the same driveability level, from the cold start, including the fast idling. From the calibration difference between the cold enrichment and the ignition timing, a large HC reduction of approx. 40 to 50% has been obtained for the FTP mode.

(2) Development of Hyper-Atomization Fuel Injector

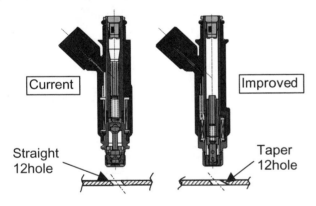

Fig. 9 Outline of Improved Fuel Injector

Fig. 9

atomization fuel injector, newly developed and adopted for the PZEV. On the basis of the conventional 12-hole injector, improvements were made to the contour of the injection holes and the main unit. For the injection hole contour, a new taper-punched hole replaced the conventional straight-punched hole. By making a thinner fuel film inside the hole, fuel atomization was promoted immediately below the hole. Compared with the conventional 85μm, the atomization level became finer to 70μm in Sauter Mean Diameter (SMD). In addition, the injector main unit has been totally improved to make it lightweight, drastically expanding the dynamic range, which improved the fuel metering accuracy in the ultra-light load operation.

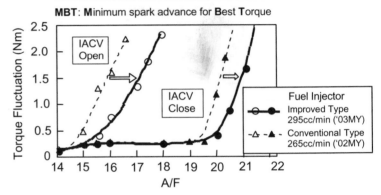

Fig.10 Lean A/F Limit of Improved Fuel Injector

(3) Improved Control of Variable Valve Timing Control System (VVT-i system)

The VVT-i system adopted for the current production model is designed to not only improve engine performance and fuel economy, but also reduce emissions through variable intake valve opening timing, which allows re-burning of the once combusted gas because it is blown back to the intake port. The conventional VVT system is mainly operated under the engine warmed-up condition. We have challenged to make the system operate even in cold condition to extract a maximum effect from the VVT. At the moment the intake valve opens, injected fuel is immediately blown back in the intake port by hot combusted gas, promoting fuel atomization. The blown-back gas with atomized fuel is taken back into the cylinder once again and burnt to promote engine-out HC reduction. In addition, as the quantity of intake port wall wetting decreases, cold driveability can be greatly improved. However, it is quite difficult to have VVT operate immediately after cold start because of the effect from engine oil viscosity, etc. To solve this problem, we have adopted a new control method: in the VVT neutral position, the overlap angle is set to approx. 6 degrees. When the oil pressure after cold start is high enough to control, the VVT is positively operated. The effects of VVT, reducing fuel enrichment, and re-burning unburned HCs in in-cylinder residual gas result in reduction of HCs and NOx.

Fig. 11 compares required fuel enrichment just after the cold start, in condition of with and without valve overlap, at the maximum retarded VVT timing. Approximately 30% fuel enrichment reduction was achieved for cold engine start. From this result, it is well understood that the wall wetting is reduced by the atomization with the fuel blown-back, which is achieved by the valve overlap obtained by the intake valve open timing set before the piston reaches the top dead center. In addition, the atomization by the fuel blow-back depends on the fuel injection timing. The effective timing for completing injection is before an intake stroke, except for the valve-overlapped timing.

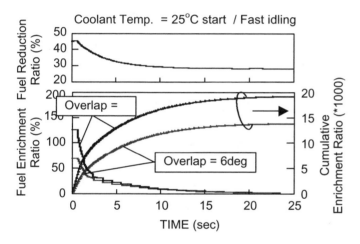

Fig.11 Required Fuel Enrichment
with/without Valve Overlap at Cold Start

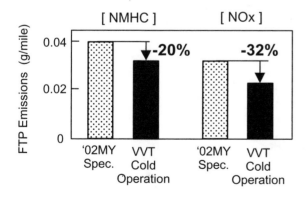

Fig.12 FTP Emissions with/without VVT
Cold Operation

(4) New VVT Control Method in Cold Condition

In cold condition, under the conventional control method, VVT hydraulic control tends to be uncertain since the engine oil viscosity is quite high. The VVT control overshoots or miss-sets to the targeted position. To solve such defect and enable to correctly set the targeted position under a very high viscosity engine oil condition, a new hydraulic control logic that observes the actual VVT position was developed. This system gives high hydraulic pulses until the deviation from the target within the pre-determined value has been developed.

Fig. 12 verifies the emission reduction when VVT is operated in cold condition. As stated earlier, the fuel atomization and the in cylinder residual gas achieved the reduction of HCs by about 20% and NOx by about 30%. Fig. 13 shows the engine torque with the cold VVT operation. The torque increased by about 7% by advancing the VVT position from the most retarded position to a prescribed position. This contributes much to the acceleration performance in cold condition.

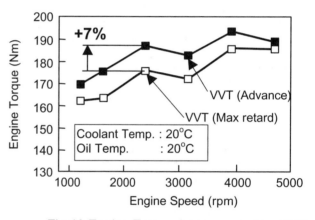

Fig.13 Engine Torque Improvement by VVT
Advance Timing (at Cold WOT)

DEVELOPMENT OF IMPROVED CATALYST WARM-UP PERFORMANCE AND EMISSION REDUCTION IN COLD CONDITION THROUGH IMPROVEMENT OF CATALYST

(1) The Heat Capacity Reduction of Exhaust Manifold

Fig.14 shows the HC emission sensitivity with the thermal energy supplied to the close-coupled catalyst from the exhaust gas in cold condition. The thermal energy depends on the engine speed, ignition timing and A/F. HC reduction can be achieved by increasing engine speed, increasing exhaust gas amount and retarding the ignition timing. In addition, it is possible to reduce HCs by changing the A/F from stoichiometric to lean, if supplied thermal energy to the catalyst remains constant. In other words, it is important to have an exhaust manifold with high heat insulation to fully utilize the thermal energy of exhaust gas supplied from the engine. In addition, as shown in Fig. 15, the catalyst activating temperature lowers if the atmosphere in the catalyst is oxidation atmosphere, which greatly improves the catalyst conversion performance. In other words, it is possible to reduce the heat energy supply by about 50% by changing A/F from stoichiometric to lean.

Fig. 16 shows the outline diagram of the exhaust manifold for ULEV and PZEV. Large improvements from the ULEV are as follows: shorter branches, more compact, and a double wall structure to prevent heat loss from the exhaust gas. In addition, by locating the A/F sensor at the union of the branches, great improvements were made on the gas flow to the A/F sensor, and the high temperature exhaust gas enabled earlier activation. Since the new exhaust manifold is designed to minimize the heat loss, the close-coupled catalyst is easily exposed to high temperature exhaust gas during high power conditions. Therefore, by balancing both requirements of quick warm-up and less deterioration at high speed driving, the close-coupled catalyst position was optimized. In addition, to retain the structural reliability of the substrate and efficient catalyst conversion performance, a space was provided in front of the catalytic substrate for uniform exhaust gas flow into the catalyst. While designing, we first analyzed the gas flow from respective branches to the catalyst using an analytical simulation (CFD) device, then confirmed the effect of the improved gas flow on the actual vehicle.

Fig. 17 shows that the space provided in front of the substrate lessens the temperature distribution in the center and circumference of the catalyst front area to within 20 to 30°C, while the current ULEV exhaust manifold shows 70 to 80°C temperature distribution. The same trend, also observed in the higher speed/load range, is important in retaining the structural reliability of the ultra thin-walled ceramic substrate (2-mil (50μm)/ 900-cpsi). We obtained full structural reliability in both a 150k miles durability test using an actual vehicle, and a reliability endurance test conducted simulating the actual market driving conditions. In addition, to keep the ceramic substrate in to the catalyst canning, alumina mat is used as the retainer. To prevent the alumina from scattering, an organic solvent (binder) was used. Since vaporized solvent is detected as HCs in the emission test of a new vehicle, efforts were made for zero-binder as much as possible.

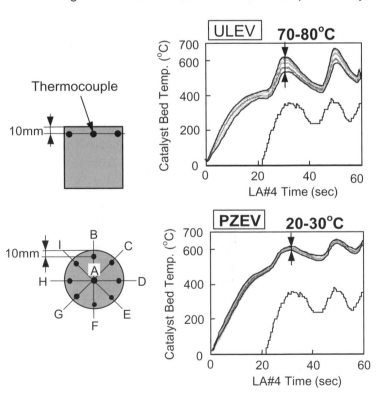

ULEV ('02MY) **PZEV ('03MY)**

Fig. 16 Exhaust Manifold with Close-Coupled Catalyst

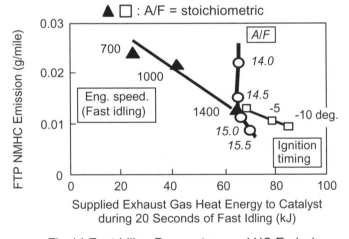

Fig. 14 Fast Idling Parameters and HC Emission

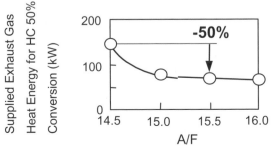

Fig. 15 Effect of A/F on Catalyst Activation

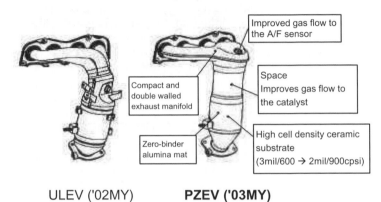

Fig. 17 Difference in the Thermal Distribution at ULEV and PZEV Exhaust Manifold

Figs. 18 and 19 show the exhaust gas mixing performance from the four cylinders on an actual vehicle. The data was collected by inserting 12 sampling tubes into the upper surface of the catalyst. The fuel injection amount of a specific cylinder was increased once every 10 cycles, and with an ultra-high response HC analyzer, the HC concentration was measured. Fig. 18 displays each cylinder's HC concentration distribution over a cycle. Fig. 19 shows the same distribution of the HC concentration deviation on the basis of the result in Fig. 18. From this analysis, we verified an improved exhaust gas mixing with the mixing space provided by the new exhaust manifold.

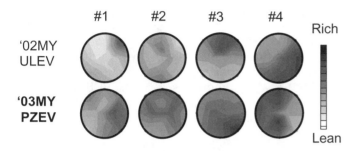

Fig.18 HC Concentration Distribution on Actual Engine

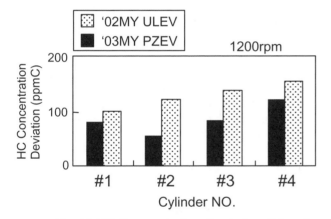

Fig.19 HC Concentration Deviation (Mean)

(2) Improvement of Close-Coupled Catalyst

Fig. 20 shows the effect on the emissions by the ceramic substrate cell density for the close-coupled catalyst. Cold NMHC emission was measured while varying the surface volume ratio of the ceramic substrate. 4-mil (100μm)/400-cpsi was used as the basis for the specific surface volume ratio of 100%. It shows that HCs can be reduced in cold condition by making the ceramic substrate wall thinner or by raising the cell density. However, as this graph illustrates, raising the cell density more than required tends to just increase the thermal capacity of the substrate, suppressing the HC reduction effect. Also, the same evaluation on the NOx was made for the PZEV specification decision. As the result, we have selected 2-mil (50μm)/900-cpsi for the best balance. Next, we

will explain the improvements of the catalyst. As stated, for quicker catalyst activation at lower temperature during the engine warm-up following a cold start, we developed a compact, low heat mass exhaust manifold, a thin-walled ceramic substrate, and improved the engine control system. In addition, we worked on improving the catalyst itself. The basic concepts of the improvements are as follows:

1. Optimization of the total precious metal loading amount and adopting the front high loading, same as conventional catalyst.

2. Enhancements of the high temperature resistance through the improvement of oxygen storage capacity (OSC), low temperature catalyst activation and optimization of conversion performance after warmed up.

3. Decreased manufacturing variability: coating thickness, precious metal amount, and precious metal distribution, which reduces emissions variability.

Fig. 21 shows the effect for catalytic temperature rise with front high loading. Eight times the nominal loading rate of platinum (Pt) is additionally loaded at the front portion (20 mm) of catalyst, improving the quick activation performance and allowing the temperature to exceed around 140°C in the latter half of the FTP fast idling. This is a suitable technology for the PZEV system as it shows much higher activation at lower temperatures found in an early stage after cold starting. To use limited precious metal resources as effectively as possible, the high loading, especially on the front portion of catalyst where it is sensitive for the deterioration, makes a great contribution to improve the durability. For the PZEV system, precious metal amount and the length of the front high loading portion are optimized on the basis of the 150k mile emission durability. [4]

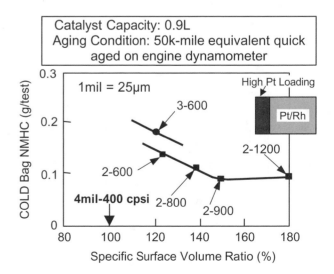

Fig.20 The Ceramic Substrate Cell Density and HC Emissions

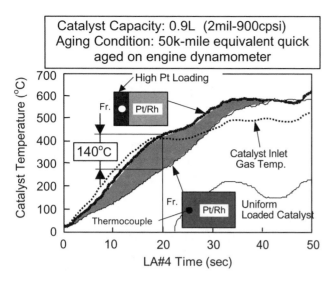

Catalyst Capacity: 0.9L (2mil-900cpsi)
Aging Condition: 50k-mile equivalent quick aged on engine dynamometer

Fig.21 Effect of Front High Loading Catalyst

(3) Improvement of 3-way Catalytic Function

Since the PZEV regulation requires 15year/150k miles emission warranty, it is important to consider the deterioration of the catalyst. When we made a detailed analysis on the cause of deterioration for catalysts in a high-temperature durability test, we found that both the OSC for moderating A/F variation and the specific surface area for the oxygen storage material (promoter of CZ system zirconia (ZrO_2) solid solution with ceria (CeO_2)) decreased significantly at the same time. From this result, we paid attention to the thermal resistance of ceria, which has an OSC function to enhance the converting performance of 3-way catalyst.

Our concept is positioning alumina (Al_2O_3) particles, which are hard-to-react, among CZ particles to suppress the high-temperature-induced growth of CZ particles in the subsidiary catalyst. Accordingly, a new subsidiary catalyst of ACZ-type (Al_2O_3-CeO_2-ZrO_2) was developed. Fig. 22 compares the emissions with CZ- and ACZ-type catalysts. HC and NOx emissions were reduced by about 10% and 20%, respectively, with ACZ-type. [5]

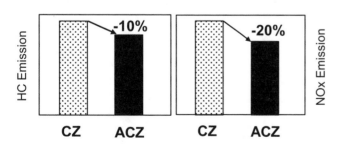

Fig.22 Effect of ACZ-type Catalyst

(4) Improvement of Underfloor Catalyst (UFC)

In previous section, we have reported the low-temperature activation of the close-coupled catalyst. Since even a close-coupled catalyst needs several seconds to activate after a cold start, part of the engine out emissions go through the close-coupled catalyst without being converted. Particularly, since fuel supply is high in the cold starting, it is no exaggeration to say that the emission amount from this period is almost equivalent to the PZEV standards. Therefore, the most effective measure is to suppress the go through HCs. Fig. 23 shows the schematic diagram of an HC adsorbing type 3-way catalyst. The catalyst temporarily adsorbs HCs emitted immediately after a start and converts them gradually as they are desorbed and the catalyst is activated. The catalyst is coated with HC adsorbent as the bottom layer, close to the substrate, and with normal 3-way catalyst for the upper layer. The adsorbent coating amount is determined balancing the required HC adsorption amount and thermal characteristics of the catalyst. The HC adsorbent itself is almost identical to the one used "Toyota-HCAC-System" adopted on 2001 Model Year Prius. [6] Fig. 24 shows the tailpipe HC emission behavior of the newly developed HC adsorbing type 3-way catalyst and the conventional 3-way catalyst. Temporarily, the catalyst adsorbs about 70% or more of HCs. Some of the adsorbed HCs are converted during the HC desorbing period, but the other HCs are emitted without conversion, thus the HC reduction for the total FTP mode remains 20 to 30%. However, as shown in Fig. 25, variation in the amount of entering HCs to the HC Adsorbing type 3-way Catalyst do not much affect the tailpipe HC amount within the PZEV operation area. Therefore, the catalyst system is suitable from the viewpoint of variability reduction.

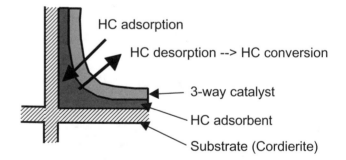

Fig.23 Image Diagram of HC Adsorbing type 3-way Catalyst

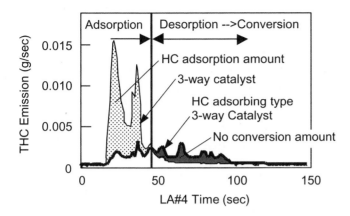

Fig.24 Comparison of HC Emission Behavior
with/without HC Adsorbent

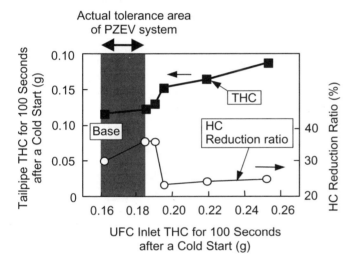

Fig.25 HC Emission Sensitivity to UFC Inlet HC

DEVELOPMENT OF EXHAUST EMISSION VARIABILITY REDUCTION

As a countermeasure for reducing exhaust emission variability in cold condition, we have added intake air amount feedback control for stable engine speed, and early lean A/F feedback (F/B) control to stabilize the exhaust gas amount during fast idling. In addition, to prevent decrease of the catalyst conversion performance from variance in the A/F among the cylinders after warmed-up, we first optimized the A/F sensor installation position on the exhaust manifold to make a drastic improvement of gas exposure for the sensor from each cylinder. Next, considering the mixed exhaust gas exposure, the sub-O_2 sensor position has moved downward from just downstream of the close coupled catalyst, to in front of the under floor catalyst. By the mixing effect, the exhaust gas from each cylinder has equal exposure to the sub-O_2 sensor. Moreover, an improved O_2 sensor tip and optimized sub-feedback control, for effective anti-variability, was adopted.

(1) Improvement of sub-O_2 sensor control λ point

Fig. 26 shows the hydrogen concentration in the exhaust gas downstream of the close-coupled catalyst. Although the engine out hydrogen concentration decreases after passing through the catalyst, the hydrogen still remains at 50 to 100ppm in the under floor catalyst inlet gas. Moreover, hydrogen tends to increase when variability exists in the A/F per cylinder. This is caused by increased concentration of hydrogen from the A/F rich cylinders because the hydrogen generation is correlated with A/F. If hydrogen is present in the gas, the controlled A/F by the O_2 sensor shifts leaner from stoichiometric because of the faster diffusion of hydrogen than oxygen. This causes emission variability, particularly for NOx.

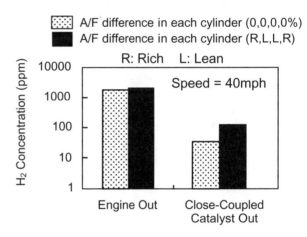

Fig.26 Hydrogen (H_2) Concentration in Exhaust Gas

Fig. 27 shows a sectional view of the improved O_2 sensor. A catalyst layer, applied to the top of the coating layer, converts the hydrogen in the exhaust gas, eliminating the lean shift and realizing stoichiometric controlled A/F. Fig. 28 shows the sensor characteristics in a model gas evaluation using the H_2-O_2-N_2 mixed gas. The improved sensor's switching point is nearer the stoichiometric point compared with the conventional sensor's leaner switching point.

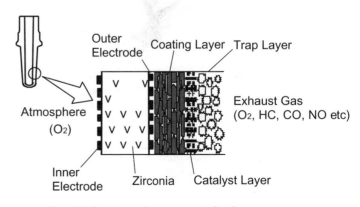

Fig.27 Outline of Improved O_2 Sensor

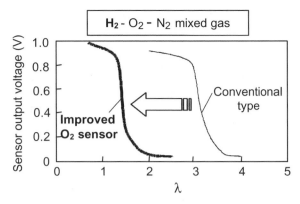

Fig.28 H_2 Sensitivity of Improved O_2 Sensor
(Model Gas Characteristics)

(2) Improvement of Sub-feedback Control

As stated in the preceding section, we have improved the sub-O_2 sensor tip to eliminate the effect of hydrogen in the exhaust gas. As the result, gas sensitivity has improved to make the controlled A/F stabilize. In addition, to regulate the controlled A/F appropriately at stoichiometric to enhance the catalyst conversion performance, we have improved the sub-feedback control based on the conventional model (control using the A/F sensor and sub-O_2 sensor). Since the improved O_2 sensor accurately senses the oxygen concentration in the emission control system, we optimized the A/F correction gain to have the sub-O_2 sensor output quickly converged to a target voltage.

Fig. 29 shows the result of the NOx emission fluctuation due to the A/F variance in the individual cylinders with the improved O_2 sensor and the improved sub-feedback control. As the result, we were able to realize a drastic reduction of the emission variability and an average emission reduction in NOx.

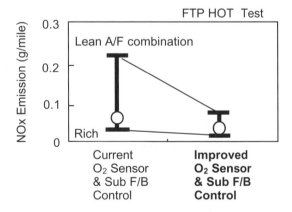

Fig.29 NOx Emission and Variability Reduction

(3) Improvement of Catalyst Deterioration Suppressing Control

To achieve the 150k miles long distance warranty, it is very effective to suppress deterioration of the catalyst. In addition to improving the catalyst for anti-deterioration, we adopted a new catalyst deterioration suppressing control system. In order to investigate the effect of lean A/F for catalyst deterioration, we conducted engine dynamometer aging tests with 4 combinations of temperature and A/F as shown in Fig. 30. Fig. 30 shows the results of the CO-NOx conversion rate from our studies, indicating the catalyst aging temperature and aging A/F as the parameter. Though dependent on the type of precious metal, loading quantity, etc., the figure indicates that the degree of deterioration tends to increase as the catalyst temperature increases and the A/F becomes leaner. According to the actual driving pattern, fuel-cut (F/C) can exist under the condition of high-speed, high-load driving with high catalyst temperature. The catalyst deterioration suppressing control estimates the catalyst temperature and, under a high temperature condition, it inhibits the fuel-cut. This catalyst deterioration suppression largely contributes to the catalyst durability.

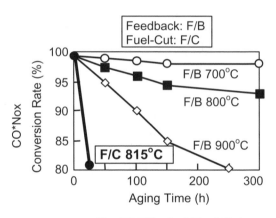

Fig.30 Effect of Fuel-Cut

SUMMARY OF EMISSION AFTER 150K MILES DURABILITY

Lastly, Fig. 31 compares the exhaust emissions between the 150k miles-aged PZEV with the 100k miles-aged ULEV. The PZEV emission control system we have introduced, incorporating the various low-emission technologies including newly developed and improved ones, have fully cleared the PZEV standard with the production parts variation within the design tolerances. Classifying the reduction ratio from the ULEV level for each emission reduction technology adopted for the PZEV, the largest improvement on HCs was the combustion improvement during cold start and warm up. Regarding NOx, the largest improvement was from the catalytic system and the emission variability reduced through the improved sub-feedback control.

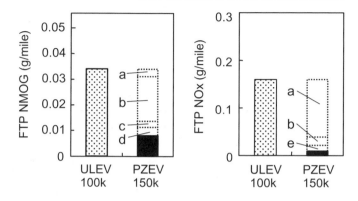

a: Improvement of exhaust manifold and catalyst performance
b: Combustion improvement, rapid warm-up control and VVT improvement
c: Cold lean A/F F/B control (during fast idling)
d: HC adsorbing/converting 3-way catalyst
e: Improvement of O_2 sensor and sub-F/B control

Fig.31 ULEV and PZEV Emissions
(Useful Life Aged)

CONCLUSION

We have made a technological development on the subject of realizing the PZEV tailpipe emission standards by using various low-emission technologies accumulated thus far.

1. Regarding the important concept of cold lean burning and a ignition timing retard control for meeting the PZEV standards, we have succeeded in compatibly achieving the low-emission and the acquisition of driveability with a newly developed IACV-system, improved injector and VVT control.

2. We have developed a highly reliable emission control system capable of meeting cold emission standards (50°F) and long-distance durability through the adoption of a compact exhaust manifold, optimized catalyst and the newly developed HC adsorbing type 3-way catalyst.

3. With regard to the mass sale vehicles, we have been able to provide a low cost PZEV emission system capable of enduring 150k miles. This was accomplished by reducing the variability among the vehicles caused by the variance in operating conditions and the manufacturing tolerance of the emission control devices, and by improving the existing engine control.

ACKNOWLEDGMENTS

The present PZEV system development has been achieved jointly with related departments within the company and outside. The authors wish to express their appreciation to a number of persons who took part in the development.

REFERENCES

1. H.Kitagawa, T.Mibe, K.Okamatsu and Y.Yasui:"L4-Engine Development for a Super Ultra Low Emissions Vehicle" SAE Paper 2000-01-0887

2. K.Nishizawa,S.Mitsuishi,K.Mori,andS.Yamamoto: "Development of Second Generation of Gasoline P-ZEV Technology" SAE Paper 2001-01-1310

3. T.Inoue, M.Kusada, H.Kanai,S. Hino and Y.Hyodo: "Improvement of a Highly Efficient Hybrid Vehicle and Integrating Super Low Emissions" SAE Paper 2001-01-2930

4. T.Takada, H.Hirayama, T.Itoh, and T.Yaegashi: "Study of Divided Converter Catalytic System Satisfying Quick Warm up and High Heat Resistance" SAE Paper 960797

5. T.Kanazawa,J.Suzuki,T.Takada,T.Suzuki,A.Morikawa,and A.Suda: "Development of Three-Way Catalyst Using Composite Alumina-Ceria-Zirconia" JSAE 20025084

6. T.Kanazawa and K.Sakurai: "Development of the Automotive Exhaust Hydrocarbon Adsorbent" SAE2001-01-0660

L4-Engine Development for a Super Ultra Low Emissions Vehicle

Hiroshi Kitagawa, Toshihiro Mibe, Kazunori Okamatsu and Yuji Yasui
Tochigi R&D Center, Honda R&D Co., Ltd.

Copyright © 2000 Society of Automotive Engineers, Inc.

ABSTRACT

HONDA has developed technology to fulfill the strictest emissions standards to date. That is, we have developed the technology necessary to satisfy the state of California's SULEV (LEV-II) regulations. We were able to move from the conventional 600-cell, 4.3mil three-way catalyst to a 1200-cell, 2.0mil catalyst through the application of new canning technology. We were also able to achieve early catalyst light-off and improved conversion performance without increasing the precious metal content. However, it was not possible to satisfy the SULEV standards using only these improvements to the catalyst. It was also necessary for us to develop new emission control technology for the various stages of engine operation: cold, warm-up and post warm-up. Specifically, we developed technology that dramatically increases catalyst light-off speed by controlling intake air and ignition timing. Further, we developed technology to reduce exhaust emissions to the ultimate degree possible when the engine is cold and the catalyst is inactive. Also, in order to further improve the conversion performance of the catalyst once it is active, we developed a Secondary O2 Feedback Control System. This feedback system is an extremely high precision system that utilizes an adaptive sliding mode controller to identify various catalyst models in real time and predict the post-catalyst HEGO sensor output. The above technologies were used in combination to achieve the SULEV emissions standards.

INTRODUCTION

In recent years, it has become increasingly important to address environmental issues. In particular, it has become urgent that the air pollution produced by automobiles be decreased. Accordingly, emissions regulations throughout the world have become more stringent with the state of California introducing the new and more rigorous LEV-II regulations.

Table 1. LEV-II Emission Standards (g/mile)

Durability	Category	NMOG	CO	NOx	HCHO
50 km ile	LEV	0.075	3.4	0.05	0.015
	ULEV	0.040	1.7	0.05	0.008
120 km ile	LEV	0.090	4.2	0.07	0.018
	ULEV	0.055	2.1	0.07	0.011
	SULEV	**0.010**	**1.0**	**0.02**	**0.004**
150 km ile (Option)	LEV	0.090	4.2	0.07	0.018
	ULEV	0.055	2.1	0.07	0.011
	SULEV	0.010	1.0	0.02	0.004

Table 1 indicates Emissions Standards by category as set forth by the LEV-II Regulations. The LEV-II Regulations have revised even the durability requirements so that emissions through 120,000 miles will now also be regulated. What is notable about the new regulations is that in addition to the NMOG standards, NOx standards have been become much more rigorous.

Figure 1 indicates the LEV-I ULEV 100,000 mile and SULEV 120,000 mile NOx standards. The standards for the new SULEV category are very stringent compared to the LEV-I ULEV standards requiring an NMOG level 2/11 and a NOx level 1/15 that currently required.

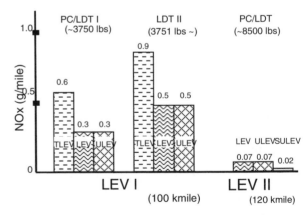

Figure 1.   Comparison of LEV-I and LEV-II NOx Regulations

In September 1997, Honda became the first company in the world to sell a gasoline-powered vehicle that satisfied California's (LEV-I) ULEV standards. The engine system powering this vehicle is characterized by achieving ultra low emissions using only an underfloor catalyzer due to the application of high precision fuel control. This fuel control is attained through the use of modern control and adaptive learning control techniques.

Now we have developed technology to even further reduce emissions. The main technologies developed are: (1) a three-way catalyst with increased cell density and thinner walls as well as the related canning technology; (2) early light-off control of the three-way catalyst; (3) optimized fuel control at and immediately following start-up; and, (4) predictive fuel control that adapts to changes in three-way catalyst characteristics and calculates the future value of the HEGO sensor.

It is the adoption of these technologies that has made it possible for Honda to sell gasoline-powered vehicles that fulfill SULEV standards earlier than any other company in the world. This paper describes these technologies in detail.

SULEV SYSTEM DESIGN

KEY AREAS OF IMPROVEMENT FOR THE SULEV – 25°C Modal Mass Emissions tests were performed with the Honda Accord ULEV (50,000 miles). Figures 2 and 3 show the accumulated emissions for NMHC and NOx, respectively. As you can see from these results, both of these components in the emissions exceed the SULEV standard during fast idle, that is, during the time from start-up to initial acceleration. It is obvious then that the degree to which emissions can be controlled before the engine warms up is critical. After fast idle, during LA-4 mode operation, the emissions level gradually increases. As indicated in Figure 2, the cumulative NMHC level "A" reached in the period from initial start to the end of the hill climb is 0.005g, which is approximately 50% of the standard (even without considering the longer useful life or compliance margins).

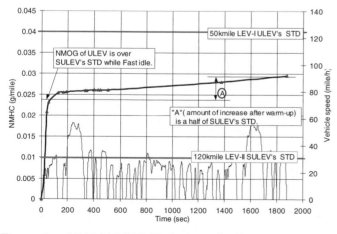

Figure 2.   ULEV NMHC Emissions (In-House Data)

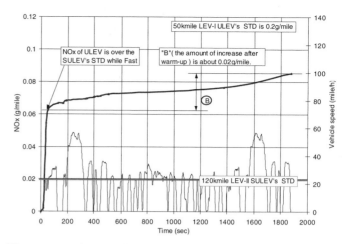

Figure 3.   ULEV NOx-Emission (In-House Data)

The cumulative NOx level "B" emitted from initial acceleration at the bottom of the hill to the time when the vehicle crested the top is approximately 0.02 g/mile as shown in Figure 3. This level is equivalent to that of the SULEV standards (again, without considering the longer useful life or compliance margins). Therefore, in order to satisfy the SULEV emissions standards, the (Accord) ULEV emissions performance had to be improved not only during the warm-up period after a cold start when the catalyst is inactive but also after warm-up and throughout the entire spectrum. What's more, this improvement had to be effective even after 120,000 miles once the catalyst has begun to deteriorate.

SUMMARY OF SULEV – In order to achieve the super ultra low emissions levels of the SULEV, we enhanced our existing ULEV technology by developing the following.

Catalyst with High Cell Density and Thinner Walls – By applying a 1200-cell catalyst with 2.0mil catalyst support, we are able to improve early catalyst light-off and conversion performance. Further, by reducing the thickness of the walls of the catalyst (2.0mil) and applying a construction using only an underfloor catalyst, we can limit the backpressure increase and minimize engine power reduction.

Reduction of Exhaust Emissions From Start Through Warm-Up

Quick Warm-Up System – Intake air volume control and ignition timing control are used to achieve early light-off of the three-way catalyst and to improve exhaust gas conversion performance during warm-up.

Lean Air/Fuel Ratio Control – A lean air/fuel ratio condition is enabled by VTEC (Variable Valve Timing and Lift Electronic Control) due to the high swirl ratio that is generated when only one valve is open. As the Quick Warm-Up system is implemented to stablize combustion, a lean air/fuel ratio condition can be attained even earlier. Further fuel control optimization is attained and cold engine emissions reduced.

Exhaust Emissions Control Using Three-Way Catalyst Light-off Conditions

New Secondary Oxygen Feedback System – Conversion performance is optimized by the constant identification of changes to the three-way catalyst conditions as affected by engine operating conditions, catalyst temperature and deterioration. This system calculates the future value of the HEGO sensor and determines the optimum fuel feedback target level, resulting in even higher fuel feedback system precision.

OVERALL SULEV SYSTEM CONFIGURATION – Figure 4 shows the overall configuration of the system that resulted in the achievement of SULEV standards.

The current Accord ULEV system is comprised of components (a)-(h).

The following are the new components applied to the system to achieve SULEV standards:

(i) 1200-cell, 2.0 mil tri-metal underfloor catalyst, 1.7 liter, 330 g/ft^3 (Pt: Pd: Rh = 1:15:1)

(j) Quick Warm-up System

(k) Lean Air/Fuel Ratio Control at Cold Start

(l) New Secondary O2 Feedback Control System: PRISM (PRediction and Identification type Sliding Mode control)

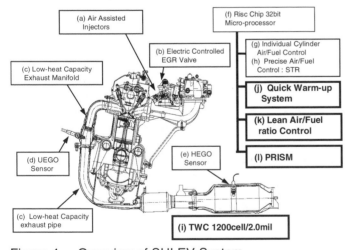

Figure 4.　Overview of SULEV System

1200-CELL, 2.0MIL THREE-WAY CATALYST

Application of the 1200-cell catalyst is crucial to achieving early catalyst light-off and high conversion performance and thereby to satisfying the SULEV standards. In this section, we describe catalyst wall thickness selection, the related canning technology and emissions performance.

CATALYST WALL THICKNESS SELECTION – It is known that increasing the cell density of the catalyst and reducing wall thickness are effective methods for improving the conversion performance of a catalyst. Figure 5 shows the relationship between cell density, surface area and bulk density. That is, surface area increases with cell density, while bulk density decreases with reductions in wall thickness.

A 1200-cell, 2.0mil catalyst has 20% less bulk density than does a 2.5mil construction. Also, backpressure increases with cell number. However, through the use of a thin walled catalyst, we were able to minimize power loss to merely 2kw of horsepower and 5Nm of torque.

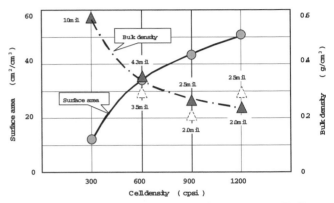

Figure 5.　Relationship Between Cell Density, Bulk Density and Surface Area

CANNING TECHNOLOGY – As the new 1200-cell, 2.0mil catalyst has low isostatic strength, it is not possible to use conventional canning techniques. (Isostatic strength is the static breaking strength of radical pressure.)

Figure 6 shows the two types of canning constructions. The conventional method uses clamshell type canning (Figure 6-(a)) wherein the catalyst and holding seal mat are sandwiched into an upper and lower case. As this construction has a fixed inside case diameter, variations in catalysts and holding seal mats lead to significant surface pressure differences being generated during the canning process. Therefore, catalysts with an isostatic strength of 15 kg/cm^2 or more are required for this type of canning.

The new technique employs a rolling construction (Figure 6-(b)) in which the catalyst and holding seal mat are fixed after being rolled into a case.

This innovative new technique has made it possible for the surface pressure generated during canning to be evenly dispersed outward about the circumference of the product. It has also made it possible to change the inside case diameter to match that of the catalyst, thereby making it possible to change the catalyst diameter. In addition to this rolling type canning method, holding seal mat selection, and improvements to the production method have contributed to the reduction of surface pressure generated during canning.

Figure 7 shows the surface pressures that are generated in the catalyst during canning. The rolling type construction exhibits a much more even surface pressure distribution than that of the clam shell type with the surface pressure consistently being less than 5 kg/cm^2.

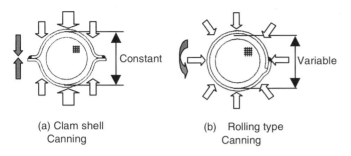

(a) Clam shell Canning   (b) Rolling type Canning

Figure 6.   Clam Shell Type Canning and Rolling Type Canning

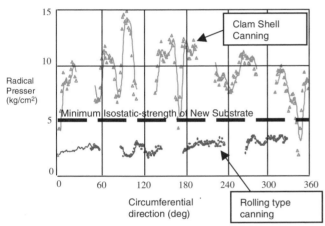

Figure 7.   Radical Pressure of Clam Shell Type and Rolling Type Canning

EMISSIONS PERFORMANCE – Figure 8 shows an emissions comparison of the 98 Accord ULEV catalyst and a 1200-cell, 2.0mil catalyst over the FTP. We installed a 600-cell, 4.3mil catalyst (100K aged) and a 1200-cell, 2.0mil catalyst (with 120K aged) in the same vehicle and took measurements. NMOG decreased by 31% and NOx by 48% with the new catalyst. Both the improved conversion brought about by increased cell density and the early catalyst light-off resulting from decreased wall thickness contribute to these reductions. However, at this point we were still far from attaining our goal of compliance with the SULEV standards.

QUICK WARM-UP SYSTEM

Figure 9 presents an overview of the Quick Warm-up System (QWS) operation. The horizontal axes represent time, and the vertical axes represent the RACV (rotary air control valve) opening, engine speed and ignition timing, respectively, from the top.

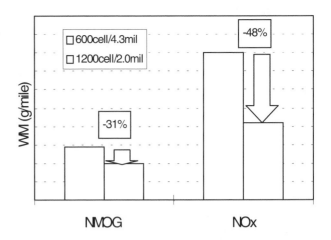

Figure 8.   Comparison of Emissions from 600 Cell and 1200 Cell Catalysts

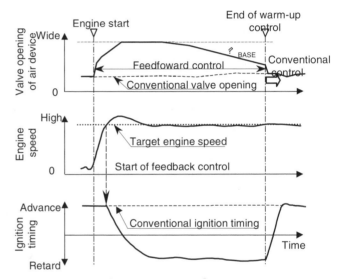

Figure 9.   Basic Quick Warm-up System

While the engine is idling after start-up, the QWS uses the RACV (by means of feed-forward control) to supply large quantities of intake air. The amount of intake air is gradually reduced as the engine warms up. When the engine speed exceeds a preset level, ignition timing PI (proportional + integral) feedback control begins. The ignition timing is controlled to maintain target engine speed. Normally, as indicated in the diagram, when intake air is increased, ignition timing is controlled to retard timing to achieve a timing that will maintain the target engine speed.

Figure 10 shows the increased temperature effect of the Quick Warm-up System. At initial acceleration in the FTP cycle, the exhaust and three-way catalyst temperatures are higher than those of the ULEV by approximately 240°C and 190°C respectively (catalyst temperatures taken at the center of a location 40mm from the front end of the catalyst). This indicates that the early light-off of the catalyst is effective. (The improvements in the light-off performance of the high cell density, thin-walled catalyst is partly due to the increased temperature and partly due to the catalyst lighting off at lower temperatures.)

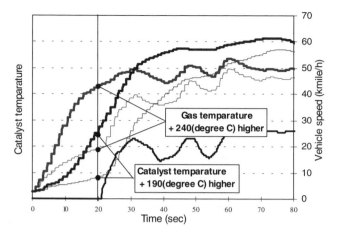

Figure 10. Increased Temperature Effect Due to QWS at Start-Up

LEAN AIR/FUEL RATIO CONTROL – Once the effects of the increased catalyst temperature at start-up brought about by the introduction of the QWS were verified, we found another area to apply exhaust emissions reduction. Figure 11 indicates engine speed and intake air pressure at a 25°C engine start-up for both the ULEV system and the SULEV system. Though the engine speed is similar for both systems, the intake air pressure of the SULEV system is higher. This is due to the increase in intake air provided by the QWS. As there is a high engine load at start-up, combustion is more stable than that of the ULEV system. Figure 12 indicates the air/fuel ratio at the corresponding times. The ULEV system utilizes VTEC to achieve a lean burn approximately 6 seconds after start-up. We used the same VTEC system for the SULEV. However, the more stable combustion and optimized fuel control at start-up enabled us to run even leaner during the initial start-up period.

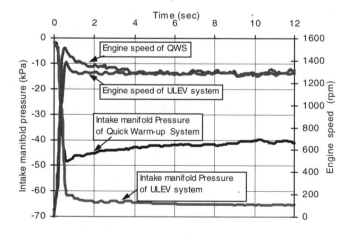

Figure 11. Comparison of QWS and ULEV Intake Manifold Pressu

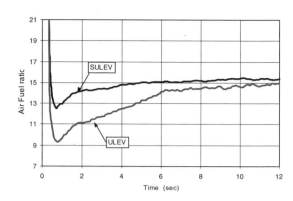

Figure 12. Comparison of SULEV and ULEV Air to Fuel Ratios

Figure 13 indicates emissions (THC and NOx) at start-up for both the ULEV and SULEV in the 25°C FPT cycle. You can see that the levels of the THC and NOx were dramatically lower in the SULEV. This is the result of the reduction in exhaust emissions due to a leaner mixture immediately following start up and to the early light-off of the catalyst enabled by the QWS.

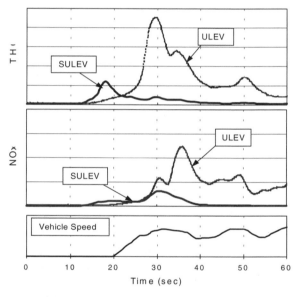

Figure 13. Emissions Reduction Effects to QWS and Lean Air/Fuel Ratio Control at Start-up (FTP cycle)

Figure 14 shows the emissions reduction effects of the QWS and lean air/fuel ratio control. In the FTP cold transient phase, the QWS and Lean Air/Fuel ratio Control reduce NMOG by approximately 70% and NOx by approximately 36%.

PRISM

In developing this new control system, the first thing we looked into was the correlation between catalytic conversion and post-catalyst O2 sensor output. We measured the catalyst conversion ratio with a warm engine maintained in a constant operating condition wherein the exhaust gas air/fuel ratio was gradually altered.

714

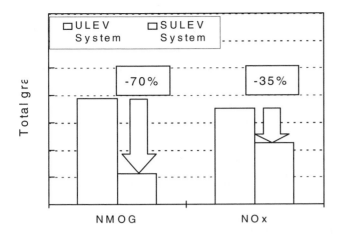

Figure 14. Emissions Reduction Effects of QWS and Lean Air/Fuel ratio Control (Total grams during FTP CT Phase)

Figure 15 shows the conversion ratio of the three-way catalyst and the voltage output (Vout) of the O2 sensors placed between the catalyst. (The horizontal axis represents voltage output ϕin of the linear air-fuel ratio sensors positioned upstream of the catalyst).

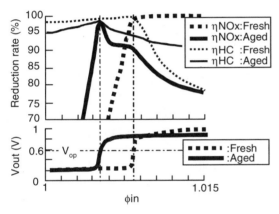

Figure 15. Reduction Rate Characteristics of the Three-way Catalyst

The ϕin for a maximum hydrocarbon and NOx conversion rate shifts toward a lower voltage (toward a reduced excess air ratio) as the three-way catalyst deteriorates. You can see that at this point the Vout value equals the oxygen sensor switching point. Based on this, we decided to construct an air-fuel ratio control an approximately 0.6V constant (Vop in the figure). The objective of this mechanism is to maintain a maximum conversion rate regardless of catalyst deterioration.

Figure 16 illustrates how the PRISM (PRediction and Identification type Sliding Mode control) control system is configured. The system is provided with an identifier for measuring the changes in the dynamic characteristics of the linear model of the catalyst and for identifying the model parameters in the on board system.

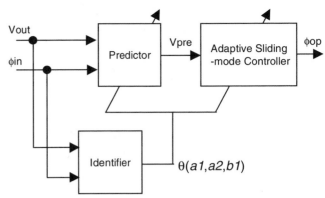

Figure 16. PRediction and Identification type Sliding Mode Control (PRISM)

Using a predictor designed to compensate for dead time, the system calculates the future value of the HEGO sensor and determines the optimum fuel feedback target level using an adaptive sliding mode controller. The Secondary O2 Feedback section of the self-tuning regulator (STR) in the ULEV was replaced by this system.

Figure 17 shows the overall configuration of the fuel control system. The PRISM control system calculates the fuel feedback ϕop target value in order to ensure that the HEGO sensor output (Vout) converges at the optimum value (Vop in Figure 15).

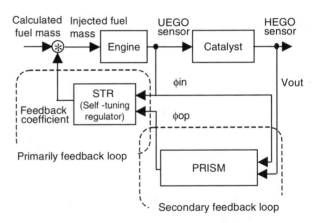

Figure 17. Overall Configuration of Double Feedback loop

Here, the fuel is controlled so that the air/fuel ratio control (pre-catalyst) using a self-tuning regulator (STR) conforms ϕin to ϕop. In order for this to be accomplished a double feedback loop is constructed wherein Vout converges with Vop.

The identifier is a delayed auto regressive (DARX) model with dead time dm (control cycle) as shown in Figure 16.

$$VO2\ (k+1) = a1 \times VO2\ (k-1) + b1 \times KACT\ (k-dm) \quad (1)$$

$$VO2\ (k) = Vout\ (k) - VO2\_TARGET \quad (2)$$

$$KACT\ (k) = \phi op\ (k) - KACT\_BASE \quad (3)$$

a1, a2, b1:   model parameters

VO2 (k):   deviation between Vout and Vop

VO2_TARGET: desired Vop value

KACT (k):   deviation from base value of ϕin

k: control time

dm: dead time

Model parameters a1, a2 and b1 are defined as vectors symbolized by θ. The VO2 and KACT sampling data are defined as a component by the Greek symbol ξ (θ^T is the transposed matrix of θ.)

$$\theta^T = [\, a1 \; a2 \; b1] \qquad (4)$$

$$\xi^T (k) = [VO2(k) \; VO2(k) \; KACT(k-dm)] \qquad (5)$$

The use of equations (4) and (5) provides a different expression of the plant as shown below.

$$VO2(k+1) = \theta^T \xi(k) \qquad (6)$$

In order to consistently identify status changes in the three-way catalyst, θ (plant model parameter) is successively calculated on board using equation (6). The model parameters (expressed by a discrete system) estimated by the identifier are defined as a1(k), a2(k) and b1(k), and θ is used for the estimation vector which is defined by these parameters. This is applied to equation (6) to produce the estimation plant model of equation (8).

$$\theta(k)^T = [\, a1(k) \; a2(k) \; b1(k)] \qquad (7)$$

$$VO2\text{-hat}(k) = \theta(k)^T \xi(k-1) \qquad (8)$$

Equation (9) is used to define the difference between the actual plant and the estimated plant as the estimated error.

$$e\text{-id}(k) = VO2(k) - VO2\_hat(k) \qquad (9)$$

We used recursive weighted least square estimation (RWLS) which minimizes the variance of this error to create the identification algorithm.

The adaptive sliding mode controller for calculating ϕop for the control target model in equation (1) is expressed by the arithmetic equations listed below.

$$\phi_{op} = Usl(k) + KACT\_BASE \qquad (10)$$

$$Usl(k) = Ueq(k) + Urah(k) + Uadp(k) \qquad (11)$$

$$VO2p(k) = Vpre(k) - VO2\_TARGET \qquad (12)$$

$$\sigma p(k) = VO2p(k) + S \times VO2P(K-1) \qquad (13)$$

$$Ueq(k) = \frac{-1}{b1(k)}\{(a1(k)-1)+S\}VO2p(k) + (a2(k)-S)VO2p(k-1) \qquad (14)$$

$$Urch(k) = \frac{-1}{b1(k)} F \cdot \sigma p(k) \qquad (15)$$

$$Uadp(k) = \frac{-1}{b1(k)} G \cdot \sum_{i=0}^{k} \sigma p(i) \qquad (16)$$

Ueq (Equivalent Control Input): Input to restrict status to switching line

Urch (Target Achievement Law Input): Input to place status on switching line

Uadp (Adaptive Law Input): Input to control modelling errors and disturbances while placing status on switching line

S: switching function parameter

F: target achievement law gain

G: adaptive law gain

The fundamental principle of the sliding mode is to apply the control input in such a way that the switching function $\sigma p(k)$ is confined to zero. If σp moves from zero, the control input continuously works toward the switching point, as shown by the control input in Figure 18. When confined to this line, it begins to slide toward zero.

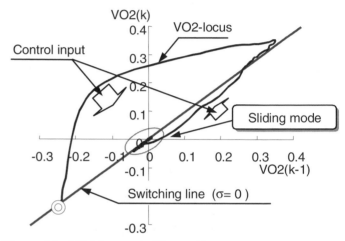

Figure 18. VO2-Locus in Phase Plane

Figure 19 shows the control status of PRISM at the second hill in the FTP first phase. You can see that the post-catalyst HEGO sensor output is set to the target voltage.

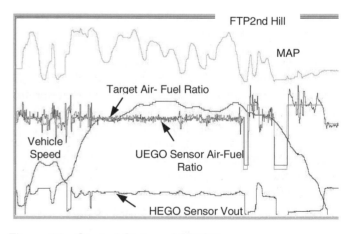

Figure 19. Control Status of PRISM

Figure 20 shows the emissions measurement of the 98 Accord ULEV secondary O2 feedback system and the PRISM control system. The same catalyst was used in both vehicles so it is clear that the PRISM control system is particularly effective in reducing NMOG and NOx levels.

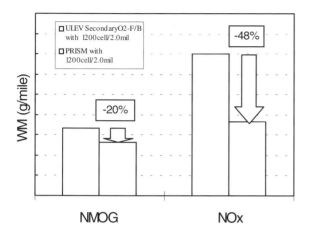

Figure 20. Comparison of ULEV's Secondary O2 Feedback and PRISM Emissions

VEHICLE TEST RESULTS – SULEV   regulation standards were achieved through the combined application of the above described technology. We measured the emissions of an Accord SULEV vehicle with 120K aged components over the FTP. The Modal Mass Emissions results from this test are indicated in Figures 21 and 22. Figure 21 indicates NMHC results and Figure 22 indicates NOx results.

Both of these emissions components are kept to extremely low levels during the first 20 seconds after start-up due to the emissions reductions brought about by lean burn at start-up; and, early catalyst light-off brought about by the application of the Quick Warm-up System. Further, the fuel feedback increase is strictly controlled after catalyst light-off with the PRISM system.

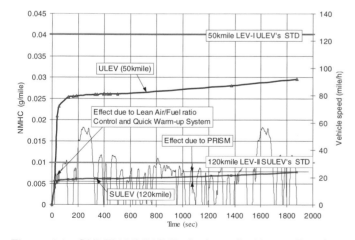

Figure 21. SULEV NMHC Emissions (In-House Data)

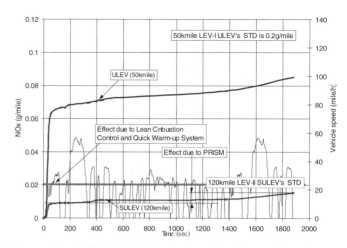

Figure 22. SULEV NOx Emissions (In-House Data)

Table 2 indicates emissions achievement values. As you can see, we have achieved SULEV standards with a more than sufficient margin.

Table 2.   Accord SULEV Emissions (g/mile) (In-House Data)

	Durability	NMOG	CO	Nox
SULEV STD	120kmile	0.010	1.000	0.020
Achieved Values	120kmile	0.0084	0.0155	0.0161

CONCLUSION

1. Rolling-type canning technology makes it possible to use a mass produced 1200-cell, 2.0mil catalyst with low isostatic strength. This results in improved catalyst conversion performance and light-off properties and thus to a decrease in emissions.

2. The Quick Warm-up System dramatically increases the heating up of the catalyst and ultimately to the speed of catalyst light-off.

3. Early lean air/fuel ratio control after start-up is made possible due to the QWS, resulting in cold engine emissions reductions.

4. Use of the PRISM (PRediction and Identification type Sliding Mode control) as the new secondary oxygen feedback system enables optimum fuel feedback thereby lowering the emissions.

Each of the technologies developed to achieve SULEV standards had a high degree of difficulty. However, the resulting SULEV Accord system construction is simple and requires no special hardware for subsequent processing. Further, the implementation of these technologies makes it possible to achieve the most rigorous super ultra low emissions standards in the world, at the same time maximizing engine output, all for minimal additional cost.

REFERENCES

1. Kisshi, Kikuchi, Seki, Kato, and Fujimori : Development of High Performance L4 Engine ULEV System, SAE Paper 980415 (1998)

2. Maki, H., Hasegawa, Y., Akazaki, S., Komoriya, I., Nishimura, Y. and Hirota, T.: Real Time Control Using STR in Feedback System, SAE Paper 950007 (1995)

3. Shimasaki, Y., et al.: "Study on Conformity Technology with ULEV Using EHC System" SAE Paper 960342 (1996)

4. Fujii, S.: "Digital Adaptive Control", Computrol, No.27, p28~ , Corona, July 1989

5. Suzuki, T.: " Basic of Adaptive Control", Computrol, No.32, p7~ , Corona, July 1990

6. Nonami, Den: Sliding Mode Control, Tokyo, Coronasha, 1994, p263

7. Yasui, Akazaki, Ueno and Iwaki: Adaptive Sliding Mode Control for Secondary O2 Feedback, Honda R&D Technical Review, Vol.11, No.1, p.65~71 (1999)

8. Yasui, Akazaki, Ueno and Iwaki: Adaptive Sliding Mode Control for Secondary O2 Feedback, A Second Report, Honda R&D Technical Review, Vol.11 No.2, p.65~71 (1999)

9. Ueno, Akazaki, Yasui and Iwaki: A Quick Warm-up System during Engine Start-up Period Using Adaptive Control of Intake Air and Ignition Timing, Honda R&D Technical Review, Vol.11, No.1, p.73~78 (1999)

ABOUT THE EDITOR AND AUTHORS

Editor

Dr. Fuquan (Frank) Zhao is a Research Executive of Technical Affairs at DaimlerChrysler Corporation. In this position, he is engaged in many facets of corporate powertrain research and development activities, including combustion system development, emissions reduction, fuel economy improvements, engine system control, engine design, energy management systems, alternative propulsion systems, fuel effects, and strategic planning for future products and technologies. Dr. Zhao received his B.S. in Mechanical Engineering from Jilin University of Technology (China) in 1985. He obtained his M.S. in Mechanical Engineering from the University of Hiroshima (Japan) in 1989 and his Ph.D. there in 1992. His previous experience includes Assistant Professor in Mechanical Engineering at Wayne State University, Research Fellow at Imperial College of Science, Technology and Medicine (UK), and Postdoctoral Fellow at Wayne State University and the University of Hiroshima. He received several awards both internally and externally. He is currently the chair of the combustion committee for the SAE Fuels & Lubricants Activity. Dr. Zhao is the principal author of more than 100 technical publications on various subjects related to engine combustion and emissions control technologies. He is the principal author of the book *"Automotive Gasoline Direct-Injection Engines"* published by SAE in 2002, and the leading editor of the book *"Homogeneous Charge Compression Ignition (HCCI) Engines"* published by SAE in 2003.

Overview Authors

Dr. Michael Akard received a bachelor of science in chemistry from the University of New Mexico in 1990 and a Ph.D. in analytical chemistry from the University of Michigan in 1994. He is the author of 11 publications in journals ranging from Analytical Chemistry to the Journal of Chromatographic Science. He has made 18 presentations at PITTCON, SAE and Anachem conferences. He has one U.S. patent for a method and a product. He currently works at Horiba Instruments, Inc. as an Analytical Product Specialist in the Ann Arbor, Michigan office.

Dr. Paul Andersen received a Bachelor's Degree in Chemical Engineering from the University of Illinois (Urbana) and a Ph.D. in Chemical Engineering from Northwestern University. Since joining Johnson Matthey in 1992, Paul has worked on the development of various emission control catalysts including Three Way Catalysts, NOx Adsorber Catalysts, SCR Catalysts, and Oxidation Catalysts. Paul is currently the Technical Director in Johnson Matthey's North American Technical Center.

Prof. Choongsik Bae has worked in the Department of Mechanical Engineering, KAIST (Korea Advanced Institute of Science and Technology) since 1998. He received his B.S. and M. Sci. degree in Aerospace Engineering from Seoul National University (Korea). He worked as researcher for Korea Aerospace Research Institute and as a Teaching Associate for KAIST before beginning his doctoral study at Imperial College in London, UK. He received his Ph.D. degree in Mechanical Engineering from Imperial College in 1993 and worked as a Research Associate. He then joined Chungnam National University and soon moved to

KAIST. He has been involved in various studies of experimental engine works, and research on the fuel spray, flow and combustion in SI and CI engines. He received SAE's Arch T. Colwell Merit Award in 1997. He was honored as one of the Presidential Researchers in Korea (2000), leading Engine Laboratory at KAIST, which was selected as one of the National Research Laboratories. He also serves as a General Secretary of Combustion Engineering Research Center at KAIST, vitalizing the efforts in research as well as education.

Dr. Todd Ballinger graduated from the University of Pittsburgh in 1993 with a Ph.D. in physical chemistry/surface science. After conducting post-doctoral research in catalysis and surface science at Texas A&M University and the Naval Research Laboratory, he joined Johnson Matthey, Catalytic Systems Division, in 1995 as a staff scientist. At Johnson Matthey, he has been involved in the development of advanced three-way catalysts and hydrocarbon trap/catalysts for automotive catalytic converters, as well as the development of advanced catalyst systems for achieving very low emissions from vehicles.

Mark Borland is a Senior Engineering Specialist in the Advanced Engine Systems Development Group at DaimlerChrysler Corporation. In his 13 years there, he has worked on the development of emission systems and control algorithms for both gasoline- and diesel-powered vehicles. Most recently, he was part of the team responsible for the development of the software and hardware concepts for meeting the tailpipe emissions on DaimlerChryler's first production PZEV vehicle. Mr. Borland earned a B.S. in Mechanical Engineering from the University of Wisconsin–Milwaukee and an M.S. in Applied Statistics from Oakland University in Rochester, Michigan.

Prof. Wai K. Cheng is a Professor of Mechanical Engineering at MIT and Associate Director of the Sloan Automotive Lab. His research interest is in engine performance and emissions. He has made major contributions to the mixture preparation process in spark ignition engines, and has authored more than 70 technical publications. He had SAE's Teetor Award and Oral Presentation Award in the past, and is a Fellow of the Society.

Dr. James A. Eng is a Staff Research Engineer at General Motors Research and Development. Dr. Eng received his Ph.D. from Princeton University working on hydrocarbon emissions and post-flame oxidation mechanisms from homogeneous charge SI engines. During the past 10 years he has worked in the areas of chemical kinetics, understanding the effects of fuels on engine performance and emissions, cold-start hydrocarbon emissions, and HCCI combustion.

Timothy Gernant is an OBD Calibration Engineer at Ford Motor Company. He joined Ford upon graduation from the University of Michigan with a Bachelor's Degree in Mechanical Engineering in 1995. He continued with his education at the University of Michigan and received a Master's Degree in Automotive Engineering in 2000. Mr. Gernant holds multiple patents relating to OBDII diagnostic systems.

Kathleen Grant has been an OBDII Calibration Engineer at Ford Motor Company for 10 years. She is an electrical engineering graduate from Wayne State University and has worked with diagnostic systems and engine controls for over 20 years. Kathleen has been granted U.S. patents relating to OBDII diagnostic systems. Previously she was a Project

Engineer at General Motors. Most recently Kathleen has been calibrating OBD functionality on PZEV emission level vehicles.

Christopher Hadre is the core design and release engineer for vapor canisters with DaimlerChrysler Advanced Evaporative Systems Group, and has been a specialist in evaporative systems development for over 5 years. Chris graduated with a BASc. in Mechanical Engineering from the University of Windsor in 1998.

Prof. Matthew J. Hall is a Professor in the Department of Mechanical Engineering at the University of Texas at Austin. He has over 20 years of experience in automotive and vehicle research. He received B.S. and M.S. degrees in Mechanical Engineering from the University of Wisconsin-Madison. He received his Ph.D. from Princeton University. He was a post-doc at the Combustion Research Facility of Sandia National Laboratories and at the University of California, Berkeley. His primary research interests center around combustion processes, with an emphasis on internal combustion engines. His focus is on experimental measurements studying engine performance, emissions, and flows, with a specialization in optical diagnostic techniques and sensors. He has received several awards from SAE, including the Ralph R. Teetor Award (1993), the Arch T. Colwell Merit Award (1985), the Horning Award (1987), and the Myers Award (1998).

Dr. David Lafyatis has a Ph.D. in Chemical Engineering from the University of Delaware. He completed a Post Doctoral Fellowship at the Rijksuniversiteit Gent where he studied reaction kinetics in a transient reactor. He has worked at Johnson Matthey for 9 years in exhaust aftertreatment technology, and is currently a Technical Program Manager.

Prof. Ronald D. Matthew obtained his Bachelor's Degree in Mechanical Engineering from the University of Texas followed by three Graduate Degrees from the University of California at Berkeley, culminating in 1977 with a Ph.D. specializing in combustion. He joined the faculty of the Department of Mechanical Engineering at the University of Texas in 1980 where he established their combustion and engines research program. He is the head of the General Motors Foundation Combustion Sciences and Automotive Research Laboratories on the U.T. campus. He is also the faculty advisor for U.T.'s student branch of the Society of Automotive Engineers, and has been since he founded U.T.'s student branch in 1980. He has been involved in research in the area of combustion, engines, emissions, and alternative fuels for over 25 years. His research includes experimental work and numerical modeling of fundamental combustion processes and combustion within engines. His present research is focused primarily on controlling HC emissions from PFI SI engines, the spark ignition process, engine friction, and alternative diesel fuels. He has received several awards from SAE, including the Ralph R. Teetor Award (1979), the Arch T. Colwell Merit Award (1992), the Excellence in Engineering Education (Triple E) Award (2002), the Phil Myers Award (2002), and the Faculty Advisor Award (1990, 1997, 2002). In 1996 and again in 1998, U.T.'s body of work on fractal engine modeling was nominated for the ComputerWorld Award and selected for inclusion in the Smithsonian's National Museum of American History Permanent Research Collection on Information, Technology, and Society. Dr. Matthews was elected a Fellow of SAE in 2002.

Kimiyoshi Nishizawa is a Senior Manager of Engine Engineering Department No. 2 at Nissan Motor Co., Ltd. He graduated from Tokyo University, Japan. His first job at Nissan was to design and develop components for gasoline engines. Later he became an engine system engineer focusing on emissions reduction. His recent work focused on developing the PZEV system of the Nissan Sentra CA for the U.S. market, launched in February 2000; the U-LEV system of the Nissan Bluebird Sylphy for the Japan market, launched in August 2000; and, most recently, developing emission systems to meet Japan U-LEV standards and applying it to more than 80% of Nissan vehicles sold in Japan.

Dr. Stephen Russ is a Technical Leader for Engine Combustion in Ford's V-Engine Engineering Division. He joined Ford after receiving his Ph.D. in Mechanical Engineering from the University of Minnesota in 1993. Dr. Russ worked in the Ford Research Laboratory for 6 years researching engine combustion, emissions formation and advanced diagnostics. For the past 5 years he has been leading the development of several Ford V-engine programs. He has authored 20 SAE technical papers and has organized and chaired SAE technical sessions on engine combustion and emissions. He has been awarded 11 U.S. patents for various engine technologies. Dr. Russ was selected to participate in the 1999-2000 SAE Industrial Lectureship Program and has given invited lectures at several universities.

Prof. Tariq Shamim is an Associate Professor of Mechanical Engineering at the University of Michigan-Dearborn. He is a graduate of the University of Michigan-Ann Arbor, where he received his Ph.D. in Mechanical Engineering and a Master's Degree in Aerospace Engineering. He received another Master's Degree in Mechanical Engineering from the University of Windsor, Canada and a Bachelor's Degree in Mechanical Engineering from the N.E.D. University, Karachi, Pakistan. His research and teaching interests are in the area of computational thermo-fluids with major emphasis on combustion, emission control, fuel cell, and thermal spray. His research is supported by the National Science Foundation, Department of Energy, Department of Defense, and the automotive industry. He is actively involved with several professional organizations including SAE, ASME, and Combustion Institute.

Jenny Spravsow has a Bachelor's Degree in Mechanical Engineering from Lawrence Tech University in Southfield, MI and a Master's Degree in Mechanical Engineering from Oakland Univeristy in Rochester, MI. She has been employed by DaimlerChrysler since 1998, currently holding the position of Product Development Engineer in the Advanced Evaporative Systems Group.

Glenn Zimlich is an OBD Calibration Technical Expert at Ford Motor Company. He joined Ford in 1990 after completing a Bachelor of Mechanical Engineering Degree at the University of Detroit. Glenn completed his Master's of Mechanical Engineering Degree from the University of Detroit in 1993. He has received multiple U.S. patents relating to on-board diagnostics.